NONLINEAR OPTICS AND PHOTONICS

Nonlinear Optics and Photonics

Guang S. He

State University of New York at Buffalo

UNIVERSITY PRESS

Great Clarendon Street, Oxford, OX2 6DP,
United Kingdom

Oxford University Press is a department of the University of Oxford.
It furthers the University's objective of excellence in research, scholarship,
and education by publishing worldwide. Oxford is a registered trade mark of
Oxford University Press in the UK and in certain other countries

© Guang S. He 2015

The moral rights of the author have been asserted

First Edition published in 2015

Impression: 1

All rights reserved. No part of this publication may be reproduced, stored in
a retrieval system, or transmitted, in any form or by any means, without the
prior permission in writing of Oxford University Press, or as expressly permitted
by law, by licence or under terms agreed with the appropriate reprographics
rights organization. Enquiries concerning reproduction outside the scope of the
above should be sent to the Rights Department, Oxford University Press, at the
address above

You must not circulate this work in any other form
and you must impose this same condition on any acquirer

Published in the United States of America by Oxford University Press
198 Madison Avenue, New York, NY 10016, United States of America

British Library Cataloguing in Publication Data

Data available

Library of Congress Control Number: 2014945198

ISBN 978-0-19-870276-4

Printed in Great Britain by
Clays Ltd, St Ives plc

*To
my wife Lan,
my daughter Katherine and her husband Stephen,
and my grandchildren
Emma, Joshua, Kate, Colette, and Adelyn*

Preface

In this book, the author intends to provide a comprehensive presentation on most of the major topics in nonlinear optics and photonics, and to emphasize equally the principles, experiments, techniques and applications.

Chapters 1 to 10 present the fundamentals of modern nonlinear optics, and could therefore be adopted as a textbook with problems given at the end of each chapter. Chapters 11 to 17 cover the advanced topics of techniques and applications of nonlinear optics and photonics, which may also be beneficial to some researchers and experts. This book also includes 16 color pages of selected photographs to illustrate some typical nonlinear optical effects and phenomena.

Part of this book (Chapters 2–5 and 18) is basically the same as that originally written by the author in a previous book (G. S. He and S. H. Liu, *Physics of Nonlinear Optics*, World Scientific, 1999); Chapters 6–11 are rewritten and updated; Chapters 12–17 are entirely new chapters dealing with the latest progress and achievements in nonlinear optics and photonics.

As the author of this book, I am grateful to Prof. P. N. Prasad, the executive director of the Institute of Lasers, Photonics and Biophotonics, the State University of New York at Buffalo, for his long-term (almost three decades) support of my research activity. I am also thankful to Dr. C. Y-C Lee (Air Force Office of Scientific Research) for his encouragement of my research efforts.

In my research career, I have benefited from the collaboration and interaction with many talented colleagues and outstanding scientists. For this reason, I would like to thank Dr. L.-S. Tan, Prof. H. Ågren, Prof. M. Samoc, Prof. R. W. Boyd, Prof. Y. R. Shen, Prof. I. C. Khoo, Prof. Y. Cui, Prof. D. L. Andrews, Prof. M. T. Swihart, Dr. A. Baev, and Prof. S. H. Liu. I also want to express my gratitude to Dr. Y. Shen and Prof. Q. D. Zheng for writing some important sections, and to Prof. S. Y. Tian who critically read the entire manuscript and provided many necessary corrections. Finally, the advice received from S. Adlung and the assistance from Dr. J. Zhu and J. White are greatly appreciated.

The International Systems of Units (SI) is adopted throughout this book for all formulations.

Guang S. He
Buffalo, New York
September 2014
Email: gshe@buffalo.edu
gshe41@gmail.com

Contents

1 **Introduction**   1
   1.1 Conventional optics and nonlinear optics   1
   1.2 Major topics of nonlinear optics and photonics   3
   1.3 Characterization of intense coherent optical field   7
     1.3.1 Intensity and brightness   7
     1.3.2 Spatial and temporal coherence   8
     1.3.3 Photon mode and degeneracy   9
   1.4 Two theoretical regimes   10
     1.4.1 Semiclassical theory   10
     1.4.2 Quantum theory of radiation   11
   1.5 Applicability of the two theories in nonlinear optics and photonics   13

2 **Nonlinear Polarization of an Optical Medium**   18
   2.1 Optical field-induced electric polarization in a medium   18
   2.2 Various mechanisms causing nonlinear polarization in a medium   22
   2.3 Manipulation of nonlinear susceptibilities   23
   2.4 Basic properties of nonlinear susceptibilities   25
     2.4.1 Relative magnitude of various orders of susceptibilities   25
     2.4.2 Spatial-symmetry restrictions on susceptibilities   25
     2.4.3 Resonance enhancement of susceptibilities   26
     2.4.4 Permutation symmetry of susceptibilities   26
     2.4.5 Complex conjugation and time-reversal symmetry of susceptibilities   27
   2.5 Nonlinear coupled-wave equations   27
   2.6 Complex expressions of optical wave fields   30

3 **Second-Order Nonlinear (Three-Wave) Frequency Mixing**   33
   3.1 Second-harmonic generation (SHG)   33
     3.1.1 Quantum description of the mechanism of SHG   33
     3.1.2 Semiclassical description of SHG   35
     3.1.3 Nonlinear crystals for SHG   39
     3.1.4 SHG devices   42
   3.2 Optical sum- and difference-frequency generation   43
     3.2.1 Optical sum-frequency generation   43
     3.2.2 Optical difference-frequency generation   45
     3.2.3 Experimental setups for optical sum- and difference-frequency generation   47
   3.3 Optical parametric amplification and oscillation   47
     3.3.1 General description   47
     3.3.2 Solutions of coupled-wave equations   49
     3.3.3 Experimental devices   51

3.4	Special second-harmonic generation	55
	3.4.1 Other materials for SHG	55
	3.4.2 SHG from surfaces and interfaces	55

4 Third-Order Nonlinear (Four-Wave) Frequency Mixing — 59

4.1	Various FWFM processes	59
4.2	Third-harmonic generation (THG)	62
	4.2.1 Basic theoretical descriptions	62
	4.2.2 Phase-matching methods for THG	64
	4.2.3 Resonance enhancement of THG	65
	4.2.4 Materials and devices for THG and third-order sum-frequency generation	68
4.3	Raman-enhanced FWFM	70
	4.3.1 Coherent Stokes- and anti-Stokes ring emission	70
	4.3.2 Raman-enhanced FWFM using two incident beams with a small crossing angle	73
4.4	Special non-resonant four-photon parametric interaction effects	74
	4.4.1 Continuum generation via four-photon parametric interaction	74
	4.4.2 Frequency-degenerate four-photon parametric interaction	76
4.5	Second-harmonic generation (SHG) via third-order nonlinear processes	78
	4.5.1 Electric field-induced SHG	78
	4.5.2 SHG in optical fibers	79

5 Induced Refractive-Index Changes — 83

5.1	Description of refractive index in linear optics	83
5.2	Description of refractive index in nonlinear optics	85
5.3	Two-beam induced refractive-index changes	87
5.4	Two-photon resonance enhanced refractive-index change	88
5.5	Raman resonance enhanced refractive-index change	90
5.6	Mechanisms of induced refractive-index changes	92
	5.6.1 Various mechanisms of induced refractive-index changes	92
	5.6.2 Molecular reorientation contribution	93
	5.6.3 Optical electrostriction contribution	95
	5.6.4 Temporal responses of induced refractive-index changes	96
5.7	Second-order nonlinearity induced refractive-index change (optical Pockels effect)	99

6 Self-Focusing, Self-Phase Modulation, and Spectral Self-Broadening — 105

6.1	Basic theory of the self-focusing effect	105
	6.1.1 General description	105
	6.1.2 Induced waveguide model of self-trapping	108
	6.1.3 Theory of steady-state self-focusing	109
	6.1.4 Another empirical formula for steady-state self-focusing	113
	6.1.5 Dynamic self-focusing process	114
6.2	Direct observation of self-focusing effect	116
6.3	Self-phase modulation and spectral self-broadening	119
	6.3.1 Self-phase modulation and frequency chirp of intense short light pulses	119
	6.3.2 Spectral self-broadening of intense short light pulses	123
	6.3.3 Beat-frequency enhanced spectral self-broadening	125

	6.4	Optical coherent continuum generation	127
		6.4.1 White-light continuum generation with ultrashort laser pulses	127
		6.4.2 Experimental observation of coherent continuum generation	129
		6.4.3 Applications of coherent continuum generation	134
7	**Stimulated Scattering of Intense Coherent Light**	136
	7.1	Introduction to light scattering	136
		7.1.1 Origins of light scattering	136
		7.1.2 Classification of light scattering	137
		7.1.3 Differences between spontaneous and stimulated scattering	140
	7.2	Theory of stimulated Raman scattering (SRS)	142
		7.2.1 Quantum-electrodynamical description of Raman scattering	142
		7.2.2 Probabilities of spontaneous and stimulated Raman scattering	146
		7.2.3 Gain coefficient and threshold condition	148
	7.3	Experimental studies of SRS	151
		7.3.1 Raman media and experimental setups	151
		7.3.2 Experimental properties of SRS	153
		7.3.3 Four-wave frequency mixing (FWFM) in SRS experiments	154
		7.3.4 Spin-flip, electronic, and rotational SRS effects	160
	7.4	Stimulated Brillouin scattering (SBS)	168
		7.4.1 Fundamental description of Brillouin scattering	168
		7.4.2 Equations of interaction between light and acoustic field	171
		7.4.3 Solution of coupled equations and gain coefficient of SBS	173
		7.4.4 Materials and experimental setups for SBS studies	177
		7.4.5 Major issues of experimental studies on SBS	179
	7.5	Stimulated Kerr scattering (SKS)	183
		7.5.1 Stimulated Rayleigh-wing scattering	183
		7.5.2 Discovery of a super-broadband stimulated scattering	185
		7.5.3 Physical model of Rayleigh–Kerr and Raman–Kerr scattering	187
		7.5.4 Cross-section of Kerr scattering	188
		7.5.5 Exponential gain of SKS	192
		7.5.6 Experimental studies of forward SKS	194
		7.5.7 Experimental studies of backward SKS	198
	7.6	Stimulated Rayleigh–Bragg scattering (SRBS)	200
		7.6.1 Early studies of stimulated thermal Rayleigh scattering in a linearly absorbing medium	200
		7.6.2 Finding of frequency-unshifted backward stimulated scattering in a two-photon absorbing medium	202
		7.6.3 Physical model of SRBS	203
		7.6.4 Threshold requirement of SRBS	204
		7.6.5 Experimental results of SRBS in multi-photon absorbing media	205
	7.7	Stimulated Mie scattering (SMS)	210
		7.7.1 Spontaneous and stimulated Mie scattering	210
		7.7.2 SMS in metallic nanoparticle suspensions	211
		7.7.3 SMS in semiconductor nanoparticle suspensions	215
8	**Optical Phase Conjugation**	222
	8.1	Definitions and features of PCWs	222
		8.1.1 Introduction to optical phase conjugation	222

		8.1.2	Definitions of backward PCWs	223

- 8.1.2 Definitions of backward PCWs — 223
- 8.1.3 Special capability of optical PCWs — 225
- 8.2 PCW generation via nonlinear wave mixing — 226
  - 8.2.1 Backward PCW generation via degenerate four-wave mixing (FWM) — 226
  - 8.2.2 Holographic model of backward PCW generation via degenerate FWM — 229
  - 8.2.3 Forward PCW generation via FWM and three-wave mixing — 233
  - 8.2.4 Experimental studies of backward PCW generation via FWM — 236
- 8.3 PCW generation via backward stimulated scattering — 243
  - 8.3.1 Findings of phase-conjugation behavior of backward stimulated scattering — 243
  - 8.3.2 Experimental studies on phase-conjugation properties of different types of backward stimulated scattering — 244
  - 8.3.3 Theoretical explanations: quasi-collinear FWM model — 248
  - 8.3.4 Theoretical treatment in unfocused-beam approximation — 250
- 8.4 PCW generation via backward stimulated emission (lasing) — 254
  - 8.4.1 Mechanism of generating phase-conjugate backward stimulated emission — 254
  - 8.4.2 Experimental studies on PCW generation via backward stimulated emission — 256
- 8.5 Applications of optical phase conjugation — 260
  - 8.5.1 Applications in special laser-device systems — 260
  - 8.5.2 Applications in high-speed and long-distance optical fiber communication systems — 262

9 Nonlinear and Ultrahigh Resolution Laser Spectroscopy — 274

- 9.1 Major mechanisms of spectral broadening — 274
  - 9.1.1 Doppler broadening of the gaseous medium — 275
  - 9.1.2 Collision broadening of the gaseous medium — 275
  - 9.1.3 Transit-time broadening — 276
  - 9.1.4 Second-order (transverse) Doppler broadening — 277
  - 9.1.5 Recoil broadening — 277
  - 9.1.6 Influence from laser spectral linewidth — 278
- 9.2 Saturation spectroscopy — 278
  - 9.2.1 General description of the saturated-absorption effect — 278
  - 9.2.2 Basic theoretical considerations — 281
  - 9.2.3 Experimental setups and results — 283
  - 9.2.4 Crossover resonances in saturation spectroscopy — 287
- 9.3 Two-photon absorption spectroscopy — 289
  - 9.3.1 General description — 289
  - 9.3.2 Theoretical considerations of 2PA — 291
  - 9.3.3 Experimental studies — 293
- 9.4 Coherent Raman spectroscopy — 295
  - 9.4.1 General description — 295
  - 9.4.2 Coherent anti-Stokes Raman spectroscopy (CARS) — 296
  - 9.4.3 Raman-induced Kerr effect spectroscopy (RIKES) — 301
  - 9.4.4 Raman gain spectroscopy (RGS) and inverse Raman spectroscopy (IRS) — 303
- 9.5 Laser polarization spectroscopy — 306
  - 9.5.1 Doppler-free saturated-absorption polarization spectroscopy — 306
  - 9.5.2 CARS polarization spectroscopy — 308
  - 9.5.3 Polarization labeling molecular spectroscopy — 310

		Contents	xiii

9.6 Laser cooling and trapping spectroscopy 312
    9.6.1 Principles of laser cooling and trapping 312
    9.6.2 Techniques for laser cooling and trapping 314
    9.6.3 Ultrahigh resolution spectroscopy using laser cooling and trapping techniques 316

10 Optical Coherent Transient Effects 323

10.1 Coherent transient interaction of intense light with a resonant medium 323
10.2 Self-induced transparency 324
    10.2.1 Definition of 2π-pulse and self-induced transparency 324
    10.2.2 Shape and speed of the 2π-pulse 327
    10.2.3 Experimental studies of self-induced transparency 330
10.3 Photon echo effect 334
    10.3.1 Concept of photon echo 334
    10.3.2 Theoretical description of photon echo 335
    10.3.3 Experimental studies of photon echo 339
10.4 Optical nutation effect 342
    10.4.1 Conceptual description 342
    10.4.2 Optical Bloch equation 344
    10.4.3 Solution for optical nutation effect 347
    10.4.4 Experimental studies of optical nutation 349
10.5 Optical free induction decay effect 352

11 Optical Bistability 359

11.1 Basic consideration of a nonlinear F-P etalon 359
    11.1.1 Background of optical bistability studies 359
    11.1.2 Theory of steady-state optical bistability 360
    11.1.3 Dynamic response of a nonlinear F-P etalon 364
11.2 Design of optical bistability experiments 365
    11.2.1 Influences of spatial and spectral structures of the incident laser beam 365
    11.2.2 Standard setup for experimental studies 367
11.3 Experimental studies of optical bistability 368
    11.3.1 Early observations of optical bistable effects 368
    11.3.2 Nonlinear materials for optical bistable devices 370
    11.3.3 Semiconductor bistable devices 372
    11.3.4 Optical waveguide bistable devices 373
    11.3.5 Transient and thermal optical bistability 375
11.4 Recent development of bistability studies 378

12 Optical Temporal Solitons 386

12.1 Conditions for temporal soliton formation 386
    12.1.1 Group velocity and group velocity dispersion (GVD) 386
    12.1.2 Refractive index and GVD of silica glass 387
    12.1.3 Balance between GVD and self-phase modulation in a nonlinear medium 388
12.2 Basic properties of temporal solitons 390
    12.2.1 Wave equation governing pulse propagation in a nonlinear dispersive medium 390
    12.2.2 Solitary solutions of nonlinear wave equation in optical fiber systems 391

	12.2.3 Experimental observation of temporal solitons in optical fibers	392
	12.2.4 Soliton-like pulse formation in $n_2 < 0$ nonlinear media with positive GVD	394
	12.2.5 Long-distance transmission of temporal solitons in fibers	395
12.3	Pulse narrowing and self-frequency shift of temporal solitons	396
	12.3.1 Pulse narrowing of higher-order temporal solitons through a shorter fiber	396
	12.3.2 Self-frequency shift of temporal solitons due to Raman gain	398
12.4	Fiber soliton lasers	401
	12.4.1 Principles of soliton lasers	401
	12.4.2 Original version of soliton lasers	401
	12.4.3 Rare earth-doped fiber soliton lasers	403
	12.4.4 Fiber Raman soliton lasers	405

13 Optical Spatial Solitons 410

13.1	Definition of spatial bright and dark solitons	410
13.2	Formation of bright spatial solitons	411
	13.2.1 Nonlinear materials for spatial soliton formation	411
	13.2.2 Spatial soliton formation in generalized Kerr-type nonlinear media	412
	13.2.3 Spatial soliton formation in second-order nonlinear crystals	413
	13.2.4 Spatial soliton formation in liquid crystals	415
	13.2.5 Spatial soliton formation in photorefractive crystals	416
	13.2.6 Formation of spiraling spatial solitons	417
13.3	Formation of dark spatial solitons	419
13.4	Spatial soliton interactions and applications	423
	13.4.1 General features of spatial soliton interactions	423
	13.4.2 Soliton interactions in Kerr-type media	424
	13.4.3 Soliton interactions in second-order nonlinear crystals	426
	13.4.4 Soliton interactions in PR media	427

14 Multi-Photon Nonlinear Optical Effects 432

14.1	Description of multi-photon absorption (MPA)	432
	14.1.1 Introduction to MPA studies	432
	14.1.2 Mechanisms of MPA	434
	14.1.3 Formulations of MPA-induced light attenuation	436
	14.1.4 Theoretical expression of 2PA cross-section	438
14.2	Highly multi-photon absorbing materials	440
	14.2.1 Need for highly multi-photon active materials	440
	14.2.2 Basic structures of multi-photon active chromophores	440
	14.2.3 Features of novel multi-photon active materials	441
14.3	Characterizations of MPA materials	444
	14.3.1 Selection of excitation wavelengths	444
	14.3.2 Measurement of MPA cross-section at discrete wavelengths	445
	14.3.3 Saturation effect of MPA in the sub-picosecond regime	449
	14.3.4 Measurements of MPA spectra	450
	14.3.5 Characterization of MPA-induced fluorescence emission	456
14.4	Multi-photon pumped (MPP) frequency upconversion lasing	458
	14.4.1 General features of MPP lasing materials and devices	459
	14.4.2 Two-photon pumped (2PP) cavity lasing	461
	14.4.3 Three- to five-photon pumped lasing	462

14.5	MPA-based optical limiting, stabilization, and reshaping	466
	14.5.1 Principles of optical limiting	466
	14.5.2 MPA-based optical limiting	466
	14.5.3 MPA-based optical stabilization	470
	14.5.4 MPA-based optical reshaping	474
14.6	3D data storage and microfabrication based on multi-photon excitation (MPE)	476
	14.6.1 Common features of MPE for data storage and microfabrication	476
	14.6.2 3D data storage in two-photon active materials	479
	14.6.3 Two-photon polymerization-based 3D microfabrication	481

15 Nonlinear Photoelectric Effects 489

15.1	Introduction to photoelectric effects	489
	15.1.1 One-photon photoemission effect	489
	15.1.2 Electronic band structures of solids	490
	15.1.3 One-photon induced photoconductivity in semiconductors	491
	15.1.4 Image-potential states (IPSs) of an electron at a metal surface	491
15.2	Multi-photon photoemission (MPPE) effects	493
	15.2.1 Early observations of MPPE phenomena	493
	15.2.2 Resonance-enhanced MPPE effects	494
	15.2.3 MPPE studies on clean and/or adsorbing metal surfaces	497
15.3	Multi-photon photoconductivity (MPPC) effects	500
	15.3.1 Mechanisms of multi-photon induced photoconductivity	500
	15.3.2 Observations of MPPC effects in semiconductors and dielectric media	502
	15.3.3 2PPC-based spectroscopic studies on semiconductors	502
	15.3.4 MPPC-based autocorrelation measurements of ultrashort laser pulses	504
	15.3.5 Other related studies	507

16 Fast and Slow Light 513

16.1	Definitions of light speeds	513
	16.1.1 Phase velocity of a monochromatic light	513
	16.1.2 Group velocity of a quasi-monochromatic light pulse	514
16.2	Group velocity in a resonant medium	517
	16.2.1 Complex refractive index of an absorbing medium	517
	16.2.2 Group refractive index of an absorbing medium	519
	16.2.3 Group velocity of a light pulse in an absorbing medium	520
	16.2.4 Group velocity in a gain medium	522
16.3	Fast/slow light propagation in a resonant medium	524
	16.3.1 Features of light pulse propagation in a resonant medium	524
	16.3.2 Light propagation versus causality and special relativity	526
	16.3.3 Methods of creating fast and slow light propagation	526
16.4	Studies on fast light effects	529
	16.4.1 Fast light in linear absorbing media	529
	16.4.2 Fast light in double-line gain media	532
	16.4.3 Fast light in induced absorption systems	535
	16.4.4 Backward motion of a pulse peak inside a fast light medium	537

16.5 Studies on slow light effects	540
16.5.1 Slow light based on electromagnetically induced transparency (EIT)	540
16.5.2 Slow light based on absorption saturation or coherent population oscillations	544
16.5.3 Light pulses halted (stored) in an EIT medium	546
16.5.4 Slow light effect in a Raman gain medium	548
16.5.5 Slow light effect in a Brillouin gain medium	550
16.5.6 Slow/fast light effects in a semiconductor absorber/amplifier or a fiber amplifier	552

17 Terahertz Nonlinear Photonics — 559

17.1 THz generation via optical rectification	559
17.1.1 Principle of generating THz radiation in a second-order nonlinear crystal	559
17.1.2 Experiments on THz generation in second-order nonlinear materials	562
17.1.3 THz generation via four-wave mixing in plasmas	563
17.2 THz detection using nonlinear optical methods	565
17.2.1 THz detection via electro-optic (EO) sampling	565
17.2.2 THz detection via FWM	566
17.3 Nonlinear optical applications of strong THz fields	570
17.3.1 Nonlinear phase modulation with intense single-cycle THz pulses	570
17.3.2 Strong-field THz-induced nonlinear absorption in semiconductors	575

18 Detailed Theory of Nonlinear Susceptibilities — 579

18.1 Density matrix and interaction energy	579
18.1.1 Basic equations of the density matrix	579
18.1.2 Expression for interaction energy	581
18.2 Expressions for susceptibilities based on density matrix solutions	584
18.2.1 Solutions of density matrix equations	584
18.2.2 Explicit formulations of various-order susceptibilities	586
18.3 Properties of nonlinear susceptibilities	590
18.3.1 Local-field corrections	590
18.3.2 Spatial symmetry	592
18.3.3 Permutation symmetry and time-reversal symmetry of susceptibilities	595
18.4 Resonance enhancement of nonlinear susceptibilities	598
18.4.1 Introduction	598
18.4.2 Resonance enhancement of the first- and second-order susceptibilities	598
18.4.3 Resonance enhancement of the third-order susceptibility	599
18.5 Quantum-mechanical expressions for nonlinear susceptibilities	602
18.5.1 Validity of quantum-mechanical expressions for nonlinear susceptibilities	602
18.5.2 Born–Oppenheimer approximation for nonlinear susceptibilities of a molecular medium	603

Appendices

Appendix 1 Physical Constants Commonly Used in Nonlinear Optics	605
Appendix 2 Numerical Estimates and Conversion of Units	606
Appendix 3 Tensor Elements of the Linear Susceptibility for Crystals and other Media	609

Appendix 4	Tensor Elements of the Second-Order Susceptibility for Various Crystal Classes	610
Appendix 5	Tensor Elements of the Susceptibility of Second-Harmonic Generation for Various Crystal Classes	613
Appendix 6	Tensor Elements of the Third-Order Susceptibility for Crystals and other Media	616
Appendix 7	Tensor Elements of the Nuclear Third-Order Susceptibility in the Born–Oppenheimer Approximation	620
Appendix 8	Derivation of Formulae for Self-Induced Transparency of a 2π-Pulse	622
Index		625

1
Introduction

The invention of the laser has had a revolutionary impact on the development of modern optics. Based on the interaction of laser radiation with matter, a great number of new optical effects and phenomena have been discovered, which are essentially different from those already known in conventional optics. The comprehensive studies on these new optical effects and their applications are the foundation of nonlinear optics and photonics.

1.1 Conventional optics and nonlinear optics

The principle of the laser was proposed by Schawlow and Townes in 1958,[1] and the first laser device (a ruby solid laser) was demonstrated by Maiman in 1960.[2] In 1961, Franken et al. observed second-harmonic generation in a quartz crystal irradiated by the 694.3-nm output beam from a ruby laser.[3] This observation and a series of other new optical effects (such as stimulated scattering, third-order harmonic generation, self-focusing, and so on, discovered shortly thereafter) symbolized the formation of nonlinear optics. In this sense, nonlinear optics is an outgrowth of the laser technology.

In conventional optics, which was established primarily in the 19th century, many basic mathematical equations or formulae often manifested a linear feature. The following are several typical examples showing such a feature of conventional optics.

First, in order to interpret the refraction, reflection, dispersion, scattering, as well as birefringence of light propagation in an optical medium, we should consider an important physical quantity, the light induced electric polarization of the medium. In the regime of conventional optics, the electric polarization vector **P** was simply assumed to be linearly proportional to the electric field strength **E** of an applied monochromatic optical wave, i.e.,

$$\mathbf{P} = \varepsilon_0 \chi \mathbf{E}, \tag{1.1-1}$$

where ε_0 is the free-space permittivity and χ is the susceptibility of a given medium. With this linear assumption, Maxwell's equations lead to a set of linear differential equations involving only the terms proportional to the first power of the field **E**. As a result, the refractive index of the medium is a material constant, which is independent of the light intensity, and there will be no coupling between the light waves of different frequencies in a transparent medium; in other words, no light emission at any new frequencies would be expected.

Second, in conventional optics the attenuation of a light beam propagating through an absorptive medium can be described as

Nonlinear Optics and Photonics. First Edition. Guang S. He. © Guang S. He 2015.
Published in 2015 by Oxford University Press.

$$\frac{dI(z)}{dz} = -\alpha I(z), \tag{1.1-2}$$

where I is the beam intensity, z is the propagation distance, and α is a material parameter independent of the light intensity. The physical meaning of Eq. (1.1-2) is that the decrease of the beam intensity over a unit propagation length is linearly proportional to the local intensity itself. From Eq. (1.1-2) we obtain a well-known exponential attenuation expression

$$I(z) = I(0)e^{-\alpha z}. \tag{1.1-3}$$

This expression implies that for a given propagation length of $z = l$, the transmitted intensity $I(l)$ is linearly proportional to the initial intensity $I(0)$, and the transmittance of $T = I(l)/I(0)$ is not dependent on $I(0)$.

The third example is related to the conventional description of light scattering. In this case, the scattering light intensity from a small volume of scattering medium is assumed to be simply proportional to the local intensity of an incident light beam, i.e.,

$$I_{\text{scat}}(x, y, z) \propto I_0(x, y, z). \tag{1.1-4}$$

Here I_{scat} is the scattering light intensity, I_0 is the incident light intensity, and (x, y, z) refers to the spatial location of the small scattering volume being considered.

All these and some other simple linear assumptions or conclusions given by conventional optics were widely accepted, and verified by the most experimental observations and measurements, based on the use of ordinary light sources. However, this situation has been radically changed since the beginning of the 1960s.

Shortly after the discovery of optical second-harmonic generation, several other coherent frequency-mixing effects (including optical sum-frequency, difference-frequency, and third-harmonic generation) were sequentially observed. Researchers immediately realized that all these new optical effects facilitated by laser radiation could be readily interpreted, if the linear term on the right-hand side of Eq. (1.1-1) is replaced by a power series

$$\mathbf{P} = \varepsilon_0 [\chi^{(1)} \mathbf{E} + \chi^{(2)} \mathbf{E}^2 + \chi^{(3)} \mathbf{E}^3 + \cdots]. \tag{1.1-5}$$

Here, $\chi^{(1)}$, $\chi^{(2)}$, and $\chi^{(3)}$ are the first-order (linear), second-order (nonlinear), and third-order (nonlinear) susceptibility, and so on. They are material coefficients and in general are tensors. Substituting Eq. (1.1-5) into Maxwell's equations leads to a set of nonlinear differential equations that involve high-order power terms of electric field strength; these terms are responsible for various observed coherent optical frequency-mixing effects.[4] Based on Eq. (1.1-5), it can also be derived that upon the action of laser radiation the refractive index of a medium is no longer a constant even at a given wavelength, instead there will be an induced refractive-index change, depending on the intensity (in a third-order nonlinear medium) or the amplitude (in a second-order nonlinear medium) of the incident laser beam(s). This nonlinear assumption can be used to well explain the observed self-focusing effect of a laser beam and some other related nonlinear optical effects.

In the same period, researchers also found that the attenuation of an intense laser beam propagating in an absorptive optical medium did not follow the linear description given by Eq. (1.1-2) or Eq. (1.1-3). For instance, in a one-photon absorptive medium, if the intensity of the incident beam is high enough, the attenuation coefficient α no longer remains constant and may vary depending on the incident light intensity owing to the absorption saturation effect.

Moreover, if there is a two-photon absorption process involved, the attenuation of an intense laser beam must be described as

$$\frac{dI(z)}{dz} = -\alpha I(z) - \beta I^2(z), \tag{1.1-6}$$

where β is the two-photon absorption coefficient. In a more general case, if we further extend our consideration to include multi-photon (three-photon or more) absorption processes, then Eq. (1.1-6) should be generalized to the following form:

$$\frac{dI}{dz} = -\alpha I - \beta I^2 - \gamma I^3 - \cdots. \tag{1.1-7}$$

Here, γ is the three-photon absorption coefficient, and so on.

Lastly, the Raman and Brillouin scattering behavior of an intense laser beam propagating in a transparent medium does not obey the simple linear relation expressed by Eq. (1.1-4) with increasing the input intensity. Once the input laser intensity exceeds a certain threshold level, the ordinary (spontaneous) scattering suddenly becomes stimulated scattering with a high directionality and brightness.

Based on these comparisons described above, we can conclude that the main concern in conventional optics is the propagation and interaction with matter of the light from ordinary incoherent light sources, wherein the intensities of light beams are relatively low so that even a simple linear approximation provides a satisfactory theoretical explanation for the related optical effects and phenomena. In this sense, conventional optics may also be called *linear optics*. In contrast, *nonlinear optics* mainly deals with the interaction of intense coherent light with matter. In the latter case, the monochromatic and directional brightness of a laser beam can be so high that a great number of new effects and phenomena could be generated, which are best explained using some nonlinear approximations. In general, the contents of nonlinear optics are much more comprehensive than those of linear optics and require more complicated and sophisticated theoretical treatments.

The nonlinear expression of electric polarization **P** given by Eq. (1.1-5) can be used to well describe various nonlinear frequency-mixing effects as well as other effects related to the induced refractive-index change. Hence the study of these types of optical effects was originally termed "nonlinear optics." However, it should be pointed out that not all nonlinear optical effects could be well interpreted by Eq. (1.1-5). For instance, optical transient coherent effects are well described by the Bloch equations instead of Eq. (1.1-5). The stimulated Raman scattering, multi-photon absorption, and nonlinear photoelectric effects can be perfectly elucidated by the theory of quantum electrodynamics (quantum theory of radiation), rather than the nonlinear polarization theory.

1.2 Major topics of nonlinear optics and photonics

From the contemporary point of view, all optical effects and phenomena that are related to the interaction of intense coherent light with matter are the subjects of nonlinear optics or nonlinear photonics. The following are the major topics of the modern nonlinear optics and photonics.
Fundamental studies:

- Nonlinear electric susceptibility theory
- Three-wave frequency mixing in $\chi^{(2)}$ media

4   *Introduction*

- Four-wave frequency mixing in $\chi^{(3)}$ media
- Induced refractive-index changes
- Self-focusing, self-phase modulation, and spectral self-broadening
- Stimulated scattering
- Nonlinear spectroscopic effects
- Coherent transient effects
- Multi-photon processes

Special techniques and applications:

- Optical phase conjugation
- Optical bistability
- Optical solitons
- Multi-photon techniques
- Nonlinear photoelectric effects
- Fast and slow light
- Terahertz nonlinear photonics
- Nonlinear opto-thermal, -mechanical, -chemical, and -biological effects

The following are the introductory summaries of these major subjects.

In conventional optics, the linear susceptibility χ introduced in Eq. (1.1-1) is a key parameter that characterizes the reaction of the optical medium to an input light field; whereas in nonlinear optics, the nonlinear susceptibilities of $\chi^{(n)}$ ($n = 2, 3, \ldots$) introduced in Eq. (1.1-5) play the key roles. The basic theory of the nonlinear susceptibilities indicates that in the approximation of dipole-moment transition, the even-order nonlinear susceptibilities $\chi^{(2)}, \chi^{(4)}, \ldots$ are nonzero only in those anisotropic crystals of non-centrosymmetry. In contrast, the odd-order nonlinear susceptibilities of $\chi^{(3)}, \chi^{(5)}, \ldots$ are nonzero in all types of materials including the crystal of centrosymmetry and isotropic media.

In a second-order ($\chi^{(2)}$) nonlinear medium, one can expect to observe the second-harmonic generation, sum-frequency generation, difference-frequency generation, parametric amplification/oscillation, and optical rectification. All these effects are the result of nonlinear three-wave frequency mixing processes.

In a third-order ($\chi^{(3)}$) nonlinear medium, one can anticipate the third-harmonic generation, third-order sum-frequency generation, and third-order parametric interaction. All these effects are the results of the nonlinear four-wave frequency mixing processes.

Upon the action of intense coherent light in a third-order nonlinear medium, the four-wave mixing could also lead to intensity-dependent refractive-index change. Similarly, special three-wave mixing in a second-order nonlinear crystal may lead to an amplitude-dependent refractive-index change.

In practice, the laser beam induced intensity-dependent refractive-index changes often play an important role in various different nonlinear optical processes. For instance, when a laser beam with a Gaussian-type transverse intensity distribution propagates in a third-order nonlinear medium, the laser beam may be self-focused or self-trapped.[5] In this case, the induced refractive-index change $\Delta n(r) \propto I(r)$, where r is the transverse radius of the beam, and the nonlinear medium behaves like a focusing lens or light waveguide. In another case, when the input is an intense pulsed laser beam, the intensity-dependent refractive-index change is a function of time, i.e., $\Delta n(t) \propto I(t)$, then the laser pulses themselves will experience a temporal phase modulation and consequently a spectral broadening. These are the so-called self-modulation effect and spectral self-broadening effect, respectively.

Laser frequency conversion can be achieved not only by using three- or four-wave frequency mixing methods, but can also be implemented through various stimulated scattering processes. Stimulated Raman scattering (1962) and stimulated Brillouin scattering (1964) are among the earliest observed nonlinear optical effects.[6,7] The (discrete) frequency-shift range is about 10^1–10^4 cm^{-1} for the former depending on the types of Raman transitions, and 10^{-2}–1 cm^{-1} for the latter depending on the physical state (gas, liquid, or solid) of the scattering medium. The later reported stimulated Kerr scattering in Kerr liquids could produce a continuous frequency-shift range of more than 400–700 cm^{-1} depending on the pump intensity level. In contrast, the more recently reported stimulated Rayleigh–Bragg scattering in multi-photon absorbing liquids and stimulated Mie scattering in nanoparticle-suspended liquids manifest no frequency shift, owing to the reflection feedback from a standing wave induced Bragg grating inside the scattering medium. The studies on various stimulated scattering effects allow us to have much extended and in-depth knowledge of the interaction between the intense coherent light and the scattering medium. In addition, the backward stimulated scattering is one of the most effective approaches to generate backward optical phase-conjugate waves.

The above-mentioned nonlinear optical effects of multi-wave frequency mixing and stimulated scattering are commonly produced in transparent optical media, i.e., the input laser frequency and the newly generated coherent emission frequencies are far from the linear absorption frequencies of the optical materials. However, if the input is a short pulsed laser field or a rapidly switching on/off laser field, and the frequency of the laser field is rather close to the linear absorption frequency, then the medium transient reaction may lead to the appearance of several interesting effects, such as self-transparency of a 2π pulse, photon echo, optical nutation, and optical free induction decay. Studies on these optical coherent transient effects enable researchers to have some unconventional spectroscopic approaches to investigate the molecular transition dynamics, spectral broadening properties, and relevant relaxation constants of a given resonant medium.

Regarding high-resolution spectroscopy, the resolving power of all traditional spectroscopic techniques is severely restricted by two factors: one is the limitation of resolution capability of the utilized key dispersion elements (prisms, grating, or Fabry–Perot etalon); the other is the limitation imposed by the Doppler broadening effect of a gas sample. These two factors make the spectral resolving power of traditional spectroscopic methods not greater than 10^6 in the order of magnitude. By contrast, these two limitations can be overcome by newly developed laser nonlinear spectroscopic methods, with which no traditional dispersion elements are needed and the Doppler broadening influence can be avoided. For instance, the Doppler-free saturation absorption spectroscopy and two-photon absorption spectroscopy can readily offer a resolving power higher than 10^8–10^{10}. Moreover, the later developed laser cooling and trapping spectroscopy can even remove the second-order (transverse) Doppler broadening by cooling atoms to an extremely low temperature (lower than 10^{-4}–10^{-6} K) and the spectral resolving power can be higher than 10^{11}–10^{13}.

Multi-photon absorption and multi-photon excitation are among the most important subjects in nonlinear optics and photonics. Although the theory of two-photon absorption was first proposed in 1931 by M. Göppert-Mayer,[8] the experimental observation became available only after the invention of lasers.[9] Moreover, with an applied coherent light of high photon degeneracy, not only two-photon absorption is available, the multiple (three, four, five, ...) photon absorption can also take place in appropriate materials. Multi-photon excitation processes can be rigorously elucidated by the quantum theory of radiation with introduction of the concept of intermediate state and the associated representation of a virtual energy level. Studies on multi-photon absorption, multi-photon photoemission, multi-photon conductivity, multi-photon ionization of atoms, multi-photon molecular dissociation, and multi-photon opto-chemical reactions are of great significance for both fundamental research and various technological applications.

From the history of nonlinear optics, we can find that since the advent of the laser, the most nonlinear optical effects and phenomena have been discovered in experiments, and for a long time the research momentum in these areas remains strong steadily, because many nonlinear optical effects have their special potentials for scientific and technological applications. The following are some distinctive examples of nonlinear optical applications.

It is known that the optical phase conjugation refers to a special nonlinear optical technique that can generate an optical output beam with a wave front reversed to that of an input signal beam. The unique feature of a phase-conjugate beam is that after this beam passed through an optically disturbed propagation medium, the aberration influence from the medium can be removed and the original information carried by the input signal beam can be finally restored. The major methods to generate optical phase-conjugate waves are four- or three-wave mixing, backward stimulated scattering, and backward stimulated emission (lasing), respectively. The phase conjugation technique can be particularly useful in building up high-brightness laser oscillator and/or amplifier systems, laser weapon and tracking systems, and high-speed and long-distance optical fiber communication systems. The same technique has also provided a physical approach to generate the stimulated scattering or stimulated emission (lasing) from a randomly distorted gain medium.

Optical bistability usually refers to the bistable behavior of an active or passive optical element. A commonly used optical bistable device is a Fabry–Perrot etalon with a nonlinear medium between two reflection mirrors or surfaces. Upon the incidence of a laser beam, the transmittance or reflectivity of this device may manifest a nonlinear dependence on the input light intensity. The key requirements for an efficient optical bistable device are the high nonlinear refractive-index coefficient of the used medium and the fast response time of the device. Various optical bistable devices could be highly useful for optical switching, optical limiting, optical differential amplification, optical counting in future optical logic circuits, optical data processing, and telecommunication systems.

When a short or ultrashort laser pulse propagates along an optical glass fiber, the group velocity dispersion (GVD) effect may cause considerable broadening of the pulse duration; on the other hand, the same pulse may also experience a frequency-chirping influence induced by the self-phase modulation effect. The influences from these two effects can be counterbalanced by properly choosing the input laser wavelength and controlling the GVD property of the fiber material. Under this condition, the laser pulse duration and shape can remain unchanged or periodically reproducible along the propagation distance in the fiber; this corresponds to the formation of an optical temporal soliton. The optical temporal-soliton technique has been successfully utilized to increase the fiber communication distance, to compress the optical pulse width, and to build up various fiber soliton lasers offering very stable and ultrashort output pulses.

In the recent decades, researchers have put more efforts in developing novel multi-photon active materials and exploring their applications in different areas of photonics. One of the most direct applications is the multi-photon absorption based optical limiting, stabilization, and reshaping. In addition, the multi-photon pumped lasing becomes an alternative approach to achieve coherent frequency up-conversion without phase-matching requirement. Furthermore, based on the nonlinear nature of the dependence of multi-photon absorption on the local laser intensity inside a multi-photon active medium, researchers have developed high-density three-dimensional

(3D) optical data storage, high-precision 3D micro-fabrication, and high-resolution 3D up-conversion microscopy.

There is another important topic in the area of nonlinear photonics, i.e., the study of multi-photon photoelectric effects. Historically, it is known that the photoelectric effect was discovered in 1887 by Hertz;[10] it was one of the experimental facts that inspired Einstein in 1905 to postulate the concept of photons—the fundamental particles constituting an optical field.[11] According to his hypothesis, the emission of a photoelectron from a metal surface will take place by absorbing one incident photon if the photon energy $h\nu$ is greater than the work function of the metal. For this reason, if the light wavelength is longer than a certain value, no photoelectric effect will occur. This inference was in agreement with the experimental observations at that time and later. However, shortly after the invention of lasers, researchers found that the photoemission could be easily observed from a metal surface irradiated by a pulsed laser beam, even though the photon energy was much lower than the work function of a given metal sample. This type of new findings could be well interpreted by multi-photon excitation processes. Now the study of multi-photon photoemission becomes a very useful technique to investigate the spectral structure of the image-potential states and excitation dynamic of various metal surfaces. On the other hand, the studies on the multi-photon excited photoconductivity in semiconductors have provided a new approach to measure the exciton spectrum and to detect the ultrashort laser pulses without needing a second-harmonic crystal and phase-matching requirement.

Finally, another new topic in photonics is the fast and slow light propagation in absorptive or gain media. This issue becomes more interesting and attractive because on the one hand, the studies on which can help researchers to have an in-depth understanding about the special relativity and causality in physics; on the other hand, the fast and slow light techniques could be useful for light speed control, optical buffers and delay lines, optical computing and telecommunications.

1.3 Characterization of intense coherent optical field

1.3.1 Intensity and brightness

It is well known that the laser radiation is generated based on the stimulated emission from a population inversion system, whereas the ordinary light is based on spontaneous emission from conventional light sources. Consequently, these two emission mechanisms lead to a huge quantitative difference of the parameters characterizing the properties of light radiation.

The following are the commonly adopted photometric parameters to characterize a quasi-directional and quasi-monochromatic light field.

- **Intensity** is defined as

$$I = \frac{P}{S}, \quad (1.3\text{-}1)$$

  where P is the total light power (in units of watt) and S is the cross-section of the light beam (in units of cm^2). The unit of the intensity is W/cm^2.

- **Spectral intensity** is defined as

$$I(\nu) = \frac{P}{S\Delta\nu}, \quad (1.3\text{-}2)$$

where $\Delta\nu$ is the spectral width of the light radiation (in units of hertz). The unit of the spectral intensity is W/(cm² Hz).

- **Brightness** is defined as

$$B = \frac{P}{S\Omega} = \frac{I}{\Omega}, \qquad (1.3\text{-}3)$$

where Ω is the divergent solid angle of the light beam (in units of steradian). The unit of the brightness is W/(cm² sr).

- **Spectral brightness** is defined as

$$B(\nu) = \frac{P}{S\Omega\Delta\nu} = \frac{I(\nu)}{\Omega}. \qquad (1.3\text{-}4)$$

The unit of the spectral brightness is W/(cm² sr Hz).

The spectral width and divergent angle of a laser beam can be significantly smaller in many orders of magnitudes than the light emission from any ordinary light sources; thereby the spectral brightness of the former is incomparably greater than the latter.

1.3.2 Spatial and temporal coherence

The spatial coherence refers to the phase relationship between the field vibrations examined at two separate transverse positions of a light beam. The spatial or transverse coherent length is defined by

$$d_{\text{coh}} = \frac{\lambda}{\theta}, \qquad (1.3\text{-}5)$$

where λ is the light wavelength and θ is the plane divergent angle of the beam (in units of radian). The physical meaning of d_{coh} is that there is a certain and predicable phase relationship of light vibrations examined at any two transverse positions of which the distance is equal to or smaller than d_{coh}. Furthermore, the quantity defined by

$$S_{\text{coh}} = d_{\text{coh}}^2 = \frac{\lambda^2}{\theta^2} = \frac{\lambda^2}{\Omega}, \qquad (1.3\text{-}6)$$

is the so-called coherent cross-section of the light beam, within S_{coh} all light vibrations are coherent each other.

The temporal coherence refers to the phase relationship between the field vibrations at the same spatial position but examined at two different time moments. The coherent time is defined by

$$t_{\text{coh}} = \frac{1}{\Delta\nu}, \qquad (1.3\text{-}7)$$

where $\Delta\nu$ is the spectral width (or spread range) of the light beam. The physical meaning of t_{coh} is that within a time interval of $\Delta t \leq t_{\text{coh}}$, the field vibrations examined at any two moments always have certain and predictable phase relation. For an ideal monochromatic wave, $\Delta \nu \Rightarrow 0$ and $t_{coh} \Rightarrow \infty$.

The light emitted from ordinary light sources usually exhibits very large divergence angle (Ω) and spectral width ($\Delta \nu$), so that its spatial and temporal coherence properties are very poor. For this reason, sometimes it is called incoherent light characterized by its low spectral brightness. In contrast, the laser light manifests extremely small divergent angle and spectral width; hence, it is usually called highly coherent light.

1.3.3 Photon mode and degeneracy

According to the quantum theory of radiation, any type of optical field can be viewed as an ensemble of a large number of photons that are distributed in different fundamental modes, each mode corresponds to a quantum statistic state (e.g., a monochromatic plane wave), and the average number of photons occupying a single mode is termed photon degeneracy. For a quasi-directional and quasi-monochromatic light beam, the total photon number passing through a give beam section of S within a given time interval of Δt is

$$F = \frac{P\Delta t}{h\nu}, \qquad (1.3\text{-}8)$$

where P is the beam power and $h\nu$ is the energy of a single photon. On the other hand, the mode number associated with the above F photons is given by

$$M = \frac{\Delta t}{t_{coh}} \cdot \frac{S}{S_{coh}} = \frac{\Delta t}{(1/\Delta \nu)} \cdot \frac{S}{(\lambda^2/\Omega)}. \qquad (1.3\text{-}9)$$

Here $\Delta \nu$ and Ω are the spectral width and solid divergent angle of the light beam, respectively. Assuming the light beam is non-polarized, there should be two independent polarization states, so that the photon degeneracy $\bar{n}$ can be finally determined by

$$\bar{n}(\nu) = \frac{F}{2M} = \frac{P}{(2h\nu/\lambda^2)S\Omega\Delta\nu}. \qquad (1.3\text{-}10)$$

The photon degeneracy is a dimensionless quantity.

On comparing Eq. (1.3-4) with Eq. (1.3-10) one can see that there is only a difference of factor $(2h\nu/\lambda^2)^{-1}$ between the spectral brightness $B(\nu)$ and the photon degeneracy $\bar{n}(\nu)$; therefore, they can be viewed as two equivalent quantities.

According to conventional optics, the brightness of a light beam cannot be increased by passing it through any types of optical imaging or transmission systems. It can also be expressed in terms of quantum statistics that the total number of modes for a given photon ensemble cannot be compressed by any ordinary optical systems; therefore, the photon degeneracy cannot be increased by any types of ordinary optical devices. However, these two equivalent conclusions are no longer valid for lasers and nonlinear optical devices. It is well known that the spectral brightness or photon degeneracy of a weak optical signal can be dramatically increased, based on the coherent amplification through a lasing medium, a stimulated scattering medium, or an optical parametric amplifier system. For a laser oscillator system, the number of the total lasing modes can be drastically restricted by choosing an appropriate cavity design and mode selection technique. As a result, the photon degeneracy of the output laser beam can be extremely high.

Table 1.1 *Characteristics of radiation from ordinary light source and lasers.*

Parameters	Sun	Gas Lasers	Solid Lasers	Q-Switched or mode-locked lasers
Monochromaticity ($\Delta \nu/\nu$)	White light	10^{-8}–10^{-13}	10^{-3}–10^{-8}	10^{-2}–10^{-6}
Directionality Ω(sr)	6.8×10^{-5} (on the Earth)	10^{-5}–10^{-7}	10^{-6}–10^{-8}	10^{-6}–10^{-8}
Brightness (W/cm² sr)	$\sim 10^3$	10^4–10^8	10^7–10^{11}	10^{12}–10^{17}
Spectral brightness (W/cm² sr Hz)	$\sim 10^{-12}$	10^{-2}–10^2	10–10^3	10^4–10^7
Photon degeneracy	$\leq 10^{-2}$	10^8–10^{12}	10^{11}–10^{13}	10^{14}–10^{17}

In Table 1.1, we list the typical parameters of light radiation from the strongest ordinary light source (the Sun) and from laser devices. From this table one can see that the spectral brightness or photon degeneracy of the radiation from the high peak-power laser devices can be 10^{15}–10^{19} times greater than that of the radiation from an ordinary light source (like the Sun). In the sense of radiation potentials in interacting with matter and creating various new effects, the difference between the ordinary light and the laser light is mostly similar to that between the conventional weapons and the strategic nuclear weapons.

1.4 Two theoretical regimes

Two major theoretical approaches can be employed in nonlinear optics as well as in laser physics. The first is the semiclassical theory, and the second is the quantum electrodynamical theory (or the quantum theory of radiation). These two theories are parallel and can be complementarily utilized to elucidate various nonlinear optical effects, though they have their own advantages and disadvantages.

1.4.1 Semiclassical theory

The most essential feature of the semiclassical theory is that the medium composed of atoms or molecules is described by the theory of quantum mechanics, while the light radiation is described by the classical Maxwell's equations. The key issue of semiclassical theory in nonlinear optics is to give the expressions of macroscopic nonlinear electric polarization of the optical media. For this purpose, the density matrix method, which is a special approach adopted in both quantum mechanics and statistical physics, is used to derive the expressions for various orders of electric susceptibilities, $\chi^{(1)}$, $\chi^{(2)}$, $\chi^{(3)}$,..., and the corresponding various orders of polarization components, $\mathbf{P}^{(1)}$, $\mathbf{P}^{(2)}$, $\mathbf{P}^{(3)}$,.... Substituting the appropriate nonlinear polarization components into the generalized wave equations, we are able to predict many possible nonlinear optical responses of the medium to the input intense optical field(s). Figure 1.1 shows a schematic diagram to demonstrate the framework of the semiclassical theory in nonlinear optics.

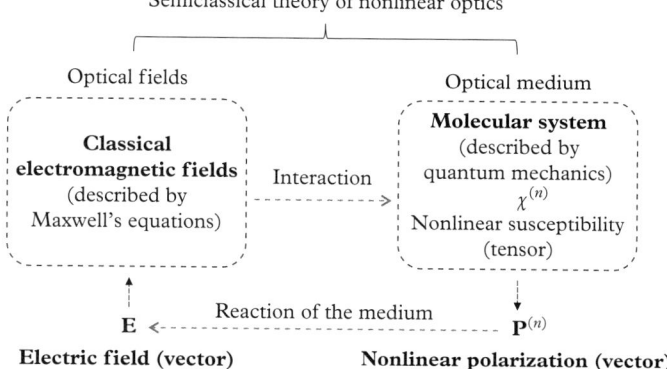

Figure 1.1 *Schematic diagram showing the interaction of optical fields with a medium within the semiclassical theoretical regime.*

According to the classical electromagnetic theory of light, the intensity of a quasi-parallel and quasi-monochromatic light beam is equal to the magnitude of the Poynting vector of a plane monochromatic electromagnetic wave, i.e.,

$$I = \frac{1}{2}\varepsilon_0 c n_0 |E(\nu)|^2, \qquad (1.4\text{-}1)$$

where c is the speed of light in the free space, n_0 is the linear refractive index of the medium, and $E(\nu)$ is the electric field strength of the light wave. For light radiation from ordinary light sources, the values of $E(\nu)$ are extremely small in comparison with the internal electric field of an atom or a molecule (see Appendix 2); therefore all nonlinear terms of $\mathbf{P}^{(2)}$, $\mathbf{P}^{(3)}$,..., can be neglected. On the contrary, laser devices may provide an electric field comparable in order of magnitude to the atomic or molecular internal electric field; thereby the nonlinear electric polarization contributions cannot be neglected any longer.

1.4.2 Quantum theory of radiation

By contrast, the quantum theory of radiation that is a part of quantum electrodynamics treats the medium and optical fields as a combined and quantized system. In other words, both the medium and the optical fields should be described in the way of quantum mechanics.[12,13] As a result, the wave function of the combined system is expressed as the product of the eigenfunction of a molecular system and the eigenfunction of the quantized photon field. In this case, the key issue is to determine the probability of state changes of the combined system due to interaction between the photon fields and the medium. Usually, the state changes of the combined system are related to the transition of molecular system from its initial eigenstate to the final state and the simultaneous changes of the photon numbers among different photon modes. Figure 1.2 shows a schematic diagram to demonstrate the framework of the quantum electrodynamical theory in nonlinear optics and photonics.

Since the eigenvalue is +1 for the photon creation operator and −1 for the photon annihilation operator, the photon number change associated with each quantum transition of the combined system can only be either +1 or −1. However, for many nonlinear optical effects, a multiple photon

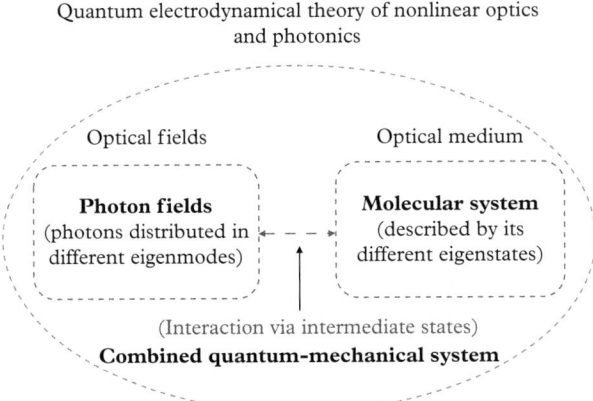

Figure 1.2 *Schematic diagram showing the interaction of photon fields with a medium within the quantum electrodynamical regime.*

number change can be involved. In these cases, we must visualize that multiple photon number changes take place through several steps. For each step, the photon number change is only +1 or −1, and two sequential steps are connected by an intermediate state that is a peculiar but very useful concept introduced in the quantum theory of radiation.

As an example, Fig. 1.3 schematically shows how we can interpret the second-order sum-frequency generation by using the concept of the intermediate state. We assume that in the very beginning, the input optical field contains only two photon modes of frequency ν_1 and ν_2 with the photon degeneracy $\bar{n}_{\nu 1} \gg 1$ and $\bar{n}_{\nu 2} \gg 1$, respectively; while the molecule stays in its ground state, and the energy separation between its ground state and the first excited state is greater than $h\nu_1$, $h\nu_2$, and $h(\nu_1 + \nu_2)$, so that one-photon absorption in these three frequencies is not possible. Then in the first step, there is an annihilation of a photon of frequency ν_1 while the molecule is excited to an intermediate state that characterizes the whole combined system, in this intermediate state the molecule is not located in its any specific excited state but possibly in its all excited states with a certain probability distribution. In the next step, there is another annihilation of a photon of frequency ν_2 while the molecule is excited to another intermediate state in which the molecule may associate to its all excited states with a different probability distribution. In the final step, there is a creation of new photon of frequency $(\nu_1 + \nu_2)$ while the molecule returns to its ground state.

It is important to notice that in an intermediate state the molecular energy uncertainty is so great (determined by the energy spread of all excited states), then according to the uncertainty principle, the staying time of molecule in an intermediate state approaches to zero (at least much shorter than the period of light). In this sense, the above-mentioned three steps actually take place simultaneously, and we can also say that the real elementary process is the annihilation of one $h\nu_1$ photon and one $h\nu_2$ photon, and the simultaneous creation of one $h(\nu_1 + \nu_2)$ photon. The other key point is that in the beginning and end of this elementary process, the molecular quantum state actually remains unchanged, so that the conservation of energy and momentum should hold only among the changed photon fields.

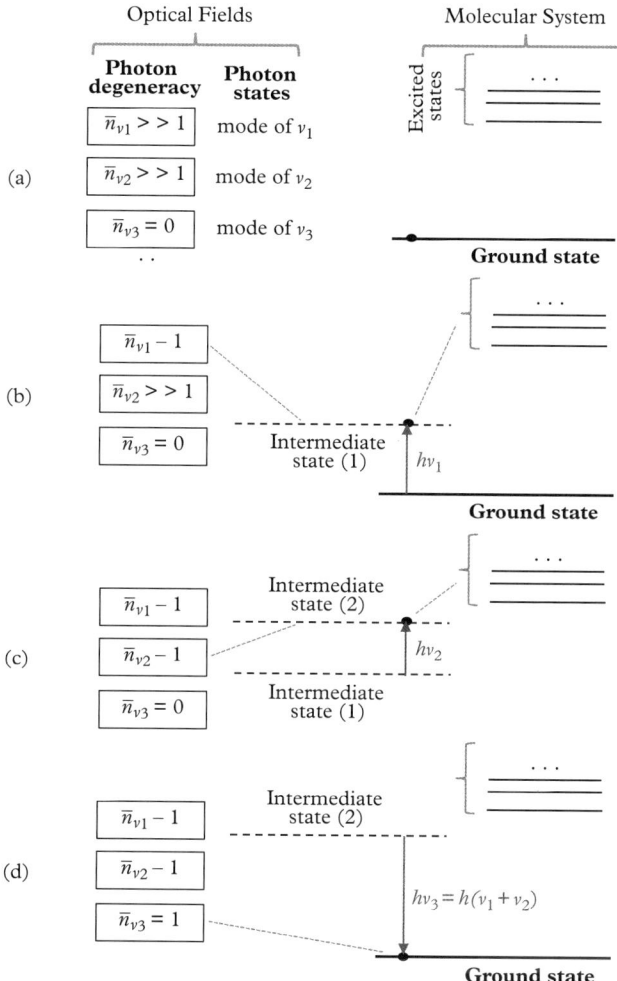

Figure 1.3 *Schematic diagram showing the elementary process of sum-frequency generation via two intermediate states introduced in the quantum theory of radiation.*

1.5 Applicability of the two theories in nonlinear optics and photonics

As mentioned at the beginning of Section 1.4, both the semiclassical and the quantum theory can be complementarily employed to describe various new optical effects and phenomena that arise from the interaction of intense coherent light with matter. The semiclassical theory of nonlinear polarization processes gives a satisfactory and quantitative description of various (three- and four-wave) nonlinear frequency-mixing effects as well as other nonlinear optical effects related to the

14 Introduction

induced refractive-index change. In addition, this theory also gives a farfetched description of two-photon absorption (cf. Section 9.3.2) as well as Raman gain (cf. Section 8.2.4 in He and Liu)[14] by using the imaginary part of the corresponding resonant $\chi^{(3)}$. However, the semiclassical theory cannot explain the microscopic origins of all these effects; it is unable to interpret the nonlinear photoelectric effects and to establish the relation between the spontaneous and the stimulated Raman scattering. Sometimes, it may also cause confusion that the same third-order nonlinear susceptibility $\chi^{(3)}$ is used to formally describe many effects (such as third-harmonic generation, refractive-index changes, two-photon absorption, Raman gain, CARS spectroscopic effect, etc.) although the physical mechanisms of these effects could be entirely different. Even for the same issue like intensity-dependent refractive-index changes, the measured effective $\chi^{(3)}$ values for a given medium can be drastically different, depending on the laser pulse duration and the specific contributing mechanism. In this sense, generally talking about the magnitude of $\chi^{(3)}$ values is not truly meaningful.

On the other hand, at least in principle, the quantum theory of radiation can give a natural and consistent explanation of the most nonlinear optical effects and phenomena, including three- and four-wave frequency mixing (three- and four-photon parametric interactions), stimulated Raman and stimulated Kerr scattering, multi-photon absorption, multi-photon photoemission and photoconductivity, and some nonlinear spectroscopy effects. In all these cases, the elementary nonlinear processes can be clearly illustrated using an energy-level diagram involving the multi-step quantum transitions via virtual energy levels representing the relevant intermediate states. However, the original mathematical derivation in the regime of quantum electrodynamics can be rather lengthy sometimes.

Figure 1.4 schematically shows the applicability of the two theories adopted in nonlinear optics and photonics.

Throughout this book, we intend to use these two theoretical approaches in a complementary way. In all possible cases, we will briefly describe the principles, mechanisms, and the microscopic

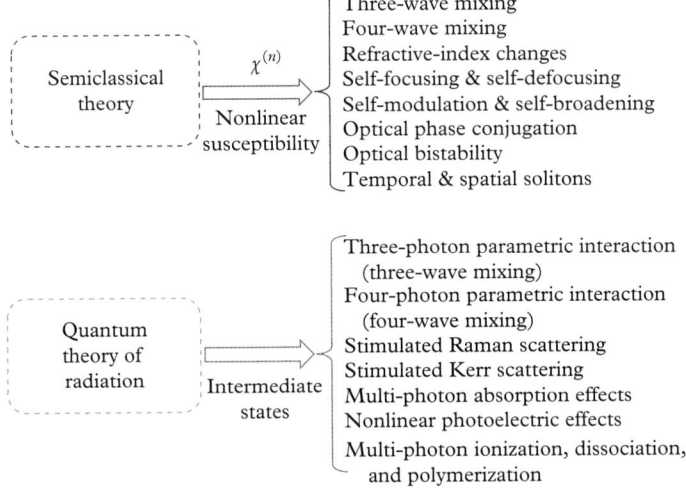

Figure 1.4 *Schematic diagram showing the applicability of two theoretical regimes in nonlinear optics and photonics.*

pictures of various nonlinear optical processes from the viewpoint of the quantum theory of radiation; in the meantime, we will also use the semiclassical approach of nonlinear polarization to present those formulae widely adopted in nonlinear optics. Only in some individual cases, such as stimulated Raman scattering and stimulated Kerr scattering, the mathematic treatment of quantum theory of radiation will be presented to give a rigorous quantitative description.

Since the beginning of this new century, the term "photonics" has more often appeared in the scientific literature and books. Although there is no standard definition of this term, it may be acceptable that the term *photonics* connotes the development of modern optics with emphasizing the photon nature of light radiation and laser applications in various scientific and technological areas.

To date, a considerable number of books on nonlinear optics have been published, which contain different features and may concentrate on different major issues.[15-49] Nevertheless, most of these books emphasized the three- and four-wave frequency-mixing effects as well as some nonlinear effects related to the refractive-index change, while only presenting the semiclassical theory of nonlinear polarization.

By contrast, this book covers many more subjects, including major topics that have been active only in recent decades, presents the descriptions from two different theoretical viewpoints, and emphasizes those nonlinear optical techniques of great potential for applications. For all these reasons, the new book is entitled *Nonlinear Optics and Photonics*.

PROBLEMS

1. For a quasi-parallel light beam from an ordinary light source, if the beam power is 100 mW, beam diameter is 1 cm, plane divergence angle is 0.1 radian, central wavelength is ~550 nm with a spectral bandwidth of ~50 nm, please calculate the corresponding intensity, spectral intensity, brightness, spectral brightness, and photon degeneracy, respectively.
2. For a laser beam of 550 nm wavelength from a laser source, if the beam power is 100 mW, beam diameter is 1 cm, plane divergence angle is 0.0005 radian, spectral bandwidth is ~0.1 nm, please repeat the calculations mentioned above.
3. For a pulsed laser beam from a solid laser source, if the pulse duration is 1 ns, pulse energy is 10 mJ, and the diameter of the focused beam spot is 50 μm, please calculate the focused light intensity I and the corresponding local electric field strength E, then compare the latter with the internal field of the hydrogen atom (cf. Appendix 2).
4. What is the basic difference between the semiclassical theory and the quantum theory of radiation?
5. What is the difference between a real excited state of the molecule and an intermediate state of the combined system including the molecule and photon field?

REFERENCES

1. A. L. Schawlow and C. H. Townes, *Phys. Rev.* **112**, 1940 (1958).
2. T. H. Maiman, *Nature* **187**, 493 (1960); T. H. Maiman, *Br. Commun. Electron.* **7**, 674 (1960).
3. P. A. Franken, A. E. Hill, C. W. Peters, and G. Weinreich, *Phys. Rev. Lett.* **7**, 118 (1961).

4. J. A. Armstrong, N. Bloembergen, J. Ducuing, and P. S. Pershan, *Phys. Rev.* **127**, 1918 (1962).
5. R. Y. Chiao, E. Garmire, and C. H. Townes, *Phys. Rev. Lett.* **13**, 479 (1964); erratum, **14**, 1056 (1965).
6. G. Eckhardt, R. W. Hellwarth, F. J. McClung, S. E. Schwarz, D. Weiner, and E. J. Woodbury, *Phys. Rev. Lett.* **9**, 455 (1962).
7. R. Y. Chiao, C. H. Townes, and B. P. Stoicheff, *Phys. Rev. Lett.* **12**, 592 (1964).
8. M. Göppert-Mayer, *Ann. Phys.* **401** (3), 273 (1931).
9. W. Kaiser and C. G. B. Garrett, *Phys. Rev. Lett.* **7**, 229 (1961).
10. H. Hertz, *Ann. Phys.* **267** (8), 983 (1887).
11. A Einstein, *Ann. Phys*, **322** (6), 132 (1905).
12. E. Fermi, *Rev. Mod. Phys.* **4**, 87 (1932).
13. W. Heitler, *The Quantum Theory of Radiation*, 3rd ed. (Oxford University Press, London, 1954).
14. G. S. He and S. H. Liu, *Physics of Nonlinear Optics* (World Scientific, Singapore, 1999).
15. N. Bloembergen, *Nonlinear Optics*, 4th ed. (World Scientific, Singapore, 1996).
16. P. N. Butcher, *Nonlinear Optical Phenomenon* (Ohio State University Press, Columbus, OH, 1965).
17. G. C. Baldwin, *An Introduction to Nonlinear Optics*, 2nd ed. (Plenum Press, New York, 1974).
18. S. A. Akhmanov and R. V. Khokhlov, *Problem of Nonlinear Optics* (Gordon & Breach, New York, 1972).
19. F. Zernike and J. E. Midwinter, *Applied Nonlinear Optics* (Wiley, New York, 1973).
20. V. S. Letokhov and V. P. Chebotayev, *Nonlinear Laser Spectroscopy* (Springer-Verlag, Berlin, 1977).
21. D. C. Hanna, M. A. Yuratich, and D. Cotter, *Nonlinear Optics of Free Atoms and Molecules* (Springer-Verlag, Berlin, 1979).
22. Y. R. Shen, *The Principles of Nonlinear Optics* (Wiley, New York, 1984).
23. J. F. Reintjes, *Nonlinear Optical Parametric Processes in Liquid and Gases* (Academic Press, New York, 1984).
24. B. Ya. Zel'dovich, N. F. Pilipetsky, and V. V. Shkunov, *Principles of Phase Conjugation* (Springer-Verlag, Berlin, 1985).
25. M. Schubert and B. Wilhelmi, *Nonlinear Optics and Quantum Electronics* (Wiley, New York, 1986).
26. M. D. Levenson and S. Kano, *Introduction to Nonlinear Laser spectroscopy* (Academic Press, Boston, 1988).
27. N. B. Delone, *Fundamentals of Nonlinear Optics of Atomic Gases* (Wiley, New York, 1988).
28. D. N. Klyshko, *Photonics and Nonlinear Optics* (Gordon & Breach, New York, 1988).
29. G. P. Agrawal, *Nonlinear Fiber Optics*, 4th ed. (Academic Press, New York, 2007).
30. V. S. Butylkin, A. E. Kaplan, Yu. G. Khronopulo, and E. I. Yakuborich, *Resonant Nonlinear Interactions of Light with Matter* (Springer-Verlag, New York, 1989).
31. P. N. Butcher and D. Cotter, *The Elements of Nonlinear Optics* (Cambridge University Press, New York, 1990).
32. P. N. Prasad and D. J. Williams, *Introduction to Nonlinear Optical Effects in Molecules and Polymers* (Wiley, New York, 1991).
33. R. W. Boyd, *Nonlinear Optics*, 3rd ed. (Academic Press, New York, 2008).
34. P. Yeh, *Introduction to Photorefractive Nonlinear Optics* (Wiley, New York, 1993).
35. V. P. Bespalov and G. A. Pasmanik, *Nonlinear Optics and Adaptive Laser Systems* (Nova Science, New York, 1993).
36. I. C. Khoo, *Liquid Crystals: Physical Properties and Nonlinear Optical Phenomena* (Wiley, New York, 1994).
37. C. L. Tang and L. K. Cheng, *Fundamentals of Optical Parametric Processes and Oscillators* (Harwood Academic, Amsterdam, 1995).
38. E. G. Sauter, *Nonlinear Optics* (Wiley, New York, 1996).
39. W. Demtroder, *Laser Spectroscopy: Basic Concepts and Instrumentation* (Springer-Verlag, Berlin, 1996).
40. D. L. Mills, *Nonlinear Optics: Basic Concepts* (Springer, 1998).

41. A I. Maimistov and A. M. Basharov, *Nonlinear Optical Waves* (Springer, 1999).
42. D. L. Andrews and P. Allcock, *Optical Harmonics in Molecular Systems: Quantum Electrodynamical Theory* (Wiley-VCH, Weinheim, 2002).
43. Y S. Kivshar and G. P. Agrawal, *Optical Solitons* (Academic, 2003).
44. R. L. Sutherland, *Handbook of Nonlinear Optics*, 2nd ed. (Marcel Dekker, New York, 2003).
45. J. V. Moloney and A. C. Newell, *Nonlinear Optics* (Westview Press, New York, 2004).
46. P. Mandel, *Nonlinear Optics* (Wiley-VCH. Weinheim, 2010).
47. G. New, *Introduction to Nonlinear Optics* (Cambridge University Press, New York, 2011).
48. P. E. Powers, *Fundamentals of Nonlinear Optics* (CRC Press, New York, 2011).
49. G. E. Stegeman and R. A. Stegeman, *Nonlinear Optics: Phenomena, Materials and Devices* (Wiley, New York, 2012).

2
Nonlinear Polarization of an Optical Medium

A considerable number of nonlinear optical effects, including three- and four-wave frequency mixing, intense light induced refractive-index changes, self-focusing, self-phase modulation, and optical phase-conjugation, can be conveniently and quantitatively described by using the nonlinear (electric) polarization theory within the framework of the semiclassical regime.

In nonlinear polarization theory, the key issue is to establish the relationship between the optical electromagnetic fields and the induced electric polarization of the medium, and more specifically is to specify the nonlinear relations between the Fourier components of the induced polarization field and the corresponding Fourier components of the interacting optical fields. The connection between them is the nonlinear susceptibilities of the medium, which characterize the nonlinearity of the medium and are the functions of the various possible frequency combinations of the involved Fourier components of the optical fields. For this purpose, we need to introduce the definition and tensor formulation of various orders of nonlinear susceptibilities of the medium and to derive the nonlinear coupled wave equations, which can be used to describe various coherent wave-mixing processes.[1–10]

2.1 Optical field-induced electric polarization in a medium

Under the interaction of an applied optical field, the atoms or molecules of the medium may respond in the following two possible ways: (i) a real transition of a certain amount of atoms or molecules from one of their eigenstates to another, and (ii) a perturbation or distortion reaction on the normal distribution or motion of the internal electric charges within an atoms or molecule. For resonant interaction, the first type of response may become predominant owing to one- or two-photon absorption when the frequency of the applied optical field exactly matches the corresponding absorptive frequency. For near-resonant or nonresonant interaction, wherein the optical field frequency is only close to or far from the absorptive frequency and the number of atoms or molecules that make a real quantum transition is negligible, the second type of response of the medium may become dominant. In this chapter, we shall mainly describe the basic behavior of the second type of response of the medium.

As mentioned above, the applied optical field may cause the distortion on the internal distribution or motion of the electric charges (electrons, nuclei, or ions) within a molecular system, in comparison to the normal situation in which the external optical field is absent. Such interaction between the optical field and the medium leads to a field-induced electric dipole moment that in turn will act as a new source to emit a secondary electromagnetic wave. This is the fundamental physical picture given by the semiclassical theory to explain the secondary-wave emission and the refractive-index behavior of the optical medium.

In order to consider the macroscopic behavior of a given medium, we need to introduce a macroscopic physical quantity, the electric polarization vector **P**, which is defined as the sum of the field-induced molecular electric dipole moment vector per unit volume in the medium. Assuming that the number of molecules in a unit volume is N and the induced dipole moment of the ith molecule is $\mathbf{p}_i$, then, **P** is determined by

$$\mathbf{P}(t) = \sum_{i=1}^{N} \mathbf{p}_i(t). \tag{2.1-1}$$

Since the applied field is an electromagnetic wave varying at optical frequency, $\mathbf{p}_i$ and **P** are also functions of time.

From Eq. (2.1-1), one can see that the electric polarization of a medium is determined by two factors: one is the field-induced dipole moment of each individual molecule of the medium, and the other is the statistically averaged property of a great number of molecules. For a given applied field, the induced molecular dipole moment is determined by the microscopic structure of molecules or their quantum eigenfunctions; whereas the result of summation of $\mathbf{p}_i$ depends on the macroscopic symmetry of the medium or the average property of an ensemble of molecules.

Since the electric polarization response of a medium is induced by an applied optical field, a certain relation between the electric polarization $\mathbf{P}(t)$ and the applied optical electric field strength $\mathbf{E}(t)$ may be established. As an arbitrary function of time, $\mathbf{P}(t)$ can be generally expressed by the following series:

$$\mathbf{P}(t) = \mathbf{P}^{(1)}(t) + \mathbf{P}^{(2)}(t) + \mathbf{P}^{(3)}(t) + \cdots + \mathbf{P}^{(n)}(t) + \cdots. \tag{2.1-2}$$

Here, it is assumed that the nth term in the above expression is related to the nth power of the field function $\mathbf{E}(t)$ and can be expressed in the form of a multiple integral[5,9]

$$\mathbf{P}^{(n)}(t) = \varepsilon_0 \int_{-\infty}^{\infty} dt_1 \int_{-\infty}^{\infty} dt_2 \cdots \int_{-\infty}^{\infty} dt_n R^{(n)}(t; t_1, t_2 \cdots, t_n) \mathbf{E}(t_1) \mathbf{E}(t_2) \cdots \mathbf{E}(t_n), \tag{2.1-3}$$

where $R^{(n)}(t; t_1, t_2 \cdots, t_n)$ is called the nth-order polarization response function of the medium and is an $(n + 1)$th-rank tensor. Furthermore, if we consider that the response behavior of the medium should not depend on choosing the specific time t, Eq. (2.1-3) can be simplified as

$$\mathbf{P}^{(n)}(t) = \varepsilon_0 \int_{-\infty}^{\infty} dt_1 \int_{-\infty}^{\infty} dt_2 \cdots \int_{-\infty}^{\infty} dt_n R^{(n)}(t_1, t_2 \cdots, t_n) \mathbf{E}(t-t_1) \mathbf{E}(t-t_2) \cdots \mathbf{E}(t-t_n). \tag{2.1-4}$$

Since the polarization of the medium is induced by the applied optical field, it is required by the causality condition that $R^{(n)}$ must be zero when any temporal variable among t_1 to t_n takes a negative value. In addition, **E** and **P** are both measurable and *real* quantities, therefore, $R^{(n)}$ must also be a real function.

So far we have given general and phenomenological relations between $\mathbf{E}(t)$ and $\mathbf{P}(t)$ that are arbitrary functions of time t. According to the principle of Fourier transform, any functions of

20 Nonlinear Polarization of an Optical Medium

time t can be generally expressed in terms of Fourier integrals. Thus we can present $\mathbf{E}(t)$ and $\mathbf{P}^{(n)}(t)$ in the form:

$$\left.\begin{array}{l} \mathbf{E}(t) = \displaystyle\int_{-\infty}^{\infty} d\omega \mathbf{E}(\omega) e^{-i\omega t}, \\[1em] \mathbf{P}^{(n)}(t) = \displaystyle\int_{-\infty}^{\infty} d\omega \mathbf{P}^{(n)}(\omega) e^{-i\omega t} \end{array}\right\}. \tag{2.1-5}$$

Here $\mathbf{E}(\omega)$ and $\mathbf{P}^{(n)}(\omega)$ are the monochromatic Fourier components of $\mathbf{E}(t)$ and $\mathbf{P}^{(n)}(t)$, respectively. Moreover, we shall establish a general relation between the Fourier component $\mathbf{P}^{(n)}(\omega)$ of the nth-order polarization $\mathbf{P}^{(n)}(t)$ and the Fourier component $\mathbf{E}(\omega)$ of the applied field $\mathbf{E}(t)$. For this purpose, substituting the Fourier integral of $\mathbf{E}(t)$ in Eq. (2.1-5) into Eq. (2.1-4) leads to

$$\mathbf{P}^{(n)}(t) = \varepsilon_0 \int_{-\infty}^{\infty} d\omega_1 \cdots \int_{-\infty}^{\infty} d\omega_n \chi^{(n)}(\omega_1, \cdots, \omega_n) \mathbf{E}(\omega_1) \cdots \mathbf{E}(\omega_n) \exp\left[-it \sum_{m=1}^{n} \omega_m\right], \tag{2.1-6}$$

where

$$\chi^{(n)}(\omega_1, \cdots \omega_n) = \int_{-\infty}^{\infty} dt_1 \cdots \int_{-\infty}^{\infty} dt_n \cdot R^{(n)}(t_1, \cdots, t_n) \cdot \exp\left[i \sum_{m=1}^{n} \omega_m t_m\right]. \tag{2.1-7}$$

Here $\chi^{(n)}(\omega_1, \cdots, \omega_n)$ is called the nth-order *susceptibility* of the medium and it is a $(n+1)$th-rank tensor. It is essential to introduce this physical quantity through which we can establish a simple relation between the nth-order Fourier-component $\mathbf{P}^{(n)}(\omega)$ and the related Fourier-components $\mathbf{E}(\omega)$ of the optical field. To explain this feature, we can take the second-order susceptibility $\chi^{(2)}$ as an example.

According to Eq. (2.1-6) the second-order temporal component $\mathbf{P}^{(2)}(t)$ can be written as

$$\mathbf{P}^{(2)}(t) = \varepsilon_0 \int_{-\infty}^{\infty} d\omega_1 \int_{-\infty}^{\infty} d\omega_2 \chi^{(2)}(\omega_1, \omega_2) \mathbf{E}(\omega_1) \mathbf{E}(\omega_2) e^{-i(\omega_1 + \omega_2)t}. \tag{2.1-8}$$

Taking a transform of $\omega_1 + \omega_2 \Rightarrow \omega$ in the above equation leads to

$$\mathbf{P}^{(2)}(t) = \varepsilon_0 \int_{-\infty}^{\infty} d\omega \int_{-\infty}^{\infty} d\omega_2 \chi^{(2)}(\omega - \omega_2, \omega_2) \mathbf{E}(\omega - \omega_2) \mathbf{E}(\omega_2) e^{-i\omega t}. \tag{2.1-9}$$

In reality, $\mathbf{E}(\omega_2)$ represents the amplitude function of a quasi-monochromatic coherent optical field of frequency ω_2 with a narrow spectral linewidth, thus $\mathbf{E}(\omega_2)$ can be expressed in the form of a δ-function:

$$\mathbf{E}(\omega_2) = \mathbf{E}(\omega')\delta(\omega_2 - \omega'). \tag{2.1-10}$$

Substituting Eq. (2.1-10) into Eq. (2.1-9) and completing the integration over ω_2 leads to

$$\mathbf{P}^{(2)}(t) = \varepsilon_0 \int_{-\infty}^{\infty} d\omega \chi^{(2)}(\omega - \omega', \omega') \mathbf{E}(\omega - \omega') \mathbf{E}(\omega') e^{-i\omega t}. \quad (2.1\text{-}11)$$

On the other hand, according to the second equality of Eq. (2.1-5), $\mathbf{P}^{(2)}(t)$ can also be expressed in the form of a Fourier integral

$$\mathbf{P}^{(2)}(t) = \int_{-\infty}^{\infty} d\omega \mathbf{P}^{(2)}(\omega) e^{-i\omega t}. \quad (2.1\text{-}12)$$

Comparing Eq. (2.1-12) to Eq. (2.1-11), we obtain

$$\mathbf{P}^{(2)}(\omega) = \varepsilon_0 \chi^{(2)}(\omega - \omega', \omega') \mathbf{E}(\omega - \omega') \mathbf{E}(\omega'). \quad (2.1\text{-}13)$$

Letting $\omega - \omega' = \omega_1$, $\omega' = \omega_2$, and $\omega = \omega_1 + \omega_2$, we can finally establish a simple relation between the Fourier component of the second-order polarization and the corresponding monochromatic components of the optical field, i.e.,

$$\mathbf{P}^{(2)}(\omega = \omega_1 + \omega_2) = \varepsilon_0 \chi^{(2)}(\omega_1, \omega_2) \mathbf{E}(\omega_1) \mathbf{E}(\omega_2), \quad (2.1\text{-}14)$$

where $\chi^{(2)}$ denotes the second-order susceptibility that is a third-rank tensor.

Following a similar procedure, the general form of the Fourier component of the nth-order polarization can be obtained as

$$\mathbf{P}^{(n)}(\omega) = \varepsilon_0 \chi^{(n)}(\omega_1, \omega_2, \cdots, \omega_n) \mathbf{E}(\omega_1) \mathbf{E}(\omega_2) \cdots \mathbf{E}(\omega_n). \quad (2.1\text{-}15)$$

Here, the following algebraic summation of frequencies is fulfilled:

$$\omega = \omega_1 + \omega_2 + \cdots + \omega_n = \sum_{m=1}^{n} \omega_m. \quad (2.1\text{-}16)$$

In principle, we can consider the relationship between $\mathbf{P}^{(n)}(t)$ and $\mathbf{E}(t)$ as being arbitrary functions in the time domain; the connection is the polarization response function $R^{(n)}$ of the medium (see Eq. (2.1-3)). On the other hand, we can also consider the relationship between $\mathbf{P}^{(n)}(\omega)$ and $\mathbf{E}(\omega)$ as being the Fourier components of the polarization and applied field in the frequency domain; the connection in this case is the susceptibility function $\chi^{(n)}$ of the medium (see Eq. (2.1-15)). Although $\chi^{(n)}$ and $R^{(n)}$ are related by Eq. (2.1-7), $R^{(n)}$ in general cannot be theoretically determined by using a macroscopic or phenomenological method. However, the values and properties of $\chi^{(n)}$ can be theoretically determined by knowing the microscopic response for each individual molecule as well as the statistical feature of a large number of molecules. The details of the theoretical determination of $\chi^{(n)}$ will be given in Chapter 18 within the framework of semiclassical theory.

In practice, it is very convenient to use the nonlinear susceptibility, $\chi^{(n)}$, which is independent of time, to describe many nonlinear optical processes. In these cases, the real input optical waves acting on the medium are usually from various laser sources and can be recognized as ideal monochromatic Fourier components of $\mathbf{E}(\omega_n)$ with different frequencies of ω_n. For this reason, the Fourier components of the nonlinear polarization $\mathbf{P}^{(n)}(\omega)$ can be easily determined via $\chi^{(n)}$ and $\mathbf{E}(\omega_n)$, without involving any time factor.

2.2 Various mechanisms causing nonlinear polarization in a medium

It is already known that there are various physical mechanisms that may cause nonlinear polarization responses in the medium.

(1) *Electron-cloud distortion*: This denotes the applied optical field-induced perturbation change of the outer-shell electron cloud of an atom, ion, or molecule, in comparison to the undisturbed status without applying an external field. This mechanism features a very fast response time approximately less than 10^{-15}–10^{-16} s.

(2) *Intramolecular motion*: This denotes the polarization contribution from an optical field-induced relative motion (vibration, rotation, etc.) between the nuclei (or ions) within a molecule. The response time of this mechanism is about 10^{-12}–10^{-14} s.

(3) *Molecular reorientation*: This denotes the additional electric polarization contribution from an optical field-induced reorientation of anisotropic molecules in a liquid. The response time of this process is dependent on the rotational viscosity of molecules in the liquid and is about 10^{-12}–10^{-13} s.

(4) *Induced acoustic motion*: This denotes the polarization contribution from an optical field-induced macroscopic acoustic motion related to the so-called optical electrostriction interaction. The response time of this mechanism is $\sim 10^{-9}$–10^{-10} s depending on the phase state of the medium (solid, liquid, or gas).

(5) *Induced population change*: The distribution of molecules in their different eigenstates may be changed owing to one-photon or two-photon absorption or Raman transition for resonant interaction. That may give rise to an additional contribution to the polarization of the medium. The response time of this contribution is strongly dependent on the dynamic properties of molecular transition and relaxation for a given medium and, in general, is slower than that of mechanisms (1) (2), and (3).

In different conditions of the applied optical field or for different media, the relative contributions from various mechanisms can be significantly different. It is essentially important to determine which mechanism is dominant compared to others for a specific case. Among various possible mechanisms mentioned above, only the first mechanism (electron-cloud distortion) can contribute to any kinds of nonlinear optical processes; whereas the contributions from other mechanisms are conditional. For instance, the second mechanism (intramolecular motion) can give a contribution only for a medium composed of molecules (not atoms); the third mechanism (molecular reorientation) is applicable only for a liquid composed of anisotropic molecules; the contribution from the mechanism of population change should be considered only for strong resonant interaction, and so on. Another criterion is that if the duration of the incident pulsed laser field is considerably shorter than the response time of a considered mechanism, the contribution from this mechanism can be excluded. For most optical frequency-mixing effects, such as second-harmonic and third-harmonic generation, sum-frequency generation, optical parametric oscillation, and four-photon parametric interaction, the electron-cloud distortion is the only mechanism in consideration because only this response is fast enough to follow the variation at optical frequency. If the duration of the incident pulsed laser field is considerably longer than the period of intramolecular motion, the contribution from the second mechanism may become important. For example, some Raman resonance-enhanced four-wave-mixing effects, including the CARS (coherent anti-Stokes Raman spectroscopy) effect and Raman-enhanced refractive index change, are really based on this mechanism. If the duration of the applied laser pulsed field is much longer than the response time of molecular reorientation for a given liquid composed of anisotropic molecules, the third mechanism may become the major origin for nonlinear effects such as the stimulated Kerr scattering and the orientational Kerr effect related refractive index change. Finally, if the duration of the incident optical field is longer than the period of the induced hypersonic sounds, the fourth mechanism may become a major contribution for nonlinear optical effects such as stimulated Brillouin scattering, self-focusing, and optical breakdown damage.

In summary, for a given investigated nonlinear optical process or phenomenon, it often is the primary task to determine what is the major mechanism dominating the relevant process.

2.3 Manipulation of nonlinear susceptibilities

According to Eq. (2.1-15), the Fourier component of the linear polarization for a medium is

$$\mathbf{P}^{(1)}(\omega) = \varepsilon_0 \chi^{(1)}(\omega) \mathbf{E}(\omega). \tag{2.3-1}$$

This relation means that in the first-order approximation, the induced electric polarization of the medium at a given frequency is linearly proportional to the input optical field of the same frequency. In other words, a monochromatic incident light wave of frequency ω can only induce a secondary wave radiation at the same frequency ω. Based on Eq. (2.3-1) the propagation properties of an ordinary light beam, such as reflection, refraction, diffraction, birefringence, scattering, as well as dispersion of refractive index can be explained very well. For an anisotropic medium, $\chi^{(1)}(\omega)$ is a second-rank tensor having $3 \times 3 = 9$ elements. Based on the matrix form of a second-rank tensor, Eq. (2.3-1) can be expressed as

$$\begin{bmatrix} P_x^{(1)}(\omega) \\ P_y^{(1)}(\omega) \\ P_z^{(1)}(\omega) \end{bmatrix} = \varepsilon_0 \begin{bmatrix} \chi_{xx}^{(1)}(\omega) & \chi_{xy}^{(1)}(\omega) & \chi_{xz}^{(1)}(\omega) \\ \chi_{yx}^{(1)}(\omega) & \chi_{yy}^{(1)}(\omega) & \chi_{yz}^{(1)}(\omega) \\ \chi_{zx}^{(1)}(\omega) & \chi_{zy}^{(1)}(\omega) & \chi_{zz}^{(1)}(\omega) \end{bmatrix} \begin{bmatrix} E_x(\omega) \\ E_y(\omega) \\ E_z(\omega) \end{bmatrix}. \tag{2.3-2}$$

Here, $P_x^{(1)}(\omega)$, $E_x(\omega)$, etc., are the components of the vectors $\mathbf{P}^{(1)}(\omega)$ and $\mathbf{E}(\omega)$ in Cartesian coordinates, respectively. Moreover, Eq. (2.3-2) can also be equivalently rewritten in a summation form

$$P_i^{(1)}(\omega) = \varepsilon_0 \sum_j \chi_{ij}^{(1)}(\omega) E_j(\omega). \quad (i,j = x, y, z) \tag{2.3-3}$$

In Appendix 3, the non-vanishing components of $\chi^{(1)}$ are given for various media with different symmetry properties. In particular, for an isotropic medium, there is only one nonzero element, and $\chi^{(1)}$ becomes a scalar coefficient.

24 Nonlinear Polarization of an Optical Medium

In a similar manner, according to Eq. (2.1-15) the Fourier component of the second-order nonlinear polarization can be written as

$$\mathbf{P}^{(2)}(\omega = \omega_1 + \omega_2) = \varepsilon_0 \chi^{(2)}(\omega_1, \omega_2)\mathbf{E}(\omega_1)\mathbf{E}(\omega_2). \tag{2.3-4}$$

The physical meaning of this expression is that in the second-order approximation, the radiation at a new frequency of $\omega = \omega_1 + \omega_2$ can be induced by two incident monochromatic waves with frequencies of ω_1 and ω_2. The second-order susceptibility $\chi^{(2)}$ is a third-rank tensor having $3 \times 9 = 27$ elements. Based on the matrix form of a third-rank tensor, Eq. (2.3-4) can be expressed as

$$\begin{bmatrix} P_x^{(2)}(\omega) \\ P_y^{(2)}(\omega) \\ P_z^{(2)}(\omega) \end{bmatrix} = \varepsilon_0 \begin{bmatrix} \chi_{xxx}^{(2)} & \chi_{xxy}^{(2)} & \cdots & \chi_{xzz}^{(2)} \\ \chi_{yxx}^{(2)} & \chi_{yxy}^{(2)} & \cdots & \chi_{yzz}^{(2)} \\ \chi_{zxx}^{(2)} & \chi_{zxy}^{(2)} & \cdots & \chi_{zzz}^{(2)} \end{bmatrix} \begin{bmatrix} E_x(\omega_1)E_x(\omega_2) \\ E_x(\omega_1)E_y(\omega_2) \\ E_x(\omega_1)E_z(\omega_2) \\ E_y(\omega_1)E_x(\omega_2) \\ E_y(\omega_1)E_y(\omega_2) \\ E_y(\omega_1)E_z(\omega_2) \\ E_z(\omega_1)E_x(\omega_2) \\ E_z(\omega_1)E_y(\omega_2) \\ E_z(\omega_1)E_z(\omega_2) \end{bmatrix}. \tag{2.3-5}$$

Also, the Cartesian components of $\mathbf{P}^{(2)}(\omega)$ can be expressed in the following form

$$P_i^{(2)}(\omega = \omega_1 + \omega_2) = \varepsilon_0 \sum_{jk} \chi_{ijk}^{(2)}(\omega_1, \omega_2) E_j(\omega_1) E_k(\omega_2). \qquad (i,j,k = x,y,z) \tag{2.3-6}$$

Finally, based on Eq. (2.1-15) the Fourier component of the third-order nonlinear polarization can be expressed as

$$\mathbf{P}^{(3)}(\omega = \omega_1 + \omega_2 + \omega_2) = \varepsilon_0 \chi^{(3)}(\omega_1, \omega_2, \omega_3)\mathbf{E}(\omega_1)\mathbf{E}(\omega_2)\mathbf{E}(\omega_3). \tag{2.3-7}$$

The physical meaning of the above expression is that in the third-order approximation, the radiation at a new frequency of $\omega = \omega_1 + \omega_2 + \omega_3$ can be generated by intense optical fields containing three frequency components of ω_1, ω_2, and ω_3. The third-order susceptibility $\chi^{(3)}$ is a fourth-rank tensor having $3 \times 27 = 81$ elements. In its matrix form, Eq. (2.3-7) can be written as

$$\begin{bmatrix} P_x^{(3)}(\omega) \\ P_y^{(3)}(\omega) \\ P_z^{(3)}(\omega) \end{bmatrix} = \varepsilon_0 \begin{bmatrix} \chi_{xxxx}^{(3)} & \cdots & \chi_{xzzz}^{(3)} \\ \chi_{yxxx}^{(3)} & \cdots & \chi_{yzzz}^{(3)} \\ \chi_{zxxx}^{(3)} & \cdots & \chi_{zzzz}^{(3)} \end{bmatrix} \begin{bmatrix} E_x(\omega_1)E_x(\omega_2)E_x(\omega_3) \\ E_x(\omega_1)E_x(\omega_2)E_y(\omega_3) \\ \cdots \\ \cdots \\ \cdots \\ E_z(\omega_1)E_z(\omega_2)E_y(\omega_3) \\ E_z(\omega_1)E_z(\omega_2)E_z(\omega_3) \end{bmatrix}. \tag{2.3-8}$$

Equation (2.3-8) can also be rewritten in the following summation form

$$P_i^{(3)}(\omega) = \varepsilon_0 \sum_{jkl} \chi_{ijkl}^{(3)}(\omega_1, \omega_2, \omega_3) E_j(\omega_1) E_k(\omega_2) E_l(\omega_3). \qquad (i,j,k,l = x,y,z) \tag{2.3-9}$$

From Eqs. (2.3-4)–(2.3-9) one can see that the various-order nonlinear susceptibilities of a medium are the key parameters to describe the nonlinear coupling between different incident waves, and to interpret the generation of coherent light emission with new frequencies.

2.4 Basic properties of nonlinear susceptibilities

In principle, one could first consider the induced electric dipole moment of a given molecular system, and then determine the macroscopic polarization of the medium based on a statistical average over a large number of molecules. In practice, however, such a task is extremely difficult because of the complexity of theoretical calculations as well as the lack of the specific knowledge about the intermolecular interaction and the constitutional structure of a given medium. For this reason, it is important to find out some common features of the nonlinear susceptibilities, based on which we could simplify the mathematical treatments and to extract some useful conclusions.

Based on the theoretical analyses described in Chapter 18, it can be proved that various orders of susceptibilities of a medium possess the following basic properties.

2.4.1 Relative magnitude of various orders of susceptibilities

Assuming that the major mechanism contributing to the nonlinear polarization is the electron-cloud distortion and there is a nonresonant interaction between the optical field and the medium, the relative ratio between the successive orders of susceptibilities can be roughly estimated as

$$|\chi^{(n)}|/|\chi^{(n-1)}| \approx 1/|E_0|, \qquad (2.4\text{-}1)$$

where $|E_0|$ is the magnitude of average electric field strength inside an atom (it is about 10^{11} V/m for a hydrogen atom). According to Eq. (2.1-15), the ratio between the successive orders of polarization can be roughly estimated by

$$|P^{(n)}|/|P^{(n-1)}| \approx |E|\,|\chi^{(n)}|/|\chi^{(n-1)}| \approx |E|/|E_0|. \qquad (2.4\text{-}2)$$

Here, $|E|$ is the magnitude of an applied optical field. For the incident light field provided by ordinary light sources, the ratio of $|E|/|E_0|$ is so small that all nonlinear polarization terms can be neglected. However, if the applied field is a laser radiation of high spectral brightness, the ratio of $|E|/|E_0|$ can no longer be completely neglected (see Appendix 2), the second-order and third-order (or even higher order) nonlinear polarization contributions may play vital roles. This is one of the basic reasons why various nonlinear optical effects can only be observed by using laser radiation. Here, we have considered only the contribution from the mechanism of electron-cloud distortion. If we also consider the additional contributions from other mechanisms mentioned in Section 2.2, the magnitude of the effective nonlinear (mainly the third-order) susceptibility can be further increased. That is the reason why some third-order nonlinear effects (such as CARS, self-focusing, optical phase conjugation, and optical bistability) could be observed even in a modest laser intensity level.

2.4.2 Spatial-symmetry restrictions on susceptibilities

The susceptibility tensors for a given medium must remain unchanged upon the symmetric operation allowed for this medium. This requirement reduces the number of the independent and nonzero elements of the susceptibility tensors. In respect to this matter, the most important conclusion is that for all centrosymmetric crystals and isotropic media (gases, liquids, and amorphous solids), all tensor elements of even-order susceptibilities ($\chi^{(2)}$, $\chi^{(4)}$, ...) must be zero in the so-called electric-dipole moment approximation. Consequently, no second-order nonlinear optical effects can be observed in these sorts of materials in the same approximation. In contrast, the odd-order susceptibility tensors ($\chi^{(1)}$, $\chi^{(3)}$, $\chi^{(5)}$, ...) will not vanish for any media. The independent and nonzero components of the second-order susceptibility for noncentrosymmetric crystals are given in Appendix 4; the independent nonzero components of the third-order susceptibility for different point-group crystals and isotropic media are given in Appendix 6.

2.4.3 Resonance enhancement of susceptibilities

It can be proved by the theories given in Chapter 18 that, when one or a combination of several frequency components of the applied optical field approaches a resonant frequency of the medium, the magnitude of the tensor elements for appropriate orders of susceptibilities can be significantly increased. The resonance mechanisms can be one-photon or two-photon absorption resonance, Raman resonance, Brillouin resonance, and so on. On the other hand, sometimes an exact resonance may lead to a rapid depletion of either the incident optical pump wave or the newly generated coherent wave. In practice, one has to find a compromise between the enhancement of nonlinear susceptibility and the attenuation of the useful optical wave. For this reason, a quasi-resonance (or near-resonance) condition is often employed by tuning the frequency of the incident laser field to be close (but not equal) to the resonant frequency of the medium.

2.4.4 Permutation symmetry of susceptibilities

From Eqs. (2.3-4), (2.3-6), (2.3-7), and (2.3-9) one can see that the elements of various-order nonlinear susceptibilities are functions of their Cartesian subscripts (i, j, k, ...) and frequencies of the coupled light waves (ω_1, ω_2, ω_3, ...). It can be proved that in general (including resonant interactions), the following permutation-symmetry relations among the tensor elements for second-order and third-order susceptibilities hold:

$$\chi^{(2)}_{ijk}(\omega_1, \omega_2) = \chi^{(2)}_{ikj}(\omega_2, \omega_1), \tag{2.4-3}$$

and

$$\chi^{(3)}_{ijkl}(\omega_1, \omega_2, \omega_3) = \chi^{(3)}_{ikjl}(\omega_2, \omega_1, \omega_3) = \chi^{(3)}_{iljk}(\omega_3, \omega_1, \omega_2) = \cdots. \tag{2.4-4}$$

The above two relations imply that the elements of nonlinear susceptibility tensors remain unchanged for all simultaneous interchanges of positions of the subscripts (j, k, l, ...) and the corresponding frequency arguments (ω_1, ω_2, ω_3, ...). This is the so-called intrinsic permutation symmetry of the nonlinear susceptibility tensors.

In the case of non-resonant interaction, the range of the permutation symmetry can be further expanded. To show this we can formally introduce a new frequency argument $\omega' = -(\omega_1 + \omega_2 + \cdots)$ as an additional parameter of the susceptibility elements, then in addition to Eqs. (2.4-3) and (2.4-4) the following relations are valid:

$$\chi^{(2)}_{ijk}[\omega' = -(\omega_1 + \omega_2); \omega_1, \omega_2] = \chi^{(2)}_{jik}(\omega_1; \omega', \omega_2) = \chi^{(2)}_{kji}(\omega_2; \omega_1, \omega'), \tag{2.4-5}$$

and

$$\chi^{(3)}_{ijkl}[\omega' = -(\omega_1 + \omega_2 + \omega_3); \omega_1, \omega_2, \omega_3] = \chi^{(3)}_{jikl}(\omega_1; \omega', \omega_2, \omega_3) = \cdots. \tag{2.4-6}$$

These two relations plus Eqs. (2.4-3) and (2.4-4) imply that the elements of the nonlinear susceptibility tensors are invariant for arbitrary simultaneous interchanges of the all subscripts (i, j, k, l, ...) and corresponding frequency arguments (ω', ω_1, ω_2, ω_3, ...). This property is called the overall permutation symmetry of the nonlinear susceptibility tensors, which holds only for non-resonant interaction.

2.4.5 Complex conjugation and time-reversal symmetry of susceptibilities

It can be proved that various orders of susceptibilities are complex quantities in general, so that they can be expressed by two parts: the real part and the imaginary part. From the general phenomenological definition of the nth-order susceptibility given by Eq. (2.1-7) one can find that, since the polarization response function $R^{(n)}$ must be a real quantity, the manipulation of complex conjugation for $\chi^{(n)}$ leads to

$$[\chi^{(n)}(\omega_1, \omega_2, \cdots, \omega_n)]^* = \chi^{(n)}(-\omega_1, -\omega_2, \cdots, -\omega_n). \tag{2.4-7}$$

This relation represents the complex conjugation symmetry.

Furthermore, if we consider the nonresonant case, the various orders of susceptibilities can be approximately viewed as real quantities, then we obtain

$$[\chi^{(n)}(\omega_1, \omega_2, \cdots, \omega_n)]^* = \chi^{(n)}(\omega_1, \omega_2, \cdots, \omega_n). \tag{2.4-8}$$

On the other hand, the relation expressed by Eq. (2.4-7) should always hold, thus it turns out from Eqs. (2.4-7) and (2.4-8) that

$$\chi^{(n)}(\omega_1, \omega_2, \cdots, \omega_n) = \chi^{(n)}(-\omega_1, -\omega_2, \cdots, -\omega_n). \tag{2.4-9}$$

The above relation implies that the susceptibility tensors are invariant for simultaneous change of the sign for all frequency arguments. This property is termed the time-reversal symmetry, which holds only for non-resonance interaction.

By using the permutation symmetry, complex-conjugation symmetry and time-reversal symmetry of various-order susceptibilities, the number of independent and nonzero elements of nonlinear susceptibilities can be further reduced. This may greatly simplify the analyses or calculations of the related nonlinear optical processes.

2.5 Nonlinear coupled-wave equations

According to the classical Maxwell's electromagnetic theory of light, for non-magnetic optical dielectric media, the spatiotemporal variations of an optical field obey the following equations:

$$\left. \begin{array}{l} \nabla \times \mathbf{H} = \varepsilon_0 \dfrac{\partial \mathbf{E}}{\partial t} + \dfrac{\partial \mathbf{P}}{\partial t} \\ \nabla \times \mathbf{E} = -\mu_0 \dfrac{\partial \mathbf{H}}{\partial t} \end{array} \right\}, \tag{2.5-1}$$

where **H** is the magnetic field intensity and ε_0 and μ_0 are the free-space permittivity and permeability, respectively. In general, **E**, **H**, and **P** could be arbitrary spatiotemporal functions that can be expressed in the following Fourier integral forms:

$$\left.\begin{aligned}\mathbf{E}(t,r) &= \int_{-\infty}^{\infty} \mathbf{E}(\omega,r) e^{-i\omega t} d\omega \\ \mathbf{H}(t,r) &= \int_{-\infty}^{\infty} \mathbf{H}(\omega,r) e^{-i\omega t} d\omega \\ \mathbf{P}(t,r) &= \int_{-\infty}^{\infty} \mathbf{P}(\omega,r) e^{-i\omega t} d\omega \end{aligned}\right\}. \quad (2.5\text{-}2)$$

Here, r represents the spatial variables. The physical meaning of these expressions is that any given optical field and induced polarization field can always be viewed as a certain superposition of infinite monochromatic Fourier components. Substituting Eq. (2.5-2) into Eq. (2.5-1) leads to

$$\left.\begin{aligned}\nabla \times \mathbf{H}(\omega,r) &= -\varepsilon_0 i\omega \mathbf{E}(\omega,r) - i\omega \mathbf{P}^{(1)}(\omega,r) - i\omega \mathbf{P}'(\omega,r) \\ \nabla \times \mathbf{E}(\omega,r) &= \mu_0 i\omega \mathbf{H}(\omega,r) \end{aligned}\right\}. \quad (2.5\text{-}3)$$

To obtain the above equations the following assumptions are used:

$$\left.\begin{aligned}\mathbf{P}(t,r) &= \mathbf{P}^{(1)}(t,r) + \mathbf{P}'(t,r) \\ \mathbf{P}(\omega,r) &= \mathbf{P}^{(1)}(\omega,r) + \mathbf{P}'(\omega,r) \end{aligned}\right\}. \quad (2.5\text{-}4)$$

Here, $\mathbf{P}^{(1)}$ and $\mathbf{P}'$ are the linear and nonlinear parts of the corresponding function. Taking $\nabla \times$ operation on both sides of the second equation of Eq. (2.5-3) and substituting $\nabla \times \mathbf{H}(\omega,r)$ from the first equation lead to

$$\nabla \times \nabla \times \mathbf{E}(\omega,r) = \varepsilon_0 \mu_0 \omega^2 \mathbf{E}(\omega,r) + \mu_0 \omega^2 [\mathbf{P}^{(1)}(\omega,r) + \mathbf{P}'(\omega,r)]. \quad (2.5\text{-}5)$$

Considering that

$$\left.\begin{aligned}\mathbf{P}^{(1)}(\omega,r) &= \varepsilon_0 \chi^{(1)}(\omega) \mathbf{E}(\omega,r) \\ \varepsilon(\omega) &= \varepsilon_0 [1 + \chi^{(1)}(\omega)] = \varepsilon_0 \varepsilon_r(\omega) \end{aligned}\right\}, \quad (2.5\text{-}6)$$

where $\varepsilon(\omega)$ is the linear dielectric constant at frequency ω and $\varepsilon_r(\omega)$ is the linear relative dielectric constant, Eq. (2.5-5) can be finally written as

$$\nabla \times \nabla \times \mathbf{E}(\omega,r) - \mu_0 \omega^2 \varepsilon(\omega) \mathbf{E}(\omega,r) = \mu_0 \omega^2 \mathbf{P}'(\omega,r). \quad (2.5\text{-}7)$$

Eq. (2.5-7) is the fundamental equation for $\mathbf{E}(\omega, r)$ and $\mathbf{P}'(\omega, r)$, and it is called the *nonlinear wave equation*.

According to Eqs. (2.3-4) and (2.3-7) given in Section 2.3, the monochromatic Fourier component $\mathbf{P}'(\omega, r)$ of the nonlinear polarization in Eq. (2.5-7) can be written as

$$\begin{aligned}\mathbf{P}'(\omega,r) &= \mathbf{P}^{(2)}(\omega,r) + \mathbf{P}^{(3)}(\omega,r) + \cdots \\ &= \varepsilon_0 [\chi^{(2)}(\omega_1,\omega_2)\mathbf{E}(\omega_1)\mathbf{E}(\omega_2) + \chi^{(3)}(\omega'_1,\omega'_2,\omega'_3)\mathbf{E}(\omega'_1)\mathbf{E}(\omega'_2)\mathbf{E}(\omega'_3) + \cdots]. \end{aligned} \quad (2.5\text{-}8)$$

Now let us explain the physical meaning of the nonlinear wave equation. For a weak incident optical field, $\mathbf{P}'(\omega, r) \Rightarrow 0$, Eq. (2.5-7) becomes the ordinary (linear) wave equation:

$$\nabla \times \nabla \times \mathbf{E}(\omega, r) - \mu_0 \omega^2 \varepsilon(\omega) \mathbf{E}(\omega, r) = 0. \qquad (2.5\text{-}9)$$

In this case, the polarization response of a medium to a given monochromatic component $\mathbf{E}(\omega, r)$ of the applied field is only represented by the dielectric constant $\varepsilon(\omega)$; the other frequency components of the field have no influence on either $\mathbf{P}^{(1)}(\omega, r)$ or $\mathbf{E}(\omega, r)$. In contrast, if the applied field is an intense laser light, the second- and/or third-order polarization components expressed by Eq. (2.5-8) may no longer be neglected. Then Eq. (2.5-7) becomes a nonlinear differential equation, and the term $\mathbf{P}'(\omega, r)$ can be recognized as a physical source that can emit coherent radiation at a new frequency.

For example, if the incident field involves two monochromatic components of $\mathbf{E}(\omega_1, r)$ and $\mathbf{E}(\omega_2, r)$, through the second-order polarization process one may expect a new monochromatic radiation $\mathbf{E}(\omega = \omega_1 + \omega_2, r)$ contributed by

$$\mathbf{P}^{(2)}(\omega = \omega_1 + \omega_2, r) = \varepsilon_0 \chi^{(2)}(\omega_1, \omega_2) \mathbf{E}(\omega_1, r) \mathbf{E}(\omega_2, r). \qquad (2.5\text{-}10)$$

Substituting the above expression into Eq. (2.5-7) we will find that three Fourier components, i.e., $\mathbf{E}(\omega, r)$, $\mathbf{E}(\omega_1, r)$, and $\mathbf{E}(\omega_2, r)$ are involved in the nonlinear wave equation of $\mathbf{E}(\omega, r)$. In order to solve all of these three functions, it is necessary to write down two other similar wave equations involving $\mathbf{P}^{(2)}(\omega_1, r)$ and $\mathbf{P}^{(2)}(\omega_2, r)$, respectively. Then we can, in principle, solve these three equations coupled each to other. Furthermore, if there are three frequency components (ω_1, ω_2, and ω_3) involved in the applied optical field, through the third-order polarization process we may expect a new radiation at frequency $\omega = \omega_1 + \omega_2 + \omega_3$ contributed by

$$\mathbf{P}^{(3)}(\omega = \omega_1 + \omega_2 + \omega_3, r) = \varepsilon_0 \chi^{(3)}(\omega_1, \omega_2, \omega_3) \mathbf{E}(\omega_1, r) \mathbf{E}(\omega_2, r) \mathbf{E}(\omega_3, r). \qquad (2.5\text{-}11)$$

Substituting the above expression into Eq. (2.5-7) leads to a nonlinear differential equation that involves four field functions. To solve them it is necessary to write down three other similar equations involving $\mathbf{P}^{(3)}(\omega_1, r)$, $\mathbf{P}^{(3)}(\omega_2, r)$, and $\mathbf{P}^{(3)}(\omega_3, r)$, respectively. In other words, Eq. (2.5-7) actually represents a set of coupled differential equations which can also be termed nonlinear coupled-wave equations.

It is usually difficult to solve the nonlinear coupled-wave equations. Therefore, theorists have to use various approaches to obtain either approximate analytical solutions or numerical solutions. We discuss here how to simplify Eq. (2.5-7) by utilizing both the plane-wave approximation and the slowly-varying amplitude approximation. For laser radiation with high monochromaticity and high directionality, the incident field can be viewed as a superposition of several linearly polarized monochromatic plane waves propagating along the z-axis. Each of them can be expressed as

$$\mathbf{E}(\omega, z) = \mathbf{a}_0 A(\omega, z) e^{ikz}, \qquad (2.5\text{-}12)$$

where $\mathbf{a}_0$ is the unit vector along the polarization direction, $A(\omega, z)$ is the scalar amplitude function, $k = \dfrac{2\pi}{\lambda} n_0(\omega)$ is the magnitude of the wavevector of the plane wave, λ is the wavelength in free space, and $n_0(\omega)$ is the linear refractive index at ω. Moreover, considering that the nonlinear polarization is still a much weaker contribution compared to the linear polarization, the relative amplitude change (increase or decrease) of each involved monochromatic wave can be nearly neglected in a propagation length shorter than a wavelength, so that the second-order spatial derivative and the square of the first-order spatial derivative of each amplitude function can be ignored approximately. This is the so-called slowly-varying amplitude approximation, and it is widely applied to the theoretical analyses or calculations for many nonlinear optical processes.

Substituting Eq. (2.5-12) into Eq. (2.5-7) and using the slowly-varying amplitude approximation leads to the following equation:

$$\frac{\partial A(\omega, z)}{\partial z} = \frac{ik}{2\varepsilon(\omega)}[\mathbf{a}_0 \cdot \mathbf{P}'(\omega, z)]e^{-ikz}. \qquad (2.5\text{-}13)$$

Actually, the amplitude $A(\omega, z)$ not only is a function of the propagation distance z, but also can be a function of the transverse variable x and y, thus we use here the form of partial derivative. Moreover, it is allowed that the wave front may deviate from an ideal plane wave to a small extent in an arbitrary way, therefore $A(\omega, z)$ can be a complex function. On the other hand, as mentioned earlier, $\mathbf{P}'(\omega, z)$ contains the product of amplitude of other monochromatic components to the second-order or third-order power. Thus Eq. (2.5-13) represents a set of equations of multi-wave coupling and can be termed the *nonlinear coupled-wave equation(s)* in quasi-plane wave and slowly-varying amplitude approximations. In comparison with Eq. (2.5-7), Eq. (2.5-13) is obviously simplified, but it is still difficult to obtain a rigorous analytical solution. In Chapters 3 and 4, we will discuss some specific cases in which an approximate analytical solution could be obtained.

2.6 Complex expressions of optical wave fields

Now let us try to find the best mathematical representation of the optical field functions. As we mentioned before in this chapter, in principle, the electric field $\mathbf{E}(t)$ and the electric polarization $\mathbf{P}(t)$ both should be two real quantities. However, if we treat the Fourier components of these two quantities, $\mathbf{E}(\omega)$ and $\mathbf{P}(\omega)$, also as real functions, it will lead to a tedious mathematical formulation. On the contrary, if we treat $\mathbf{E}(\omega)$ and $\mathbf{P}(\omega)$ as two complex functions, the related mathematical work can be greatly simplified. This treatment will bring no contradiction, because the final measurable quantity is the light intensity of $I(\omega) \propto |\mathbf{E}(\omega)|^2 = \mathbf{E}(\omega) \cdot \mathbf{E}^*(\omega)$ but not $\mathbf{E}(\omega)$ itself. In addition, based on the complex expressions of $\mathbf{E}(\omega)$ and $\mathbf{P}(\omega)$, sometimes a meaningful insight can be obtained. There are two examples to support the statement of such advantages. First, a negative frequency argument can be formally introduced into the optical field functions, which is necessary to describe the optical difference-frequency generation as well as other more complicated frequency-mixing processes. Second, a complex optical field function can be employed to describe a distorted wave front as well as the optical phase conjugation properties (see Chapter 8).

According to Eq. (2.1-5), the monochromatic Fourier component can be expressed as an inverse-Fourier integral

$$\mathbf{E}(\omega) = \frac{1}{2\pi} \int_{-\infty}^{\infty} dt \mathbf{E}(t) e^{i\omega t}. \qquad (2.6\text{-}1)$$

As we recognize that $\mathbf{E}(t)$ is a real function and $\mathbf{E}(\omega)$ is a complex function, taking a complex-conjugation operation on the latter leads to

$$E^*(\omega) = \frac{1}{2\pi} \int_{-\infty}^{\infty} dt \mathbf{E}(t) e^{-i\omega t}. \qquad (2.6\text{-}2)$$

Comparing the above two expressions we obtain

$$\mathbf{E}^*(\omega) = \mathbf{E}(-\omega). \qquad (2.6\text{-}3)$$

The physical meaning of Eq. (2.6-3) is that in formulating a nonlinear frequency-mixing process, the term of $\mathbf{E}^*(\omega) = \mathbf{E}(-\omega)$ plays a frequency-deduction role.

By using the complex representation of Fourier components of the field, it is easy to give the complete expressions of all involved nonlinear polarization components. For instance, in the second-order optical sum-frequency process $(\omega_1 + \omega_2 \Rightarrow \omega_3)$ created by $\mathbf{E}(\omega_1)$ and $\mathbf{E}(\omega_2)$, the three involved nonlinear polarization components can be written as

$$\left.\begin{aligned}\mathbf{P}^{(2)}(\omega_3 = \omega_1 + \omega_2) &= \varepsilon_0 \chi^{(2)}(\omega_1, \omega_2) \mathbf{E}(\omega_1) \mathbf{E}(\omega_2) \\ \mathbf{P}^{(2)}(\omega_1 = \omega_3 - \omega_2) &= \varepsilon_0 \chi^{(2)}(\omega_3, -\omega_2) \mathbf{E}(\omega_3) \mathbf{E}^*(\omega_2) \\ \mathbf{P}^{(2)}(\omega_2 = \omega_3 - \omega_1) &= \varepsilon_0 \chi^{(2)}(\omega_3, -\omega_1) \mathbf{E}(\omega_3) \mathbf{E}^*(\omega_1)\end{aligned}\right\}. \qquad (2.6\text{-}4)$$

Similarly, for the third-order sum-frequency process $(\omega_1+\omega_2+\omega_3 \Rightarrow \omega_4)$ created by $\mathbf{E}(\omega_1)$, $\mathbf{E}(\omega_2)$, and $\mathbf{E}(\omega_3)$, the four involved nonlinear polarization components are

$$\left.\begin{aligned}\mathbf{P}^{(3)}(\omega_4 = \omega_1 + \omega_2 + \omega_3) &= \varepsilon_0 \chi^{(3)}(\omega_1, \omega_2, \omega_3) \mathbf{E}(\omega_1) \mathbf{E}(\omega_2) \mathbf{E}(\omega_3) \\ \mathbf{P}^{(3)}(\omega_1 = \omega_4 - \omega_2 - \omega_3) &= \varepsilon_0 \chi^{(3)}(\omega_4, -\omega_2, -\omega_3) \mathbf{E}(\omega_4) \mathbf{E}^*(\omega_2) \mathbf{E}^*(\omega_3) \\ \mathbf{P}^{(3)}(\omega_2 = \omega_4 - \omega_1 - \omega_3) &= \varepsilon_0 \chi^{(3)}(\omega_4, -\omega_1, -\omega_3) \mathbf{E}(\omega_4) \mathbf{E}^*(\omega_1) \mathbf{E}^*(\omega_3) \\ \mathbf{P}^{(3)}(\omega_3 = \omega_4 - \omega_1 - \omega_2) &= \varepsilon_0 \chi^{(3)}(\omega_4, -\omega_1, -\omega_2) \mathbf{E}(\omega_4) \mathbf{E}^*(\omega_1) \mathbf{E}^*(\omega_2)\end{aligned}\right\}. \qquad (2.6\text{-}5)$$

Substituting Eqs. (2.6-4) or (2.6-5) into the nonlinear coupled-wave equations of Eq. (2.5-7) or Eq. (2.5-13), at least in principle, we could obtain the complete solutions of the corresponding nonlinear optical processes.

PROBLEMS

1. How many mechanisms may contribute to the intense coherent light induced nonlinear electric polarization responses of an optical medium? What is the response-time range for each of them?
2. Please give the general expression of $P_y^{(2)}(\omega_1 + \omega_2)$ (total nine terms) for the second-order sum-frequency generation.
3. Please write the specific expressions of $P_i^{(2)}(\omega = \omega_1 + \omega_2)$ ($i = x, y, z$) for the 4mm crystals (cf. Appendix 4).
4. If there is only one monochromatic intense light wave $\mathbf{E}(\omega_0)$ incident upon a third-order nonlinear medium, how many nonlinear polarization responses with different possible frequency components may be expected? Please write the vector expressions of these possible third-order nonlinear polarization components $\mathbf{P}(\omega')$.

REFERENCES

1. J. A. Armstrong, N. Bloembergen, J. Ducuing, and P. S. Pershan, *Phys. Rev.* **127**, 1918 (1961).
2. P. A. Franken and J. F. Ward, *Rev. Mod. Phys.* **35**, 23 (1963).
3. P. S. Pershan, in *Progress in Optics*, Vol. V, edited by E. Wolf (North-Holland, Amsterdam, 1966), pp. 85–141.
4. J. Ducuing, in *Quantum Optics*, edited by R. J. Glauber (Academic Press, New York, 1969), pp. 421–472.
5. P. N. Butcher, *Nonlinear Optical Phenomena* (Ohio State University Press, Columbus, 1965).
6. N. Bloembergen, *Nonlinear Optics* (Benjamin, New York, 1965).
7. Y. R. Shen, *The Principles of Nonlinear Optics* (Wiley, New York, 1984).
8. M. Schubert and B. Wilhelmi, *Nonlinear Optics and Quantum Electronics* (Wiley, New York, 1986).
9. P. N. Butcher and D. Cotter, *The Elements of Nonlinear Optics* (Cambridge University Press, Cambridge, 1990).
10. R. W. Boyd, *Nonlinear Optics*, 3rd ed. (Academic Press, New York, 2008).

3
Second-Order Nonlinear (Three-Wave) Frequency Mixing

In this chapter, we briefly discuss the second-order nonlinear three-wave frequency mixing processes, which generally include optical second-harmonic generation, optical sum- and difference-frequency generation, optical parametric amplification and oscillation. The most essential feature for these processes is the coupling interaction among three coherent monochromatic light waves in a second-order nonlinear crystal of no centrosymmetry. Since there is no net energy and momentum exchange between the light (photon) fields and the nonlinear medium, the conservation of energy and momentum holds only among the three interacting waves and requires that certain phase-matching conditions should be fulfilled. In practice, the second-harmonic and/or sum-frequency generation is one of the most efficient techniques for laser frequency up-conversion, while the second-order parametric oscillation or amplification is a highly useful technique to produce frequency-tunable coherent emission.

3.1 Second-harmonic generation (SHG)

3.1.1 Quantum description of the mechanism of SHG

Optical second-harmonic generation (SHG) or frequency doubling means that a monochromatic coherent optical wave of frequency v can induce a new coherent wave emission at frequency $2v$ in a second-order nonlinear medium.[1]

As we mentioned in Section 1.4, the quantum theory of radiation, which is within the framework of quantum electrodynamics, can give a concise and clear description of the essential mechanisms or physical pictures about various optical frequency-mixing effects and many other nonlinear optical effects. The major advantage of using this theory is that it can reveal not only the coherent wave property of the optical field, but also the quantum nature of the field when there is an exchange of energy and momentum either among different components of optical fields or between the optical field and the medium. The basic starting point of this theory is that we must treat the field and the molecule of a medium as a combined and quantized system. Any interaction between the optical field(s) and the molecule is always accompanied by (i) a quantum transition of the molecule among its different eigenstates, and (ii) a simultaneous change of photon distribution among different modes (photon states). In order to give a rigorous and complete physical description for these processes, the key concept of *intermediate state* must be introduced, which is a quantum state charactering the whole system involving both optical fields and the molecule.

Nonlinear Optics and Photonics. First Edition. Guang S. He. © Guang S. He 2015.
Published in 2015 by Oxford University Press.

In such an intermediate state of the whole system, the photon number (degeneracy) in a mode has changed by +1 or −1 while the molecule has left its original state (mostly ground state); however, at this moment the molecule is not certainly located in any specific excited state, but correlated with all possible excited states with a certain occupying probability for each. In this case, it is convenient to introduce a virtual energy level in the energy diagram to represent such an intermediate state. In the intermediate state represented by a virtual state-level, the energy range of all molecular eigenstates that may be occupied by an "excited" molecule is nearly unlimited; according to the uncertainty principle the staying time of this molecule in that intermediate state is infinitely short. Relying on the concept of intermediate state and its representation of a virtual energy level, many nonlinear optical effects can be interpreted very clearly.

As an example, the schematic energy-level diagram describing SHG is shown in Fig. 3.1. In this case, the elementary process of SHG occurs via multiple quantum transitions or multiple steps. In the first step, there is an annihilation of one photon from the fundamental light field while a molecule of the medium left its initial (ground) state to an intermediate state; in the second step, there is the annihilation of another fundamental photon while the excited molecule is situated in another intermediate state; in the final step, this excited molecule is returning to its initial state while there is the creation of a new photon of doubled frequency. Since the staying time of the molecule in each intermediate state is extremely short (near the response time of electron-cloud distortion), the above-mentioned three steps actually occur instantaneously and simultaneously. Here we have to visualize that the elementary SHG process undergoes three quantum transitions via two intermediate states, because in the quantum theory of radiation for each quantum transition of the whole system the photon number in a given mode can only change either by −1 (i.e., annihilation of a photon) or +1 (i.e., creation of a photon). This rule holds for any types of optical processes involving emission, absorption, or molecular scattering of light. We should keep in mind that for all those phenomena (such as nonlinear frequency mixing, Raman scattering, and multi-photon absorption), the observable elementary process is actually an instantaneous single event even though which may undergo multiple quantum transitions via one or multiple intermediate states. With such an understanding, for instance, the elementary process of SHG can also be simply described as the annihilation of two fundamental photons and the simultaneous creation of one second-harmonic photon. (Also see Figure 1.3.)

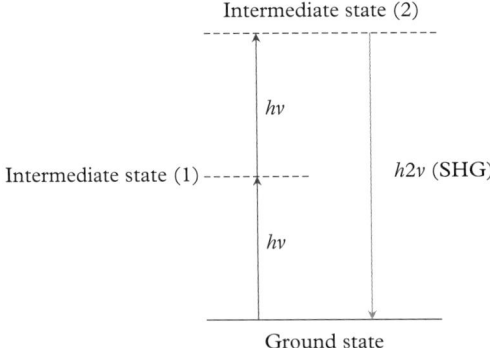

Figure 3.1 *Quantum transition mechanism of SHG via intermediate states.*

It should be noted that at the beginning and end of this elementary process, the energy and momentum of the molecule are unchanged, therefore the conservation of energy and momentum should hold only among the annihilated photons and created photon(s). Assuming the frequencies of the fundamental and second harmonic waves are v_1 and v_2, the conservation of energy and momentum requires

$$\left.\begin{array}{c} hv_2 = h2v_1 \\ \mathbf{k}_2 = \mathbf{k}_1 + \mathbf{k}_1 \end{array}\right\}. \tag{3.1-1}$$

Here, h is Planck constant and $\mathbf{k}_1$ and $\mathbf{k}_2$ are the wavevectors of these two waves. For collinear interaction, $\mathbf{k}_1$ and $\mathbf{k}_2$ are in the same direction, and from the second relation in Eq. (3.1-1) we obtain the following requirement:

$$\frac{2\pi}{\lambda_2} n(v_2) = \frac{4\pi}{\lambda_1} n(v_1), \tag{3.1-2}$$

where λ_1 and λ_2 are the wavelengths of these two waves in free space. From the first equality of Eq. (3.1-1) we know that $\lambda_1 = 2\lambda_2$, hence Eq. (3.1-2) leads to the following phase-matching requirement for the refractive index of the nonlinear medium:

$$n(2v_1) = n(v_1). \tag{3.1-3}$$

From the above discussion it is clear that the phase-matching condition is a natural result of the conservation of energy and momentum from the viewpoint of the quantum theory of radiation.

Finally, it is interesting to point out that the above-mentioned basic feature of the elementary process is not related to either the specific molecular structure or the macroscopic symmetric property of a given medium. In other words, for a given incident fundamental wave, we can use entirely different second-order nonlinear materials to obtain the coherent second-harmonic output with the same properties (wavelength, spectral linewidth, beam divergence and temporal behavior), except that the conversion efficiency may be different for those materials.

3.1.2 Semiclassical description of SHG

With the fundamental knowledge described in Chapter 2, we already know that in electric-dipole moment approximation, the second-order nonlinear optical effects can occur in the non-centrosymmetrical crystals only. Based on the concept of nonlinear polarization, which is within the framework of semiclassical theory, the source of second-harmonic field is the following nonlinear polarization component:

$$\mathbf{P}^{(2)}(2\omega) = \varepsilon_0 \chi^{(2)}(\omega, \omega) \mathbf{E}(\omega) \mathbf{E}(\omega). \tag{3.1-4}$$

Here, $\mathbf{E}(\omega)$ is the incident fundamental optical field and $\chi^{(2)}(\omega, \omega)$ is the second-order susceptibility tensor of a given nonlinear medium for SHG.

Assuming both the incident fundamental field $\mathbf{E}(\omega)$ and the frequency-doubled field $\mathbf{E}(2\omega)$ are linearly-polarized monochromatic plane waves propagating along z-axis and can be described as

$$\left.\begin{array}{c} \mathbf{E}_1(\omega, z) = \mathbf{a}_1 A_1(z) e^{ik_1 z} \\ \mathbf{E}_2(2\omega, z) = \mathbf{a}_2 A_2(z) e^{ik_2 z} \end{array}\right\}, \tag{3.1-5}$$

Second-Order Nonlinear (Three-Wave) Frequency Mixing

then the corresponding polarization components of these two waves should be

$$\left.\begin{array}{l} \mathbf{P}^{(2)}(\omega, z) = \varepsilon_0 \chi^{(2)}(2\omega, -\omega) \mathbf{E}_2 \mathbf{E}_1^* = \varepsilon_0 \chi^{(2)}(2\omega, -\omega) \mathbf{a}_2 \mathbf{a}_1 A_2(z) A_1^*(z) e^{i(k_2-k_1)z} \\ \mathbf{P}^{(2)}(2\omega, z) = \varepsilon_0 \chi^{(2)}(\omega, \omega) \mathbf{E}_1 \mathbf{E}_1 = \varepsilon_0 \chi^{(2)}(\omega, \omega) \mathbf{a}_1 \mathbf{a}_1 A_1^2(z) e^{2ik_1 z} \end{array}\right\}. \quad (3.1\text{-}6)$$

Here, $A_1(z)$ and $A_2(z)$ are the amplitude functions, $\mathbf{a}_1$ and $\mathbf{a}_2$ are the unit vectors along the light polarization direction, and k_1 and k_2 are the absolute values of wavevectors for these two waves, respectively.

To consider the amplitude change of these two waves along the z-axis, substituting the second equality of Eq. (3.1-6) into the coupled-wave equation of Eq. (2.5-13) leads to

$$\begin{aligned}\frac{\partial A_2(z)}{\partial z} &= \frac{ik_2 \varepsilon_0}{2\varepsilon(2\omega)} \left[\mathbf{a}_2 \cdot \chi^{(2)}(\omega, \omega) \mathbf{a}_1 \mathbf{a}_1\right] A_1^2(z) e^{i(2k_1-k_2)z} \\ &= \frac{ik_2 \varepsilon_0}{2\varepsilon(2\omega)} \left[\mathbf{a}_2 \cdot \chi^{(2)}(\omega, \omega) \mathbf{a}_1 \mathbf{a}_1\right] A_1^2(z) e^{i\Delta kz}, \end{aligned} \quad (3.1\text{-}7)$$

where

$$\Delta k = 2k_1 - k_2 = \frac{4\pi}{\lambda_1} [n(\omega) - n(2\omega)]. \quad (3.1\text{-}8)$$

Here, λ_1 is the wavelength of the fundamental wave and Δk is the phase mismatch factor that is proportional to the refractive index difference at frequency ω and 2ω. This factor plays a very important role in the processes of nonlinear frequency mixing. In general, the refractive index of a medium is always a function of frequency (dispersion effect), i.e., $n(\omega) \neq n(2\omega)$; only in some special cases we may have $n(\omega) = n(2\omega)$.

Now let us first consider the case of phase mismatching, i.e., $\Delta k \neq 0$. Comparing the expression of $\mathbf{P}^{(2)}(2\omega, z)$ in Eq. (3.1-6) with the expression of $\mathbf{E}_2(2\omega, z)$ in Eq. (3.1-5), one can see that the phase velocity of the polarization wave at 2ω is different from the phase velocity of the optical wave at the same frequency. Therefore, the energy from the $\mathbf{E}_1(\omega, z)$ wave cannot transfer to the $\mathbf{E}_2(2\omega, z)$ wave effectively. In this case, we can assume that for the fundamental wave $\mathbf{E}_1(\omega, z)$, the amplitude change along the z-axis can be nearly neglected, i.e.,

$$\left.\begin{array}{l} A_1(z) \approx A_1(0) \\ A_2(0) \approx 0 \end{array}\right\}, \quad (3.1\text{-}9)$$

where $A_1(0)$ is the initial amplitude of the fundamental wave at the incident surface of the nonlinear medium. Under this condition, Eq. (3.1-7) leads to the following solution:

$$A_2(z) = \frac{\varepsilon_0 k_2}{\Delta k 2\varepsilon(2\omega)} \left[\mathbf{a}_2 \cdot \chi^{(2)}(\omega, \omega) \mathbf{a}_1 \mathbf{a}_1\right] A_1^2(0) \left(e^{i\Delta kz} - 1\right). \quad (3.1\text{-}10)$$

The intensity change of the second-harmonic wave along the z direction is

$$I_2(z) \propto A_2 A_2^* \propto I_1^2(0) \left(\frac{\sin \Delta kz/2}{\Delta k/2}\right)^2. \quad (3.1\text{-}11)$$

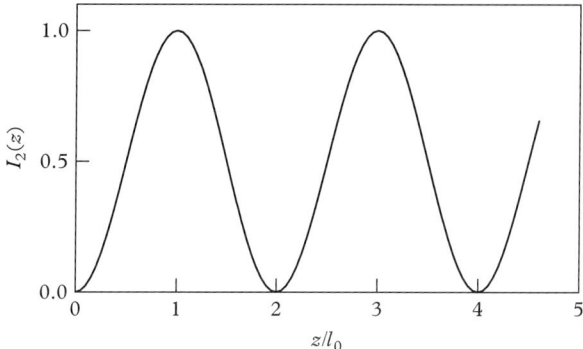

Figure 3.2 *Variation of normalized second-harmonic intensity versus z in condition of $\Delta k \neq 0$.*

Here, $I_1(0) \propto A_1^2(0)$ is the intensity of the incident fundamental wave. From Eq. (3.1-11) one can see that the intensity of the second harmonic is proportional to the square of the intensity of the fundamental wave, and varies periodically along the z-axis, as shown in Fig. 3.2. The period of the intensity variation is $2\pi/\Delta k$.

It can be seen in Fig. 3.2 that when the propagation distance of the fundamental wave is approaching a characteristic length,

$$l_0 = \frac{\pi}{|\Delta k|} = \frac{\lambda_1}{4|n(\omega) - n(2\omega)|}, \qquad (3.1\text{-}12)$$

the intensity of the second-harmonic wave reaches its first maximum that is inversely proportional to $(\Delta k)^2$; l_0 is defined as the effective interaction length or coherent length of SHG under the condition of $\Delta k \neq 0$. In the range of $l_0 \leq z \leq 2l_0$, the intensity of second harmonic is getting smaller. A larger Δk value leads to a smaller peak intensity and more rapid variation. Owing to this reason, for a second-order nonlinear medium with a thickness much larger than l_0, the efficiency of SHG is extremely low. In order to significantly increase the conversion efficiency, it is necessary to realize $\Delta k \Rightarrow 0$ or $n(\omega) \Rightarrow n(2\omega)$. For most transparent media, the normal dispersion effect makes $n(2\omega) > n(\omega)$. However, fortunately, we can use the birefringence effect of some non-centrosymmetrical crystals to compensate the dispersion effect and make the phase-matching condition of $n(2\omega) = n(\omega)$ possible.

Under the phase-matching condition of $\Delta k \Rightarrow 0$, the coupling between the fundamental wave and the second harmonic wave may become so strong that the depletion of $A_1(z)$ along the z-axis can no longer be neglected. Therefore, it is necessary to write down the following two coupled wave equations based on Eq. (2.5-13):

$$\left. \begin{array}{l} \dfrac{\partial A_1(z)}{\partial z} = \dfrac{i\varepsilon_0 k_1}{2\varepsilon(\omega)} [\mathbf{a}_1 \cdot \chi^{(2)}(2\omega, -\omega) \mathbf{a}_2 \mathbf{a}_1] A_2(z) A_1^*(z) \\[2mm] \dfrac{\partial A_2(z)}{\partial z} = \dfrac{i\varepsilon_0 k_2}{2\varepsilon(2\omega)} [\mathbf{a}_2 \cdot \chi^{(2)}(\omega, \omega) \mathbf{a}_1 \mathbf{a}_1] A_1^2(z) \end{array} \right\}. \qquad (3.1\text{-}13)$$

Noticing that for the SHG process, we have

$$\mathbf{a}_1 \cdot \chi^{(2)}(2\omega, -\omega) \mathbf{a}_2 \mathbf{a}_1 = 2\mathbf{a}_2 \cdot \chi^{(2)}(\omega, \omega) \mathbf{a}_1 \mathbf{a}_1 = 2\chi_e^{(2)}, \qquad (3.1\text{-}14)$$

where $\chi_e^{(2)}$ is termed the effective second-order susceptibility value for SHG, then Eq. (3.1-13) can be simplified as

$$\left.\begin{array}{l}\dfrac{\partial A_1(z)}{\partial z} = i\dfrac{2\pi \chi_e^{(2)}}{\lambda_1 n(\omega)} A_2 A_1^* \\[2mm] \dfrac{\partial A_2(z)}{\partial z} = i\dfrac{2\pi \chi_e^{(2)}}{\lambda_1 n(2\omega)} A_1^2 \end{array}\right\}. \qquad (3.1\text{-}15)$$

The solutions of this coupled-wave equation are[2-4]

$$\left.\begin{array}{l}A_1(z) = A_1(0)\text{sech}\,[\zeta A_1(0)z] \\ A_2(z) = iA_1(0)\tanh\,[\zeta A_1(0)z]\end{array}\right\}, \qquad (3.1\text{-}16)$$

where ζ is a constant coefficient determined by

$$\zeta = \dfrac{2\pi \chi_e^{(2)}}{\lambda_1 n(\omega)}. \qquad (3.1\text{-}17)$$

Based on Eq. (3.1-16) we can further obtain the intensity change of these two waves along the propagation direction:

$$\left.\begin{array}{l}I_1(z) = I_1(0)\text{sech}^2[\zeta A_1(0)z] \\ I_2(z) = I_1(0)\tanh^2[\zeta A_1(0)z]\end{array}\right\}. \qquad (3.1\text{-}18)$$

Figure 3.3 shows the normalized intensity change of these two waves along the z-axis under the condition of phase matching of $\Delta k \Rightarrow 0$. In this case, we can also define an effective interaction length by

$$l_0' = [\zeta A_1(0)]^{-1} = \left[\dfrac{2\pi \chi_e^{(2)}}{\lambda_1 n(\omega)} A_1(0)\right]^{-1}. \qquad (3.1\text{-}19)$$

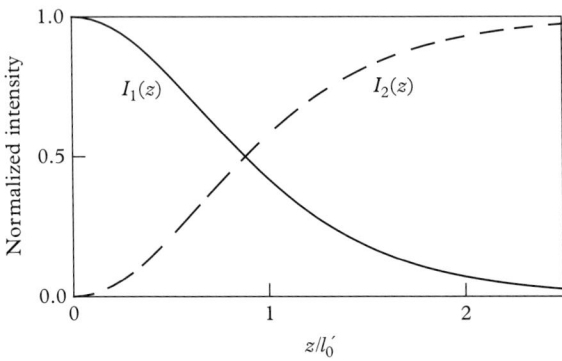

Figure 3.3 *Normalized intensities of the fundamental wave and second-harmonic wave versus z in the phase-matching condition.*

From Fig. 3.3 we can see that when the fundamental wave is propagating over a distance of $z = l'_0$, more than 58% of its power is coupled into the second-harmonic wave. For this reason, l'_0 can be considered as a rough criterion for choosing the optimized thickness of a nonlinear crystal for phase-matched SHG.

The power transfer efficiency from the fundamental wave to the second-harmonic wave is determined by

$$\eta = \frac{I_2(z)}{I_1(z)} = \left\{ \frac{\tanh[\zeta A_1(0)z]}{\operatorname{sech}[\zeta A_1(0)z]} \right\}^2. \qquad (3.1\text{-}20)$$

If only consider the range of $\zeta A_1(0) \ll 1$, we have

$$\eta \approx [\zeta A_1(0)z]^2. \qquad (3.1\text{-}21)$$

One can see that the efficiency of SHG is proportional to the initial intensity of the fundamental wave and the square of the effective second-order susceptibility value of $\chi_e^{(2)}$.

3.1.3 Nonlinear crystals for SHG

Based on the principle described above, the requirements for nonlinear crystals of SHG are as follows.

(1) *No centrosymmetry*: In the electric dipole approximation, isotropic media and centrosymmetrical crystals cannot be used to generate efficient second-order nonlinear effects. Therefore, the media for SHG should be the crystals having no centrosymmetry. This requirement is the same as that for the piezoelectric effect; thus all SHG crystals are piezoelectric crystals although the physical mechanisms for these two effects are not correlated to each other.

(2) *Phase matching*: It has been proved theoretically that only when the phase-matching condition of $n(2v) = n(v)$ is fulfilled the SHG can be effectively obtained. Usually $n(2v) > n(v)$ due to the normal dispersion, one has to choose a special direction within a given crystal so that the dispersion effect of refractive index can be compensated by the birefringence effect of an anisotropic crystal. The angle between this special direction and the optical axis of the crystal is termed the phase-matching angle. Depending on the specific polarization states of the two interacting waves, two methods can be employed to satisfy this requirement. The first method is called type I phase matching, in which the incident fundamental wave contains only one polarization component (e.g., *o*-ray), whereas the second-harmonic wave contains the other polarization component (e.g., *e*-ray). The second method is called type II phase matching, in which the fundamental field is composed of two polarization components (*o*-ray and *e*-ray) while the second-harmonic field still contains only one polarization component (*e*-ray). For negative uniaxial crystals ($n^o > n^e$), the two methods for phase-matching can be expressed as

$$\left. \begin{array}{ll} n^o(v) + n^o(v) = 2n^e(2v) & \text{(type I)} \\ n^o(v) + n^e(v) = 2n^e(2v) & \text{(type II)} \end{array} \right\}. \qquad (3.1\text{-}22)$$

For positive uniaxial crystals, the roles of ordinary and extraordinary ray are reversed.

(3) *Larger effective second-order susceptibility value*: Since the efficiency of SHG is proportional to the square of the effective second-order susceptibility value of $\chi_e^{(2)}$, the crystals possessing a larger $\chi_e^{(2)}$ value are preferred. This requirement is especially important for SHG by using a laser beam of continuous-wave (cw) or quasi-continuous wave as the incident fundamental beam.

(4) *Higher optical damage threshold*: To ensure a higher efficiency and to withstand a higher incident optical power, all crystals for SHG should possess the best optical quality and high transmissivity for both the fundamental and second-harmonic waves. The attenuation coefficients of the crystals for these two waves are usually kept less than 10^{-2} cm^{-1}. Depending on the material properties and the parameters of the input fundamental laser beam, various mechanisms may cause optical damages of the nonlinear crystal, such as opto-thermal effect, optical electrostriction effect, self-focusing effect, plasma generation near the crystal surface, and so on.

It should be pointed out that even phase matching can be fulfilled in a special direction within the nonlinear crystals, owing to the birefringence effect the fundamental beam and the second-harmonic beam will be still separated gradually along their propagation directions. This is the so-called walk-off effect that will limit the effective interaction length and reduce the conversion efficiency. For this reason, it will be better if the phase-matching requirement can be fulfilled in a direction perpendicular to the optical axis ($\phi_0 = 90°$) of a uniaxial crystal. Under this arrangement, the separation effect of two beams can be eliminated. Some piezoelectric crystals can be used to achieve $\phi_0 = 90°$ phase matching. For these crystals, their dispersion and birefringence are dependent sensitively on the temperature of the crystal, so that properly adjusting the temperature can ensure a phase matching in the direction of $\phi_0 = 90°$. This method is termed the temperature-tuning phase matching in distinction from the commonly used angle-tuning phase matching.

Figure 3.4 schematically shows how to determine the phase-matching angle in commonly used negative uniaxial crystals. From this figure we can see that the refractive index of the ordinary ray (n^o) is independent of the direction, while the refractive index of extraordinary ray (n^e) is a function of the angle ϕ between the propagation direction and the optical axis; for any direction of $\phi \neq 0$, $n^e < n^o$ (see Fig. 3.4(a)). Owing to normal dispersion effect, we usually have $n^o(2\nu) > n^o(\nu)$;

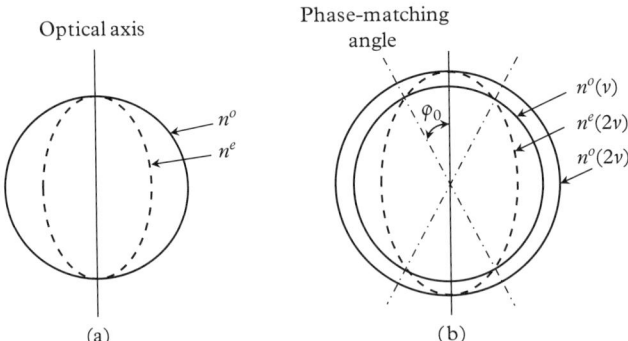

Figure 3.4 *Birefringence (a) and angular phase matching (b) of a negative uniaxial crystal.*

Table 3.1 Data of nonlinear crystals for SHG.

Crystal	Crystal Class	Transparent Range (μm)	d_{il} (10^{-12} m/V)	Phase Matching	Damage Threshold (MW/cm^2)	Efficiency (%)
KH$_2$PO$_4$ (KDP)	$\bar{4}2m$ (a)	0.2–1.5	d_{36} = 0.39	φ-tuning	~500	20–30
KD$_2$PO$_4$ (KD*P)	$\bar{4}2m$ (a)	0.2–1.5	d_{36} = 0.37	φ-tuning	~1000	20–40
NH$_4$H$_2$PO$_4$ (ADP)	$\bar{4}2m$ (a)	0.2–1.2	d_{36} = 0.47	φ-tuning	~500	20–30
KNbO$_3$ (KN)	$mm2$ (b)	0.4–5	d_{31} = –12.8	T-tuning	>500	30–60
β–BaB$_2$O$_4$ (BBO)	$3m$ (a)	0.19–3	d_{22} = 2.3	φ-tuning	~1500	40–60
LiB$_3$O$_5$ (LBO)	$mm2$ (b)	0.16–2.6	d_{31} = 0.85	φ-tuning	~2000	40–60
KTiOPO$_4$ (KTP)	$mm2$ (b)	0.3–4.3	d_{24} = 3.3	φ-tuning	>500	40–60
LiNbO$_3$ (LN)	$3m$ (a)	0.33–5	d_{31} = –4.3 d_{22} = 2.1	T-tuning	~100	30–50
LiIO$_3$	6 (a)	0.3–5.5	d_{31} = 4.4	φ-tuning	~100	30–50
CsH$_2$AsO$_4$ (CDA)	$\bar{4}2m$ (a)	0.26–1.43	d_{36} = 0.40	φ-tuning T-tuning	~500	30–50
RbH$_2$AsO$_4$ (RDA)	$\bar{4}2m$ (a)	0.26–1.46	d_{36} = 0.39 (0.694 μm)	φ-tuning T-tuning	~350 (0.694 μm)	30–50
RbH$_2$PO$_4$ (RDP)	$\bar{4}2m$ (a)	0.22–1.4	d_{36} = 0.4	φ-tuning	~300	30–50
AgGaSe$_2$	$\bar{4}2m$ (a)	0.7–18	d_{36} = 33 (10.6 μm)	φ-tuning	~10 (10.6 μm)	5–10
Ag$_3$AsS$_3$	$3m$ (a)	0.6–13	d_{15} = 11 d_{22} = 18 (10.6 μm)	φ-tuning	~30 (10.6 μm)	5–10
CdGeAs$_2$	$\bar{4}2m$ (a)	2.4–18	d_{36} = 235 (10.6 μm)	φ-tuning	~40 (10.6 μm)	10–20
Te	32 (c)	3.8–32	d_{11} = 649 (10.6 μm)	φ-tuning	~50 (10.6 μm)	10–15

(a) Negative uniaxial crystal, (b) negative biaxial crystal, and (c) positive uniaxial crystal. φ-tuning: angle tuning; T-tuning: temperature tuning.

but for a special direction of ϕ_0, we may find $n^e(2\nu) = n^o(\nu)$ (see Fig. 3.4(b)), i.e., an angle-tuning type I phase matching can be achieved.

To date, three types of nonlinear crystals have commonly been used for SHG, which can be briefly described as follows.

(A) *Crystals for angle-tuning phase matching*: KDP, ADP, KD*P, KTP, LiIO$_3$, BBO, LBO, etc. (see Table 3.1). These crystals are mainly used for SHG in visible and near ultra-violet (UV) spectral range when the input fundamental wave is in the near infra-red (IR) or visible spectral range. The advantages of these crystals are their high optical damage threshold and conversion efficiency.

(B) *Crystals for temperature-tuning phase matching*: LiNbO$_3$, Ba$_2$NaNb$_5$O$_{15}$, CDA, CD*A, KNbO$_3$, etc. These crystals possess a larger nonlinear susceptibility and temperature effect of refractive index, therefore can be used for temperature-tuning phase matching. They have very good transmission property in the 0.4–5 μm spectral range and can generate second harmonic radiation in visible and near IR range.

(C) *Semiconductor crystals for SHG in IR range*: Ag$_3$AsS$_3$, AgGaSe$_2$, CdGeAs$_2$, Te, CdSe, etc. The nonlinear susceptibilities of these crystals are higher by an order of magnitude than the other two types of crystals mentioned above. These semiconductor crystals possess a good transmission property in the near- and mid-IR spectral range, therefore, are suitable for SHG in IR range.

Based on the intrinsic permutation symmetry of nonlinear susceptibility tensor expressed by Eq. (2.4-3), for SHG process we have

$$\chi^{(2)}_{ijk}(\omega,\omega) = \chi^{(2)}_{ikj}(\omega,\omega), \tag{3.1-23}$$

so that $\chi^{(2)}_{ijk}(\omega,\omega)$ can be simplified and replaced by a shorthand notation d_{il} according to the following relationship:

$$d_{il} \Rightarrow \chi^{(2)}_{ijk}(\omega,\omega) \quad \begin{cases} l = 1,2,3,4,5,6 \\ jk = xx, yy, zz, yz, zx, xy \end{cases}. \tag{3.1-24}$$

Table 3.1 summarizes the measured values of d_{il} and other related data of commonly used crystals for SHG.[5-7] In this table, unless noted otherwise, the incident fundamental wave is assumed to be a 1064-nm pulsed laser beam with ~10-ns pulse duration.

3.1.4 SHG devices

To date, SHG still is the most useful and well-developed technique for frequency up-conversion of a laser beam or other coherent optical emission. The mostly used fundamental-wave sources are Nd-doped solid lasers and Ti:sapphire lasers. In earlier years, the power (or energy) conversion efficiency from the fundamental wave to the second-harmonic wave was 10–20% for most experimental and commercial devices. Now this efficiency has been raised to larger than 30–50%. For some optimum experimental devices, the conversion efficiency of SHG can be up to ~80%.[8,9]

In principle, there are three basic designs for SHG devices, which are shown schematically in Fig. 3.5, featuring with (a) a single-pass configuration, (b) an external cavity configuration, and (c) an intracavity configuration, respectively. The design shown in Fig. 3.5(a) is the most commonly

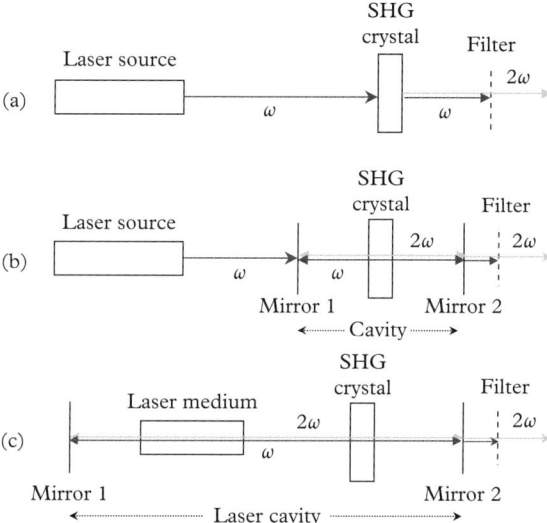

Figure 3.5 *Fundamental designs for SHG devices.*

employed geometry, which offers the advantage of simplicity and compactness; however, in this case, the conversion efficiency is usually limited to 30–50%. In the case shown in Fig. 3.5(b), the nonlinear crystal is placed in an external resonant cavity consisting of two mirrors; mirror 1 has a high reflectivity for the 2ω beam and low reflectivity for the ω beam, while mirror 2 possesses a high reflectivity for the ω beam and low reflectivity for the 2ω beam; so that one can expect a higher intracavity intensity for the ω beam and a higher output coupling for the 2ω beam.[10] Based on this design the highest conversion efficiency can be reached.[8,9] The benefit of using an external cavity may be related to the increase of effective interaction length as well as the constructive interference enhancement of the intracavity intensity. In the case shown in Fig. 3.5(c) the nonlinear crystal is placed inside the optical resonant cavity of the laser source itself. This design is mainly suitable for low-gain laser system in which the intracavity intensity of the lasing beam of ω is much higher than the output laser-beam intensity. Placing the SHG crystal inside the laser cavity may lead to a higher conversion efficiency.[11] It is obvious that in the cases of Fig. 3.5(b) and (c), the advantage of simplicity and compactness no longer exists.

Since the efficiency of SHG is determined by many factors, such as the fundamental-wave intensity, effective second-order susceptibility value, effective interaction length, as well as the damage threshold for a given nonlinear crystal, there is no simple rule to determine which configuration is the best for a specific case, instead it is often a technical choice.

3.2 Optical sum- and difference-frequency generation

3.2.1 Optical sum-frequency generation

When an intense coherent optical field consisting of two monochromatic components of (angular) frequency ω_1 and ω_2 passes through a second-order nonlinear crystal, a new coherent radiation at the sum frequency of $\omega_3 = \omega_1 + \omega_2$ may be obtained.[12] The optical sum-frequency generation in this case arises from the following nonlinear polarization component:

$$\mathbf{P}^{(2)}(\omega_3, r) = \varepsilon_0 \chi^{(2)}(\omega_1, \omega_2)\mathbf{E}(\omega_1, r)\mathbf{E}(\omega_2, r). \tag{3.2-1}$$

Assume the two incident monochromatic components are linearly polarized plane waves propagating along the z-axis, i.e.,

$$\left. \begin{array}{l} \mathbf{E}(\omega_1, z) = \mathbf{a}_1 A_1(z)e^{ik_1 z} \\ \mathbf{E}(\omega_2, z) = \mathbf{a}_2 A_2(z)e^{ik_2 z} \end{array} \right\}, \tag{3.2-2}$$

where $\mathbf{a}_1$ and $\mathbf{a}_2$ are the unit vectors of these two waves along their polarization directions, then Eq. (3.2-1) becomes

$$\mathbf{P}^{(2)}(\omega_3, z) = \varepsilon_0 \chi^{(2)}(\omega_1, \omega_2)\mathbf{a}_1 \mathbf{a}_2 A_1(z) A_2(z) e^{i(k_1 + k_2)z}. \tag{3.2-3}$$

The corresponding sum-frequency field can be written as

$$\mathbf{E}(\omega_3 = \omega_1 + \omega_2, z) = \mathbf{a}_3 A_3(z)e^{ik_3 z}. \tag{3.2-4}$$

Comparing Eq. (3.2-4) with Eq. (3.2-3) we can see that, if $k_3 \neq (k_1 + k_2)$, the phase-velocity of the polarization wave of $\mathbf{P}^{(2)}(\omega_3, z)$ is different from that of the radiation wave of $\mathbf{E}(\omega_3, z)$; therefore, the power transfer among the three optical waves cannot take place effectively. Only in the condition of

$$\Delta k = k_3 - (k_1 + k_2) \Rightarrow 0, \tag{3.2-5}$$

the phase matching can be achieved and the sum-frequency wave can be effectively generated. In general, owing to dispersion effect we always have $k_3 \neq (k_1 + k_2)$. However, as mentioned in Section 3.1, we can use the birefringence effect of the refractive index to compensate the dispersion effect in a second-order nonlinear crystal.

Under phase-matching condition, we can write down the complete expressions for the three related nonlinear polarization components of $\mathbf{P}^{(2)}(\omega_3, z)$, $\mathbf{P}^{(2)}(\omega_1, z)$, and $\mathbf{P}^{(2)}(\omega_2, z)$ (see Eq. (2.6-4)). Substituting them into the coupled-wave equations of Eq. (2.5-13), the solutions of these three waves could be finally obtained.

The phase-matching condition for optical sum-frequency generation can also be derived within the regime of quantum theory of radiation by considering the conservation of energy and momentum among the three interacting optical waves. Based on Fig. 1.3, the energy-level diagram for sum-frequency generation is shown in Fig. 3.6. In the present case, the elementary process can occur through three quantum transitions. First, there is an annihilation of one photon from the ω_1 wave while a molecule of the nonlinear medium is excited to an intermediate state (1), then another photon from the ω_2 wave is annihilated while the molecule is further excited to another intermediated state (2), finally without any delay this molecule returns to the ground state with emitting a sum-frequency photon. As the molecular staying time in these two intermediate states are infinitely short, the whole process takes place instantaneously as a single real physical event.

Since at the beginning and end of this elementary process the quantum states of the molecule are actually unchanged, the conservation of energy and momentum holds only between the annihilated photons and the created photon. Thus, we have

$$\left. \begin{array}{l} h\nu_3 = h\nu_1 + h\nu_2 \\ \mathbf{k}_3 = \mathbf{k}_1 + \mathbf{k}_2 \end{array} \right\}. \tag{3.2-6}$$

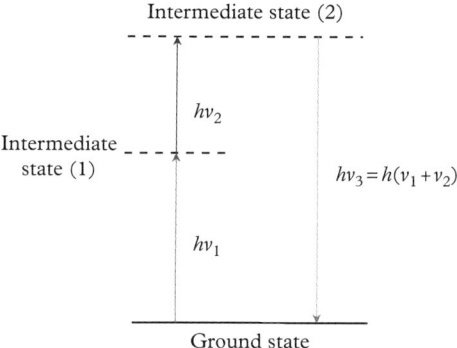

Figure 3.6 *Quantum transition diagram describing the optical sum-frequency generation in a second-order nonlinear crystal.*

Here $\mathbf{k}_1$, $\mathbf{k}_2$, and $\mathbf{k}_3$ are wavevectors of these three waves. For collinear interaction, $\mathbf{k}_1$ and $\mathbf{k}_2$ are in the same direction; then from the second relation in Eq. (3.2-6) we obtain the following refractive-index matching requirement:

$$\frac{2\pi}{\lambda_3} n(\nu_3) = \frac{2\pi}{\lambda_1} n(\nu_1) + \frac{2\pi}{\lambda_2} n(\nu_2), \tag{3.2-7}$$

where λ_1, λ_2, and λ_3 are the wavelengths of these three waves in free space. As $\nu = c/\lambda$ (where c is the light speed in free space), Eq. (3.2-7) can be rewritten as

$$\nu_3 n(\nu_3) = \nu_1 n(\nu_1) + \nu_2 n(\nu_2). \tag{3.2-8}$$

This result is exactly the same as required by Eq. (3.2-5) based on the semi-classical consideration.

Finally, it should be pointed that the SHG and sum-frequency generation are essentially the same kind of second-order nonlinear processes; in other words, SHG can be viewed as a degenerate ($\omega_1 = \omega_2$) sum-frequency generation. Therefore, all crystal materials suitable for the former can also be used for the latter. In practice, researchers sometimes utilize two nonlinear crystals: the first is for SHG, the second is for the mixing between the second harmonic (2ω) wave generated by the first crystal and the transmitted (residual) fundamental (ω) wave. After passing through the second crystal, a sum-frequency wave at frequency of 3ω will be obtained provided that the phase-matching requirements for these two crystals can be fulfilled, respectively.

3.2.2 Optical difference-frequency generation

Assuming that the incident optical field contains two different frequency components of ω_1 and ω_2 with $\omega_1 > \omega_2$, a new radiation wave at the difference-frequency of $\omega_3 = \omega_1 - \omega_2$ can be generated in a second-order nonlinear crystal.[13,14] The corresponding polarization source of the new radiation is

$$\mathbf{P}^{(2)}(\omega_3, r) = \varepsilon_0 \chi^{(2)}(\omega_1, -\omega_2) \mathbf{E}(\omega_1, r) \mathbf{E}^*(\omega_2, r). \tag{3.2-9}$$

46   Second-Order Nonlinear (Three-Wave) Frequency Mixing

Moreover, if the two incident waves can be described by Eq. (3.2-2), the above expression can be rewritten as

$$\mathbf{P}^{(2)}(\omega_3, z) = \varepsilon_0 \chi^{(2)}(\omega_1, -\omega_2)\mathbf{a}_1\mathbf{a}_2 A_1(z)A_2^*(z)e^{i(k_1-k_2)z}. \tag{3.2-10}$$

The corresponding difference-frequency radiation field is

$$\mathbf{E}(\omega_3 = \omega_1 - \omega_2, z) = \mathbf{a}_3 A_3(z)e^{ik_3 z}. \tag{3.2-11}$$

Comparing Eq. (3.2-11) with Eq. (3.2-10) one can see that, if $k_3 \neq (k_1 - k_2)$, the phase velocity of the polarization wave of $\mathbf{P}^{(2)}(\omega_3, z)$ is different from that of the radiation wave of $\mathbf{E}(\omega_3, z)$; therefore, the energy transfer among the three optical waves cannot take place effectively. Only in the case of

$$\Delta k = k_3 - (k_1 - k_2) \Rightarrow 0, \tag{3.2-12}$$

the phase matching can be achieved and the difference-frequency wave can be effectively generated. From the above requirement, we obtain the refractive-index matching condition (for collinear interaction):

$$\nu_3 n(\nu_3) = \nu_1 n(\nu_1) - \nu_2 n(\nu_2). \tag{3.2-13}$$

Owing to the dispersion effect, this requirement cannot be fulfilled automatically, however, in a special direction or a certain temperature of the nonlinear crystal it may be satisfied as we mentioned before for SHG.

Under phase-matching condition, the solutions of nonlinear coupled wave equations will show us that, while the high-frequency wave of $\mathbf{E}(\omega_1, z)$ is getting weaker along the propagation direction z, both the new radiation wave of $\mathbf{E}(\omega_3, z)$ and the low-frequency incident wave of $\mathbf{E}(\omega_2, z)$ are growing up along the same direction. This implies that the optical power flows from the high-frequency wave to other two low-frequency waves. From the viewpoint of quantum theory of radiation, here we are dealing with an elementary process involving the annihilation of one high-frequency photon ($h\nu_1$) and the simultaneous creation of two low-frequency photons ($h\nu_3$ and $h\nu_2$). The principle of difference-frequency generation is actually the same as that of the optical parametric amplification that will be discussed in detail in the next section of this chapter.

Optical difference-frequency generation is a special technique of frequency down-conversion; in principle, it can be used to generate coherent radiation from IR to sub-millimeter wavelength range. The requirements of nonlinear crystals for optical difference-frequency generation are basically the same as that for SHG.

In an extreme case, when the frequency difference between the two incident monochromatic wave is very small (i.e., $\omega_2 \Rightarrow \omega_1$) we have

$$\mathbf{P}^{(2)}(\omega_3 \Rightarrow 0) = \varepsilon_0 \chi^{(2)}(\omega_1, -\omega_2)\mathbf{E}(\omega_1)\mathbf{E}^*(\omega_2 \Rightarrow \omega_1). \tag{3.2-14}$$

In this case, one may observe the generation of an induced low-frequency or even dc electric field in a second-order crystal. This is the so-called optical rectification effect, which can be utilized to generate laser excitation-based coherent terahertz radiation and will be further discussed in Chapter 17.

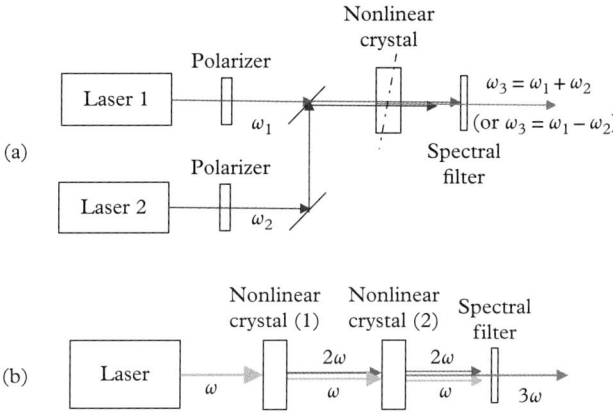

Figure 3.7 *Optical layouts of optical sum-frequency generation based on (a) the output from two different laser source and (b) the fundamental wave and its second-harmonic wave.*

3.2.3 Experimental setups for optical sum- and difference-frequency generation

Figure 3.7 shows the two basic layouts for optical sum- or deference-frequency generation experiment. In Fig. 3.7(a), there are two laser sources with different output frequencies and one second-order crystal for either sum-frequency mixing or difference-frequency generation. To reach a higher efficiency, it is better to ensure a collinear interaction between the three waves of different frequencies. The newly generated coherent emission beam of frequency ω_3 can be separated from other two beams via passing through a spectral dispersion element (prism or grating), a dichromatic reflector, or an appropriate spectral filter.

In practice, the setup shown in Fig. 3.7(b) is more commonly used for efficient frequency up-conversion, based on only one laser source with output frequency of ω. In this configuration, two second-order nonlinear crystals are adopted: the first one is for SHG of 2ω, whereas the second one is for sum-frequency generation of 3ω. The efficiency of energy transfer from the beam of ω to the sum-frequency beam of 3ω could be up to 20–30%.

3.3 Optical parametric amplification and oscillation

3.3.1 General description

Now we discuss the so-called optical parametric amplification in a second-order nonlinear crystal.[15] As indicated in Section 3.2.2, the principle of optical difference-frequency generation is essentially the same as that of optical parametric amplification except that, for the former the key issue is the generation of a difference-frequency wave, but for the latter the key issue is the amplification of a low-frequency incident wave. Here we assume that the incident optical fields contain two monochromatic waves: a strong pump beam of ω_p and a weak signal beam of ω_s with $\omega_p > \omega_s$. As a result of second-order polarization within a nonlinear crystal in phase matching condition,

Second-Order Nonlinear (Three-Wave) Frequency Mixing

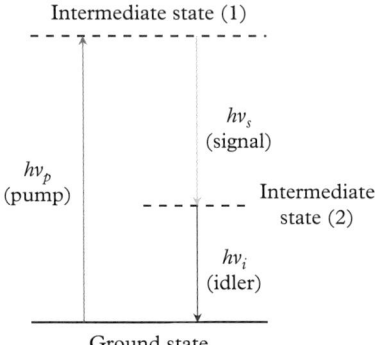

Figure 3.8 *Quantum transition description of the optical parametric amplification.*

the ω_s wave can be amplified, meanwhile a new coherent wave at the idler frequency of $\omega_i = \omega_p - \omega_s$ will be simultaneously generated.

From the viewpoint of quantum theory of radiation, the mechanism of the above process can be clearly interpreted. Figure 3.8 shows the diagram describing the quantum transition behavior of an elementary optical parametric amplification process. This elementary process can take place through three quantum transitions. First, there is an annihilation of one photon from the ω_p wave while a molecule of the nonlinear medium is excited to an intermediate state (1), without any delay this molecule can make a quantum transition to another intermediate state (2) with the creation of a photon of frequency ω_s, then the molecule returns to its ground state with a simultaneous creation of a new photon of frequency $\omega_i = \omega_p - \omega_s$. As the molecular staying time at each intermediate state is infinitely short and all related quantum transitions occur simultaneously, so that this elementary process is an instantaneous real physical event.

Since at the beginning and end of this elementary process the quantum state of the molecule is unchanged, the conservation of energy and momentum should hold only between the annihilated pump photon and the created new photons. From this consideration, we obtain the following relations:

$$\left.\begin{array}{c} h\nu_p = h\nu_s + h\nu_i \\ \mathbf{k}_p = \mathbf{k}_s + \mathbf{k}_i \end{array}\right\}. \qquad (3.3\text{-}1)$$

Assuming these three waves are propagating along the same direction (collinear interaction), from the second equality of Eq. (3.3-1) we obtain the following phase-matching requirement:

$$\nu_p n(\nu_p) = \nu_s n(\nu_s) + \nu_i n(\nu_i). \qquad (3.3\text{-}2)$$

Based on the above quantum picture, the decrease in the number of pump photons should be equal to the increase in the number of signal photons and in the number of the newly generated idler photons. In other words, if the decrease of the photon number of the pump beam inside a nonlinear crystal is $-\Delta m$, then the increase of the photon number for the signal wave (as well as for the idle wave) should be $+\Delta m$. This relationship can be expressed as:

$$-\frac{\partial I(\nu_p, z)}{h\nu_p \partial z} = \frac{\partial I(\nu_s, z)}{h\nu_s \partial z} = \frac{\partial I(\nu_i, z)}{h\nu_i \partial z}. \tag{3.3-3}$$

Here $I(\nu, z)$ is the intensity of the wave connected with frequency ν. This relation is equivalent to the so-called Manley–Rowe relations obtained from the theory of electronics.[16]

If we place a second-order nonlinear crystal inside an optical resonant cavity and provide only a strong pump wave, once the phase-matching condition of Eq. (3.3-1) is satisfied, the optical oscillation at two frequencies (ν_p and ν_i) may be achieved. This is termed optical parametric oscillation effect.[17] In this case, the initial signal wave is from the spontaneous thermal radiation or due to the intrinsic fluctuation of the quantized radiation field. Compared to the optical parametric amplification, the advantages of optical parametric oscillation are (i) only one pump beam is needed, (ii) the parametric oscillating frequency can be continuously tuned by changing the angle or temperature of the nonlinear crystal, and (iii) the spatial and temporal properties of the oscillating beams can be controlled by using appropriate optical cavity designs.

3.3.2 Solutions of coupled-wave equations

We consider the optical parametric amplification as a three-wave mixing process occurring in a second-order nonlinear crystal, and assume the three waves are linearly polarized and propagating along z-axis:

$$\left. \begin{array}{l} \mathbf{E}(\omega_p, z) = \mathbf{a}_p A_p(z) e^{ik_p z} \\ \mathbf{E}(\omega_s, z) = \mathbf{a}_s A_s(z) e^{ik_s z} \\ \mathbf{E}(\omega_i = \omega_p - \omega_s, z) = \mathbf{a}_i A_i(z) e^{ik_i z} \end{array} \right\}. \tag{3.3-4}$$

Accordingly, the nonlinear polarization sources of these waves can be expressed as the following forms:

$$\left. \begin{array}{l} \mathbf{P}^{(2)}(\omega_p, z) = \varepsilon_0 \chi^{(2)}(\omega_s, \omega_i) \mathbf{a}_s \mathbf{a}_i A_s(z) A_i(z) e^{i(k_s + k_i)z} \\ \mathbf{P}^{(2)}(\omega_s, z) = \varepsilon_0 \chi^{(2)}(\omega_p, -\omega_i) \mathbf{a}_p \mathbf{a}_i A_p(z) A_i^*(z) e^{i(k_p - k_i)z} \\ \mathbf{P}^{(2)}(\omega_i, z) = \varepsilon_0 \chi^{(2)}(\omega_p, -\omega_s) \mathbf{a}_p \mathbf{a}_s A_p(z) A_s^*(z) e^{i(k_p - k_s)z} \end{array} \right\}. \tag{3.3-5}$$

Substituting Eq. (3.3-5) into the nonlinear coupled-wave equations of Eq. (2.5-13) leads to

$$\left. \begin{array}{l} \dfrac{\partial A_p(z)}{\partial z} = \dfrac{ik_p}{2n^2(\omega_p)} [\mathbf{a}_p \cdot \chi^{(2)}(\omega_s, \omega_i) \mathbf{a}_s \mathbf{a}_i] A_s A_i e^{-i\Delta k z} \\ \dfrac{\partial A_s(z)}{\partial z} = \dfrac{ik_s}{2n^2(\omega_s)} [\mathbf{a}_s \cdot \chi^{(2)}(\omega_p, -\omega_i) \mathbf{a}_p \mathbf{a}_i] A_p A_i^* e^{i\Delta k z} \\ \dfrac{\partial A_i(z)}{\partial z} = \dfrac{ik_i}{2n^2(\omega_i)} [\mathbf{a}_i \cdot \chi^{(2)}(\omega_p, -\omega_s) \mathbf{a}_p \mathbf{a}_s] A_p A_s^* e^{i\Delta k z} \end{array} \right\}, \tag{3.3-6}$$

where $\Delta k = k_p - (k_s + k_i)$ is the phase mismatch factor. According to Eqs. (2.4-5) and (2.4-9), the following relation holds:

$$\chi_e^{(2)} = \mathbf{a}_p \cdot \chi^{(2)}(\omega_s, \omega_i) \mathbf{a}_s \mathbf{a}_i = \mathbf{a}_s \cdot \chi^{(2)}(\omega_p, -\omega_i) \mathbf{a}_p \mathbf{a}_i = \mathbf{a}_i \cdot \chi^{(2)}(\omega_p, -\omega_s) \mathbf{a}_p \mathbf{a}_s. \tag{3.3-7}$$

Here, $\chi_e^{(2)}$ is the effective nonlinear susceptibility value of the crystal for this process. Assuming the phase-matching condition of $\Delta k \Rightarrow 0$ is satisfied, the initial boundary conditions of the three waves are

$$\left.\begin{aligned}A_p(z=0) &= A_p(0)\\ A_s(z=0) &= A_s(0) \ll A_p(0)\\ A_i(z=0) &= A_i(0) = 0\end{aligned}\right\}, \qquad (3.3\text{-}8)$$

and the depletion of $A_1(z)$ in a short propagation distance can be nearly neglected in small-signal approximation, then Eq. (3.3-6) can be simplified as

$$\left.\begin{aligned}\frac{\partial A_s(z)}{\partial z} &= \frac{ik_s}{2n^2(\omega_s)} \chi_e^{(2)} A_p(0) A_i^*(z)\\ \frac{\partial A_i(z)}{\partial z} &= \frac{ik_i}{2n^2(\omega_i)} \chi_e^{(2)} A_p(0) A_s^*(z)\end{aligned}\right\}. \qquad (3.3\text{-}9)$$

The solutions of the above equations, which satisfy the boundary condition of Eq. (3.3-8), can be obtained as[3,4]

$$\left.\begin{aligned}A_s(z) &= A_s(0)\cosh(\gamma_0 z)\\ A_i^*(z) &= iA_s(0)\sqrt{\frac{\lambda_s n(\omega_s)}{\lambda_i n(\omega_i)}} \sinh(\gamma_0 z)\end{aligned}\right\}, \qquad (3.3\text{-}10)$$

where

$$\gamma_0 = \frac{\pi \chi_e^{(2)}}{\sqrt{\lambda_s \lambda_i n(\omega_s) n(\omega_i)}} A_p(0). \qquad (3.3\text{-}11)$$

Assuming that the following relation is fulfilled:

$$\gamma_0 z \gg 1,$$

the hyperbolic functions in Eq. (3.3-10) can be approximately replaced by the exponential functions, i.e.,

$$\left.\begin{aligned}A_s(z) &\approx \frac{1}{2} A_s(0) e^{\gamma_0 z}\\ A_i^*(z) &\approx \frac{1}{2} iA_s(0) \sqrt{\frac{\lambda_s n(\omega_s)}{\lambda_i n(\omega_i)}} \cdot e^{\gamma_0 z}\end{aligned}\right\}. \qquad (3.3\text{-}12)$$

From the above final solutions we can see that under the condition of a weak input signal and neglecting the depletion of the strong pump wave, both the signal wave and the idler wave experience an exponential gain; the gain coefficient is γ_0 that is proportional to $\chi_e^{(2)}$ and the initial amplitude $A_p(0)$ of the pump wave.

In order to achieve the optical parametric oscillation, we can place a second-order nonlinear crystal inside an optical cavity consisting of two mirrors: the first mirror having a high transmissivity for the input pump beam and high reflectivity for both the signal and idler beams, and the second mirror having a modest reflectivity for both the signal and idler beams. In this case, the net gain for a round-path within the cavity is determined by

$$\exp[(\gamma_0 - 2\alpha_0)l],$$

where α_0 is the linear attenuation coefficient and l is the optical path length of the crystal. One can see that the attenuation length is $2l$, while the effective gain length is only l because the backward signal and idler beams reflected from the output mirror cannot get gain from the forward pump beam. Assuming the reflectivity of the output mirror is R, the threshold condition for optical parametric oscillation is determined by

$$R \exp[(\gamma_0 - 2\alpha_0)l] \geq 1. \tag{3.3-13}$$

The above expression can be rewritten as

$$\gamma_0 \geq \frac{1}{l}\left(\ln\frac{1}{R} + 2\alpha_0 l\right), \tag{3.3-14}$$

Finally, substituting the γ_0 expression of Eq. (3.3-11) into Eq. (3.3-14) leads to the following threshold condition:

$$A_p(0) \geq \frac{\sqrt{\lambda_s \lambda_i n(\omega_s) n(\omega_i)}}{\pi \chi_e^{(2)}} \frac{1}{l}\left(\ln\frac{1}{R} + 2\alpha_0 l\right). \tag{3.3-15}$$

Since $A_p(0) \propto \sqrt{I_p(0)}$, Eq. (3.3-15) implies that for a given nonlinear medium and interaction length l, only when the input pump beam intensity is higher than a certain threshold the optical parametric oscillation can be achieved.

3.3.3 Experimental devices

The requirements of nonlinear crystals for optical parametric amplification or oscillation are essentially the same as that for SHG. In other words, the nonlinear materials must be non-centrosymmetrical crystals, highly transparent for pump, signal and idler beams, able to fulfill the phase matching by using angle tuning or temperature tuning. In principle, all commonly used SHG crystals mentioned in Section 3.1.3 as well as those listed in Table 3.1 can be employed for optical parametric amplification and oscillation purposes.

Figure 3.9 shows the typical experimental setup for optical parametric amplification (OPA) measurement. In this case, a strong input pump beam and a weak input signal beam are launched to a second-order nonlinear crystal in phase-matching manner, while at the output end of the crystal there will be three beams, i.e., the depleted pump beam, the amplified signal beam, and he newly generated idler beam. These three beams can be separated by a dispersive element (prism or grating) or by appropriate monochromatic filters with narrow spectral pass-band.

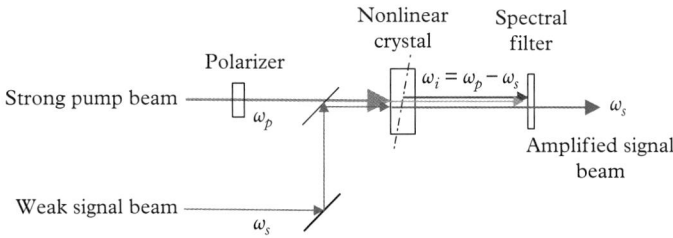

Figure 3.9 *Experimental setup for optical parametric amplification (OPA) measurement.*

Table 3.2 Experimental data for OPA performances.

Crystal	Pump Wavelength (μm)	OPA Wavelength (μm)	Tuning Method	Pump Condition	Signal Gain (dB)	Conversion Efficiency (%)
LiNbO$_3$	1.064	1.9		70 MW/cm^2 8 ns	21.3	22.5 [18]
LiNbO$_3$	1.064	λ_i = 1.5–1.58 λ_s = 3.3–3.65	φ-tuning	7 ns 10 Hz	12.6	18 [19]
Poled-LiNbO$_3$	0.782	λ_s = 1.55 λ_i = 1.58		100 ns 100 Hz	4.1	[20]
MgO:LiNbO$_3$	0.778	λ_s = 0.9–1.1 λ_i = 2.6–4.5	φ-tuning	10 GW/cm^2 2 ps		18 [21] (quantum)
β-BaB$_2$O$_4$	0.355	λ_s = 0.43–2	φ-tuning	2.7 GW/cm^2 15 ps	28.5	30 [22]
β-BaB$_2$O$_4$	0.3547	λ_s = 0.42–2.8		424 MW/cm^2 30 ps	14	61 [23]
β-BaB$_2$O$_4$	0.412	0.48–0.82	φ-tuning	13 GW/cm^2 140 fs		14–19 [24]
β-BaB$_2$O$_4$	0.790	λ_s = 1.1–1.6	φ-tuning	50 fs 1 kHz	15.2	45 [25]
LiB$_3$O$_5$	0.532	λ_s = 0.72–2	T-tuning	3.1 GW/cm^2 35 ps	30	20 [26]
LiB$_3$O$_5$	0.53	λ_s = 0.65–2.5	T-tuning φ-tuning	4.4 GW/cm^2 25 ps	31.8	24 [27]
KTiOPO$_4$	~0.8	λ_s = 1.02–1.2 λ_i = 2.6–3.7	φ-tuning	41 GW/cm^2 1.7 ps		25 [28]
KNbO$_3$	0.8–0.82	λ_s = 1.053 λ_i = 3–4	φ-tuning	150 GW/cm^2 200 fs	54.8	15 [29]
RbTiOAsO$_4$	~0.8	λ_s = 1.053 λ_i = 3–4	φ-tuning	200 GW/cm^2 100 fs	45.5	3 [30]

As an example, Table 3.2 summarizes some experimental data on OPA performances. For a pulsed laser pump and small signal input, the gain of parametric amplification can be as high as 30–50 dB, the energy transfer efficiency from the pump pulse to the signal pulse can be more than 20–40%.

Although the principles for optical parametric amplification and for optical parametric oscillation are exactly the same, the latter is more useful from the viewpoint of applications. This is because only one pump beam is needed and the continuous tuning of wavelengths of the output beams can be obtained. To explain the frequency-tuning ability of an optical parametric oscillator, we may return to Eqs. (3.3-1) and (3.3-2), and rewrite the phase-matching requirements for an optical parametric oscillation (OPO) process as

$$\left.\begin{array}{l} \nu_p = \nu_s + \nu_i \\ \nu_p n(\nu_p) = \nu_s n(\nu_s) + \nu_i n(\nu_i) \end{array}\right\}. \qquad (3.3\text{-}16)$$

Here, ν_p, ν_s, and ν_i are the frequencies of the pump, signal and idler wave; $n(\nu_p)$, $n(\nu_s)$, and $n(\nu_i)$, respectively, are their refractive indices. Under appropriate arrangement for the angle (or temperature) of a given nonlinear crystal, the above two requirements can be satisfied and the oscillation at two different frequencies (ν_s and ν_i) can be achieved. Based on this working condition, if we slightly change the angle or temperature of the crystal, the refractive-index relation between these three waves is changed, therefore the oscillating frequencies can be continuously tuned to different values. For some nonlinear crystals that possess a larger electro-optical or magneto-optical effect of refractive index, the oscillating frequency could also be tuned by applying an external electric or magnetic dc field.

Figure 3.10(a) shows the typical experimental setup for OPO operation, in which the pump beam is a frequency-doubled laser beam, the OPO crystal is placed in an external cavity, and the frequency tuning relies on the change of orientation angle of the crystal. In Fig. 3.10(b), a strong pump beam of high intensity simply passes through the nonlinear crystal only once to generate simultaneous signal and idler beams together without the need of a cavity, such a device is termed optical parametric generator (OPG).

For the updated OPO devices pumped with a high-peak power pulsed laser beam, the conversion efficiency from the pump pulse energy to the output pulse energy can be higher than 50–70%. Listed in Table 3.3 are some experimental data for the typical pulsed OPO devices.

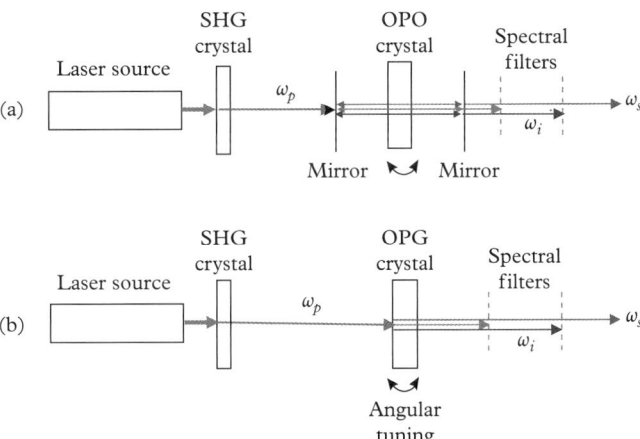

Figure 3.10 *Experimental setups for (a) optical parametric oscillation (OPO) and (b) optical parametric generation (OPG).*

Table 3.3 Experimental data for OPO performances.

Crystal	Pump Wavelength (μm)	OPO Wavelength (μm)	Tuning Method	Pump Condition	Conversion Efficiency (%)
LiNbO$_3$	0.6943		φ-tuning	150 MW/cm^2 30 ns	45 [31]
LiNbO$_3$	1.054	1.35–2.11	φ-tuning	150 MW/cm^2 1.3 ps	15 [32]
MgO:LiNbO$_3$	0.532	0.966–1.185	T-tuning	300 W (cw)	38.3 [33]
MgO:LiNbO$_3$	0.532	1.0–1.12	T-tuning	100 μs	84 [34]
Poled-LiNbO$_3$	1.047	λ_s = 1.67–2.81	T-tuning	2 kW 2.4 ps	43 [35]
Poled-LiNbO$_3$	1.064	1.45–4	T-tuning	10 ns	70 [36]
β-BaB$_2$O$_4$	0.532	λ_s = 0.7–1 λ_i = 1.2–1.8	φ-tuning	2 GW/cm^2 85 ps	30 [37]
β-BaB$_2$O$_4$	0.360	0.406–3.17	φ-tuning	2 GW/cm^2 25 ps	30 [38]
β-BaB$_2$O$_4$	0.355	0.430–2	φ-tuning	2 00 mJ/pulse 10 Hz	61 [39]
LiB$_3$O$_5$	0.3457	0.435–1.922	φ-tuning	45 MW/cm^2 10 ns	22 [40]
LiB$_3$O$_5$	0.532	0.9–1.23	T-tuning	5 GW/cm^2 55 ps	50 [41]
LiB$_3$O$_5$	0.5145	λ_s = 0.932–0.946	T-tuning	3.4 W (cw)	28 [42]
KTiOPO$_4$	0.532	1.06–1.09	φ-tuning	650 MW/cm^2 85 ps	30 [43]
KTiOPO$_4$	0.780	λ_s = 1.2–1.38 λ_t = 0.6–0.68	φ-tuning	2.5 W 125 fs, 90 MHz	55 [44]
KTiOPO$_4$	0.527	1.01–1.1	φ-tuning	30 ps 76 MHz	44 [45]
KTiOAsO$_4$	0.780	λ_s = 1.29–1.44 λ_t = 1.83–1.91	φ-tuning	1.2 W 90 fs	10~15 [46]
ZnGeP$_2$	3	4–10	φ-tuning	1 GW/cm^2 100 ps	17.6 [47] (quantum)

3.4 Special second-harmonic generation

3.4.1 Other materials for SHG

So far we have only considered inorganic piezoelectric crystals as the major second-order nonlinear materials for highly efficient SHG, OPA, OPO, and OPG. Actually, researchers have also paid considerable attention to exploiting organic bulk crystals for SHG. Of all the reported organic materials for second-order nonlinear optical applications, the crystals of urea,[48,49] NPP (*N*-(4-nitrophenyl)-L-prolinol),[50,51] and DAST (4-dimethylamino-*N*-methylstilbazolium tosylate)[52,53] can offer a superior performance for SHG or OPO. Compared to inorganic crystals, the organic compounds may have a much higher second-order nonlinearity, but sometimes might be limited by the available crystal size and lower optical damage threshold.

SHG is useful not only for coherent optical frequency-upconversion, but also for fundamental studies of science. We know that besides the inorganic and organic bulk crystals, other materials and substances, which are not totally disordered or not in perfect centrosymmetry, can also be employed for observing the second-harmonic signal by using IR laser beam excitation. For example, polycrystal, crystal powder, microcrystal, liquid crystal, and crystal film or waveguide options can also produce strong SHG signals. In addition, the anisotropic polymer bulk or film, dc-field poled polymer film, as well as the periodically poled polymer waveguide can also offer a considerable SHG yield. Even some kinds of wood, paper, fabrics, human skin, some medical and biological samples can also generate easily observed second-harmonic signal induced by high peak power IR laser pulses. Through these special SHG studies, researchers could acquire a meaningful knowledge of the sample structures and constitution properties, which may not be available by using other technical approaches.

As will be explained in Chapter 18 by the detailed nonlinear-susceptibility theory, even for isotropic or centrosymmetrical media in which the second-order susceptibility vanishes in the approximation of electric-dipole interaction, there is still a relatively weak second-order nonlinearity if we consider the contribution in the electric-quadrupole and magnetic-dipole approximation. In fact, the observation of a very weak second-harmonic signal from a calcite crystal of centrosymmetry was among the first observations of nonlinear optical effects.[54]

If we apply a dc electric field to an isotropic or centrosymmetric medium, a considerably enhanced second-harmonic signal can be observed.[54] In this case, the observed second-harmonic signal is mainly due to the third-order nonlinearity that we shall discuss in the next chapter.

3.4.2 SHG from surfaces and interfaces

At the surface of a medium or the interface between two different media, the local symmetry property may be different from that of the isotropic or centrosymmetric bulk medium; consequently, a local second-order nonlinearity may arise at the surface or interface. Let us consider two special circumstances. The first occurs at the interface between two isotropic or centrosymmetrical media. At the boundary, the microscopic molecular interaction field for each medium undergoes a sudden change and the centrosymmetry within this thin interface layer may be destroyed. In this case, the thin interface layer, which may involve only a single or few molecular monolayers, can be recognized as a quasi-macroscopic second-order nonlinear medium that may produce a stronger SHG signal than that from the bulk substrates. The other circumstance occurs when a group of chosen molecules are adsorbed on the surface of an isotropic or centrosymmetrical medium. Through the interaction between the adsorbed molecular monolayer and the surface layer of the substrate, the adsorbed molecules may manifest a regular orientation on the surface. As a result, the centrosymmetry is locally destroyed and a SHG signal might be observed due to the local second-order nonlinearity.

As an example, let us further consider the second circumstance. In this case, the second-order susceptibility in the surface region for SHG process can be written as

$$\chi^{(2)}_{xyz} = N_s L_{xyz} \langle T^{x'y'z'}_{xyz} \rangle \alpha^{(2)}_{x'y'z'}, \qquad (3.4\text{-}1)$$

where N_s is the adsorbed molecule number per unit area on the substrate surface, $\alpha^{(2)}_{x'y'z'}$ is the second-order hyperpolarizability of a single adsorbed molecule, $\langle T^{x'y'z'}_{xyz} \rangle$ is a matrix coefficient characterizing the transfer from the coordinates of the adsorbed molecule to the laboratory coordinates, and L_{xyz} is a correction factor of the local-field near the surface. Here, $\alpha^{(2)}_{x'y'z'}$ is determined by the structure of the adsorbed molecule as well as the interaction between the adsorbed molecules and the substrate, and L_{xyz} is related to the regularity (roughness) of the surface as well as the surface-field property.

Since the 1980s, the observation of SHG from surfaces or interfaces has become a useful approach to investigate the surface nonlinearity. The investigated systems include solid–liquid, solid–gas, solid–vacuum, and liquid–gas interfaces, and so on.[55-58] The adsorbate layer can be prepared by various methods including dye solution evaporation,[59] Langmuir–Blodgett films,[60] evaporated films,[56] and an electrolytic process.[61,62] Through these studies of SHG from surfaces or interfaces, the following information can be obtained:[63,64]

(1) the second-order susceptibility values of surface-adsorbed molecular layers, molecular symmetry, and orientation behavior on the surface;
(2) the two-photon transition enhancement of surface SHG;
(3) the symmetry of atomic layer distribution on a solid surface and their phase-change properties;
(4) the rough-surface and local field-enhanced SHG.

PROBLEMS

1. Similar to Fig. 1.3, sketch a diagram showing the elementary process of the second-harmonic generation via two intermediate states.
2. For the second-harmonic generation (SHG) of a laser beam with frequency ω_0, we could use many crystals of entirely different molecular structures, but why do we always get the SHG exactly at the same frequency $2\omega_0$?
3. For SHG in positive uniaxial crystals, please indicate the two ways to meet the phase-matching requirements (cf. Eq. (3.1-22)).
4. Similar to Fig. 1.3, sketch a diagram showing the elementary process of the second-order parametric generation (i.e., the annihilation of one pump photon and the simultaneous creations of one signal photon and one idler photon).
5. For a second-order parametric amplification processes, the initial pump photon number is 1000, the initial signal photon number is 50, after the amplifier the retained pump photon number is 250. How many signal photon number and idler photon number should be there after passing through the amplifier?

REFERENCES

1. P. A. Franken, A. E. Hill, C. W. Peters, and G. Weinreich, *Phys. Rev. Lett.* **7**, 118 (1961).
2. D. A. Kleinman, *Phys. Rev.* **128**, 1761 (1962).
3. J. Ducuing, in *Quantum Optics*, edited by R. J. Glouber (Academic Press, New York, 1969), pp. 421–472.
4. R. L. Byer, in *Nonlinear Optics*, edited by P. G. Harper and B. S. Wherrett (Academic Press, New York, 1977), pp. 47–160.
5. D. N. Nikogosyan, *Soviet J. Quantum Electron.* **7**, 1 (1979).
6. R. C. Eckardt, H. Masuda, Y. X. Fan, and R. L. Byer, *IEEE J. Quantum Electron.* **26**, 922 (1990).
7. D. A. Roberts, *IEEE J. Quantum Electron.* **28**, 2057 (1992).
8. Z. Y. Ou, S. F. Pereira, and H. J. Kimble, *Opt. Lett.* **17**, 640 (1992).
9. R. Paschotta, P. Kurz, R. Henking, S. Schiller, and J. Mlynek, *Opt. Lett.* **19**, 1325 (1994).
10. A. Ashkin, G. D. Boyd, and J. M. Dziedzic, *IEEE J. Quantum Electron.* **2**, 109 (1966).
11. J. E. Geusic, H. J. Levinstein, S. Singh, R. G. Smith, and L. G. Van Uitert, *Appl. Phys. Lett.* **12**, 306 (1968).
12. M. Bass, P. A. Fraken, A. E. Hill, C. W. Peters, and G. Weinreich, *Phys. Rev. Lett.* **8**, 18 (1962).
13. K. E. Neihuhr, *Appl. Phys. Lett.* **2**, 136 (1963).
14. A. W. Smith and N. Braslou, *J. Appl. Phys.* **34**, 2105 (1963).
15. C. C. Wang and G. W. Racette, *Appl. Phys. Lett.* **6**, 169 (1965).
16. J. M. Manley and H. E. Rowe, *Proc. IRE* **44**, 904 (1956); **47**, 2115 (1959).
17. J. A. Giordmaine and R. C. Miller, *Phys. Rev. Lett.* **14**, 973 (1965).
18. R. A. Baumgartner and R. L Byer, *IEEE J. Quantum Electron.* **QE-15**, 432 (1979).
19. M. J. T. Milton, T. J. McIlveen, D. C. Hanna, and P. T. Woods, *Opt. Commun.* **93**, 186 (1992).
20. M. L. Bortz, M. A. Arbore, and M. M. Fejer, *Opt. Lett.* **20**, 49 (1995).
21. S. Lin and T. Suzuki, *Opt. Lett.* **21**, 579 (1996).
22. J. Y. Huang, J. Y. Zhang, Y. R. Shen, C. Chen, and B. Wu, *Appl. Phys. Lett.* **57**, 1961 (1990).
23. F. Huang and L. Huang, *IEEE J. Quantum Electron.* **30**, 2601 (1994).
24. S. R. Greenfield and M. R. Wasielewski, *Opt. Lett.* **20**, 1394 (1995).
25. K. R. Wilson and V. V. Yakovlev, *J. Opt. Soc. Am. B* **14**, 444 (1997).
26. J. Y. Huang, Y. R. Shen, C. Chen, and B. Wu, *Appl. Phys. Lett.* **58**, 1579 (1991).
27. S. Lin, J. Y. Huang, J. Ling, C. Chen, and Y. R. Shen, *Appl. Phys. Lett.* **59**, 2805 (1991).
28. D. E. Gragson, D. S. Alavi, and G. L. Richmond, *Opt. Lett.* **20**, 1991 (1995).
29. V. Petrov and F. Noack, *Opt. Lett.* **21**, 1576 (1996).
30. V. Petrov, F. Noack, and R. Stolzenberger, *Appl. Opt.* **36**, 1164 (1997).
31. J. Falk and J. E. Murray, *Appl. Phys. Lett.* **14**, 245 (1969).
32. R. Laenen, H. Graener, and A. Laubereau, *Opt. Commun.* **77**, 226 (1990).
33. D. C. Gerstenberger and R. W. Wallace, *J. Opt. Soc. Am. B* **10**, 1681 (1993)
34. G. Breitenbach, S. Schiller, and J. Mlynek, *J. Opt. Soc. Am. B* **12**, 2095 (1995).
35. S. D. Butterworth, V. Pruneri, and D. C. Hanna, *Opt. Lett.* **21**, 1345 (1996).
36. L. E. Myers, R. C. Eckardt, M. M. Fejer, R. L. Byer, W. R. Bosenberg, and J. W. Pierce, *J. Opt. Soc. Am. B* **12**, 2102 (1995).
37. L. J. Bromley, A. Guy, and D. C. Hanna, *Opt. Commun.* **67**, 316 (1988).
38. S. Burdulis, R. Grigonis, A. Piskarskas, G. Sinkevicius, V. Sirutkaitis, A. Fix, J. Nolting, and R. Wallenstein, *Opt. Commun.* **74**, 398 (1990).
39. A. Fix, T. Schroder, and R. Wallenstein, *J. Opt. Soc. Am. B* **10**, 1744 (1993).
40. Y. Wang, Z. Xu, D. Deng, W. Zheng, B. Wu, and C. Chen, *Appl. Phys. Lett.* **59**, 531 (1991).
41. M. Ebrahimzadeh, G. J. Hall, and A. I. Ferguson, *Appl. Phys. Lett.* **60**, 1421 (1992).
42. G. Robertson, M. J. Padgett, and M. H. Dunn, *Opt. Lett.* **19**, 1735 (1994).
43. L. J. Bromley, A. Guy, and D. C. Hanna, *Opt. Commun.* **70**, 350 (1989).

44. W. S. Pelouch, P. E. Powers, and C. L. Tang, *Opt. Lett.* **17**, 1070 (1992).
45. Ch. Grasser, D. Wang, R. Beigang, and R. Wallenstein, *J. Opt. Soc. Am. B* **10**, 2218 (1993).
46. P. E. Powers, S. Ramakrishna, C. L. Tang, and L. K. Cheng, *Opt. Lett.* **18**, 1171 (1993).
47. K. L. Vodopyanov, *J. Opt. Soc. Am. B* **10**, 1723 (1993).
48. M. J. Rosker and C. L. Tang, *J. Opt. Soc. Am. B* **2**, 691 (1985).
49. A. J. Henderson, M. Ebrahimzaden, and M. H. Dunn, *J. Opt. Soc. Am. B* **7**, 1402 (1990).
50. I. Ledoux, J. Zyss, A. Migus, D. Hulin, and A. Antonetti, *J. Appl Phys.* **64**, 3309 (1988).
51. S. X. Dou, D. Josse, and J. Zyss, *J. Opt. Soc. Am. B* **10**, 1708 (1993).
52. S. R. Marder, J. W. Perry, and W. P. Schaefer, *Science* **245**, 626 (1989).
53. S. R. Marder, J. W. Perry, and C. P. Yakymyshyn, *Chem. Mater.* **6**, 1137 (1994).
54. R. W. Terhune, P. D. Maker, and C. M. Savage, *Phys. Rev. Lett.* **8**, 404 (1962).
55. T. F. Heinz, H. W. K. Tom, and Y. R. Shen, *Phys. Rev. A* **28**, 1883 (1983).
56. G. T. Bord, Th. Rasing, J. R. R. Leite, and Y. R. Shen, *Phys. Rev. A* **30**, 519 (1984).
57. T. F. Heinz, M. M. T. Loy, and W. A. Thomson, *Phys. Rev. Lett.* **54**, 63 (1985).
58. G. Berkovic, Th. Rasing, and Y. R. Shen, *J. Opt. Soc. Am. B* **4**, 945 (1987).
59. T. F. Heinz, C. K. Chen, D. Ricard, and Y. R. Shen, *Phys. Rev. Lett.* **48**, 478 (1982).
60. G. Morowsky, R. Steinhoff, L. F. Chi, J. Hutter, and G. Wagniere, *Phys. Rev. B* **38**, 6274 (1988).
61. C. H. Lee, R. K. Chang, and N. Bloembergen, *Phys. Rev. Lett.* **18**, 167 (1967).
62. C. K. Chen, T. F. Heinz, D. Ricard, and Y. R. Shen, *Phys. Rev. Lett.* **46**, 1010 (1981).
63. N. Bloembergen, *Appl. Phys. B* **68**, 289 (1999).
64. Y. R. Shen, *Appl. Phys. B* **68**, 295 (1999).

4

Third-Order Nonlinear (Four-Wave) Frequency Mixing

Coherent light emission with new frequencies can be generated upon the action of three incident coherent waves in a third-order nonlinear optical medium. This is the so-called four-wave frequency mixing (FWFM) effect or equivalently the four-photon parametric interaction effect, which can take place in any types of transparent nonlinear media, including centrosymmetric or isotropic materials. The effective occurring of this effect requires that the phase-matching condition must be fulfilled. FWFM is a very useful physical approach to produce laser frequency conversion or frequency variable coherent emission in third-order nonlinear materials. In addition, some types of FWFM processes may also lead to an induced refractive index change that shall be discussed in the next chapter.

4.1 Various FWFM processes

In Chapter 3, we have given a concise but very clear description about the elementary mechanism of various three-wave mixing processes in terms of the quantum theory of radiation. In those cases, three-photon parametric interaction takes place in a second-order nonlinear medium. In order to reach high efficiency, the phase-matching requirement should be satisfied.

We now consider various four-wave mixing processes in a third-order nonlinear medium. The common feature of these processes is the parametric interaction between four photons in a third-order nonlinear medium that is transparent at the frequencies of the interacting four photons. Figure 4.1 depicts some major four-wave mixing effects in the quantum-transition scheme.

Figure 4.1(a) shows the third-order nonlinear sum-frequency generation, i.e., three waves of different frequencies of ν_1, ν_2, and ν_3 interact with a nonlinear medium to generate a new coherent radiation at the sum frequency ν_4. In this case, the elementary process can be viewed as an event of four steps. The first step is the annihilation of an incident photon of frequency ν_1 while a molecule undergoes a transition from its initial (ground) state to an intermediate state (i); the second step in the sequential transition of this molecule to another intermediate state (ii), accompanied by the annihilation of another incident photon of ν_2; the next step is the sequential annihilation of an incident photon of ν_3 while the molecule is excited to an new intermediate state (iii); finally without any delay the molecule returns to the ground state while there is the generation of a new photon with the sum-frequency of $\nu_4 = (\nu_1 + \nu_2 + \nu_3)$. As the time in which the molecule stays in each intermediate state is extremely short (near to the response time of electron-cloud distortion), these multiple-step processes actually occur simultaneously. With this understanding and for the

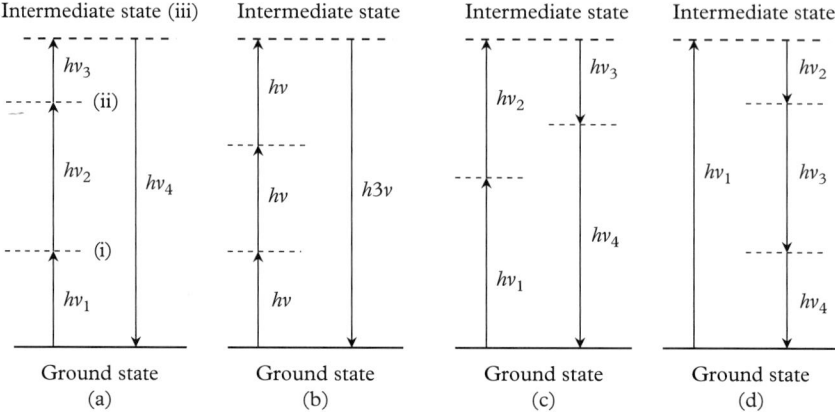

Figure 4.1 *Quantum-transition description of some typical four-wave mixing effects: (a) sum-frequency generation; (b) third-harmonic generation; (c, d) four-photon parametric interaction. The horizontal dashed lines represent the involved intermediate states characterizing both the molecule and the photon fields.*

simplicity of statement, we can also say that the above-mentioned process is actually a single-step process with the annihilation of three incident photons and the simultaneous creation of a new sum-frequency photon.

Since at the beginning and the end of the elementary process of four-wave sum-frequency mixing shown in Fig. 4.1(a), the physical states of the molecule are unchanged; therefore, the conservation of energy and momentum should be maintained only between the annihilated photons and the created photon. Thus, we have the following phase-matching requirement:

$$\left.\begin{array}{l} v_4 = v_1 + v_2 + v_3 \\ \mathbf{k}_4 = \mathbf{k}_1 + \mathbf{k}_2 + \mathbf{k}_3 \end{array}\right\}, \tag{4.1-1}$$

where $\mathbf{k}_i$ is the wavevector of the ith wave. If $\mathbf{k}_1$, $\mathbf{k}_2$, and $\mathbf{k}_3$ are in the same direction (collinear interaction), the second relation in Eq. (4.1-1) leads to the following phase-matching requirement for refractive indices:

$$v_4 n(v_4) = v_1 n(v_1) + v_2 n(v_2) + v_3 n(v_3). \tag{4.1-2}$$

This type of nonlinear sum-frequency generation requires three incident waves ($v_1 \neq v_2 \neq v_3$), or requires two incident waves ($v_1 = v_2$ and v_3). The contribution to the corresponding third-order nonlinear polarization comes solely from the mechanism of electron-cloud distortion. If one of the incident waves is tunable, the output sum-frequency emission will also be tunable.

Figure 4.1(b) shows the third-harmonic generation (THG) in a third-order nonlinear medium. In this case, a single monochromatic coherent optical wave of frequency $v_1 = v$ may induce a new coherent optical wave at the frequency of $v_2 = 3v$. The elementary process of THG is the same as the sum-frequency process just described above. The conservation of energy and momentum requires that

$$\left.\begin{array}{l} v_2 = 3v_1 \\ \mathbf{k}_2 = 3\mathbf{k}_1 \end{array}\right\}. \tag{4.1-3}$$

Here, k_1 and k_2 are the wavevectors of these two waves. From the second relation of Eq. (4.1-3) we obtain the following refractive-index requirement for collinear interaction:

$$\frac{2\pi}{\lambda_2} n(\nu_2) = \frac{6\pi}{\lambda_1} n(\nu_1), \tag{4.1-4}$$

where λ_1 and λ_2 are the wavelengths of these two waves in free space. As $\lambda_1 = 3\lambda_2$, Eq. (4.1-4) leads to the phase-matching condition for THG:

$$n(3\nu_1) = n(\nu_1). \tag{4.1-5}$$

It is clear that the phase-matching condition is a logical result of the conservation of energy and momentum. Similar to the third-order sum-frequency generation, the major contribution to the corresponding third-order nonlinear polarization for THG is from the mechanism of electron-cloud distortion.

Figures 4.1(c) and (d) show other two types of four-wave mixing processes, i.e., the often called four-photon parametric interaction processes. The process shown in Fig. 4.1(c) involves the annihilation of two incident photons at frequencies ν_1 and ν_2 and the simultaneous creation of two new photons at frequencies ν_3 and ν_4. The conservation of energy and momentum among these four photons requires

$$\left. \begin{array}{c} \nu_3 + \nu_4 = \nu_1 + \nu_2 \\ k_3 + k_4 = k_1 + k_2 \end{array} \right\}. \tag{4.1-6}$$

The optical power will transfer from the waves of ν_1 and ν_2 into the waves of ν_3 and ν_4. Lastly, the process shown in Fig. 4.1(d), which can be viewed as a reverse process of that shown in Fig. 4.1(a), features the annihilation of one incident photon of frequency ν_1 and the creation of three new photons at frequencies ν_2, ν_3, and ν_4. The conservation of energy and momentum among these four photons requires:

$$\left. \begin{array}{c} \nu_2 + \nu_3 + \nu_4 = \nu_1 \\ k_2 + k_3 + k_4 = k_1 \end{array} \right\}. \tag{4.1-7}$$

In this case, the optical power flows from the incident ν_1 wave into the simultaneously created waves of ν_1, ν_2, and ν_3. For the last two types of four-photon parametric processes mentioned above, the main contribution to the nonlinear polarization is still the mechanism of electron-cloud distortion. Nevertheless, if a resonance (such as Raman-type) enhancement is involved, the contribution from other mechanisms (such as intramolecular motion) may play a significant role.

Finally, one can conclude that for all the above mentioned four-wave frequency mixing or four-photon parametric effects, (i) the phase-matching requirement should be met, (ii) optical power transfer and redistribution among different waves must obey the conservation of the total optical power, and (iii) the change of photon numbers among different waves should maintain a certain ratio. For example, in the case shown in Fig. 4.1(c), according to the last two requirements we obtain the following relations for changes of intensities and photon numbers among these four waves:

$$\left. \begin{array}{c} -\left(\dfrac{\partial I_1(z)}{\partial z} + \dfrac{\partial I_2(z)}{\partial z} \right) = \left(\dfrac{\partial I_3(z)}{\partial z} + \dfrac{\partial I_4(z)}{\partial z} \right) \\ -\dfrac{\partial I_1(z)}{h\nu_1 \partial z} = -\dfrac{\partial I_2(z)}{h\nu_2 \partial z} = \dfrac{\partial I_3(z)}{h\nu_3 \partial z} = \dfrac{\partial I_4(z)}{h\nu_4 \partial z} \end{array} \right\}, \tag{4.1-8}$$

where $I_i(z)$ is the intensity of the wave with frequency ν_i, h is Planck's constant, and it is assumed that all four waves are nearly propagating along the same direction of z. The second relation in Eq. (4.1-8) implies that the decrease of the photon numbers of ν_1 and ν_2 waves is equal to the increase of the photon numbers of ν_3 and ν_4 waves. Eq. (4.1-8) can be recognized as the Manley–Rowe relations in the third-order nonlinear frequency-mixing processes.

In the following sections of this chapter, we shall further discuss several typical four-wave mixing or four-photon parametric interaction effects in detail because they are more important for both fundamental studies and applications.

4.2 Third-harmonic generation (THG)

4.2.1 Basic theoretical descriptions

We discuss here the third-harmonic generation (THG) or frequency-tripling effect based on the third-order nonlinear polarization process. The first optical THG was observed in 1962 in a calcite crystal possessing centrosymmetry.[1] Based on the nonlinear polarization theory, the source of a third-harmonic field is the following nonlinear polarization component:

$$\mathbf{P}^{(3)}(3\omega) = \varepsilon_0 \chi^{(3)}(\omega,\omega,\omega) \mathbf{E}(\omega) \mathbf{E}(\omega) \mathbf{E}(\omega). \tag{4.2-1}$$

Here, $\mathbf{E}(\omega)$ is the incident fundamental optical field and $\chi^{(3)}(\omega,\omega,\omega)$ is the third-order susceptibility tensor of a given nonlinear medium for THG.

Assuming both the incident fundamental filed $\mathbf{E}(\omega)$ and the frequency-tripled field $\mathbf{E}(3\omega)$ are linearly-polarized monochromatic plane waves propagating along the z-axis and can be specifically written as

$$\left. \begin{array}{l} \mathbf{E}_1(\omega,z) = \mathbf{a}_1 A_1(z) e^{ik_1 z} \\ \mathbf{E}_2(3\omega,z) = \mathbf{a}_2 A_2(z) e^{ik_2 z} \end{array} \right\}, \tag{4.2-2}$$

the corresponding polarization components of these two waves should be

$$\left. \begin{array}{l} \mathbf{P}^{(3)}(\omega,z) = \varepsilon_0 \chi^{(3)}(3\omega,-\omega,-\omega) \mathbf{E}_2 \mathbf{E}_1^* \mathbf{E}_1^* \\ \mathbf{P}^{(3)}(3\omega,z) = \varepsilon_0 \chi^{(3)}(\omega,\omega,\omega) \mathbf{E}_1 \mathbf{E}_1 \mathbf{E}_1 \end{array} \right\}. \tag{4.2-3}$$

Substituting Eq. (4.2-2) into Eq. (4.2-3) leads to

$$\left. \begin{array}{l} \mathbf{P}^{(3)}(\omega,z) = \varepsilon_0 \chi^{(3)}(3\omega,-\omega,-\omega) \mathbf{a}_2 \mathbf{a}_1 \mathbf{a}_1 A_2(z) A_1^*(z) A_1^*(z) e^{i(k_2-2k_1)z} \\ \mathbf{P}^{(3)}(3\omega,z) = \varepsilon_0 \chi^{(3)}(\omega,\omega,\omega) \mathbf{a}_1 \mathbf{a}_1 \mathbf{a}_1 A_1(z) A_1(z) A_1(z) e^{i3k_1 z} \end{array} \right\}. \tag{4.2-4}$$

Here, $A_1(z)$ and $A_2(z)$ are the amplitude functions, $\mathbf{a}_1$ and $\mathbf{a}_2$ are the unit vectors of light polarization, and k_1 and k_2 are the absolute values of wavevectors for the fundamental wave and the third-harmonic wave, respectively.

To consider the amplitude change of these two waves along the z-axis, substituting the second equality of Eq. (4.2-4) into the coupled-wave equation of Eq. (2.5-13) leads to

$$\begin{aligned} \frac{\partial A_2(z)}{\partial z} &= \frac{ik_2 \varepsilon_0}{2\varepsilon(3\omega)} [\mathbf{a}_2 \cdot \chi^{(3)}(\omega,\omega,\omega) \mathbf{a}_1 \mathbf{a}_1 \mathbf{a}_1] A_1^3(z) e^{i(3k_1-k_2)z} \\ &= \frac{ik_2 \varepsilon_0}{2\varepsilon(3\omega)} \chi_e^{(3)} A_1^3(z) e^{i\Delta k z}. \end{aligned} \tag{4.2-5}$$

Here,

$$\chi_e^{(3)} = \mathbf{a}_2 \cdot \chi^{(3)}(\omega,\omega,\omega)\mathbf{a}_1\mathbf{a}_1\mathbf{a}_1 \qquad (4.2\text{-}6)$$

is the effective third-order susceptibility value of the medium for THG and

$$\Delta k = 3k_1 - k_2 = \frac{6\pi}{\lambda_1}[n(\omega) - n(3\omega)] \qquad (4.2\text{-}7)$$

is the phase mismatch factor that is proportional to the refractive-index difference at frequencies ω and 3ω. In general, the refractive index of a medium is always a function of frequency (dispersion effect), i.e., $n(\omega) \neq n(3\omega)$; only in some special cases we may have $n(\omega) = n(3\omega)$.

Now, let us first consider the situation in phase-mismatched condition, i.e., $\Delta k \neq 0$. In that case, comparing the expression of $\mathbf{P}^{(3)}(3\omega,z)$ in Eq. (4.2-4) to the expression of $\mathbf{E}_2(3\omega,z)$ in Eq. (4.2-2), we can see that the phase velocity of the polarization wave at 3ω is different from that of the optical wave at the same frequency. Therefore, the energy from $\mathbf{E}_1(\omega,z)$ wave cannot transfer to $\mathbf{E}_2(3\omega,z)$ wave effectively. In this case, we can assume that for the fundamental wave $\mathbf{E}_1(\omega,z)$ the amplitude change along the z-axis can be nearly neglected, i.e.,

$$\left.\begin{array}{l} A_1(z) \approx A_1(0) \\ A_2(0) \approx 0 \end{array}\right\}, \qquad (4.2\text{-}8)$$

where $A_1(0)$ is the initial amplitude of the fundamental wave at the incident surface of the nonlinear medium. Under this condition, Eq. (4.2-5) leads to a solution:

$$A_2(z) = \frac{\varepsilon_0 k_2}{2\Delta k \varepsilon(3\omega)} \chi_e^{(3)} A_1^3(0)(e^{i\Delta kz} - 1). \qquad (4.2\text{-}9)$$

The intensity change of the third-harmonic wave along the z-direction is

$$I_2(z) \propto A_2 A_2^* \propto |\chi_e^{(3)}|^2 I_1^3(0)\left(\frac{\sin \Delta kz/2}{\Delta k/2}\right)^2. \qquad (4.2\text{-}10)$$

Here, $I_1(0) \propto A_1^2(0)$ is the intensity of the incident fundamental wave. From Eq. (4.2-10), we can see that the intensity of the third-harmonic is proportional to the cube of the intensity of fundamental wave, and it varies periodically along the z-axis as that shown in Fig. 3.1 for second-harmonic generation under the condition of phase mismatch, the period of the intensity variation is $2\pi/\Delta k$. A larger Δk leads to a smaller peak intensity and more rapid variation. Similar to Eq. (3.1-9), the coherent length of THG can be written as

$$l_0 = \frac{\pi}{|\Delta k|} = \frac{\lambda_1}{6|n(\omega) - n(3\omega)|}. \qquad (4.2\text{-}11)$$

In order to significantly increase the conversion efficiency, it is necessary to fulfill phase-matching requirement, i.e., to realize $\Delta k \Rightarrow 0$ or $n(\omega) \Rightarrow n(3\omega)$. For most transparent media, the normal dispersion effect makes $n(3\omega) > n(\omega)$; however, under certain circumstances we could make the phase matching be fulfilled by employing some special methods that shall be discussed later.

Considering that, in general, the third-order nonlinear susceptibility is much smaller than the second-order susceptibility even with the phase-matching condition of $\Delta k \Rightarrow 0$, the power-transfer efficiency from the fundamental wave to the third-harmonic wave is only in the order of $\leq 10^{-2}$–10^{-1} for most practical cases. Therefore, the undepleted-fundamental wave approximation can be applicable to a phase-matched case. This implies that the interaction length z_0 between the fundamental beam and the nonlinear medium could be much shorter than the coherent length l_0, i.e., $z_0 \ll l_0$. Thus within the range of $z \leq z_0$ we have

$$\sin \frac{\Delta k z}{2} \approx \frac{\Delta k z}{2}. \qquad (4.2\text{-}12)$$

In this approximation, Eq. (4.2-10) becomes

$$I_2(z_0) \propto \left|\chi_e^{(3)}\right|^2 I_1^3(0) z_0^2. \qquad (4.2\text{-}13)$$

Eq. (4.2-13) is obtained in the plane-wave approximation, from which one can conclude that under the phase-matching condition, the intensity of third-harmonic signal is proportional to the square of the effective third-order susceptibility value of the nonlinear medium, the cube of the intensity of the fundamental, and the square of the interaction length, respectively. In practice, we can utilize resonant or near-resonant enhancement of the third-order susceptibility to achieve a higher value of $\chi_e^{(3)}$, and can also use a focused fundamental beam to produce a higher local intensity of $I_1(0)$. However, in case of a focused-beam configuration, the final expression of third-order harmonic intensity is more complicated than Eq. (4.2-13).[2]

4.2.2 Phase-matching methods for THG

Based on the theoretical consideration mentioned above, we know that phase matching is the primary requirement for THG as well as for other four-wave mixing processes. In general, $n(\omega) \neq n(3\omega)$ owing to the refractive-index dispersion effect; therefore we have to utilize some special methods to realize the phase-matching requirement for a given THG material. The following are two examples of such special methods.

1. *Birefringence compensation*: In an anisotropic transparent crystal, the normal dispersion leads to $n(3\omega) > n(\omega)$; however, we can choose a special angle to compensate this dispersion effect by birefringence. This technique is essentially the same as what we utilized for second-harmonic generation. For instance, in a negative uniaxial crystal the phase-matching can be met via the following possible combinations involving different light polarization states:

$$\left. \begin{array}{ll} n^o(\omega) + n^o(\omega) + n^o(\omega) = 3n^e(3\omega) & (\text{typeI}) \\ n^o(\omega) + n^o(\omega) + n^e(\omega) = 3n^e(3\omega) & (\text{typeII}) \\ n^o(\omega) + n^e(\omega) + n^e(\omega) = 3n^e(3\omega) & (\text{typeII}) \end{array} \right\}. \qquad (4.2\text{-}14)$$

2. *Dispersion compensation in a mixed system*: Suppose a system combines two components: part A showing normal dispersion and part B showing anomalous dispersion at frequencies of ω and 3ω, i.e., $n_a(\omega) < n_a(3\omega)$ and $n_b(\omega) > n_b(3\omega)$, where n_a and n_b are the refractive indices of these two components when they exist separately. The phase matching of the mixed system can be achieved by properly choosing the ratio of these two components so that

Third-harmonic generation (THG)

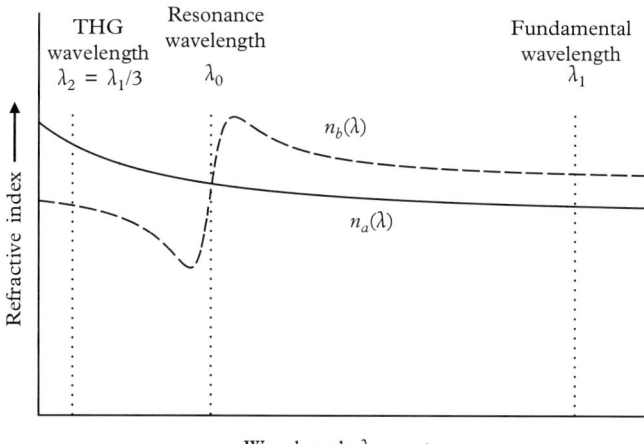

Figure 4.2 *THG phase matching via dispersion compensation of a two-component mixture.*

$$\frac{c_a}{c_a + c_b}n_a(\omega) + \frac{c_b}{c_a + c_b}n_b(\omega) = \frac{c_a}{c_a + c_b}n_a(3\omega) + \frac{c_b}{c_a + c_b}n_b(3\omega), \qquad (4.2\text{-}15)$$

where c_a and c_b are the mole concentrations of these two components in the mixture. The principle of this method is schematically illustrated in Fig. 4.2, where the normal dispersion of n_a is expressed by a solid curve while the anomalous dispersion of n_b is presented by a dashed curve. In this case, it is assumed that there is no resonant absorption for component A in the considered spectral range; whereas there is a one-photon absorption around some position of λ_0 between λ_1 and λ_2 for component B. By properly adjusting the percentage ratio of these two components, the phase matching could be realized between a given fundamental wavelength λ_1 and the corresponding third-harmonic wavelength $\lambda_2 = \lambda_1/3$.

4.2.3 Resonance enhancement of THG

From Eq. (4.2-13) we know that under the phase-matching condition the third-harmonic signal intensity is proportional to the square of the effective third-order susceptibility value. For this consideration, the resonant enhancement of the third-order susceptibility of nonlinear media is highly desirable. One of the most useful mechanisms is the resonant enhancement based on a two-photon absorptive transition. In this case, as shown in Fig. 4.3, a two-photon absorptive transition is assumed to be allowed between the ground state and a real exited state t. If we can tune the frequency ω of the fundamental wave and ensure $2\omega \Rightarrow \omega_0$, where ω_0 is the resonant frequency of the two-photon absorptive transition, a significant enhancement of the third-order nonlinearity can be expected.

Based on the theory of nonlinear susceptibilities described in Chapter 18, it turns out that the third-order nonlinear susceptibility of the medium with an allowed two-photon transition can be expressed as follows:

$$\chi^{(3)}(\omega, \omega, \omega) = \chi^{(3)}_{NR}(\omega, \omega, \omega) + \chi^{(3)}_R(\omega, \omega, \omega). \qquad (4.2\text{-}16)$$

66   Third-Order Nonlinear (Four-Wave) Frequency Mixing

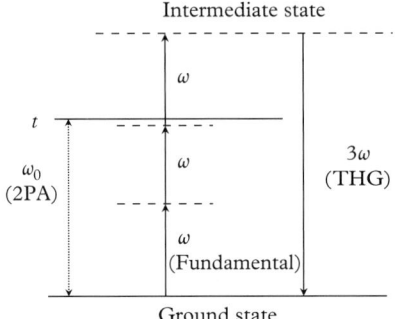

Figure 4.3 *Two-photon transition enhancement for THG.*

Here, $\chi_{NR}^{(3)}$ is the non-resonant contribution that is a real quantity; $\chi_{R}^{(3)}$ is the resonant contribution that is a complex quantity expressed by (cf. Eq. (18.4-7))

$$\chi_R^{(3)} = K \frac{1}{(\omega_0 - 2\omega) - i\Gamma}, \qquad (4.2\text{-}17)$$

where 2Γ is the full spectral width of the two-photon transition at ω_0 and K is a real tensor coefficient. In the range $(\omega_0 - 2\omega) \approx 0$, $\chi_R^{(3)}$ can be significantly larger than $\chi_{NR}^{(3)}$; therefore, from Eqs. (4.2-6) and (4.2-17) we have

$$\left|\chi_e^{(3)}\right|^2 \propto \left|\chi_R^{(3)}\right|^2 \propto \frac{1}{(\omega_0 - 2\omega)^2 + \Gamma^2}. \qquad (4.2\text{-}18)$$

The two-photon enhancement of the effective third-order susceptibility is schematically shown in Fig. 4.4. The full-width at half-maximum (FWHM) of the resonant curve is 2Γ, while the total spectral tuning range within which the resonant contribution is considerably larger than the non-resonant contribution is about $\sim 8\Gamma$.

In practice, the same enhancement principle can also be used for third-order optical sum-frequency generation. In that case, two coherent monochromatic waves of ω_1 and ω_2 are incident on a third-order nonlinear medium possessing a two-photon transition at ω_0. If the frequency (ω_1) of one beam meets the two-photon resonance requirement, i.e., $2\omega_1 \Rightarrow \omega_0$, a significant enhancement of the third-order nonlinearity is expected, which can be expressed as

$$\left|\chi_e^{(3)}(\omega_1,\omega_1,\omega_2)\right|^2 \propto \frac{1}{(\omega_0 - 2\omega_1)^2 + \Gamma^2}, \qquad (4.2\text{-}19)$$

where $\chi_e^{(3)}(\omega_1,\omega_1,\omega_2)$ is the effective third-order susceptibility value describing the sum-frequency generation process. Moreover, if the frequency ω_2 of the other incident beam is continuously tunable over a large spectral range, the output frequency $(2\omega_1 + \omega_2)$ of the sum-frequency signal will also be continuously tunable over a large spectral scale. This type of resonance-enhanced and tunable third-order optical sum-frequency generation is schematically

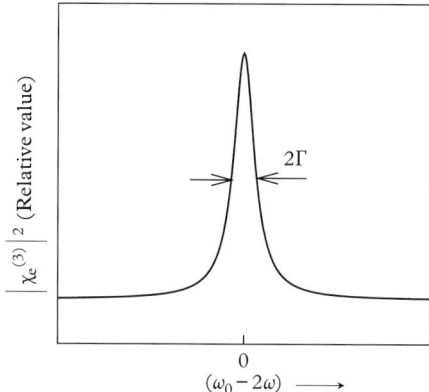

Figure 4.4 *Two-photon transition enhancement of the effective third-order susceptibility value.*

shown in Fig. 4.5. The major advantage of this technique is that when we continuously change the frequency ω_2, the two-photon resonance condition of $(\omega_0 - 2\omega_1) \Rightarrow 0$ always remains unchanged.

So far, we have discussed the advantages of utilizing a two-photon transition resonance to enhance the third-order nonlinearity for a given medium. However, it should be noted that under some circumstances, the excessive two-photon absorption may lead to an undesirable depletion of the incident beam. For this reason a near-resonance, as shown in Figs. 4.3 and 4.5, is often employed to ensure both a small attenuation owing to real two-photon absorption and a large enhancement of the effective third-order susceptibility value.

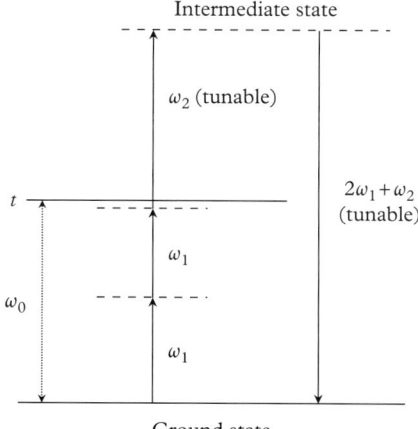

Figure 4.5 *Two-photon transition enhanced tunable third-order sum-frequency generation.*

4.2.4 Materials and devices for THG and third-order sum-frequency generation

Based on the considerations we discussed above, the requirements of nonlinear materials for THG as well as for third-order sum-frequency generation (SFG) can be summarized as: (i) low losses for both incident wave(s) and the newly-generated third-harmonic wave or sum-frequency wave, (ii) an achievable phase matching, and (iii) a higher effective third-order susceptibility value. Regarding the last requirement, a resonance enhancement of the third-order nonlinearity for a given medium is highly desirable.

Figure 4.6(a) shows a typical experimental setup for THG. The fundamental beam from a high-power laser source is focused into a third-order nonlinear medium to provide a higher local intensity. After passing through the nonlinear medium the transmitted fundamental beam is blocked by a spectral filter (or a spectrometer device) while the third-harmonic output can be isolated and detected. It should be noted that a lens with short focal length may generate a higher local light intensity near the focal plane, but the effective interaction length (i.e., the focal depth) becomes shorter. In practice, a lens with an appropriate focal length is chosen to reach a compromise between the local intensity and the effective interaction length. That rule is applicable to most nonlinear optical experiments when a focused incident laser beam is used. Figure 4.6(b) shows a typical setup for third-order SFG performance. In that case, two laser sources are employed to provide the ω_1 and ω_2 waves separately. In order to obtain a tunable sum-frequency output, at least one laser frequency should be tunable. If the nonlinear medium is a gas or liquid system contained in a sealed pipe or cell, the output window of the pipe or cell should be made of a special material having high transmission for the optical signal of THG or the third-order SFG.

To date, several types of nonlinear materials have been utilized for THG and third-order SFG, which can be briefly described as follows.

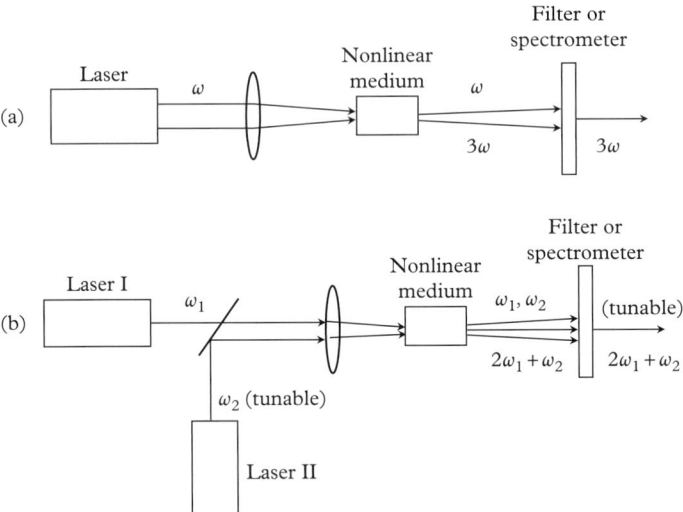

Figure 4.6 *Experimental setups for (a) THG and for (b) tunable third-order SFG.*

A. *Metal vapors*: It is found that metal vapors are among the most suitable nonlinear media for THG and third-order SFG because of their low losses for ultraviolet (UV) emission, higher third-order nonlinearity, achievable phase-matching, and applicable resonant enhancement.[3–14] Based on the anomalous dispersion of a properly chosen one-photon absorption line in the near infrared (IR) or visible spectral range of metal vapors, one could find a negative dispersion for the given fundamental wave, i.e., $[n(3\omega)-n(\omega)] < 0$; this effect can be compensated by adding a proper rare gas possessing positive dispersion, i.e., $[n'(3\omega)-n'(\omega)] > 0$. In addition, it is easy to find an allowed two-photon transition in the UV range to achieve a resonance enhancement of the nonlinearity of metal vapor systems. The commonly used systems are alkali-metal vapors with a rare gas added as the buffer medium, such as a Rb–Xe mixture,[4,6–8] or a Na–Xe mixture.[6,8,10,12] The other metal vapor–rare gas systems are Sr–Xe,[5] Mg–He,[9] Mg–Kr,[11] Tl vapor,[13] and Zn–Ar.[14] Using near-IR or visible laser emission as the fundamental wave, one can obtain THG or third-order SFG at UV or vacuum UV (VUV, 200–100 nm) spectral ranges. The conversion efficiency from the fundamental to the third-harmonic can be up to 2–10%.[6,7,10]

B. *Rare gases*: They are among the nonlinear media for the earliest THG studies.[2] Rare gases have a superior transparency in the whole visible and UV (up to extreme UV (XUV, ≤100 nm)) spectral ranges, a high chemical and physical stability, as well as a high breakdown threshold. The often used systems for fundamental studies of THG are Xe,[15–17] Kr,[15,18] and Ar.[19] In general, the conversion efficiency of THG is quite low owing to the lack of phase matching or resonance enhancement.

C. *Molecular gases*: The molecular gases, such as SF_6, BCl_3, CO, CD_4, and DCl–CF_4, have been employed for THG studies using the ~10.6 μm output from CO_2 lasers as the fundamental wave.[20–23] In general, the THG efficiency is relatively low ($\leq 10^{-2}$).

D. *Dye solutions*: It is well known that many organic dyes have a strong one-photon absorption band in the visible spectral range providing a negative dispersion between frequencies 3ω and ω, which can be compensated by the normal dispersion of an appropriate transparent solvent. Thus, a dye solution can be employed for THG purposes.[24–26] In addition, a strong two-photon absorption enhancement of the third-order nonlinearity is also expected.[27,28] A conversion efficiency of 1% has been reported in a dye-solution with phase-matching.[26]

E. *Crystal materials*: There are two essentially different approaches to get third-harmonic emission in crystal materials. One is based on second-order nonlinearity of crystals with no centrosymmetry; in this case, we can mix a second-harmonic beam of 2ω with a fundamental beam of ω to generate a sum-frequency wave of 3ω. The other approach is the direct generation of a third-harmonic wave based on the third-order nonlinearity of a crystal sample.[1,29–33] In both cases, the shortest wavelength of the 3ω wave is limited by the transparency edge of crystals in the UV range. The conversion efficiency from the ω wave to the 3ω wave is much higher for the first approach than the second.

F. *Other materials*: In addition to the nonlinear media mentioned above, there are various other materials which have been used for THG studies, such as liquid crystals,[34] transparent liquid droplet,[35] optical fibers,[36–39] optical bulk glasses,[40] laser-induced plasma in the air,[41–43] and metal nanostructures.[44,45]

Table 4.1 lists the typical experimental data for THG performance on some third-order nonlinear materials mentioned above.

70 Third-Order Nonlinear (Four-Wave) Frequency Mixing

Table 4.1 Experimental data for THG performances in third-order nonlinear media.

Nonlinear Media	λ_0 (nm)	$\lambda_0/3$ (nm)	Input Intensity (W/cm^2)	Phase Matching	Resonance Enhancement	Efficiency
Rb–Xe	1064	354.7	10^{10}	yes		10% [6]
Rb–Xe	1064	354.7	10^{11}	yes		2.8% [7]
Na–Xe	1064	354.7	2×10^{11}	yes		2.7% [6]
Mg–He	430.8	143.6		yes	yes	0.2% [9]
Xe, Kr	212–226	70.7–75.3	20–60 kW			[15]
CD$_4$	10.2 (μm)	3.4 (μm)	7.8×10^8		yes	0.68% [22]
CO–O$_2$ (liquid)	9.36 (μm)	3.12 (μm)	3×10^{10}	yes		4% [23]
PMC dye solution	1054	351.3	2.5×10^{11}	yes		1% [26]
β-BaB$_2$O$_4$	1054	351.3	5×10^{10}	yes		0.8% [30]
KTP	1618	539.3	3×10^{10}	yes		2.4% [33]
SiO$_2$–Ge fiber	532	177				0.3% [36]
Laser-induced plasma	1064	354.7	10^9			3% [41]

In summary, the optical THG or third-order SFG is an effective technique to produce intense and tunable coherent radiation in the short-wavelength (up to VUV and XUV) range of the optical spectrum and in the media of third-order nonlinearity. Coherent short wavelength UV emission is highly useful in many research and application areas such as:

1. studies of UV absorptive and fluorescence spectroscopy;
2. studies of UV ionization and dissociation of atoms and molecules;
3. high-density optical data storage and processing;
4. high-resolution UV photo-ablation and photo-lithography;
5. coherent UV-excited photo-chemical and photo-biological studies.

4.3 Raman-enhanced FWFM

4.3.1 Coherent Stokes- and anti-Stokes ring emission

In this section, we shall consider several special third-order nonlinear optical processes, which are called Raman-enhanced FWFM effects or Raman-enhanced four-photon parametric interaction

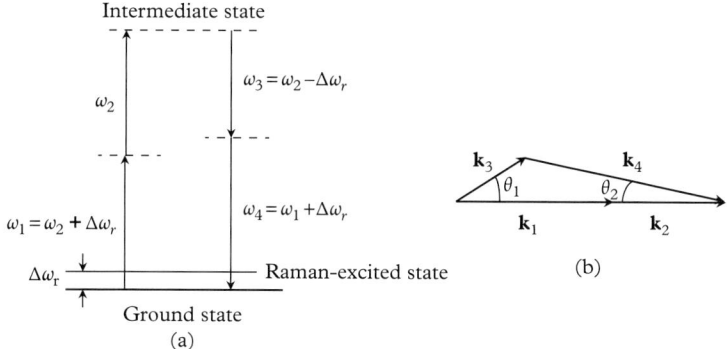

Figure 4.7 (a) Raman-enhanced four-wave mixing and (b) noncollinear phase-matching.

effects. The features of these effects are the noncollinear phase-matching requirement and the Raman-resonance enhanced third-order nonlinearity. In these cases, the general quantum description shown in Fig. 4.1(c) is applicable except that a Raman medium is adopted as the third-order nonlinear medium.

Suppose two coherent light waves of frequencies ω_1 and ω_2 ($\omega_1 > \omega_2$) are incident upon a Raman medium that possesses a characteristic Raman mode frequency of $\Delta\omega_r$, and the frequency difference between these two waves is nearly equal to the Raman-mode frequency, i.e., $\omega_1 - \omega_2 \approx \Delta\omega_r$. If we further assume that the two incident waves are propagating along the same direction in the Raman medium, the quantum-transition processes can be schematically shown in Fig. 4.7(a). Here the elementary process is related to the annihilation of two incident photons and the simultaneous creation of two new photons. In this case, the frequencies of the two generated photons should be

$$\left. \begin{array}{l} \omega_3 = \omega_2 - \Delta\omega_r \\ \omega_4 = \omega_1 + \Delta\omega_r \end{array} \right\}, \tag{4.3-1}$$

and the phase-matching (or wavevector-matching) condition

$$\mathbf{k}_1 + \mathbf{k}_2 = \mathbf{k}_3 + \mathbf{k}_4 \tag{4.3-2}$$

can be fulfilled in the way shown in Fig. 4.7(b).

It should be pointed out that for all processes discussed in this section, at the beginning and end of the elementary process, the quantum state of molecules of the medium remains unchanged; in other words, no real Raman transition of molecules from ground state to a Raman-excited state should be considered. Such a real Raman transition between two real states is the issue of Raman scattering or stimulated Raman scattering process (see Chapter 7); once it occurs, the excited molecule will stay in the Raman-excited state for certain time period.

From Fig. 4.7(b) one can see that the newly generated two coherent beams will propagate along two specific directions; the angles between these two beams and the incident beams are determined by the $\Delta\omega_r$ value as well as the dispersion property of the refractive index of the medium. For common Raman media in liquid or solid phase, the angles θ_1 and θ_2 are usually in the range 1–4°. One can further imagine that if we rotate the triangle shown in Fig. 4.7(b) by 360°

utilizing ($k_1 + k_2$) as the rotation axis, the phase-matching condition is always fulfilled. Therefore, we expect that the k_3 and k_4 waves will be observed in the form of conical emission at well-defined angles of θ_1 and θ_2.

The nonlinear polarization components contributing to the generation of k_3 and k_4 waves can be written as

$$\left. \begin{array}{l} P^{(3)}(\omega_3) = \varepsilon_0 \chi^{(3)}(\omega_2, \omega_2, -\omega_1) E_2(\omega_2) E_2(\omega_2) E_1^*(\omega_1) \\ P^{(3)}(\omega_4) = \varepsilon_0 \chi^{(3)}(\omega_1, \omega_1, -\omega_2) E_1(\omega_1) E_1(\omega_1) E_2^*(\omega_2) \end{array} \right\}, \quad (4.3\text{-}3)$$

Similar to the derivation of Eq. (4.2-13), under phase-matching condition the intensities of k_3 and k_4 waves can be expressed as

$$\left. \begin{array}{l} I_3(\omega_3) \propto \left|\chi_e^{(3)}\right|^2 I_2^2(\omega_2) I_1(\omega_1) z_0^2 \\ I_4(\omega_4) \propto \left|\chi_e^{(3)}\right|^2 I_1^2(\omega_1) I_2(\omega_2) z_0^2 \end{array} \right\}, \quad (4.3\text{-}4)$$

where $\chi_e^{(3)} = \chi_e^{(3)}(\omega_2, \omega_2, -\omega_1) = \chi_e^{(3)}(\omega_1, \omega_1, -\omega_2)$ is the effective nonlinear susceptibility value describing the above four-wave mixing process, and z_0 is the effective interaction length. Recalling the discussion of two-photon resonance given in Section 4.2.3, under the Raman-resonance condition of

$$\omega_1 - \omega_2 \Rightarrow \Delta\omega_r, \quad (4.3\text{-}5)$$

we have

$$\left|\chi_e^{(3)}\right|^2 \propto \left|\chi_R^{(3)}\right|^2 \propto \frac{1}{[\Delta\omega_r - (\omega_1 - \omega_2)]^2 + \Gamma^2}. \quad (4.3\text{-}6)$$

Here, $\left|\chi_R^{(3)}\right|$ is the resonant part of the third-order susceptibility (cf. Eqs. (18.4-11) and (18.4-12)), 2Γ is the full spectral width of a given Raman transition. From Eq. (4.3-6), one can see that when $\omega_1 - \omega_2 \Rightarrow \Delta\omega_r$, the efficiency of this four-wave mixing process can be significantly enhanced.

Figure 4.8 shows the experimental setup for observing the coherent ring emission based on the Raman-enhanced four-wave mixing. Two collinear laser beams pass through a Raman medium,

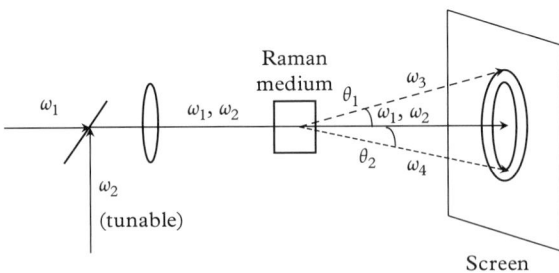

Figure 4.8 Stokes and anti-Stokes coherent ring emission generated by Raman-enhanced four-wave mixing.

and the frequency (ω_2) of one laser beam is tunable. By tuning ω_2 until the Raman resonance is reached, two strong forward coherent emission rings can be observed on a screen at frequencies ω_3 and ω_4 and with angles of θ_1 and θ_2, respectively. It is customary to call the beam of $\omega_3 = \omega_2 - \Delta\omega_r$ the coherent Stokes emission (not scattering), and the beam of $\omega_4 = \omega_1 + \Delta\omega_r$ the coherent anti-Stokes emission (not scattering).

In practice, we can use only one laser beam of ω_1 to pump a Raman medium to generate the first-order Stokes stimulated Raman scattering of $\omega_2 = \omega_1 - \Delta\omega_r$. Since the latter is basically collinear with the former, we may observe the same ring-like coherent emission, as shown in Fig. 4.8. It is obvious that in this case, the stimulated scattering process takes place first without phase-matching requirement, then the four-photon parametric process follows with a defined phase-matching requirement.

4.3.2 Raman-enhanced FWFM using two incident beams with a small crossing angle

To continue the discussion of Raman-enhanced four-wave mixing described in the above subsection, we consider another experiment in which the two incident laser beams of ω_1 and ω_2 are not collinear but have a small intersection angle θ within the Raman medium. Here, it is assumed that the frequency difference between these two beams meets the Raman resonance requirement, and the crossing angle θ can be adjusted to a special value θ_0 so that the phase-matching requirement could be satisfied in the ways shown in Fig. 4.9(a) and (b), respectively.

The elementary quantum process shown in Fig. 4.9(a) involves the annihilation of two photons in the ω_1 wave and the simultaneous creation of two photons in the ω_2 and $\omega_4 = \omega_1 + \Delta\omega_r$ waves, respectively. On the other hand, the process shown in Fig. 4.9(b) is related to the annihilation of two photons in the ω_2 wave and the simultaneous generation of two photons in the ω_1 and $\omega_3 = \omega_2 - \Delta\omega_r$ waves, respectively. If the considered Raman frequency shift ($\Delta\omega_r$) is not too great and the refractive-index dispersion effect is not too large, the four angles indicated in Figs. 4.9(a) and (b) are approximately the same.

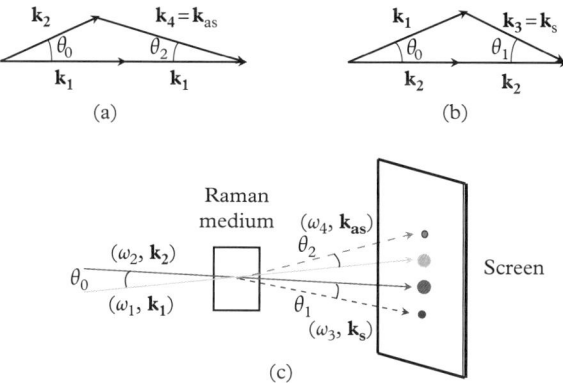

Figure 4.9 *(a, b) Phase-matching conditions for Raman-enhanced four-wave mixing with two incident beams having a crossing angle θ_0; (c) spatial location of coherent Stokes and anti-Stokes emission.*

The experimental behavior of this special four-wave mixing is schematically shown in Fig. 4.9(c). By smoothly adjusting the angle between the two incident beams until it reaches the phase-matched angle θ_0, a coherent Stokes emission beam and a coherent anti-Stokes emission beam can be observed simultaneously in two symmetrical positions on a screen, provided that the two incident beams are strong enough and the Raman resonance is fulfilled.

In summary, studies of various Raman-enhanced four-wave mixing processes are of great importance because they could:

(i) provide an effective technique to generate frequency-shifted coherent light emission in the direction different from the incident beams;

(ii) offer a clear physical insight to understand and explain the generation of high-order Stokes and anti-Stokes coherent emission components, which can be observed in some stimulated Raman scattering experiments and sometimes may become quite complicated;[46–52]

(iii) provide a novel spectral technique to measure Raman spectra of various nonlinear media with a higher sensitivity and other advantages over traditional Raman spectroscopy.[53–55]

4.4 Special non-resonant four-photon parametric interaction effects

4.4.1 Continuum generation via four-photon parametric interaction

Here the continuum means the coherent light emission with a continuous spectral distribution over a very broad spectral range. It can be generated via two mechanisms: one is the four-photon parametric interaction we are going to discuss here; the other is the light-induced phase modulation we will discuss in Chapter 6.

We now consider a single, intense, and monochromatic coherent light wave interacting with a transparent non-resonant medium that does not possess two-photon or Raman enhancement. In this case, the general picture of the third-order nonlinear process shown in Fig. 4.1(c) can be simplified and redrawn in Fig. 4.10(a).

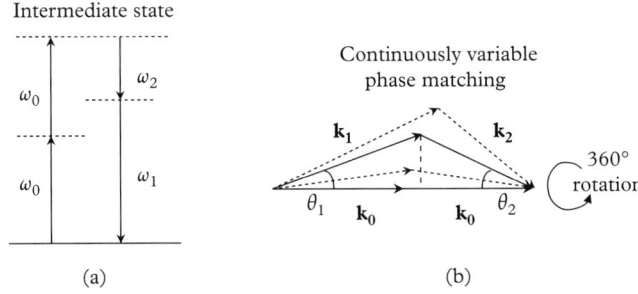

Figure 4.10 *(a) One incident beam-induced four-photon parametric interaction and (b) the phase-matching condition.*

The elementary process that we see here is the annihilation of two photons from the incident wave of ω_0 and the simultaneous creation of two new photons at different frequencies of ω_1 and ω_2. Since there is no energy and momentum exchange between the medium and the photon field, the conservation of energy and momentum should be maintained only among these four photons, i.e.,

$$\left.\begin{array}{l}2\omega_0 = \omega_1 + \omega_2 \\ 2\mathbf{k}_0 = \mathbf{k}_1 + \mathbf{k}_2\end{array}\right\}. \qquad (4.4\text{-}1)$$

The phase matching can be achieved in the way shown in Fig. 4.10(b). Here the special feature is that not only one but also many possible combinations of $(\mathbf{k}_1 + \mathbf{k}_2)$ can fulfill the requirements imposed by Eq. (4.4-1). From the first requirement of Eq. (4.4-1) we have

$$\frac{2}{\lambda_0} = \frac{1}{\lambda_1} + \frac{1}{\lambda_2}, \qquad (4.4\text{-}2)$$

where $\lambda_0, \lambda_1, \lambda_2$ are the wavelengths in free space of these three waves, respectively. Furthermore, from the geometrical consideration of the wavevector matched triangle we have

$$\left.\begin{array}{l}\dfrac{n(\lambda_1)}{\lambda_1}\cos\theta_1 + \dfrac{n(\lambda_2)}{\lambda_2}\cos\theta_2 = \dfrac{2n(\lambda_0)}{\lambda_0} \\ \dfrac{n(\lambda_1)}{\lambda_1}\sin\theta_1 = \dfrac{n(\lambda_2)}{\lambda_2}\sin\theta_2\end{array}\right\}. \qquad (4.4\text{-}3)$$

It can be seen from Fig. 4.10(b) that when the phase-matched triangle rotate around the $\mathbf{k}_0$-direction over 360°, the same phase-matching condition always holds. For this reason, we expect to see a ring-type coherent emission with continuously changed frequencies and propagation angles.

In order to know the spectral and spatial structure of the newly generated coherent ring emission, for a given input wavelength of λ_0 one can assume an arbitrary value of the blue-shifted wavelength $\lambda_1 < \lambda_0$, then get the corresponding value of the red-shifted wavelength λ_2 through Eq. (4.4-2). Substituting values of such a pair of wavelengths into Eq. (4.4-3) leads to a solution of θ_1 and θ_2,

$$\left.\begin{array}{l}\theta_1 = \arccos\left\{\dfrac{\lambda_0\lambda_1}{4n(\lambda_0)n(\lambda_1)}\left[\left(\dfrac{2n(\lambda_0)}{\lambda_0}\right)^2 + \left(\dfrac{n(\lambda_1)}{\lambda_1}\right)^2 - \left(\dfrac{n(\lambda_2)}{\lambda_2}\right)^2\right]\right\} \\ \theta_2 = \arcsin\left[\dfrac{\lambda_2 n(\lambda_1)}{\lambda_1 n(\lambda_2)}\sin\theta_1\right]\end{array}\right\}, \qquad (4.4\text{-}4)$$

provided that the refractive-index values at different wavelengths are known for a given nonlinear medium. When we change the allowed values of the combinations of λ_1 and λ_2, the corresponding values of θ_1 and θ_2 will also be changed. That implies a wavelength–angle dependence for the coherent emission. From Eq. (4.4-2) it is known that the wavelength of the blue-shifted emission should not be shorter than $\lambda_0/2$.

For example, if we choose fused silica glass as the nonlinear medium and 600 nm as the wavelength of the incident laser beam, the calculated wavelength–angle dependence in the visible spectral range is shown in Fig. 4.11 for the four-photon interaction discussed here. In calculation we utilized the following refractive index-dispersion formula for fused quartz glass:[56]

$$n^2(\lambda) - 1 = \frac{0.6961663\lambda^2}{\lambda^2 - (0.0684043)^2} + \frac{0.4079426\lambda^2}{\lambda^2 - (0.1162414)^2} + \frac{0.8974794\lambda^2}{\lambda^2 - (9.896161)^2}, \qquad (4.4\text{-}5)$$

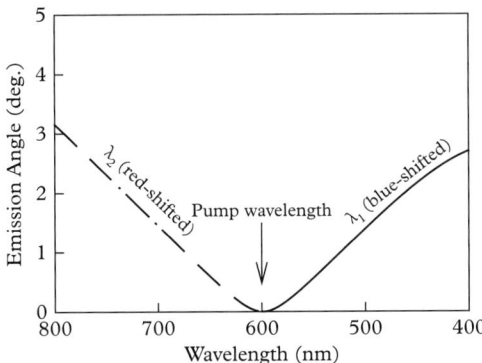

Figure 4.11 *Wavelength–angle dependence of coherent ring emission based on one-beam induced four-photon parametric interaction.*

where the value of wavelength λ is in units of μm. From Fig. 4.11 we can see that within a small angle range ($\leq 2°$), the Stokes and anti-Stokes spectral components have a nearly symmetric wavelength–angle distribution.

In practice, it is relatively easy to observe the coherent emission with a continuous spectral distribution in both sides of the input laser spectral line by using focused high peak-power ultrashort ($\leq 10^{-11}$ s) laser pulses to pass through many types of transparent optical media (optical glasses, crystals, water, organic solvents, etc.). Usually, the observed continuum emission consists of two components.[57–59] One is the conical or non-collinear component, which features the dependence of wavelength on the propagation angle, as shown in Fig. 4.11. The other is a collinear component propagating along the same direction as the incident beam, and there is no wavelength–angle dependence. The generation mechanism for the former is the four-photon parametric interaction discussed here; whereas for the latter is the intense-light induced self-phase modulation that will be discussed in Chapter 6.

4.4.2 Frequency-degenerate four-photon parametric interaction

In this subsection, we consider another type of the forward four-photon parametric interaction process in which all four photons have the same frequency but propagate along different directions. In this case, we assume that two intense laser beams with the same frequency are incident on the third-order nonlinear medium with a small intersection angle θ, as shown in Fig. 4.12.

Under this arrangement, the elementary process of the four-photon parametric interaction can be described as the annihilation of two photons from the two pump beams and the simultaneous creation of two new photons of the same frequency but propagating along other directions. The phase-matching condition in this case can be written as

$$\mathbf{k}_1(\omega) + \mathbf{k}_2(\omega) = \mathbf{k}_3(\omega) + \mathbf{k}_4(\omega). \qquad (4.4\text{-}6)$$

This condition is indicated by a parallelogram shown in Fig. 4.12(a). One can find that if we rotate the bottom half of the parallelogram over 360°, the phase-matching requirement is always met.

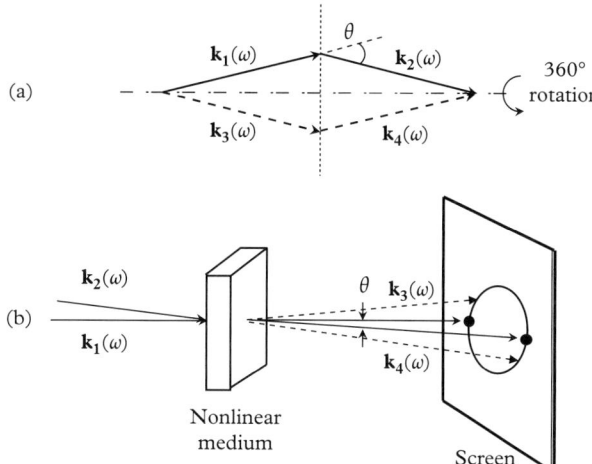

Figure 4.12 *Two-beam excited coherent ring emission via degenerate four-photon parametric interaction.*

Therefore, we could observe a coherent emission ring with a conical angle equal to θ, as shown schematically in Fig. 4.12(b). In practice, using two beams from a picosecond or sub-picosecond high-power laser source to excite some transparent nonlinear media, one can easily generate such a type of coherent ring emission.

Based on the principle described above, we can further design a three-beam based degenerate four-photon parametric amplification experiment, as shown in Fig. 4.13. Under this arrangement two strong laser beams (beam 1 and beam 2) of same frequencies pass through a $\chi^{(3)}$ nonlinear medium with a small intersection angle (θ_{12}) at the horizontal plane, while the third (weak) laser beam (beam 3) of the same frequency is incident nearly on a vertical plane at a small intersection angle (θ_{32}) with respect to beam 2. Carefully adjust the latter angle and the spatial overlapping between these three beams until beam 3 is exactly located at a circle that passes through beams 1 and 2 and has an angular diameter of θ_{12} on a screen, a newly generated coherent emission beam 4 can be observed at the location symmetrical to that of beam 3 along the circle. Meanwhile,

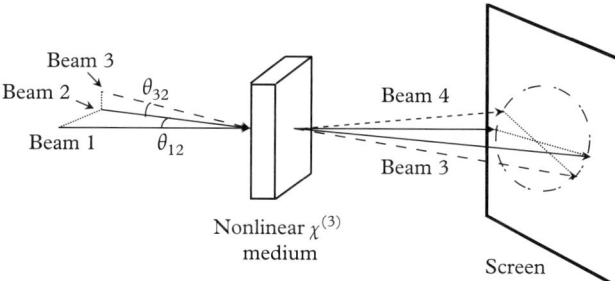

Figure 4.13 *Forward degenerate four-photon parametric amplification experiment.*

after passing through the nonlinear medium, the intensity of beam 3 becomes stronger. These observations can be well understood if we consider the annihilation of two photons from beams 1 and 2, and the simultaneous creation of two new photons contributing to beams 3 and 4. One can find the similarity between the process we discussed here and the second-order nonlinear optical parametric amplification described in Section 3.3. In the present case, we are dealing with four-photon optical parametric amplification processes in which beams 1 and 2 are the strong pump waves, beam 3 is the input weak signal wave, and beam 4 is the idler wave.

In summary, the principles and techniques described in this subsection can be applied to investigate (i) the third-order nonlinearity of non-resonant media, (ii) the dynamic response of the nonlinear media by controlling the temporal delay of one laser beam, and (iii) the relative contributions from different possible nonlinear polarization mechanisms. Regarding the last issue, for instance, in some forward degenerate four-wave mixing experiments the fourth beam may be generated by a pure (simultaneous) four-photon parametric interaction process based on the electronic nonlinearity, or otherwise by the diffraction of the induced holographic gratings formed via various possible mechanisms of the light-induced refractive-index change. In the latter case, how to distinguish the major contributing mechanism from all other possible mechanisms is an interesting but sometimes rather difficult research issue.

4.5 Second-harmonic generation (SHG) via third-order nonlinear processes

4.5.1 Electric field-induced SHG

For centrosymmetric or isotropic media, the second-order nonlinear processes are forbidden in electric-dipole moment approximation. Nevertheless, in the same approximation the SHG is allowed in these media if a strong dc electric field is applied.[1] We can call this effect the dc electric field-induced SHG (EFISHG). It is a third-order nonlinear effect and can be described by the following electric polarization component

$$\mathbf{P}^{(3)}(2\omega) = \chi^{(3)}(0,\omega,\omega)\mathbf{E}(0)\mathbf{E}(\omega)\mathbf{E}(\omega), \tag{4.5-1}$$

where $\mathbf{E}(0)$ and $\mathbf{E}(\omega)$ are the applied dc electric field and the incident fundamental optical wave, respectively. Assuming that these two fields have the same linear polarization (say along x-axis direction), then Eq. (4.5-1) can be simplified to a scalar form

$$P^{(3)}(2\omega) = \chi_e^{(3)}(0,\omega,\omega)E(0)E(\omega)E(\omega), \tag{4.5-2}$$

where $\chi_e^{(3)}(0,\omega,\omega) = \chi_{xxxx}^{(3)}(0,\omega,\omega)$ is the effective third-order susceptibility value related to EFISHG. For a liquid medium consisting of the same polar molecules, this value can be expressed as[60–62]

$$\chi_e^{(3)} = Nf\left[\gamma_e(0,\omega,\omega) + \frac{\vec{\mu}_0 \cdot \vec{\beta}(\omega,\omega)}{5k_BT}\right]. \tag{4.5-3}$$

Here, N is the number of molecules in a unit volume, f is a local-field correction factor, $\gamma_e(0,\omega,\omega)$ is the average molecular third-order polarizability due to electron-cloud distortion, k_B is the

Boltzmann constant, T is the absolute temperature of the medium, $\vec{\mu}_0$ is the permanent electric dipole moment of the polar molecule, and $\vec{\beta}(\omega,\omega)$ is the vector part of the molecular second-order polarizability defined by $\beta_i = \sum_j \beta_{ijj}$. The first term in Eq. (4.5-3) represents the contribution from the mechanism of electron-cloud distortion of molecules; the second term represents the contribution from the partial reorientation of the polar molecules along the dc field. For nonpolar molecules ($\vec{\mu}_0 = 0$) or molecules with inversion symmetry ($\vec{\beta} = 0$), the second contribution vanishes while the first contribution remains.

It should be noted that for non-resonant nonlinear media, if the dispersion effect of the third-order polarizability of molecules is negligible, the value of $\gamma_e(0,\omega,\omega)$ measured with the EFISHG method should be nearly the same as the value of $\gamma_e(\omega,\omega,\omega)$ measured with the THG method. Based on this fact one can separate the second contribution from the first by comparing the result from EFISHG to that from THG. Consequently, the useful information about molecular second-order polarizability could be extracted.[60–64]

4.5.2 SHG in optical fibers

In the early days of nonlinear optical studies, people thought that it would be difficult to observe SHG in optical glass fibers because the core materials were isotropic. However, at the beginning of the 1980s researchers did observe a weaker SHG signal from single-mode and multi-mode optical fibers with lengths of several tens to hundred meters.[65,66] Later, a research group in Europe reported the observation of strong SHG signal from much shorter (0.5–1 m) single-mode Ge- and P-doped optical fiber samples.[67,68] In the latter case, a long irradiation time with the input fundamental 1.06 μm laser beam was needed for the fiber sample. After several hours' irradiation of the fundamental beam passing through the sample, the intensity of SHG could reach a higher level and the conversion efficiency could be as high as $\geq 5\%$. This surprising result stimulated the research interests in this specific issue. Soon after, a research group at AT&T reported another experimental arrangement for producing SHG in a short single-mode fiber, and they could give a physical explanation of the observed phenomenon.[69,70]

According to the second arrangement, the fiber sample should be simultaneously excited with both the fundamental 1.06-μm laser beam and the 532-nm second-harmonic beam. After a certain period of time the input 532-nm beam was removed, then a quite strong SHG signal could be still observed in the output end of the fiber sample. The needed pre-irradiation time depended on the intensity levels of the two input beams and could be several minutes to hours. The conversion efficiency could be ~0.24%. According to the proposed theoretical model,[69,70] the observed SHG in fibers arises from two sequential nonlinear processes. The first is a third-order optical rectification process induced by simultaneous excitation of the ω wave and the 2ω wave; the corresponding third-order nonlinear polarization component (ignoring its vector property) is

$$P_{dc}^{(3)}(\omega' = 0) = \varepsilon_0 \chi_e^{(3)}(\omega,\omega,-2\omega)|E(\omega)E(\omega)E^*(2\omega)|\cos \Delta kz, \quad (4.5\text{-}4)$$

where $\Delta k = 2k(\omega) - k(2\omega)$ is the phase-mismatch factor and z is the propagation distance along the fiber core. The above process creates a periodic internal dc field with a period of $2\pi/\Delta k$ within the core of the fiber sample. On the other hand, there could be some kinds of static or laser irradiation-induced color centers, defects, or trapping centers in the doped-core region. A partial deviation from centrosymmetry could take place due to charge-separation or reorientation of these centers upon the action of the internal dc field (like a poling field). This will give rise to SHG described by the following second-order polarization component (ignoring its vector property)

$$P^{(2)}(2\omega) = \varepsilon_0 \chi_e^{(2)}(\omega,\omega)|E(\omega)E(\omega)|e^{i2k(\omega)z}. \tag{4.5-5}$$

Here it can be further assumed that the magnitude and spatial distribution of $\chi_e^{(2)}(\omega,\omega)$ are determined by the induced internal dc field described by Eq. (4.5-4). This assumption implies that the spatial variation period of the magnitude of $\chi_e^{(2)}$ is $\pi/\Delta k$ that just equals the coherent length l_0 of SHG for $\Delta k \neq 0$ (see Eq. (3.1-9) and Fig. 3.1). Therefore, the second-harmonic signal can continuously grow along the fiber over a length much longer than the coherent length of SHG.

Another possibility is that once the induced dc field described by Eq. (4.5-4) is built up, there could be a dc-field induced SHG, as we just described in Section 4.5.1.

The above-mentioned physical model may also be used to qualitatively explain the experiments using only a single fundamental laser beam.[67,68] Under that condition, it can be assumed that there is a very weaker initial SHG signal within the fiber core due to electric quadrupole contribution or static local violation of centrosymmetry. This initial second-harmonic wave and the strong fundamental wave may interact with each other to create the internal dc field, and then the same processes as described above could take place.

PROBLEMS

1. Similar to Fig. 1-3, sketch a diagram showing the elementary process of the four-wave mixing (or four-photon parametric interaction) through three intermediate states, as specified in Fig. 4.1(c).
2. For a medium transparent in visible range, if its longest absorption wavelength is ~400 nm (corresponding to the transition from the ground state to the lowest excited state) and the shortest absorption wavelength is ~150 nm (corresponding to the transition from the ground state to the highest excited state), please calculate the energy spread range ($h\Delta\nu$) of the molecular excited states and the corresponding life time (Δt) of an intermediate state, based on the uncertainty relation, $\Delta\nu\Delta t \approx 1$.
3. For third-harmonic generation (THG) in quartz glass, if the fundamental wavelength is 1500 nm, and the refractive-index curve is shown in Fig. 12.1(a), please calculate the rough value of the coherent length of THG.
4. For coherent Stokes and anti-Stokes ring emission shown in Figs. 4.7 and 4.8, assume the Raman resonance frequency of the medium is 900 cm^{-1} and the wavelength of the one input laser beam is $\lambda_1 = 530$ nm, please determine the wavelength ($\lambda_2 < \lambda_1$) of another input laser beam and the wavelengths (λ_3 and λ_4) of the two emission rings. If the refractive-indices are $n(\lambda_1) = 1.503$, $n(\lambda_2) = 1.500$, $n(\lambda_3) = 1.498$, and $n(\lambda_4) = 1.508$, please calculate the phase-matching angles θ_1 and θ_2.
5. For the non-resonant four-photon parametric interaction shown in Fig. 4.10 and satisfying Eq. (4.2-2), if the input pump laser wavelength is $\lambda_0 = 600$ nm, the intensity of the measured angular emission at a wavelength of $\lambda_1 = 500$ nm is 1 mW/cm^2, please determine the corresponding wavelength λ_2 and predict the emission intensity at that wavelength.

REFERENCES

1. R. W. Terhune, P. D. Maker, and C. M. Savage, *Phys. Rev. Lett.* **8**, 404 (1962).
2. J. F. Ward and G. H. New, *Phys. Rev.* **185**, 57 (1969).
3. S. E. Harris and R. B. Miles, *Appl. Phys. Lett.* **19**, 385 (1971).

4. J. F. Young, G. C. Bjorklund, A. H. Kung, R. B. Miles, and S. E. Harris, *Phys. Rev. Lett.* **27**, 1551 (1971).
5. T. Hodgson, P. P. Sorokin, and J. J. Wynne, *Phys. Rev. Lett.* **32**, 343 (1974).
6. D. M. Bloom, G. W. Bekkers, J. F. Young, and S. E. Harris, *Appl. Phys. Lett.* **26**, 687 (1975).
7. H. Puell, K. Spannev, W. Falkenstein, W. Kaiser, and C. R. Vidal, *Phys. Rev. A* **14**, 2240 (1976).
8. Y. Ohasi, Y. Ishibashi, T. Kobayashi, and H. Inaba, *Jpn. J. Appl. Phys.* **15**, 1817 (1976).
9. S. C. Wallace and G. Zdasiuk, *Appl. Phys. Lett.* **28**, 449 (1976).
10. V. Mitev, L. Pavlov, K. Stamenov, and A. Juazapavicius, *Dokl. Bolg. Acad. Nauk.* **30**, 1001 (1977).
11. H. Junginger, H. B. Puell, H. Scheingraber, and C. R. Vidal, *IEEE J. Quantum Electron.* **16**, 1132 (1980).
12. S. Dimov, L. Pavlov, K. Stamenov, Yu. I. Heller, and A. K. Popov, *Appl. Phys. B* **30**, 35 (1983).
13. F. Sh. Ganikhanov, N. I. Koroteev, V. B. Morozov, M. V. Rychev, S. N. Sazanov, and V. G. Tunkin, *Pis'ma Zh. Tech. Fiz.* **14**, 1570 (1988).
14. S. Zhao, P. Zhang, G. Zhang, and W. Zhao, *Appl. Opt.* **28**, 4521 (1989).
15. K. Miyazaki, H. Sakai, and T. Sato, *Appl. Opt.* **28**, 699 (1989).
16. W. Ubachs, K. S. E. Eikema, and W. Hogervorst, *Appl. Phys. B* **57**, 411 (1993).
17. B. Wellegehausen, K. Mossavi, A. Egbert, B. N. Chichkov, and H. Welling, *Appl. Phys. B* **63**, 451(1996).
18. G. Schilling, W. E. Ernst, and N. Schwentner, *IEEE J. Quantum Electron.* **29**, 2702 (1993).
19. R. Eramo and M. Matera, *Appl. Opt.* **33**, 1691 (1994).
20. H. Kildal and T. F. Deutsch, *IEEE J. Quantum Electron.* **QE12**, 429 (1976).
21. M. H. Kang, V. T. Nguyen, T. Y. Chang, T. C. Damen, and E. G. Burkhardt, *Appl. Phys. Lett.* **33**, 303 (1978).
22. J. E. Decker, F. Yergeau, Y. Beaudoin, M. M. Denariez-Roberge, and S. L. Chin, *IEEE J. Quantum Electron.* **25**, 1747 (1989).
23. S. R. J. Bruek and H. Kildal, *Appl. Phys. Lett.* **33**, 928 (1978).
24. P. P. Bey, J. F. Giuliani, and H. Rabin, *Phys. Rev. Lett.* **19**, 819 (1967).
25. W. Leupacher, A. Penzkofer, B. Runde, and K. H. Drexhage, *Appl. Phys. B* **44**, 133 (1987).
26. C. Schwan, A. Penzkofer, N. J. Mark, and K. H. Drexhage, *Appl. Phys. B* **57**, 203 (1993).
27. G. S. He, *Chin. Phys. Lasers*, **14**, 162 (1987).
28. G. S. He, M. Yoshida, J. D. Bhawalkar, and P. N. Prasad, *Appl. Opt.* **36**, 1155 (1997).
29. W. L. Glab and J. P. Hessler, *Appl. Opt.* **26**, 3181 (1987).
30. P. Qiu and A. Penzkofer, *Appl. Phys. B* **45**, 225 (1988).
31. I. V. Tomov, B. Van Wonterghem, and P. M. Rentzepis, *Appl. Opt.* **31**, 4172 (1992).
32. H. Kildal, R.F. Beglev, M. M. Choy, and R. L. Byer, *J. Opt. Soc. Am.* **62**, 1398 (1972).
33. J. P. Feve, B. Boulanger, and Y. Guillien, *Opt. Lett.* **25**, 1373 (2000).
34. S. A. Kkopyan, S. M. Arakelyan, R. V. Kochikyan, S. Ts. Nersisyan, and Yu. S. Chilingaryan, *Sov. J. Quantum Electron.* **7**, 814 (1977).
35. W. P. Acker, D. H. Leach, and R. K. Chang, *Opt. Lett.* **14**, 402 (1989).
36. V. I. Borisov, V. M. Kulichkov, V. I. Lebedev, V. P. Minkovich, and N. P. Minkovich, *Zh. Prikl. Spektrosk.* **52**, 163 (1990).
37. D. L. Nicacio, E. A. Gouveia, N. M. Borges, and A. S. Gouveia, *Appl. Phys. Lett.* **62**, 2179 (1993).
38. G. Qin, M. Liao, C. Chaudhari, X. Yan, T. Kito, T. Suzuki, and Y. Ohishi, *Opt. Lett.* **35**, 58 (2010).
39. W. Gao, K. Ogawa, X. Xue, M. Liao, D. Deng, T. Cheng, T. Suzuki, and Y. Ohishi, *Opt. Lett.* **38**, 2566 (2013).
40. F. Miyaji, K. Tadanaga, and S. Sakka, *Appl. Phys. Lett.* **60**, 2060 (1992).
41. A. B. Fedotov, S. M. Gladkov, N. I. Koroteev, and A. M. Zheltikov, *J. Opt. Soc. Am. B* **8**, 363 (1991).
42. H. Xiong, H. Xu, Y. Fu, Y. Cheng, Z. Xu, and S. L. Chin, *Phys. Rev. A* **77**, 043802 (2008).
43. S. Suntsov, D. Abdollahpour, D. G. Papazoglou, and S. Tzortzakis, *Phys. Rev. A* **81**, 033817 (2010).

44. T. V. Konstantinova, P. N. Melentiev, A. E. Afanasiev, A. A. Kuzin, P. A. Starikov, A. S. Baturin, A. V. Tausenev, A. V. Konyashchenko, and V. I. Balykin, *Quantum Electron.* **43**, 379 (2013).
45. T. V. Konstantinova, P. N. Melentiev, A. E. Afanasiev, A. A. Kuzin, P. A. Starikov, A. S. Baturin, A. V. Tausenev, A. V. Konyashchenko, and V. I. Balykin, *JETP* **117**, 21 (2013).
46. E. Garmire, F. Pandarese, and C. H. Townes, *Phys. Rev. Lett.* **11**, 160 (1963).
47. H. J. Zeiger, P. E. Tannenwald, S. Kern, and R. Herendeen, *Phys. Rev. Lett.* **11**, 419 (1963).
48. R. Chiao and B. P. Stoicheff, *Phys. Rev. Lett.* **12**, 290 (1964).
49. E. Garmire, *Phys. Lett.* **17**, 251 (1965).
50. P. D. Maker and R. W. Terhune, *Phys. Rev.* **137**, A80 (1965).
51. G. S. He, F. X. Zhou, D. Liu, and S. H. Liu, *Chin. Phys. Lasers*, **13**, 81 (1986).
52. G. S. He, D. Liu, and S. H. Liu, *Opt. Commun.* **70**, 145 (1989).
53. Eli Yablonovitch, N. Bloembergen, and J. J. Wynne, *Phys. Rev. B* **3**, 2067 (1971).
54. M. D. Levenson, C. Flytzanis, and N. Bloembergen, *Phys. Rev. B* **6**, 3962 (1972).
55. D. Heiman, R. W. Hellwarth, M. D. Levenson, and G. Martin, *Phys. Rev. Lett.* **36**, 189 (1976).
56. I. H. Malitson, *J. Opt. Soc. Am.* **55**, 1205 (1965).
57. R. R. Alfano and S. L. Shapiro, *Phys. Rev. Lett.* **24**, 584 (1970).
58. R. R. Alfano and S. L. Shapiro, *Phys. Rev. Lett.* **24**, 592 (1970).
59. G. S. He, G. C. Xu, Y. Cui, and P. N. Prasad, *Appl. Opt.* **32**, 4507 (1993).
60. B. F. Levine and C. G. Bethea, *J. Chem. Phys.* **63**, 2666 (1975).
61. J. C. Oudar and D. S. Chemla, *J. Chem. Phys.* **66**, 2664 (1977).
62. K. D. Singer and A. F. Garito, *J. Chem. Phys.* **75**, 3572 (1981).
63. S. J. Lalama, K. D. Singer, A. F. Garito, and K. N. Desai, *Appl. Phys. Lett.* **39**, 940 (1981).
64. C. C. Teng and A. F. Garito, *Phys. Rev. B* **28**, 6766 (1983).
65. Y. Fujii, B. S. Kawasaki, K. O. Hill, and D. C. Johnson, *Opt. Lett.* **5**, 48 (1980).
66. Y. Sasaki and Y. Ohmori, *Appl. Phys. Lett.* **39**, 446 (1980).
67. U. Osterberg and W. Margulis, *Opt. Lett.* **11**, 516 (1986).
68. U. Osterberg and W. Margulis, *Opt. Lett.* **12**, 57 (1987).
69. R. H. Stolen and H. W. K. Tom, *Opt. Lett.* **12**, 585 (1987).
70. H. W. K. Tom, R. H. Stolen, G. D. Aumiller, and W. Pleibel, *Opt. Lett.* **13**, 512 (1988)

5
Induced Refractive-Index Changes

In conventional optics, the refractive index of a material at a given wavelength is a constant independent of the light intensity. Since the advent of laser, it is found that the refractive index can be a variable upon the action of intense coherent light. Bases on the third-order nonlinear polarization mechanisms, the induced refractive-index change is proportional to the intensity of the incident laser beam, whereas in a second-order nonlinear medium it can be proportional to the amplitude of the intense coherent light field.

Intense coherent light induced refractive-index change is the fundamental process involved in several major nonlinear optical effects, including self-focusing, self-phase modulation, optical phase conjugation, optical bistability, and optical solitons. It is also the physical basis to develop novel devices and techniques for all-optical switching and steering, real-time holography, optical data storage and processing.

5.1 Description of refractive index in linear optics

It is well known that under the action of a monochromatic optical field of angular frequency ω, the electric field vector **E**, electric displacement vector **D**, and electric polarization vector **P** in a dielectric medium are related by the following equation:

$$\mathbf{D}(\omega) = \varepsilon_0 \mathbf{E}(\omega) + \mathbf{P}(\omega), \tag{5.1-1}$$

where ε_0 is the free-space permittivity. In the scope of conventional optics, the electric polarization vector can be written as

$$\mathbf{P}(\omega) = \varepsilon_0 \chi^{(1)}(\omega) \mathbf{E}(\omega), \tag{5.1-2}$$

where $\chi^{(1)}(\omega)$ is the linear susceptibility of the medium. Substituting Eq. (5.1-2) into Eq. (5.1-1) leads to

$$\mathbf{D}(\omega) = \varepsilon_0 \mathbf{E}(\omega) + \varepsilon_0 \chi^{(1)}(\omega) \mathbf{E}(\omega) = \varepsilon_0 \varepsilon_r \mathbf{E}(\omega). \tag{5.1-3}$$

Here,

$$\varepsilon_r(\omega) = [1 + \chi^{(1)}(\omega)] \tag{5.1-4}$$

is the relative dielectric constant of the medium. From Eq. (5.1-4), one can see that the dielectric constant is determined by the linear susceptibility.

Nonlinear Optics and Photonics. First Edition. Guang S. He. © Guang S. He 2015.
Published in 2015 by Oxford University Press.

In general, $\varepsilon_r(\omega)$ is a second-rank tensor, but for isotropic media it is a scalar. For non-resonant interaction, it is a real quantity, whereas for resonant interaction it is a complex quantity.

For nonmagnetic optical media, the refractive index is defined by

$$n(\omega) = \sqrt{\text{Re}\{\varepsilon_r(\omega)\}} = \sqrt{1 + \text{Re}\{\chi^{(1)}(\omega)\}}. \tag{5.1-5}$$

Here, $\text{Re}\{\varepsilon_r(\omega)\}$ is the real part of the dielectric constant at frequency ω.

For non-resonant interaction we have (see Chapter 18, Eq. (18.2-21))

$$\chi_{ij}^{(1)}(\omega) = \frac{1}{\varepsilon_0 \hbar} \sum_{a,b} N_a (p_i)_{ab} (p_j)_{ba} \frac{2\omega_{ba}}{\omega_{ba}^2 - \omega^2}, \tag{5.1-6}$$

where $\hbar$ is Planck's constant divided by 2π, $(p_i)_{ab} = (p_i)_{ba}$ is the dipole-transition matrix element between the molecular state (a) and state (b), ω_{ba} is the transition frequency between these two states, N_a is the population density of state (a), and finally the subscripts a and b represent all allowed eigenstates of the molecular system, respectively. In obtaining Eq. (5.1-6), we assumed that the light frequency ω considered here is much lower than any molecular electronic transition frequency.

For isotropic media the linear susceptibility becomes a real scalar,

$$\chi^{(1)}(\omega) = \frac{1}{\varepsilon_0 \hbar} \sum_{a,b} N_a (p_0)_{ba}^2 \frac{2\omega_{ba}}{\omega_{ba}^2 - \omega^2}, \tag{5.1-7}$$

where $(p_0)_{ba}$ is the transition matrix element and p_0 is the component of the dipole-moment operator in the direction of the light polarization. From Eq. (5.1-5), the refractive index is determined by

$$n^2(\omega) - 1 = \frac{1}{\varepsilon_0 \hbar} \sum_{a,b} N_a (p_0)_{ba}^2 \frac{2\omega_{ba}}{\omega_{ba}^2 - \omega^2}, \text{ (normal dispersion)} \tag{5.1-8}$$

Since it is assumed that $\omega < \omega_{ba}$, from Eq. (5.1-8) one can see that $n(\omega)$ becomes greater when the light frequency ω increases. This is the normal dispersion behavior of the refractive index in the transparent spectral range of an isotopic medium and can be schematically shown in Fig. 5.1 by the dashed-line curve.

Next, let us consider the resonant interaction in an isotropic medium, i.e., the light frequency is very close to an absorptive transition frequency. In this case, the linear susceptibility contains two parts, i.e., the non-resonant contribution and the resonant contribution. The former is a real and the latter is a complex, therefore the linear susceptibility can be expressed as

$$\chi^{(1)}(\omega) = \chi_{nonr}^{(1)}(\omega) + \chi_{res}^{(1)}(\omega) = \chi_{nonr}^{(1)}(\omega) + \text{Re}\{\chi_{res}^{(1)}(\omega)\} + \text{Im}\{\chi_{res}^{(1)}(\omega)\}. \tag{5.1-9}$$

Here the non-resonant contribution $\chi_{nonr}^{(1)}(\omega)$ is still given by Eq. (5.1-7), while the resonant contribution can be written as (cf. Eq. (18.4-2))

$$\chi_{res}^{(1)}(\omega) = \frac{(N_o - N_t)(p_0)_{to}^2}{\varepsilon_0 \hbar} \frac{1}{\omega_{to} - \omega - i\Gamma}. \tag{5.1-10}$$

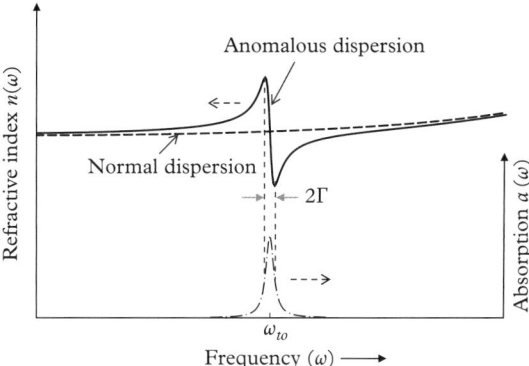

Figure 5.1 *Refractive index as a function of frequency: the normal dispersion (dashed-line curve) and the anomalous dispersion (solid-line curve). The linear absorption lineshape is shown in the bottom by the dash-dotted curve.*

where $(p_0)_{to}$ is the dipole-transition matrix element between the ground state (o) and high exited state (t), ω_{to} is the resonant absorption frequency between these two states, 2Γ is the full spectral width of this transition, and N_o and N_t are, respectively, the molecular population densities of these two states.

It is well known that the imaginary part of Eq. (5.1-10) determines the linear absorption cross-section, while the real part determines the resonant contribution to the linear refractive index. Substituting the real part of Eq. (5.1-10) into Eq. (5.1-9) we finally have

$$n^2(\omega) - 1 = \chi^{(1)}_{nonr}(\omega) + \frac{(N_o - N_t)(p_0)^2_{to}}{\varepsilon_0 \hbar} \cdot \frac{(\omega_{to} - \omega)}{(\omega_{to} - \omega)^2 + \Gamma^2}. \text{ (anomalous dispersion)} \qquad (5.1\text{-}11)$$

The above equation describes the anomalous dispersion behavior in the spectral range close to an (one-photon) absorption line. Based on Eq. (5.1-11) we can obtain the curve shown in Fig. 5.1 by the solid-line.

From Eqs. (5.1-8) and (5.1-11) one can see that for both resonant and non-resonant cases, the refractive index is only a function of frequency, and is independent of the intensity (or amplitude) of the incident light field. This conclusion is valid only for a weak incident light field.

5.2 Description of refractive index in nonlinear optics

Under the action of an intense coherent light field, the contribution of nonlinear polarization to refractive index can no longer be neglected. For simplicity, we still assume that the medium is isotropic and only the third-order polarization contribution should be considered.[1,2] The incident light is a monochromatic wave linearly polarized along the x-axis and can be expressed as

$$\mathbf{E}(\omega) = \mathbf{a}_x E_0(\omega), \qquad (5.2\text{-}1)$$

where $\mathbf{a}_x$ is a unit vector along the x-axis direction and $E_0(\omega)$ is the light amplitude function. In this case, the overall polarization vector can be written as

$$\begin{aligned} \mathbf{P}(\omega) &= \mathbf{P}^{(1)}(\omega) + \mathbf{P}^{(3)}(\omega) \\ &= \varepsilon_0[\chi^{(1)}(\omega)\mathbf{E}(\omega) + \chi^{(3)}(\omega,-\omega,\omega)\mathbf{E}(\omega)\mathbf{E}^*(\omega)\mathbf{E}(\omega)]. \end{aligned} \tag{5.2-2}$$

For isotropic media, $\chi^{(1)}(\omega)$ is a scalar quantity, and the only involved nonzero element of third-order susceptibility tensor is $\chi^{(3)}_{xxxx}$ (see Appendix 6). Substituting Eqs. (5.2-1) and (5.2-2) to Eq. (5.1-1) leads to

$$\begin{aligned} D(\omega) &= \varepsilon_0\{E_0(\omega) + \chi^{(1)}(\omega)E_0(\omega) + \chi_e^{(3)}(\omega,-\omega,\omega)E_0(\omega)E_0^*(\omega)E_0(\omega)\} \\ &= \varepsilon_0\{[1 + \chi^{(1)}(\omega)] + \chi_e^{(3)}(\omega,-\omega,\omega)|E_0(\omega)|^2\}E_0(\omega), \end{aligned} \tag{5.2-3}$$

where $\chi_e^{(3)}(\omega,-\omega,\omega) = \mathbf{a}_x \cdot \chi^{(3)}(\omega,-\omega,\omega)\mathbf{a}_x\mathbf{a}_x\mathbf{a}_x = \chi^{(3)}_{xxxx}$ is the effective third-order susceptibility value for induced refractive-index change and $D(\omega)$ is the electric displacement function. Comparing Eq. (5.2-3) to Eq. (5.1-3), we obtain a generalized expression for the relative dielectric constant of a third-order nonlinear medium

$$\begin{aligned} \varepsilon_r(\omega) &= [1 + \chi^{(1)}(\omega)] + \chi_e^{(3)}(\omega,-\omega,\omega)|E_0(\omega)|^2 \\ &= \varepsilon_r^0(\omega) + \chi_e^{(3)}(\omega,-\omega,\omega)|E_0(\omega)|^2, \end{aligned} \tag{5.2-4}$$

where ε_r^0 is the linear relative dielectric constant of the medium. If we assume in Eq. (5.2-4) the nonlinear contribution is much small than the linear contribution, the generalized expression of refractive index should be

$$\begin{aligned} n(\omega) &= \sqrt{\text{Re}\{\varepsilon_r(\omega)\}} \\ &\approx n_0(\omega) + \frac{1}{2n_0(\omega)}\text{Re}\{\chi_e^{(3)}(\omega,-\omega,\omega)\}|E_0(\omega)|^2. \end{aligned} \tag{5.2-5}$$

Here, $n_0(\omega)$ is the linear refractive index of the medium. The physical meaning of Eq. (5.2-5) is that under the action of an intense incident optical field, the refractive index for a third-order nonlinear medium is no longer a constant. The induced refractive-index change is proportional to the intensity of the applied optical field and can be written as

$$\Delta n(\omega) = n(\omega) - n_0(\omega) = n_2|E_0(\omega)|^2, \tag{5.2-6}$$

where n_2 is called the nonlinear refractive-index coefficient of the medium and is defined by

$$n_2(\omega) = \frac{1}{2n_0(\omega)}\text{Re}\{\chi_e^{(3)}(\omega,-\omega,\omega)\}. \tag{5.2-7}$$

Since the light intensity I is proportional to $|E_0|^2$ (cf. Eq. (1.4-1)), one can give an equivalent expression of the induced refractive-index change:

$$\Delta n(\omega) = \frac{2n_2}{\varepsilon_0 c n_0(\omega)} \cdot I_0(\omega) = \bar{n}_2 \cdot I_0(\omega). \tag{5.2-8}$$

In most non-resonant cases, the induced refractive-index change is very small, i.e., usually $\Delta n \leq 10^{-4}$–10^{-5}. However, it can be significantly increased by resonance enhancement. The specific enhancement can be realized via one-photon transition resonance, two-photon transition resonance, and Raman-transition resonance, respectively.

Let us first consider the enhanced refractive-index change by one-photon transition resonance. According to the theory of nonlinear susceptibility described in Chapter 18, the one-photon resonant contribution to the third-order nonlinear susceptibility matrix element of an isotropic medium is given by (cf. Eq. (18.4-5))

$$\chi_{e,r}^{(3)}(\omega,-\omega,\omega) = K \frac{1}{(\omega_{to}-\omega)-i\Gamma}, \qquad (5.2\text{-}9)$$

where ω_{to} is the one-photon transition frequency from the ground state (o) to an excited state (t), 2Γ is the full spectral width of this transition, and K is a real coefficient determined by

$$K = \frac{(N_o - N_t)(p_0)_{to}}{\varepsilon_0 6 \hbar^3} \sum_{b,c}(p_0)_{ob}(p_0)_{bc}(p_0)_{ct} \left[\frac{2}{(\omega_{bo}-\omega)\omega_{co}} + \frac{2}{(\omega_{bo}+\omega)\omega_{co}} + \frac{2}{(\omega_{bo}-\omega)(\omega_{co}-2\omega)} \right]. \qquad (5.2\text{-}10)$$

Here, the definitions of N_o, N_t, and $(p_0)_{to}$ are the same as those indicated in Eqs. (5.1-7) and (5.1-8), and the subscripts b and c represent all possible eigenstates of the molecular system. Substituting the real part of $\chi_{e,r}^{(3)}$ from Eq. (5.2-9) into Eq. (5.2-7) leads to

$$n_2 = \frac{K}{2n_0(\omega)} \frac{\omega_{to}-\omega}{(\omega_{to}-\omega)^2 + \Gamma^2}. \text{ (one-photon resonance)} \qquad (5.2\text{-}11)$$

One can see that $n_2(\omega)$ changes its sign when we scan ω from the low-frequency side to the high-frequency side of ω_{to}, and at exact resonance position of $\omega = \omega_{to}$ we expect $n_2 = 0$.

In practice, when the incident laser frequency ω is close to the peak absorption frequency ω_{to}, the loss of input light signal will increase rapidly owing to linear absorption, which may also cause undesirable opto-heating effect. These facts are unfavorable for some applications.

5.3 Two-beam induced refractive-index changes

Now let us consider the refractive-index change in an isotropic third-order nonlinear medium induced by two light beams.[2-5] Assume that there are two incident monochromatic waves polarized along the x-axis but having different frequencies, i.e.,

$$\left.\begin{array}{l}\mathbf{E}(\omega_1) = \mathbf{a}_x E_0(\omega_1) \\ \mathbf{E}(\omega_2) = \mathbf{a}_x E_0(\omega_2)\end{array}\right\}. \qquad (5.3\text{-}1)$$

The polarization components at frequencies of ω_1 and ω_2 are

$$\left.\begin{array}{l}\mathbf{P}(\omega_1) = \varepsilon_0[\chi^{(1)}(\omega_1)\mathbf{E}(\omega_1) + \chi^{(3)}(\omega_1,-\omega_1,\omega_1)\mathbf{E}(\omega_1)\mathbf{E}^*(\omega_1)\mathbf{E}(\omega_1)+ \\ \qquad \chi^{(3)}(\omega_1,-\omega_2,\omega_2)\mathbf{E}(\omega_1)\mathbf{E}^*(\omega_2)\mathbf{E}(\omega_2)] \\ \mathbf{P}(\omega_2) = \varepsilon_0[\chi^{(1)}(\omega_2)\mathbf{E}(\omega_2) + \chi^{(3)}(\omega_2,-\omega_2,\omega_2)\mathbf{E}(\omega_2)\mathbf{E}^*(\omega_2)\mathbf{E}(\omega_2)+ \\ \qquad \chi^{(3)}(\omega_2,-\omega_1,\omega_1)\mathbf{E}(\omega_2)\mathbf{E}^*(\omega_1)\mathbf{E}(\omega_1)]\end{array}\right\}. \qquad (5.3\text{-}2)$$

By substituting the above expressions into Eq. (5.1-1) and following the same procedures as indicated in the previous section, the refractive-index changes can be written as

$$\left.\begin{aligned}\Delta n(\omega_1) &= \frac{1}{2n_0(\omega_1)}\text{Re}\{\chi_e^{(3)}(\omega_1,-\omega_1,\omega_1)\}|E_0(\omega_1)|^2+ \\ &\quad \frac{1}{2n_0(\omega_1)}\text{Re}\{\chi_e^{(3)}(\omega_1,-\omega_2,\omega_2)\}|E_0(\omega_2)|^2 \\ \Delta n(\omega_2) &= \frac{1}{2n_0(\omega_2)}\text{Re}\{\chi_e^{(3)}(\omega_2,-\omega_2,\omega_2)\}|E_0(\omega_2)|^2+ \\ &\quad \frac{1}{2n_0(\omega_2)}\text{Re}\{\chi_e^{(3)}(\omega_2,-\omega_1,\omega_1)\}|E_0(\omega_1)|^2\end{aligned}\right\}. \quad (5.3\text{-}3)$$

In the expression of $\Delta n(\omega_1)$, the first term represents the refractive-index change induced by the ω_1 beam itself; the second term represents the coupled refractive-index change induced by the ω_2 beam. Similarly there are two terms in $\Delta n(\omega_2)$. In other words, there is an additional refractive-index change at the frequency of one beam induced by another beam of different frequency, and this additional change is proportional to the intensity of the other beam. If we only consider non-resonant interaction and neglect the dispersion effect, the effective third-order susceptibility elements are real and become

$$\chi_e^{(3)}(\omega_1,-\omega_1,\omega_1) \approx \chi_e^{(3)}(\omega_2,-\omega_2,\omega_2) \approx \chi_e^{(3)}(\omega_1,-\omega_2,\omega_2) \approx \chi_e^{(3)}(\omega_2,-\omega_1,\omega_1).$$

Then Eq. (5.3-3) is simplified to

$$\left.\begin{aligned}\Delta n(\omega_1) &= \frac{1}{2n_0(\omega_1)}\chi_e^{(3)}(\omega_1,-\omega_1,\omega_1)[|E_0(\omega_1)|^2 + |E_0(\omega_2)|^2] \\ \Delta n(\omega_2) &= \frac{1}{2n_0(\omega_2)}\chi_e^{(3)}(\omega_2,-\omega_2,\omega_2)[|E_0(\omega_2)|^2 + |E_0(\omega_1)|^2]\end{aligned}\right\}. \quad (5.3\text{-}4)$$

Moreover, if $|E_0(\omega_1)|^2 \gg |E_0(\omega_2)|^2$, we have

$$\Delta n(\omega_2) = \frac{1}{2n_0(\omega_2)}\chi_e^{(3)}(\omega_2,-\omega_2,\omega_2)E_0(\omega_1)^2. \quad (5.3\text{-}5)$$

The physical meaning of Eq. (5.3-5) is that the refractive-index change for a weaker beam of ω_1 can be effectively controlled by another strong beam of ω_2.

5.4 Two-photon resonance enhanced refractive-index change

The study of two-beam (or two-frequency) induced refractive-index change is of importance not only because it provides the way to control the refractive-index change of one beam (or one frequency) by another beam (or another frequency). This study has also revealed the mechanisms that can be used to achieve a resonance-enhanced refractive-index change, and at the same time to maintain a small energy losses and opto-heating effect.

Let us now consider the two-photon resonance enhanced refractive-index changes. For this purpose, from Eq. (5.3-3) we first obtain the general expression of coupled refractive-index changes owing to two-wave interaction with matter:

$$\left.\begin{array}{l}\Delta n'(\omega_1) = \dfrac{1}{2n_0(\omega_1)} \text{Re}\{\chi_e^{(3)}(\omega_1, -\omega_2, \omega_2)\} |E_0(\omega_2)|^2 \\[2mm] \Delta n'(\omega_2) = \dfrac{1}{2n_0(\omega_2)} \text{Re}\{\chi_e^{(3)}(\omega_2, -\omega_1, \omega_1)\} |E_0(\omega_1)|^2\end{array}\right\}. \quad (5.4\text{-}1)$$

It shall be proved that when the frequency sum of the two incident light waves is close to a two-photon absorption frequency $\omega_{t'o}$ of the medium, i.e., $(\omega_1 + \omega_2) \Rightarrow \omega_{t'o}$, the coupled refractive-index change for these two waves can be significantly enhanced.[4-6] Based on the general expression of third-order susceptibility given in Chapter 18 and neglecting the non-resonant part, the resonant contribution from two-photon transition to $\chi_e^{(3)}$ can be expressed as (cf. Eq. (18.4-7)):

$$\left.\begin{array}{l}\chi_{e,r}^{(3)}(\omega_1, -\omega_2, \omega_2) = K' \dfrac{1}{[\omega_{t'o} - (\omega_1 + \omega_2)] - i\Gamma'} \\[2mm] \chi_{e,r}^{(3)}(\omega_2, -\omega_1, \omega_1) = K' \dfrac{1}{[\omega_{t'o} - (\omega_1 + \omega_2)] - i\Gamma'}\end{array}\right\}, \quad (5.4\text{-}2)$$

where $2\Gamma'$ is the full spectral width of a given two-photon absorptive transition, and K' is a real coefficient determined by

$$K' = \dfrac{N_o - N_{t'}}{\varepsilon_0 6 \hbar^3} \left| \sum_b \left[\dfrac{(p_0)_{ob}(p_0)_{bt'}}{\omega_{bo} - \omega_1} + \dfrac{(p_0)_{ob}(p_0)_{bt'}}{\omega_{bo} - \omega_2} \right] \right|^2. \quad (5.4\text{-}3)$$

Here, N_o and $N_{t'}$ are the molecular population densities in the ground state (o) and a real excited state (t'), subscript b represents all possible eigenstates, $(p_0)_{bo} = (p_0)_{ob}$ is the electric dipole matrix element related to the quantum transition from state (o) to state (b), and ω_{bo} is the corresponding transition frequency.

From Eq. (5.4-2), one can see that under the two-photon resonance condition, $\chi_{e,r}^{(3)}$ is a complex quantity: its imaginary part determines the attenuation of the incident beams due to two-photon absorption (see Section 9.3.2), and the real part determines the enhanced refractive-index change. Substituting the real part of Eq. (5.4-2) into Eq. (5.4-1) leads to

$$\left.\begin{array}{l}\Delta n'(\omega_1) = \dfrac{K'}{2n_0(\omega_1)} \cdot \dfrac{\omega_{t'o} - (\omega_1 + \omega_2)}{[\omega_{t'o} - (\omega_1 + \omega_2)]^2 + \Gamma'^2} \cdot |E_0(\omega_2)|^2 \\[2mm] \Delta n'(\omega_2) = \dfrac{K'}{2n_0(\omega_2)} \cdot \dfrac{\omega_{t'o} - (\omega_1 + \omega_2)}{[\omega_{t'o} - (\omega_1 + \omega_2)]^2 + \Gamma'^2} \cdot |E_0(\omega_1)|^2\end{array}\right\}. \quad (5.4\text{-}4)$$

Assuming $n_0(\omega_1) \approx n_0(\omega_2) \approx n_0$, Eq. (5.4-4) can be rewritten as

$$\left.\begin{array}{l}\Delta n'(\omega_1) = n'_2 |E_0(\omega_2)|^2 \\ \Delta n'(\omega_2) = n'_2 |E_0(\omega_1)|^2\end{array}\right\}. \quad (5.4\text{-}5)$$

Here, the nonlinear refractive-index coefficient n'_2 under two-photon resonance condition is only a function of $[\omega_{t'o} - (\omega_1 + \omega_2)]$:

$$n'_2 = \dfrac{K'}{2n_0} \cdot \dfrac{\omega_{t'o} - (\omega_1 + \omega_2)}{[\omega_{t'o} - (\omega_1 + \omega_2)]^2 + \Gamma'^2}. \quad (5.4\text{-}6)$$

The tuning behavior of n'_2 as a function of $[\omega_{t'o} - (\omega_1 + \omega_2)]$ is illustrated in Fig. 5.2, from which one can see that when $(\omega_1 + \omega_2)$ approaches $\omega_{t'o}$ from the low-frequency side, the refractive-index

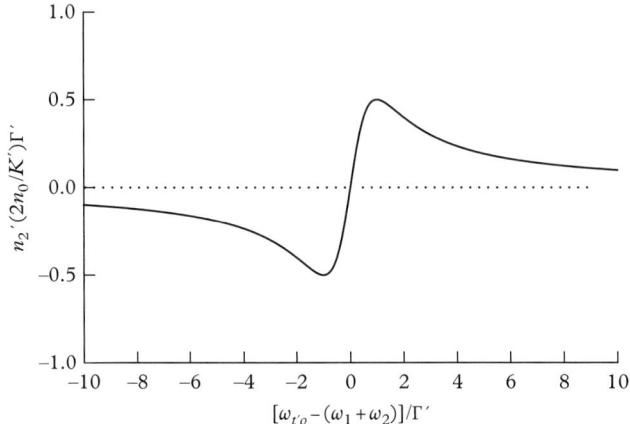

Figure 5.2 *Resonant enhancement curve of the nonlinear refractive-index coefficient n'_2 near a two-photon absorption frequency $\omega_{t'o}$.*

increment is positive; whereas when $(\omega_1 + \omega_2)$ approaches $\omega_{t'o}$ from the high-frequency side, the refractive-index increment is negative. This means that by tuning $(\omega_1 + \omega_2)$ we can control both the magnitude and the sign of the resonance-enhanced refractive index change. The refractive-index change reaches its maximum when $\omega_{t'o} - (\omega_1 + \omega_2) = \pm\Gamma'$, and a considerable resonant contribution remains over a relatively broad tuning range (roughly $\pm 8\Gamma'$).

It should be noted that in comparison with one-photon absorption, the two-photon absorption probability is much smaller and the attenuation influence on the incident light waves over the spectral range mentioned above can be tolerated.

The two-photon resonance mechanism is also applicable to the case where only one laser beam of frequency ω is incident. Now the resonant enhancement condition is $2\omega \Rightarrow \omega_{t'o}$, and the enhanced refractive-index change will be

$$\left.\begin{array}{l}\Delta n(\omega) = n_2 \cdot |E_0(\omega)|^2 \\ n_2 = \dfrac{K'}{2n_0(\omega)} \cdot \dfrac{\omega_{t'o} - 2\omega}{(\omega_{t'o} - 2\omega)^2 + \Gamma'^2}\end{array}\right\}, \qquad (5.4\text{-}7)$$

where

$$K' = \dfrac{N_o - N_{t'}}{\varepsilon_0 6\hbar^3} \left|\sum_b \left[\dfrac{2(p_0)_{ob}(p_0)_{bt'}}{\omega_{bo} - \omega}\right]\right|^2. \qquad (5.4\text{-}8)$$

5.5 Raman resonance enhanced refractive-index change

We next consider another type of resonance enhancement, i.e., the Raman-enhanced refractive-index change.[2, 5, 7–9] In this case, the frequency difference $(\omega_1 - \omega_2) > 0$ of the two incident beams is very close to a Raman-transition frequency ($\Delta\omega_r$) of the nonlinear medium, i.e., $|\omega_1 - \omega_2| \Rightarrow \Delta\omega_r$. Under this condition, from the general expression of Raman-enhanced third-order susceptibility derived in Chapter 18 we can write the resonant contribution to the third-order susceptibility element (cf. Eqs. (18.4-8) and (18.4-9)) as

$$\left. \begin{aligned} \chi_{e,r}^{(3)}(\omega_1, -\omega_2, \omega_2) &= K_r \frac{1}{[\Delta\omega_r - (\omega_1 - \omega_2)] - i\Gamma_r} \\ \chi_{e,r}^{(3)}(\omega_2, -\omega_1, \omega_1) &= K_r \frac{1}{[\Delta\omega_r - (\omega_1 - \omega_2)] - i\Gamma_r} \end{aligned} \right\}, \tag{5.5-1}$$

where $2\Gamma_r$ is the full spectral width of the considered Raman transition, and K_r is a real coefficient determined by

$$K_r = \frac{N_o - N_t}{\varepsilon_0 6\hbar^3} \left| \sum_b \left[\frac{(p_0)_{ob}(p_0)_{bt}}{\omega_{bo} - \omega_1} + \frac{(p_0)_{ob}(p_0)_{bt}}{\omega_{bo} + \omega_2} \right] \right|^2. \tag{5.5-2}$$

Here, N_o and N_t are the molecular population densities in the ground state and the excited Raman state, respectively.

From Eq. (5.5-1) we can see that the Raman-enhanced third-order nonlinear susceptibility element is a complex quantity: its imaginary part determines the attenuation (or gain) of one beam upon the action of the other beam, while the real part determines the enhanced refractive-index change. Substituting the real part of Eq. (5.5-1) into Eq. (5.4-1) leads to

$$\left. \begin{aligned} \Delta n'(\omega_1) &= \frac{K_r}{2n_0(\omega_1)} \cdot \frac{\Delta\omega_r - (\omega_1 - \omega_2)}{[\Delta\omega_r - (\omega_1 - \omega_2)]^2 + \Gamma_r^2} \cdot |E_0(\omega_2)|^2 \\ \Delta n'(\omega_2) &= \frac{K_r}{2n_0(\omega_2)} \cdot \frac{\Delta\omega_r - (\omega_1 - \omega_2)}{[\Delta\omega_r - (\omega_1 - \omega_2)]^2 + \Gamma_r^2} \cdot |E_0(\omega_1)|^2 \end{aligned} \right\}. \tag{5.5-3}$$

Assuming $n_0(\omega_1) \approx n_0(\omega_2) \approx n_0$, we obtain

$$\left. \begin{aligned} \Delta n'(\omega_1) &= n'_2 |E_0(\omega_2)|^2 \\ \Delta n'(\omega_2) &= n'_2 |E_0(\omega_1)|^2 \end{aligned} \right\}, \tag{5.5-4}$$

where

$$n'_2 = \frac{K_r}{2n_0} \cdot \frac{\Delta\omega_r - (\omega_1 - \omega_2)}{[\Delta\omega_r - (\omega_1 - \omega_2)]^2 + \Gamma_r^2}. \tag{5.5-5}$$

From the above expression, we can see that the nonlinear refractive-index coefficient n'_2 is a function of $[\Delta\omega_r - (\omega_1 - \omega_2)]$. The tuning curve of n'_2 is illustrated in Fig. 5.3 under the Raman resonance condition.

It is shown in Fig. 5.3 that when the frequency difference $(\omega_1 - \omega_2)$ of the two incident beams approaches the Raman-transition frequency $\Delta\omega_r$ from the low-frequency side, the refractive-index increment is positive; whereas when $(\omega_1 - \omega_2)$ approaches $\Delta\omega_r$ from the high-frequency side, the refractive-index increment is negative. The maximum refractive-index change is achieved at the positions of $[\Delta\omega_r - (\omega_1 - \omega_2)] = \pm\Gamma$. The considerable resonance-enhanced refractive-index change can be expected over a broad spectral range of $\sim(\pm 8\Gamma)$. Here we find another very useful approach to control both the magnitude and the sign of the resonance-enhanced refractive-index change by tuning $(\omega_1 - \omega_2)$ nearby $\Delta\omega_r$.

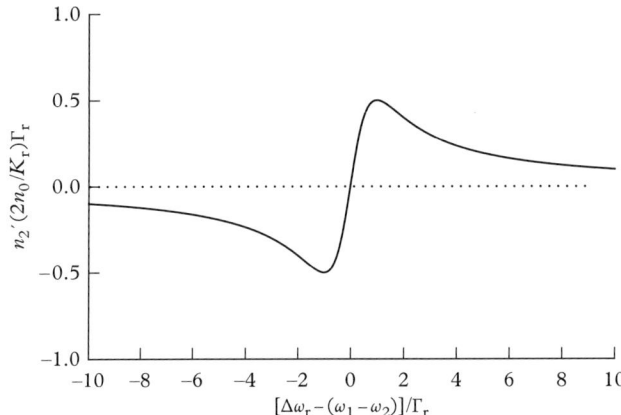

Figure 5.3 *Resonant tuning curve of the nonlinear refractive-index coefficient n_2' near a Raman transition frequency $\Delta\omega_r$.*

5.6 Mechanisms of induced refractive-index changes

5.6.1 Various mechanisms of induced refractive-index changes

In Section 5.2 we obtained the general formulation for the intense-light induced refractive-index change in a third-order nonlinear optical medium. This formulation is essentially applicable only for steady-state cases in which the duration of the applied optical field is much longer than the response-time of the dominant nonlinear polarization mechanism.

In Section 2.2 we already discussed various mechanisms contributing to the nonlinear polarization, including electron-cloud distortion, intramolecular motion, molecular reorientation, electrostriction, induced population change, and opto-thermal effect. All these mechanisms have their own characteristic response-time, and might be partially responsible for the intense light-induced refractive-index change.[10-13] Sometimes it is a research task to determine which mechanism is dominant for a nonlinear medium under the given experimental conditions.

In Sections 5.4 and 5.5 we focused our discussions on two types of resonance enhanced refractive-index changes: one is the two-photon transition enhancement, the other is the Raman transition enhancement. In the first case, the dominant mechanism is the electron-cloud distortion, which is fast enough to respond to the two-photon transition. In the second case, the dominant mechanism is the intramolecular motion that is solely responsible for the Raman transition.

However, in many practical cases, the employed nonlinear materials are transparent and non-resonant at the incident laser wavelength(s). In those situations, either the molecular reorientation or the electrostriction could be the major mechanism for observed refractive-index changes. Strictly speaking, the explicit expressions for the third-order nonlinear susceptibility given in Chapter 18 are not applicable to those cases when the molecular reorientation or electrostriction becomes the major source of the nonlinear polarization. Nevertheless, in the following parts of this section we shall prove that a similar formula for intense light induced refractive-index change can be obtained for these two mechanisms.

In cases of resonant interactions, another two mechanisms may also give rise to induced refractive-index changes: one is the population change effect and the other is the opto-thermal effect. It is known that the refractive index is a function of the population distribution among different eigenstates of a given molecular system; any resonant (one-photon, two-photon, or Raman) transition will change the original population distribution as well as the related refractive index of the medium.

Furthermore, the population change is often accompanied by the radiationless transition that causes the local temperature increase of the medium. Since the refractive index of the medium is also dependent on the temperature of the medium owing to the change of density, an additional refractive-index change induced by the opto-thermal effect might be observable.

5.6.2 Molecular reorientation contribution

Now let us consider a liquid composed of anisotropic and nonpolar molecules. In the absence of an external electric field, the orientations of molecules are randomly distributed, therefore the liquid is macroscopically isotropic. Under the action of an intense optical field linearly polarized, the anisotropic and nonpolar molecules in the liquid have a tendency to reorientate themselves along the polarization direction of the optical field. Because of such a light-induced reorientation of molecules, the liquid medium becomes partially anisotropic. In other words, there is an induced anisotropic refractive-index change. This is the well-known (reorientational) optical Kerr effect in liquid. The typical Kerr-type liquids include CS_2 and C_6H_6.

For simplicity, it is assumed that the molecule in a Kerr liquid possesses two-dimensional anisotropy, namely, there is a rotation axis for molecular symmetry. Along this axis, the molecular polarizability is $\alpha_{//}$; in the direction perpendicular to that axis the molecular polarizability is $\alpha_\perp$, and we assume here $\alpha_{//} > \alpha_\perp$. These assumptions are basically applicable to the liquid of carbon disulfide (CS_2).

We assume further that the incident optical field is linearly polarized, i.e., $\mathbf{E} = \mathbf{a}_0 E_0$, where E_0 is the amplitude function and $\mathbf{a}_0$ is a unit vector along the polarization direction of the applied optical field. Let $\mathbf{p}$ denote the induced molecular electric-dipole moment; θ is the angle between the vector $\mathbf{E}$ and the molecular symmetric axis (x-axis), as shown in Fig. 5.4. In this case, the potential energy function for the interaction between the optical field and an anisotropic molecule is

$$\begin{aligned} U(\theta) &= -\mathbf{p} \cdot \mathbf{E} = -[p_x E_x + p_y E_y] = -[\alpha_{//} E_x^2 + \alpha_\perp E_y^2] \\ &= -[\alpha_{//}\cos^2\theta + \alpha_\perp \sin^2\theta] E_0^2 \\ &= -[\alpha_\perp + (\alpha_{//} - \alpha_\perp)\cos^2\theta] E_0^2. \end{aligned} \tag{5.6-1}$$

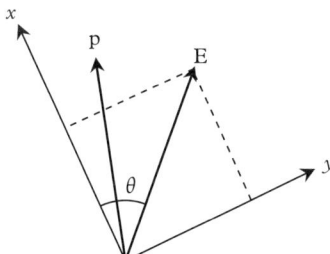

Figure 5.4 *Orientation of the field vector* $\mathbf{E}$ *and the induced dipole moment vector* $\mathbf{p}$ *in a molecular coordinate system.*

From the above expression, one can see that a smaller angle θ corresponds to a lower interaction potential. This implies that the molecule having a smaller θ is more stable under an applied optical field than the molecule having a larger θ. For this reason, molecules prefer to reorientate themselves along the polarization direction of the optical field. On the other hand, the rotational viscosity and thermal collision inside the liquid will damp the optical field-induced molecular reorientation. Under the condition of thermal equilibrium, the distribution function of molecular orientations can be expressed as[14]

$$F(\theta) = \frac{e^{-U(\theta)/k_B T}}{\int_0^\pi e^{-U(\theta)/k_B T} \cdot 2\pi \sin\theta \, d\theta}$$

$$\approx \frac{1}{4\pi}\left[1 + \frac{E_0^2}{6k_B T}(\alpha_{//} - \alpha_\perp)(3\cos^2\theta - 1)\right]. \tag{5.6-2}$$

Here, k_B is the Boltzmann constant, T is the absolute temperature, and it is assumed that the second term in the brackets is much smaller than the first. From Eq. (5.6-2), we find that all liquid molecules obey an axially symmetric orientational distribution along the polarization direction of the optical field: numbers of molecules of $|\theta| \approx 0°$ are much greater than those molecules of $|\theta| \approx 90°$. As a result of that symmetric distribution, the macroscopic polarization vector $\mathbf{P}$ of the medium points to the same direction as the optical field vector $\mathbf{E}$ does. In other words, $\mathbf{P}$ is determined by the summation of components of the molecular dipole moment along $\mathbf{E}$ over a great number of molecules. For each molecule the projection of $\mathbf{p}$ in the direction of $\mathbf{E}$ can be written as

$$p(\theta) = \mathbf{p} \cdot \mathbf{a}_0 = (\alpha_{//}\cos^2\theta + \alpha_\perp \sin^2\theta)E_0, \tag{5.6-3}$$

and the macroscopic electric polarization vector is given by[14]

$$\mathbf{P} = \mathbf{a}_0 N \int_0^\pi p(\theta) F(\theta) 2\pi \sin\theta \, d\theta$$

$$= \left[\frac{N(2\alpha_\perp + \alpha_{//})}{3} + \frac{2N(\alpha_{//} - \alpha_\perp)^2}{45 k_B T}E_0^2\right]\mathbf{E}. \tag{5.6-4}$$

Here, N is the molecular density of the liquid. Substituting Eq. (5.6-4) into Eq. (5.1-1) leads to

$$\mathbf{D} = \varepsilon_0 \mathbf{E} + \mathbf{P}$$

$$= \varepsilon_0\left[1 + \frac{N(2\alpha_\perp + \alpha_{//})}{3\varepsilon_0} + \frac{2N(\alpha_{//} - \alpha_\perp)^2}{45\varepsilon_0 k_B T}E_0^2\right]\mathbf{E} \tag{5.6-5}$$

$$= \varepsilon_0[\varepsilon_r + \Delta\varepsilon]\mathbf{E},$$

where

$$\varepsilon_r = 1 + \frac{N(2\alpha_\perp + \alpha_{//})}{3\varepsilon_0} \tag{5.6-6}$$

is the linear relative dielectric constant of the liquid, and the nonlinear contribution to the dielectric constant is

$$\Delta\varepsilon = \frac{2N(\alpha_{//} - \alpha_\perp)^2}{45\varepsilon_0 k_B T}E_0^2. \tag{5.6-7}$$

From Eqs. (5.6-6) and (5.6-7) we can see that both ε_r and $\Delta\varepsilon$ are real quantities because there is no resonant interaction. Finally, we can obtain the following expression for refractive-index change induced by a linearly polarized optical field

$$\Delta n \approx \frac{1}{2n_0}\Delta\varepsilon = \frac{N(\alpha_{//}-\alpha_\perp)^2}{45\varepsilon_0 n_0 k_B T} E_0^2. \tag{5.6-8}$$

Comparing Eq. (5.6-8) to Eq. (5.2-6) that is derived from the more general consideration, we find the similarity in the light intensity-dependence of induced refractive-index changes. For the molecular reorientation process, a nominal nonlinear refractive-index coefficient can be given by

$$n_2^{\text{Kerr}} = \frac{N(\alpha_{//}-\alpha_\perp)^2}{45\varepsilon_0 n_0 k_B T}. \tag{5.6-9}$$

It should be noted that (i) in the above theoretical derivation we did not consider the damping effect due to rotational viscosity inside the liquid, (ii) the result is not dependent on the frequency of the optical field, and (iii) if the incident beam is of a circular polarization, the molecules of a Kerr liquid have a tendency to reorientate along a plane perpendicular to the propagation direction of the beam.

Experimental studies using laser pulses of duration longer than $\sim 10^{-12}$ s and shorter than $\sim 10^{-8}$ s have shown that in many Kerr liquids (such as CS_2, benzene, toluene, etc.), the observed refractive-index change is mostly produced by the molecular reorientation mechanism. The reorientational optical Kerr effect can be applied to many novel optical devices, including the optical Kerr gate (shutter), the optical bistable Fabry–Perot interferometer, the optical phase-conjugate reflector, the real-time holography, etc. Among all known Kerr media, CS_2 liquid is the best because of its fast response time and large n_2^{Kerr} value.

5.6.3 Optical electrostriction contribution

All types of transparent optical media (solids, liquids, gases, or plasmas) are composed of elementary particles (molecules, ions, or atoms). Under the action of an applied optical field, these particles encounter a ponderomotive force (or electrostriction force) that is proportional to the gradient of the square of the field amplitude. The interaction between the applied optical field and a great number of particles gives rise to an induced density change of the medium. Since the dielectric constant of a medium is a function of the density of particles, the electrostriction-induced density change can also lead to a refractive-index change. The same mechanism will also give rise to the stimulated Brillouin scattering that will be discussed later in Chapter 7.

First, let us give an expression for density-variation ($\Delta\rho$) induced dielectric-constant change

$$\Delta\varepsilon = \frac{\partial\varepsilon}{\partial\rho}\Delta\rho = \frac{\gamma}{\rho_0}\Delta\rho. \tag{5.6-10}$$

Here, ρ_0 is the average density of the medium without an applied optical field, and

$$\gamma = \rho_0 \frac{\partial\varepsilon}{\partial\rho} \tag{5.6-11}$$

is known as the electrostrictive coefficient. In the presence of an optical field, the density change of the medium follows an equation given in Chapter 7 by Eq. (7.4-17), and the steady-state solution of that equation is

$$\Delta\rho = \frac{\varepsilon_0 \gamma}{2v_a^2} E_0^2, \qquad (5.6\text{-}12)$$

where v_a is the acoustic velocity in the medium and E_0 is the amplitude of the optical field. Substituting Eq. (5.6-12) into Eq. (5.6-10) leads to

$$\Delta\varepsilon = \frac{\varepsilon_0 \gamma^2}{2\rho_0 v_a^2} E_0^2. \qquad (5.6\text{-}13)$$

Thus the induced refractive-index change should be

$$\Delta n = \frac{1}{2n_0}\Delta\varepsilon = \frac{\varepsilon_0 \gamma^2}{4 n_0 \rho_0 v_a^2} E_0^2. \qquad (5.6\text{-}14)$$

Once again, we find that this expression in its final form is analogous to Eq. (5.2-6) derived from the more general consideration. In the present case of electrostrictive mechanism, a nominal nonlinear refractive-index coefficient is given by

$$n_2^{es} = \frac{\varepsilon_0 \gamma^2}{4 n_0 \rho_0 v_a^2}. \qquad (5.6\text{-}15)$$

Numerous experimental studies have shown that when intense laser pulses with a duration longer than $\sim 10^{-9}$ s pass through various transparent optical media (including solids, non-Kerr liquids, gases, and plasmas), the observed refractive-index changes are mostly due to the optical electrostrictive process, provided that there is no spectral resonance-enhancement or opto-thermal effect involved.

5.6.4 Temporal responses of induced refractive-index changes

In experimental studies of various induced refractive-index changes Δn, a primary issue is to clarify what is the dominant mechanism responsible to the observed Δn. The most effective approach to address this issue is to consider the different temporal responses of Δn produced by different mechanisms.

Common sense tells us that if the temporal response of a dominant nonlinear polarization mechanism is much faster than the variation period of the amplitude of an applied optical field, the induced refractive-index change can vary instantaneously with the variation of the applied field. On the contrary, if the temporal response of a dominant nonlinear polarization mechanism is slower than the variation of the applied optical field, we will see a retarded refractive-index change. Furthermore, if the input light pulse duration is much shorter than the response times of some nonlinear polarization mechanisms, then the contributions from those mechanisms to the observed Δn can be surely ignored.

Precisely speaking, there are two parameters used to characterize the temporal behavior of the induced refractive-index change. One is the rise time (τ_{rise}) defined as the time interval between the start of an applied rectangular light pulse and the time at which the refractive-index change reaches its maximum, as shown schematically in Fig. 5.5. The other is the relaxation (or decay) time (τ_{relax}) defined as the time interval between the end of an applied rectangular light pulse and the time at which the refractive-index change (Δn) vanishes. Simply speaking, the first parameter determines how fast the induced refractive-index change can occur upon the action of a laser pulse, and the second parameter determines how long the induced refractive-index change can

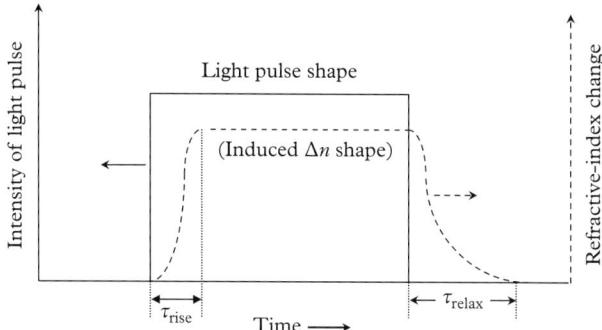

Figure 5.5 *Temporal response of refractive-index change induced by an intense rectangular light pulse.*

last when the applied laser pulse is gone. These two parameters can be measured for a given medium by properly designed experiments.

If Δn is mainly caused by the electron-cloud distortion, we have $\tau_{rise} \approx \tau_{relax} \leq 10^{-15}$–$10^{-16}$ s. In another case, if Δn is mainly due to intramolecular motion (Raman enhancement), we have $\tau_{rise} \approx \tau_{relax} \leq 10^{-12}$–$10^{-14}$ s.

If the molecular reorientation in liquid is the major contribution to Δn, τ_{rise} and τ_{relax} can be considerably different because in that case τ_{rise} is determined by the molecular rotational inertia, but τ_{relax} is determined by the average thermal collision interval or the so-called Debye time. For CS_2 liquid the former is measured to be $\sim(1.6$–$1.7) \times 10^{-13}$ s, and the latter is measured to be $\sim(1.5$–$2) \times 10^{-12}$ s.[15-17]

An experimental result of temporary response of the induced refractive-index change in CS_2 excited by an ultrashort laser pulse is shown in Fig. 5.6, based on a Kerr-gate setup.[15] In Fig. 5.6, the upper curve is the intensity of the transmitted signal through the Kerr gate of CS_2 versus the delay time of a probe pulse, which represents the induced refractive-index change as a function of time; the lower curve is the cross-correlation between the pump and probe pulses, which indicates the zero-point of the delay-time scale as well as the pulse-width (~ 0.15 ps) used. From Fig. 5.6 we can clearly see that the delay time between the peak of the pump pulse and the peak of the induced refractive-index change is ~ 0.17 ps, whereas the decay time of the induced refractive-index change is ~ 2 ps.

Furthermore, if the optical electrostrictive mechanism becomes the major origin of Δn, the determinations of τ_{rise} and τ_{relax} are rather complicated. In such a case, there is an induced acoustic motion inside the medium and the process is no longer localized; as a result, the temporary behavior of the induced refractive-index change may depend on the acoustic velocity as well as the interaction size between the light beam and the induced acoustic wave. Nevertheless, the shortest limits of τ_{rise} and τ_{relax} seem determined by the inertia of molecular translation, or in other words determined by the shortest period of the hypersonic sound induced by the electrostriction mechanism. The latter is measured approximately to be 10^{-9}–10^{-10} s using the backward stimulated Brillouin scattering technique.

In the cases of precise resonant interactions, the one- or two-photon absorption or Raman-transition will take place and the population distribution among different energy levels of molecules may change considerably. Therefore, one may observe a refractive-index change mainly due to the population change. Under these circumstances both τ_{rise} and τ_{relax} may vary over a broad range,

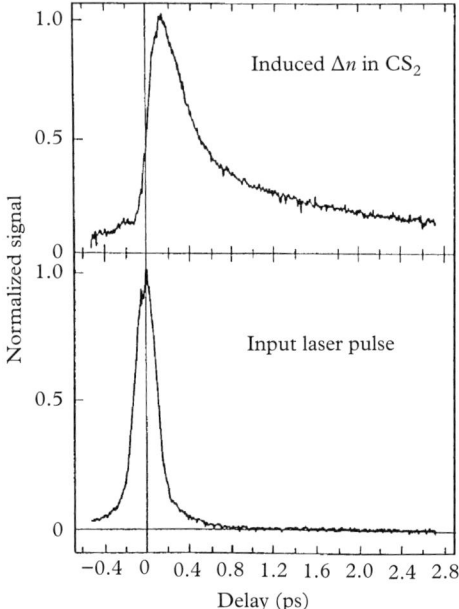

Figure 5.6 *Transmission through CS_2 optical Kerr gate versus delay time (top) and cross-correlation of the pump and probe pulses making t = 0 (bottom). (Reproduced with permission from Greene and Farrow.[15] Copyright 1982, AIP Publishing LLC).*

depending on the specific energy-level structure and transition property as well as the intensity and pulse duration of the applied optical field. Nevertheless, in general, we have $\tau_{rise} \ll \tau_{relax}$ because the former is determined by a fast build-up time of population change, and the latter determined by a relatively slow population relaxation process after the optical excitation. In other words, a molecule can leave the ground state to an excited state very fast through absorptive or Raman transition; however, this molecule may need a longer time for passing through many steps to finally return to the ground state.

In many precisely resonant interaction cases, the opto-thermal effect may take place. The absorbed optical energy will partially transfer to the thermal energy through radiationless transitions, resulting in a local temperature change. For most optical media, the refractive index is sensitive to temperature change; hence, we could see a refractive-index change due to the opto-thermal effect. In this case, τ_{rise} is mainly determined by the time interval between a moment while molecule left the ground state to a real excited state and the moment at which a radiationless transition takes place. On the other hand, τ_{relax} is determined by the thermal diffusion that is a slow process. In general, the opto-thermal process is of less scientific significance because of its relatively slow response time and extremely long relaxation time. Unfortunately, it is often unavoidable and sometimes becomes the dominant source for the observed refractive-index changes in resonant media.

Table 5.1 *Estimated values of τ_{rise} and τ_{relax} for different mechanisms causing refractive-index changes.*

Mechanism	Rise time (s)	Relaxation time (s)	Resonant enhancement
Electron-cloud distortion	$\leq 10^{-15}$–10^{-16}	$\leq 10^{-15}$–10^{-16}	yes
			no
Intramolecular motion	$\leq 10^{-12}$–10^{-14}	$\leq 10^{-12}$–10^{-14}	yes
Molecular reorientation	$\leq 10^{-12}$–10^{-13}	$\leq 10^{-11}$–10^{-12}	no
Electrostriction	$\geq 10^{-9}$–10^{-10}	$\geq 10^{-9}$–10^{-10}	no
			yes
Population change	10^{-10}–10^{-13}	$\geq 10^{-8}$–10^{-10}	yes
Opto-thermal effect	10^{-8}–10^{-10}	$\geq 10^{-3}$–10^{-6}	yes

As a summary, Table 5.1 lists the roughly estimated values of τ_{rise} and τ_{relax} for different mechanisms induced refractive-index changes.

In conclusion, the mechanism of electron-cloud distortion is solely responsible for the one- or two-photon resonance enhanced Δn, and it also contributes to the non-resonant interaction induced Δn for input pulse with any duration. The intramolecular motion is responsible for the Raman-enhanced Δn. The molecular reorientation is responsible only for the induced Δn in Kerr liquids. The mechanism of electrostriction mainly contributes to the induced Δn in non-resonant transparent media excited by ns-laser pulses; however, the same mechanism could also provide an enhanced Δn if there are two light beams and their small frequency difference and propagation directions satisfy the Brillouin resonance requirement.[18–20] Finally, to avoid or reduce the opto-thermal effect, one could employ the two-photon near resonance enhancement or near Raman resonance enhancement rather than the one-photon absorption enhancement.

5.7 Second-order nonlinearity induced refractive-index change (optical Pockels effect)

In this chapter, so far we only considered the refractive-index change based on the third-order nonlinearity of the medium. In these cases, the induced refractive-index change is proportional to the square of the amplitude of the optical field or the intensity of the incident optical beam. In this section we shall prove that under special conditions the second-order nonlinearity can also provide a light-induced refractive-index change.[21,22] In this case, the medium must be a second-order nonlinear crystal, the induced Δn will be proportional to the amplitude of the applied optical field, and the major contribution to the induced refractive-index change comes from the electron-cloud distortion.

We now examine a special three-wave mixing process occurring in a second-order nonlinear crystal when two input coherent monochromatic waves of frequencies ω and 2ω pass through the crystal in a phase-matching manner (Fig. 5.7). For simplicity, we assume that the medium is a negative uniaxial crystal and the phase matching can be fulfilled in two possible ways.

100 Induced Refractive-Index Changes

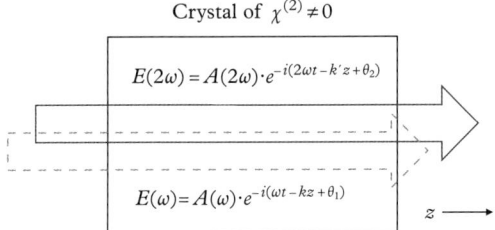

Phase matching: $k' = 2k$ or $n(2\omega) = n(\omega)$

Figure 5.7 *Refractive-index change induced in a second-order nonlinear crystal by coupling a fundamental wave of frequency ω and its second-harmonic wave of 2ω under phase-matching condition.*

Let us first examine the case of type-I ($o, o \to e$) phase matching. The two incident plane waves are of the forms

$$\left. \begin{array}{l} \mathbf{E}(\omega) = \mathbf{a}^o E(\omega) = \mathbf{a}^o A(\omega) e^{i(kz-\theta_1)} \\ \mathbf{E}(2\omega) = \mathbf{a}^e E(2\omega) = \mathbf{a}^e A(2\omega) e^{i(2kz-\theta_2)} \end{array} \right\}, \quad (5.7\text{-}1)$$

where $\mathbf{a}^o$ and $\mathbf{a}^e$ are the unit vectors of polarization of the ordinary ray and the extraordinary ray, k and $2k$ are the absolute values of wavevectors of these two waves, and θ_1 and θ_2 are the initial phase factors of the two waves, respectively. Since these two waves are plane waves, the amplitudes $A(\omega)$ and $A(2\omega)$ are real functions here. The corresponding second-order polarization components are

$$\left. \begin{array}{l} \mathbf{P}(\omega) = \varepsilon_0[\chi^{(1)}(\omega)\mathbf{E}(\omega) + \chi^{(2)}(2\omega,-\omega)\mathbf{E}(2\omega)\mathbf{E}^*(\omega)] \\ \mathbf{P}(2\omega) = \varepsilon_0[\chi^{(1)}(2\omega)\mathbf{E}(2\omega) + \chi^{(2)}(\omega,\omega)\mathbf{E}(\omega)\mathbf{E}(\omega)] \end{array} \right\}. \quad (5.7\text{-}2)$$

Based on Eq. (5.7-2) we can give the expressions for Fourier components of the electric displacement vector at frequencies ω and 2ω. For instance, the component at ω is

$$\begin{aligned} \mathbf{D}(\omega) &= \varepsilon_0 \mathbf{E}(\omega) + \mathbf{P}(\omega) \\ &= \varepsilon_0[\mathbf{E}(\omega) + \chi^{(1)}\mathbf{E}(\omega) + \chi^{(2)}(2\omega,-\omega)\mathbf{E}(2\omega)\mathbf{E}^*(\omega)]. \end{aligned} \quad (5.7\text{-}3)$$

Substituting Eq. (5.7-1) into Eq. (5.7-3) leads to

$$\mathbf{D}(\omega) = \varepsilon_0[\mathbf{a}^o + \chi^{(1)}(\omega)\mathbf{a}^o + \chi^{(2)}(2\omega,-\omega)\mathbf{a}^e \mathbf{a}^o A(2\omega) e^{i\delta\theta}] E(\omega), \quad (5.7\text{-}4)$$

where the phase-difference factor $\delta\theta$ is determined by

$$\delta\theta = 2\theta_1 - \theta_2. \quad (5.7\text{-}5)$$

Taking a scalar-product operation on both sides of Eq. (5.7-4) by $\mathbf{a}^o$ leads to

$$\begin{aligned} D^o(\omega) &= \varepsilon_0[1 + \mathbf{a}^o \cdot \chi^{(1)}(\omega)\mathbf{a}^o + \mathbf{a}^o \cdot \chi^{(2)}(2\omega,-\omega)\mathbf{a}^e \mathbf{a}^o A(2\omega) e^{i\delta\theta}] E(\omega) \\ &= \varepsilon_0 \varepsilon^o(\omega) E(\omega). \end{aligned} \quad (5.7\text{-}6)$$

Here, $D^o(\omega)$ is the component of vector $\mathbf{D}(\omega)$ in the $\mathbf{a}^o$ direction, and ε^o is the complex relative dielectric constant in the same direction. We thus obtain the refractive index at frequency ω for the ordinary ray in the crystal medium

$$\begin{aligned} n^o(\omega) &= \sqrt{\text{Re}\{\varepsilon^o(\omega)\}} \\ &= \text{Re}[\varepsilon_r^o(\omega) + \mathbf{a}^o \cdot \chi^{(2)}(2\omega,-\omega)\mathbf{a}^e\mathbf{a}^o A(2\omega)e^{i\delta\theta}]^{1/2} \\ &\approx n_0^o(\omega) + \text{Re}\left[\frac{1}{2n_0^o(\omega)}\mathbf{a}^o \cdot \chi^{(2)}(2\omega,-\omega)\mathbf{a}^e\mathbf{a}^o A(2\omega)e^{i\delta\theta}\right], \end{aligned} \quad (5.7\text{-}7)$$

where $n_0^o(\omega)$ is the linear refractive index for the ordinary ray

$$n_0^o(\omega) = \sqrt{\text{Re}\{\varepsilon_r^o(\omega)\}} = \sqrt{1 + \mathbf{a}^o \cdot \text{Re}\{\chi^{(1)}(\omega)\}\mathbf{a}^o}. \quad (5.7\text{-}8)$$

Following the same procedure we can also obtain the expression of $n^e(2\omega)$, the refractive index of the extraordinary ray at frequency 2ω. Specifically, the induced refractive-index changes of the two coupling waves are

$$\left.\begin{aligned} \Delta n^o(\omega) &= \text{Re}\left[\frac{1}{2n_0^o(\omega)}\mathbf{a}^o \cdot \chi^{(2)}(2\omega,-\omega)\mathbf{a}^e\mathbf{a}^o A(2\omega)e^{i\delta\theta}\right] \\ \Delta n^e(2\omega) &= \text{Re}\left[\frac{1}{2n_0^e(2\omega)}\mathbf{a}^e \cdot \chi^{(2)}(\omega,\omega)\mathbf{a}^o\mathbf{a}^o \frac{A^2(\omega)}{A(2\omega)}e^{i\delta\theta}\right] \end{aligned}\right\}. \quad (5.7\text{-}9)$$

Based on Eq. (5.7-9) one can see that here the nominal refractive-index changes are complex in general: the real parts of which determine the real refractive-index change while the imaginary parts determine the energy transfer between the two incident waves. If we further assume that, the nonlinear crystal is transparent for both incident waves and no spectral resonant enhancement is involved, then $\chi^{(2)}(2\omega,-\omega)$ and $\chi^{(2)}(\omega,\omega)$ can be recognized as real quantities. Replacing the exponential term in Eq. (5.7-9) by triangular functions and only taking the real parts leads to

$$\left.\begin{aligned} \Delta n^o(\omega) &= \frac{1}{2n_0^o(\omega)}\mathbf{a}^o \cdot \chi^{(2)}(2\omega,-\omega)\mathbf{a}^e\mathbf{a}^o A(2\omega) \cos\delta\theta \\ \Delta n^e(2\omega) &= \frac{1}{2n_0^e(2\omega)}\mathbf{a}^e \cdot \chi^{(2)}(\omega,\omega)\mathbf{a}^o\mathbf{a}^o \frac{A^2(\omega)}{A(2\omega)} \cos\delta\theta \end{aligned}\right\}. \quad (5.7\text{-}10)$$

Using the following relationship

$$\mathbf{a}^o \cdot \chi^{(2)}(2\omega,-\omega)\mathbf{a}^e\mathbf{a}^o = 2\mathbf{a}^e \cdot \chi^{(2)}(\omega,\omega)\mathbf{a}^o\mathbf{a}^o = 2\chi^{(2)}_{e,(I)}, \quad (5.7\text{-}11)$$

where $\chi^{(2)}_{e,(I)}$ is termed the effective second-order nonlinear susceptibility value under the type-I phase-matching condition, Eq. (5.7-10) is simplified to

$$\left.\begin{aligned} \Delta n^o(\omega) &= \frac{\chi^{(2)}_{e,(I)}}{2n_0^o(\omega)}2A(2\omega)\cos\delta\theta \\ \Delta n^e(2\omega) &= \frac{\chi^{(2)}_{e,(I)}}{2n_0^e(2\omega)}\frac{A^2(\omega)}{A(2\omega)}\cos\delta\theta \end{aligned}\right\}. \quad (5.7\text{-}12)$$

102 Induced Refractive-Index Changes

Since these refractive-index changes result from the nonlinear interaction between the two incident waves, the phase-matching should be always satisfied, i.e.,

$$n_0^o(\omega) + \Delta n^o(\omega) = n_0^e(2\omega) + \Delta n^e(2\omega), \qquad (5.7\text{-}13)$$

It is already assumed that $n_0^o(\omega) = n_0^e(2\omega)$, and the dynamic phase-matching requires that

$$\Delta n^o(\omega) = \Delta n^e(2\omega). \qquad (5.7\text{-}14)$$

Substituting Eq. (5.7-12) into Eq. (5.7-14) we obtain the following special requirement for the amplitude relation between the two incident waves

$$A(\omega) = \sqrt{2} A(2\omega). \qquad (5.7\text{-}15)$$

One can see from Eq. (5.7-12) that when $\delta\theta = \pm\dfrac{\pi}{2}$, $\Delta n^o(\omega) = \Delta n^e(2\omega) = 0$, there is no induced refractive-index change, but there is energy transfer occurring between the two waves. However, when

$$\delta\theta = 2\theta_1 - \theta_2 = 0 \text{ or } \pi, \qquad (5.7\text{-}16)$$

there is no energy transfer between the two waves, but we obtain the maximum refractive-index change:

$$\left.\begin{aligned}
\Delta n^o(\omega) &= \pm\dfrac{\sqrt{2}\chi^{(2)}_{e,(I)}}{2 n_0^o(\omega)} A(\omega) \\
\Delta n^e(2\omega) &= \pm\dfrac{\chi^{(2)}_{e,(I)}}{n_0^e(2\omega)} A(2\omega) \\
\Delta n^o(\omega) &= \Delta n^e(2\omega)
\end{aligned}\right\}. \qquad (5.7\text{-}17)$$

The physical meaning of Eq. (5.7-17) is that if the special amplitude relation of Eq. (5.7-15) and the phase relation of Eq. (5.7-16) are fulfilled, the induced refractive-index change for each wave is linearly proportional to the amplitude of the wave itself. Therefore, we could term this special effect the optical frequency Pockels effect.[21,22] It is seen from Eq. (5.7-17) that when $\delta\theta = 0$, there is a positive maximum refractive-index change; when $\delta\theta = \pi$, there is a negative maximum change.

Similarly, for the case of type-II ($o, e \rightarrow e$) phase matching in a negative uniaxial crystal, the incident optical fields can be expressed as

$$\left.\begin{aligned}
\mathbf{E}^o(\omega) &= \mathbf{a}^o E^o(\omega) = \mathbf{a}^o A(\omega) e^{-i(\omega t - k^o z + \theta_1)} \\
\mathbf{E}^e(\omega) &= \mathbf{a}^e E^e(\omega) = \mathbf{a}^e A(\omega) e^{-i(\omega t - k^e z + \theta_1)} \\
\mathbf{E}^e(2\omega) &= \mathbf{a}^e E^e(2\omega) = \mathbf{a}^e A(2\omega) e^{-i[2\omega t - (k^o + k^e)z + \theta_2]}
\end{aligned}\right\}. \qquad (5.7\text{-}18)$$

Here, we assume that two components of the fundamental wave have the same amplitude functions and the same initial phase factors. Through the same derivation procedure, the maximum refractive-index changes for the three waves can be obtained as

$$\Delta n^o(\omega) = \pm \frac{\sqrt{2}}{2n_0^o(\omega)} \chi_{e,(II)}^{(2)} A(\omega)$$

$$\Delta n^e(\omega) = \pm \frac{\sqrt{2}}{2n_0^e(\omega)} \chi_{e,(II)}^{(2)} A(\omega)$$

$$\Delta n^e(2\omega) = \pm \frac{1}{n_0^e(2\omega)} \chi_{e,(II)}^{(2)} A(2\omega)$$

$$\Delta n^e(2\omega) = \frac{1}{2}[\Delta n^o(\omega) + \Delta n^e(\omega)] \approx \Delta n^o(\omega) \approx \Delta n^e(\omega)$$

(5.7-19)

Here,

$$\chi_{e,(II)}^{(2)} = \mathbf{a}^e \cdot \chi^{(2)}(\omega,\omega)\mathbf{a}^o\mathbf{a}^e = \frac{1}{2}\mathbf{a}^o \cdot \chi^{(2)}(2\omega,-\omega)\mathbf{a}^e\mathbf{a}^e$$

$$= \frac{1}{2}\mathbf{a}^e \cdot \chi^{(2)}(2\omega,-\omega)\mathbf{a}^e\mathbf{a}^o$$

(5.7-20)

is the effective second-order susceptibility value under the type-II phase-matching condition, and it is a scalar coefficient.

It should be indicated that the optical frequency Pockels effect is the responsible mechanism for the formation of spatial solitons in the second-order nonlinear optical crystals, which will be discussed later in Chapter 13.

The discussion of the optical Pockels effect implies that for general three-wave mixing (three-photon parametric interaction) processes, we should consider not only the energy transfer between the coupling waves but also the possible refractive-index changes depending on the initial amplitude relation as well as the phase relation. From Eq. (5.7-9), the general expression of complex refractive-index changes of the ω wave and 2ω wave, one can see that the energy transfer and real refractive-index changes of these two waves can be controlled by changing the initial phase relation and the amplitude ratio between these two waves.[23–25] The same conclusion is also applicable to general four-wave mixing (four-photon parametric interaction) processes.

PROBLEMS

1. According to Eq. (5.2-8), if the induced refractive-index change is $\Delta n = 10^{-4}$ measured at the laser intensity level of $I = 500$ MW/cm^2 in a transparent medium of $n_0 = 1.45$, calculate the values of n_2 and $\chi_e^{(3)}$.
2. For two-beam Raman enhanced refractive-index change described by Fig. 5.3, if the full linewidth $(2\Gamma_r)$ is 4 cm^{-1}, and the two-beam frequency detuning value changed from 20 cm^{-1} to 2 cm^{-1}, what is the rough ratio of the induced refractive-index changes in these two cases?
3. In what conditions the electron-distortion will be the only mechanism contributing to the laser induced refractive-index change?
4. In a transparent solid medium, if the input laser pulsewidth is ≥ 1 ns, what mechanisms can be the dominate contributing to the induced refractive-index change?
5. If the input laser wavelength is very close to a linear absorption line of the medium, how many mechanisms may contribute to the induced refractive-index change?

REFERENCES

1. P. D. Maker, R. W. Terhune, and C. M. Savage, *Phys. Rev. Lett.* **12**, 507 (1964).
2. P. D. Maker and R. W. Terhune, *Phys. Rev.* **137**, A801 (1965).
3. R. Y. Chiao, P. L. Kelley, and E. Garmire, *Phys. Rev. Lett.* **17**, 1158 (1966).
4. P. F. Ziao and G. C. Bjorklund, *Phys. Rev. Lett.* **36**, 584 (1976).
5. G. S. He, *Chin. Phys. Lasers*, **14**, 162 (1987).
6. G. S. He, M. Yoshida, J. D. Bhawalkar, and P. N. Prasad, *Appl. Opt.* **36**, 1155 (1997).
7. G. S. He, Y. Cheng, F. Zhou, B. Yu, and S. H. Liu, *Chin. J. Lasers*, **10**, 455 (1983).
8. G. S. He, Y. Cheng, F. Zhou, B. Yu, F. Wang, M. Liu, and S.H. Liu, *Chin. J. Lasers*, **11**, 96 (1984).
9. G. H. He, D. Liu, and S. H. Liu, *Opt. Commun.* **70**, 145 (1989).
10. Y. R. Shen, *Phys. Lett.* **20**, 378 (1966).
11. R. W. Hellwarth, *Phys. Rev.* **152**, 156 (1966).
12. R. G. Brewer, J. R. Lifsitz, E. Garmire. R. Y. Chiao, and C. H. Townes, *Phys. Rev.* **166**, 326 (1968).
13. O. Svelto, in *Progress in Optics*, Vol. XII, edited by E. Wolf (North-Holland, Amsterdam, 1974), pp. 3–54.
14. A. Yariv, *Quantum Electronics*, 3rd ed. (Wiley, New York, 1988). pp. 482–484.
15. B. I. Greene and R. C. Farrow, *J. Chem. Phys.* **77**, 4779 (1982).
16. C. Kapouzos, W. T. Lotshaw, D. McMorrow, and G. A. Kenney-Wallace, *J. Phys. Chem.* **91**, 2028 (1987).
17. S. Ruhman, L. R. Williams, A. G. Joly, B. Kohler, and K. A. Nelson, *J. Phys. Chem.* **91**, 2237 (1987).
18. N. A. Andreev, V.-I. Bespalov, A. M. Kiselev, A. Z. Matveev, G. A. Pasmanik, and A. A. Shilov, *JETP Lett.* **32**, 625 (1980).
19. M. D. Skeldon, P. Narum, and R. W. Boyd, *Opt. Lett.* **12**, 343 (1987).
20. A. M. Scott and K. D. Ridlex, *IEEE J. Quantum Electron.* **25**, 438 (1989).
21. D. Liu, G. S. He, and S. H. Liu, *Chin. Phys. Lasers* **13**, 234 (1986).
22. G. S. He, D. Liu, and S. H. Liu, *Chin. Quantum Electron.* **3**, 116 (1986).
23. R. DeSalvo, D. J. Hagan, M. Sheik-Bahae, G. Stegeman, and E. W. Van Stryland, *Opt. Lett.* **17**, 28 (1992).
24. G. Assanto, G. Stegeman, M. Sheik-Bahae, and E. W. Van Stryland, *Appl. Phys. Lett.* **62**, 1323 (1993).
25. D. J. Hagan, Z. Wang, G. Stegeman, and E. W. Van Stryland, *Opt. Lett.* **19**, 1305 (1994).

6
Self-Focusing, Self-Phase Modulation, and Spectral Self-Broadening

The intense coherent light induced refractive-index change of a nonlinear medium may lead to several reactions on the incident light itself. First, the laser beam with a nonuniform transverse intensity distribution can induce a transversely nonuniform refractive-index change in the nonlinear medium, which in turn may affect the shape and spatial structure of the laser beam itself. This is related to the self-focusing, self-defocusing, or self-trapping effect.[1-4] Secondly, if the incident beam is consisting of short or ultrashort laser pulses, the pulsed change of light intensity may cause a fast variation of the phase encountered by the pulsed beam itself. This is the so-called self-phase modulation effect. Finally, according to the principle of Fourier-transform, a fast phase modulation will cause a broadening of the frequency spectrum. This is the so-called spectral self-broadening effect.

In experiments, some other nonlinear optical effects such as stimulated scattering, optical breakdown, and optical damage could be facilitated by a self-focusing process, because the latter may lead to an anomalously high local light-intensity. In addition, the spectral self-broadening effect is one of the most effective techniques for generating coherent continuum emission that is useful for many applications.

6.1 Basic theory of the self-focusing effect

6.1.1 General description

In conventional optics, the geometric structure and propagation behavior of a light beam can be controlled by various optical elements such as lenses and prisms. These elements are made by high-quality optical materials, which have a uniform refractive-index distribution. The function of such an optical element is ensured by a regular variation of thickness, or equivalently, by the regular variation of optical path. For instance, a convex lens can be used to focus an incident light beam, whereas a concave lens can make an incident beam more divergent, and so on. If an ordinary quasi-parallel light beam passes through an optical plate, the beam structure (size and divergence angle) will not be changed because the thickness or optical path of the plate is the same over different sections of the incident beam. In nonlinear optics, however, even the medium is equivalent to a parallel plate; the spatial structure of an intense laser beam may be changed due to the nonlinear interaction between the medium and the laser beam. The earliest theoretical considerations of the spatial structure change of an intense coherent light beam propagating in a nonlinear medium were reported during 1962–1964.[5-7]

Now let us consider a quasi-parallel and linearly polarized laser beam passing through a centrosymmetric or isotropic optical medium; the electric field of the incident beam can be written as

$$\mathbf{E} = \mathbf{a}E = \mathbf{a}E_0(x, y, z)e^{i(kz-\omega t)}. \quad (6.1\text{-}1)$$

Here, **a** is a unit vector in the direction of light polarization and $E_0(x, y, z)$ is the amplitude function depending on the spatial coordinates. The third-order nonlinear refractive-index change is expressed by (cf. Eq. (5.2-6))

$$\Delta n(x, y, z) = n_2 |E_0(x, y, z)|^2,$$

where n_2 is the nonlinear refractive-index coefficient determined by (cf. Eq. (5.2-7))

$$n_2 = \frac{1}{2n_0} \chi_e^{(3)}(\omega, -\omega, \omega).$$

Here, $\chi_e^{(3)}$ is the effective third-order susceptibility value of the medium. For simplicity, it can be further assumed that the transverse intensity distribution of the beam is of circular symmetry, and then we have

$$\Delta n(r, z) = n_2 |E_0(r, z)|^2.$$

Here, r is the radial variable in the beam section.

Usually, the transverse intensity decreases with the increase of r for most laser beams. In these cases, we can expect four types of possible changes of a quasi-parallel incident laser beam propagating through a nonlinear optical medium, as shown schematically in Fig. 6.1.

In Fig. 6.1(a), it is assumed that the induced refractive-index change is negligible and the beam size increases gradually with the propagation distance owing to the diffraction-divergence of the incident beam. In Fig. 6.1(b), it is assumed that n_2 is positive and the incident beam is focused inside the medium due to the induced refractive-index change. In this case, the induced refractive-index change makes the medium like a focusing lens; thereby this phenomenon is specifically termed the self-focusing effect. In Fig. 6.1(c), it is assumed that n_2 is positive and once the propagating beam is focused onto a small spot, the beam size will maintain unchanged over a longer propagation distance within the nonlinear medium; in other words, the beam is finally confined in a waveguide channel formed by the induced refractive-index change; here we term this phenomenon the self-trapping effect. Finally, in Fig. 6.1(d), it is assumed that n_2 is negative and the induced refractive-index change makes the medium like a negative lens, so that the propagating beam becomes more divergent after passing through the nonlinear medium, this phenomenon is termed the self-defocusing effect.

If we assume that the nonlinear medium is transparent for the incident laser beam, then the change of the beam's size will lead to a corresponding change of the local intensity at the beam center. Figure 6.2 shows schematically the central-intensity change of a beam propagating within a nonlinear medium under the four possible conditions indicated in Fig. 6.1.

One can further imagine that if the laser beam has a randomly nonuniform transverse intensity distribution and the nonlinear medium possesses a positive n_2 value, then after passing a certain

Basic theory of the self-focusing effect 107

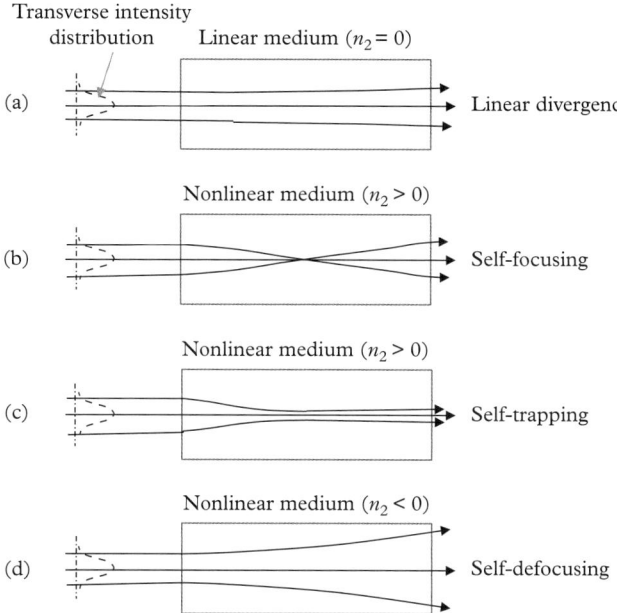

Figure 6.1 *Four types of structure changes of a quasi-parallel laser beam propagating through a nonlinear optical medium: (a) linear divergence, (b) self-focusing, (c) self-trapping, and (d) self-defocusing.*

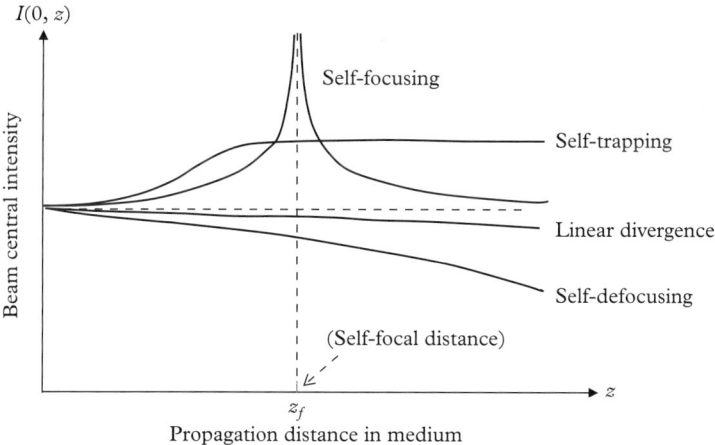

Figure 6.2 *Central-intensity change of a laser beam vs. the propagation distance in a medium under the four possible conditions shown in Fig. 6.1.*

propagation distance the whole beam may break into several spatially separated small filaments; each filament might last over a longer distance and with a much higher local intensity. That is one of the mechanisms responsible for the early observed optical damages inside solid lasing materials or propagation media; these damages were often observed in the form of one or several long tracks of damage spots. Since then, researchers had realized that to maintain a uniform transverse intensity-distribution is one of the effective measures to avoid the optical damages in a lasing or propagation medium.

In Chapter 5, we have discussed various mechanisms of induced refractive-index changes and the corresponding polarity of n_2 values. These discussions and conclusions are applicable to the topics described in this chapter. For instance, it is known that if the major contribution for induced refractive-index change arises from non-resonant electronic nonlinearity, n_2 usually will be positive. However, if the resonant electronic nonlinearity (one- or two-photon resonance) or Raman resonance is the predominant origin for the induced refractive-index change, n_2 can be either positive or negative depending on resonance-tuning conditions. Furthermore, if molecular reorientation or optical electrostriction is the main origin of refractive-index changes, n_2 is usually positive. Finally, if the opto-thermal mechanism becomes the major source for the induced refractive-index change, a self-defocusing could be easily observed for those cases wherein a thermal-expansion induced refractive-index decrease plays the major role.

6.1.2 Induced waveguide model of self-trapping

In 1964, Chiao et al. proposed a simple but conceptually important model to describe under what condition a thin laser beam could be self-trapped in a nonlinear medium of $n_2 > 0$ without further divergence.[6] This model is based on the consideration that when the original divergence of a laser beam is offset by the confinement function of a waveguide induced by the beam itself, the self-trapping effect will take place.

In a most simple case, shown in Fig. 6.3, we assume that a quasi-parallel laser beam with a nearly uniform transverse intensity distribution and a diameter of d_0 is incident into a third-order nonlinear medium of $n_2 > 0$. In the region covered by the laser beam within the medium, there is a refractive-index increase of $\Delta n = n_2 |E_0|^2$, so that there will be an induced waveguide that in turn can confine the light beam itself.

The condition of critical angle for internal total-reflection in the boundary of the induced waveguide is determined by

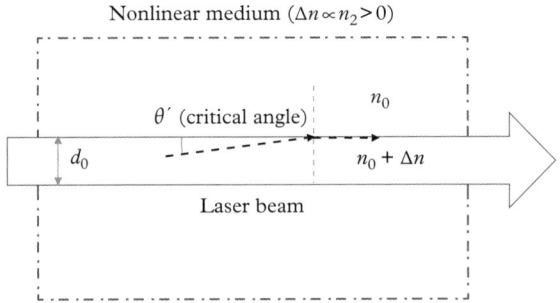

Figure 6.3 *Self-trapping of a laser beam through an induced waveguide in the nonlinear medium of $n_2 > 0$.*

$$\cos\theta' = \frac{n_0}{n_0 + \Delta n}. \tag{6.1-2}$$

Here, n_0 is the linear refractive index of the medium. When θ' is quite small, the function $\cos\theta'$ can be replaced by $(1 - \theta'^2/2)$; thus we have

$$\theta'^2 \approx \frac{2\Delta n}{n_0} = \frac{2n_2}{n_0}|E_0|^2.$$

Assuming the original divergence angle of the incident beam is θ_0, the self-trapping condition should be $\theta' \geq \theta_0$, i.e.,

$$|E_0|^2 \geq \frac{n_0 \theta_0^2}{2n_2}. \tag{6.1-3}$$

The physical meaning of this expression is that the intensity requirement for the incident beam to be self-trapped is proportional to the square of the divergence angle and inversely proportional to the nonlinear refractive-index coefficient n_2.

Moreover, if the input laser beam is a plane wave, the divergence angle is determined by diffraction, i.e.,

$$\theta_{dif} = \frac{1.22\lambda}{n_0 d_0},$$

then the self-trapping condition will be

$$|E_0|^2 \geq \frac{(1.22\lambda)^2}{2n_0 n_2 d_0^2}. \tag{6.1-4}$$

This means that the beam with a smaller diameter is easily to be self-trapped in a medium with a larger n_2 value.

6.1.3 Theory of steady-state self-focusing

Let us consider a quasi-parallel and linearly polarized laser beam propagating in a transparent and isotropic nonlinear medium of $n_2 > 0$, the electric field of the incident beam is expressed by Eq. (6.1.1). The Maxwell wave equation for non-absorptive medium is

$$\nabla^2 \mathbf{E} - \mu_0 \frac{\partial^2}{\partial t^2}\mathbf{D} = 0. \tag{6.1-5}$$

Here, μ_0 is the free-space permeability. From Eqs. (5.2-3) and (5.2-4) we know that

$$\mathbf{D} = \varepsilon_0 \left[\varepsilon_r + \chi_e^{(3)}|E_0|^2 \right] \mathbf{E}, \tag{6.1-6}$$

where ε_0 is the free-space permittivity, ε_r is the linear dielectric constant, and $\chi_e^{(3)} = \chi_e^{(3)}(\omega, -\omega, \omega)$ is the effective third-order nonlinear susceptibility value of the medium. Substituting Eq. (6.1-6) into Eq. (6.1-5) leads to the following scalar equation:

$$\nabla^2 E - \frac{\varepsilon_r}{c^2}\frac{\partial^2}{\partial t^2}E - \frac{\chi_e^{(3)}}{c^2}\frac{\partial^2}{\partial t^2}(|E_0|^2 E) = 0. \tag{6.1-7}$$

Here, c is the light speed in free space. If assume $\chi_e^{(3)} \Rightarrow 0$, Eq. (6.1-7) reduces to an ordinary linear wave equation. Substituting Eq. (6.1-1) into Eq. (6.1-7) leads to

$$(\nabla_x^2 + \nabla_y^2) E_0 + 2ik\frac{\partial E_0}{\partial z} + \frac{\chi_e^{(3)}}{\varepsilon_r} k^2 |E_0|^2 E_0 = 0, \tag{6.1-8}$$

where k is the absolute value of wavevector and $\nabla_x^2 = \frac{\partial^2}{\partial x^2}, \nabla_y^2 = \frac{\partial^2}{\partial y^2}$. To obtain Eq. (6.1-8) we have assumed that (i) the amplitude of the incident optical field is a slow-varying function so that the term of second-order z derivative of E_0 can be ignored, and (ii) the temporal variation of the amplitude is much slower than the response time of the induced refractive-index change so that both the first- and the second-order temporal derivatives can be neglected (steady-state assumption). For simplicity, it is also assumed that the amplitude and phase-front of the incident beam are of circular symmetry, i.e.,

$$E_0(r, z) = A(r, z) e^{ikS(r,z)}. \tag{6.1-9}$$

Here, $A(r, z)$ is the real amplitude function, $S(r, z)$ is the phase function that represents the deviation of the real wavefront from an ideal plane wavefront, r is the radial variable, and z is the propagation distance in a given nonlinear medium. Substituting Eq. (6.1-9) into Eq. (6.1-8) results in

$$\left. \begin{array}{l} 2\dfrac{\partial S}{\partial z} + \left(\dfrac{\partial S}{\partial r}\right)^2 = \dfrac{1}{k^2 A}\left(\dfrac{\partial^2 A}{\partial r^2} + \dfrac{1}{r}\dfrac{\partial A}{\partial r}\right) + \dfrac{\chi_e^{(3)}}{\varepsilon_r} A^2 \\[2mm] \dfrac{\partial (A^2)}{\partial z} + \dfrac{\partial (A^2)}{\partial r}\dfrac{\partial S}{\partial r} + A^2 \left(\dfrac{\partial^2 S}{\partial r^2} + \dfrac{1}{r}\dfrac{\partial S}{\partial r}\right) = 0 \end{array} \right\}. \tag{6.1-10}$$

It can be proved that when $\chi_e^{(3)} = 0$ spherical waves (plane wave is a special case of them) are the solution of the above equations. Thus we can assume that when $\chi_e^{(3)} \neq 0$ the solution of Eq. (6.1-10) still has the form of a spherical wave but the distribution of transverse intensity and wavefront will vary with z. Based on this consideration one can try the following trial solution:[8]

$$\left. \begin{array}{l} S(r, z) = \beta(z)\dfrac{r^2}{2} + \phi(z) \\[2mm] A^2(r, z) = \dfrac{A_0^2}{\alpha^2(z)} e^{-\frac{r^2}{a_0^2 \alpha^2(z)}} \end{array} \right\}. \tag{6.1-11}$$

Here, we assumed a Gaussian function for the transverse-intensity distribution, A_0 is the initial amplitude value in the beam center, and a_0 is the radius value wherein the intensity decreases to the ($1/e$) value of the central intensity. The initial condition of this solution is

$$\left. \begin{array}{l} \beta(0) = \dfrac{1}{R}, \phi(0) = 0, \alpha(0) = 1 \\[2mm] A^2(r, 0) = A_0^2 e^{-r^2/a_0^2} \end{array} \right\}. \tag{6.1-12}$$

Here, R is the initial radius of curvature of the wavefront. The physical meaning of the trial solution expressed by Eq. (6.1-11) is that for an incident laser beam with a spherical wavefront and Gaussian intensity distribution, while it propagates in a nonlinear medium, its wavefront still retains the spherical form but with a radius of curvature varying with z through the factor $\beta(z)$; also, its transverse-intensity distribution remains a Gaussian form but with a beam size varying with z via the factor $\alpha(z)$. These assumptions are only valid for paraxial rays.

Substituting Eq. (6.1-11) into the first equation of Eq. (6.1-10) and using the paraxial approximation

$$e^{-r^2/(a_0^2\alpha^2)} \approx 1 - \frac{r^2}{a_0^2\alpha^2}, \tag{6.1-13}$$

we obtain equations for $\beta(z)$ and $\phi(z)$ separately:

$$\left.\begin{array}{l} \dfrac{d\beta}{dz} + \beta^2 = \dfrac{1}{k^2 a_0^4 \alpha^4} - \dfrac{\chi_e^{(3)}}{\varepsilon_r} \dfrac{A_0^2}{a_0^2 \alpha^4} \\[8pt] 2\dfrac{d\phi}{dz} = -\dfrac{2}{k^2 a_0^2 \alpha^2} + \dfrac{\chi_e^{(3)}}{\varepsilon_r} \dfrac{A_0^2}{\alpha^2} \end{array}\right\}. \tag{6.1-14}$$

Then substituting Eq. (6.1-11) into the second equation of Eq. (6.1-10), we get an equation for $\alpha(z)$

$$\frac{d^2\alpha}{dz^2} = \left(\beta^2 + \frac{d\beta}{dz}\right)\alpha.$$

By utilizing the first equation of Eq. (6.1-14), the above equation becomes

$$\frac{d^2\alpha}{dz^2} = \frac{1}{k^2 a_0^4 \alpha^3} - \frac{\chi_e^{(3)}}{\varepsilon_r} \frac{A_0^2}{a_0^2 \alpha^3}. \tag{6.1-15}$$

From Eqs. (6.1-14) and (6.1-15) we see that once a solution for $\alpha(z)$ is obtained the functions of $\beta(z)$ and $\phi(z)$ can also be solved. Therefore, we should solve Eq. (6.1-15) first.

Multiplying the two sides of Eq. (6.1-15) by $2\dfrac{d\alpha}{dz}$ and integrating with z leads to

$$\left(\frac{d\alpha}{dz}\right)^2 = \left(\frac{\chi_e^{(3)}}{\varepsilon_r} \frac{A_0^2}{a_0^2} - \frac{1}{k^2 a_0^4}\right)\frac{1}{\alpha^2} + C. \tag{6.1-16}$$

Here, C is an integration constant. Using the initial condition of Eq. (6.1-12), it is readily to obtain the final solution for $\alpha(z)$:[8]

$$\left.\begin{array}{l} \alpha^2(z) = Cz^2 + \dfrac{2}{R}z + 1 \\[8pt] C = \dfrac{1}{R^2} - \dfrac{1}{a_0^2}\left(\dfrac{\chi_e^{(3)}}{\varepsilon_r} A_0^2 - \dfrac{1}{k^2 a_0^2}\right) \end{array}\right\}. \tag{6.1-17}$$

Returning to the expression of amplitude $A(r, z)$ in Eq. (6.1-11), we can find that the characteristic spot size $a_0\alpha(z)$ varies with the propagation length z within the nonlinear medium.

112 Self-Focusing, Self-Phase Modulation, and Spectral Self-Broadening

When $\alpha(z) \Rightarrow 0$, $A^2(0, z) \Rightarrow \infty$, this corresponds to the focus point of a self-focused beam. Letting $\alpha(z) \Rightarrow 0$ in the first part of Eq. (6.1-17), the focal distance is obtained as

$$z_f = \left[-\frac{2}{R} \pm \sqrt{\left(\frac{2}{R}\right)^2 - 4C} \right] \frac{1}{2C}. \tag{6.1-18}$$

The above solution can be rewritten as

$$\frac{1}{z_f} = -\frac{1}{R} \pm \frac{1}{a_0} \sqrt{\frac{\chi_e^{(3)}}{\varepsilon_r} A_0^2 - \frac{1}{k^2 a_0^2}}. \tag{6.1-19}$$

This solution shall be discussed further for the following three specific cases.

6.1.3.1 Plane incident wave

We have $R \Rightarrow \infty$ here and, as self-focusing requires a positive value of z_f, the solution should be

$$\frac{1}{z_f} = \frac{1}{a_0} \sqrt{\frac{\chi_e^{(3)}}{\varepsilon_r} A_0^2 - \frac{1}{k^2 a_0^2}}. \tag{6.1-20}$$

The first term in the square root of the above expression represents the converging effect due to self-focusing, while the second term represents the diverging effect due to diffraction. When these two tendencies are equal, we have $z_f \Rightarrow \infty$ (self-trapping) and the beam will propagate within the nonlinear medium without any divergence. Thus, we can define a critical amplitude condition by

$$A_{00}^2 = \frac{\varepsilon_r}{\chi_e^{(3)}} \frac{1}{k^2 a_0^2} = \frac{n_0}{2n_2} \frac{1}{k^2 a_0^2}. \tag{6.1-21}$$

The relation between A_0^2 and optical power P is (cf. Eq. (1.4-1))

$$P \approx \pi a_0^2 \frac{n_0 \varepsilon_0 c}{2} A_0^2, \tag{6.1-22}$$

where c is the light speed in free space. Accordingly, the critical power for self-trapping is

$$P_0 \approx \pi a_0^2 \frac{n_0 \varepsilon_0 c}{2} A_{00}^2 = \frac{\varepsilon_0 c \lambda^2}{16\pi n_2}. \tag{6.1-23}$$

Here, λ is the wavelength in free space. Based on Eqs. (6.1-22) and (6.1-23) we can rewrite Eq. (6.1-20) as

$$\frac{1}{z_f} = \frac{2}{a_0^2 n_0} \sqrt{\frac{n_2}{\varepsilon_0 c \pi}} \sqrt{P - P_0}. \tag{6.1-24}$$

In this case the focal distance is

$$z_f = \frac{a_0^2 n_0}{2} \sqrt{\frac{\varepsilon_0 c \pi}{n_2}} \bigg/ \sqrt{P - P_0} = k a_0^2 \bigg/ \sqrt{(P/P_0) - 1}. \tag{6.1-25}$$

The physical meaning of the above expression is that a Gaussian beam with a smaller beam size and higher power will have a shorter self-focusing distance.

6.1.3.2 Convergent spherical incident wave

We have $R < 0$ here and Eq. (6.1-19) becomes

$$\frac{1}{z_f} = \frac{1}{|R|} \pm \frac{1}{ka_0^2}\sqrt{(P/P_0) - 1}. \tag{6.1-26}$$

The original converging center moves toward the entrance side owing to self-focusing. For a loosely converging incident beam with $|R| > ka_0^2/\sqrt{(P/P_0) - 1}$, the second term in Eq. (6.1-26) should take the plus sign, and there is only one focal point. Whereas, for a tightly converging incident beam with $|R| < ka_0^2/\sqrt{(P/P_0) - 1}$, the second term in Eq. (6.1-26) can take both the plus and minus signs. This means that in addition to the first focal point shifted to the entrance direction, there will be a second focal point behind the first.

6.1.3.3 Divergent spherical incident wave

We have $R > 0$ here and the second term in Eq. (6.1-19) can only take the plus sign, i.e.,

$$\frac{1}{z_f} = -\frac{1}{R} + \frac{1}{ka_0^2}\sqrt{(P/P_0) - 1}. \tag{6.1-27}$$

If we expect that the beam becomes convergent, $z_f > 0$, the following requirement should be satisfied:

$$R > ka_0^2/\sqrt{(P/P_0) - 1}. \tag{6.1-28}$$

This expression implies that to observe self-focusing at a given optical power level, the divergence of the incident beam should not be too large or the R value cannot be too small.

Figure 6.4 shows schematically the propagation behavior of a Gaussian beam in a nonlinear medium of $n_2 > 0$ for (a) a plane incident wave, (b) a loosely focused incident wave, (c) a tightly focused incident wave, and (d) a slightly divergent incident wave, respectively.

In summary, if one wants to avoid self-focusing in experiments, the following measures can be taken: (i) use a laser beam with a uniform transverse intensity distribution, (ii) have a larger beam size, or (iii) let the beam enter the nonlinear medium with a slight divergence.

6.1.4 Another empirical formula for steady-state self-focusing

In the above subsection we have described the analytical solutions of the nonlinear wave equation of Eq. (6.1-8), based on the so-called paraxial approximation (or aberrationless approximation).[8] This approximate analytical theory can provide a straightforward physical picture about the basic behavior of steady-state self-focusing and self-trapping under several simple incident conditions. However, it could not give a more detailed and rigorous description of self-focusing or self-trapping behavior under various real experimental conditions. For instance, it can only predict the appearance of the second self-focal point, but con not predict what will happen after the second focal point. This kind of questions can only relay on the numerical solutions of the nonlinear wave equations. In the experimental section (Section 6.2), we will discuss the multi-focus behavior of the self-focused ultrashort pulsed laser beam.

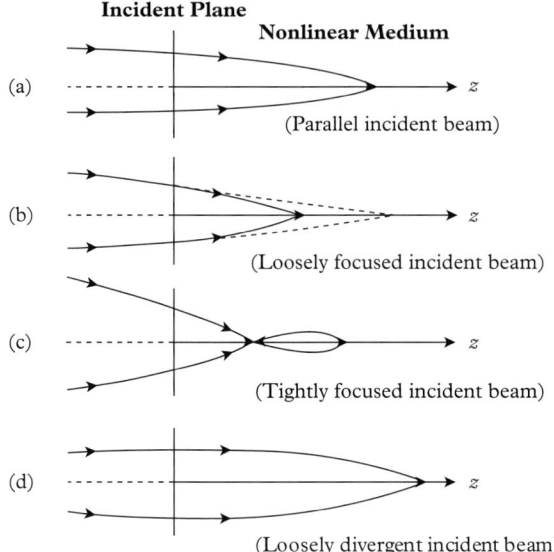

Figure 6.4 *Self-focusing configurations under various incident conditions: (a) a parallel beam, (b) a loosely focused beam, (c) a tightly focused beam, and (d) a slightly divergent beam.*

Independently, Kelley in 1965 gave an empirical formula of self-focal distance for a parallel incident beam with a Gaussian transverse intensity distribution:[9]

$$z_f = \frac{a_0^2 n_0}{2}\sqrt{\frac{\varepsilon_0 c \pi}{n_2}} \bigg/ \left(\sqrt{P} - \sqrt{P_0}\right). \qquad (6.1\text{-}29)$$

This formula was based on a phenomenological consideration and verified by the numerical solution of the nonlinear wave equation of Eq. (6.1-8).

Comparing Eq. (6.1-29) to Eq. (6.1-25), one can find that the difference between these two formulae is only in the denominator. When $P \gg P_0$ the difference between these two formulae becomes very small. Some later numerical calculations further indicated that certain correction factors depending on the specific range of (P/P_0) should be applied to the empirical formula of Eq. (6.1-29).[4,10]

6.1.5 Dynamic self-focusing process

In the preceding subsections, the steady-state self-focusing process is described where we neglected both the first- and the second-order temporal derivatives; thereby all formulae describing the self-focusing behavior are independent of time t. However, if the pulse duration is quite short (less than 10^{-10}–10^{-11} s) and the variation of the pulse profile is not too slow, then the first-order temporal derivative of the amplitude function can no longer be neglected. In this case, the nonlinear wave equation of Eq. (6.1-7) reduces to

$$\left(\nabla_x^2 + \nabla_y^2\right) E_0 + 2ik\left(\frac{\partial E_0}{\partial z} + \frac{n_0}{c}\frac{\partial E_0}{\partial t}\right) + \frac{\chi_e^{(3)}}{\varepsilon_r} k^2 |E_0|^2 E_0 = 0. \tag{6.1-30}$$

In obtaining this equation, we assumed that second-order spatial derivatives with respect to z and the second-order temporal derivative (as well as the product of two first-order temporal derivatives) could still be neglected. Comparing Eq. (6.1-30) to Eq. (6.1-8), we find that these two equations are similar except that there is an additional term of first-order temporal derivative involved in the former.

To solve the time-dependent nonlinear wave equation of Eq. (6.1-30), we first make a transformation on the variable t by

$$t = \zeta + \frac{z}{c/n_0}, \tag{6.1-31}$$

where c/n_0 is the light speed in the nonlinear medium and ζ is a newly introduced temporal variable. Using the following derivative rule of a compound function

$$\begin{aligned}\frac{\partial}{\partial z} E_0\left(z, t = \zeta + \frac{n_0 z}{c}\right) &= \frac{\partial E_0(z,t)}{\partial z} + \frac{\partial E_0(z,t)}{\partial t}\frac{\partial t}{\partial z} \\ &= \frac{\partial E_0(z,t)}{\partial z} + \frac{\partial E_0(z,t)}{\partial t}\frac{n_0}{c},\end{aligned} \tag{6.1-32}$$

and comparing the above expression to the terms inside the second parentheses of Eq. (6.1-30), we find that after making the transformation expressed by Eq. (6.1-31), the nonlinear wave equation of Eq. (6.1-30) can be simplified as[1,2]

$$\left(\nabla_x^2 + \nabla_y^2\right) E_0(\zeta) + 2ik\frac{\partial E_0(\zeta)}{\partial z} + \frac{\chi_e^{(3)}}{\varepsilon_r} k^2 |E_0|^2 E_0(\zeta) = 0. \tag{6.1-33}$$

Comparing the dynamic wave equation of Eq. (6.1-33) to the steady-state wave equation of Eq. (6.1-8), we see that they are formally the same except that in the former case the amplitude function E_0 involves a parametric variable ζ. Therefore, all solutions for Eq. (6.1-8) derived previously can be formally applied to this subsection. For a quasi-parallel incident beam with a Gaussian transverse-intensity distribution, the analytical solution in the paraxial approximation can be written as

$$z_f(t) = K \left/ \sqrt{P\left(\zeta = t - \frac{n_0 z_f}{c}\right) - P_0}\right., \tag{6.1-34}$$

where

$$K = \frac{a_0^2 n_0}{2}\sqrt{\frac{\varepsilon_0 c \pi}{n_2}} \tag{6.1-35}$$

is a real factor that can be experimentally determined. Similarly, based on the empirical steady-state formula of Eq. (6.1-29) we have

$$z_f(t) = K \left/ \left(\sqrt{P\left(\zeta = t - \frac{n_0 z_f}{c}\right)} - \sqrt{P_0}\right)\right.. \tag{6.1-36}$$

Both Eq. (6.1-34) and Eq. (6.1-36) can be used to describe the dynamic behavior of the moving foci of a pulsed laser beam propagating in a nonlinear medium.

In the dynamic case of a self-focused pulsed laser beam, the self-focused point will be moving following the input power variation. In the beginning, the pulse power is lower and the self-focal distance from the incident plan is longer; when the pulse power reaches its maximum value, this self-focal distance becomes shortest; after that, the pulse power is getting lower and that distance increases again. This simple picture is suitable if the refractive-index change (Δn) is an instantaneous function of the optical power variation, such as in the case when the electron-cloud distortion is the major mechanism leading to the induced Δn. Otherwise, the moving focus behavior will be much complicated if the response time of Δn is much slower than the pulse power variation. In the latter case, the moving focus behavior of a pulsed laser beam was thoroughly investigated by Shen and Loy.[11,12]

6.2 Direct observation of self-focusing effect

The early experimental evidence of the self-focusing or self-trapping was mostly obtained by indirect observations, such as to check optical damage trace in glass or crystal samples, or to measure the anomalous apparent decrease of the stimulated scattering threshold.[1,3] The reason to do so was that the investigated samples were usually high quality optical media, in a side direction of which the spatial profile of a laser beam was hard to observe or to record in virtue of scattering. Of course, in a few cases, one might measure the near-field patterns of a laser beam at the entry and exit positions of a solid or liquid sample to get some clues about the possible self-focusing behavior, but it was still difficult to know what had really happened inside the nonlinear medium during the nonlinear propagation process.

In the early part of this century, some direct observations of self-focusing and self-trapping have been reported. For instance, the self-focusing induced multi-focus structure and long filament formation are observed in a fused-silica sample through the plasma emission generated by a tightly focused femtosecond laser beam.[13,14] Another experimental approach is to directly observe the self-focusing induced filaments in a transparent liquid or gas medium, provided that the induced filament is "visible" through its secondary emission.[15–17]

The most recent demonstration is the direct visible observation on the self-focusing behavior and multi-focus structure of an ultrashort pulsed IR laser beam propagating in a liquid medium that is doped with three-photon absorbing and upconversion fluorescent dye molecules.[18] In this case, the dye molecules play the role of an indicator to directly manifest the spatial profile of a self-focused laser beam with a much higher image contrast, ensured by the cubic intensity dependence of the three-photon excited fluorescence emission.

Figure 6.5 shows the optical layout used for such an experimental demonstration. The input laser beam was from a Ti:sapphire laser pumped optical parametric generator with the output wavelength of ~1.3 μm, pulsewidth of ~160 fs, and repetition rate of 1 kHz. The tested sample was an APSS dye solution of 0.02 M concentration in the solvent of DMSO. The APSS molecules are highly three-photon absorbing around ~800-nm wavelength and able to emit ~540-nm fluorescence that is proportional to the cube of the local IR laser intensity. To avoid the possible thermal effect, the input laser beam was chopped to lower the pulse repetition to 50 Hz.

As shown schematically in Fig. 6.5(a), the laser beam of low intensity was pre-focused via an $f = 10$ cm lens onto the center of a 5-cm long dye solution cell, at the geometric focus position one could see a visible green spot due to three-photon induced fluorescence from the dye molecules. When the input laser intensity is properly increased, one could see that the focal point moved

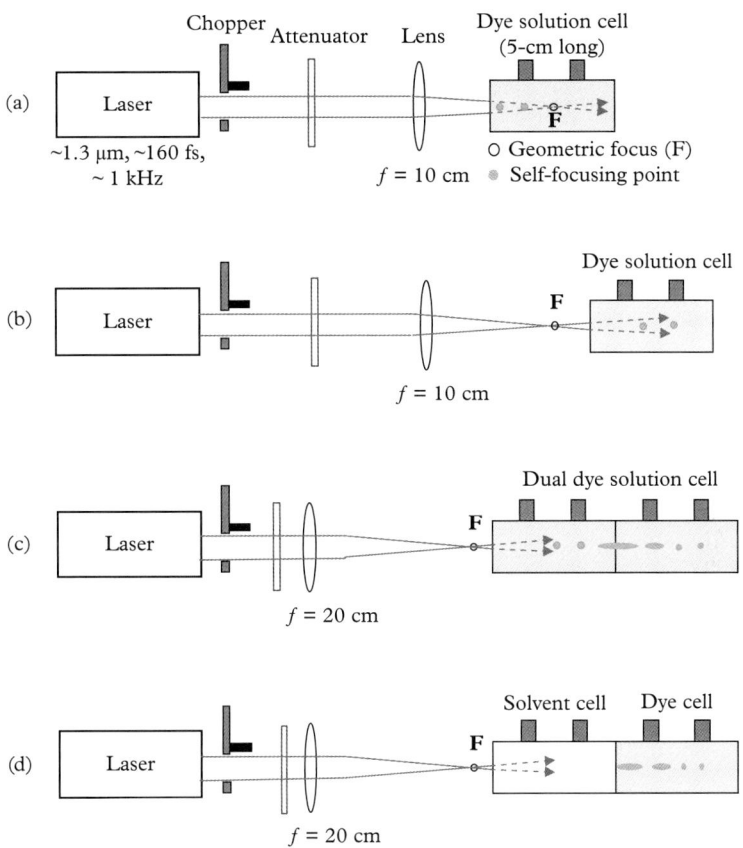

Figure 6.5 *Optical layouts for observing self-focusing and multi-focus structure of a pulsed laser beam. (a) The incident beam is pre-focused onto the center of the dye cell. (b) The beam is pre-focused in air at the position 1.5 cm before the entry window of the dye cell. (c) The beam is pre-focused in air at the position 1.5 cm before the entry window of the first dye cell connected to the second dye cell. (d) The first dye cell is replaced by a solvent cell.*

toward the entry window of the dye cell; with further increasing the input laser intensity one could see the appearance of multiple foci due to self-focusing processes within the dye solution.

Figure 6.6, which shows photographs taken at different input laser pulse energy levels, clearly demonstrates the focal point moving and the multiple foci generation following the increase of the input laser intensity.

For the situation shown in Fig. 6.5(b), the input laser beam was pre-focused via the same $f = 10$ cm lens at the point in air 1.5 cm before the front window of the dye solution cell. The experimental results are shown in Fig. 6.7. At low input intensity levels, no visible spot could be seen because of the initial beam divergence, whereas by properly raising the input intensity levels one could see the appearance of a single focus and then multiple foci structure. This is qualitatively in agreement with that predicted by the theoretical conclusions described in Section 6.1.3.

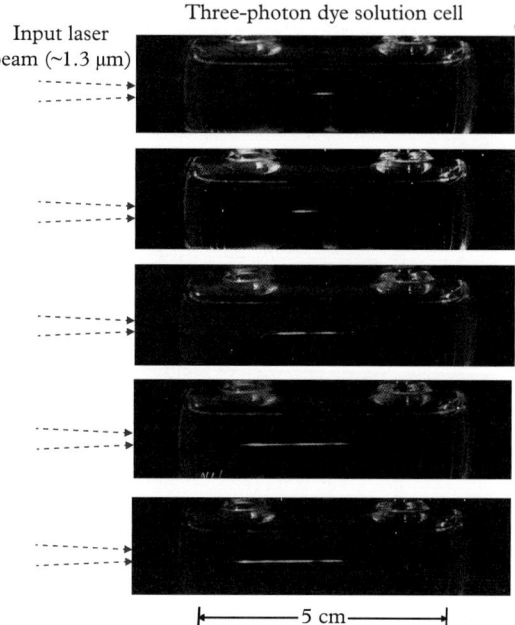

Figure 6.6 *Photographs showing the moving focal point and multi-focus structures of a self-focused IR laser beam propagating in a three-photon absorbing dye solution cell. The input beam was pre-focused at the center of the cell. From the top to the bottom, the input pulse energy was varied from 1.5 to 7.8 µJ. (After He et al.[18] Copyright 2009, IEEE) (See also Color fig. 1.).*

To obtain the result shown in Fig. 6.7(a), the input beam was pre-focused in air at the position 6.5 cm before the front window of the dye solution cell, there is no visible fluorescence spot can be seen even at an input energy level up to ∼9 µJ, because the beam divergence at the entry window of the cell was too large.

To obtain the result shown in Fig. 6.7(b), another 5-cm long dye cell is placed in the front of the second dye cell, at the same input level one can see a clear multi-focus trace across both two cells, which indicates that the previously divergent laser beam gets refocused in the first dye cell and remains self-focused in the second dye cell.

Now one question arises: what is the dominant contributor to the induced refractive-index change? To answer this question, the setup shown in Fig. 6.5(d) can be employed, in which the first dye cell is replaced by a cell filled with the pure solvent (DMSO). Under the same input conditions, it was shown that in the second dye cell, the same self-focusing behavior still could be seen as that when the first cell was filled with the dye solution. The test results are shown in Fig. 6.7(c), which sufficiently reveal that the major contribution to the induced Δn is from the solvent and not from the dye molecules. In other words, the latter only played the role of an indicator of the self-focused beam's spatial profile.

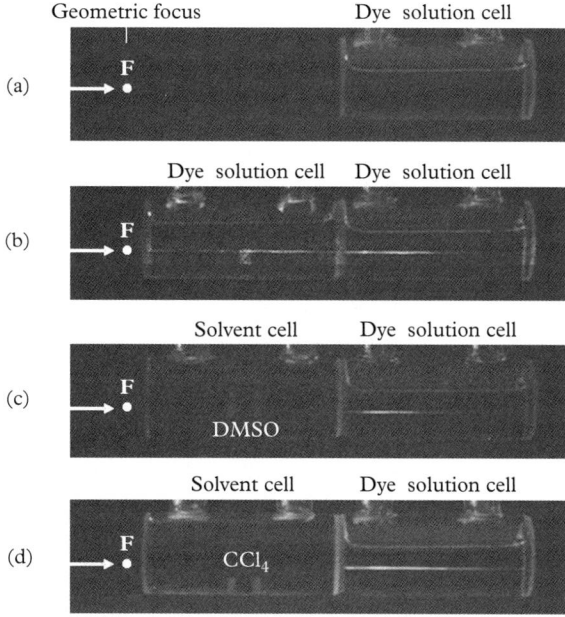

Figure 6.7 Photographs showing (a) a divergent laser beam passed through a dye solution cell 6.5 cm from the geometric focal point in air, (b) another dye cell is placed in the front of the second dye cell, (c) the first cell is filled with the pure solvent of DMSO, and (d) the first cell is filled with the pure solvent of CCl$_4$. The input laser pulse energy is ~9 µJ. (After He et al.[18] Copyright 2009, IEEE).

It is obvious that this type of experimental design can be used to comparing the self-focusing capability of different liquids or solvents filled in the first cell. As an example, shown in Fig. 6.7(d) is the result obtained by filling the pure solvent of CCl$_4$ into the first cell, which even manifests a stronger self-focusing capability.

6.3 Self-phase modulation and spectral self-broadening

6.3.1 Self-phase modulation and frequency chirp of intense short light pulses

In Section 6.2 we have considered the impact of the induced refractive index change on the beam size and the wavefront structure of an input laser beam. In this section, we shall consider the connection between an induced dynamic refractive-index change and the corresponding phase change experienced by the intense light pulse itself.

For simplicity, we consider here only the temporal behavior of the induced refractive-index change in a third-order nonlinear medium. Assume that there is a quasi-plane input laser wave with a short pulsewidth; the time-dependent refractive-index change can be written as

$$\Delta n(t) = n_2 E_0^2(t), \tag{6.3-1}$$

where n_2 is the nonlinear refractive-index coefficient that is assumed to be positive throughout this section, and $E_0(t)$ is the real amplitude function of the incident intense light pulse. It is also assumed that the pulse duration is much longer than both the rise-time and the decaying-time of the induced Δn (see Section 5.6), so that Δn can instantaneously follow the change of the pulse intensity. Under these conditions the phase variation experienced by the beam itself will be

$$\Delta\phi(t) = \int_0^L \frac{\omega_0}{c} \Delta n(t) dz = \frac{\omega_0}{c} L n_2 E_0^2(t). \tag{6.3-2}$$

This is the so-called self-phase-modulation effect. Here ω_0 is the initial circular frequency of the light pulse, and L is the thickness of the nonlinear medium. According to the principle of Fourier transform, a fast changed phase factor results in a corresponding frequency change, i.e.,

$$\Delta\omega(t) = \omega(t) - \omega_0 = -\frac{\partial}{\partial t}\Delta\phi(t) = -\frac{\partial}{\partial t}\left[\frac{\omega_0}{c} L n_2 E_0^2(t)\right]. \tag{6.3-3}$$

The reason for taking a negative sign here is that a fast increase in phase (or in optical path) implies that the medium seems suddenly thicker and consequently, according to the principle of the Doppler effect the instantaneous frequency will become lower. For simplicity, we assume that the input light pulse exhibits a Gaussian temporal profile:

$$E_0(t) = A_0 \exp\left[-(t-t_0)^2/\tau^2\right], \tag{6.3-4}$$

where A_0 is the peak amplitude, t_0 is the time position of the pulse peak, and τ is the half-width at the 1/e point of the peak amplitude. Substituting Eq. (6.3-4) into Eq. (6.3-2) leads to

$$\Delta\phi(t) = \Delta\phi_{max} \exp\left[-2(t-t_0)^2/\tau^2\right]. \tag{6.3-5}$$

Here,

$$\Delta\phi_{max} = \left[\frac{\omega_0 n_2 L}{c} A_0^2\right] \tag{6.3-6}$$

is the maximum induced phase change in units of radians, which is proportional to n_2 (the non-linearity of the medium), L (the nonlinear interaction length), and A_0^2 (the peak intensity of the light pulse). Substituting Eq. (6.3-5) into Eq. (6.3-3) leads to an explicit expression of the frequency-shift as a function of time:

$$\Delta\omega(t) = \Delta\phi_{max} \frac{4(t-t_0)}{\tau^2} \exp\left[-2(t-t_0)^2/\tau^2\right]. \tag{6.3-7}$$

Letting $\frac{\partial}{\partial t}[\Delta\omega(t)] = 0$, we can determine that the maximum frequency shift occurs at the positions of $(t-t_0) = \pm\tau/2$, and the maximum value of frequency shift is

$$|\Delta\omega_{max}| = \Delta\phi_{max}\frac{2}{\tau}e^{-0.5} = \Delta\phi_{max}\frac{1.21}{\tau}. \tag{6.3-8}$$

From Eq. (6.3-8) we see that the maximum frequency shift is proportional to the maximum phase shift $\Delta\phi_{max}$, and is inversely proportional to the half-width τ of the input light pulse.

As an example, Fig. 6.8 shows (a) the amplitude function of an incident pulse with a Gaussian temporal profile of $\tau = 1$ ps, (b) the induced phase-change function with an assumed $\Delta\phi_{max} = 10\pi$, and (c) the frequency chirp curve, i.e., $\Delta\omega(t)$ as a function of time t.

From Fig. 6.8 we see that the leading part of the pulse leads to a red frequency shift (Stokes shift), whereas the trailing part of the pulse leads to a blue frequency shift (anti-Stokes shift). As we mentioned this kind of frequency shift is caused by the Doppler effect: during the period of leading part of the pulse the nonlinear medium is getting thicker, then it is getting thinner during the period of trailing part of the pulse. At the pulse-peak position, the frequency shift is zero. The maximum frequency shift occurs at the positions of two inflection points of the curve shown in Fig. 6.8(b). In Fig. 6.8(c) we see a very important feature of the frequency-chirped output pulse: the red-shifted and the blue-shifted spectral components occupy the leading half and trailing half of the output pulse, respectively, while the original unshifted spectral component is located in the center of the output pulse.

Assume that the input pulse with a pulsewidth of $(\Delta t)_0$ is unchirped, i.e., different portions of this pulse have the same spectral structure with a central frequency of ω_0 and a smaller spectral width of $\delta\omega$, as shown schematically in Fig. 6.9(a). After passing through a nonlinear medium of $n_2 > 0$, the output frequency-chirped pulse is shown in Fig. 6.9(b), which indicates red-shifted spectral components in the leading half and blue-shifted spectral components in the trailing half of the output pulse. If let this frequency-chirped pulse further pass through a second linear medium of positive group velocity dispersion (GVD), the temporal profile of the output pulse will

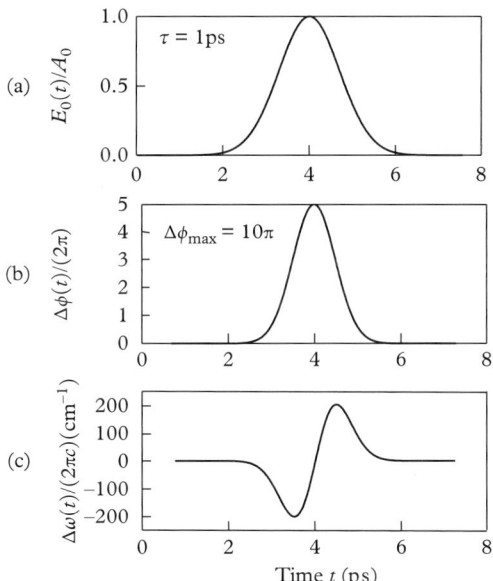

Figure 6.8 *(a) Normalized amplitude profile of the input pulse, (b) the induced phase-change function, and (c) the frequency chirp curve of the output pulse.*

122   Self-Focusing, Self-Phase Modulation, and Spectral Self-Broadening

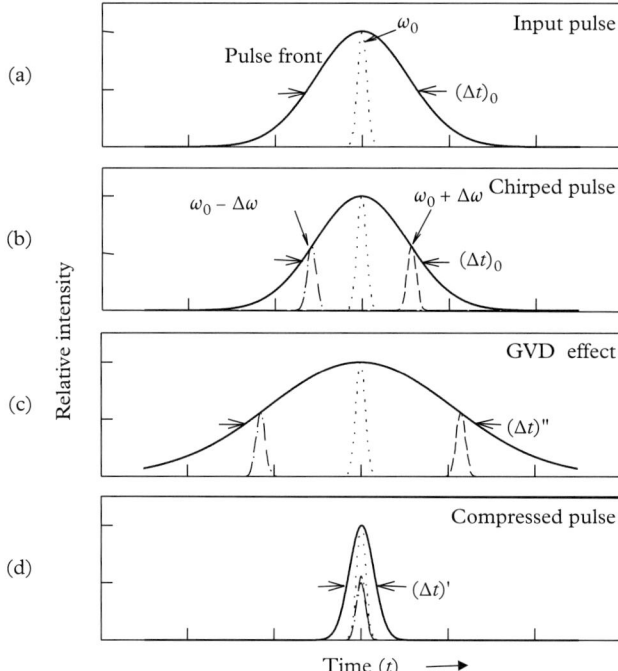

Figure 6.9 (a) Profile of an input unchirped pulse, (b) output frequency-chirped pulse after passing through a nonlinear medium, (c) pulse-expanding of a chirped pulse after passing through a linear medium of positive-dispersion, and (d) pulse-compressing of a chirped pulse after passing through a linear medium of negative-dispersion.

be expanded because the red-shifted spectral components propagate faster than the blue-shifted components, as shown schematically in Fig. 6.9(c).

In a contrary case, if let the same frequency-chirped pulse pass through a second linear medium of negative GVD, the temporal profile of the output pulse will be compressed because the red-shifted components propagate slower than the blue-shifted components. By properly choosing the negative GVD property and propagation length of the second medium, it can be achieved that all different spectral components finally overlap each other, as shown in Fig. 6.9 (d). Because of that process, the pulsewidth of the final output pulse can be significantly compressed in comparison with the original pulsewidth. The narrowest pulsewidth $(\Delta t)'$ of the compressed pulse is limited by the uncertainty principle, i.e.,

$$(\Delta t)' \geq 2\pi |\Delta\omega_{\max}|^{-1}, \qquad (6.3\text{-}9)$$

where $\Delta\omega_{max}$ is the maximum frequency shift range of the chirped pulse.

In practice, a single-mode optical fiber is commonly employed as a nonlinear medium to provide an efficient self-phase modulation and frequency chirping because of the longer interaction length and higher local intensity within the fiber core.[19] On the other hand, a pair of gratings or two pairs of prisms are often employed as the second transmission system with an equivalent negative GVD property.[20,21]

6.3.2 Spectral self-broadening of intense short light pulses

In the preceding subsection, we know that an intense and short light pulse may induce a fast self-phase modulation in the nonlinear medium. According to the Fourier transform principle, a fast phase change will lead to corresponding spectral broadening. In this case, the amplitude function of the spectral distribution for the output light pulse can be determined by taking an inverse-Fourier transform (cf. Eq. (2.6-1)):

$$E_0(\omega) = \frac{1}{2\pi} \int_{-\infty}^{\infty} \left[E_0(t) e^{i\Delta\phi(t)} e^{-i\omega_0 t} \right] e^{i\omega t} dt, \qquad (6.3\text{-}10)$$

where $\Delta\phi(t) \propto |E_0(t)|^2$ is given by Eq. (6.3-2). The above expression can be written in another form

$$E_0(\Delta\omega) = \frac{1}{2\pi} \int_{-\infty}^{\infty} \left[E_0(t) e^{i\Delta\phi(t)} \right] e^{i\Delta\omega t} dt. \qquad (6.3\text{-}11)$$

Here $\Delta\omega = \omega_0 - \omega$ is the frequency shift due to self-phase modulation. Finally, the spectral intensity distribution can be expressed as

$$I(\Delta\omega) \propto |E_0(\Delta\omega)|^2. \qquad (6.3\text{-}12)$$

As an example, Fig. 6.10 shows the computed self-broadened spectral intensity distribution of a quasi-monochromatic light pulse passed through a nonlinear medium for different values of $\Delta\phi_{max}$ and pulse duration. Here it is still assumed that the amplitude profile of the input light pulse is a Gaussian curve with a half-width τ at the $1/e$ point. From Fig. 6.10 we can see the following features:

1. For the same pulse duration the spectral broadening range is proportional to the maximum induced phase change $\Delta\phi_{max}$.

2. For the same value of $\Delta\phi_{max}$ the spectral broadening range is inversely proportional to the pulse duration τ.

3. There is a quasi-periodic modulation structure of the spectrum where the period number on each side of the central frequency position is determined by an integer closest to $(\Delta\phi_{max}/2\pi)$.

The first two features are exactly predicted by Eq. (6.3-8). In particular, one can find that the maximum spectral shift value shown in Fig. 6.10(a) is in agreement with the corresponding value shown in Fig. 6.8(c). The third feature can also be explained based on Fig. 6.8(c) from which one can see that for a given frequency shift value there are two waves originated from different moments. These two waves of the same frequency will interfere with each other constructively or destructively depending on their relative phase-difference. Because of this interference, we will observe a quasiperiodic modulation structure of the broadened spectrum.[22]

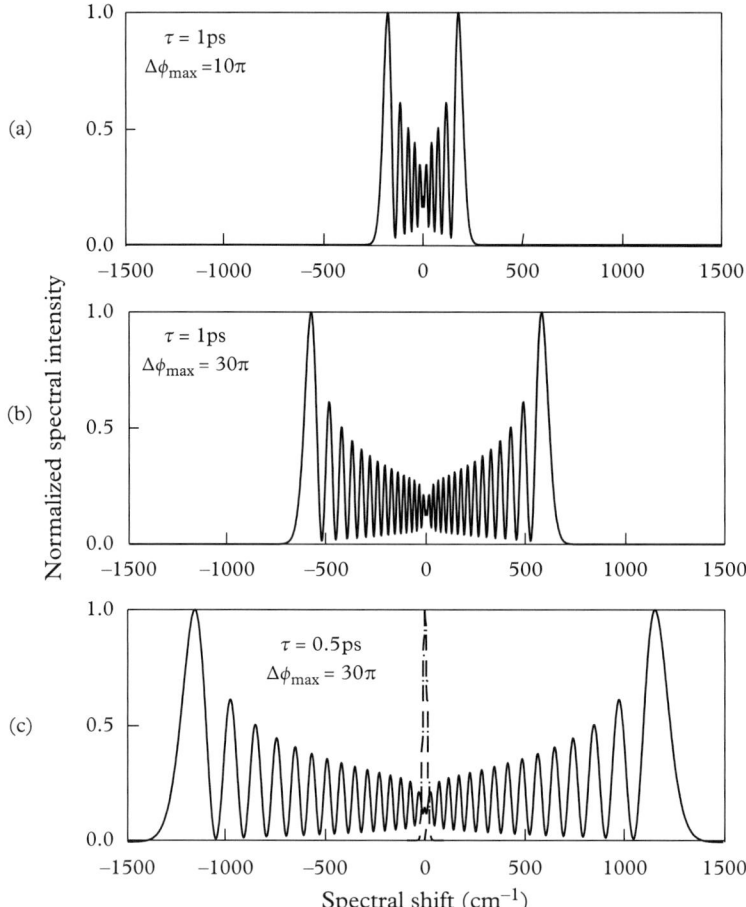

Figure 6.10 *Spectral self-broadening in different conditions: (a) $\tau = 1$ ps and $\Delta\phi_{max} = 10\pi$, (b) $\tau = 1$ ps and $\Delta\phi_{max} = 30\pi$, and (c) $\tau = 0.5$ ps and $\Delta\phi_{max} = 30\pi$. The dash-dotted curve in (c) is the original spectral line with a Fourier-transform limited linewidth.*

So far in this section we have assumed that the input pulsed light has a uniform transverse intensity distribution and thereby $E_0(t)$ is only the function of time. On such a simplified assumption, one would expect a regular and fully modulated spectral broadening like that shown in Fig. 6.10. In practice, however, the real input pulsed laser beam usually exhibits a Gaussian-like transverse intensity distribution, and the induced phase change $\Delta\phi(t, r)$ is also dependent on the radial variable r. In this case, the observed spectral self-broadening appearance will not be the same as that shown in Fig. 6.10. Instead, there will be two major differences: (i) a spectral maximum will be located at the $\Delta\omega = 0$ position and (ii) the periodic modulation structure may not be as sharp as that shown in Fig. 6.10.

The other fact is that in obtaining Fig. 6.8 and Fig. 6.10, it is assumed that the induced Δn and $\Delta\phi$ are instantaneous functions of $E_0(t)$, which is true when the major contribution to the induced Δn is from the electron-cloud distortion mechanism. In this case, as the red-shifted spectral components are created by the leading part of the input light pulse whereas the blue-shifted spectral components are created by the trailing part of the same pulse, the overall self-broadening spectral distribution should be symmetric if the input temporal profile of the input pulse is also symmetric. However, if the rise- or decay-time of Δn is slower than the variation time of the input pulse, the temporal profile of Δn or $\Delta\phi$ may become asymmetric, consequently one could observed an asymmetric spectral broadening. The asymmetric spectral broadening, which was observed in CS_2 liquid by the early studies,[23–25] can be explained by this consideration, wherein the major contribution to the induced Δn comes from the reorientational Kerr effect that exhibits a much slower relaxation time than the rise time of the induced Δn.

6.3.3 Beat-frequency enhanced spectral self-broadening

So far, we have considered the spectral self-broadening behavior of a quasi-monochromatic input light pulse that possesses only one frequency component with a narrower spectral linewidth. Now we shall consider another interesting situation in which the input light pulse possesses two discrete frequency components with a spectral spacing much greater than their linewidths. In this particular case, the electric field of the input light pulse with a linear polarization status can be expressed as

$$\tilde{E}(t) = E_{01}(t)e^{-i\omega_1 t} + E_{02}(t)e^{-i\omega_2 t}, \tag{6.3-13}$$

where ω_1 and ω_2 are the central frequencies of these two spectral components, and E_{01} and E_{02} are their time-dependent amplitude functions. Assuming $E_{01}(t) = E_{02}(t) = E_0(t)$, the above expression can be simplified as

$$\begin{aligned}\tilde{E}(t) &= E_0(t)\left[1 + e^{-i\Delta\omega' t}\right]e^{-i\omega_1 t} \\ &= \tilde{E}_0(t)e^{-i\omega_1 t}.\end{aligned} \tag{6.3-14}$$

Here $\Delta\omega' = \omega_2 - \omega_1$, and

$$\tilde{E}_0(t) = E_0(t)\left[1 + e^{-i\Delta\omega' t}\right] \tag{6.3-15}$$

is the time-dependent amplitude function of the total field of the input pulse.

After passing through a nonlinear medium of length L the induced phase-change will be

$$\begin{aligned}\Delta\phi(t) &= \frac{\omega_0}{c}L\Delta n(t) = \frac{\omega_0}{c}Ln_2\left|\tilde{E}_0(t)\right|^2 \\ &= \frac{\omega_0}{c}Ln_2 2E_0^2(t)(1 + \cos\Delta\omega' t),\end{aligned} \tag{6.3-16}$$

where $\omega_0 = (\omega_1 + \omega_2)/2$ is the average frequency of the light pulse. The complex field function of the output pulse after passing through the nonlinear medium is

$$\tilde{E}'(t) = \tilde{E}(t)e^{i\Delta\phi(t)} = E_0(t)\left[e^{-i\omega_1 t} + e^{-i\omega_2 t}\right]e^{i\Delta\phi(t)}. \tag{6.3-17}$$

The spectral broadening behavior of the ω_1 line is determined by the following inverse-Fourier transform:

$$E_0(\omega) = \frac{1}{2\pi} \int_{-\infty}^{\infty} \left[E_0(t) e^{-i\omega_1 t} e^{i\Delta\phi(t)} \right] e^{i\omega t} dt. \qquad (6.3\text{-}18)$$

This expression can be further simplified as

$$E_0(\Delta\omega) = \frac{1}{2\pi} \int_{-\infty}^{\infty} \left[E_0(t) e^{i\Delta\phi(t)} \right] e^{i\Delta\omega t} dt, \qquad (6.3\text{-}19)$$

where $\Delta\omega = \omega - \omega_1$. Thus the final spectral intensity distribution around the ω_1 line will be

$$I(\Delta\omega) \propto |E_0(\Delta\omega)|^2 = \left| \frac{1}{2\pi} \int_{-\infty}^{\infty} \left[E_0(t) e^{i\Delta\phi(t)} \right] e^{i\Delta\omega t} dt \right|^2. \qquad (6.3\text{-}20)$$

Here, $\Delta\phi(t)$ is given by Eq. (6.3-16). It is obvious that for the ω_2 line we will see the same spectral broadening behavior.

On the other hand, from Eq. (6.3-15) we know that the temporal variation of the intensity of the input light pulse is determined by

$$I_0(t) \propto \left| \tilde{E}_0(t) \right|^2 \propto E_0^2 (1 + \cos \Delta\omega' t), \qquad (6.3\text{-}21)$$

where $\Delta\omega'$ is the frequency difference between the two spectral components contained in the input light pulse and $E_0(t)$ is the amplitude function of each component. Eq. (6.3-21) implies that there is an intensity modulation at the beat-frequency of $\Delta\omega' = |\omega_2 - \omega_1|$.

As an example, Fig. 6.11 shows the normalized intensity variation of the half of the input pulse as a function of time t. It is assumed here that the two spectral components possess the same Gaussian amplitude profile with $\tau = 40$ ps at the $1/e$ point, and their frequency difference is $\Delta\omega'/(2\pi c) = 41.67$ cm^{-1}. From Fig. 6.11 we see that the instantaneous refractive-index change will depend not only on the pulse envelope but also on the modulation period. The latter for our example is equal to 0.8 ps. Under the conditions described above and based on Eq. (6.3-20), the theoretical spectral-intensity distributions are shown in Fig. 6.12 for different values of $\Delta\phi_{\max}$. From Fig. 6.12 we can see the following features.

1. At a smaller value of $\Delta\phi_{\max} \leq 2\pi$, there are only several discrete side bands with a frequency spacing equal to $\Delta\omega' = \omega_2 - \omega_1$;
2. At a moderate value of $\Delta\phi_{\max} \approx 10\pi$, both the overall spectral broadening range and the spectral width of each side band are remarkably increased;
3. At a higher value of $\Delta\phi_{\max} \geq 30\pi$, the overall spectral broadening range is further stretched to both the far-long-wavelength side and the far-short-wavelength side, and more importantly, a quasi-continuous spectral distribution appears.

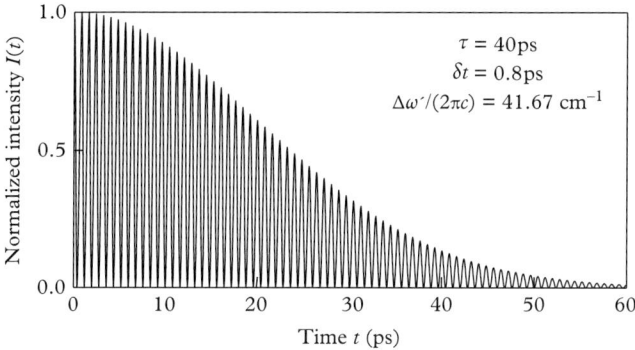

Figure 6.11 *Normalized intensity variation of the input light pulse possessing a Gaussian amplitude profile of $\tau = 40$ ps and two spectral components separated by 41.67 cm^{-1}. The intensity-modulation period is 0.8 ps.*

Comparing Fig. 6.12(d) to Fig. 6.10(b) we can also see that at the same level of $\Delta\phi_{max} = 30\pi$, the spectral broadening range for the former is greater than that for the latter, although the duration of the input pulse envelope for the former is 40 times longer than that for the latter.

At the beginning of this subsection, it is assumed that the linewidth of the two incident spectral components is negligible comparing to the spacing between these two lines. In a real case, the spectral linewidth of each component might be comparable to the spacing between these two spectral lines, then the discrete spectral structures may become blurred even in a moderate level of $\Delta\phi_{max}$.

Some early studies indicated that the spectral broadening behavior of a laser beam propagating in a nonlinear medium was essentially dependent on the spectral structure of the input laser pulse: a multi-axial-mode structure (or multi-frequency-component structure) of the pump pulse was positively favorable for observing a greater spectral broadening.[22,26] Also, it was known that the continuum emission could be generated in a single-mode fiber pumped by nanosecond laser pulses with a broader spectral band.[27,28] This type of experimental results, obtained by using laser pulses even with an envelope of nanoseconds, can be partially explained based on the mechanism of beat-frequency enhanced spectral self-broadening, the frequency beating occurred between different spectral components of the input laser pulses.

6.4 Optical coherent continuum generation

6.4.1 White-light continuum generation with ultrashort laser pulses

The term ultrashort laser pulse is generally defined by the pulse duration less than several picoseconds. Since the pulsewidth is so short and the peak power is so high, even a loosely focused ultrashort laser beam can generate the coherent continuum emission with a superbroad spectral distribution by passing through a transparent optical medium. Because the spectral broadening range can readily cover the entire visible spectral region, this nonlinear optical effect often refers

128 Self-Focusing, Self-Phase Modulation, and Spectral Self-Broadening

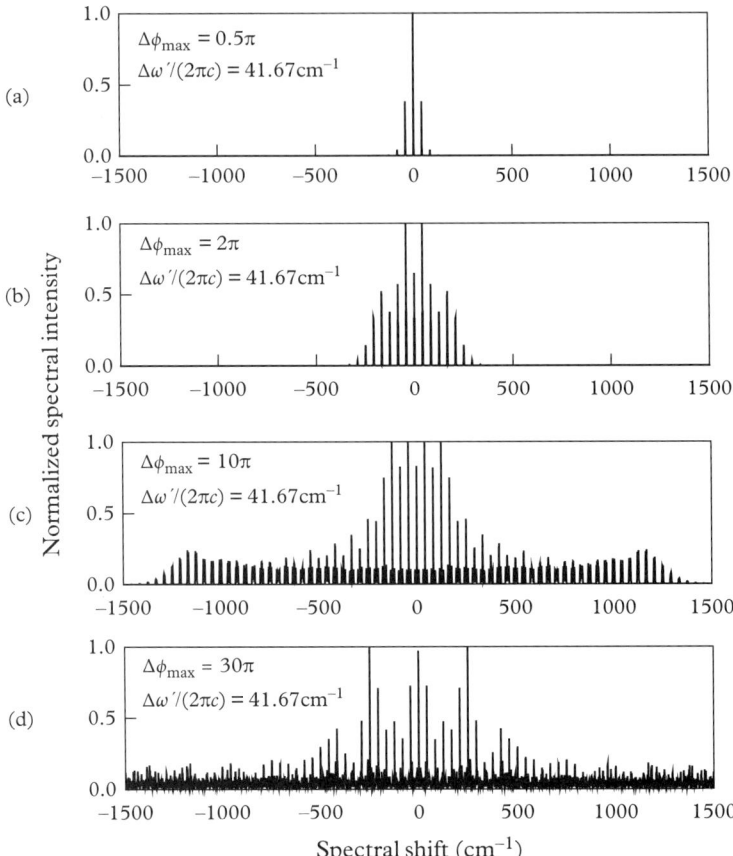

Figure 6.12 *Beat-frequency enhanced spectral broadening at different values of $\Delta\phi_{max}$. The other parameters of the input pulse are the same as indicated in Fig. 6.8.*

to white-light continuum generation (WLCG). Unlike ordinary white light emitted from incoherent light sources, the different spectral components of WLCG retain certain predictable phase relations. Such coherent continuum emission is generated through the third-order nonlinearity of a transparent medium via two major mechanisms: one is the self-broadening due to self-phase modulation; the other is the four-wave frequency mixing (FWFM) or the so-called four-photon parametric interaction. Since the spectral and spatial structures of the coherent emission generated via these two mechanisms are essentially different, thereby the emission components generated through these two mechanisms can be readily identified or separated.

The first clear observation of coherent optical continuum generation was reported in 1970 by Alfano and Shapiro.[29,30] In their initial work, the 532-nm and picosecond laser pulses of high peak power were used to pump a number of transparent glass and crystal samples (BK-7 glass, quartz, calcite, etc.). The forward coherent emission with a super-broad band spectrum ranging from 400 to 700 nm was observed on both Stokes and anti-Stokes sides of the pump spectral line. It was found that there were two different spectral components contributing to the observed continuum generation: one was the axial spectral component propagating collinearly with the input pump beam; the other was the off-axis spectral component having a propagation angle with respect to the pump beam. The former was characterized by a quasi-periodic modulation structure due to self-phase modulation (see Section 6.3.2); the latter was characterized by an angle-dependent spectral distribution due to non-resonant four-photon parametric interaction (see Section 4.4.1).

Depending on the experimental conditions (such as the input-pulse duration, peak power, interaction length, and types of the tested samples), the mechanisms contributing to the observed coherent continuum emission may not be limited to only these two mechanisms mentioned above. In some cases, other mechanisms (such as stimulated scattering and other resonant interactions) may also give considerable contributions to the observed spectra of coherent continuum generation.[31–36]

6.4.2 Experimental observation of coherent continuum generation

To experimentally distinguish various contributions from different possible mechanisms to the observed superbroad band coherent emission, a simple but quite useful experimental design has been reported.[37] This optical layout of such setup design is schematically shown in Fig. 6.13, where the input laser pulse is focused onto the center of a 5- or 10-cm long tested sample that can be a liquid cell or a solid rod. When the local intensity in the focal area reaches a higher level (e.g., $\geq$5–10 GW/cm^2), very strong continuum emission will be observed in the forward direction.

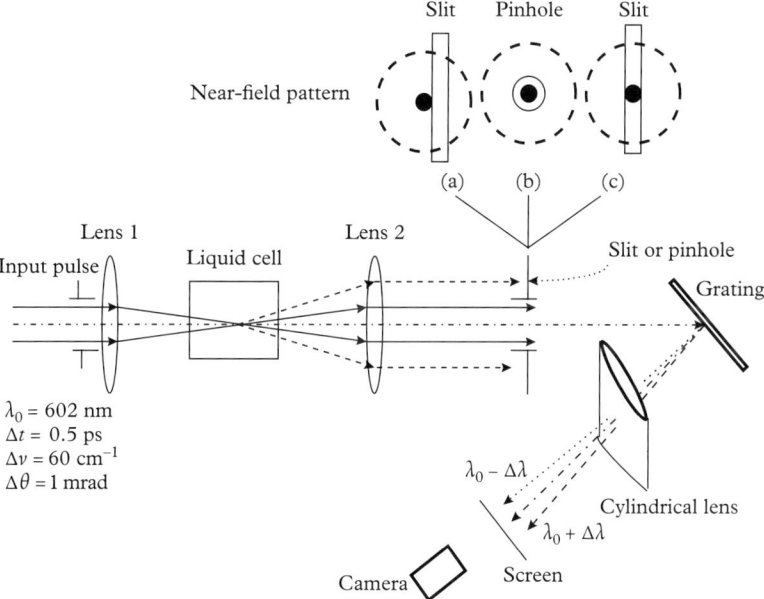

Figure 6.13 *Experimental setup for observing the collinear and non-collinear coherent emission from a nonlinear medium pumped by a laser beam of ultrashort pulses.*

130 *Self-Focusing, Self-Phase Modulation, and Spectral Self-Broadening*

After passing through a lens, the forward coherent emission is re-collimated and at the near-field pattern position of the continuum output, one could see a bright central spot with white-light like color (collinear emission), and the central spot is surrounded by a series of emission rings (non-collinear emission) that correspond to different emission angle and have different colors.

In this case, if put a small pinhole or a narrow slit in the near-field position, one could choose different portions of the forward beam to be reflected from a grating, and then be focused by a cylindrical lens on an observation screen. The special feature of this setup is that along the horizontal direction on the screen one could observe the spectral distribution, while in the vertical direction one can see the dependence of the spectrum on the propagation angle of emission.

For three specially chosen positions of the pinhole or slit, as shown in Fig. 6.13, the observed spectra of continuum generation in a 5-cm-long heavy water (D_2O) are shown in Fig. 6.14, respectively. The photographs demonstrated in Fig. 6.14(a) is the non-collinear (off-axial) coherent emission component, which was obtained when a $\sim$1.3-mm slit was placed in the position a

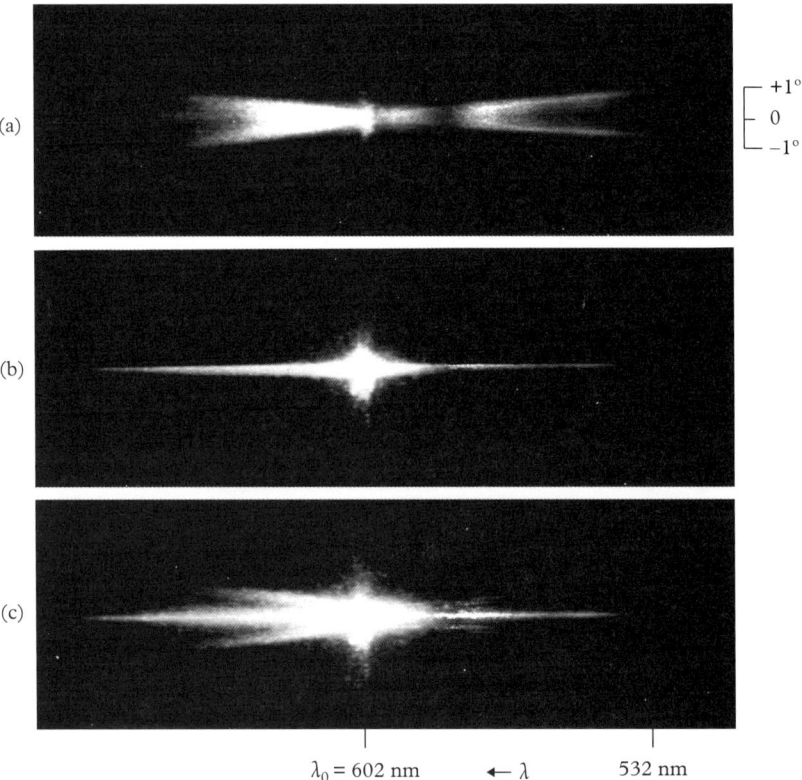

Figure 6.14 *Spectral and spatial structures of continuum generation from heavy water at a pump intensity level of $I_0 \approx 10$ GW/cm^2: (a) non-collinear emission due to four-photon parametric interaction, (b) collinear emission due to self-phase modulation, and (c) overall coherent emission. (After He et al.[37] Copyright 1993, Optical Society of America) (See also Color fig. 3.).*

shown in Fig. 6.13. One can see an obvious angle-dependence of the off-axis coherent emission on the propagation angle; the spectral distributions on both sides of the excitation wavelength are nearly symmetric, and the observed maximum deviation angle is about ±1°, corresponding to a maximum wavelength shift range from ∼500 nm to ∼700 nm. This behavior of the non-collinear coherent emission is in semi-quantitative agreement with the theoretical prediction shown in Fig. 4.11 based on the mechanism of non-resonant four-photon parametric interaction. For this mechanism, the electron-cloud distortion is the only origin of nonlinearity and the phase-matching requirements can be expressed by (cf. Eq. (4.4-1)):

$$\left.\begin{array}{l} 2\omega_0 = \omega_1 + \omega_2 \\ 2\mathbf{k}_0 = \mathbf{k}_1 + \mathbf{k}_2 \end{array}\right\}, \tag{6.4-1}$$

where $\mathbf{k}_0$ is the wavevector of the pump beam, while $\mathbf{k}_1$ and $\mathbf{k}_2$ are the wavevectors of Stokes-shifted and anti-Stokes-shifted emission beams. For most commonly used optical media transparent in visible spectral range, there is no dramatic difference in their dispersion behavior, and therefore we do not expect any obvious difference in the angle-dependence of the spectrum of non-resonant four-photon parametric emission.

Figure 6.14(b) shows the collinear (axial) spectral component when an ∼0.8-mm pinhole was placed in the position b shown in Fig. 6.13. In this case, only the central part of the forward continuum beam propagating along the original direction of the input pump beam was displayed on the screen. The spectrum obtained in that condition shows nearly symmetric distributions on both sides of the excitation wavelength. Furthermore, a quasi-periodic modulation structure among the axial spectral components was observed by utilizing a grating spectrograph with an ∼10 cm^{-1} spectral resolution. All these observations verified that the self-phase-modulation is the major mechanism contributing to the axial or collinear emission in the coherent continuum output.

Finally, Fig. 6.14(c) shows the overall spectral structure when the slit was placed in the position c indicated in Fig. 6.13. In this case, both the collinear and non-collinear spectral components were recorded simultaneously.

Under the same experimental conditions, among the five measured transparent liquid samples, heavy water showed the strongest continuum generation, water showed strong continuum emission, carbon tetrachloride showed a moderate intensity of continuum generation, benzene showed very weak continuum generation, and carbon disulfide showed no continuum generation.

For instance, Fig. 6.15 shows the observed spectra of a 10-cm-long benzene (C_6H_6) sample at the same pump level ($I_0 \approx 10$ GW/cm^2). In Fig. 6.15(a), an apparent unsymmetrical distribution on the two sides of the pump line was probably due to the fact that the non-collinear emission in benzene was so weak, that the intensity of the red-shifted part was lower than the exposure threshold of the photographic film of low sensitivity for dark-red beams.

On the other hand, two stronger spots at the wavelength position of ∼568 nm can be clearly seen in Fig. 6.15(a). These stronger spectral spots result from a Raman resonance-enhanced four-wave mixing process, as generally discussed in Section 4.3. In such a particular case, the phase-matching and frequency-resonance requirements are

$$\left.\begin{array}{l} 2\omega_0 = \omega_s + \omega_{as} \\ \omega_0 - \omega_s = \omega_{as} - \omega_0 \approx \Delta\omega_r \\ 2\mathbf{k}_0 = \mathbf{k}_s + \mathbf{k}_{as} \end{array}\right\}, \tag{6.4-2}$$

where $\Delta\omega_r$ is the Raman-mode frequency of 992 cm^{-1} for benzene and ω_s and ω_{as} are the frequencies of coherent Raman emission components due to FWFM. These resonance-enhanced

132 Self-Focusing, Self-Phase Modulation, and Spectral Self-Broadening

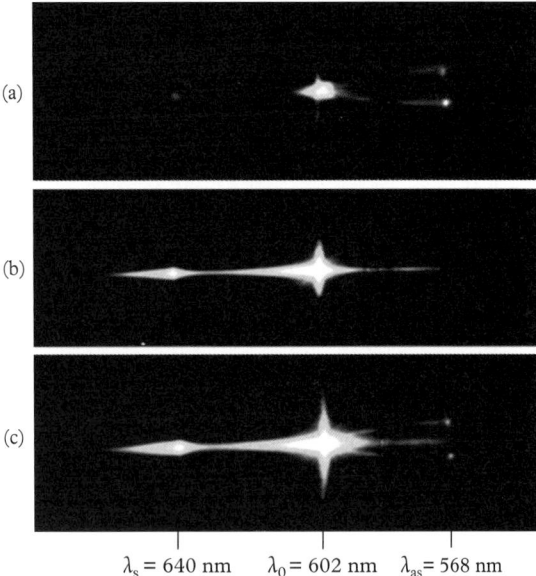

Figure 6.15 *Spectral and spatial structures of coherent emission from benzene at the pump intensity of $I_0 \approx$ 10 GW/cm^2: (a) non-collinear emission, (b) collinear emission, and (c) overall coherent emission. (After He et al.[37] Copyright 1993, Optical Society of America) (See also Color fig. 4.).*

components will add to the relatively weak spectral emission of the non-resonant four-photon parametric emission.

Figure 6.15(b) shows the spectral feature of the collinear emission from the benzene sample: the self-broadening due to self-phase-modulation is much weaker than that from heavy water; however, there is a very strong spectral spot at the Stokes-shifted Raman line position of ~640 nm for the same 992 cm^{-1} mode. This strong emission is due to stimulated Raman scattering that does not require phase-matching (see Section 7.3). Moreover, we can see from Fig. 6.15(b) that there are red-shifted broadband contributions added on the Stokes sides of both the pump line and the stimulated Raman scattering line, respectively. These additional coherent red-shifted broad band components are due to stimulated Rayleigh–Kerr scattering and stimulated Raman–Kerr scattering, respectively,[38,39] which will be described in detail in Section 7.5.

Differences between Fig. 6.14 for heavy water and Fig. 6.15 for benzene can be explained if one considers the competition among various mechanisms that may cause the spectral broadening of the input laser pulses. For liquids such as heavy water and water, the nonlinearity from the molecular reorientation (optical Kerr effect) is negligible, and the electron-cloud distortion is the only major contribution to the nonlinearity of the medium; therefore, one would expect a stronger non-resonant four-photon parametric emission as well as a stronger collinear spectral super-broadening owing to self-phase modulation. On the other hand, benzene is a liquid exhibiting a much larger Raman scattering cross-section and a stronger optical reorientational Kerr effect, so that we would

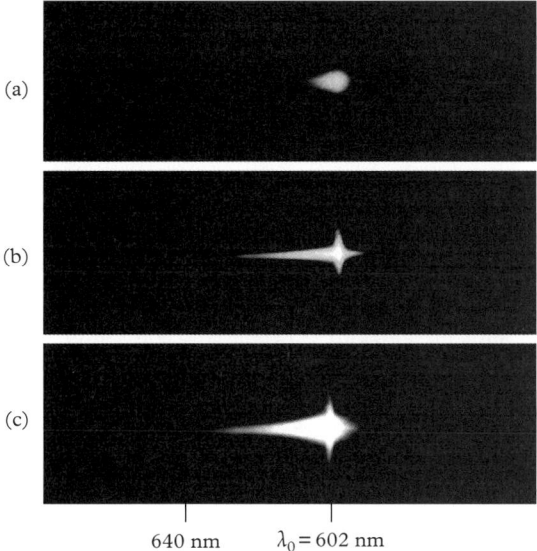

Figure 6.16 *Spectral and spatial structures of coherent emission from carbon disulfide at the pump intensity of $I_0 \approx 10$ GW/cm^2: (a) non-collinear emission, (b) collinear emission, and (c) overall coherent emission. (After He et al.[37] Copyright 1993, Optical Society of America) (See also Color fig. 5.).*

expect a strong stimulated Raman scattering as well as a strong stimulated Raman–Kerr scattering, the latter is inherently related to the optical Kerr effect.

As a quite different example, Fig. 6.16 shows spectral photographs of the forward coherent emission from a 10-cm-long carbon disulfide (CS$_2$) sample, obtained under the same experimental condition. Figure 6.16(a) shows that there is no angle-dependent four-photon parametric emission at all, whereas, Fig. 6.16(b) shows that there is only a strong red-shifted broadband coherent emission added in the pump line, which is also caused by the stimulated Rayleigh–Kerr scattering characterized by a monotonically declining intensity distribution.

The dramatic difference between Fig. 6.16 and Fig. 6.14 can be explained by the fact that carbon disulfide is the best liquid exhibiting a large reorientational optical Kerr effect and the strongest stimulated Rayleigh–Kerr scattering, as proved by numerous experiments performed with laser pulses of various pulsewidth (from 10^{-8} to 10^{-13} s). In the case shown in Fig. 6.16, the stimulated Rayleigh–Kerr scattering becomes the dominant process that extracted the most pump energy, and all other possible nonlinear processes were suppressed.

From the above discussions, we realize that the spectral-broadening behavior of intense light pulses in nonlinear media can be obviously different or quite complicated, depending sensitively on the type of the medium and the parameters (pulse duration, spectral width, peak power, etc.) of the pump pulses. Nevertheless, this kind of study is important and meaningful because through them researchers could gain a better understanding about various possible mechanisms and their relative contributions to the observed spectral broadening for a given nonlinear medium.

6.4.3 Applications of coherent continuum generation

The white-light continuum generation and superbroad band coherent emission are highly useful for fundamental research as well as for various applications.[40] The continuum generation or superbroad band coherent emission can be specially employed in the following research areas.

1. Transient absorptive spectral measurement and transmission measurement over a broad (or superbroad) spectral range without the need of time-consuming spectral scanning, in other words, all data covering the entire measured spectral range can be obtained within a short time (even one pulse) period;
2. Coherent Raman gain or inverse-Raman spectroscopic measurements, which need a broad-band (or superbroad-band) coherent emission source to acquire Raman spectral data without the need of spectral scanning;
3. Wavelength-division multiplexing optical telecommunications, which need a coherent light source of a broad spectral band, so that these systems could more efficiently transmit or receive a tremendous amount of signals;
4. Multi-wavelength based holography, optical data storage, and multi-color optical display.

PROBLEMS

1. In a transparent third-order nonlinear medium of $n_0 = 1.45$ and $\chi_e^{(3)} = 2 \times 10^{-21} m^2/V^2$, the divergent angle is $\theta_0 = 0.001$ radian for an input quasi-parallel laser beam; please use Eq. (6.1-3) and Eq. (1.4-1) to calculate the critical (threshold) light intensity for realizing self-trapping.
2. In a transparent third-order nonlinear medium of $n_0 = 1.45$ and $\chi_e^{(3)} = 2 \times 10^{-21} m^2/V^2$, if the input plane Gaussian laser beam has a wavelength of $\lambda = 600$ nm and beam size (radius) of $a_0 = 1$ mm, please use Eq. (6.1-23) to calculate the critical power for realizing self-trapping. If the input laser power is ten times higher than the calculated critical power, calculate the self-focusing distance by using Eq. (6.1-25).
3. For an intense laser pulse propagating in a third-order nonlinear medium of $n_2 < 0$, please sketch a diagram like Fig. 6.8 to show the induced self-phase change and frequency chirp behavior.
4. Please describe the two major mechanisms contributing to the white-light continuum generation.

REFERENCES

1. S. A. Akhmanov, R. V. Khokhlov, and A. P. Sukhorukov, in *Laser Handbook*, Vol. 2, edited by F. T. Arecchi and E. O. Schulz-Dubois (North-Holland, Amsterdam, 1972).
2. O. Svelto, in *Progress in Optics*, Vol. 12, edited by E. Wolf (North-Holland, Amsterdam, 1974), pp. 1–51.

3. Y. R. Shen, *Prog. Quantum Electron.* **4**, 1 (1975).
4. J. H. Marburger, *Prog. Quantum Electron.* **4**, 35 (1975).
5. G. A. Askar'yan, *Sov. Phys. JETP* **15**, 1088 (1962).
6. R. Y. Chiao, E. Garmire, and C. H. Townes, *Phys. Rev. Lett.* **13**, 479 (1964).
7. V. I. Talanov, *Izv. VUZ-Radiofizika* **7**, 564 (1964).
8. S. A. Akhmanov, A. P. Sukhorukov, and R. V. Khokhlov, *Sov. Phys. JETP* **23**, 1025 (1966).
9. P. L. Kelley, *Phys. Rev. Lett.* **15**, 1005 (1965).
10. E. L. Dawes and J. H. Marburger, *Phys. Rev.* **179**, 862 (1969).
11. Y. R. Shen and M. M. T. Loy, *Phys. Rev. A* **3**, 2099 (1971).
12. M. M. T. Loy and Y. R. Shen, *IEEE J. Quantum Electron.* **9**, 409 (1973).
13. Z. Wu, H. Jiang, L. Luo, H. Guo, H. Yang, and Q. Gong, *Opt. Lett.* **27**, 448 (2002).
14. Z. Wu, H. Jiang, Q. Sun, H. Yang, and Q. Gong, *Phys. Rev. A* **68**, 063820 (2003).
15. A. Brodeur and S. L. Chin, *J. Opt. Soc. Am. B* **16**, 637(1999).
16. G. Heck, J. Sloss, and R. J. Levis, *Opt. Commun.* **259**, 216 (2006).
17. C. P. Hauri, R. B. Lopez-Martens, C. I. Blaga, K. D. Schultz, J. Cryan, R. Chirla, P. Colosimo, G. Doumy, A. M. March, C. Roedig, E. Sistrunk, J. Tate, J. Wheeler, L. F. DiMauro, and E. P. Power, *Opt. Lett.* **32**, 868 (2007).
18. G. S. He, A. P. Zhang, Q. Zheng, H.-Y. Qin, P. N. Prasad, S. He, and H. Agren, *IEEE J. Quantum Electron.* **45**, 816 (2009).
19. A. M. Johnson and C. V. Shank, in *The Supercontinuum Laser Sources*, edited by R. R. Alfano (Springer-Verlag, New York, 1989).
20. E. B. Treasy, *Phys. Lett.* **28A**, 34 (1968); E. B. Treasy, *IEEE J. Quantum Electron.* **5**, 454 (1969).
21. R. L. Fork, O. E. Martinez, and J. P. Gordon, *Opt. Lett.* **9**, 150 (1984).
22. F. Shimizu, *Phys. Rev. Lett.* **19**, 1097 (1967).
23. A. C. Cheung, D. M. Rank, R. Y. Chiao, and C. H. Townes, *Phys. Rev. Lett.* **20**, 786 (1968).
24. T. K. Gustafson, J. P. Taran, H. A. Haus, J. R. Lifsitz, and P. L. Kelley, *Phys. Rev.* **177**, 306 (1969).
25. R. Polloni, C. A. Sacchi, and O. Svelto, *Phys. Rev. Lett.* **23**, 690 (1969).
26. N. Bloembergen and P. Lallemand, *Phys. Rev. Lett.* **16**, 81 (1966).
27. C. Lin and R. H. Stolen, *Appl. Phys. Lett.* **28**, 216 (1976).
28. I. Ilev, H. Kumagai, K. Toyoda, and I. Koprinkov, *Appl. Opt.* **35**, 2548 (1996).
29. R. R. Alfano and S. L. Shapiro, *Phys. Rev. Lett.* **24**, 584 (1970).
30. R. R. Alfano and S. L. Shapiro, *Phys. Rev. Lett.* **24**, 592 (1970).
31. N. G. Bondarenko, I. V. Eremina, and V. I. Talanov, *JETP Lett.* **12**, 85 (1970).
32. N. N. Il'ichev, V. V. Korobkin, V. A. Korshunov, A. A. Malyutin, T. G. Okroashvili, and P. P. Pashinin, *JETP Lett.* **15**, 133 (1972).
33. G. E. Busch, R. P. Jones, and P. M. Rentzepis, *Chem. Phys. Lett.* **18**, 178(1973).
34. R. L. Fork. C. V. Shank, C. Hirlimann, R. Yen, and W. J. Tomlinson, *Opt. Lett.* **8**, 1 (1983).
35. P. B. Corkum, P. P. Ho, R. R. Alfano, and J. T. Manassah, *Opt. Lett.* **10**, 624 (1985).
36. I. Golub, *Opt. Lett.* **15**, 305 (1990).
37. G. S. He, G. C. Xu, Y. Cui, and P. N. Prasad, *Appl. Opt.* **32**, 4507(1993).
38. G. S. He and P. N. Prasad, *Phys. Rev. A* **41**, 2687 (1990).
39. G. S. He, R. Burzynski, and P. N. Prasad, *J. Chem. Phys.* **93**, 7647 (1990).
40. R. R. Alfano, ed., *The Supercontinuum Laser Sources* (Springer-Verlag, New York, 1989).

7

Stimulated Scattering of Intense Coherent Light

When a high-intensity laser beam passes through a transparent and homogeneous optical medium, the induced scattering light may manifest the properties of stimulated radiation characterized by threshold requirement, high directionality and high brightness. This is the stimulated scattering of intense coherent light. Discoveries and studies of various types of stimulated scattering effects have provided a much deeper and more comprehensive understanding of interactions between the intense light and matter.[1] On the other hand, devices based on stimulated Raman scattering, stimulated Brillouin scattering, and stimulated spin-flip Raman scattering can be employed as highly efficient tunable frequency shifters of coherent light. Stimulated Kerr scattering can be used to generate a superbroad-band coherent light and to investigate the molecular anisotropy and intermolecular interaction in condensed phase. In addition, the recently developed Stimulated Rayleigh–Bragg scattering and Stimulated Mie scattering can significantly reduce the threshold requirement, and provide a highly efficient approach to generate no-frequency-shift optical phase conjugation waves.

7.1 Introduction to light scattering

7.1.1 Origins of light scattering

When a directional light beam passes through any type of transmission medium except the vacuum, a certain amount of energy of the incident light will convert into scattered light that may propagate along all other directions. This phenomenon is known as light scattering effect.

There are various types of light scattering effects caused by different physical mechanisms. Generally speaking, the light scattering arises from either the inhomogeneity of refractive-index distribution of the medium, or the spatiotemporal fluctuation of electric polarization response of the medium. More specifically, there are three major origins that give rise to most known light scattering phenomena.

7.1.1.1 Macroscopic inhomogeneity of refractive index

Many light-scattering phenomena seen daily are due to the macroscopic inhomogeneity of the refractive index of the medium. For example, if we mix two transparent liquids of different refractive indices without rapid stirring, at the very beginning the mixed liquid becomes highly scattering for an incident light beam because of the random inhomogeneous refractive-index distribution within the mixture. Gradually, the light scattering becomes weaker, and will finally disappear when a macroscopic homogeneous refractive-index distribution is reestablished in the mixture. The other example is the so-called critical opalescence phenomenon. It occurs when the temperature of a transparent medium (liquid or solid) is very close to its phase-transition point (boiling point or melting point). Under these circumstances, the macroscopic spatial distribution of the refractive index is highly inhomogeneous. As a result, strong light scattering can be observed.

7.1.1.2 *Contained impurity scattering centers*

Most light scattering phenomena seen daily are related to the existence of impurity particles (dust, aerosol, droplets, etc.) in the air or water. In those cases, the refractive indices or dielectric constants of the particles are obviously different from that of the pure air or water; therefore, these particles or suspensions play the roles of scattering centers. Under these circumstances, the behavior of light scattering depends on the relative sizes of scattering centers in comparison with the wavelength of the incident light, and also depends on the refractive-index difference between the particles and the host medium.

7.1.1.3 *Molecular motion related fluctuations*

Even for transparent and macroscopically homogeneous media (such as neat water or air), there are still intrinsic light scattering processes associated with the molecular motion. The well-known example is Rayleigh scattering in a pure gas or liquid medium. In this case, because of the random thermal translation and collisions of molecules, there is a fluctuation of the number of molecules (or atoms) within a small volume with a size much shorter than the light wavelength. Since the refractive index of the medium is a function of the number of molecules in a unit volume, there will be refractive-index fluctuation on a small scale. That leads to molecular Rayleigh scattering even in a pure and clear medium. Another example is Raman scattering, which arises from a temporal modulation of molecular polarizability through the intramolecular (vibrational or vibration–rotational) motion. In addition, there is one more example, the so-called Rayleigh-wing scattering, which can be observed only in a Kerr liquid and is related to fluctuations of molecular orientations within a small volume. In all these three examples of light scattering, the scattering centers are the molecules themselves.

7.1.2 Classification of light scattering

There is a variety of light scattering phenomena that can be classified according to different criteria. In the preceding subsection, we have mentioned three major categories of light scattering based on general physical origins. We can also classify them based on the types of scattering centers; thus we can specify the light scattering from impurity particles, from micro-structures, from crystal lattices, and from molecules, atoms, electrons, excitons, phonons, and so on. On the other hand, in some cases where the spatial inhomogeneity of refractive index (or the spatial electric-polarization fluctuation of the medium) is the major origin of light scattering, the frequency of scattered light is the same as that of the incident light. This is the frequency-unshifted light scattering. Whereas when the temporal fluctuation of the polarization response of the medium is the major origin of light scattering, the frequency of the scattered light can be different from that of the incident light, i.e., frequency-shifted light scattering.

Here, in this subsection, we shall mainly consider those light scattering phenomena, of which the spontaneous scattering behavior can be transformed into stimulated scattering behavior.

7.1.2.1 Class I: Light scattering in a pure medium

Even for a pure transparent medium without any impurities and defects, there are still some intrinsic mechanisms that may lead to the following types of light scattering.

7.1.2.1.1 Molecular Rayleigh scattering This scattering effect is caused by the local fluctuations of molecular density in a neat and transparent medium. Assume that the medium is composed of identical molecules (or atoms), and the macroscopic refractive-index distribution of the medium is homogeneous in the thermal-equilibrium condition. Even so, considering a small volume with a size much larger than the molecule dimension but much smaller than the light wavelength, we will find that the number of molecules involved in such a small volume fluctuates randomly because of the molecular translational motion and thermal collisions. This implies a random fluctuation of the local density or refractive index inside the medium on a scale much smaller than the light wavelength. Thus we can observe the so-called Rayleigh scattering in a pure medium. For this process, the frequency of the scattered light is the same as that of the incident light, but the intensity of the scattered light is inversely proportional to the fourth power of the wavelength of the incident light beam, i.e.,

$$I_s(\lambda) \propto I_0(\lambda) \frac{1}{\lambda^4}, \quad \text{(Rayleigh scattering)} \quad (7.1\text{-}1)$$

where $I_0(\lambda)$ is the incident light intensity. This relationship is part of the conclusion given by Rayleigh's theory by assuming that each molecule of the scattering medium is a classical dipole oscillator, and the overall scattering intensity is determined by the sum of each dipole's contribution.[2] From the viewpoint of modern statistical physics, such a classical assumption is inadequate. The true reason of Rayleigh scattering is due to the molecular density fluctuation,[3] and in general, the scattering intensity is essentially determined by the isothermal compressibility, not the molecular density of the medium.[4] Nevertheless, the wavelength dependence law of Eq. (7.1-1) has been proved correct by experimental measurements as well as by the rigorous theoretical treatments.

7.1.2.1.2 Molecular Raman scattering This effect is caused by the temporal modulation of molecular polarizability of the medium.[5] In this case each individual molecule is a scattering center and its polarization response (polarizability) is periodically modulated by intramolecular vibration or rotation. As a result of this periodic modulation, the frequency of the scattered light is shifted with respect to the frequency of the incident light. Raman scattering can be observed in all types of media (gases, liquids, and solids) consisting of molecules; the frequency shift of scattering is determined by

$$\nu_0 - \nu' = \Delta \nu_r, \quad (7.1\text{-}2)$$

where ν_0 and ν' are the frequencies of the incident light and the scattered light, $\Delta \nu_r$ is the frequency of intramolecular motion (vibrational and/or rotational). For the majority of Raman scattering studies, the molecular vibrational motion is the main mechanism, and, consequently, the frequency shift is in the range 10^2–10^3 cm^{-1}.

7.1.2.1.3 Brillouin scattering This is a light scattering effect caused by spontaneous thermal acoustic vibrations inside the medium.[6] In any type of optical media (gases, liquids, and solids), there are always spontaneous thermal vibrations, which can be visualized as a series of elastic acoustic waves with different frequencies and propagation directions. Each of these acoustic waves may induce a periodic spatiotemporal modulation of the density (and hence the refractive-index) of the medium, and gives rise to a frequency-shifted light scattering. The unique feature of this effect is that the frequency shift of the scattering light is dependent on the scattering angle, i.e.,

$$\Delta \nu = \nu_0 - \nu' = 2\nu_0 \frac{n_0 v_a}{c} \sin \frac{\theta}{2}, \quad (7.1\text{-}3)$$

where n_0 is the linear refractive index of the medium, v_a is the speed of the acoustic wave in the medium, and θ is the scattering angle between the incident light beam and the considered scattered light beam. Usually, the spontaneous elastic vibrations of a medium are very weak, and the frequency shift is quite small (≤ 1 cm^{-1}); hence the observation of Brillouin scattering was difficult before the advent of laser.

7.1.2.2 Class II: Mie scattering in a medium containing impurity particles

For most real transparent optical media in gas, liquid or solid phase, they cannot be absolutely pure and always contain more or less impurities, which can be particles, dust, aerosol, droplets, micro-structural defects, etc. Since the dielectric constants or refractive-indices of these impurity particles are different from that of the host medium, the partial energy of an incident directional light beam will spread itself over a broad solid-angle range through diffraction or reflection at the boundary between the impurity particles and the host medium. In that sense these impurity particles play the role of scattering centers. Tyndall was one of the earliest scientists devoted to investigate this kind of light scattering in the 19th century.[7] In the beginning of the 20th century, Mie contributed a complete theory based on the solution of Maxwell's equations to describe the scattering behavior of a single metallic (or dielectric) microsphere (or many of them separated by a distance much larger than the wavelength) suspended in a neat homogeneous medium.[8] For this reason, the light scattering caused by impurity particles suspended in a neat homogeneous medium can be termed Tyndall–Mie scattering or simply *Mie scattering*, which could be much stronger than Rayleigh scattering from the host medium, depending on the specific particle's parameters (such as density, size, shape, and refractive index). Although there is no frequency shift between an incident monochromatic light and the corresponding Mie scattering light, there is a predicable wavelength dependence of the Mie scattering cross-section of an individual particle, which is numerically determined by the parameter of $q = 2\pi r n_s/\lambda$, where r is the radius of the particle, λ is the incident light wavelength in vacuum, and n_s is the refractive index of the surrounding medium.[9]

The intensity of Mie scattering can be written as

$$I_s(\lambda) \propto I_0(\lambda) N_p \sigma_M(\lambda), \quad (7.1\text{-}4)$$

where N_p is the density of impurity particles and σ_M is the Mie scattering cross-section for each particle. Furthermore, there is also a phenomenological expression for wavelength dependence of σ_M,

$$\sigma_M(\lambda) = b\lambda^\eta, \quad \text{(Mie scattering)} \quad (7.1\text{-}5)$$

where b is an arbitrary proportional constant and η is a phenomenologically introduced power factor. As an example, for polystyrene microspheres suspended in water, the numerically calculated curve of σ_M as a function of $q^{-1} = \lambda/(2\pi r n_s)$ is plotted in Fig. 7.1(a), while the curve of power

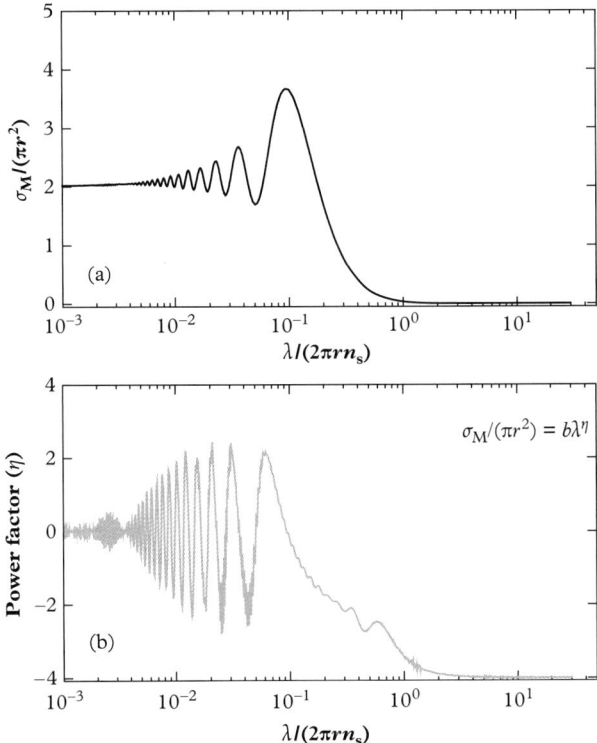

Figure 7.1 (a) Calculated Mie scattering cross-section of a polystyrene microsphere (of radius r) suspended in water (of refractive index n_s) as a function of $q^{-1} = \lambda/(2\pi r n_s)$; (b) Calculated curve of power factor η vs. q^{-1}. (After He et al.[10] Copyright 2009, AIP Publishing LLC).

factor η versus q^{-1} is plotted in Fig. 7.1(b), from which one can see that when $q \ll 1$ we have $\eta \Rightarrow -4$, when $q \gg 1$ we have $\eta \Rightarrow 0$. Between these two extreme situations, there is no simple analytical relationship; the η value varies between -4 and 0 in an oscillating way.[9,10]

Finally, it is worth noticing that even for those nanoparticles whose sizes approximately fulfil the condition of $q \ll 1$ and the λ^{-4} law, their other scattering properties described by Mie theory are generally different from molecular Rayleigh scattering of a neat transparent medium.[11]

7.1.3 Differences between spontaneous and stimulated scattering

In 1962, when researchers investigated the spectral structure of a ruby laser using a nitrobenzene liquid cell as the Q-switched element, they observed an unexpected ~767-nm lasing line added to the previously known 694.3-nm lasing line of ruby.[12] This newly observed stimulated emission line could not be explained based on the fluorescence property of the lasing medium (ruby). Shortly after that observation, the researchers found that the frequency difference between these two stimulated emission lines was just equal to the frequency of the strong Raman vibrational mode of nitrobenzene liquid. The researchers then realized that they had actually observed a stimulated Raman scattering from the nitrobenzene-filled liquid cell pumped by the intense 694.3-nm ruby lasing line.[13] Since then, researchers have used intense laser beams to excite various optical media and successively discovered other stimulated scattering effects, including (but not limited to) stimulated Brillouin scattering, stimulated Rayleigh scattering, and even, as recently reported, stimulated Mie scattering.

Generally speaking, the differences between spontaneous scattering and stimulated scattering in a scattering medium are quite similar to that between spontaneous emission and stimulated emission in a lasing medium. In experiments, one can easily distinguish stimulated scattering from the corresponding spontaneous scattering based on the following features.

1. *Threshold of stimulated scattering*: If the intensity of an input pump light beam is lower than a certain threshold, there is only an extremely weak spontaneous scattering that needs to be detected by a highly sensitive detector system. However, if the intensity of the pump light beam is higher than a certain threshold, a very strong and highly directional coherent scattering signal can be readily observed by human eyes on a paper screen or an IR sensitive paper card.

2. *Spatial structure of stimulated scattering*: The spontaneous scattering in a scattering medium spreads over all 4π solid angle, though there may be some angle dependence of the scattering intensity distribution. For most of stimulated scattering observation, in order to reach a higher local intensity the incident pump light is focused into the scattering medium. In this case the stimulated scattering can only be generated along the forward direction, the backward direction or both of them, depending on the specific experimental conditions. Moreover, the beam size and divergence angle of the stimulated scattering are nearly the same as those of the incident pump beam.

3. *Mechanism comparisons between the stimulated and spontaneous scattering*: Although all types of stimulated scattering processes originate from the corresponding spontaneous scattering, the basic mechanism for stimulated scattering is not necessarily the same as that for spontaneous scattering. For example, the spontaneous Brillouin scattering is caused by spontaneous thermal acoustic waves in the medium, but the stimulated Brillouin scattering is actually based on the interaction between the incident intense coherent light and the electrostriction-induced acoustic waves. Also, the spontaneous Rayleigh-wing scattering is caused by spontaneous fluctuations of molecular orientations via thermal collisions, whereas the stimulated Kerr scattering is related to the intense light-induced molecular reorientation in a Kerr liquid. For stimulated Rayleigh–Bragg scattering as well as stimulated Mie–Bragg scattering, the formation of a standing-wave Bragg grating inside the medium plays the key role in providing a positive feedback.

It should be also noted that stimulated scattering and stimulated emission (lasing) have the same physical properties, such as high brightness, high photon degeneracy, high directionality, and a controllable spectral structure. The only difference between these two stimulated processes is that the generation of stimulated emission requires population inversion in a lasing medium, whereas the generation of stimulated scattering does not require such population inversion in a scattering medium.

7.2 Theory of stimulated Raman scattering (SRS)

7.2.1 Quantum-electrodynamical description of Raman scattering

We assume here that the scattering medium is composed of identical and independent molecules, each of them possesses a discrete eigenenergy level structure, and the frequency of the incident excitation light field is not equal or close to any absorption frequencies of molecular transitions among their different energy levels. In this sense the molecules cannot absorb the incident photons; however the former can scatter the latter through a two-step quantum transition process. As shown in Fig. 7.2, the molecular ground state is denoted by energy level a, a lower (vibrational) excited state is denoted by energy level c, and the spacing between these two real energy levels is $h\Delta v_r$.

Figure 7.2(a) shows the two-step elementary process of the Stokes-shifted Raman scattering. In the first step, there is the annihilation of an incident photon of frequency v_0, at the same time a molecule in state a is excited to an intermediate state; in the second step, this molecule returns to Raman excited state c, while there is the creation of a scattered photon at the frequency $v_0 - \Delta v_r$. In this case the observed scattering spectral line is shifted to the low-frequency side. In other words, there is a net energy transfer from the incident photon to the molecule by the amount of $h\Delta v_r$, here Δv_r is termed Raman mode frequency of the medium.

Figure 7.2(b) shows the elementary process of anti-Stokes-shifted Raman scattering. In the first step, there is the annihilation of an incident photon of frequency v_0, and a molecule in excited state c is further excited to an intermediate state; in the second step, this molecule returns to

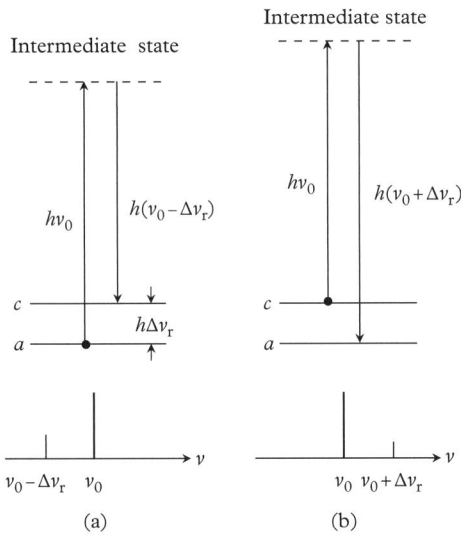

Figure 7.2 *Quantum-transition diagrams of (a) Stokes-shifted Raman scattering and (b) anti-Stokes-shifted Raman scattering. The corresponding scattering spectra are also shown.*

ground state *a*, while there is the creation of a scattered photon of frequency $v_0 + \Delta v_r$. In this case the observed scattering spectral line is shifted to the high-frequency side.

The intermediate state, schematically depicted by a dashed energy level line in Fig. 7.2, is the key concept uniquely introduced in quantum electrodynamics. Also adequate for many other nonlinear optical effects (such as multi-photon absorption), quantum electrodynamics can give a rigorous and concise description of the elementary Raman scattering process.[14] In this scheme, one must treat optical fields and the medium as a combined and quantized system. For this purpose, the light fields are recognized as a photon ensemble with a certain photon-population distribution among their different photon states or electromagnetic modes. Assuming the incident optical field is a quasi-parallel and quasi-monochromatic laser beam, the average photon number $\bar{n}$ (photon degeneracy) in a given mode of this beam can be estimated using Eq. (1.3-10). Referring to Fig. 7.2, the elementary two-step process of Raman scattering may take place in two possible ways.

In the first way (way I):

1. One photon in an arbitrary mode of the incident field is annihilated, $\bar{n}(v_0) \to \bar{n}(v_0) - 1$, while one molecule makes a transition from its initial real energy level to an intermediate state (denoted by a dashed energy level line);
2. One photon is created in a mode of the scattering field, $\bar{n}(v') \to \bar{n}(v') + 1$, while the excited molecule returns to another real energy level.

In the second way (way II):

1. One photon is created in a mode of the scattering field, $\bar{n}(v') \to \bar{n}(v') + 1$, while one molecule makes a transition from its initial real energy level to an intermediate state;
2. One photon in an arbitrary mode of the incident field is annihilated, $\bar{n}(v_0) \to \bar{n}(v_0) - 1$, while this molecule returns to another real energy level.

In both of these two equivalent ways, between the two elementary steps the molecule is in a special intermediate state, which means that this molecule may stay in its all possible eigenstates (except states *a* and *c*). In such an intermediate state the energy location range of the excited molecule is entirely uncertain, therefore according to uncertainty principle, the staying time of the molecule in this intermediate state is infinitely short. In other words, the two steps of the elementary scattering process occur simultaneously, thus we should say that the molecular Raman scattering is actually a single instantaneous elementary process.

If we denote the Hamiltonian operator of the interaction between the optical field and a molecule by H', the overall wave function of the system combining the optical field and the molecular system is Ψ satisfying the following wave equation:

$$i\hbar \frac{\partial \Psi}{\partial t} = (H_0 + H')\Psi. \tag{7.2-1}$$

Here, H_0 is the Hamiltonian operator of the combined system if the interaction between the field and the molecule is neglected. In this case the unperturbed eigenfunction, Φ_n, which is determined by H_0, can be expressed as the product of the eigenfunction of optical field and eigenfunction of a molecule. If we consider the existence of the interaction Hamiltonian (H'), the wave function Ψ can be expanded by the unperturbed eigenfunction Φ_n in the following way:

$$\Psi(t) = \sum_n A_n(t)\Phi_n \exp(-iE_n t/\hbar). \tag{7.2-2}$$

Here, the expansion coefficient $A_n(t)$ represents the amplitude of probability to find the perturbed system in its unperturbed state Φ_n at time t with a given eigenvalue E_n. The temporal variation of $A_n(t)$ is governed by the following equation:

$$\dot{A}_n(t) = -\frac{i}{\hbar} \sum_m H'_{nm} A_m(t) \exp[i(E_n - E_m)t/\hbar], \tag{7.2-3}$$

where

$$H'_{nm} = \int \Phi_n^* H' \Phi_m d\tau \tag{7.2-4}$$

is the matrix element of the interaction Hamiltonian determined by unperturbed eigenfunctions with different eigenvalues.

For an elementary Raman scattering process, the combined system at the beginning is described by its unperturbed quantum number of a', in the end is described by another unperturbed quantum number of c'. Between these two-step the system is described by an intermediate state where all unperturbed states (except states a' and c') are involved, one of them is denoted by b'. Based on Eq. (7.2-3), the probability amplitudes of these eigenstates satisfy the following equations:

$$\left. \begin{aligned} \dot{A}_{a'}(t) &= -\frac{i}{\hbar} \sum_{b'} H'_{a'b'} A_{b'}(t) \exp[i(E_{a'} - E_{b'})t/\hbar] \\ \dot{A}_{b'}(t) &= -\frac{i}{\hbar} H'_{b'a'} A_{a'}(t) \exp[i(E_{b'} - E_{a'})t/\hbar] \\ \dot{A}_{c'}(t) &= -\frac{i}{\hbar} \sum_{b'} H'_{c'b'} A_{b'}(t) \exp[i(E_{c'} - E_{b'})t/\hbar] \end{aligned} \right\}. \tag{7.2-5}$$

The initial conditions of these equations are

$$\left. \begin{aligned} A_{a'}(0) &= 1 \\ A_{c'}(0) &= 0 \end{aligned} \right\}. \tag{7.2-6}$$

To solve Eq. (7.2-5), we can examine the process in such a short time period that the probability amplitude $A_{a'}(t)$ does not change so much, i.e., $A_{a'}(t) \approx A_{a'}(0) \approx 1$. Thus we can first find the solution for $A_{b'}(t)$ as

$$A_{b'}(t) = \frac{H'_{b'a'}}{E_{a'} - E_{b'}} \exp\left[i(E_{b'} - E_{a'})t/\hbar\right]. \tag{7.2-7}$$

Substituting the above solution into the third equation of Eq. (7.2-5) leads to

$$\dot{A}_{c'}(t) = -\frac{i}{\hbar} \sum_{b'} \frac{H'_{c'b'} H'_{b'a'}}{E_{a'} - E_{b'}} \exp\left[i(E_{c'} - E_{a'})t/\hbar\right]. \tag{7.2-8}$$

Here, the summation is over all possible eigenstates. Completing the integration with respect to t and noticing the second initial condition indicated by Eq. (7.2-6), we obtain

$$A_{c'}(t) = \frac{K_{c'a'}}{E_{a'} - E_{c'}} \left\{ \exp\left[i(E_{c'} - E_{a'})t/\hbar\right] - 1 \right\}, \qquad (7.2\text{-}9)$$

where

$$K_{c'a'} = \sum_{b'} \frac{H'_{c'b'} H'_{b'a'}}{E_{a'} - E_{b'}}. \qquad (7.2\text{-}10)$$

Furthermore, the probability of finding the system in the state c' at time t can be expressed as

$$|A_{c'}(t)|^2 = 2|K_{c'a'}|^2 \frac{[1 - \cos(E_{c'} - E_{a'})t/\hbar]}{(E_{a'} - E_{c'})^2}. \qquad (7.2\text{-}11)$$

It has been assumed that the time period we are considering here is so short that the change of $A_{d'}(t)$ is very small. However, on the other hand, this time interval is still much longer than the period of the optical wave, i.e., $t \gg \hbar/E_{c'}$ and $t \gg \hbar/E_{a'}$. Under this condition, in Eq. (7.2-11) we can let $t \to \infty$. Noticing that

$$\lim_{t \to \infty} \frac{[1 - \cos(E_{c'} - E_{a'})t/\hbar]}{(E_{a'} - E_{c'})^2} = \frac{\pi}{\hbar} t \delta(E_{c'} - E_{a'}),$$

we obtain

$$w_{c'a'} = |A_{c'}(t)|^2 / t = |K_{c'a'}|^2 \frac{2\pi}{\hbar} \delta(E_{c'} - E_{a'}). \qquad (7.2\text{-}12)$$

The physical meaning of $w_{c'a'}$ is the probability for a molecule to scatter a photon into an arbitrary mode of the scattering field within a unit time interval. As shown in Fig. 7.2(a), the molecule is initially in its eigenstate a (with eigenenergy E_a), and after scattering it returns to another eigenstate c (with eigenenergy E_c), owing to energy conservation of the combined system we have

$$E_{c'} - E_{a'} = (E_c + h\nu') - (E_a + h\nu_0) = 0, \qquad (7.2\text{-}13)$$

where ν' is the frequency of the scattered light. If the interaction volume between the incident light and the medium is V, within a unit frequency interval the total mode numbers of the scattering light field are determined by

$$g(\nu') = \frac{8\pi}{c^3} \nu'^2 V.$$

Then the overall probability of a molecule to scatter a photon into one of all possible modes of the scattering light field is

$$w(t) = \int_0^\infty w_{c'a'} g(\nu') d\nu' = \frac{8\pi \nu'^2 V}{\hbar^2 c^3} |K_{c'a'}|^2 = \frac{8\pi \nu'^2 V}{\hbar^2 c^3} \left| \sum_{b'} \frac{H'_{c'b'} H'_{b'a'}}{E_{a'} - E_{b'}} \right|^2. \qquad (7.2\text{-}14)$$

Raman scattering always exhibits a limited spectral width; however, as a δ-function is involved in the integral of Eq. (7.2-12), the integral limit can be extended to infinity. According to Eq. (7.2-13) the Raman scattering frequency follows the following relation:

$$h\nu' = h\nu_0 - (E_c - E_a) = h(\nu_0 - \Delta\nu_r), \qquad (7.2\text{-}15)$$

which corresponds to Stokes Raman scattering.

7.2.2 Probabilities of spontaneous and stimulated Raman scattering

From Eq. (7.2-14) one can see that to calculate the probability of an elementary Raman scattering process, the key issue is to find the explicit expressions for $H'_{c'b'}$ and $H'_{b'a'}$, which are the matrix elements of the interaction Hamiltonian H' and determined by unperturbed eigenfunctions of the combined system. According to quantum electrodynamical theory, these two matrix elements for the two possible ways of an elementary Raman process can be finally expressed as[14]

$$\left.\begin{array}{l} H'_{b'a'} = -\dfrac{e}{\mu}\sqrt{\dfrac{\hbar c^2}{4\pi\varepsilon_0 V \nu_0}}\sqrt{\bar{n}(\nu_0)}\displaystyle\int \varphi_b^* p_0^e e^{i(\mathbf{k}_0\cdot\mathbf{r})}\varphi_a d\tau \\[2ex] H'_{c'b'} = -\dfrac{e}{\mu}\sqrt{\dfrac{\hbar c^2}{4\pi\varepsilon_0 V \nu'}}\sqrt{\bar{n}(\nu')+1}\displaystyle\int \varphi_c^* p^e e^{-i(\mathbf{k}'\cdot\mathbf{r})}\varphi_b d\tau \end{array}\right\}, \quad \text{(Way I)} \qquad (7.2\text{-}16)$$

and

$$\left.\begin{array}{l} H'_{b'a'} = -\dfrac{e}{\mu}\sqrt{\dfrac{\hbar c^2}{4\pi\varepsilon_0 V \nu'}}\sqrt{\bar{n}(\nu')+1}\displaystyle\int \varphi_b^* p^e e^{-i(\mathbf{k}'\cdot\mathbf{r})}\varphi_a d\tau \\[2ex] H'_{c'b'} = -\dfrac{e}{\mu}\sqrt{\dfrac{\hbar c^2}{4\pi\varepsilon_0 V \nu_0}}\sqrt{\bar{n}(\nu_0)}\displaystyle\int \varphi_c^* p_0^e e^{i(\mathbf{k}_0\cdot\mathbf{r})}\varphi_b d\tau \end{array}\right\}. \quad \text{(Way II)} \qquad (7.2\text{-}17)$$

Here, e is the electron charge and $\mu = m_0 c^2$, where m_0 is the electron rest mass; $\bar{n}(\nu_0)$ and $\bar{n}(\nu')$ are the photon degeneracy values of the incident field and scattering field, respectively; p_0^e and p^e are components of the electron momentum operator (multiplied by the speed of light in vacuum) in the polarization directions of incident light and scattering light, respectively; $\mathbf{k}_0$ and $\mathbf{k}'$ are the wavevectors of the incident photon and the scattered photon, and φ_a, φ_b, and φ_c are the molecular eigenfunctions without applied photon field. In its final calculation of Eq. (7.2-14), the summation over quantum number b' of the combined system will be converted to the summation over quantum number b of the molecular system, where b denotes one of any molecular eigenstates (except a and c).

Substituting Eqs. (7.2-16) and (7.2-17) into Eq. (7.2-14) and summing up, these two contributions lead to

$$w(t) = \dfrac{8\pi}{V} c\sigma_r [\bar{n}(\nu')+1]\bar{n}(\nu_0), \qquad (7.2\text{-}18)$$

where σ_r is a coefficient having a dimension of area, which is called the differential cross-section of Raman scattering, determined by

$$\sigma_{\mathrm{r}} = \frac{e^4 v'}{(4\pi\varepsilon_0)^2 \mu^4 v_0} \left\{ \sum_b \left[\frac{(p^e)_{cb}(p_0^e)_{ba}}{h(v_{ba} - v_0)} + \frac{(p_0^e)_{cb}(p^e)_{ba}}{h(v_{ba} + v')} \right] \right\}^2. \tag{7.2-19}$$

Here, v_{ba} is the molecular transition frequency from eigenstate a to eigenstate b. In obtaining Eq. (7.2-18) we have used the electric-dipole approximation so that each exponential factor in Eqs. (7.2-16) and (7.2-17) can be replaced by 1 (see Section 18.1.2).

Furthermore, we have the following relation for matrix elements of the electron momentum operator (multiplied by the speed of light) determined by molecular eigenfunctions:

$$(p_0^e)_{ba} = im_0 2\pi v_{ba}(p_0)_{ba} \frac{c}{e}, \tag{7.2-20}$$

where $(p_0)_{ba} = (er_0)_{ba}$ is the matrix element of the component of the molecular electric-dipole moment operator in the polarization direction of the incident light. If assume that $v_{ba} \approx v_0$, $v_{bc} \approx v'$, the differential cross-section of Raman scattering can be simplified as

$$\sigma_{\mathrm{r}} = \left(\frac{2\pi}{c}\right)^4 \frac{v_0 v'^3}{(4\pi\varepsilon_0)^2 \hbar^2} \left\{ \sum_b \left[\frac{(p_0)_{ab}(p)_{bc}}{v_{ba} - v_0} + \frac{(p)_{ab}(p_0)_{bc}}{v_{ba} + v'} \right] \right\}^2. \tag{7.2-21}$$

Here, $(p)_{cb}$ is the matrix element of the component of the molecular electric-dipole moment operator in the polarization direction of the scattered light. In addition, here we also use the identity of $(p_0)_{ba} = (p_0)_{ab}$, $(p)_{cb} = (p)_{bc}$, etc.

For a given scattering medium, the value of σ_{r} can be theoretically calculated if related molecular eigenfunctions are known. However, it can be more easily determined by experimental measurement. Assume that the molecule number in eigenstate a within a unit volume is denoted by N_a, then according to Eq. (7.2-18), the overall probability for $N_a V$ molecules to scatter one photon from $\bar{n}(v_0)$ incident photons into an arbitrary scattering mode can be written as

$$W(t) = 8\pi N_a c \sigma_{\mathrm{r}} \left[\bar{n}(v_0) + \bar{n}(v_0)\bar{n}(v') \right]. \tag{7.2-22}$$

If the incident beam is produced by an ordinary light source, i.e., $\bar{n}(v_0) \ll 1$ (see Section 1.3), then $\bar{n}(v') \ll 1$ also. Based on this consideration, the second term of Eq. (7.2-22) can be neglected, and we have

$$W(t) = 8\pi N_a c \sigma_{\mathrm{r}} \bar{n}(v_0). \tag{7.2-23}$$

This corresponds to the situation of spontaneous Raman scattering, where the coefficient 8π implies the scattering field occupies all the 4π solid angle and contains two possible independent polarization states. Because $\bar{n}(v_0) \ll 1$ and the value of σ_{r} is small, the ordinary spontaneous Raman scattering signal is very weak. Therefore, a highly sensitive detector and longer recording time are needed for ordinary Raman measurements.

In contrast, if the incident light is a high-intensity laser beam, its photon degeneracy can be $\bar{n}(v_0) \geq 10^{10}$, which makes it possible that $\bar{n}(v') \gg 1$. In this case the second term in Eq. (7.2-22) may become much larger than the first term, and we can rewrite Eq. (7.2-22) as

$$W(t) = 2\Omega N_a c \sigma_{\mathrm{r}} \bar{n}(v_0)\bar{n}(v'). \tag{7.2-24}$$

This corresponds to the situation of stimulated Raman scattering (SRS), where Ω is the solid angle occupied by the stimulated scattering light. Comparing Eq. (7.2-24) to Eq. (7.2-23) one finds that as $\bar{n}(\nu_0)\bar{n}(\nu') \gg 1$, the probability of stimulated scattering can be dramatically greater than that of spontaneous scattering.[15]

We can also see from Eq. (7.2-24) that, the probability for molecules to scatter a photon into a given mode is proportional to the number of photons occupying this mode already. This situation is exactly the same as that for stimulated emission in a lasing medium. In the later case the probability of emitting a photon into a given mode is proportional to the number of photons already existing in that mode. In that sense, the generation and amplification of SRS in a scattering medium are the same as those of stimulated emission in a lasing medium. Specifically, if the gain of the stimulated scattering (or emission) is considerably greater than the losses due to various attenuation mechanisms, a strong one-pass (without a cavity) or multi-pass (with a cavity) stimulated amplification (photon avalanche) can be achieved.

7.2.3 Gain coefficient and threshold condition

In this subsection we shall give an explicit gain expression of SRS. The quantity $W(t)$ given by Eq. (7.2-24) represents the probability of $N_a V$ molecules to scatter one photon into the solid angle Ω per second under the action of $\bar{n}(\nu_0)$ photons in a given mode of the incident light. On the other hand, within a unit time interval there are many pump photon modes passing through the medium, but the total photon numbers per second are simply determined by $P_0/h\nu_0$, where P_0 is the power of the incident pump beam. After taking account of that fact, Eq. (7.2-24) can be modified as

$$\frac{d}{dt}\left[\frac{P'}{h\nu'}\right] = 2\Omega N_a c\sigma_r \frac{P_0}{h\nu_0}\bar{n}(\nu'). \tag{7.2-25}$$

Here, P' is the power of the stimulated scattering beam. Further assuming that the beam sections of the incident light and stimulated scattering light are nearly the same, then Eq. (7.2-25) becomes

$$\frac{dI'}{dt} = 2\Omega N_a c\sigma_r I_0 \bar{n}(\nu')\frac{\nu'}{\nu_0}, \tag{7.2-26}$$

where I' and I_0 are the overall intensities of the stimulated scattering beam and incident light beam, respectively. If $I(\nu')$ and $\Delta\bar{\nu}'$ are the monochromatic intensity and spectral width of the stimulated scattering beam, the overall scattering intensity is $I' \approx I(\nu')\Delta\bar{\nu}'$. Moreover, remembering that the monochromatic intensity is related to the photon degeneracy by (cf. Eqs. (1.3-2) and (1.3-10))

$$\bar{n}(\nu') = \frac{I(\nu')}{2h\nu'\Omega/\lambda'^2},$$

then Eq. (7.2-26) can be further transformed to

$$\frac{dI(\nu')}{dt} = N_a c\sigma_r \frac{\lambda'^2}{h\nu_0 \Delta\bar{\nu}'} I_0 \cdot I(\nu'). \tag{7.2-27}$$

For simplicity, we consider here only the steady-state stimulated scattering process. In that case all related physical quantities are not explicit functions of time t. Therefore, in Eq. (7.2-27) we can let $dt = dz/c$ and obtain

$$\frac{dI(v',z)}{dz} = N_a \sigma_r \frac{\lambda'^2}{h\nu_0 \Delta \bar{v}'} I_0 I(v',z). \tag{7.2-28}$$

Here, z is the spatial variable along the propagation direction of the stimulated scattering beam. The solution of the above equation is

$$I(v',z) = I(v',0) e^{Gz}, \tag{7.2-29}$$

where the exponential gain coefficient G is determined by

$$G(v') = N_a \sigma_r \frac{\lambda'^2}{h\nu_0 \Delta \bar{v}'} I_0. \tag{7.2-30}$$

To obtain Eq. (7.2-29) we have employed the small-signal approximation, i.e., the depletion of the input pump beam within the medium is negligible and I_0 can be approximately recognized as a constant. From Eq. (7.3-30) we find that the exponential gain coefficient G of SRS is proportional to the molecule density N_a in the initial level, the differential cross-section σ_r, and the overall pump intensity I_0; whereas G is inversely proportional to the spectral width $\Delta \bar{v}'$ of the SRS.

Until now, we have not considered the specific spectral structure of stimulated scattering. For simplicity, we can consider two extreme situations. In the first, we assume the spectral width of the pump light is much smaller than that of the spontaneous Raman scattering (most experimental cases belong to this situation). Under this condition, the spectral distribution of stimulated scattering is mainly determined by the spectral structure of spontaneous scattering, and Eq. (7.2-30) can be modified to

$$G(v') = N_a \sigma_r \frac{\lambda'^2}{h\nu_0} I_0 S(v'). \tag{7.2-31}$$

Here, $S(v')$ is a normalized lineshape function, which has the same dimension as $(1/\Delta \bar{v}')$ does. For example, if the spontaneous Raman line exhibits a Lorentzian profile, $S(v')$ can be expressed as

$$S(v') = \frac{2\Gamma}{4\pi^2 [\Delta v_r - (v_0 - v')]^2 + \Gamma^2}, \tag{7.2-32}$$

where $2\Gamma/(2\pi)$ is the full width at the half-maximum position of the Raman line on the frequency scale.

In the second extreme situation, we assume that the spectral width of the pump laser beam is much larger than that the spontaneous Raman spectral width of the medium. In this condition, the spectral distribution of stimulated scattering will be mainly determined by the spectral structure of the input pump light, and Eq. (7.2-30) should be modified to

$$G'(v') = N_a \sigma_r \frac{\lambda'^2}{h\nu_0} I_0 S'(v'), \tag{7.2-33}$$

where $S'(v')$ is a normalized spectral distribution function identical to that of the pump light. If it is further assumed that the pump light also exhibits a Lorentzian spectral profile, we then have

$$S'(\nu') = \frac{2\Delta\nu_0}{4\pi^2[\Delta\nu_r - (\nu_0 - \nu')]^2 + (\Delta\nu_0)^2}, \quad (7.2\text{-}34)$$

where $(2\Delta\nu_0)/2\pi$ is the full spectral width of the pump light. According to Eqs. (7.2-31) and (7.2-33), the ratio of peak gain at $\nu_0 - \nu' = \Delta\nu_r$ for these two extreme situations is determined by

$$\frac{G'_{max}}{G_{max}} = \frac{S'(\nu' = \nu_0 - \Delta\nu_r)}{S(\nu' = \nu_0 - \Delta\nu_r)} = \frac{\Gamma}{\Delta\nu_0} \ll 1. \quad (7.2\text{-}35)$$

This relation implies that under the same overall pump intensity (I_0) condition, it is much easier to observe SRS by using a pump laser beam with a spectral linewidth narrower than that of the spontaneous Raman line.

So far in this subsection we have assumed that the initial level of scattering molecules is the ground state a, and the final level is an excited level c, as shown in Fig. 7.2(a). That corresponds to the Stokes-shifted SRS. If the molecule population in level c is not negligible, through analogous procedures we can obtain the expression of exponential gain coefficient for anti-Stokes-shifted SRS:

$$G_{as}(\nu') = N_c \sigma_r \frac{\lambda'^2}{h\nu_0 \Delta\bar{\nu}'} I_0 = \frac{N_c}{N_a} G(\nu'), \quad (7.2\text{-}36)$$

where N_c is the molecule density in the level c, and G is given by Eq. (7.2-30). Usually we have $N_a \gg N_c$, and therefore $G \gg G_{as}$. This implies that the Stokes stimulated scattering is more easily observed than the anti-Stokes stimulated scattering in general. Nevertheless, in some cases, when the population in excited state c is considerably increased through Stokes stimulated scattering processes, the real anti-Stokes SRS signal can also be generated.

As mentioned early, one of the most essential features of stimulated scattering is the threshold requirement. This feature can be explained by the following two considerations. First, there may be some mechanisms (such as Mie scattering and absorption from impurities) that might bring an attenuation influence on the initial stimulated scattering signal, therefore the gain of simulated scattering must be high enough to overcome various losses. Second, the probability for molecules to scatter a photon into a given mode is proportional to the number of photons existing in that mode; hence a photon accumulation process in a given scattering mode is needed for the onset of initial stimulated scattering signals.

With adopting an optical resonant cavity, the threshold condition of stimulated scattering can be written as

$$e^{(G-\alpha)L}R > 1, \quad (7.2\text{-}37)$$

where α and L are the overall attenuation coefficient and gain length of the Raman medium, and R is the reflectivity of mirrors of the cavity. Without using an optical cavity, the threshold requirement for generation of a single-pass stimulated scattering is

$$e^{(G-\alpha)L} \gg 1. \quad (7.2\text{-}38)$$

From Eqs. (7.2-37) and (7.2-38), we can see that the threshold requirement can be lowered by using a cavity. For this reason, an optical cavity is mainly suitable for low-gain stimulated scattering systems, whereas single-pass scheme is most suitable for high-gain stimulated scattering systems.

7.3 Experimental studies of SRS

7.3.1 Raman media and experimental setups

To date, more than a thousand research articles on SRS experimental studies have been reported. Depending on pump laser wavelengths and Raman transition frequencies of the media, the spectral range covered by SRS can extent from UV to mid-IR. In some special cases, the SRS output can be tunable. In this sense SRS is a highly efficient approach to generate frequency-shifted intense coherent radiation. The conversion efficiency from the pump laser intensity to SRS intensity can be up to 30–50%.

A medium used for SRS should fulfill the following basic requirements: (i) it should be transparent at the wavelengths of both pump beam and SRS beam; (ii) it should possess a larger differential cross-section of Raman scattering; (iii) it should withstand a higher intensity of the input pump laser beam.

The commonly used media for SRS can be summarized as follows.

1. *Liquids*: Most fundamental studies of SRS have been done in various transparent organic solvents, such as benzene (C_6H_6), carbon disulfide (CS_2), nitrobenzene, carbon tetrachloride (CCl_4), acetone, dimethyl sulfoxide (DMSO), as well as in liquid gases, etc. These liquids generally exhibit a relatively large Raman cross-section for their vibrational Raman modes and can withstand higher pump intensity without breakdown or bubbling.

2. *Solids*: Optical crystals (such as $CaCO_3$, $Ba(NO_3)_2$, $Pb(NO_3)_2$, $NaNO_3$, KNO_3, etc.), optical fibers, optical glasses, and optical organic solid materials are often used for experimental studies of SRS working at their vibrational Raman modes.

3. *Gases*: Highly efficient SRS can be achieved in molecular gas systems (such as H_2, D_2, N_2, CH_4, SF_6, etc.) working at their vibrational, vibration–rotational, or pure rotational Raman modes. A higher pressure is often needed to get a higher exponential gain because the latter is proportional to the molecular density. Advantages of using gas systems as Raman media are their low losses at both pump and SRS wavelengths, long gain length, and high breakdown threshold. In addition, highly efficient SRS based on electronic transitions can also be achieved in metal vapor systems.

4. *Semiconductors*: A special SRS, the so-called spin-flip SRS, can be achieved in some semiconductor crystals placed in an external magnetic field. In this case, the Raman transition takes place between two sub-levels of the Landau level of conduction electrons in a semiconductor crystal. The unique feature of this type of SRS is that the frequency shift of the stimulated scattering can be continuously tuned by changing the intensity of the applied magnetic field.

Frequency shift (Δv_r) values (in units of cm^{-1}) of SRS are determined by the corresponding Raman-transition frequencies of scattering media. In general, we have $\Delta v_r \approx 10\text{--}10^2$ cm^{-1} for pure rotational transition, $\Delta v_r \approx 10^2\text{--}10^3$ cm^{-1} for vibrational transition, $\Delta v_r \approx 8 \times 10^3\text{--}3 \times 10^4$ cm^{-1} for electronic transition, and $\Delta v_r \approx 0\text{--}10^2$ cm^{-1} for spin-flip transition, respectively.

As shown in Fig. 7.3, there are three types of experimental devices that can be used to generate SRS. For the first type of setup shown in Fig. 7.3(a), the scattering medium is placed inside the cavity of a laser oscillator that is the pump source. The feature of this configuration is that the intracavity intensity of the pump laser beam can be much higher than that of the output laser beam, and a multi-pass exponential gain of SRS can be obtained. This configuration is mainly

152   Stimulated Scattering of Intense Coherent Light

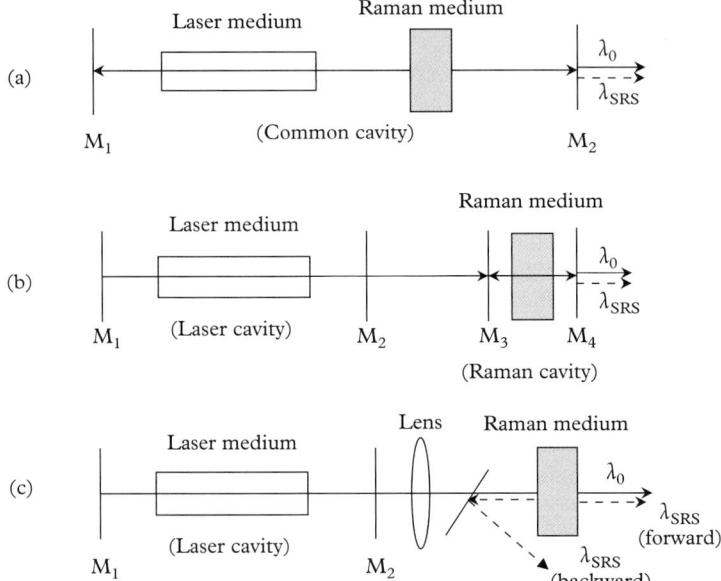

Figure 7.3 *Schematic experimental setups of SRS: (a) Raman medium is placed inside a laser cavity, (b) Raman medium is placed in an external cavity, and (c) Raman medium is excited by a focused single-pass laser beam.*

suitable for those experiments when a cw laser or low-peak power pulses laser has to be used as a pump source, or the scattering medium possesses a relatively small Raman cross-section.

For the second type of experimental setup shown in Fig. 7.3(b), the scattering medium is placed in a separate external cavity. The optical axis of the external cavity can be either collinear with that of the laser cavity, or has a certain crossing angle with the latter. In both cases, there still is a multi-pass stimulated scattering amplification within the second cavity. The advantages of using this configuration are the lowered threshold for SRS generation and easy control of the spectral and spatial structures of stimulated scattering.

For the third type of SRS setup shown in Fig. 7.3(c), the scattering medium is simply placed outside the pump laser source, and the observed SRS is based on a single-pass amplification of the initial Raman scattering signal. Obviously, this setup is suitable for high-peak-power laser pulses pumped Raman systems. To obtain a higher local intensity, the pump laser beam is usually focused through a lens, or compressed by using a reverse beam expander.

When the input pump laser beam is a quasi-parallel or unfocused beam, the gain length of SRS is determined by the path length of the Raman medium. Whereas, if the pump beam is a focused laser beam, the effective gain length is mainly determined by the focal depth within which a higher local pump intensity is maintained. When a focused pump beam is used, by choosing proper beam diameter and focusing length the effective gain length can be several millimeters to several centimeters for liquid-cell and crystal-rod samples, or more than several tens of centimeters for high-pressure gas samples. In particular, the effective gain length can be drastically increased when the Raman medium is in a form of channel waveguide or optical fiber. In that case the high local intensity of a focused pump laser beam can be maintained within the whole pass-length of the waveguide materials, and consequently the conversion efficiency from pump beam to SRS beam can be significantly increased. For most experimental devices of SRS, the overall conversion efficiency from the pulsed pump intensity (or energy) to the SRS pulsed intensity (or energy) ranges from 10^{-2} to 10^{-1}, and under optimum conditions it can be higher than 30–50%.

7.3.2 Experimental properties of SRS

In general, a complete experimental study of SRS in a given scattering medium involves the following major issues.

1. *Threshold measurement*: The generation of SRS exhibits a sharp threshold feature. For a given Raman medium and experimental condition, no SRS signal can be observed until the input pump intensity is higher than a certain threshold level. For most experiments using common Raman media and a single-pass device the pump-intensity threshold ranges from 10^7 to 10^8 W/cm^2.

2. *Spatial structure*: The spatial structure of SRS output beam can be completely determined by measuring its near-field and far-field patterns. In most of experiments without using an optical cavity, the output SRS beam is collinear with the transmitted pump laser beam, and the near- and far-field patterns are basically the same for both the SRS beam and the pump beam, provided that the latter is focused onto the sample. On the other hand, if the Raman medium is in an external cavity, as shown in Fig. 7.3(b), the near- and far-field distributions may depend on the cavity design and specific experimental conditions.

3. *Temporal behavior*: Owing to the threshold requirement of SRS, the pulse duration of SRS is always shorter than that of the input pump pulse. When the pump level is just slightly higher than the threshold value, the duration of a SRS pulse might be much shorter than that of the pump pulse.

4. *Spectral structure*: For most experimental SRS studies, once the input pump laser intensity is higher than the threshold level, a Stokes-shifted SRS spectral line will be first observed, which corresponds to the Raman transition with a larger cross-section for the given Raman medium. If increase the pump intensity further, one may observe the additional second or even the third SRS components on the longer wavelength side with the same frequency shift, they have the same propagation direction as the pump beam and are termed the second- and third-order Stokes SRS. The generation of this type of multi-order SRS is based on the cascaded stimulated scattering process; in other words, the strong first-order Stokes SRS can play a role of pump beam and induce the second-order Stokes SRS, while the second-order SRS can generate the third-order SRS, and so on. Under high pump condition and in some suitable media, SRS with several different Raman modes or SRS at anti-Stokes frequency position can also be observed.

5. *Steady-state and transient SRS*: It is known that the molecular Raman transition can take place between their vibrational states or rotational states. If the pump laser pulse duration (τ_p) is much longer than the phase-relaxation time of the excited Raman mode, one may expect a steady-state or quasi-steady-state SRS process. On the other hand, if the duration of the pump pulse is comparable to or even shorter than the above dephasing time, one will expect a transient SRS process. In this chapter we mainly consider the steady-state stimulated scattering process.

6. *Relation between forward and backward SRS*: For single-pass SRS device shown in Fig. 7.3(c), the stimulated scattering may be observed in one of the following three possible manners: (i) nearly symmetric forward and backward SRS output, (ii) a stronger forward output and a weaker backward output, and (iii) only a forward SRS output.

Theoretically, if the pump source is a cw laser or a pulsed laser with a pulse-length ($l_0 = c\tau_p$) drastically longer than the gain length (L) of the Raman medium, the forward and backward SRS output should be basically symmetric because the interaction length for both beams is equally determined by L. Whereas, if l_0 is shorter than L, we should observe a stronger forward output than the backward output, because the gain length for the former is still determined by L, but for the latter it is determined by l_0. Furthermore, if l_0 is greatly shorter than L, no backward SRS can be observed.

7.3.3 Four-wave frequency mixing (FWFM) in SRS experiments

In Chapter 4 we have described various (including Raman-enhanced) four-wave frequency mixing (FWFM) or four-photon parametric effects. The most distinctive feature of these effects is that at the beginning and end of an elementary process, the molecular status of the nonlinear medium does not change, i.e., there is no exchange of energy and momentum between the molecular system and optical fields. Consequently, the conservation of energy and momentum should be satisfied only among all involved optical waves, and that leads to the phase-matching requirement. If the dispersion effect of refractive index of the medium within a considered spectral range is not negligible, the phase-matching requirement cannot be satisfied along the same direction in an isotropic medium; in other words, FWFM usually is not a collinear process.

In contrast, SRS is not a FWFM effect. As mentioned earlier in this chapter, at the beginning and end of an elementary process of Raman scattering, the quantum states of the molecule system have been changed because of the exchange of energy and momentum between the Raman medium and the optical fields. In this case, the conservation of energy and momentum applies to the whole system involving both molecules and optical fields; therefore, no phase-matching requirement is needed. That is the reason why SRS is collinear with the pump beam and can also be observed in the backward direction.

Although SRS and FWFM are two essentially different nonlinear effects, sometimes researchers may confuse them as in many SRS experiments they can be observed sequentially or simultaneously. Nevertheless, the appearances of these two types of effects are distinguishable in liquids and solids: the SRS emission appears as a coherent beam that is collinear with the transmitted pump laser beam, whereas the FWFM emission appears as a pair (or several pairs) of conical ring-shape beams with discrete phase-matching angles.[16-20] However, in gaseous Raman media, because of their small refractive-index dispersion, the phase-matching angle may be either smaller than the divergence angle of a collimated pump beam or within the convergence angle of a focused pump beam, so that a nearly collinear FWFM emission might be observed in forward direction too.

There are two major reasons that can explain why SRS processes are often accompanied by FWFM processes. First, a Raman medium efficient for the former is also efficient for the latter. Second, during SRS processes a Raman-enhanced refractive-index change may lead to a Raman-enhanced self-focusing, which may significantly increase the local intensity of the existing optical waves. As a result of this type of self-focusing, both high-order SRS and high-order FWFM can be observed.

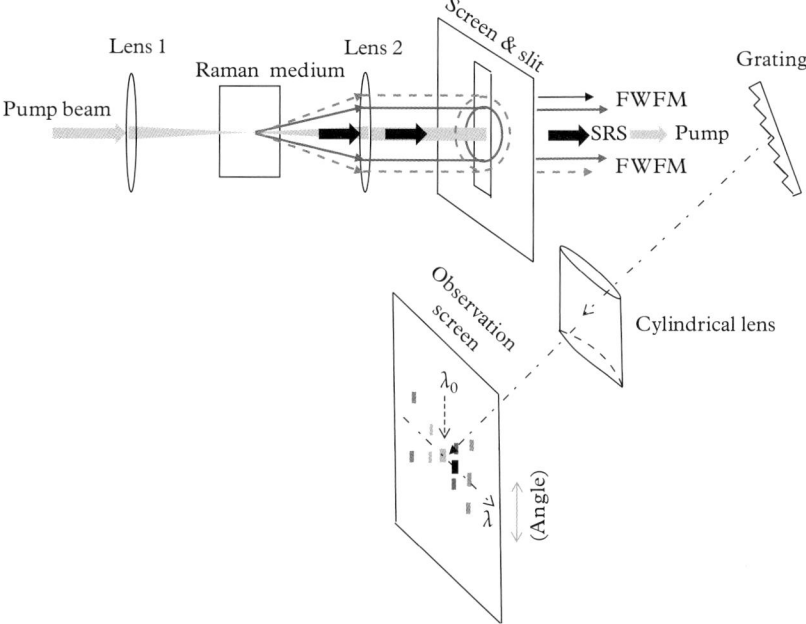

Figure 7.4 *Schematic setup for simultaneously measuring the spectral and spatial structures of coherent Raman emissions (SRS plus FWFM).*

The most effective method to distinguish FWFM from SRS is to measure the spectra of the coherent Raman emissions (including both SRS and FWFM signals) as a function of the propagation angle at different pump intensity levels. Such an experimental setup is shown schematically in Fig. 7.4, which is basically similar to Fig. 6.13. In present case, a focused pump beam passes through a Raman medium once; the transmitted pump beam and the forward coherent Raman output are re-collimated by a lens. If place a screen to observe the near-field pattern of the output beams, we will see a strong central spot that contains the transmitted pump beam and collinear SRS components. When the pump intensity further increases, we may see additional coherent Raman emission rings that are generated by FWFM processes. If arrange the overall output beams pass through a vertically placed slit in the near-field central position and then let the selected beams' parts be reflected from a dispersion grating, we can finally separate different coherent emission components on both the spectral scale and the spatial (angular) scale. By measuring the spatial structures of different spectral components, the sequence of their appearance as well as the dependence of their appearance on the pump intensity, one can identify the contributions from SRS and from FWFM, respectively.

Although the specific behavior and appearances of SRS and the accompanied FWFM may vary with different Raman media or under different pump conditions, some common features among them should remain. For clarity, in Fig. 7.5 we depict schematically three typical spatial structures of spectra of the forward coherent Raman beams, which may be observed in SRS experiments using a liquid (or solid) sample and the setup shown in Fig. 7.4.

The spectral structure shown in Fig. 7.5(a) is usually observed when the pump intensity is slightly higher than the threshold for SRS. In this case there is only a pump line and Stokes SRS

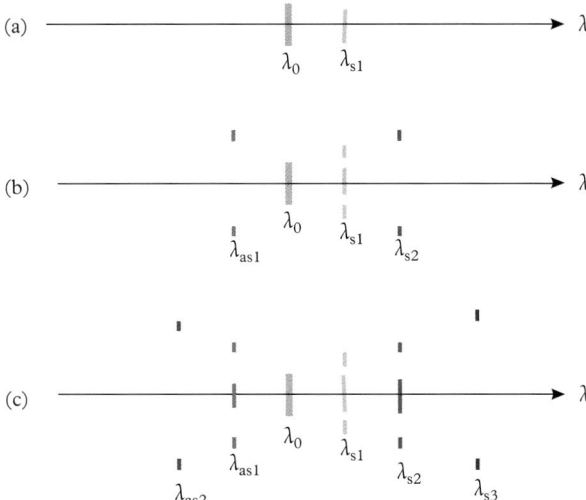

Figure 7.5 Examples of three typical spectral structures of forward coherent Raman emissions recorded in the observation screen shown in Fig. 7.4: (a) the transmitted pump line and the first-order SRS line (at a lower pump level), (b) the same as above but with the additional first-order Stokes ring, second-order Stokes ring, and first-order anti-Stokes ring (at a moderate pump level), (c) the same as above but with the additional second-order SRS line, third-order Stokes ring, and second-order anti-Stokes ring (at a high pump level). Different vertical positions correspond to different propagation angles inside the Raman medium.

line in the forward axial direction. The frequency shift corresponds to a characteristic Raman mode of the medium with a larger Raman cross-section.

When the pump intensity increases to a moderate level, a spectral structure shown in Fig. 7.5(b) may be observed, which is featuring additional off-axial spectral components on both sides of the pump line. These off-axial spectral components corresponding to ring-shape coherent emissions are resulted from various Raman-enhanced FWFM processes (see Section 4.3). The spacing between two adjacent spectral components is equal to Raman-shift frequency of the medium. In the present case, the phase matching conditions for generating the first-order Stokes (s_1), second-order Stokes (s_2), and first-order anti-Stoles (as_1) ring emissions are shown in Fig. 7.6(a) and (b), and can be written as

$$\left. \begin{array}{l} \mathbf{k}_0(\text{axial}) + \mathbf{k}_0(\text{axial}) = \mathbf{k}_{s1}(\text{ring}) + \mathbf{k}_{as1}(\text{ring}) \\ \mathbf{k}_0(\text{axial}) + \mathbf{k}_{s1}(\text{axial}) = \mathbf{k}_{s2}(\text{ring}) + \mathbf{k}_{as1}(\text{ring}) \end{array} \right\}. \quad (7.3\text{-}1)$$

Here, $\mathbf{k}_0$ (axial) and $\mathbf{k}_{s1}$ (axial) are the wavevectors of the pump beam and the collinear first-order Stokes SRS beam, while $\mathbf{k}_{s1}$ (ring), $\mathbf{k}_{s2}$ (ring), and $\mathbf{k}_{as1}$ (ring) are the wavevectors of the three ring-shape beams.

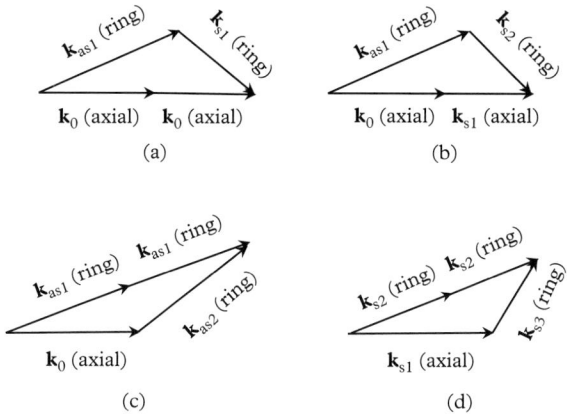

Figure 7.6 *Phase-matching conditions for generating a pair of FWFM emission rings under different combinations. Here the Raman medium is assumed a liquid or solid with non-negligible refractive-index dispersion effect.*

At a high pump level we can see multiple high-order Stokes SRS lines along the axial direction based on the cascaded SRS processes, as well as multiple pairs of high-order Stokes/anti-Stokes rings based on the cascaded FWFM processes. These possibly observed structures are shown schematically in Fig. 7.5(c), where we can see the second-order axial SRS line as well as more additional pairs of ring components: the third-order Stokes ring and the second-order anti-Stokes ring. The phase-matching conditions for generating the latter pair of ring beams are shown in Fig. 7.6(c) and (d), and can be written as

$$\left. \begin{array}{l} 2\mathbf{k}_{as1}(\text{ring}) = \mathbf{k}_0(\text{axial}) + \mathbf{k}_{as2}(\text{ring}) \\ 2\mathbf{k}_{s2}(\text{ring}) = \mathbf{k}_{s1}(\text{axial}) + \mathbf{k}_{s3}(\text{ring}) \end{array} \right\}. \qquad (7.3\text{-}2)$$

It should be noted that the really observed spectral structures might be either simpler or more complicated than that shown in Fig. 7.5, depending on the specific experimental condition and the Raman medium used. We describe in what follows the experimental observation on spectral-spatial structures of the coherent Raman output from two typical Raman media: one is calcite ($CaCO_3$) crystal, another is benzene (C_6H_6) liquid, based on a setup shown in Fig. 7.4. In addition, we will also briefly mention the high-order SRS behavior in the gas of hydrogen (H_2).

7.3.3.1 Calcite crystal sample

Calcite is one of the best crystal materials for SRS studies because of its large cross-section of Raman mode at frequency of 1086 cm^{-1}, superior optical quality, and high optical damage threshold. In the following experiment, the crystal sample was 1-cm long with its optical axis along the propagation direction of a pulsed pump laser beam of 532 nm wavelength and ∼8 ns pulse duration. The experimental setup is the one shown in Fig. 7.4.

Figure 7.7 shows the photographs of coherent Stokes emission spectra of the calcite sample at three different pump levels. To obtain these photographs the transmitted pump beam was greatly

158 Stimulated Scattering of Intense Coherent Light

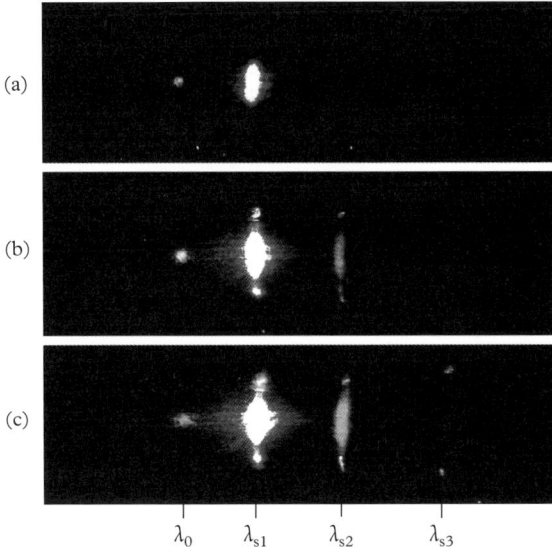

Figure 7.7 *Spatial structures of Stokes spectral components of SRS and FWFM output from a 1-cm-long calcite sample pumped with $\lambda_0 = 532$-nm laser pulses at various intensity levels: (a) $I_0 = 550$ MW/cm^2, (b) $I_0 = 900$ MW/cm^2, and (c) $I_0 = 1.4$ GW/cm^2. $\lambda_{s1} = 565$ nm, $\lambda_{s2} = 602$ nm, and $\lambda_{s3} = 644$ nm (See also Color fig. 6.).*

attenuated and blue-green-cut glass filters were used. At a low pump level as shown in Fig. 7.7(a), we can only see the first-order Stokes component ($\lambda_{s1} \approx 565$ nm) along the axial direction; obviously it is due to a pure stimulated scattering process without any phase-matching requirement. At a moderate pump level as shown in Fig. 7.7(b), we can see the second-order Stokes SRS component ($\lambda_{s2} \approx 602$ nm) in the same axial direction, which is due to a pure cascaded stimulated scattering process without phase-matching requirement. In the meantime, we can also see the additional ring emissions at λ_{s1} and λ_{s2} wavelength positions; obviously they are generated by FWFM processes under phase-matching conditions. Finally, at a high pump level, as shown in Fig. 7.7(c), we see an additional ring emission at $\lambda_{s3} \approx 644$ nm, which is also a product of the cascaded FWFM process.

Next, Fig. 7.8 shows the coherent anti-Stokes emission spectra from the same sample at three different pump levels. To obtain these photographs the transmitted pump beam was moderately attenuated and orange-red-cut glass filters were used. At the low pump level, shown in Fig. 7.8(a), we can only see the relatively weak first-order anti-Stokes SRS component ($\lambda_{as1} \approx 503$ nm) along the axial direction. At the moderate pump level, shown in Fig. 7.8(b), we can see the ring-shape emissions at both λ_{as1} and $\lambda_{as2} \approx 477$ nm wavelength positions, which are generated from phase-matched FWFM processes. Finally, at a high pump level, shown in Fig. 7.8(c), both the anti-Stokes SRS components and FWFM components become stronger.

In conclusion, the spectral structures of coherent Raman output from the calcite sample are basically similar to that illustrated in Fig. 7.5.

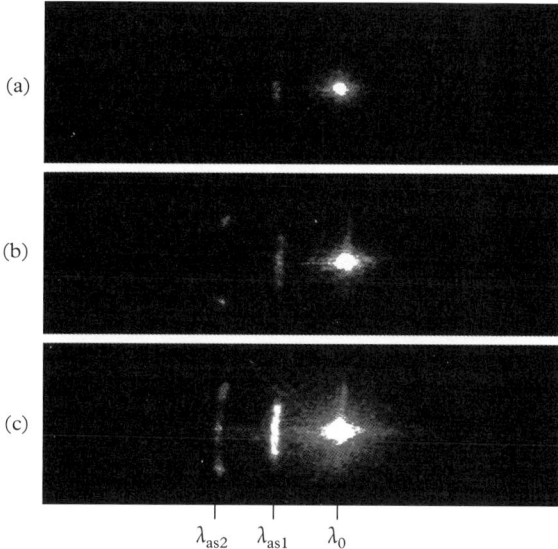

Figure 7.8 *Spatial structures of anti-Stokes spectral components of SRS and FWFM output from a 1-cm-long calcite under the same conditions as described in the caption of Fig. 8.7; $\lambda_0 = 532$ nm, $\lambda_{as1} = 503$ nm, and $\lambda_{as2} = 477$ nm (See also Color fig. 7.).*

7.3.3.2 Liquid benzene sample

Benzene is the most commonly used liquid sample for SRS and coherent Raman studies with its strong Raman vibration mode of 992 cm^{-1}. In the experiment exemplified here, the sample was placed in a 10-cm-long liquid cell pumped with a focused pulsed dye laser beam through an $f_1 = 20$ cm focusing lens. The wavelength, spectral width, pulse duration, and beam size of the pump laser are $\lambda \approx 615$ nm, $\Delta\lambda \approx 0.6$ nm, $\Delta t = 6$ ns, and $\phi = 2$ mm, respectively.

At a pump intensity level of $I_0 \approx 300$ MW/cm^2, the spatial/spectral structures of forward coherent emissions are shown in Fig. 7.9, which is digitally recorded by a CCD array. Here we can see that there are two axial Stokes SRS components and one anti-Stokes SRS component along the direction of the transmitted pump beam. We can also see one ring component in the second-order Stokes wavelength (λ_{s2}) position and two rings in the λ_{as1} and λ_{as2} positions, respectively. The angular diameter of the λ_{s2} ring is measured to be $\sim 4.3°$. All these coherent emission rings are generated by FWFM processes under phase-matching conditions. One can find that the measured spatial structures of SRS and FWFM components for benzene sample are quite similar to that schematically shown in Fig. 7.5, but not exactly the same.

7.3.3.3 Hydrogen gas sample

Hydrogen is one of the best gaseous media for SRS generation with its very strong vibrational Raman mode of 4155 cm^{-1}. It is easy to generate high-order stimulated coherent Raman in high-pressure H$_2$ gas, pumped by ns- and mJ-laser pulses. In this case, the observed high-order coherent Raman emissions are the results of cascaded SRS and FWFM processes as we mentioned above.

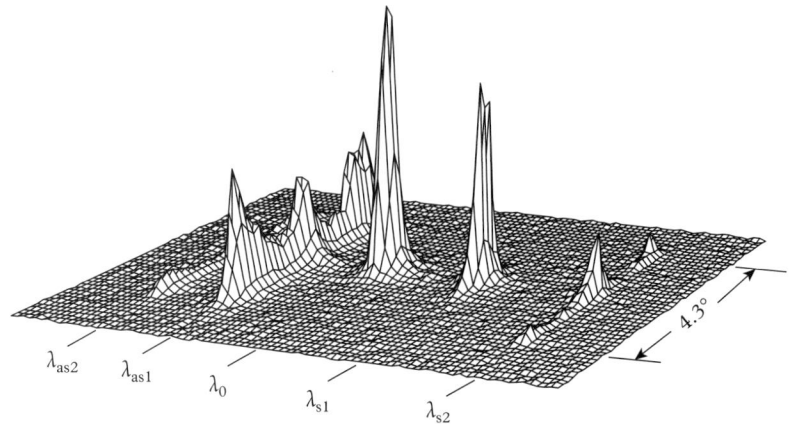

Figure 7.9 *Spatial and spectra structures of the forward SRS and FWFM output from a 10-cm-long benzene liquid cell at a pump level of $I_0 \approx 300$ MW/cm². The SRS frequency shift for benzene is 992 cm⁻¹. Since different spectral components have been attenuated via properly chosen spectral filters, this figure does not indicate the real intensity ratio among different spectral components.*

As an example, shown in Fig. 7.10 are the anti-Stokes coherent Raman emission spectra from a 45-cm long H_2 gas cell of ∼19 bar pressure, recorded by a CCD array at different pump level of the 1064-nm and ∼8-ns pulsed laser beam focused by an f = 50 cm lens into the center of the cell. In this experiment the Stokes components could not be recorded because they are out of the spectral response range of the CCD camera used.

Under the experimental condition mentioned above, no backward high-order coherent Raman emission could be observed, which implies that the observed forward high-order emissions are still due to phase-matched FWFM processes. Since the refractive-index dispersion for gaseous media is much smaller than liquids or solids, the phase-matching requirement may be fulfilled either along the axial direction or within a quite small angle range. As regards this aspect, there is another experimental result that indicated the phase-matching dependence of high-order coherent Raman emission on the refractive-index dispersion.[21] In such an experiment, an 80-cm long H_2-gas cell was pumped by 266-nm and 8-ns laser pulses with ∼100 mJ energy, up to ninth-order anti-Stokes coherent Raman emissions could be generated in the vacuum UV spectral range, where the refractive-index dispersion is no longer negligible when the pressure of H_2-gas is high enough. Shown in Fig. 7.11 is the variation of measured near-field pattern for the sixth-order anti-Stokes (∼160 nm) emission following the increase of gas pressure. One can see that when the dispersion effect is considerably large, the ring-shape of FWFM emission still can be observed even in a gaseous Raman medium.

7.3.4 Spin-flip, electronic, and rotational SRS effects

Although most SRS studies are related to vibrational transitions of the Raman media, there are several special SRS effects that are related to spin-flip transitions in semiconductors, electronic transitions in atomic metal vapors, and rotational transitions in molecular gas systems, respectively.

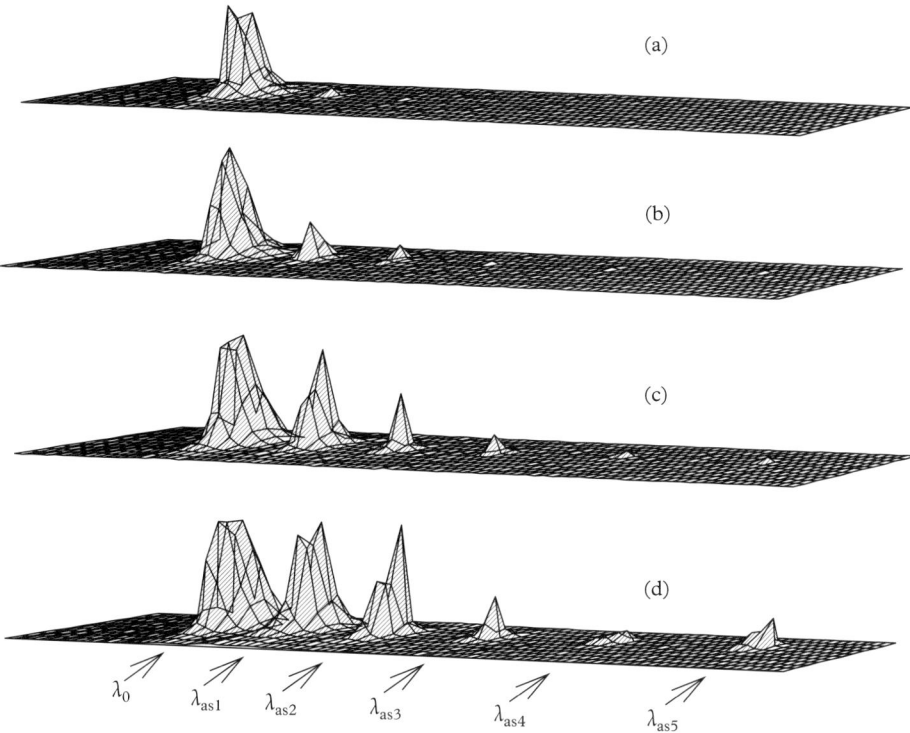

Figure 7.10 *Spectra of forward anti-Stokes coherent Raman emissions from a 45-cm long high-pressure H_2 cell at various pump levels: (a) $I_0 = 0.8$ GW/cm^2, (b) $I_0 = 1.1$ GW/cm^2, (c) $I_0 = 1.3$ GW/cm^2, and (d) $I_0 = 2$ GW/cm^2.*

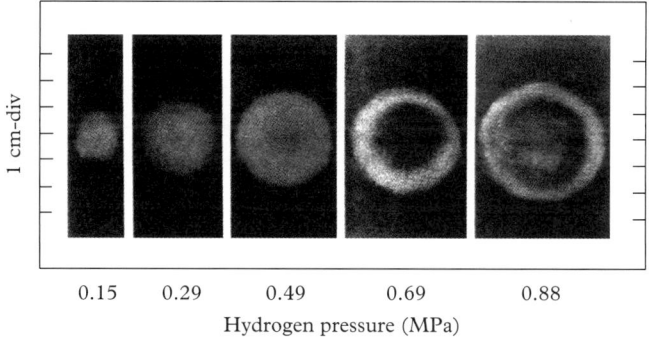

Figure 7.11 *Variation of the near-field pattern of the sixth-order anti-Stokes FWFM emission ($\lambda_{as6} \approx 160$ nm) with increasing hydrogen pressure. The pump wavelength: $\lambda_0 \approx 266$ nm, beam size: ~8 mm. (Reproduced with permission from Moriwaki et al.[21] Copyright 1993, AIP Publishing LLC).*

7.3.4.1 Spin-flip SRS in semiconductor crystals

This special SRS effect was first observed in a semiconductor crystal by Patel and Shaw in 1970.[22] The unique feature of this effect is that the Raman medium is a semiconductor sample placed in an intense dc magnetic field. In this case, the scattering centers are the conduction electrons, and the Raman transition takes place between two Zeeman-splitting sub-levels of those electrons. It is more special that the frequency interval of Zeeman-splitting sub-levels is directly dependent on the applied magnetic field. Therefore, with a frequency-fixed pump laser source, one can generate a frequency-tunable SRS output by changing the intensity of the magnetic field.

The typical material for spin-flip SRS study is the semiconductor crystal of n-type InSb. With an applied magnetic field, the energy level of the conduction electron will be split into a series of equally spaced sub-levels that are referred to Landau levels labeled with different orbital quantum numbers of n, as illustrated schematically in Fig. 7.12. For simplicity, here we neglect the dependence of the energy-level structures on the wavevector of the wave function, hence different Landau levels can be schematically represented by different horizontal lines (not to scale).

Furthermore, if we consider the interaction between the electron spin and magnetic field, each Landau level should involve two sub-levels that correspond to two possible spin-states: the lower sub-level corresponds to a spin-up state, the higher sub-level corresponds to a spin-down state. For each Landau level the interval between its two sub-levels is determined by

$$\Delta E = |g^*|\beta B, \tag{7.3-3}$$

where g^* is the effective gyromagnetic ratio, β is the Bohr magneton, and B is the magnetic induction of the applied field. In the normal situation of an InSb sample, the most electrons in conduction band populate in the low sub-level of their $n = 0$ Landau level. Upon the action of an incident optical field, a conduction electron may make a Raman transition from its low spin-up state to the higher spin-down state within the same $n = 0$ Landau level, and scatter a red-shifted photon simultaneously. The frequency shift between the input photon and scattered photon is determined by

$$\Delta v = v_0 - v_s = \Delta E/h = |g^*|\beta B/h, \tag{7.3-4}$$

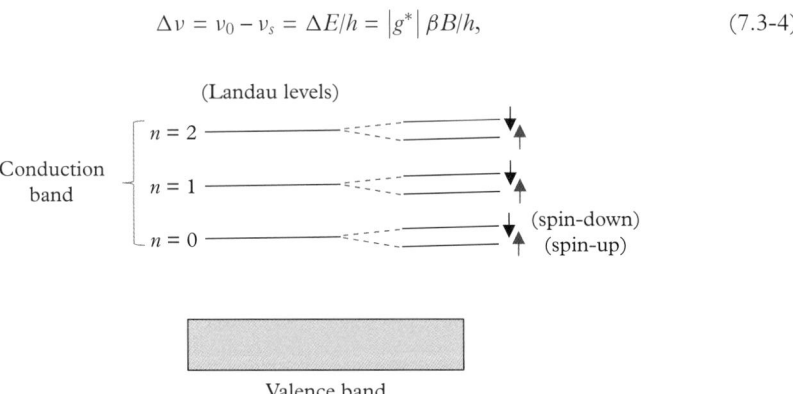

Figure 7.12 *Upon the action of an applied magnetic field, the electronic energy level in conduction band is split into a series of Landau levels, each Landau level involves two sub-levels corresponding to two different spin states.*

where ν_0 and ν_s are the frequencies of the input light and scattered light, respectively. This is the basic description of spin-flip Raman scattering effect. From Eq. (7.3-4) we can see that the frequency of the scattering light can be tuned by changing the applied dc magnetic field.

As mentioned before, the concept of intermediate state should be employed to describe any type of Raman scattering. For the spin-flip Raman scattering in a semiconductor, the contributions to the intermediate state may come from all possible one-photon transition allowed eigenstates in both conduction and valence bands. However, only the contributions from those eigenstates, the energy separations between which and the initial state of the Raman transition are closer to the photon energy of the pump light, are of practical importance due to the near-resonance enhancement. Figure 7.13 shows a schematic diagram (not to scale) that shows the two possible transition pathways with near-resonant excitation.[23]

In the situation shown in Fig. 7.13(a), it is assumed that the pump photon energy is closer to the gap between the conduction band and valence band(s), thus the contributions from the eigenstates in the valence band(s) dominate. In the first step of this process, an electron in the valence band makes a transition to the empty spin-down state of the $n = 0$ Landau level with annihilating a pump photon; then the electron in the spin-up state of the $n = 0$ Landau level moves down to the valence band with creating a red-shifted photon. Figure 7.13(b) shows another situation when the pump photon energy is closer to the spacing of two different Landau levels within the conduction band. In that case the contributions from the higher Landau levels in the conduction band are more important.

Similar to the regular SRS process, if the incident light is an intense laser beam, the initial spin-flip Raman scattering signal may experience an exponential gain due to the stimulated nature of the process. When this gain is high enough to overcome various losses in a scattering medium, the stimulated spin-flip scattering signal can be observed.

In experiments of spin-flip SRS, the most commonly used scattering medium is the n-type InSb crystal.[24–26] The carrier concentration of the sample is $n_e \approx 10^{15}$–10^{16} cm^{-3}, the working

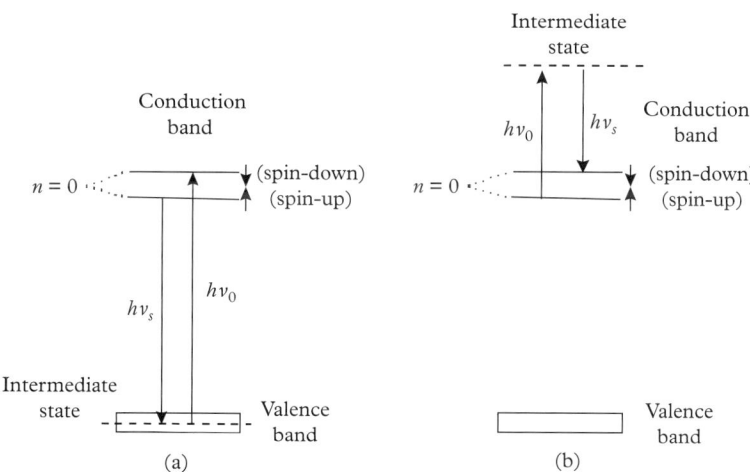

Figure 7.13 *Transition pathways of the elementary spin-flip SRS process: (a) the intermediate state is mainly related to the valence band and (b) the intermediate state is mainly related to the conduction band.*

temperature is 3–30 K, the refractive-index is ~4, and at low magnetic field $g^* \approx -50$. The pump laser sources can be CO_2 lasers with ~10 μm output wavelength, or CO lasers with ~5 μm output wavelength. At a high magnetic field level (~10 T), the Raman shift range can be 150–250 cm^{-1}, and the average tuning rate is about (22–15) cm^{-1}/T within the magnetic field range 0–8 T. In optimum conditions, the transfer efficiency from the pump light to SRS light can be higher than 30–50%. The output spin-flip SRS is tunable over a spectral range 11–14 μm when pumped with a $\lambda \approx 10$ μm laser beam, or over a spectral range 5–6 μm when pumped with a $\lambda \approx 5$ μm laser beam.

It has been shown that under proper experimental conditions, in addition to the Stokes SRS, the second-order Stokes component and anti-Stokes component could also be generated through the FWFM process; however, in this case the phase-matching requirement could be fulfilled along the same axial direction because the frequency shift is relatively small, and within such narrow spectral range the dispersion effect of the refractive index of the sample can be neglected. As an example, the measured wavelength values of above-mentioned three spectral components as a function of the applied magnetic field are shown in Fig. 7.14.[24] The sample was an InSb crystal

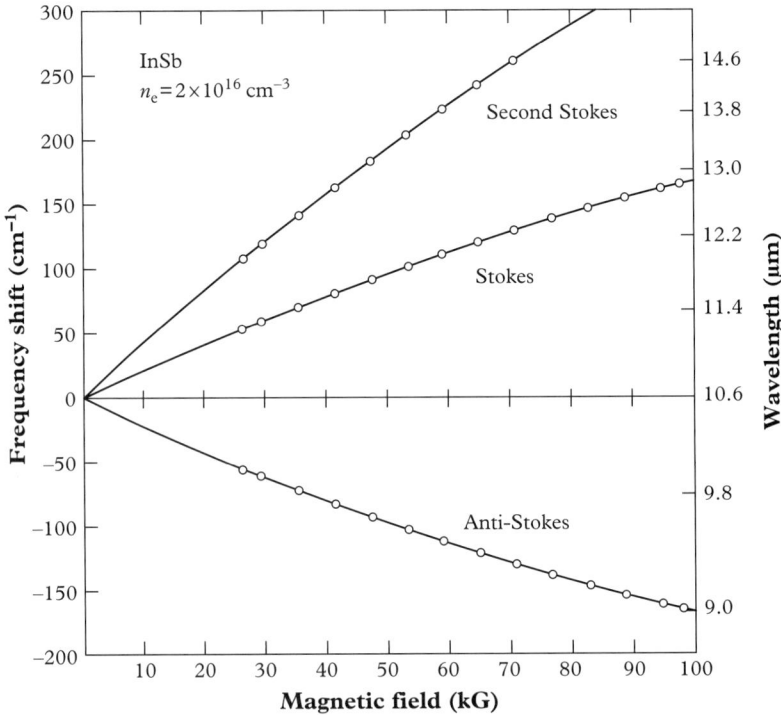

Figure 7.14 *Tuning curves for anti-Stokes, Stokes, and second-order Stokes components of the spin-flip SRS output from InSb pumped with 10.6-μm laser beam. (Reproduced with permission from Aggarwal et al.[24] Copyright 1971, AIP Publishing LLC).*

of $n_e = 2 \times 10^{16}$ cm^{-3} pumped by a $\lambda = 10.6$-μm laser beam (with peak power of the order of 1 MW) from a high-pressure pulsed CO$_2$ laser source, and the magnetic field was varied over the range 20–100 kG (2–10 T).

In addition to InSb, the spin-flip SRS can also be generated in several other semiconductor crystals including CdS, InAs, Hg$_{1-x}$Cd$_x$Te, Pb$_{1-x}$Sn$_x$Te, etc.[27–31]

The favorable feature of spin-flip SRS is its tunability over a moderate (0–10^2 cm^{-1}) frequency shift range. Therefore this technique could be useful for those applications, such as nonlinear infrared spectroscopy, tunable nonlinear optical frequency-mixing, selective photo-chemical reactions, and laser isotope separation. It seems, however, that further studies and applications of this technique have been limited by its disadvantage of requiring low-temperature operation and high magnetic field.

7.3.4.2 Electronic-transition SRS in metal vapors

For gas systems consisting of atoms, there are no vibrational and rotational eigenenergy structures, the Raman transition can only occur among their different electronic states, whose energy spacing is much greater than that of vibrational states of molecular systems. For this reason, one would expect a much greater frequency shift of electronic-transition SRS in atomic media. In practice, stimulated electronic-transition Raman scattering has been observed in a number of metal atomic systems. Metal vapors are easier to prepare for SRS studies, they exhibit relatively simple and well-known energy level structures, and most importantly, a quite number of energy levels can be chosen for resonant enhancement purposes.

It is often mentioned in this chapter that the intermediate state, introduced for describing a Raman scattering process, actually involve all possible atomic eigenstates that could give a nonzero contribution to the process. This is implied mathematically in the general expression of Raman cross-section (cf. Eq. (7.2-21)) by the summation over all related eigenstates. In particular, if the energy of the input excitation photon is quite close to one of these eigenstates, the denominator of one summed term becomes very small, and thus the contribution from this term may dominate others. This is the so-called one-photon resonance enhanced electronic SRS process, and it is shown schematically in Fig. 7.15(a), where energy levels 1 and 2 represent the initial state and final state of an electronic Raman transition, level 3 refers to one of the one-photon absorption allowed electronic excited states, and the pump photon energy is quite close to the energy interval between levels 1 and 3. The similar resonant-enhancement mechanism can also be further extended to two-photon and three-photon excited SRS processes. In the case shown in Fig. 7.15(b), the sum energy of two pump photons is close to the spacing between the initial level 1 and a two-photon absorption (2PA)-allowed exited state 3; the stimulated scattering may occur by virtue of the annihilation of two pump photons at frequency ν_0 and the creation of one scattered photon at frequency ν_s. This is the so-called stimulated hyper-Raman scattering effect. Finally, in the case shown in Fig. 7.15(c), the stimulated scattering may occur with the annihilation of three pump photons and simultaneously creating of one scattered photon, provided that the resonant enhancement can be achieved either by choosing two-pump-photon energy close to a 2PA-allowed real level 3 or by choosing the three-pump photon energy close a three-photon absorption allowed real level 4. That is the so-called high-order stimulated hyper-Raman scattering effect. The probability of hyper Raman scattering is much smaller than that of the regular electronic Raman process but much greater than that of high-order hyper-Raman process.

The highly efficient electronic-transition SRS performances have been accomplished in a series of metal vapors including potassium (K),[32–34] thallium (Tl),[35,36] cesium (Cs),[37–41] barium

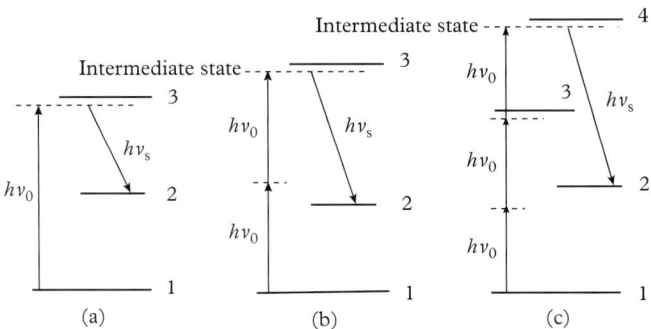

Figure 7.15 Schematic diagrams of resonance-enhanced electronic-transition SRS processes: (a) one-photon induced regular SRS, (b) two-photon induced stimulated hyper-Raman scattering, and (c) three-photon induced high-order stimulated hyper-Raman scattering. Here the dashed-line levels represent intermediate states connecting the atom and light fields.

(Ba),[42–45] lead (Pb),[36,46] bismuth (Bi),[36] thulium (Tm),[43] etc. In most cases the one-photon resonant enhancement has been applied to increase the conversion efficiency. The temperature range of those metal vapors could be from 300 °C to 1200 °C, the length of the heat-pipe oven is about 30–50 cm, and the input pump radiation can be high-power laser pulses with duration in nanoseconds or picoseconds. The frequency shift of the output electronic SRS signal ranges from ∼8000 to ∼28 000 cm^{-1}. When the input pump wavelengths are in UV or visible range, the output Stokes-shifted SRS wavelengths are in the visible or near-IR range accordingly. Under optimum conditions, the quantum conversion efficiency can be more than 30–60%, and the energy conversion efficiency can be more than 20–40%.

Figure 7.16 shows the energy levels relevant to the processes of resonance-enhanced electronic SRS in metal vapors of Ba, Tl, Pb, and Bi.[36] In the experiment, the 308-nm pump radiation was provided by a discharge-pumped XeCl laser with a pulse duration of 25 ns, pulse energy of 75 mJ, and a peak focused intensity of 3 GW/cm^2.

In addition, the stimulated hyper-Raman scattering has been observed in several metal vapor systems (such as K, Cs, Na, and Sr) with two-photon resonant enhancement.[47–51] As expected, in those cases the conversion efficiency is quite low (10^{-4}–10^{-2}) in comparison with the one-photon excited SRS. As an example of stimulated hyper-Raman scattering, Fig. 7.17 shows the energy levels relevant to the process of stimulated hyper scattering in the vapor of Cs.[48] In this case, the wavelength, pulse duration, spectral linewidth, and power level of the pump beam were 1.06 μm, 2 ns, 0.4 cm^{-1}, and 150 MW, respectively. Since two-photon energy of the pump laser is close to the $7S_{1/2}$ state, a near two-photon resonance-enhancement could be achieved for stimulated hyper-Raman scattering from the $6S_{1/2}$ state to $6P_{3/2}$ state. The wavelength of the stimulated scattering signal was 1.416 μm.

Finally, the three-photon excited high-order stimulated hyper-Raman scattering was also observed in the vapor of lithium (Li); however, in this case the measured efficiency was only about 2×10^{-6}.[52]

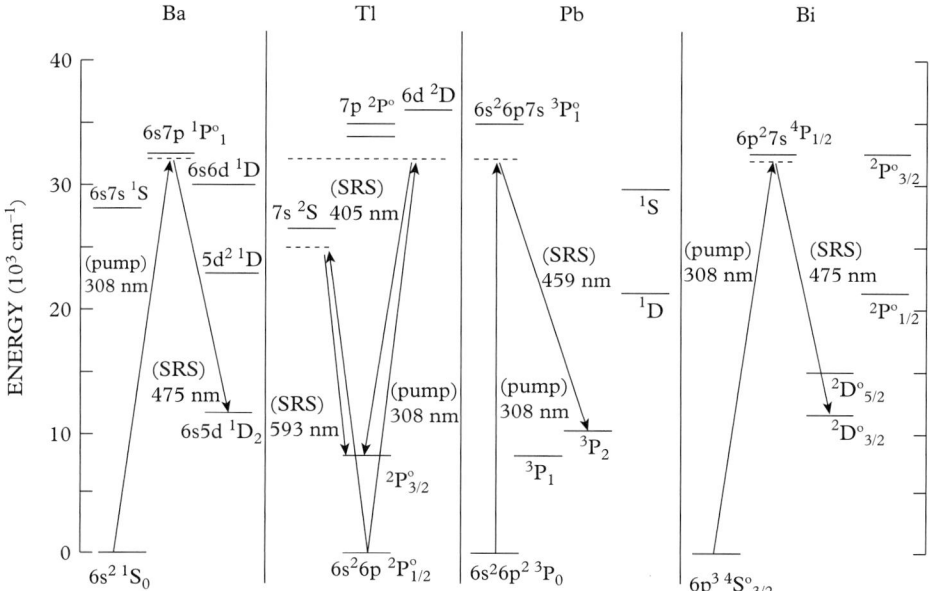

Figure 7.16 *Energy levels relevant to the process of resonance-enhanced electronic SRS in the vapors of Ba, Tl, Pb, and Bi. (Reproduced with permission from Burnham and Djeu.[36] Copyright 1978, Optical Society of America).*

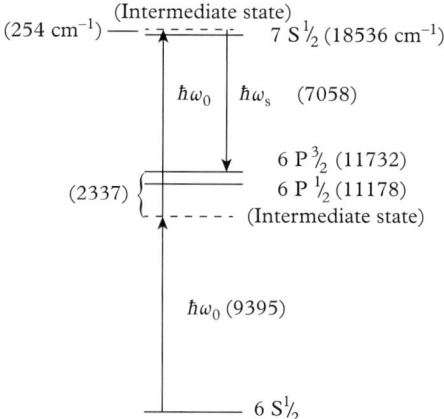

Figure 7.17 *Energy levels relevant to the process of near two-photon resonance enhanced stimulated hyper-Raman scattering in the vapor of Cs. (Reproduced with permission from Vrehen and Hikspoors.[48] Copyright 1977, Elsevier).*

7.3.4.3 Rotational transition SRS in molecular gases

It is known that the spontaneous Raman scattering of vibrational, vibration–rotational, and rotational transitions can be observed in various molecular gases. Since the advent of lasers, stimulated scattering for all these three types of Raman transitions have been achieved in a number of molecular gases. The commonly utilized gases for vibrational SRS studies and their strongest vibrational mode frequencies are: hydrogen (H_2, 4155 cm^{-1}), deuterium (D_2, 2991 cm^{-1}), methane (CH_4, 2915 cm^{-1}), sulfur hexafluoride (SF_6, 1551cm^{-1}, 2323 cm^{-1}), nitrogen (N_2, 2330 cm^{-1}), oxygen (O_2, 1550 cm^{-1}), etc. In order to get a higher SRS gain and efficiency, the pressures of these gases are usually several tens of bar. Depending on the gases used and the experimental conditions (pressure, gas-tube length, focusing geometry, and temperature), the stimulated scattering can be observed in either pure vibrational Raman transitions,[53–57] or vibration–rotational Raman transitions.[58–62] Under optimum conditions the conversion efficiency from the pump energy to the SRS energy can be higher than 50%.

The stimulated scattering in pure rotational Raman transitions can also be generated in specially optimized experimental conditions.[63–71] The well-studied gases for this purpose are H_2 and D_2. For hydrogen, the strongest rotational Raman mode frequency is 587 cm^{-1} (S(1) line) for orthohydrogen and 354 cm^{-1} (S(0) line) for parahydrogen; the natural ortho:para ratio is 3:1 in normal room-temperature conditions.[58] For deuterium, the strongest rotational Raman mode frequency is 414 cm^{-1} (S(2) line) in the temperature range 300–200 K or 170 cm^{-1} (S(0) line) in the temperature range lower than 160 K.[63] To create the priority for stimulating rotational Raman transition over vibrational transitions, the pump light should be circularly polarized and the gas pressure should be considerably lower than that suitable for vibrational SRS operation.[60,63,71] In some cases the gas might be operating in low temperature, so that the molecular population in the lower rotational level and, hence, the stimulated scattering gain can be increased.[66,69] Pumped with nanosecond laser pulses, the optimum pressures for H_2 and D_2 are around 4–8 bar and the energy conversion efficiency can be up to 50–70%.[66,67,70]

7.4 Stimulated Brillouin scattering (SBS)

7.4.1 Fundamental description of Brillouin scattering

Brillouin scattering is related to the interaction between an incident optical wave and the elastic acoustic waves in a transparent medium. The feature of this effect is that the frequency shift of the scattering light is dependent on the scattering angle and the acoustic velocity in the medium.[6]

It is well known that when a collimated monochromatic light beam is incident on a grating, there will be several diffraction (or reflection) maxima along certain directions. Moreover, if the grating is moving with a given velocity, the frequency of the diffracted (or reflected) light will shift with respect to that of the incident light because of Doppler effect. In practice, an ultrasonic cell filled with appropriate acousto-optical medium can play such a role of moving grating. In this case, the density of the medium will experience a periodical spatiotemporal modulation. When a directional and monochromatic light beam passes through this ultrasonic cell, a frequency-shifted diffraction maximum can be observed in a definite direction. The frequency shift value and direction of the diffraction maximum are determined by the properties of the ultrasonic field in the medium

Generally, there is always a spontaneous acoustic field in any optical medium, which is generated by thermal elastic motion of a great number of particles of which the medium is composed. This thermal acoustic field can be visualized as a superposition of various monochromatic and plane acoustic waves. Each of these acoustic waves may produce a periodic spatial and temporal modulation of the density of the medium and, therefore, can cause a grating diffraction effect on the incident light beam. In this case the direction and frequency shift of the diffracted light will depend on the velocity, propagation direction, and frequency of the considered acoustic wave. This is the classical picture of spontaneous Brillouin scattering of light. It is known that the thermal elastic acoustic field in an ordinary medium is very weak; therefore, the observation of Brillouin scattering is very difficult if the input light beam is from a conventional light source.

The laser technology has opened a new opportunity for the development of Brillouin scattering technique. Using lasers as intense and coherent monochromatic light sources makes the observation of spontaneous Brillouin scattering much easier than before; and most importantly, if the monochromatic intensity of the input laser beam is high enough, the scattering light exhibits a stimulated feature. This is the stimulated Brillouin scattering (SBS) effect, first observed by Chiao et al. in 1964.[72]

The essential difference between spontaneous and stimulated Brillouin scattering is that the former is caused by very weak thermal acoustic fields, whereas the latter is caused by an intense laser beam-induced electrostrictive acoustic field. Upon the electrostrictive action of an intense laser field, the induced acoustic vibration can be created in the scattering medium, which turns to play the role of an induced density-modulated grating and produces the Brillouin scattering light. If the pump intensity is so strong that the growth of both the induced electrostrictive acoustic wave and the scattered light wave can significantly overcome their losses inside the medium, one may observe the simultaneous stimulated amplification of the both fields. This is a classical description of the SBS effect.

It is different from Raman scattering that the Brillouin scattering is not related to any change of molecular microscopic states, and the energy and momentum exchange only occurs between the optical field and the elastic acoustic field. In this sense, the Brillouin scattering is essentially a parametric interaction between these two fields, and a certain phase-matching requirement must be fulfilled.

On the other hand, both the optical field and the acoustic field can be quantized; therefore we can also use a quantum theoretical model to describe the mechanism of the elementary Brillouin scattering process that can be recognized as a parametric interaction between an input photon, a scattered photon, and a phonon inside the medium. The conservation of energy and momentum can be fulfilled in the following two possible ways.

7.4.1.1 Stokes scattering generation

This process can be described as the annihilation of an incident photon and the simultaneous creation of one scattered photon and one induced phonon. In this case the conservation of energy and momentum requires that

$$\left.\begin{array}{l} \nu_0 = \nu_s + \nu_a \\ \mathbf{k}_0 = \mathbf{k}_s + \mathbf{k}_a \end{array}\right\}, \tag{7.4-1}$$

where ν_0, ν_s, and ν_a are the frequencies of the incident photon, scattered photon, and induced phonon, and $\mathbf{k}_0$, $\mathbf{k}_s$, and $\mathbf{k}_a$ are the wavevectors of these three quanta, respectively. The feature of this scattering process is that the partial energy of the input optical field is transferred to the acoustic field. The Stokes frequency shift of the scattered light is determined by the phase-matching condition, as shown in Fig. 7.18.

Based on Eq. (7.4-1) we can assume that $\nu_0 \approx \nu_s$ and $k_0 \approx k_s$ because of $\nu_a \ll \nu_0, \nu_s$; thus from Fig. 7.18(b) one can find that

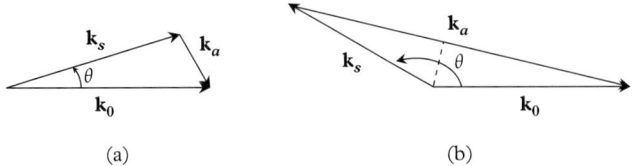

Figure 7.18 Phase-matching condition of the Stokes Brillouin process for (a) the forward scattering and (b) the backward scattering.

$$\frac{1}{2}k_a \approx k_0 \sin\frac{\theta}{2}. \quad (7.4\text{-}2)$$

Here, $k_a = 2\pi/\lambda_a = 2\pi v_a/v_a$, $k_0 = 2\pi n/\lambda_0 = 2\pi n v_0/c$; λ_0 and λ_a are the wavelengths of the incident light and the phonon; c/n and v_a are the velocities of the photon and phonon in the scattering medium; n is the refractive index of the medium, and θ is the angle between the incident light and the observed scattered light. Based on Eq. (7.4-2) and the first equation of Eq. (7.4-1) we obtain

$$\Delta v = v_0 - v_s = v_a = 2v_0\frac{nv_a}{c}\sin\frac{\theta}{2}. \quad (7.4\text{-}3)$$

From the above equation one can see that when $\theta = \pi$, the backward scattering light will have the maximum frequency shift, i.e.,

$$\Delta v_{max} = v_0 - v_s = 2v_0\frac{nv_a}{c}. \quad (7.4\text{-}4)$$

In experiments, we can accurately measure the frequency shift using a Fabry–Perot (F-P) etalon or optical heterodyne technique; therefore the velocity of the induced phonon field in a medium can be determined via Eq. (7.4-3) or Eq. (7.4-4). In addition, from Fig. 7.18 we can see that for the backward scattering light with $\theta = \pi$, the phase matching is collinear, and the induced phonon field will propagate along the same direction as the incident light.

7.4.1.2 Anti-Stokes scattering generation

This process can be described as the annihilation of one incident photon plus one existing phonon and the simultaneous creation of one scattered photon. The conservation of energy and momentum requires that

$$\left.\begin{array}{l}v_0 + v_a = v_{as}\\ \mathbf{k}_0 + \mathbf{k}_a = \mathbf{k}_{as}\end{array}\right\}, \quad (7.4\text{-}5)$$

where v_{as} and $\mathbf{k}_{as}$ are the frequency and wavevector of the anti-Stokes scattering light. The feature of this process is that the partial energy of the acoustic field in the medium is transferred to the scattering light. In this case the phase-matching condition is shown in Fig. 7.19.

The frequency shift of the anti-Stokes scattering light is determined by

$$\Delta v = v_{as} - v_0 = v_a = 2v_0\frac{nv_a}{c}\sin\frac{\theta}{2}. \quad (7.4\text{-}6)$$

When $\theta = \pi$, the backward scattering light has the maximum frequency shift. In this case, the phonon field should propagate along the opposite direction of the incident light beam.

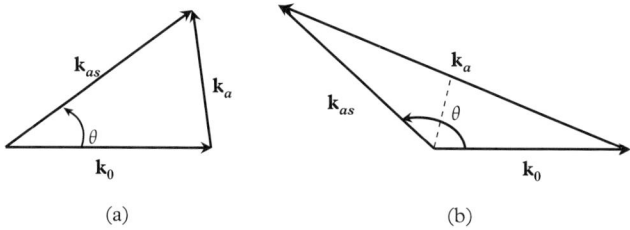

Figure 7.19 *Phase-matching condition of the anti-Stokes Brillouin process for (a) the forward scattering and (b) the backward scattering.*

7.4.2 Equations of interaction between light and acoustic field

The elementary process of SBS is related to the interaction between the intense coherent optical wave and the induced electrostrictive acoustic wave in a transparent optical medium. In order to provide a quantitative description of this process, first we need to consider the acoustic-field equation of the medium under the action of an intense optical field, and then write up Maxwell's equation taking account of opto-elastic effect, and finally solve the coupled field equations.

7.4.2.1 Acoustic-field equation of medium under action of intense light

First, let us consider the situation of a medium without applying an intense optical field. Let P denote the pressure of the elastic acoustic field in the medium, $\mathbf{u}$ denote the velocity of the elastic motion of particles of the medium along the propagation direction of the acoustic field, and ρ_0 denote the average mass density of the medium; then we apply Newton's second law to a unit volume of the medium and obtain the following dynamical equation:[73]

$$\nabla P + \Gamma \rho_0 \mathbf{u} + \rho_0 \frac{\partial \mathbf{u}}{\partial t} = 0. \tag{7.4-7}$$

Here, the first term represents the force imposed upon the unit volume of the medium owing to the pressure gradient; the second term represents a damping force proportional to the mass and velocity of the moving particle, where Γ is a damping factor; the third term represents the force proportional to the mass and acceleration. In the absence of an external optical field, these three forces keep in balance.

Now we consider the same medium in interaction with an intense applied optical field. In this case the intense optical field will exert a ponderomotive force (or electrostrictive force) on the particles of the medium and drive them to move. This force is proportional to the gradient of square of the amplitude function E of the optical field and can be expressed as[74]

$$\mathbf{f} = -\frac{\varepsilon_0}{2} \rho_0 \frac{\partial \varepsilon_r}{\partial \rho} \nabla(E^2) = -\frac{\varepsilon_0}{2} \gamma \nabla(E^2), \tag{7.4-8}$$

where

$$\gamma = \rho_0 \left(\frac{\partial \varepsilon_r}{\partial \rho} \right)_T \tag{7.4-9}$$

is called the electrostrictive coefficient, and ε_r is the relative dielectric constant of the medium. The negative sign in Eq. (7.4-8) implies that the electrostrictive force is in the counter direction of the force arising from the pressure gradient.

Considering the electrostrictive force arising from the existence of an intense optical field, Eq. (7.4-7) should be rewritten as

$$\nabla P + \Gamma \rho_0 \mathbf{u} + \rho_0 \frac{\partial \mathbf{u}}{\partial t} = \frac{\varepsilon_0 \gamma}{2} \nabla(E^2). \tag{7.4-10}$$

On the other hand, from the consideration of the conservation of mass element of the medium, we know that the elastic motion of the medium should fulfill the following continuity equation:

$$\nabla \cdot \mathbf{u} + \frac{1}{\rho_0} \frac{\partial \rho}{\partial t} = 0, \tag{7.4-11}$$

where $\rho(x, y, z, t)$ is the real mass density function of the medium with the presence of an optical field.

Taking a divergence operation on the both sides of Eq. (7.4-10) leads to

$$\nabla^2 P + \Gamma \rho_0 \nabla \cdot \mathbf{u} + \rho_0 \nabla \cdot \left(\frac{\partial \mathbf{u}}{\partial t}\right) = \frac{\varepsilon_0 \gamma}{2} \nabla^2(E^2). \tag{7.4-12}$$

Based on Eq. (7.4-11) the above equation can be rewritten as

$$\nabla^2 P - \Gamma \frac{\partial \rho}{\partial t} - \frac{\partial^2 \rho}{\partial t^2} = \frac{\varepsilon_0 \gamma}{2} \nabla^2(E^2). \tag{7.4-13}$$

In terms of the adiabatic modulus of elasticity defined by

$$\beta = \rho_0 \frac{\partial P}{\partial \rho}, \tag{7.4-14}$$

Eq. (7.4-13) becomes

$$\nabla^2 \rho - \frac{\rho_0}{\beta} \Gamma \frac{\partial \rho}{\partial t} - \frac{\rho_0}{\beta} \frac{\partial^2 \rho}{\partial t^2} = \frac{\rho_0}{\beta} \frac{\varepsilon_0 \gamma}{2} \nabla^2(E^2). \tag{7.4-15}$$

It is known in acoustics that

$$\left.\begin{array}{c} \dfrac{\beta}{\rho_0} = v_a^2 \\ \Gamma = \alpha_a v_a \end{array}\right\}, \tag{7.4-16}$$

where v_a and α_a are the velocity and attenuation coefficient of the acoustic wave in the medium. Then Eq. (7.4-15) can be rewritten in its final form:

$$\nabla^2 \rho - \frac{\alpha_a}{v_a} \frac{\partial \rho}{\partial t} - \frac{1}{v_a^2} \frac{\partial^2 \rho}{\partial t^2} = \frac{1}{v_a^2} \frac{\varepsilon_0 \gamma}{2} \nabla^2(E^2). \tag{7.4-17}$$

This is the equation describing the spatiotemporal variation of density of the medium. The term on the right-hand side of the equation represents the electrostrictive contribution from an external electric field. For optical waves, only the laser beams can be strong enough to produce a significant electrostrictive acoustic wave in the medium.

7.4.2.2 Electromagnetic field equation in medium with photoelastic effect

On the one hand, the intense optical field can create an induced elastic acoustic field through the electrostrictive mechanism. On the other hand, the induced acoustic field may produce a periodic variation of density of the medium; this density variation of the medium will react with the incident light and give rise to Brillouin scattering. In this case the dielectric constant can be expressed as

$$\varepsilon_r = \bar{\varepsilon}_r + \Delta\varepsilon_r, \tag{7.4-18}$$

where $\bar{\varepsilon}_r$ is the average relative dielectric constant when no intense optical field is applied. According to Eq. (7.4-9), under the action of an intense optical field the change of the dielectric constant of the scattering medium is determined by

$$\Delta\varepsilon_r = \rho_0^{-1}\gamma\Delta\rho. \tag{7.4-19}$$

Substituting Eq. (7.4-19) into Eq. (7.4-18) leads to

$$\varepsilon_r = \bar{\varepsilon}_r + \frac{\gamma}{\rho_0}\Delta\rho = \bar{\varepsilon}_r + \frac{\gamma}{\rho_0}(\rho - \rho_0). \tag{7.4-20}$$

If we do not consider other nonlinear optical effects described in the previous chapters, the wave equation of the optical field **E** in an isotropic and transparent medium can be written as

$$\nabla^2 \mathbf{E} - \frac{1}{c^2}\frac{\partial^2}{\partial t^2}(\varepsilon_r \mathbf{E}) = 0. \tag{7.4-21}$$

Substituting Eq. (7.4-20) into the above equation results in

$$\nabla^2 \mathbf{E} - \frac{\bar{\varepsilon}_r}{c^2}\frac{\partial^2}{\partial t^2}\mathbf{E} = \frac{1}{c^2}\frac{\gamma}{\rho_0}\frac{\partial^2}{\partial t^2}(\rho\mathbf{E}). \tag{7.4-22}$$

This is the wave equation of an intense optical field in the medium involving an induced photoelastic effect.

7.4.3 Solution of coupled equations and gain coefficient of SBS

7.4.3.1 Coupled wave equations and their solutions

Combining Eqs. (7.4-17) and (7.4-22) results in the following coupled wave equations:

$$\left. \begin{array}{l} \nabla^2 \mathbf{E} - \dfrac{\bar{\varepsilon}_r}{c^2}\dfrac{\partial^2}{\partial t^2}\mathbf{E} = \dfrac{1}{c^2}\dfrac{\gamma}{\rho_0}\dfrac{\partial^2}{\partial t^2}(\rho\mathbf{E}) \\[2ex] \nabla^2 \rho - \dfrac{\alpha_a}{v_a}\dfrac{\partial\rho}{\partial t} - \dfrac{1}{v_a^2}\dfrac{\partial^2 \rho}{\partial t^2} = \dfrac{\varepsilon_0}{v_a^2}\dfrac{\gamma}{2}\nabla^2(E^2) \end{array} \right\}. \tag{7.4-23}$$

These are the fundamental equations to describe the SBS process. For simplicity, we assume that the overall optical field is composed of two monochromatic components: a pump light of frequency ω_0 and a backward Stokes SBS of frequency ω_s; the frequency of the induced electrostrictive acoustic wave is $\omega_a = \omega_0 - \omega_s$. The optical field and the induced acoustic field can be written as

$$\left. \begin{array}{l} \mathbf{E} = \mathbf{E}_0(\mathbf{r})e^{-i\omega_0 t} + \mathbf{E}_s(\mathbf{r})e^{-i\omega_s t} \\ \rho = \rho_a(\mathbf{r})e^{-\omega_a t} \end{array} \right\}, \tag{7.4-24}$$

where $\mathbf{E}_0$, $\mathbf{E}_s$, and ρ_a are the amplitude functions of these three wave fields. Substituting the above expressions into Eq. (7.4-23) we obtain the following amplitude wave equations for the three waves:

$$\left.\begin{array}{l}[\nabla^2 + k_0^2]\mathbf{E}_0(\mathbf{r}) = -\dfrac{\omega_0^2}{c^2}\dfrac{\gamma}{\rho_0}[\mathbf{E}_s(\mathbf{r})\rho_a(\mathbf{r})] \\[2mm] [\nabla^2 + k_s^2]\mathbf{E}_s(\mathbf{r}) = -\dfrac{\omega_s^2}{c^2}\dfrac{\gamma}{\rho_0}[\mathbf{E}_0(\mathbf{r})\rho_a^*(\mathbf{r})] \\[2mm] [\nabla^2 + k_a^2 + i\alpha_a k_a]\rho_a(\mathbf{r}) = \dfrac{\varepsilon_0}{v_a^2}\dfrac{\gamma}{2}\nabla^2[\mathbf{E}_0(\mathbf{r})\cdot\mathbf{E}_s^*(\mathbf{r})]\end{array}\right\} , \qquad (7.4\text{-}25)$$

where k_0, k_s, and $k_a = \omega_a/v_a$ are the magnitudes of wavevector of these three components.

To solve Eq. (7.4-25) we further assume that these three components are plane waves, among them the incident pump light wave and the induced acoustic wave are propagating along the same z-axis direction while the backward Stokes SBS wave is propagating along −z direction, and both optical waves are linearly polarized in the same direction. Based on these assumptions the amplitude functions of these three waves can be simply expressed as

$$\left.\begin{array}{l}\mathbf{E}_0(z) = \mathbf{e}_0 A_0(z)e^{ik_0 z} \\ \mathbf{E}_s(z) = \mathbf{e}_0 A_s(z)e^{-ik_s z} \\ \rho_a = \rho_a^0(z)e^{ik_a z}\end{array}\right\}. \qquad (7.4\text{-}26)$$

Here, $\mathbf{e}_0$ is the unit vector along the polarization direction of both light fields; $A_0(z)$, $A_s(z)$, and $\rho_a^0(z)$ are the real amplitude functions of the three waves. Substituting Eq. (7.4-26) into Eq. (7.4-25) and applying the slowly-varying-amplitude approximation to the three waves, one could finally obtain the analytical solutions of Eq. (7.4-25).[75]

The first step is to obtain the solution of amplitude function ρ_a^0 for the density wave. For this purpose we substitute Eq. (7.4-26) into the third equation of Eq. (7.4-25) and neglect the terms involving the second spatial derivatives, thus we obtain a simplified equation for the density wave

$$2ik_a\dfrac{\partial \rho_a^0(z)}{\partial z} + ik_a\alpha_a\rho_a^0(z) = -\dfrac{\varepsilon_0\gamma}{2}\dfrac{(k_0+k_s)^2}{v_a^2}A_0(z)A_s^*(z)e^{i(k_0+k_s-k_a)z}. \qquad (7.4\text{-}27)$$

Since the phase mismatch factor $\Delta k = k_0 + k_s - k_a$ plays an essential role only in the exponential term, approximately we have $(k_0+k_s)^2 \approx k_a^2$, thus the above equation can be simplified as

$$\dfrac{\partial \rho_a^0(z)}{\partial z} + \dfrac{\alpha_a}{2}\rho_a^0(z) = i\dfrac{\varepsilon_0\gamma}{4}\dfrac{k_a}{v_a^2}A_0(z)A_s^*(z)e^{i\Delta k z}. \qquad (7.4\text{-}28)$$

The solution of this equation can be expressed in an integral form, i.e.,[75]

$$\rho_a^0(z) = \dfrac{ik_a}{v_a^2}\dfrac{\varepsilon_0\gamma}{4}\int_0^z A_0(z')A_s^*(z')e^{[i\Delta k z' - \alpha_a(z-z')/2]}dz' + \rho_a^0(0)e^{-\alpha_a z/2}. \qquad (7.4\text{-}29)$$

Only in range of $(z - z') \le 2/\alpha_a$ is the contribution of the exponential term to the integral in the above solution significant; whereas in normal conditions for most optical media, α_a values are quite large, i.e., $2/\alpha_a \leq 10^{-2}$ cm. Thus, within the range of $(z-z') \leq 2/\alpha_a$ the spatial variations of $A_0(z)$ and $A_s(z)$ can be ignored. Then we can move the product of these two amplitude functions outside of the integral and obtain the following solution:

$$\rho_a^0(z) = \frac{ik_a\,\varepsilon_0\gamma}{v_a^2}\frac{1}{4}A_0(z)A_s^*(z)\frac{1}{i\Delta k+(\alpha_a/2)}[e^{i\Delta kz}-e^{-(\alpha_a z/2)}]+\rho_a^0(0)e^{-\alpha_a z/2}. \qquad (7.4\text{-}30)$$

For most SBS experiments the interaction length between the optical field and the scattering medium is always much greater than the attenuation length of the phonon, i.e., $z \gg (2/\alpha_a)$, therefore the above solution can be further simplified as

$$\rho_a^0(z) = \frac{ik_a\,\varepsilon_0\gamma}{v_a^2}\frac{1}{4}A_0(z)A_s^*(z)\frac{1}{i\Delta k+(\alpha_a/2)}e^{i\Delta kz}. \qquad (7.4\text{-}31)$$

The next stage is to get the solutions for the two optical waves. For this purpose we substitute the solution of the density wave given by Eq. (7.4-31) into Eq. (7.4-25); after neglecting the terms containing the second-order spatial derivatives and the terms involving the product of two first-order spatial derivatives we obtain

$$\left.\begin{array}{l}\dfrac{\partial A_0(z)}{\partial z}=-\dfrac{\omega_0^2\varepsilon_0\gamma^2 k_a}{8c^2\beta k_0}\dfrac{1}{i\Delta k+(\alpha_a/2)}A_0(z)|A_s(z)|^2\\[2ex]\dfrac{\partial A_s(z)}{\partial z}=-\dfrac{\omega_s^2\varepsilon_0\gamma^2 k_a}{8c^2\beta k_s}\dfrac{1}{-i\Delta k+(\alpha_a/2)}A_s(z)|A_0(z)|^2\end{array}\right\}. \qquad (7.4\text{-}32)$$

Using the following relations:

$$I=\frac{1}{2}\varepsilon_0 cn_0|A|^2=\frac{1}{2}\varepsilon_0 cn_0(AA^*),$$

$$\frac{\partial}{\partial z}|A^2|=\frac{\partial}{\partial z}AA^*=A\frac{\partial A^*}{\partial z}+A^*\frac{\partial A}{\partial z},$$

we can rewrite Eq. (7.4-32) as

$$\left.\begin{array}{l}\dfrac{\partial I_0(z)}{\partial z}=-g_0 I_s(z)I_0(z)\\[2ex]\dfrac{\partial I_s(z)}{\partial z}=-g_s I_0(z)I_s(z)\end{array}\right\}, \qquad (7.4\text{-}33)$$

where g_0 and g_s are determined by

$$\left.\begin{array}{l}g_0=\dfrac{\omega_0^2\gamma^2 k_a}{4n_0 c^3\beta k_0}\dfrac{\alpha_a}{(\Delta k)^2+(\alpha_a/2)^2}\\[2ex]g_s=\dfrac{\omega_s^2\gamma^2 k_a}{4n_0 c^3\beta k_s}\dfrac{\alpha_a}{(\Delta k)^2+(\alpha_a/2)^2}\end{array}\right\}. \qquad (7.4\text{-}34)$$

To further simplify the above expressions, we rewrite the phase mismatch factor Δk by

$$\Delta k=k_0+k_s-k_a=\frac{\omega_0 n_0}{c}+\frac{\omega_s n_0}{c}-\frac{\omega_a}{v_a}\approx 2\frac{\omega_0 n_0}{c}-\frac{\omega_0-\omega_s}{v_a}$$

$$=\frac{1}{v_a}\left[\omega_s-\left(\omega_0-\frac{2\omega_0 n_0 v_a}{c}\right)\right]=\frac{1}{v_a}[\omega_s-\bar{\omega}_s], \qquad (7.4\text{-}35)$$

where $\bar{\omega}_s$ is the central frequency of the Stokes SBS, determined by Eq. (7.4-4) under a perfect phase matching condition. Noticing that the difference between ω_0 and ω_s is extremely small, we can simplify Eq. (7.4-34) as

$$g_0 = g_s = \frac{\omega_0 \gamma^2 \omega_a}{n_0^2 c^2 \beta} \frac{\alpha_a v_a}{4(\omega_s - \bar{\omega}_s)^2 + (\alpha_a v_a)^2}$$

$$= \frac{\omega_0^2 \gamma^2}{n_0 c^3 \rho_0 v_a} \frac{(\alpha_a v_a/2)}{(\omega_s - \bar{\omega}_s)^2 + (\alpha_a v_a/2)^2}. \tag{7.4-36}$$

7.4.3.2 Exponential gain and threshold condition of SBS

From Eq. (7.4-33) we see that while the intensity of the pump wave decreases along its propagation (z-axis) direction, the intensity of the backward Stokes SBS wave increases along the $-z$ direction. If the depletion of the pump beam over a short distance can be nearly neglected, the intensity $I_0(z)$ in the second equation of Eq. (7.4-33) can be replaced by its initial value, i.e.,

$$\frac{dI_s(z)}{dz} = -g_s I_0(0) I_s(z). \tag{7.4-37}$$

The solution of this equation is

$$I_s(0) = I_s(z) e^{g_s I_0(0) z} = I_s(z) e^{G_s z}. \tag{7.4-38}$$

Here, $I_s(0)$ and $I_s(z)$ are the intensities of the backward stimulated scattering beam in the incident plane and exit plane, $I_0(0)$ is the intensity of the pump beam in the incident plane, and g_s and G_s are the SBS gain factor and gain coefficient, respectively. The exponential gain coefficient is a function of the frequency ω_s, i.e.,

$$G_s(\omega_s) = g_s(\omega_s) I_0(0) = \frac{\omega_0^2 \gamma^2}{n_0 c^3 \rho_0 v_a} \frac{(\alpha_a v_a/2)}{(\omega_s - \bar{\omega}_s)^2 + (\alpha_a v_a/2)^2} I_0(0). \tag{7.4-39}$$

At the central frequency position of $\omega_s = \bar{\omega}_s$, the exponential gain coefficient reaches its maximum:

$$G_s^{max} = G_s(\omega_s = \bar{\omega}_s) = \frac{\omega_0^2 \gamma^2}{n_0 c^3 \rho_0 v_a} \frac{2}{\alpha_a v_a} I_0(0). \tag{7.4-40}$$

The full width at half maximum of the gain curve is $(\alpha_a v_a) = 1/\tau_a$, where τ_a is the lifetime of the induced phonons in the medium. According to Eq. (7.4-9), the γ value is proportional to the average density ρ_0, hence G_s^{max} is proportional to ρ_0.

Until now we have assumed that the pump light is a monochromatic wave with a spectral linewidth much smaller than $\alpha_a v_a$, so that the gain linewidth of the SBS is mainly determined by the reciprocal of lifetime of the induced phonons.

However, if the spectral linewidth of the pump laser beam is much greater than $\alpha_a v_a$, the spectral width of the gain curve will be mainly determined by the spectral linewidth of the pump beam. In such a case, Eq. (7.4-39) can be modified as

$$G_s(\omega_s) = \frac{\omega_0^2 \gamma^2}{n_0 c^3 \rho_0 v_a} S'(\omega_s) I_0(0). \tag{7.4-41}$$

Here, $S'(\omega_s)$ is a normalized line-shape function describing the spectral distribution of the pump light. In particular, if the pump light exhibits a Lorentzian spectral profile, the gain coefficient can be phenomenologically expressed as

$$G_s(\omega_s) = g_s(\omega_s)I_0(0) = \frac{\omega_0^2 \gamma^2}{n_0 c^3 \rho_0 v_a} \frac{(\Delta\omega_0/2)}{(\omega_s - \bar{\omega}_s)^2 + (\Delta\omega_0/2)^2} I_0(0). \quad (7.4\text{-}42)$$

where $\Delta\omega_0$ is the full spectral linewidth of the pump laser beam. In this case the maximum gain will be

$$G_s^{max} = G_s(\omega_s = \bar{\omega}_s) = \frac{\omega_0^2 \gamma^2}{n_0 c^3 \rho_0 v_a} \frac{2}{\Delta\omega_0} I_0(0). \quad (7.4\text{-}43)$$

Here, we have $\Delta\omega_0 \gg \alpha_a v_a$. Comparing Eq. (7.4-43) to Eq. (7.4-40) shows that at the same overall pump intensity level, the maximum gain of SBS pumped with a broader spectral linewidth is much lower than that pumped with a narrower spectral linewidth.

In analogy to the case of SRS, the threshold condition for SBS process without using an optical cavity can be expressed as

$$e^{(G_s - \alpha')L} \gg 1, \quad (7.4\text{-}44)$$

where α' is the linear attenuation coefficient for the light wave and L is the gain length of the medium.

So far, we have only considered the generation of Stokes stimulated scattering. The same derivation procedure can also apply to the generation of anti-Stokes SBS, provided that there is an existing intense acoustic wave propagating along the opposite direction with respect to the pump laser beam. It can be achieved in experiments by arranging the transmitted pump laser beam be reflected from an external mirror and reenter the same scattering medium again. In that case, the backward propagating pump laser bean will interact with the existing forward electrostrictive acoustic wave and produce a forward anti-Stokes SBS with a scattering angle of $\theta = 180°$.

7.4.4 Materials and experimental setups for SBS studies

In principle, any optical media transparent at working wavelength can be used for generating SBS. The basic requirements for good substances for SBS studies can be summarized as: (i) low losses for both optical pump and generated supersonic fields; (ii) larger electrostrictive coefficient γ and average density ρ_0; (iii) capability of withstanding high pump intensity without damage, breakdown, or boiling.

The following are three major types of optical materials commonly used for generating backward SBS.[76-97]

7.4.4.1 Solids

These include inorganic crystals, optical glasses, organic crystals, optical polymers, etc. For these solid materials, the observed frequency shifts of the backward SBS are in the range 0.3–1 cm^{-1} (in units of wavenumber); the induced acoustic frequencies are in the hypersonic range, i.e., about 3–30 GHz; the optical absorption coefficients are about 10^{-2}–10^{-3} cm^{-1}; the attenuation coefficient of the induced hypersonic waves are about 10^2–10^3 cm^{-1}; the speeds of the hypersonic waves are in the range 4000–6000 m/s; and the lifetimes of the hypersonic phonons are in the range

10^{-9}–10^{-8} s. The materials can be in the form of solid rod as well as in the form of fiber or waveguide structure.

7.4.4.2 Liquids

These include CS_2, DMSO, benzene, acetone, CCl_4, glycerol, liquid gases, etc. For these liquid media, the measured frequency shifts of the backward SBS are in the range 0.15–0.3 cm^{-1}; the induced hypersonic frequencies are in the range 4.5–9 GHz; the optical absorption coefficients are about 10^{-2} cm^{-1}; the attenuation coefficients of the induced hypersonic waves are about 10^3–10^4 cm^{-1}; the speeds of the hypersonic waves are in the range 1000–3000 m/s; and the lifetimes of the hypersonic phonons are in the range 10^{-10}–10^{-9} s. The sample material can be in a liquid cell or filled into a hollow fiber or waveguide structure.

7.4.4.3 High pressure gases

These include H_2, N_2, CH_4, CO_2, etc., in the pressure range 10^1–10^2 bar and filled into a gas cell or tube. A higher pressure is necessary because the gain coefficient of SBS is proportional to the density of the scattering medium. The frequency shifts of the backward SBS are in the range 0.01–0.05 cm^{-1}; the induced hypersonic frequencies are in the range about 0.3–1.5 GHz; the optical attenuation can be neglected in those gases; the speeds of the hypersonic wave are in the range 150–500 m/s; and the lifetimes of the induced phonons are in the range 10^{-9}–10^{-8} s.

7.4.4.4 Experimental setups

The pump sources used for generating SBS in the above media are high power pulsed laser devices, such as Q-switched Nd:YAG lasers and high power dye lasers. In order to have a high stimulated scattering gain and resolvable SBS spectral line from the pump line, the spectral linewidth of the pump laser should be as narrow as possible.

The basic optical layouts for SBS experiments are shown schematically in Fig. 7.20. Figure 7.20(a) shows the most common optical setup for backward SBS observation, where the pump source is a pulsed laser device working in a single axial mode or otherwise with a linewidth narrower than 0.1 cm^{-1}; the combination of a polarizing prism and a quarter-wave plate is adopted to prevent the backward SBS signal re-entering the pump laser cavity. The focusing lens is used to provide high pump intensity and great single-pass gain. Under this pump geometry, only backward ($\theta = 180°$) Brillouin scattering can be finally stimulated because of its longest gain length. Figure 7.20(b) shows a special setup for generating cavity-enhanced large angle ($\theta < 180°$) SBS. For the latter case, the frequency shift of the stimulated scattering can be smoothly and precisely tuned by change the scattering angle θ value; however, the interaction length between the pump beam and stimulated scattering beam will be limited by their crossing angle. Figure 7.20(c) shows another special setup for $\theta = 90°$ SBS generation, where the pump beam is focused by a positive cylindrical lens to form a high intensity focal line along the 90° direction inside the scattering medium; if necessary an external cavity can be equipped to enhance the stimulated scattering.

Finally, the setup shown in Fig. 7.20(d) is most useful for a laser oscillator-amplifier system. In an ordinary laser oscillator-amplifier system, the optical quality of the laser beam from the oscillator could become degraded after passing through an amplifier due to the aberration influence from the latter. If one utilizes a SBS medium as a phase-conjugation mirror and let the SBS beam backward pass through the amplifier, this beam will get further amplified because the frequency shift is usually smaller than the gain width of the amplifier. Most importantly, the

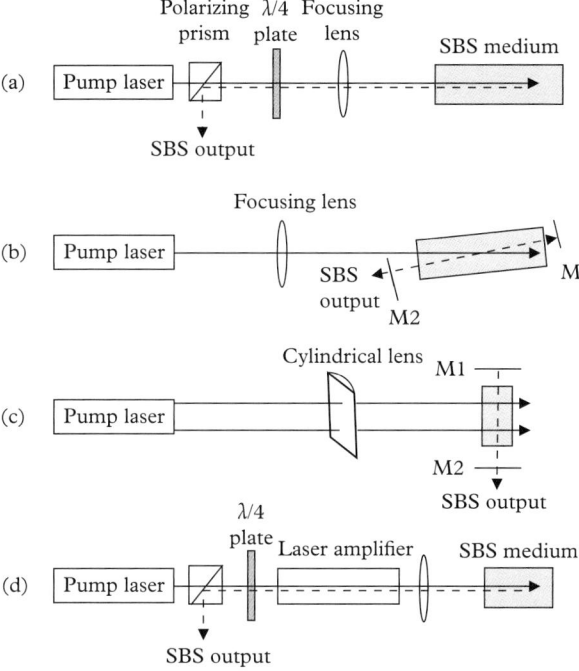

Figure 7.20 *Several types of experimental setups for generating SBS.*

aberration influence from the amplifier on this amplified beam can be finally removed due to the optical phase-conjugation property of backward SBS (see Chapter 8).

Usually, for backward SBS generated in a solid, liquid, or high pressure gas sample, the frequency shift is measured by a F-P interferometer system shown schematically in Fig. 7.21, in which the backward SBS is directed via a beam splitter into the F-P system while half of the sampled pump beam also collinearly enters the same F-P device.

As an example, Fig. 7.22 shows a photograph of the interferograms produced simultaneously by a single 532-nm pump laser pulse and the associated SBS pulse from a 5-cm long water sample. Here the full-circle rings are produced by the backward SBS beam, while the half-circle rings are due to the pump laser beam. The measured SBS frequency shift for water is ~ 0.22 cm^{-1}.

7.4.5 Major issues of experimental studies on SBS

7.4.5.1 Significance of SBS studies

SBS is a unique and efficient technique to generate intense and coherent hypersonic waves in a scattering medium; it can also be used to investigate the induced hypersonic field (such as the acoustic speed, acoustic-speed dispersion, lifetime of the phonon, etc.) as well as the opto-elastic properties of the scattering medium. The generation of SBS sometimes is accompanied by severe optical damages in the scattering medium or on the attached optical windows. Although the specific mechanisms causing these damages are not entirely clear yet, further studies will help

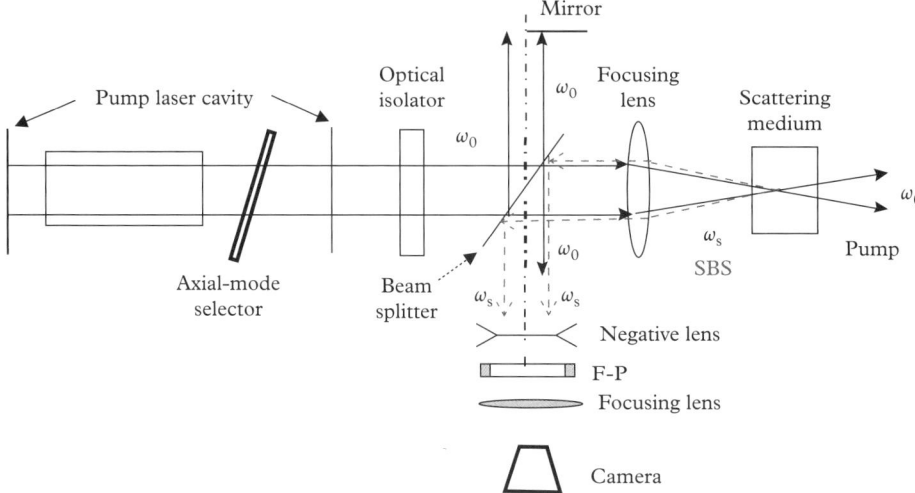

Figure 7.21 *Optical layout for the observation of backward SBS with a F-P system.*

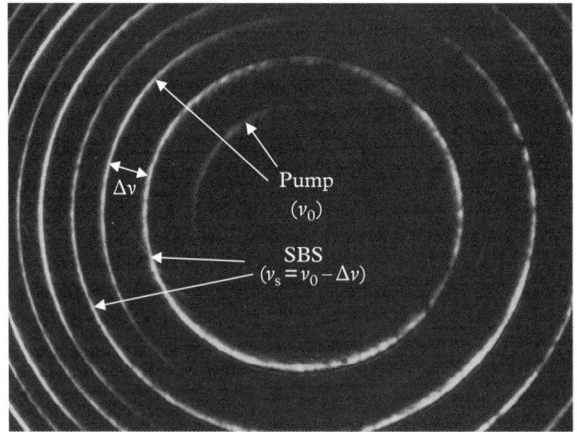

Figure 7.22 *Fabry–Perot interferograms of the backward SBS from a water sample (complete circle rings) and the 532-nm single axial-mode pump laser (half-circle rings). The free spectral range is 0.455 cm^{-1} and the measured SBS frequency shift is $\Delta v \approx 0.22$ cm^{-1}.*

researchers to provide a better understanding of this issue and to find effective solutions. In the other respect, the first discovery of the optical phase-conjugation phenomenon was based on a SBS experimental observation, and the SBS is one of the most efficient techniques to generate optical phase-conjugate waves (see Chapter 8). In addition, the most recent application of SBS is to be employed to produce slow or fast light propagation (see Chapter 16).

7.4.5.2 Difference between SRS and SBS

Essentially speaking, SRS is a microscopic scattering behavior determined by each individual molecule, which can be generated in both forward and backward directions by using laser pulses in nanosecond and picosecond regimes. In contrast, SBS is a macroscopic scattering behavior determined by the induced electrostrictive acoustic field in a continuous medium, it usually is generated in backward ($\theta = 180°$) direction with a maximum frequency shift. As the induced acoustic field exhibits certain build-up time and decay time, to reach a steady-state process the pump pulse duration should be much longer than the build-up time of the induced electrostrictive acoustic field. This is the reason for most SBS experiments the pump laser pulses' duration is longer than several nanoseconds. In some special experiments, the pump pulsewidth could be less than 1 ns but with a pulse interval ≤ 10 ns. Under some other pump conditions, the SBS generation working in its transient regime could also be observed.[85,92]

7.4.5.3 Nonlinear reflectivity of SBS

The nonlinear reflectivity of a SBS experimental device is defined as the ratio of the stimulated scattering output energy to the input pump energy.

Table 7.1 lists some reported optimum results for SBS performances in various scattering media under different conditions.

7.4.5.4 Delayed start of SBS

Most experimental SBS studies have been performed by using pump laser pulses of 10–20 ns duration. When the pump levels are not too high over the threshold, one can find an obvious time delay between the start points of the pump pulse and the SBS pulse. This delay cannot be simply explained by the pump intensity threshold requirement.

As one of many similar experimental examples, Fig. 7.23(a) shows the oscilloscope-measured temporal profiles of the 1064-nm pump pulse and the SBS pulse generated in a fluorocarbon (FC-75) liquid sample.[98] From this figure one can see that the SBS pulse started near to the peak position of the pump pulse, and lasted over the second half of the pump pulse. This is not a result of the intensity threshold requirement; instead, it is due to the accumulation requirement for the induced hypersonic field. Following the increase of pump intensity level, this delay time should become shorter. The similar result obtained under different conditions is shown in Fig. 7.23(b). In this case, the pump source was a pulse train from a mode-locked and Q-switched Nd:YAG laser, the duration of each single pulse was ~0.2 ns, the pulse spacing ~7.5 ns, the width of whole pulse train was ~45 ns, and the SF gas pressure was ~22 atm with an estimated phonon lifetime of ~18 ns.[99] From the latter figure one can see there is an obvious delay before the burst of the first SBS pulse.

The above-mentioned intrinsic delay of SBS generation essentially reflects the non-instantaneity of the interaction between the input laser field and the induced hypersonic field through the electrostrictive mechanism. The induced acoustic field is related to a collective macroscopic movement of great number of molecules; therefore, the growth of the induced acoustic wave field could be slower than the intensity change of the input pump laser field. On the other hand, the induced hypersonic field must be strong enough to provide a high enough nonlinear reflectivity for backward Stokes-shifted optical signals. For these two reasons, there should be a certain accumulation time requirement for the growth of the hypersonic field and for the burst of the SBS output signals.

Table 7.1 *Experimental data for SBS performances.*

Pump Laser Source	Pump Wavelength (μm)	Pump Conditions	Scattering Medium	Nonlinear Reflectivity (%)	Ref.
Ruby	0.6943	$\Delta \nu \leq 0.01 \text{cm}^{-1}$ 80 MW/cm^2	n-hexane ethyl ether	75 80	[80]
Nd:glass	1.064	$\Delta t = 100$ ps 10^{14} W/cm^2	laser-induced He plasma (30 atm)	30	[85]
CO$_2$	10.6	$\Delta t = 35$ ns 10^{13} W/cm^2	laser-induced H$_2$ plasma	30	[86]
Ruby	0.6943	$\Delta \nu \leq 0.01 \text{cm}^{-1}$ $\Delta t = 30$ ns	CS$_2$ CH$_4$ (80 atm)	70	[88]
XeF	0.3511	$\Delta \nu = 1$ GHz $\Delta t = 30$ ns 5 GM/cm^2	hexane isopropyl alcohol	70	[89]
Nd:YAG	1.319	$\Delta \nu = 1.6$ MHz 100 mW	GeO$_2$-doped silica fiber	65	[90]
KrF	0.248	$\Delta \nu = 0.08$ cm^{-1} $\Delta t = 27$ ns 180 mJ	SF$_6$ (9.8 atm) CCl$_2$F$_2$ (6 atm)	80 35	[91]
Iodine	1.32	$\Delta t = 1$ ns 35–55 GW/cm^2	N$_2$ (100 bar) SF$_6$ (20 bar)	50	[92]
HF	2.91	$\Delta t = 1$ μs 3 J	Xe (30–60 atm)	50	[93]
XeCl	0.308	$\Delta t = 25$ ns 150 mJ	n-hexane other liquids	30	[94]
Nd:YAG	1.064	$\Delta t = 18$ ns 14 mJ	fluorinert liquid FC75	98	[96]
Nd:YAG	1.064	$\Delta t = 14$ ns 60 mJ 20 mJ	organic crystal DLAP LAP	80 50	[97]

Δt = pump pulse duration; $\Delta \nu$ = pump spectral linewidth.

7.4.5.5 *Pulse compression effect of SBS*

Backward SBS is a very straightforward and efficient method to compress the laser pulses of nanoseconds duration. The compressed pulse usually exhibits a fast-rising edge and a slowly falling tail. The length of the tail part depends on the pump pulse shape and the pump energy level. When the pump level is not too high above the threshold level, it is easy to produce a

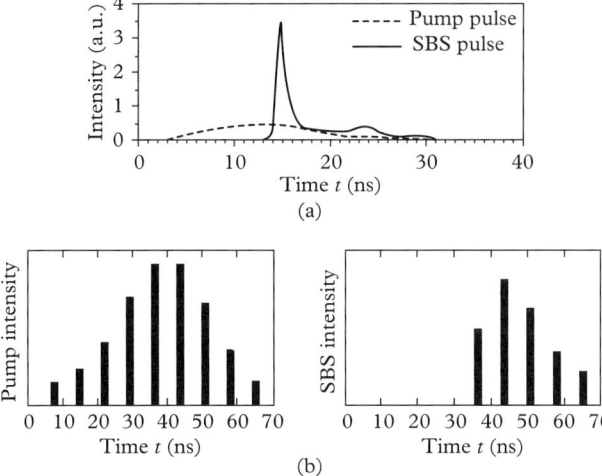

Figure 7.23 *(a) Temporal profiles of the 1064-nm pump pulse (dashed line) and the SBS pulse (solid line) generated in fluorocarbon (FC-75) liquid. (Reproduced with permission from Kmetic et al.[98] Copyright 1998, Optical Society of America); (b) Intensity distribution of the pump pulse train (left) and SBS pulse train (right) generated in SF_6 gas cell. (Reproduced with permission from Mullen.[99] Copyright 1990, IEEE).*

much narrower SBS pulse. In general experiments, the compression ratio between the pump pulse duration and the SBS pulse width can be higher than 10.[100–103]

7.5 Stimulated Kerr scattering (SKS)

7.5.1 Stimulated Rayleigh-wing scattering

The so-called stimulated Rayleigh-wing scattering (SRWS) was reported in 1965 by Mash et al.[104] They observed a broad (10–15 cm^{-1}) stimulated scattering spectrum on the Stokes side of the pump line in CS_2 and other Kerr-type liquids, excited by 694.3-nm and ~13-ns ruby laser pulses. Figure 7.24 shows the photograph of F-P interferograms obtained at two different pump levels. The authors of this report and several following papers related this effect with the laser induced reorientation of anisotropic liquid molecules, and proposed a phenomenological expression for the exponential gain factor of this stimulated scattering effect:[104–106]

$$g(\Delta\nu) \propto \frac{(2\pi\Delta\nu\tau)}{1+(2\pi\Delta\nu\tau)^2}. \tag{7.5-1}$$

Here, $\Delta\nu = \nu_0 - \nu'$ is the frequency shift of the stimulated scattering and τ is the molecular reorientation relaxation (or Debye) time of a given Kerr liquid consisting of anisotropic molecules. This expression predicts (i) a negative gain on the anti-Stokes side of the pump line, (ii) a zero-gain

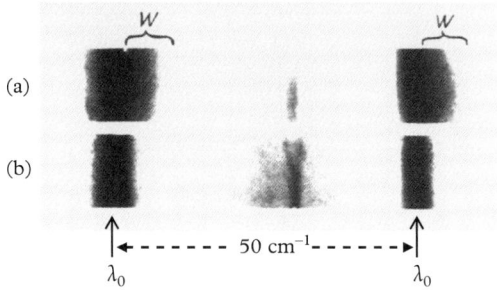

Figure 7.24 *Interferograms of backward SRWS from a CS_2 liquid cell at the pump levels of (a) $\sim$100 MW and (b) 85–90 MW. Free dispersion region of the F-P interferometer is 50 cm^{-1}. The label 'W' indicates the main spectral range predicted by the SRWS theory. (After Mash et al.[104]).*

at the pump line position, and (iii) a positive gain on the Stokes side with the maximum located at the frequency-shift position of $\Delta v_{max} = (2\pi\tau)^{-1}$. For CS_2 liquid which manifests the strongest reorientational Kerr effect, $\tau = 1.5–2$ ps, one would expect that the frequency shift of the spectral maximum will be $\Delta v_{max} = 2.7$–3.5 cm^{-1} on the Stokes side, which should be easily detected by a F-P interferometer or even a high-resolution grating spectrograph. However, in Fig. 7.24 no spectral maximum on the red-side of the pump line could be identified. Instead, one can only see a continuous spectral broadening starting from the pump line position and monotonically decaying towards the long-wavelength direction.

The aforementioned comparisons indicated that even in the beginning of SRWS studies, there were some inconsistencies between the experimental observations and the original theoretical predictions.[105,107,108] As a early effort to verify the validity of Eq. (7.5-1), Denariez and Bret performed a direct gain-curve measurement by using the pump–probe approach, based on a ruby laser working with a single-axial mode and ns-pulse duration.[106] The measured gain data as a function of the frequency shift are shown in Fig. 7.25. To obtain these results, an external signal beam of frequency v' backward passed through the nitrobenzene cell pumped by a forward laser beam of fixed frequency v_0. The input probe (signal) beam was actually a backward SBS beam generated in another liquid cell; the frequency v' of the signal beam could be varied by changing different liquids filled in the second cell. In Fig. 7.25, the fitting curves in solid-lines are based on Eq. (7.5-1), and the spectral peak on the top curve corresponds to the additional SBS amplification in nitrobenzene. Unfortunately, in the spectral region of $\Delta v = 0$–0.15 cm^{-1} that should be most essential for comparison, there was no experimental data could be used for comparison. Therefore, these results could not be regarded as strong evidence to support the theoretical prediction given by Eq. (7.5-1).

In 1970, Bol'shov et al. reported the spectral broadening properties of the pump line and the SRS line in both forward and backward directions in a CS_2 cell of 10–30 cm length, pumped by 532-nm and $\sim$10-ns laser pulses from a frequency-doubled Nd:glass laser working on a single-axial mode.[109] They observed very broad ($\geq$150 cm^{-1}) and monotonically decaying spectral broadening on the Stokes side of the pump line; at the same time they also observed a noticeable

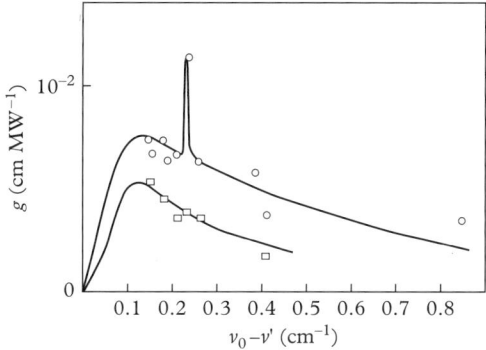

Figure 7.25 *Measured exponential gain factor of stimulated scattering amplification in nitrobenzene as a function of frequency shift Δv for parallel (top curve) and perpendicular (bottom curve) polarization of the pump beam of v_0 and signal beam of v'. (Reproduced with permission from Denariez and Bret.[106] Copyright 1968, American Physical Society).*

spectral broadening on the anti-Stokes side. These two experimental features are contrary to the predictions given by Eq. (7.5-1).

When self-focusing produced filamentation takes place in a Kerr liquid cell, the stimulated scattering spectrum may be deeply reformed due to the influence from self-phase modulation. The self-phase modulation often leads to a quasi-modulated spectral structure on both sides of the pump line.[110]

In conclusion, there are several major features of experimental results that could not be explained well either quantitatively or qualitatively by the original SRWS theory and the associated Eq. (7.5-1). Actually, according to uncertainty principle, the reorientational relaxation time (τ) of the anisotropic liquid molecules should only determine the spectral linewidth of an elementary Rayleigh-wing scattering process, not the maximum position of frequency shift. On the other hand, the original theory of SRWS failed to give a clear physical model to describe the specific mechanism of the SRWS generation.

7.5.2 Discovery of a super-broadband stimulated scattering

In 1985, He et al. reported a super-broadband (≥ 500 cm^{-1}) stimulated scattering observed on the Stokes side of the pump line and/or various-order SRS lines from a hollow glass fiber filled with a Kerr liquid such as CS$_2$ or benzene.[111,112] In that case, the liquid–core fiber system could provide higher local pump intensity and much longer interaction length. This super-broadband stimulated scattering was distinguished from all other types of stimulated scattering (including SRS, SBS, and early observed SRWS) by its special spectral distribution.

In most ordinary stimulated scattering experiments using a liquid cell, the local intensity provided by a focused pump laser beam is incompatible with the effective gain length or the focal depth. In other words, one could use a short-focal-length lens to increase the former while sacrificing the latter or, otherwise, use a long-focal-length lens to increase the latter while sacrificing the former. It is much desirable to have a special device that can provide both high local pump intensity and long effective gain length. The liquid-core hollow fiber system is one of the best approaches to achieve this goal.[113]

As an example of the early experimental results of the super-broadband stimulated scattering in a liquid-core hollow fiber system, Figure 7.26 shows the photographs of spectra for the forward output from a 7-m-long liquid-core hollow glass fiber filled with carbon disulfide (CS_2), benzene (C_6H_6), and carbon tetrachloride (CCl_4), respectively.[114] The inner diameter of the quartz glass hollow fiber was 100 μm. The parameters of the input pulsed pump laser were wavelength 532 nm (or 563 nm), pulse duration 10 ns, and spectral linewidth 0.05 cm^{-1}. The spectra were measured by a 1-m grating spectrograph with a spectral resolution better than 0.03 nm. The local pump intensity in the input end of the fiber sample could vary in the range 10^2–10^3 MW/cm^2.

For the CS_2-filled fiber system shown in Fig. 7.26(a), more than three orders of Stokes SRS are observed, which can be easily explained by cascaded SRS processes due to a higher local pump intensity and much longer gain length. In addition, most importantly, a super-broadband stimulated scattering contribution is added on the red-side of the transmitted pump line and each SRS line, respectively. The broadening range of this additional stimulated scattering contribution can reach 500–700 cm^{-1}.

For the C_6H_6-filled fiber system shown in Fig. 7.26(b), we can see a super-broadband stimulated scattering contribution added to three cascaded Stokes SRS lines. However, in this case no super-broadening is observed on the transmitted pump line.

Finally, for the CCl_4-filled fiber system shown in Fig. 7.26(c), we can see more than six orders of Stokes SRS lines; however, no spectral broadening can be observed on the pump line or any SRS line.

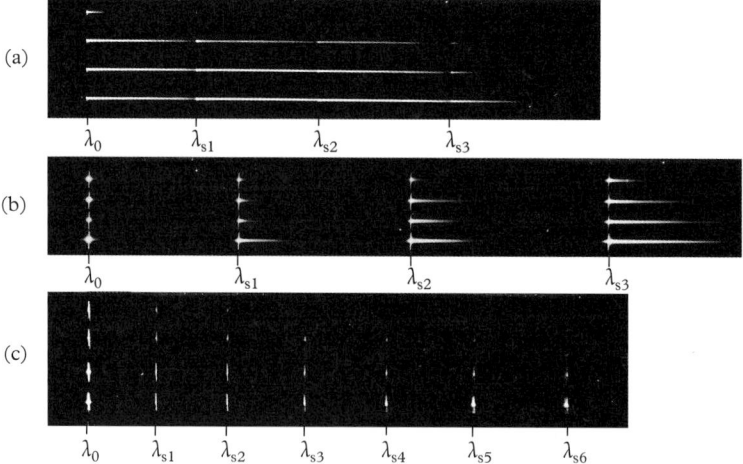

Figure 7.26 *Photographs of spectra of the forward stimulated scattering output from a 7-m-long liquid-core fiber pumped at various input intensity levels. (a) CS_2 sample, $\lambda_0 = 563$ nm, Raman shift 656 cm^{-1}; (b) C_6H_6 sample, $\lambda_0 = 532$ nm, Raman shift 992 cm^{-1}; (c) CCl_4 sample, $\lambda_0 = 532$ nm, Raman shift 460 cm^{-1}. (After He and Prasad.[114] Copyright 1990, Taylor & Francis).*

It is well known that CS_2 and C_6H_6 are typical Kerr liquids consisting of anisotropic molecules; but CCl_4 is a non-Kerr liquid consisting of isotropic molecules. Based on these facts, we can reasonably assume that the observed super-broadband stimulated scattering is inherently related to the reorientational optical Kerr effect.

7.5.3 Physical model of Rayleigh–Kerr and Raman–Kerr scattering

In a liquid consisting of nonpolar anisotropic molecules, the molecules obey a random orientational distribution if there is no applied external electric field or electromagnetic field. In the presence of an applied optical field, the anisotropic molecules may make a field-induced reorientational motion along the polarization direction of the optical field; then the liquid becomes partially anisotropic. This phenomenon is the well-known reorientational optical Kerr effect. Unlike the case of a gaseous medium, the anisotropic molecules in liquid phase cannot re-orientate themselves freely owing to the viscosity of the liquid. This means that they have to get additional energy to overcome the viscosity and to do reorientation work. If we assume the liquid is transparent at the incident light wavelength, there is no resonant absorption will occur at the wavelength of the applied optical field, then a red-shifted photon scattering process is the only way to transfer energy from the optical field to the scattering molecules. In this case, a molecule making an induced reorientation can be recognized as a scattering center, and the optical Kerr effect can be recognized as a red-shifted photon scattering process.

The microscopic picture of the elementary scattering process of optical (reorientational) Kerr effect can be described as follows: a molecule (scattering center) receives additional energy through the annihilation of an incident photon and the simultaneously creation of a red-shifted scattered photon; meanwhile the molecule exhausts this additional energy to overcome the rotational viscosity and to complete the reorientation work within the liquid. Depending on the initial and final location of the scattering molecule in its eigenstate level, the elementary scattering can occur in two possible ways, as shown schematically in Fig. 7.27.

If the scattering molecule stays in the same ground level at the beginning and end of an elementary process, as shown in Fig. 7.27(a), a red-shifted photon will be observed on the Stokes side of the pump line. This is the Rayleigh–Kerr scattering, which corresponds to the ordinary optical Kerr effect.[115] Here we assume that at the beginning and end of this elementary process, the scattering molecule has an arbitrary initial angle $\theta_0 = \theta$ and the final angle $\theta' = 0$ between the molecular axis and the polarization direction of the applied optical field.

In another case, if the induced reorientation is accompanied by a transition of the scattering molecule from the initial ground level to a Raman excited level, as shown in Fig. 7.27(b), a red-shifted photon can be observed on the Stokes side of the Raman line. This is the Raman–Kerr scattering that corresponds to the so-called Raman-induced Kerr (reorientation) effect.[116]

According to conservation of energy, for Rayleigh–Kerr scattering the energy loss of the incident photon is equal to the reorientation work; and for Raman–Kerr scattering the energy loss of the incident photon is equal to the sum of the reorientation work and the Raman excitation energy. Therefore, for these two cases the reorientation work W can be expressed as

$$\left. \begin{array}{ll} W(\Delta\theta) = h(\nu_0 - \nu') & \text{(Rayleigh – Kerr)} \\ W(\Delta\theta) = h(\nu_r - \nu') & \text{(Raman – Kerr)} \end{array} \right\}, \qquad (7.5\text{-}2)$$

where ν_0 is the frequency of the incident field, ν' is the frequency of the Kerr scattering, and ν_r is the Raman scattering frequency. Obviously, this reorientation work is an increasing function

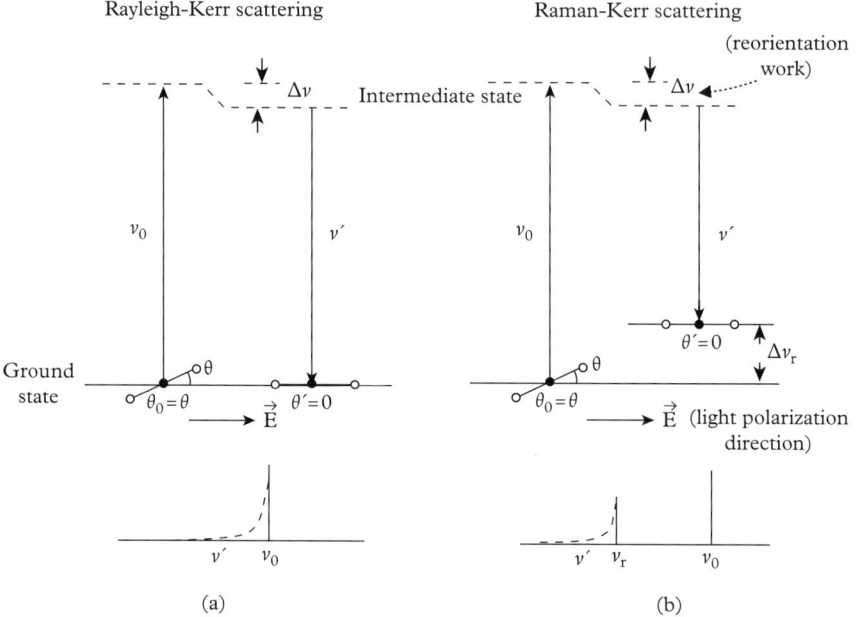

Figure 7.27 Schematic diagrams of an elementary Kerr scattering process and the induced molecular reorientation in a Kerr liquid: (a) Rayleigh–Kerr scattering and (b) Raman–Kerr scattering.

of the orientation-angle change, $\Delta\theta = \theta_0 - \theta' = \theta$; in other words, a larger $\Delta\theta$ implies a greater reorientation work $W(\Delta\theta)$ or a larger frequency redshift. Since different molecules have different initial orientation angle θ_0 with respect to the applied optical field, as a result of contributions from a great number of scattering molecules, we will see a continuous broadband spectral distribution on the red-side of the frequency of ν_0 or the frequency of ν_r, as shown in the bottom of Fig. 7.27(a) and (b).

7.5.4 Cross-section of Kerr scattering

In this subsection, we shall start from the general expression of the cross-section of Raman scattering to derive the explicit expression of the cross-section for Rayleigh–Kerr and Raman–Kerr scattering, respectively.[115,116]

From Eq. (7.2-21) we know that the differential cross-section for Raman scattering is given by

$$\sigma = \left(\frac{2\pi}{c}\right)^4 \frac{\nu_0 \nu_r^3}{(4\pi\varepsilon_0)^2 \hbar^2} \left\{ \sum_b \left[\frac{(p_0)_{ab}(p)_{bc}}{\nu_{ba} - \nu_0} + \frac{(p)_{ab}(p_0)_{bc}}{\nu_{ba} + \nu_r} \right] \right\}^2, \qquad (7.5\text{-}3)$$

where $(p_0)_{ab}$ and $(p)_{bc}$ are the matrix elements of the components of molecular dipole-moment operator $\mathbf{p}$ in the polarization directions of the incident light and scattering light, respectively; the subscript b labels all possible eigenstates contributing to the intermediate state involved in the

Raman transition from the ground state a to the excited state c. If the Raman medium is a Kerr liquid, the above expression can be further modified to describe the Raman–Kerr scattering process.

For simplicity, it is assumed that the applied optical field is linearly polarized, and the liquid consists of nonpolar molecules, each of them possesses a two-dimensional anisotropic polarizability. As shown in Fig. 7.28, $\mathbf{E}$ is the electric field vector of the applied optical field, z- and x-axes are the maximum and minimum polarizability directions for a given molecule in the liquid, and θ is the angle between $\mathbf{E}$ and the z-axis. In general, the induced molecular dipole-moment vector $\mathbf{p}$ and $\mathbf{E}$ have different directions owing to molecular anisotropy. Let $\mathbf{e}_0$ and $\mathbf{e}'$ denote the unit vectors along the directions of these two vectors, and α is the angle between $\mathbf{p}$ and the z-axis. In this case the product of matrix elements of the dipole-moment operator in Eq. (7.5-3) can be further expressed as

$$(p_0)_{ab}(p)_{bc} = (\mathbf{p} \cdot \mathbf{e}_0)_{ab}(\mathbf{p} \cdot \mathbf{e}')_{bc}$$
$$= (p_z \cos\theta + p_x \sin\theta)_{ab}(p_z \cos\alpha + p_x \sin\alpha)_{bc}. \quad (7.5\text{-}4)$$

In Eq. (7.5-4) there are two angle factors, α and θ, but, in fact, they are not independent from each other. Based on the semiclassical theory of molecular polarization, one of them can be eliminated by introducing a phenomenological parameter, the molecular polarizability tensor. In terms of semiclassical theory, the components of induced molecular dipole-moment vector $\mathbf{p}$ along the x- and z-axes can be written as

$$p_x = \alpha_{xx} E_x,$$
$$p_z = \alpha_{zz} E_z,$$

where E_x and E_z are the components of $\mathbf{E}$, whereas α_{xx} and α_{zz} are the corresponding elements of the molecular polarizability tensor. On the other hand, as shown in Fig. 7.28, E_x and E_z can be written as

$$E_x = |E| \sin\theta,$$
$$E_z = |E| \cos\theta;$$

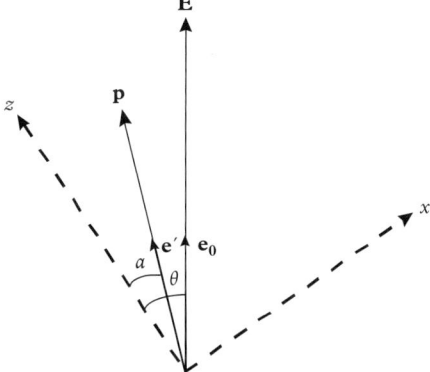

Figure 7.28 *Interaction between an applied optical field* $\mathbf{E}$ *and the induced molecular dipole moment* $\mathbf{p}$ *in a Kerr liquid.*

thus we have

$$\left.\begin{array}{l}\sin\alpha = \dfrac{p_x}{|p|} = \dfrac{\alpha_{xx}E_x}{\sqrt{(\alpha_{xx}E_x)^2+(\alpha_{zz}E_z)^2}} = \dfrac{\alpha_{xx}\sin\theta}{\sqrt{(\alpha_{xx}\sin\theta)^2+(\alpha_{zz}\cos\theta)^2}}\\[6pt] \cos\alpha = \dfrac{p_z}{|p|} = \dfrac{\alpha_{zz}E_z}{\sqrt{(\alpha_{xx}E_x)^2+(\alpha_{zz}E_z)^2}} = \dfrac{\alpha_{zz}\cos\theta}{\sqrt{(\alpha_{xx}\sin\theta)^2+(\alpha_{zz}\cos\theta)^2}}\end{array}\right\}. \quad (7.5\text{-}5)$$

Equation (7.5-5) implies that the angle α is determined by the molecular orientation angle θ and the molecular anisotropic polarizability. Only for isotropic molecules wherein $\alpha_{xx} = \alpha_{zz}$, will the angle α equal θ, otherwise $\alpha \ne \theta$. Substituting Eq. (7.5-5) into Eq. (7.5-4) results in

$$(p_0)_{ab}(p)_{bc} = [(p_z)_{ab}(p_z)_{bc}\alpha_{zz}\cos^2\theta + (p_x)_{ab}(p_x)_{bc}\alpha_{xx}\sin^2\theta + (p_x)_{ab}(p_z)_{bc}\alpha_{zz}\cos\theta\sin\theta$$
$$+ (p_z)_{ab}(p_x)_{bc}\alpha_{xx}\sin\theta\cos\theta] \times 1 \Big/ \sqrt{(\alpha_{zz}\cos\theta)^2+(\alpha_{xx}\sin\theta)^2}. \quad (7.5\text{-}6)$$

On the other hand, in the presence of an applied optical field the orientational distribution function $F(\theta)$ for molecules in a Kerr liquid is governed by (cf. Eq. (5.6-2))

$$F(\theta) = \dfrac{1}{4\pi}\left[1 + \dfrac{|E|^2}{6k_BT}(\alpha_{//}-\alpha_{\perp})(3\cos^2\theta-1)\right], \quad (7.5\text{-}7)$$

where $\alpha_{//} = \alpha_{zz}, \alpha_{\perp} = \alpha_{xx}$, k_B is the Boltzmann constant, and T is the absolute temperature. This expression means all molecules follow a symmetric orientational distribution with respect to the polarization direction of the incident light field; thus we can divide all molecules into two groups: one with $+\theta$ values and the other with $-\theta$ values. If the molecule density is N and the interaction volume is V, the overall scattering probability of all molecules having the same $|\theta|$ is determined by the summation of Eq. (7.5-3) for $NVF(\theta)$ molecules. As a result of such a summation, contributions from the first two terms in the square brackets of Eq. (7.5-6) will be doubled, while the last two terms will cancel each other. For this reason Eq. (7.5-6) can be simplified as

$$(p_0)_{ab}(p)_{bc} = (p_0)_{bc}(p)_{ab}$$
$$= [(p_z)_{ab}(p_z)_{bc}\alpha_{//}\cos^2\theta + (p_x)_{ab}(p_x)_{bc}\alpha_{\perp}\sin^2\theta] \quad (7.5\text{-}8)$$
$$\times 1 \Big/ \sqrt{(\alpha_{//}\cos\theta)^2+(\alpha_{\perp}\sin\theta)^2}.$$

Based on Eqs. (7.5-8) and (7.5-3), the final expression of cross-section for the generalized Raman scattering of an anisotropic liquid molecule can be written as

$$\sigma = \left(\dfrac{2\pi}{c}\right)^4 \dfrac{v_0 v_r^3}{(4\pi\varepsilon_0)^2}\left[\dfrac{\alpha_{zz}^{ac}\alpha_{//}\cos^2\theta + \alpha_{xx}^{ac}\alpha_{\perp}\sin^2\theta}{\sqrt{\alpha_{\perp}^2+(\alpha_{//}^2-\alpha_{\perp}^2)\cos^2\theta}}\right]^2. \quad (7.5\text{-}9)$$

Here,

$$\alpha_{zz}^{ac} = \frac{1}{h}\sum_b \left[\frac{(p_z)_{ab}(p_z)_{bc}}{v_{ba} - v_0} + \frac{(p_z)_{ab}(p_z)_{bc}}{v_{ba} + v_r}\right]$$
$$\alpha_{xx}^{ac} = \frac{1}{h}\sum_b \left[\frac{(p_x)_{ab}(p_x)_{bc}}{v_{ba} - v_0} + \frac{(p_x)_{ab}(p_x)_{bc}}{v_{ba} + v_r}\right]$$
(7.5-10)

are molecular polarizabilities characterizing the Raman transition from state a to state c. If we assume that the anisotropy of molecular polarizability is not very large so that $(\alpha_{//}^2 - \alpha_\perp^2) \ll \alpha_\perp^2$, then Eq. (7.5-9) becomes

$$\sigma = \left(\frac{2\pi}{c}\right)^4 \frac{v_0 v_r^3}{(4\pi\varepsilon_0)^2}\left[\alpha_{xx}^{ac} + (\alpha_{zz}^{ac} - \alpha_{xx}^{ac})\cos^2\theta\right]^2. \tag{7.5-11}$$

Moreover, if the second term in the square brackets is much less than the first term, Eq. (7.5-11) can be finally simplified as

$$\sigma = \left(\frac{2\pi}{c}\right)^4 \frac{v_0 v_r^3}{(4\pi\varepsilon_0)^2}\left[(\alpha_\perp^r)^2 + 2\alpha_\perp^r(\alpha_{//}^r - \alpha_\perp^r)\cos^2\theta\right], \tag{7.5-12}$$

where $\alpha_{//}^r = \alpha_{zz}^{ac}$ and $\alpha_\perp^r = \alpha_{xx}^{ac}$ are the anisotropic molecular Raman polarizabilities related to the transition from state a to state c. The first term in the brackets of Eq. (7.5-12) is independent of θ and will not vanish even when the molecules are Raman-isotropic ($\alpha_{//}^r = \alpha_\perp^r$); this term contributes to the ordinary Raman scattering of isotropic molecules in liquid. The second term in the brackets of Eq. (7.5-12) is dependent on θ and the Raman anisotropy of the molecule; it contributes to the Raman–Kerr scattering in liquid. The cross-section of the latter scattering process is

$$\sigma(\Delta\theta) = \left(\frac{2\pi}{c}\right)^4 \frac{v_0 v_r^3}{(4\pi\varepsilon_0)^2} 2\alpha_\perp^r(\alpha_{//}^r - \alpha_\perp^r)\cos^2\Delta\theta. \quad \text{(Raman – Kerr)} \tag{7.5-13}$$

Here, $\Delta\theta = \theta_0 - \theta' = \theta$ is the orientation-angle change of the scattering molecule, which is related to the frequency shift of the scattered photon through Eq. (7.5-2),

$$\Delta v = v_r - v' = h^{-1} W(\Delta\theta). \tag{7.5-14}$$

Without any specific knowledge of the relationship between $\Delta\theta$ and Δv, we can generally write

$$\Delta\theta = f(\Delta v), \tag{7.5-15}$$

where $f(\Delta v)$ is an increasing function of Δv and can be experimentally determined for a given Kerr liquid. Using the above equation we rewrite Eq. (7.5-13) as

$$\sigma(\Delta v) = \left(\frac{2\pi}{c}\right)^4 \frac{v_0 v_r^3}{(4\pi\varepsilon_0)^2} 2\alpha_\perp^r(\alpha_{//}^r - \alpha_\perp^r)\cos^2[f(\Delta v)]. \quad \text{(Raman – Kerr)} \tag{7.5-16}$$

For Rayleigh–Kerr scattering, $a = c$ and $\nu_r = \nu_0$, we have

$$\left.\begin{array}{l} \alpha_{zz}^{aa} = \alpha_{zz} = \alpha_{\parallel} = \dfrac{1}{h}\sum_b \dfrac{2\nu_{ba}(p_z)_{ab}(p_z)_{ba}}{(\nu_{ba}^2 - \nu_0^2)} \\[2mm] \alpha_{xx}^{aa} = \alpha_{xx} = \alpha_{\perp} = \dfrac{1}{h}\sum_b \dfrac{2\nu_{ba}(p_x)_{ab}(p_x)_{ba}}{(\nu_{ba}^2 - \nu_0^2)} \end{array}\right\}. \quad (7.5\text{-}17)$$

Therefore, from Eq. (7.5-9) we can obtain the expression of the cross-section for the generalized Rayleigh–Kerr scattering

$$\sigma = \left(\frac{2\pi}{c}\right)^4 \frac{\nu_0^4}{(4\pi\varepsilon_0)^2}[\alpha_{\perp}^2 + (\alpha_{\parallel}^2 - \alpha_{\perp}^2)\cos^2\theta]. \quad (7.5\text{-}18)$$

Here, the first term in the brackets is independent of θ and will not vanish even when the molecules are isotropic, i.e., $\alpha_{\parallel} = \alpha_{\perp}$; so this term represents the ordinary Rayleigh scattering without any frequency shift. Whereas, the second term in the brackets of Eq. (7.5-18) is dependent on molecular anisotropy and the initial orientation angle θ; this term represents the Rayleigh-Kerr scattering. The cross-section of this scattering process is

$$\sigma = \left(\frac{2\pi}{c}\right)^4 \frac{\nu_0^4}{(4\pi\varepsilon_0)^2}(\alpha_{\parallel}^2 - \alpha_{\perp}^2)\cos^2\theta. \quad \text{(Rayleigh – Kerr)} \quad (7.5\text{-}19)$$

This expression can also be rewritten in a form similar to Eq. (7.5-16), i.e.,

$$\sigma(\Delta\nu) \approx \left(\frac{2\pi}{c}\right)^4 \frac{\nu_0^4}{(4\pi\varepsilon_0)^2} 2\alpha_{\perp}(\alpha_{\parallel} - \alpha_{\perp})\cos^2[f(\Delta\nu)]. \quad \text{(Rayleigh – Kerr)} \quad (7.5\text{-}20)$$

Knowing the differential cross-section of Kerr scattering, we can further consider the exponential gain behavior of the stimulated Kerr scattering (SKS) in the next subsection.

7.5.5 Exponential gain of SKS

There are two mechanisms that may cause molecular reorientations in a Kerr liquid. One is molecular thermal collision that leads to a random change of molecular orientations and leads to spontaneous Rayleigh-wing scattering. The other is intense optical field-induced molecular reorientations along the polarization direction of the incident light. The latter processes are accompanied by the Kerr scattering that can produce a net exponential gain provided that the intensity of the incident light is higher than a certain threshold value.

Since we have already obtained the expressions of the cross-section for Raman–Kerr and Rayleigh–Kerr scattering, following the same procedures described in Section 7.2.3 we can give the following formula for SKS (cf. Eq. (7.2-29)):

$$I(\nu', L) = I(\nu', 0)e^{GL}, \quad (7.5\text{-}21)$$

where $I(\nu', 0)$ is the initial spectral intensity of the Kerr scattering signal, L is the gain length, and G is the exponential gain coefficient determined by (cf. Eq. (7.2-30))

$$G(\theta) = N_a F(\theta)\sigma_k \frac{\lambda'^2}{h\nu_0 \delta\bar{\nu}'} I_0. \quad (7.5\text{-}22)$$

Here, N_a is the molecular density in the ground state, $F(\theta)$ is the molecular orientation distribution function given by Eq. (7.5-7), σ_k is the cross-section of Kerr scattering determined by either Eq. (7.5-16) or Eq. (7.5-20), I_0 is the overall pump intensity, and $\delta\bar{\nu}'$ is the spectral width of an elementary Kerr scattering process. Since this elementary process is related to an optical field-induced change of molecular orientation states, the average time of a molecule staying in a certain orientation state is limited by thermal collisions and characterized by Debye time τ. This limited molecular orientation-staying time will be the major factor determining the spectral profile of an elementary Kerr scattering process, which can be written as a Lorentzian function:

$$S(\delta\nu') = \frac{[1/(4\pi\tau)]^2}{(\delta\nu')^2 + [1/(4\pi\tau)]^2}. \tag{7.5-23}$$

Here $\delta\nu' = \nu_0 - \nu'$ for Rayleigh-Kerr scattering or $\delta\nu' = \nu_r - \nu'$ for Raman-Kerr scattering. From this function we know that the full spectral width of the elementary scattering process is

$$\delta\bar{\nu}' = \frac{1}{2\pi\tau}. \tag{7.5-24}$$

For simplicity, we can approximately replace $F(\theta)$ by $1/(4\pi)$; then the gain coefficient for red-shifted stimulated Raman–Kerr scattering can be expressed as

$$G(\Delta\nu = \nu_r - \nu' \geq 0) = \frac{\pi N_a \nu_r I_0}{2\varepsilon_0^2 c^2 h \delta\bar{\nu}'} \cdot \alpha_\perp^r (\alpha_\parallel^r - \alpha_\perp^r) \cos^2[f(\Delta\nu)]. \tag{7.5-25}$$

Similarly, for the red-shifted stimulated Rayleigh–Kerr scattering, we have

$$G(\Delta\nu = \nu_0 - \nu' \geq 0) = \frac{\pi N_a \nu_0 I_0}{2\varepsilon_0^2 c^2 h \delta\bar{\nu}'} \cdot \alpha_\perp (\alpha_\parallel - \alpha_\perp) \cos^2[f(\Delta\nu)]. \tag{7.5-26}$$

Equations (7.5-25) and (7.5-26) mean that upon the pump action of I_0, the molecules complete their induced reorientation along the polarization direction of the applied optical field meanwhile stimulate red-shifted scattering photons that form a continuous broadband spectrum on the red side of the ν_0 position or ν_r position. Since different red-shifted spectral components correspond to different orientation-angle changes, the overall red-shifted stimulated scattering spectrum shows the nature of *inhomogeneous broadening*. On the other hand, at the frequency positions of ν_0 and ν_r, the stimulated scattering contributions mainly come from those molecules which have an initial orientation angle of $\theta_0 \approx 0$. The frequency shift of stimulated scattering from those molecules can almost be neglected. However, the SKS from these molecules will provide a nonzero contribution on the anti-Stokes side of ν_0 line or ν_r line due to the spectral profile of an elementary Kerr scattering process. So that, based on Eq. (7.5-23), the spectral gain distribution on the anti-Stokes side can be simply written as

$$G(\Delta\nu \leq 0) = G(0) \frac{[1/(4\pi\tau)]^2}{(\Delta\nu)^2 + [1/(4\pi\tau)]^2}. \tag{7.5-27}$$

Here, $G(0)$ is the maximum exponential coefficient in the frequency position of ν_0 or ν_r. The physical meaning of this formula is that the spectral distribution of SKS on the anti-Stokes side shows a nature of *homogeneous broadening*.

194   *Stimulated Scattering of Intense Coherent Light*

Similar to Eq. (7.2-38), the threshold condition for observing SKS can be expressed by

$$e^{[G(\Delta v)-\bar{\alpha}]L} \gg 1. \qquad (7.5\text{-}28)$$

Here $\bar{\alpha}$ is the attenuation coefficient of the scattering medium. Since the gain coefficient G is a function of frequency shift, the threshold requirements for different spectral components are different; therefore, at different pump levels the observed spectral width of SKS might be apparently different. Nevertheless, once the pump intensity is high enough, all spectral components of the Kerr scattering can be simultaneously stimulated, then we will see a complete spectral distribution of the SKS.

Comparing Eqs. (7.5-25)–(7.5-27) to Eq. (7.5-1), one can see that the theoretical conclusions based on the SKS model and SRWS model are obviously different:

i. The SKS theory predicts a monotonically decreasing broadband spectral distribution on the Stokes-side, and a narrow-band spectral broadening on the anti-Stokes side of the v_0 line (or v_r line), and the gain maximum should be located at the v_0 (or v_r) position;

ii. The SRWS theory predicts a red-shifted spectral maximum, a zero-gain at the v_0 position, and a negative gain on the anti-Stokes side of the v_0 position.

Most experimental results of broadband stimulated scattering in Kerr liquids obtained under different conditions are basically in agreement with the predictions given by the SKS theory.[107–112] Some quantitative experimental results will be discussed in the following two subsections.

7.5.6   Experimental studies of forward SKS

In this subsection we shall discuss some typical experimental results of the forward SKS in a Kerr liquid filled hollow fiber system.[115–117] In these experiments, the quantitative spectral intensity distributions of the stimulated scattering output were measured with an optical-multichannel-analyzer (OMA-III) system in conjunction with a grating spectrograph.

Figure 7.29 shows the normalized spectral intensity distributions of the transmitted 532-nm pump line and the added forward stimulated Rayleigh–Kerr scattering (SRKS) contributions from a CS_2-filled hollow glass fiber pumped at various intensity levels. The wavelength, spectral width, and pulse duration of the linearly-polarized pump laser were 532 nm, 0.85 cm^{-1}, and 10 ns, respectively. The spectral resolution of the spectrograph system was 0.4 cm^{-1}; the detector of the OMA-III system was working in its linear response range. From Fig. 7.29 one can see that: (i) the spectral broadening becomes more obvious with increasing the pump intensity, (ii) there is no spectral maximum at the position (indicated here by an arrow) predicted by the SRWS theory, (iii) there is a considerable spectral broadening on the anti-Stokes side of the pump line.

Furthermore, Fig. 7.30 shows the measured relative exponential gain curves of the forward SRKS from the same CS_2-filled hollow fiber sample at various pump intensity levels. To obtain these curves the transmitted-pump-line contribution was deducted from the forward output spectrum; then a natural logarithm operation on the spectral intensity-distribution curves was made.[115] The most remarkable feature of Fig. 7.30 is that once the pump intensity level is high enough (approaching 200–300 MW/cm^2), the relative exponential gain curves show nearly the same characteristic distribution for a given Kerr liquid.

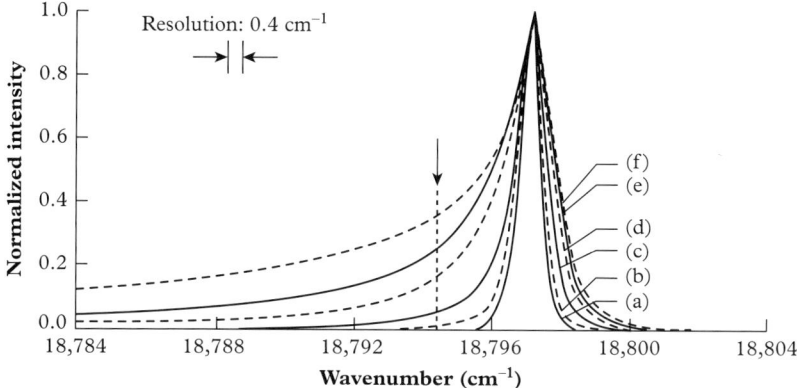

Figure 7.29 Normalized intensity distributions of the transmitted pump line ($\lambda_0 = 532$ nm) and the added SRKS contributions from a 2.5-m-long hollow fiber filled with CS_2 liquid. The arrow indicates the maximum position predicted by SRWS theory. The pump intensities were (a) 1.3 MW/cm^2, (b) 3.9 MW/cm^2, (c) 13 MW/cm^2, (d) 39 MW/cm^2, (e) 130 MW/cm^2, and (f) 390 MW/cm^2. (After He and Prasad.[115] Copyright 1990, American Physical Society).

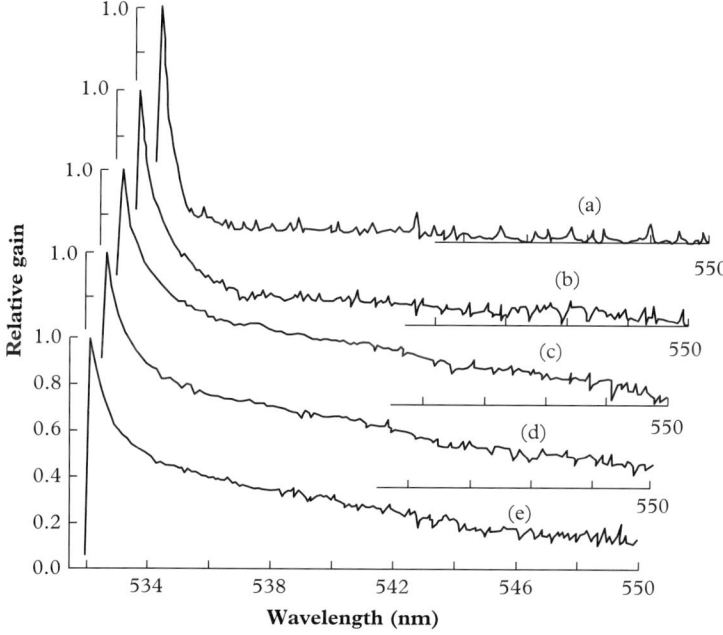

Figure 7.30 Measured relative exponential gain distributions of the forward SRKS from the same CS_2-filled hollow fiber sample at various pump intensity levels: (a) 65 MW/cm^2, (b) 91 MW/cm^2, (c) 120 MW/cm^2, (d) 200 MW/cm^2, and (e) 300 MW/cm^2. (After He and Prasad.[115] Copyright 1990, American Physical Society).

196 *Stimulated Scattering of Intense Coherent Light*

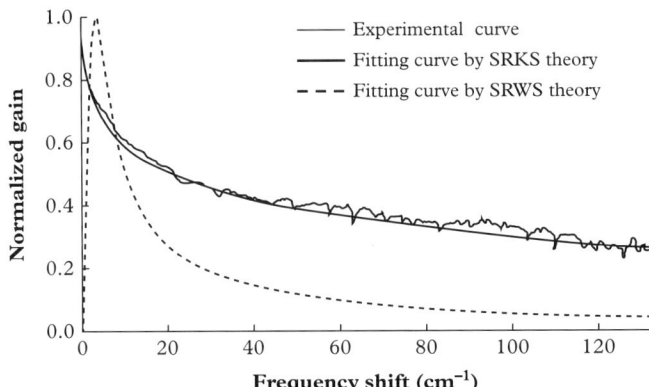

Figure 7.31 *Normalized exponential gain distribution of the forward SRKS in a CS_2-filled hollow fiber at a pump level of $I_0 = 270$ MW/cm^2. (After He and Prasad.[115] Copyright 1990, American Physical Society).*

According to Eq. (7.5-26), the normalized Stokes gain function of SRKS can be simplified as

$$G(\Delta \nu)/G(0) = \cos^2[f(\Delta \nu)], \quad (7.5\text{-}29)$$

where $f(\Delta \nu)$ is an unknown increasing function of $\Delta \nu$. As a trial function, it can be arbitrarily assumed that

$$f(\Delta \nu) = (A\Delta \nu)^B, \quad (7.5\text{-}30)$$

where A and B are two fitting parameters. For comparison, Fig. 7.31 shows the measured exponential gain curve (thin solid-line) at a pump level of 270 MW/cm^2 as well as the fitting curves given by two different theoretical models. The dashed-line curve is given by SRWS theory with a fitting parameter of τ = 2ps; whereas the thick-solid-line curve is given by SRKS theory using the following best-fitting parameters:[115]

$$A = 7.5 \times 10^9, B = 0.148. \quad (7.5\text{-}31)$$

Here, $\Delta \nu$ is in units of cm^{-1} and $f(\Delta \nu)$ is in units of degrees. It is clear that the experimental result can be well fitted by SRKS theory. On the contrary, the experimental result shown in Fig. 7.31 is totally different from that predicted by the SRWS theory.

On the other hand, the similar broadband stimulated scattering contribution has been also observed added in the various-order Stokes SRS lines, which is due to stimulated Raman–Kerr scattering in the same CS_2-filled hollow glass fiber system.[114,117] This implies that the CS_2 liquid has larger values for both $(\alpha_{//} - \alpha_{\perp})$ and $(\alpha_{//}^r - \alpha_{\perp}^r)$.

It is interesting to point out that under the same experimental conditions, the experimental results from a C_6H_6-filled hollow fiber sample are not the same as those from a CS_2-filled hollow fiber sample, although both liquids are typical Kerr media. The main difference is indicated by the broadening structures around the transmitted pump line from these two samples. The normalized spectral distributions of the transmitted 532-nm pump line from a C_6H_6-filled hollow fiber sample

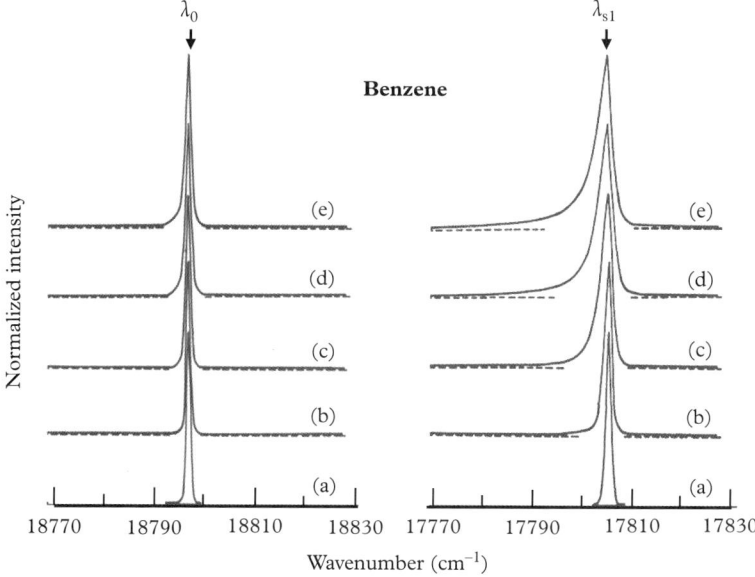

Figure 7.32 *Normalized spectral distributions around the transmitted pump line ($\lambda_0 = 532$ nm) (left column) and the first-order SRS line ($\lambda_{s1} = 562$ nm) (right column) from a 2.5-m-long benzene-filled hollow fiber at various pump intensity levels: (a) 3 MW/cm^2, (b) 9 MW/cm^2, (c) 30 MW/cm^2, (d) 90 MW/cm^2, and (e) 300 MW/cm^2. (After He et al.[116] Copyright 1990, AIP Publishing LLC).*

pumped at various pump intensity levels are shown in the left column of Fig. 7.32.[114,116] One can see that even at a high pump level (300 MW/cm^2), the spectral broadening still is nearly negligible. This fact implies that the anisotropy of polarizability for benzene molecules in ground state is smaller than that for CS$_2$ molecules in ground state. However, under the same experimental condition, a much evident spectral broadening can be easily observed on various orders of Stokes SRS lines, as shown in right column of Fig. 7.32.

From Fig. 7.32 we can see that the spectral broadening is much stronger for the SRS line than that for the pump line; this means that for C$_6$H$_6$ molecules we have $(\alpha_{//}^r - \alpha_{\perp}^r) \gg (\alpha_{//} - \alpha_{\perp})$. Based on this experimental fact, we know that for benzene liquid the anisotropy of molecular polarizability at the vibrational excited state is remarkably larger than that at the ground state. Here we are actually dealing with the Raman enhancement of the anisotropy of molecular polarizability. As a result of that effect, while molecules of benzene are being excited to a Raman excitation mode, they have a greater probability of making a reorientation motion along the polarization direction of the pump light. In that sense, this also is a Raman-enhanced optical reorientational Kerr effect.

It should be noted that the observed spectral broadening behavior mentioned above is not caused by the self-phase-modulation effect described in Chapter 6. The latter is based on the pump pulse-induced fast refractive-index change in a nonlinear medium, and consequently, the corresponding spectral broadening of the pump line should sensitively depend on the temporal structure or pulse duration of the input laser pulses. So far, we have only discussed the results from Kerr-liquid core fiber samples pump with ∼10 ns laser pulses. However, using the same samples but pumped by laser pulses with drastically different pulse duration (20 ns, 4 ps, and 0.5 ps), the observed spectral broadening behavior is basically the same as that described above.[117-120]

7.5.7 Experimental studies of backward SKS

Similar to SRS discussed in Section 7.3.2, if the spatial length of the pump laser pulse is much longer than the optical path-length of a scattering medium, one may expect nearly the same behavior for the forward and the backward SKS. Moreover, in the case of stimulated Rayleigh–Kerr scattering (SRKS), the study of backward output may provide an additional profit: the strong influence from forward transmitted pump line can be entirely eliminated. Here we shall examine some experimental results of backward SRKS generated in a 10-cm-long CS_2 liquid cell pumped by 10-ns and 532-nm laser pulses with different spectral linewidth ($\sim$0.1 and 1 cm^{-1}).[121]

Figure 7.33 shows the photographs of the backward stimulated scattering spectra generated by the 532-nm laser beam with two different spectral linewidth. When the pump linewidth is $\sim$1 cm^{-1}, as shown in Fig. 7.33(a) there is a smoothly decreasing super-broadband stimulated scattering on the red-side of the pump wavelength, which reminds us of the characteristic spectral distribution of the forward SRKS from a hollow fiber system. In contrast, pumped by the same laser wavelength but with a much narrower linewidth ($\sim$0.1 cm^{-1}), as shown in Fig. 7.33(b), the spectral broadening is not so obvious even though an overexposure was made to obtain this photograph. The other feature shown in Fig. 7.33(b) is that there is a strong and slightly broadened first-order Stokes SRS line.

In order to explain the noticeable difference between these two photographs shown in Fig. 7.33, we must consider the competition among various stimulated scattering mechanisms. It is well known that CS_2 is a good medium not only for SKS, but also for stimulated Raman and Brillouin scattering. In the present case, the spectral linewidth $\delta \tilde{v}'$ of an elementary Kerr scattering

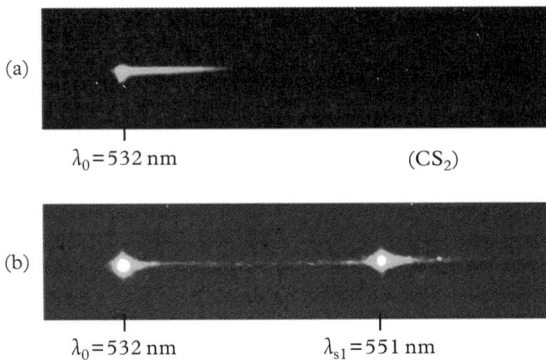

Figure 7.33 *Spectral photographs of the backward stimulated scattering from a 10-cm-long CS_2-liquid cell pumped by 532-nm laser beam with a spectral linewidth of (a) $\delta v = 1\ cm^{-1}$ and (b) $\delta v = 0.1\ cm^{-1}$, respectively. The pump intensity was 150 MW/cm^2 and the spectral resolution was $\sim$9 cm^{-1}. The second photograph was over-exposed to show the slight broadening. (After He et al.[121] Copyright 1997, AIP Publishing LLC) (See also Color fig. 8.).*

process is considerably greater than the spectral linewidth (0.1 or 1 cm^{-1}) of the 532-nm beam, therefore the SRKS behavior should not depend on the difference of the pump linewidth. On the other hand, however, it is also known that the threshold requirement for SBS could be sensitively dependent on the spectral linewidth of the pump laser, as verified by several experimental results showing that the SBS threshold rises with the increase of the pump linewidth.[122–124] Specifically, the threshold increase (or efficiency decrease) of SBS becomes more severe if the pump linewidth is much greater than the frequency shift of the backward SBS. For CS$_2$ liquid this shift is about

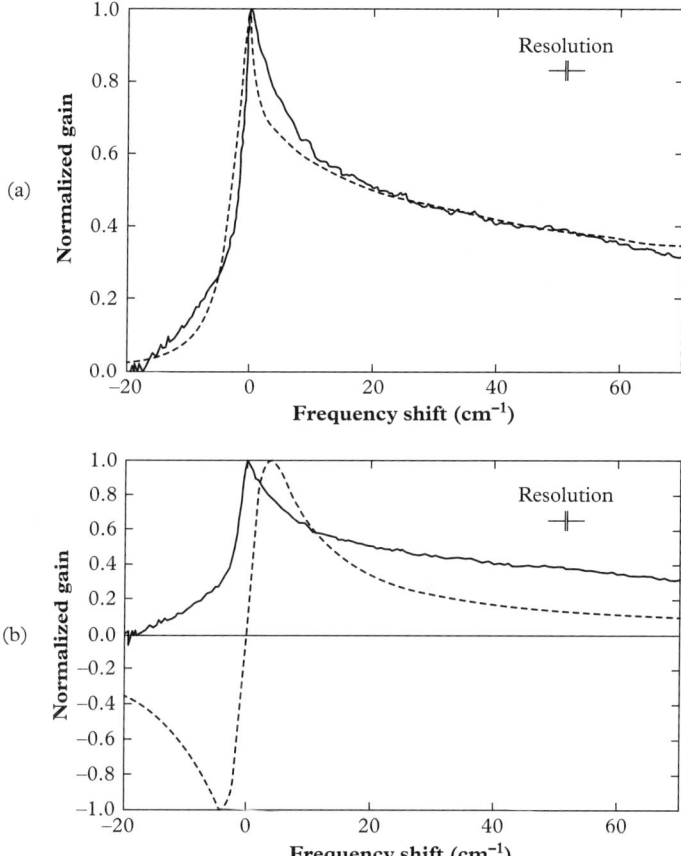

Figure 7.34 *(a) Normalized gain curves for backward stimulated scattering based on the measured data (solid line) and fitting (dashed line on the anti-Stokes side and dash-dotted line on the Stokes side) given by SRKS theory; the pump intensity $I_0 \approx 500$ MW/cm^2, linewidth 1 cm^{-1}, and the spectral resolution ~ 0.48 cm^{-1}; (b) Normalized gain curves of backward stimulated scattering based on the measured data (solid line) and fitting (dashed line) by SRWS theory with an assumed value of $\tau = 1.5$ ps. (After He et al.[121] Copyright 1997, AIP Publishing LLC).*

0.21 cm^{-1}, so that in the case of Fig. 7.33(a) the backward SBS is unlikely to take place, and the backward SRKS is the dominant process. By contrast, in case of Fig. 7.33(b) the pump linewidth is narrow enough then the SBS may take place more efficiently than SRKS. As a result of such competition, only a strong sharp line due to SBS can be observed with a much weaker broadening background due to SRKS. The observation with a F-P etalon verified that the strong sharp line shown in Fig. 7.33(b) was indeed due to the backward SBS in CS$_2$ liquid. Moreover, the observed backwards first-order SRS line shown in Fig. 7.33(b) could be generated by the narrow-line pump beam and/or the strong backward SBS beam.

Figure 7.34(a) shows the measured exponential gain curve (solid line) of backward SRKS from the 10-cm CS$_2$ cell, pumped by a 532-nm laser of 1 cm^{-1} linewidth at the intensity level of 500 MW/cm^2. In the same figure, two fitting curves (dashed line on the anti-Stokes side and dash-dotted line on the Stokes side) are given by SRKS theory with the same fitting parameter given by Eq. (7.5-31), which were used to fit the gain curve for the forward SRKR from a CS$_2$-filled hollow fiber system. In this figure we can see a fairly good agreement between the experimental results and the theoretical predictions based on the model of SRKS.

For further comparison, the same measured gain curve and the fitting curve (dashed line) given by the SRWS theory are shown in Fig. 7.34(b). Once again, one can see that there is a severe discrepancy between the experimental results and the prediction from the SRWS theory.

From the viewpoint of fundamental research, the studies of SKS reveal that the optical reorientational Kerr effect is always accompanied by a red-shifted light scattering process. On the other hand, by measuring the spectral distribution of the SKS, we can experimentally determine the molecular anisotropy as well as the relationship between the intense light-induced orientation angle change and the reorientation work done by the scattering molecule.[115,116] This kind of information and knowledge is valuable for basic research on physics of condensed matter.

From the application point of view, the stimulated Rayleigh–Kerr and Raman–Kerr scattering can be used as an efficient technical approach to generate or to amplify multi-wavelength and broadband coherent emission that is highly useful for special spectroscopic studies, for multi-wavelength coherent optical sensor systems, and for optical telecommunications based on the wavelength-division multiplexing mechanism.[118,120,125,126]

7.6 Stimulated Rayleigh–Bragg scattering (SRBS)

7.6.1 Early studies of stimulated thermal Rayleigh scattering in a linearly absorbing medium

The early theoretical analysis of so-called stimulated thermal Rayleigh scattering (STRS) was given by Herman and Gray in 1967.[127] Based on the consideration of intense light field induced density and temperature fluctuations in a linearly absorbing medium, they derived an expression of exponential gain factor for backward STRS generation,

$$g(\Delta \nu = \nu_0 - \nu') = g^{\max} \frac{-\Delta \nu (\delta \nu_0 + \delta \nu_R)}{(\Delta \nu)^2 + (\delta \nu_0 + \delta \nu_R)^2/4}. \qquad (7.6\text{-}1)$$

Here, ν_0 is the pump frequency, ν' the simulated scattering frequency, $\delta \nu_0$ the pump spectral linewidth, and $\delta \nu_R$ is the spontaneous Rayleigh scattering linewidth. From Eq. (7.6-1) one can see that the gain can be obtained only on the anti-Stokes side of the pump line; at the position of

$$\Delta \nu = -(\delta \nu_0 + \delta \nu_R)/2 \qquad (7.6\text{-}2)$$

the gain reaches its maximum value g^{max} that is determined by material properties and is proportional to the linear absorption coefficient of the medium. In liquids usually $\delta \nu_R$ is much narrower than $\delta \nu_0$, hence one would expect an anti-Stokes shift with the value equal to the half of the pump laser linewidth.

Following this theoretical prediction, almost at the same time, Rank et al. reported their experimental results of STRS in transparent liquids (CCl_4 and CS_2) added with iodine as a linear absorber.[128] The pump source was 694.3-nm and 10-ns laser pulses with a linewidth varied from 0.019 to 0.025 cm^{-1}. They measured the spectral shift by using an F-P interferometer, and claimed an anti-Stokes shift determined by the solvent of value (~ 0.01 cm^{-1}) that was close to the value of half of the pump linewidth, as predicted by Eq. (7.6-2). Their typical results are shown in Fig. 7.35. To obtain these results, the pump beam (its polarization was rotated by 90°) and the backward stimulated scattering beam were simultaneously passed through the F-P interferometer and a specially arranged four-quadrant analyzing sectors. In such a way, the pictures in quadrants I and III are the interferograms formed by the backward stimulated scattering beam, while quadrants II and IV are formed by the pump beam. Under this special arrangement, however, the experimental uncertainty of small spectral-shift might be essentially limited by the system error of the combined four-quadrant polarization analyzer. For example, taking a very careful look on Fig. 7.35(a), one can find that at a very low linear absorption coefficient value ($\alpha = 0.05$ cm^{-1}), there is only backward SBS can be seen with an obvious Stokes shift determined by the solvent. At a slightly higher value of $\alpha = 0.10$ cm^{-1}, as shown in Fig. 7.35(b), the newly generated STRS line can be seen with the SBS line together, while the former should have an anti-Stokes frequency shift according to the

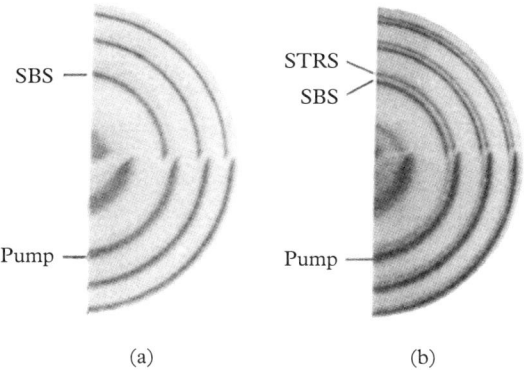

Figure 7.35 *F-P interferograms of the 694.3-nm pump line and backward STRS line from an I_2-doped CCl_4 solution, measured through a four-quadrant combined analyzer. (a) Linear absorption coefficient $\alpha = 0.05$ cm^{-1} and (b) $\alpha = 0.10$ cm^{-1}. The F-P free spectral range is 0.74 cm^{-1}. (Reproduced with permission from Rank et al.[128] Copyright 1967, American Physical Society).*

theoretical prediction. However, as shown in Fig. 7.35(b) it seems that the STRS line is also red-shifted, which is in contradiction with the prediction from the proposed STRS theory.

Later on, Bespalov et al. repeated the measurement in the same I_2-added CCl_4 (and other transparent solvents) samples pumped by 694.3-nm laser beam of 0.02 cm^{-1} linewidth, and no frequency shift was observed with a spectral resolution of 0.005 cm^{-1}, although Eq. (7.6-2) predicted a shift value of 0.01 cm^{-1}.[129] Particularly, Darée and Kaiser reconsidered the theoretical description of STRS under transient condition and gave a gain coefficient expression with no frequency shift, which was in striking contrast to the original steady-state theory mentioned above. In addition, they also re-measured the spectral behavior of STRS in the same I_2+CCl_4 sample, and showed no frequency shift within the accuracy of approximately 10^{-2} cm^{-1}.[130]

Looking back at the history of early studies of stimulated thermal scattering in linearly absorbing media, one may realize that there are two major problems or uncertainties. First, in spite of discrepancies among those different theoretical predictions, there was lack of a clear physical model (mechanism) to support their theoretical conclusions. Secondly, the original STRS theory predicted an anti-Stokes shift by a half of the pump linewidth, but all early experiments were based on a single axial-mode laser source with a very narrow spectral linewidth (usually ≤ 0.02 cm^{-1}), hence most of the frequency shift measurements of STRS were actually limited by the spectral resolution of the system used. It is quite obvious that much conclusive verification of STRS theory would be based on measurements by using pump laser with a broader spectral line (e.g., $\delta\nu_0 \geq 0.1 - 1 \text{cm}^{-1}$).

7.6.2 Finding of frequency-unshifted backward stimulated scattering in a two-photon absorbing medium

In other respects, the possible two-photon absorption (2PA) contribution to the stimulated thermal scattering in a pure organic solvents (such as benzene) was first proposed by Boissel et al. in 1978, although there was lack of specific identification of the observed stimulated backscattering in benzene pumped by 347-nm and 8-ns laser pulses.[131] The same possibility was also mentioned by Karpov et al. in 1991, in order to explain their observation in a pure liquid hexane sample, pumped by 308-nm and 8-ns laser pulses.[132]

In 2004, He et al. reported the first unambiguous observation of backward stimulated scattering in a two-photon absorbing dye-solution sample, pumped by 532-nm and 10-ns laser pulses.[133] Actually, the finding of this new stimulated scattering effect was inherently related to the characterization studies of two-photon active materials. In measuring the 2PA coefficient (β) of some two-photon active chromophore solutions by using ns-laser pulses, it was found that even at a given wavelength the measured β value for a given solution sample was not a material constant. Specifically, when the input intensity of laser pulses is higher than a certain (threshold) value, the apparent β value suddenly becomes higher with further increase of the input laser intensity. This unexpected increase of the apparent 2PA coefficient can be explained by assuming that there are some additional nonlinear processes which take partial energy from the input laser pulses. One could be a backward stimulated scattering process enhanced by 2PA. The experimental observation from backward direction of the input laser beam has verified the existence of a highly directional stimulated scattering when the input pump intensity is higher than a certain threshold value.

There are several noteworthy features of the newly observed effect, which are different from all other previously known stimulated scattering effects: (i) there is no frequency shift with a spectral resolution much narrower than the half of the pump linewidth; (ii) without two-photon absorbing dye only SBS can be observed in the pure solvent, which implies that the newly observed frequency-unshifted backward stimulated scattering is related with a two-photon excitation process; (iii) under the same experimental conditions, the pump threshold for observing frequency unshifted stimulated scattering in a two-photon absorbing dye solution is much lower than that for observing SBS in a pure solvent. All these experimental results cannot be well interpreted by any previously known theories of stimulated scattering.

In order to explain this newly observed backward stimulated scattering in a two-photon absorbing medium, He et al. have proposed a straightforward physical model, based on which the feedback mechanism of the frequency unshifted backward stimulated scattering is the reflection from an induced standing-wave Bragg grating.[133,134] This induced Bragg grating is initiated from the interference between the forward pump beam and the backward Raleigh scattering beam, such interference leads to a periodic spatial intensity modulation that further causes a periodic spatial refractive-index change. Based on this physical explanation, this special type of stimulated scattering is termed stimulated Rayleigh–Bragg scattering (SRBS).

7.6.3 Physical model of SRBS

Here let us consider a linearly transparent but two-photon (or multi-photon) absorbing medium pumped by a strong monochromatic laser beam. In our case, the backward propagating spontaneous Rayleigh scattering from the scattering medium can be recognized as original seed signals, which may interfere with the forward propagating pump beam to form a standing wave with a spatially modulated intensity distribution. This intensity modulation may further induce an intensity-dependent refractive-index change, and create a stationary Bragg grating inside the medium. A Bragg grating formed in such a way will offer a nonzero reflectivity for both the strong forward pump beam and the very weak backward scattering beam. However, the absolute value of the energy reflected from the pump beam to the scattering beam will be much greater than that from the backward scattering beam to the pump beam. As a net result, the backward scattering seed beam becomes stronger, as shown schematically in Fig. 7.36.

Moreover, a slightly stronger backward scattering seed beam will enhance the modulation depth of the induced Bragg grating and consequently increase the reflectivity of this grating, which means more energy will transfer from the pump beam to the backward scattering beam. These positive feedback processes may finally make the backward scattering signals to be stimulated.

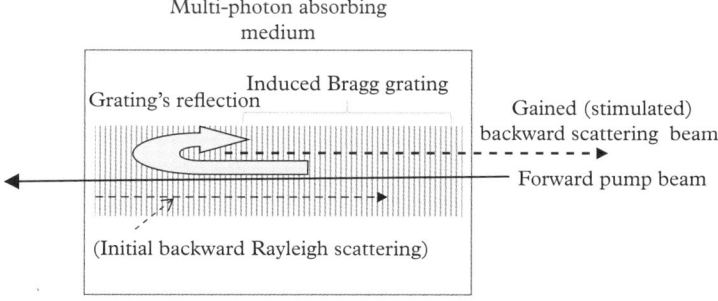

Figure 7.36 *Schematic illustration for the formation of induced stationary Bragg grating inside a multi-photon absorbing medium, as well as the energy transfer from the strong pump beam to the weak backward Rayleigh scattering beam through grating's reflection.*

Upon this assumption, a two- or multi-photon absorption (2PA or MPA) medium is desirable because of the following two reasons. First, accompanying 2PA (or MPA) there is a resonance-enhanced refractive-index change (see Section 5.3.2) that is necessary for forming an effective Bragg grating. Second, 2PA (or MPA) leads to an attenuation influence only on the strong pump beam not on the much weak backward seed scattering beam. In contrast, for a linear absorbing medium, the linear attenuation ratio is the same for both strong pump beam and weak scattering beam, which may prohibit the latter from being finally stimulated. This can explain why the effect of frequency-unshifted SRBS is much easier to be observed in a two-photon absorbing medium than in a linear absorbing medium.

Comparing SRBS with SBS, one may find that there is one thing in common: in both cases the positive feedback is based on the reflection of the pump beam from the induced grating. The difference is that for SBS the induced grating is a travelling hypersonic wave grating and the reflected beam experiences a frequency shift due to Doppler effect, whereas for SRBS the induced is a stationary (standing-wave) grating that offers reflection without any frequency shift.

According to the model described above, the key factor for SRBS generation is the induced refractive-index change, not the pump field induced heating effect. This is because the reflectivity of the induced Bragg grating is essentially determined by the intensity-dependent refractive-index changes (Δn) in a nonlinearly absorbing medium. There are various mechanisms that can possibly contribute to Δn, including population change of the absorbing molecules among their different energy levels, nonlinear refractive-index dispersion effect related with nonlinearly absorbing molecules, and stationary electrostriction effect. Different mechanisms have different temporal response features; therefore, through specially designed experiments it can be determined which mechanism is dominant under given experimental conditions.

7.6.4 Threshold requirement of SRBS

To give a mathematical description for the proposed Bragg grating model, we can write the intensity of the overall optical field inside the scattering medium as

$$I(z) = (I_p + I_s) + 2\sqrt{I_p I_s} \cos(4\pi n_0 z/\lambda_0). \tag{7.6-3}$$

Here, I_p is the intensity of the forward pump beam, I_s the intensity of backward Rayleigh scattering beam, and n_0 is the linear refractive index of the scattering medium at the pump wavelength λ_0. This periodic intensity modification will produce a refractive index change with the same spatial period ($\lambda_0/2n_0$), due to third-order nonlinear polarization effect. The spatial modulation of the intensity-dependent refractive-index change can be expressed as

$$\Delta n(z) = n_2 \Delta I(z) = 2n_2 \sqrt{I_p I_s} \cos(4\pi n_0 z/\lambda_0) = \delta n \cos(4\pi n_0 z/\lambda_0). \tag{7.6-4}$$

Here, n_2 is the nonlinear refractive-index coefficient, the value of which for a given medium is dependent on the specific mechanism of induced refractive-index change, and δn is the amplitude of the spatial refractive-index modulation.

Light induced periodic refractive-index changes inside a nonlinear medium will form an induced Bragg phase grating, which in return provides an effective reflection for both beams by the same reflectivity R. In an experiment using bulk liquid or solid sample, the laser induced Bragg grating is essentially a thick hologram grating with a cosinoidal spatial modulation, and its reflectivity is simply given by the well-known coupled wave theory of thick hologram gratings:[135]

$$R = \text{th}^2(\pi \delta n L/\lambda_0). \tag{7.6-5}$$

Here, L is the thickness of the grating or the effective gain length inside the scattering medium. From Eq. (7.6-4) we have

$$\delta n = 2n_2\sqrt{I_p I_s}. \quad (7.6\text{-}6)$$

Then Eq. (7.6-5) becomes

$$R = \text{th}^2(2\pi n_2\sqrt{I_p I_s}\, L/\lambda_0). \quad (7.6\text{-}7)$$

The threshold condition for stimulating the backward Rayleigh scattering is that for backward scattering signals the gain from grating reflection of the pump beam should be much greater than the losses due to linear attenuation; thereby the threshold requirement can be expressed as:

$$RI_p^{th} \gg I_s\{1 - \exp[-\alpha(\lambda_0)L]\}, \quad (7.6\text{-}8)$$

where $\alpha(\lambda_0)$ is the residual linear attenuation coefficient at λ_0. Under threshold condition, we can assume $R \ll 1$ and $\alpha(\lambda_0)L \ll 1$, the hyperbolic tangent function in Eq. (7.6-7) can be replaced by its arguments, then Eq. (7.6-8) can be simplified as

$$(2\pi n_2/\lambda_0)^2 L(I_p^{th})^2 \gg \alpha(\lambda_0), \quad (7.6\text{-}9)$$

or

$$I_p^{th} \gg \frac{\lambda_0}{2\pi n_2}\sqrt{\frac{\alpha(\lambda_0)}{L}}. \quad (7.6\text{-}10)$$

In obtaining Eqs. (7.6-8) and (7.6-9), we actually assume that near to the threshold level the pump attenuation due to 2PA is negligible.

The physical meaning of the above condition is that the backward SRBS is easier to be observed in a nonlinear medium possessing a larger nonlinear refractive-index coefficient n_2 and a smaller linear attenuation coefficient $\alpha(\lambda_0)$ at the pump wavelength. Two- or multi-photon absorbing media satisfy these requirements because they can manifest a resonance-enhanced refractive-index change while retain a smaller $\alpha(\lambda_0)$ value for very weak initial backward Rayleigh scattering signals.

7.6.5 Experimental results of SRBS in multi-photon absorbing media

The nonlinear medium for the first demonstration of SRBS is a two-photon absorbing dye (PRL802) solution in tetrahydrofuran (THF); this dye solution is highly transparent at 532 nm wavelength but shows an efficient 2PA at the same wavelength.[133,134] A 1-cm long quartz cuvette filled with dye solution of 0.01M concentration was pumped by 532-nm and 10-ns pulses with two significantly different spectral linewidth values, i.e., $\sim$0.08 cm^{-1} and $\sim$0.8 cm^{-1}, respectively. In both cases, once the input pump intensity is higher than a certain threshold level, a backward stimulated scattering beam with no frequency shift could be observed. When pumped with the narrow spectral line, an F-P interferometer was used to measure the spectral structure of the stimulated scattering with a spectral resolution of 0.025 cm^{-1}; when pumped with the broad spectral line, the stimulated scattering spectrum was measured by the grating spectrometer with a spectral resolution of 0.11 cm^{-1}.

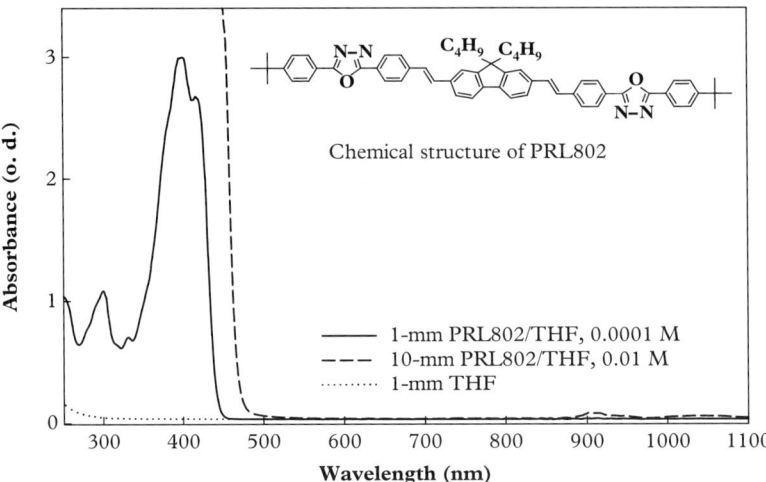

Figure 7.37 *Linear absorption spectra for solutions of PRL 802 in THF and for pure solvent THF. The chemical structure of the two-photon absorbing solute is shown in the top-right corner. (After He at al.[133] Copyright 2004, Optical Society of America).*

The chemical structure of the dye molecule and the linear absorption spectra for two samples of solutions with different concentration and path-length are shown in Fig. 7.37, from which one can see that there is no linear absorption for PRL802/THF in the spectral range from 480 to 875 nm, whereas there are strong linear absorption bands below the ~450 nm wavelength position. One therefore may expect that for the chosen pump wavelength of 532 nm, a moderate 2PA could take place.

In experiment, the input 532-nm laser beam was focused via an $f = 10$-cm lens onto the center of a 1-cm long glass cell filled with a PRL802/THF solution of 0.01M concentration. The incident angle of the input pump beam on the liquid cell was around 10–15° to avoid any reflection influence from the two optical windows of this cell. Under these experimental conditions, the 2PA induced upconversion fluorescence of ~475 nm from the sample solution could be readily seen, and once the input pump intensity was high enough (but still much lower than the threshold for generation SBS in pure THF), a highly directional backward stimulated scattering of no-frequency shift could be observed. The measured pump threshold for generating this effect in PRL802/THF solution was ~40 MW/cm^2 for pump linewidth ~0.08 cm^{-1} or ~0.8 cm^{-1}. In contrast, the pump threshold for generating SBS from a pure THF sample was measured to be ~100 MW/cm^2 for the pump linewidth of ~0.08 cm^{-1} and ~150 MW/cm^2 for the pump linewidth of ~0.8 cm^{-1}, respectively.

Figure 7.38 shows the interferograms of (a) the part of the pump beam with a spectral linewidth ~0.08 cm^{-1}, (b) the backward SRBS from the dye solution cell, (c) both the pump and SRBS beam together, and (d) both the pump and SBS beam from the pure THF sample together. All these photographs are taken by a single pulse shot.

From Fig. 7.38(a) and (b) one can see that both the backward SRBS beam and the input pump beam have nearly the same spectral width of $\Delta \approx 0.08$ cm^{-1}. Furthermore, from Fig. 7.38(c) one can see that there is no frequency shift between these two beams within the spectral resolution of

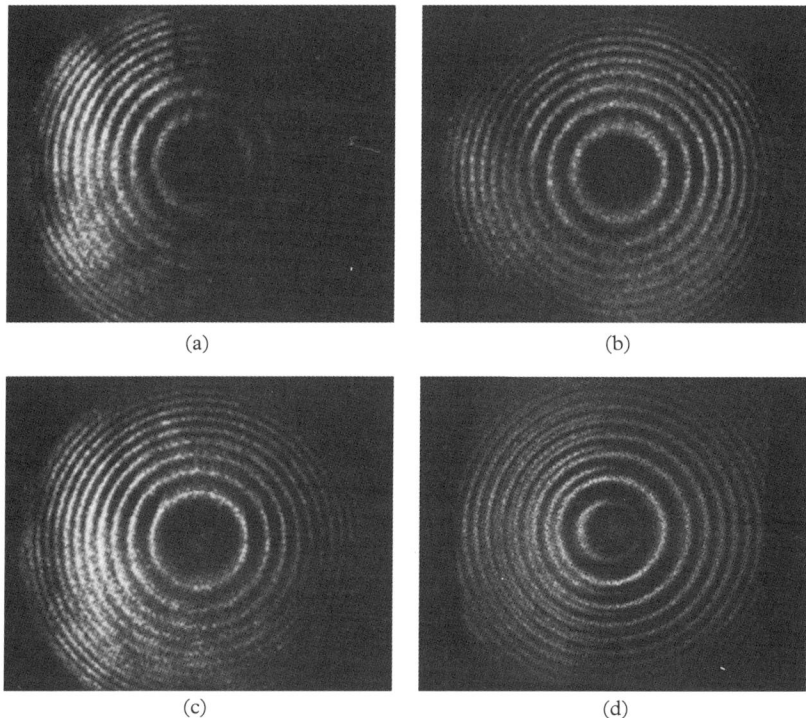

Figure 7.38 *F-P interferograms of (a) the part of the 532-nm pump beam, (b) the backward SRBS beam from a 1-cm PRL802/THF solution, (c) both beams together, and (d) the part of the pump beam and the backward SBS beam from a 1-cm pure solvent (THF) together. Pump linewidth was ~0.08 cm^{-1} and the free spectral range of the F-P interferometer was 0.5 cm^{-1}. (After He et al.^{133} Copyright 2004, Optical Society of America).*

0.025 cm^{-1}. Finally, Fig. 7.38(d) shows the interferogram of the half of the pump beam and the SBS beam from a 1-cm pure solvent (THF) sample, at a pump level of two times higher than the SBS threshold.

The measured near- and far-field patterns of the input pump beam and backward SRBS beam are shown in Fig. 7.39, respectively. The near-field patterns shown in Fig. 7.39(a) were obtained by projecting both the collimated pump beam and backward stimulated scattering beam on a ground-glass screen close to the sample position. The far-field patterns, shown in Fig. 7.39(b), were obtained by focusing these two beams via an $f = 100$ cm lens on a screen located at the focal plane. In both cases, the images on screens were recorded by a CCD camera. From Fig. 7.39 one can see that the beam size for the backward stimulated scattering is slightly larger than the pump beam in the near-field position, while the beam divergence of the stimulated scattering is slightly smaller than the pump beam, as indicated by the far-field pattern.

Under the pump condition with a broad spectral linewidth of ~0.8 cm^{-1}, the measured spectra of both pump and backward SRBS beam are shown in Fig. 7.40.^{134} Once again, there is no

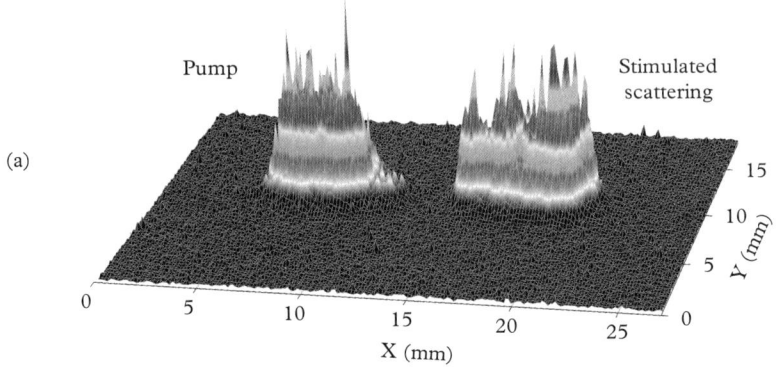

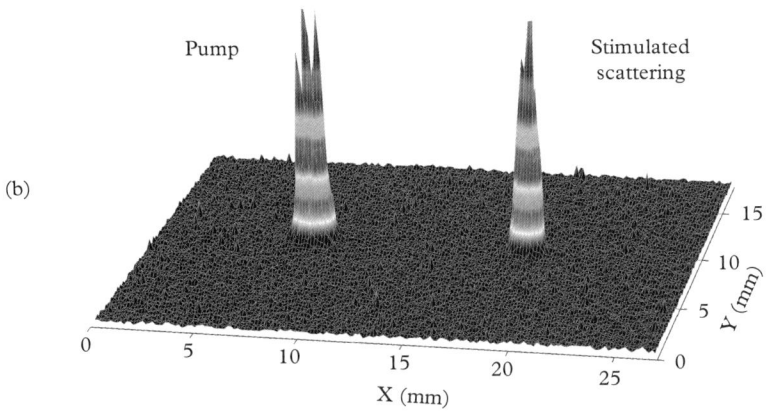

Figure 7.39 Measured near-field patterns (a) and far-field patterns (b) of the pump beam (left) and backward SRBS beam (right) at input intensity level of 160 MW/cm². (After He at al.[133] Copyright 2004, Optical Society of America).

measured frequency shift within a spectral resolution of 0.11 cm^{-1} which was much smaller than half of the pump linewidth.

Figure 7.41 shows the measured output SRBS intensity values as a function of the input pump intensity. These experimental data can be well fitted based on the theoretical model of pump reflection from the induced standing-wave Bragg grating. The theoretical curve is shown in this figure by the dashed line with the best fitting parameters: $n_2 = 2.93 \times 10^{-7}$ cm²/MW and 2PA coefficient $\beta = 9.46$ cm/GW.[134] From Fig. 7.41 one can see that at input intensity levels around 200–250 MW/cm², the conversion efficiency from the pump to the stimulated scattering is $\eta \approx 17\%$.

The same frequency unshifted SRBS effect is also observed in a three-photon absorbing dye-solution (PRL-OT04 in chloroform) pumped by 1064-nm laser pulses of duration 10–20 ns.[136] In that case, three-photon absorption enhanced refractive-index change promoted the formation of induced Bragg grating. Pumped by 1064-nm and ∼10-ns laser pulses of ∼0.8 cm^{-1} spectral linewidth, the measured pump threshold for generating SRBS in dye solution was ∼0.6 GW/cm²,

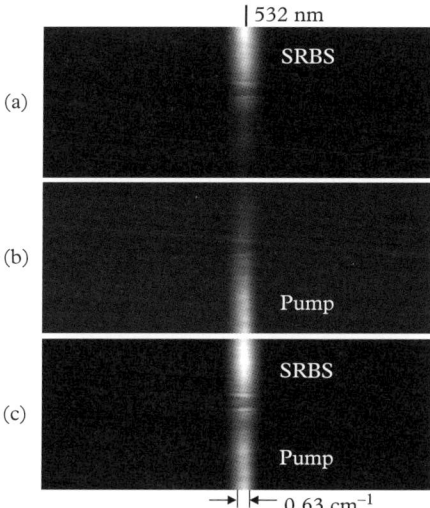

Figure 7.40 *Spectral photographs of (a) the backward SRBS beam, (b) the input 532-nm pump beam with a spectral linewidth of ~0.8 cm^{-1}, and (c) both the stimulated scattering beam and pump beam, recorded by a grating spectrometer with spectral resolution of 0.11 cm^{-1}. (After He et al.[134] Copyright 2005, American Physical Society).*

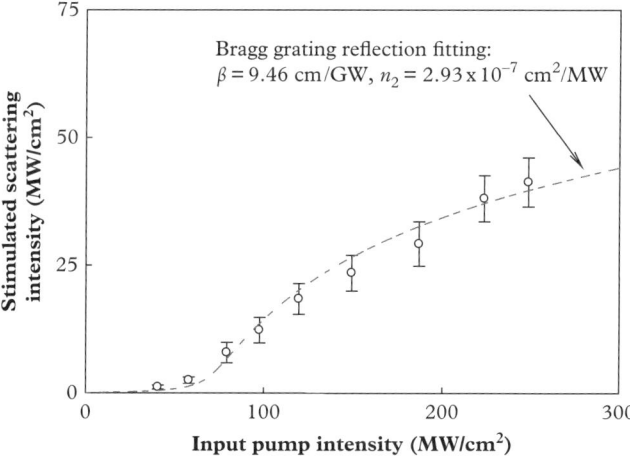

Figure 7.41 *Measured output backward SRBS intensity vs. input pump intensity; the dashed line is the best-fitting curve based on the grating reflection model. (After He et al.[134] Copyright 2005, American Physical Society).*

whereas it was 3.6 GW/cm² for generating SBS in pure solvent of chloroform. The conversion efficiency of SRBS was measured to be $\eta = 26\%$. In addition, a superior optical phase-conjugation property of the backward SRBS beam had been experimentally demonstrated by employing two different optical setups. In both cases, a specially introduced aberration influence of 4–5 mrad can be basically removed by the backward SRBS beam that retains a much smaller beam divergence of ≤0.4 mrad.

In conclusion, the preliminary results show that SRBS effect could be efficiently generated in any types of transparent multi-photon active media.[137] Comparing with other known stimulated scattering effects (such as SBS and SRS), SRBS exhibits two favorable features for applications, i.e., the low pump threshold and no frequency shift. On the other hand, there are two limitations related to this effect. First, the nonlinear reflectivity of this backward SRBS process seems to be ≤50% due to the feedback mechanism discussed in the preceding subsection. Second, so far the SRBS has been generated only by using the laser pulses in the ns-regime and with a spectral linewidth ≤1 cm⁻¹. Under these conditions, the effective gain length is essentially determined by the short one between the longitudinal coherent length and the path-length of the scattering sample. When the pump pulse duration becomes much shorter, the spectral linewidth becomes broader due to the uncertainty relationship, and the effective thickness of the induced Bragg grating will ultimately be limited by the coherent length of the pump beam. For this reason, one should not expect to generate SRBS by using sub-picosecond laser pulses or any pump linewidth notably broader than 1 cm⁻¹.

7.7 Stimulated Mie scattering (SMS)

7.7.1 Spontaneous and stimulated Mie scattering

As we mentioned in Section 7.1.2, in pure or neat optical media, spontaneous Rayleigh scattering, Raman scattering and Brillouin scattering can be observed separately or simultaneously, depending on specific experimental conditions. Pumping with monochromatic laser beam and properly choosing suitable materials of this type of media, researchers could generate various stimulated scattering, including SRS, SBS, SKS, and SRBS; in these cases the initial seed signals to be further stimulated originate from the corresponding spontaneous Raman, Brillouin, Kerr, and Rayleigh scattering, respectively.

On the other hand, in nature and daily life, there is another type of scattering phenomena, which is the scattering from impurity particles or external micro-objects suspended in a homogeneous neat medium. This type of scattering can be generally termed Mie scattering. If we say that the blue sky is a typical result of Rayleigh scattering of air molecules, then the white cloud appearance in air is the typical result of Mie scattering. For a given pure or neat optical medium and at a given incident light wavelength, the Rayleigh scattering cross-section is fixed; so that the spontaneous Rayleigh scattering capability for a given medium and at a certain wavelength is uncontrollable. By contrast, the Mie scattering capability of a given scattering system can be much easily controlled by adjusting the density, size, shape, coating and constituent of the impurity particles suspended in a neat solvent.

For a long time, researchers rarely thought the possibility to generate stimulated scattering from those impurity particles. However, the recent development of nanotechnology has brought a new opportunity for conducting thorough studies on Mie scattering properties of metallic and semiconductor nanoparticle systems.[138,139] Inspired by the finding of SRBS, in the later few years He et al. started to consider the possibility and a specific approach for generating stimulated scattering in a Mie scattering system. In doing so, they have realized that a resonant enhanced refractive-index change and effective formation of induced Bragg grating in the system are essentially important. In 2012, they formally reported the observation of stimulated Mie scattering (SMS) in metallic and semiconductor nanoparticle suspensions.[140,141] In these two cases, the initial (seed) signals are the spontaneous Mie scattering from nanoparticles suspended in a solvent, and the positive feedback is still provided by the induced Bragg grating, hence this special stimulated scattering effect can also be termed stimulated Mie–Bragg scattering (SMBS) or simply SMS. It is not surprising that metallic and semiconductor nanoparticles are the adequate candidates for realizing SMS, as both of them exhibit the property of induced refractive-index changes that might be enhanced by MPA and/or surface plasmon resonance.

7.7.2 SMS in metallic nanoparticle suspensions

Gold nanoparticles of various shapes and sizes dispersed in water or other solvents are very suitable for studying spontaneous and stimulated Mie scattering because of their high stability, good optical quality, and ease of preparation. In the first experiment that demonstrated SMS in metallic

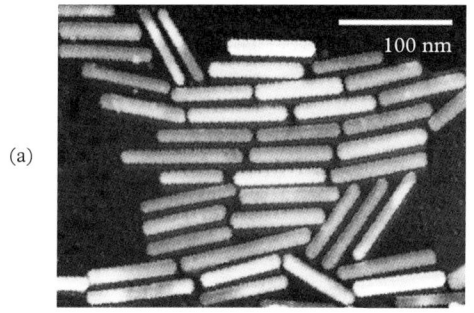

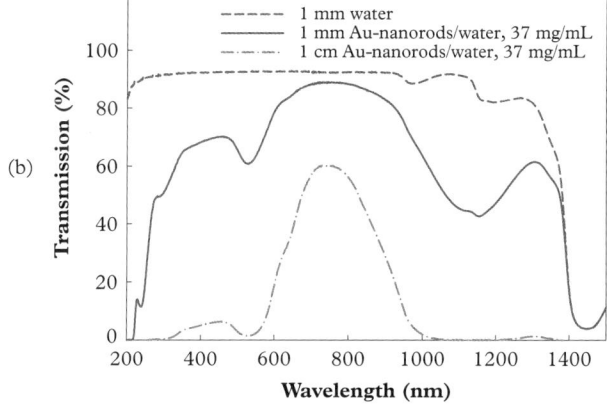

Figure 7.42 *(a) TEM image of Au nanorods used for stimulated scattering experiment. (b) Transmission spectra for 1-mm water sample and 1-mm and 1-cm Au nanorods/water samples, respectively. (After He et al.[140] Copyright 2012, American Physical Society).*

nanoparticles, Au nanorods of ~90 nm length and ~13 nm diameter dispersed in water were employed as Mie scattering centers. Their transmission electron microscope (TEM) image is shown in Fig. 7.42(a), while the transmission spectra of their suspension in water with 37 mg/mL concentration (1.6×10^{14} cm^{-3} particle density) and of two different path-lengths are shown in Fig. 7.42(b).

From Fig. 7.42(b) one can see that there are two major linear extinction bands; one is around the ~530 nm position and the other around the ~1100 nm position. These two extinction bands are caused by the surface plasmon resonance along the shorter dimension and longer dimension of the nanorods, respectively. It is noted that the general rules for generating stimulated scattering are: (i) a lower linear attenuation due to absorption, and (ii) a stronger initial spontaneous scattering signal. According to the first requirement, an ~800-nm wavelength is suitable for the choice of pump wavelength. Within this spectral range, a stronger spontaneous Mie scattering can be maintained while its contribution to the overall attenuation is still much smaller than the linear absorption.

The pump source for generating stimulated scattering was a tunable dye laser system that provided the output of ~816 nm wavelength, ~0.022 nm spectral linewidth, ~10 ns pulse duration, and 10 Hz repetition rate. The pump beam was focused via an $f = 15$ cm lens onto the center of a 2-cm-long cuvette containing the sample of Au nanorods in water of 37 mg/mL concentration. When the pump intensity is lower than a certain threshold value, only ordinary Mie scattering can be observed via an IR viewer from any direction, as shown schematically in Fig. 7.43(a). Once the pump intensity is above the threshold value, a highly directional backward stimulated scattering can be observed through a beam splitter, as shown schematically in Fig. 7.43(b). The measured pump energy (or intensity) threshold value was ~0.5 mJ (or ~2.2 GW/cm^2).

Since Mie scattering is a no-frequency-shift scattering process, after obtaining positive feedback from the induced Bragg grating's reflection, the final stimulated Mie scattering should also have no frequency shift.

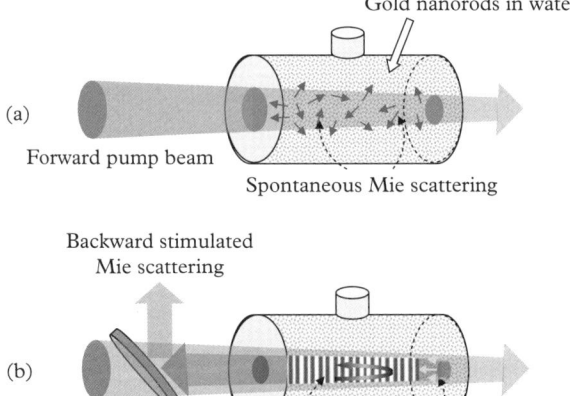

Figure 7.43 *Schematic diagrams showing (a) random Mie scattering from Au nanorods in water with below-threshold pump, (b) directional backward stimulated Mie scattering with above-threshold pump.*

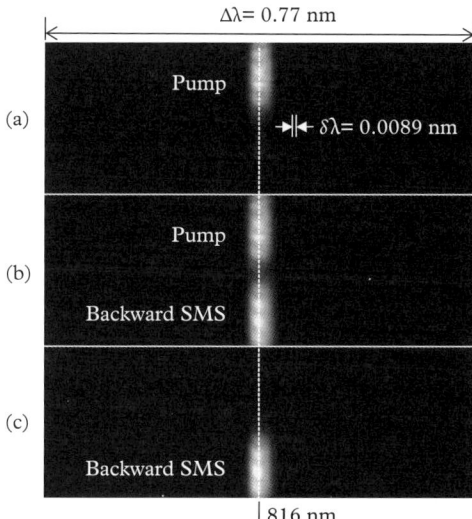

Figure 7.44 *Spectrograms for (a) the spectral line of the 816-nm pump beam alone, (b) the spectral lines of both pump and backward SMS beam together, and (c) the spectral line of the SMS beam alone. Here $\delta\lambda$ indicates the apparatus linewidth in the 816-nm range. All photos were taken upon a single-pulse exposure. (After He et al.[140] Copyright 2012, American Physical Society).*

Figure 7.44 shows photographs of the spectral lines for the pump beam alone (a), for the backward stimulated scattering beam alone (c), and for the two beams together (b), measured by a 1-m two-grating spectrograph in conjunction with a CCD-array detector. The system's resolution around the 816-nm range is $\delta\lambda \approx 0.0089$ nm. To obtain the photos shown in Figure 7.44, the pump beam and the backward stimulated scattering beam were incident with slightly shifted positions on the entrance slit of the spectrograph. With such an arrangement, the spectral lines of these two beams were separated vertically from each other, and the spectral resolution of wavelength-shift measurement should be $\delta\lambda \approx 0.0045$ nm around the 816-nm range. As shown in Fig. 7.44(b), there is no wavelength shift between the pump line and stimulated scattering line, with a spectral resolution much narrower than the half of the pump linewidth.

Figure 7.45 shows the temporal waveforms of the pump pulse and stimulated scattering pulse, measured at different pump energy levels by using a dual-channel oscilloscope in conjunction with two identical photodiodes with a 1 ns resolution. From this figure, one can see that the duration of the stimulated scattering pulse is always shorter than that of the corresponding pump pulse. This is understandable on considering the threshold requirement for pump intensity that is a function of time.

At the pump level of 1.5 mJ, the measured divergence angle of the SMS beam is ∼0.22 mrad, which is smaller than that of ∼0.36 mrad for the pump beam. In addition, the energy conversion efficiency from the pump pulse to the SMS pulse is higher than 15% at the pump levels of 3.5–4 mJ.

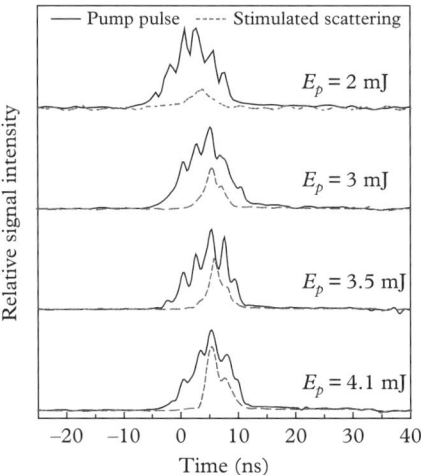

Figure 7.45 *Temporal waveforms of pump pulse and SMS pulse, measured at different pump energy (E_p) levels. All waveforms are taken upon a single-pulse. (After He et al.[140] Copyright 2012, American Physical Society).*

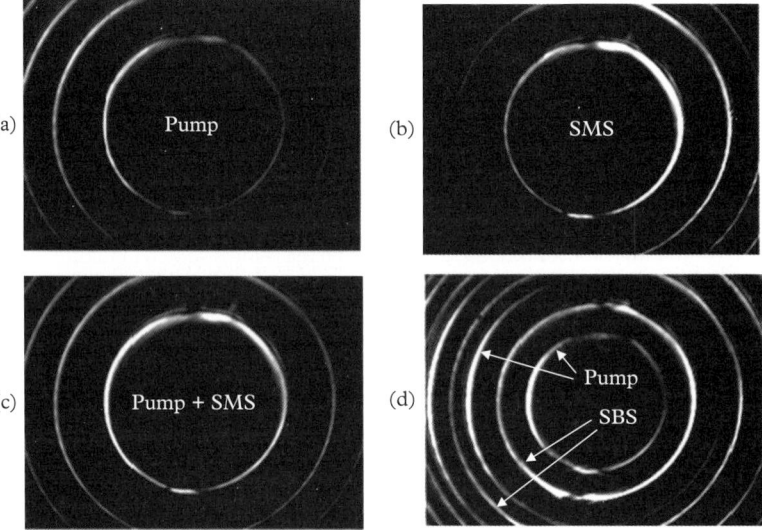

Figure 7.46 *F-P interferograms of (a) partially blocked pump beam of 1064 nm, (b) SMS beam from a sample of Au nanorods in chloroform, (c) both pump and SMS beam together, and (d) both the pump and the SBS beam from pure chloroform. Free spectral range is 0.43 cm^{-1}, all photos are taken by a single pulse exposure. (After He et al.[140] Copyright 2012, American Physical Society).*

Finally, to distinguish the SMS in a Au nanorod solution from the SBS in a pure solvent, 1064-nm laser pulses of narrow spectral linewidth (≤ 0.005 cm^{-1}) were employed to pump the Au nanorods suspended in chloroform; the spectrum structure of the generated SMS was compared to that of SBS from the pure solvent, by using a F-P interferometer. Figure 7.46 shows the recorded interferograms of (a) the input 1064-nm pump beam alone, (b) the SMS beam alone, (c) the pump and SMS beam together, and (d) the pump and the SBS beam from pure chloroform, respectively. The measured SBS's frequency shift for chloroform is ~ 0.24 cm^{-1}, whereas there is no frequency shift between the SMS beam and the pump beam.

More recently, the same SMS effect has also been observed in small (8–10 nm) metallic nanoparticles of Au, Ag, and Au/Ag alloy dispersed in toluene pumped by laser pulses of ~ 800 nm wavelength and ~ 10 ns duration.[142]

7.7.3 SMS in semiconductor nanoparticle suspensions

Most available semiconductor quantum dots (QDs) or rods (QRs) dispersed in solvents exhibit multi-photon activity and manifest resonance-enhanced refractive-index changes at appropriately suitable excitation wavelengths; therefore they are suitable candidates for generating SMS. Actually, the earliest experiments of no-frequency-shift stimulated scattering in the CdSe QRs/CHCl$_3$ and CdTe$_x$Se$_{1-x}$ QDs/CHCl$_3$ systems, respectively, have been published with somewhat ambiguous titles.[143,144]

There is a more recent example of SMS study in the medium of CdSe/CdS/ZnS double-shell structured nanocrystals (NCs) dispersed in chloroform (CHCl$_3$). The TEM image of these CdSe/CdS/ZnS NCs (abbreviated as CdSe NCs) is shown in the inset of Fig. 7.47(a), the sizes of these particles are in the range 5–6 nm.

Figure 7.47(a) shows the linear extinction spectra of a 1-mm pathlength pure solvent (CHCl$_3$) sample, and two CdSe NCs suspension samples in CHCl$_3$ with different weight concentrations. In this case, the measured extinction is determined by absorption and scattering losses. From this figure, we can see that for the spectral range beyond 700 nm, the absorption is negligible and the losses due to Mie scattering from the CdSe NCs still remain greater than the Rayleigh scattering from the solvent (the solid-line and dot-dashed-line curves are always above the dotted-line curve). Figure 7.47(b) shows the experimental setup for observing the backward stimulated scattering from a Mie scattering medium. One of the pump sources for generating stimulated scattering in CdSe NCs/CHCl$_3$ medium was a 816-nm dye laser device; the other pump source was a 1064-nm Nd:YAG laser working on a single longitudinal mode. The pulse duration and repetition rate for both 816-nm and 1064-nm pump pulses were ~ 8 ns and 10 Hz, respectively. The input pump pulses were focused via an $f = 10$ cm lens onto the center of a quartz cuvette with high optical quality or onto an ordinary glass vial with very poor optical quality; both containers were filled with scattering sample solutions. The optical pathlength of the focused pump beam passing through the scattering medium was ~ 2 cm. Two samples with different concentration values (2.0 mg/mL and 4.5 mg/mL) were tested separately for stimulated scattering generation.

At pump levels higher than the corresponding threshold value, highly directional backward SMS could be generated in the CdSe NCs/CHCl$_3$ samples by using two different pump wavelengths. Spectral measurements have once again shown that there is no frequency shift between the stimulated scattering and the pump laser. In addition, the backward SMS beam exhibits

216 *Stimulated Scattering of Intense Coherent Light*

(a)

(b)

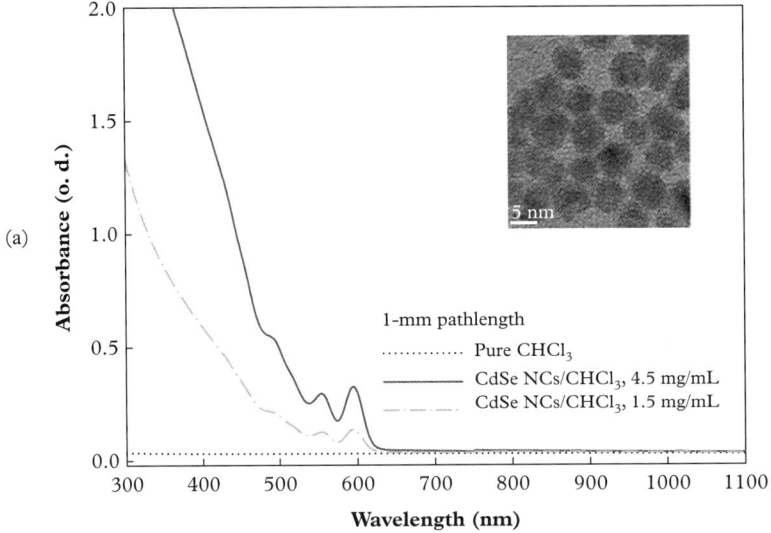
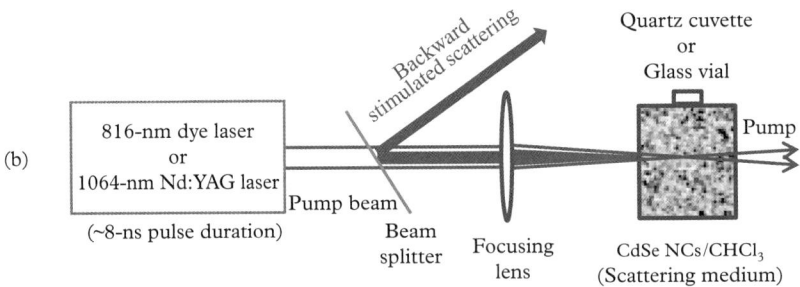

Figure 7.47 *(a) Linear extinction spectra of 1-mm $CHCl_3$ (dotted line) and 1-mm CdSe NCs/$CHCl_3$ samples with different concentrations (solid line and dot-dashed line), respectively; the TEM image of CdSe NCs is shown in the inset. (b) Experimental setup for generating backward SMS in a CdSe NCs/$CHCl_3$ sample filled in a quartz cuvette (with high optical quality) or in an ordinary glass vial (with poor optical quality). (After He et al.[141] Copyright 2012, AIP Publishing LLC).*

good phase-conjugation capability so that the aberration inferences from the glass vial containing scattering medium can be automatically removed.

Figure 7.48 shows the measured data of output SMS pulse energy as a function of the input pump energy for two samples with different concentration values. At the same pump level, the sample with a higher concentration produced much greater SMS output. Specifically, at the ~5.6 mJ pump level the conversion efficiency from the pump energy to the stimulated scattering energy for the sample with a higher (4.5 mg/mL) concentration is $\eta \approx 32\%$.

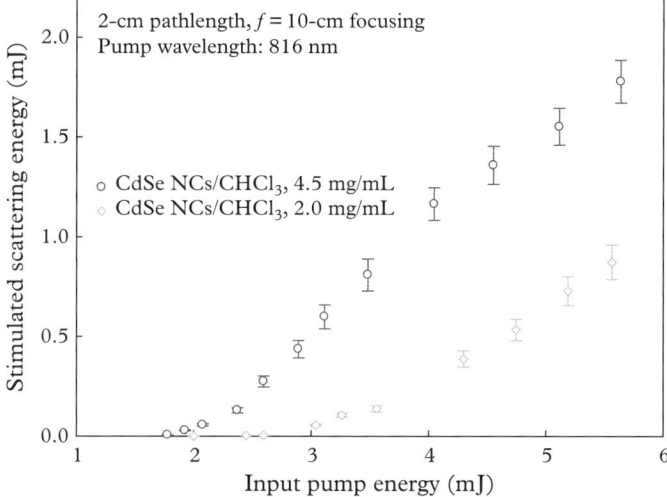

Figure 7.48 *Backward SMS energy vs. the pump pulse energy, measured in two CdSe NCs/CHCl₃ samples with different concentrations. (After He et al.[141] Copyright 2012, AIP Publishing LLC).*

Finally, it should be pointed that the specific mechanisms and dynamics of the induced Bragg grating formation are the key issues for further SMS studies.

....

PROBLEMS

1. Briefly explain the difference between spontaneous light scattering and stimulated light scattering, as well as the difference between stimulated emission (lasing) and stimulated scattering.
2. Briefly indicate the mechanism difference between the axial higher-order stimulated Raman scattering and the non-axial higher-order coherent Raman emission rings.
3. Based on Eq. (7.4-3) and assuming that $v_a = 2000$ m/s is the speed of the induced hypersonic wave and $n = 1.4$ is the refractive index of the medium, please calculate the frequency shift values (in units of cm^{-1}) of Brillouin scattering of a 530-nm light beam at the scattering angles of $\theta = 0$, 90, and 180°, respectively.
4. For a transparent Kerr liquid (like CS$_2$) consisting of anisotropic molecules, why may a linearly polarized intense light field enforce the molecules to make a reorientation movement even though there is a rotational viscosity? How may the ground-state molecules obtain the additional energy to do the reorientation work?
5. What is the common feedback (gain) mechanism to generate stimulated Rayleigh–Bragg scattering and stimulated Mie scattering? Why is there no frequency shift for these two types of stimulated scattering?

REFERENCES

1. G. S. He, in *Progress in Optics*, Vol. 53, edited by E. Wolf (Amsterdam, Elsevier, 2009), pp. 201–292.
2. L. Rayleigh, *Phil. Mag.* **47**, 375 (1899).
3. M. V. Smoluchowski, *Ann. Phys. (Berlin)* **330**, 205 (1908).
4. A. Einstein, *Ann. Phys. (Berlin)* **338**, 1275 (1910).
5. C. V. Raman and K. S. Krishnan, *Nature*, **121**, 501 (1928).
6. L. Brillouin, *Ann. Phys. (Paris)* **17**, 88 (1922)
7. J. Tyndall, *Proc. Roy. Soc. (London)* **17**, 223 (1868).
8. G. Mie, *Ann. Phys. (Berlin)* **330**, 377 (1908).
9. M. Born and E. Wolf, *The Principles of Optics* (Pergamon, New York, 1980), Chapter 13.
10. G. S. He, H.-Y. Qin, and P. N. Prasad, *J. Appl. Phys.* **105**, 023110 (2009).
11. G. S. He. J. Zhu, K.-T. Yong, A. Baev, H.-X. Cai, R. Hu, Y. Cui, X.-H. Zhang, and P. N. Prasad, *J. Phys. Chem. C* **114**, 2853 (2010).
12. E. J. Woodbury and W. K. Ng, *Proc. IRE* **50**, 2367 (1962).
13. G. Eckhardt, R. W. Hellwarth, F. J. McClung, S. E. Schwarz, D. Weiner, and E. J. Woodbury, *Phys. Rev. Lett.* **9**, 455 (1962).
14. W. Heitler, *The Quantum Theory of Radiation*, 3rd ed. (Oxford University Press, London, 1954).
15. R. W. Hellwarth, *Phys. Rev.* **130**, 1850 (1963); Y. R. Shen and N. Bloembergen, *Phys. Rev.* **137**, A1787 (1965).
16. H. J. Zeiger, P. E. Tannenwald, S. Kern, and R. Herendeen, *Phys. Rev. Lett.* **11**, 419 (1963).
17. R. W. Terhune, *Solid State Des.* **4**, No. 11, 38 (1963).
18. R. Y. Chiao and B. P. Stoicheff, *Phys. Rev. Lett.* **12**, 290 (1964).
19. P. D. Maker and R. W. Terhune, *Phys. Rev.* **137**, A801 (1965).
20. E. Garmire, *Phys. Lett.* **17**, 251 (1965).
21. H. Moriwaki, S. Wada, H. Tashiro, K. Toyoda, A. Kasai, and A. Nakamura, *J. Appl. Phys.* **74**, 2175 (1993).
22. C. K. N. Patel and E. D. Shaw, *Phys. Rev. Lett.* **24**, 451 (1970).
23. S. D. Smith, R. B. Dennis, and R. G. Harrison, *Prog. Quantum Electron.* **5**, 205 (1977).
24. R. L. Aggarwal, B. Lax, C. E. Chase, C. R. Pidgeon, D. Limbert, and F. Brown, *Appl. Phys. Lett.* **18**, 383 (1971).
25. E. D. Shaw and C. K. N. Patel, *Appl. Phys. Lett.* **18**, 215 (1971).
26. S. R. J. Brueck and A. Mooradian, *Appl. Phys. Lett.* **18**, 229 (1971).
27. J. F. Scott and T. C. Damen, *Phys. Rev. Lett.* **29**, 107 (1972).
28. R. S. Eng, A. Mooradian, and H. R. Fetterman, *Appl. Phys. Lett.* **25**, 453 (1974).
29. T. P. Sattler, B. A. Weber, and J. Nemarich, *Appl. Phys. Lett.* **25**, 491 (1974).
30. P. Norton and P. W. Kruse, *Opt. Commun.* **22**, 147 (1977).
31. K. Yasuda and J. Shirafuji, *Appl. Phys. Lett.* **34**, 661 (1979).
32. P. P. Sorokin, N. S. Shiren, J. R. Lankard, E. C. Hammond, and T. G. Kazyaka, *Appl. Phys. Lett.* **10**, 44 (1967).
33. M. Rokni and S. Yatsiv, *Phys. Lett.* **24A**, 277 (1967).
34. D. Cotter, D. C. Hanna, P. A. Karkkainen, and R. Wyatt, *Opt. Commun.* **15**, 143 (1975).
35. R. A. Weingarten, L. Levin, A. Flusberg, and S. R. Hartmann, *Phys. Lett.* **39A**, 38 (1972).
36. R. Burnham and N. Djeu, *Opt. Lett.* **3**, 215 (1978).
37. P. P. Sorokin and J. R. Lankard, *IEEE J. Quantum Electron.* **9**, 227 (1973).
38. D. Cotter, D. C. Hanna, and R. Wyatt, *Opt. Commun.* **16**, 256 (1976).
39. R. Wyatt and D. Cotter, *Opt. Commun.* **32**, 481 (1980).
40. A. L. Harris, J. K. Brown, M. Berg, and C. B. Harris, *Opt, Lett.* **9**, 47 (1984).

41. D. G. Sarkisyan, *Sov. J. Quantum Electron.* **18**, 1477 (1988).
42. N. Djeu and R. Burnham, *Appl. Phys. Lett.* **30**, 473 (1977).
43. V. S. Verkhovskii, V. M. Klimkin, V. E. Prokop'ev, V. F. Tarasenko, V. G. Sokovikov, and A. F. Fedorov, *Sov. J. Quantum Electron.* **12**, 1397 (1982).
44. S. O. Sapondzhyan and D. G. Sarkisyan, *Sov. J. Quantum Electron.* **13**, 1062 (1983).
45. Y. Manners, *Opt. Commun.* **44**, 366 (1983).
46. Y. Akiyama, Y. Matsunawa, K. Midorikawa, M. Obara, and H. Tashiro, *Appl. Phys. Lett.* **62**, 823 (1993).
47. S. Yatsiv, M. Rokni, and S. Barak, *IEEE J. Quantum Electron.* **4**, 900 (1968).
48. Q. H. F. Vrehen and H. M. J. Hikspoors, *Opt. Commun.* **21**, 127 (1977).
49. D. Cotter, D. C. Hanna, W. H. W. Tuttlebee, and M. A. Yuratich, *Opt. Commun.* **22**, 190 (1977).
50. J. Reif and H. Walther, *Appl. Phys.* **15**, 361 (1978).
51. M. A. Moore, W. R. Garrett, and M. G. Payne, *Opt. Commun.* **68**, 310 (1988).
52. F. Z. Chen, X. F. Han, and C. Y. R. Wu, *Appl. Phys. B*, **56**, 113 (1993).
53. R. W. Minck, R. W. Terhune, and W. G. Rado, *Appl. Phys. Lett.* **3**, 181 (1963).
54. T. R. Loree, R. C. Sze, and D. L. Barker, *Appl. Phys. Lett*, **31**, 37 (1977).
55. V. Wilke and W. Schmidt, *Appl. Phys.* **16**, 151 (1978).
56. T. R. Loree, R. C. Sze, D. L. Barker, and P. B. Scott, *IEEE J. Quantum Electron.* **15**, 337 (1979).
57. J. K. Brasseur, K. S. Repasky, and J. L. Carlsten, *Opt. Lett.* **23**, 367 (1998).
58. F. M. Johson, J. A. Duardo, and G. L. Clark, *Appl. Phys. Lett.* **10**, 157 (1967).
59. N. Bloembergen, G. Bret, P. Lallemand, A. Pine, and P. Simova, *IEEE J. Quantum Electron.* **3**, 197 (1967).
60. V. S. Averbakh, A. I. Makarov, and V. I. Talanov, *Sov. J. Quantum Electron.* **8**, 472 (1978).
61. P. T. Lang, W. Schartz, T. Kass, A. D. Semenov, and K. F. Renk, *Opt. Commun.* **89**, 233 (1992).
62. V. Krylov, A. Rebane, O. Ollikainen, D. Erni, V. Wild, V. Bespalov, and D. Staselko, *Opt. Lett.* **21**, 381 (1996).
63. R. W. Minck, E. E. Hagenlocker, and W. G. Rado, *Phys. Rev. Lett.* **17**, 229 (1966).
64. M. E. Mack, R. L. Carman, J. Reintjes, and N. Bloembergen, *Appl. Phys. Lett.* **16**, 209 (1970).
65. R. L. Byer and W. R. Trutna, *Opt. Lett.* **3**, 144 (1978).
66. P. Robinowitz, A. Stein, R. Brickman, and A. Kaldor, *Appl. Phys. Lett.* **35**, 739 (1979).
67. N. G. Basov, A. Z. Grasyuk, Yu. I. Karev, L. L. Losev, and V. G. Smirnov, *Sov. J. Quantum Electron.* **9**, 780 (1979).
68. G. V. Venkin, Yu. A. Il'inskii, and G. M. Mikheev, *Sov. J. Quantum Electron.* **15**, 395 (1985).
69. A. N. Bobrovskii, V. D. Bondaryuk, A. A. Kirillov, A. V. Kozhevnikov, V. A. Mishchenko, and G. D. Mylnikov. *Sov. J. Quantum Electron.* **20**, 778 (1990).
70. F. Hanson and P. Poirier, *IEEE J. Quantum Electron.* **29**, 2342 (1993).
71. M. R. Perrone, G. De Nunzio, and C. Panzera, *Opt. Commun.* **145**, 128 (1998).
72. R. Y. Chiao, C. H. Townes, and B. P. Stoicheff, *Phys. Rev. Lett.* **12**, 592 (1964).
73. G. C. Baldwin, *An Introduction to Nonlinear Optics* (Plenum Press, New York, 1969), pp. 124–132.
74. L. D. Landau and E. M. Lifshitz, *Electrodynamics of Continuous Media*, 2nd ed. (Pergamon Press, New York, 1984), Sections 15–16.
75. C. L. Tang, *J. Appl. Phys.* **37**, 2945 (1966).
76. W. Kaiser and M. Maier, in *Laser Handbook*, Vol. 2, edited by F. T. Arecchi and E. O. Schulz-Dubois (North-Holland, Amsterdam, 1972), pp. 1077–1150.
77. E. Garmire and C. H. Townes, *Appl. Phys. Lett.* **5**, 84 (1964).
78. A. S. Pine, *Phys. Rev.* **149**, 113 (1966).
79. E. E. Hagenlocker, R. W. Minck, and W. G. Rado, *Phys. Rev.* **154**, 226 (1966).
80. M. Maier, *Phys. Rev.* **166**, 113 (1968).
81. D. von der Linde, M. Maier, and W. Kaiser, *Phys. Rev.* **178**, 11 (1969).
82. D. Pohl and W. Kaiser, *Phys. Rev. B* **1**, 31 (1970).
83. E. P. Ippen and R. H. Stolen, *Appl. Phys. Lett.* **21**, 539 (1972).

84. L. M. Goldman, J. Soures, and M. J. Lubin, *Phys. Rev. Lett.* **31**, 1184 (1973).
85. T. Gruhl, H. Puell, and W. Kaiser, *Appl. Phys. Lett.* **25**, 135 (1974).
86. A. Ng, L. Pitt, D. Salzmann, and A. A. Offenberger, *Phys. Rev. Lett.* **42**, 307 (1979).
87. D. T. Hon, *Opt. Lett.* **5**, 516 (1980).
88. M. J. Damzen and M. H. R. Hutchinson, *Opt. Lett.* **8**, 313 (1983).
89. M. Slatkine, I. J. Bigio, B. J. Feldman, and R. A. Fisher, *Opt. Lett.* **7**, 108 (1982).
90. D. Cotter, *Electron Lett.* **18**, 495 (1982).
91. I. V. Tomov, R. Fedosejevs, and D. C. D. McKen, *Opt. Lett.* **9**, 405 (1984); I. V. Tomov, R. Fedosejevs, and D. C. D. McKen, *IEEE J. Quantum Electron.* **21**, 9 (1985).
92. M. A. Greiner-Mothes and K. J. Witte, *Appl. Phys. Lett.* **49**, 4 (1986).
93. M. T. Duigman, B. J. Feldman, and W. T. Whithey, *Opt. Lett.* **12**, 111 (1987).
94. H. J. Eichler, R. König, R. Menzel, H-J. Pätzold, and J. Schwartz, *J. Phys. D: Appl. Phys.* **25**, 1161 (1992).
95. H. J. Eichler, A. Haase, and R. Menzel, *IEEE J. Quantum Electron.* **31**, 1265 (1995).
96. H. Yoshida, V. Kmetic, H. Fujita, M. Nakatsuka, T. Yamanaka, and K. Yoshida, *Appl. Opt.* **36**, 3739 (1997).
97. M. Yoshimura, Y. Mori, T. Sasaki, H. Yoshida, and M. Nakatsuka, *J. Opt. Soc. Am. B* **15**, 446 (1998).
98. V. Kmetic, H. Fiedorowicz, A. A. Andreev, K. J. Witte, H. Daido, H. Fujita, M. Nakatsuka, and T. Yamanaka, *Appl. Opt.* **37**, 7085 (1998).
99. R. A. Mullen, *IEEE J. Quantum Electron.* **26**, 1299 (1990).
100. D. Neshev, I. Velchev, W. A. Majewski, W. Hogervorst, and W. Ubachs, *Appl. Phys. B* **68**, 671 (1999).
101. K. Kuwahara, E. Takahashi, Y. Matsumoto, S. Kato, and Y. Owadano, *J. Opt. Soc. Am. B* **17**, 1943 (2000).
102. A. A. Shilov, G. A. Pasmanik, O.V. Kulagin, and K. Deki, *Opt. Lett.* **26**, 1565 (2001).
103. G. Marcus, S. Pearl, and G. Pasmanik, *J. Appl. Phys.* **103**, 103105 (2008).
104. D. I. Mash, V. V. Morozov, V. S. Starunov, and I. L. Fabelinskii, *JETP Lett.* **2**, 25 (1965).
105. G. I. Zaitsev, Yu. I. Kyzylasov, V. S. Starunov, and I. L. Fabelinskii, *JETP Lett.* **6**, 35 (1967).
106. M. Denariez and G. Bret, *Phys. Rev.* **171**, 160 (1968).
107. E. J. Miller and R. W. Boyd, *Int. J. Nonlinear Opt. Phys.* **1**, 765 (1992).
108. D. Wang and G. Rivoire, *J. Chem. Phys.* **98**, 9279 (1993).
109. M. A. Bol'shov, G. V. Venkin, S. A. Zhilkin, and I. I. Nurminskii, *Sov. Phys. JETP* **31**, 1 (1970).
110. K. D. Dorkenoo, A. J. van Wonderen, and G. Rivoire, *J. Opt. Soc. Am. B* **15**, 1762 (1998).
111. G. S. He, D. Y. Tang, Z. L. Cao, F. X. Zhou, R. W. Wang, and S. H. Liu, in *Conference on Lasers and Electro-Optics, 1985 OSA Technical Digest Series* (Optical Society of America, Washington, DC, 1985), paper THM24, p. 202.
112. G. S. He, D. Y. Tang, Z. L. Cao, F. X. Zhou, R. W. Wang, and S. H. Liu, *Chin. Phys. Lett.* **2**, 477 (1985).
113. E. P. Ippen, *Appl. Phys. Lett.* **16**, 303 (1970).
114. G. S. He and P. N. Prasad, *Fiber Integrated Optics*, **9**, 11 (1990).
115. G. S. He and P. N. Prasad, *Phys. Rev. A* **41**, 2687 (1990).
116. G. S. He, R. Burzynski, and P. N. Prasad, *J. Chem. Phys.* **93**, 7647 (1990).
117. G. S. He, G. C. Xu, Y. Pang, and P. N. Prasad, *J. Opt. Soc. Am. B* **8**, 1907 (1991).
118. G. S. He and G. C. Xu, *IEEE J. Quantum Electron.* **28**, 323 (1992).
119. G. S. He, G. C. Xu, Y. Cui, and P. N. Prasad, *Appl. Opt.* **32**, 4507 (1993).
120. G. S. He, G. C. Xu, R. Burzynski, and P. N. Prasad, *Opt. Commun.* **72**, 397 (1989).
121. G. S. He, Y. Cui, and P. N. Prasad, *Sov. Phys. JETP* **85**, 850 (1997); erratum, **86**, 420 (1998).
122. I. G. Zubarev and S. I. Mikhailov, *Sov. J. Quantum Electron.* **4**, 683 (1974).
123. A. A. Filippo and M. R. Perrone, *IEEE J. Quantum Electron.* **28**, 1859 (1992).

124. G. Cook, in *Conference on Lasers and Electro-Optics, 1994 OSA Technical Digest Series* (Optical Society of America, Washington, DC, 1994), 8, p. 12.
125. X. F. Chen, Q. Li, Y. M. Hua, and Y. L. Chen, *Opt. Commun.* **111**, 137 (1994).
126. H. Z. Wang, X. G. Zheng, W. D. Mao, Z. X. Yu, and Z. L. Gao, *Phys. Rev. A* **52**, 1740 (1995).
127. R. M. Herman and M. A. Gray, *Phys. Rev. Lett.* **19**, 824 (1967).
128. D. H. Rank, C. W. Cho, N. D. Foltz, and T. A. Wiggings, *Phys. Rev. Lett.* **19**, 828 (1967).
129. V. I. Bespalov, A. M. Kubarev, and G. A. Pasmanik, *Phys. Rev. Lett.* **24**, 1274 (1970).
130. K. Darée and W. Kaiser, *Phys. Rev. Lett.* **26**, 816 (1971).
131. P. Boissel, G. Hauchecorne, F. Kerherve, and G. Mayer, *J. Phys. Lett.* **39**, 319 (1978).
132. V. B. Karpov, V. V. Korobkin, and D. A. Dolgolenko, *Sov. J. Quantum Electron.* **21**, 1235 (1991).
133. G. S. He, T.-C. Lin, and P. N. Prasad, *Opt. Express* **12**, 5952 (2004).
134. G. S. He, C. Lu, Q. Zheng, P. N. Prasad, P. Zerom, R. W. Boyd, and M. Samoc, *Phys. Rev. A* **71**, 063810 (2005).
135. H. Kogelnik, *Bell Syst. Tech. J* **48**, 2909 (1969).
136. G. S. He, Q. Zheng, and P. N. Prasad. *J. Opt. Soc. Am. B* **24**, 1166 (2007).
137. G. S. He, L.-S. Tan, Q. Zheng, and P. N. Prasad, *Chem. Rev.* **108**, 1245 (2008).
138. D. D. Evanoff and G. Chumanov, *J. Phys. Chem. B* **108**, 13957 (2004).
139. P. K. Jain. K. S. Lee, I. H. El-Sayed, and M. A. El-Sayed, *J. Phys. Chem B* **110**, 7238 (2006).
140. G. S. He, K.-T. Yong, J. Zhu, and P. N. Prasad, *Phys. Rev. A* **85**, 043839 (2012).
141. G. S. He, W.-C. Law, L. Liu, X. Zhang, and P. N. Prasad, *Appl. Phys. Lett.* **101**, 011110 (2012).
142. G. S. He, W.-C. Law, A. Baev, S. Liu, M. T. Swihart, and P. N. Prasad, *J. Chem. Phys.* **138**, 024202 (2013).
143. G. S. He, K.-T. Yong, H.-Y. Qin, Q. Zheng, P. N. Prasad, S. He, and H. Ågren, *IEEE J. Quantum Electron.* **44**, 894 (2008).
144. G. S. He, J. Zhu, K.-T. Yong, R. Hu, Y. Cui, and P. N. Prasad, *J. Chem. Phys.* **131**, 214301 (2009).

8
Optical Phase Conjugation

Optical phase conjugation is a special nonlinear optical technique that deals with the generation of optical phase conjugate waves (PCWs) and their applications. For a coherent and monochromatic light wave (signal beam) with a given wavefront shape, one can generate a new light wave with a reversed wavefront shape (in respect to the propagation direction) via nonlinear optical methods, this newly generated optical wave is termed the PCW of the input signal beam. There are three major nonlinear optical methods can be employed to generate various PCWs: nonlinear four- or three-wave mixing, backward stimulated scattering, and backward stimulated emission (lasing). The unique feature of PCWs is their capability to overcome the disturbing and aberration influences from the propagation medium. Therefore, this technique can be especially useful in high brightness laser oscillator/amplifier systems, laser aiming and targeting systems, long-distance and high bit-rate optical fiber communication systems. It also provides an effective approach to achieve lasing or stimulated scattering from a randomly disturbed gain medium.

8.1 Definitions and features of PCWs

8.1.1 Introduction to optical phase conjugation

Optical phase conjugation (OPC) is one of the new laser-based techniques.[1-5] Before the 1960s it was well known that there were two impossibilities within the regime of conventional optics. The first was that the brightness of any given light beams cannot be increased through any types of optical imaging systems or specially designed devices. The second was that a perfect and reversible optical imaging system was impossible because of the aberration influences from optical elements and propagating media. The first impossibility has been removed after the inventions of laser oscillators and amplifiers, while the second restraint can also be released by utilizing the OPC technique.

In general, a pair of optical waves is phase conjugated to each other if their complex amplitude functions are conjugated each other with respect to their phase factors. Optical PCWs can be generated through various nonlinear optical processes (such as four-wave mixing, three-wave mixing, backward stimulated scattering, and others). They can also be generated through one-photon or multi-photon pumped backward stimulated emission in a lasing medium, provided that the pump is a laser beam. In most cases, the generation of various optical PCWs is based on the coherent light induced holographic grating and the subsequent wavefront reconstruction.

8.1.2 Definitions of backward PCWs

8.1.2.1 Backward degenerate PCWs

Suppose there is an input quasi-plane and monochromatic wave with an arbitrary phase distortion deviated from an ideal plane-monochromatic wave, i.e.,

$$\mathbf{E}(z,x,y,\omega) = \mathbf{E}(z,x,y)e^{-i\omega t} = \mathbf{A}_0(z,x,y)e^{i[kz+\phi(z,x,y)]}e^{-i\omega t}. \tag{8.1-1}$$

Here, z is the longitudinal variable along the propagation direction, x and y are the radial variables along the beam section, ω is the frequency of the optical field, $k = n_0/(2\pi\lambda)$ is the magnitude of the corresponding wavevector, $\mathbf{E}(z, x, y)$ is the complex amplitude function, $\mathbf{A}_0(z, x, y)$ is the real amplitude function, and finally $\phi(z, x, y)$ is the phase-distortion function that describes the deviation of the real wavefront from an ideal plane wave.

If there is a backward propagating wave, which can be expressed as

$$\mathbf{E}'(z,x,y,\omega) = a\mathbf{E}^*(z,x,y)e^{-i\omega t} = a\mathbf{A}_0(z,x,y)e^{-i[kz+\phi(z,x,y)]}e^{-i\omega t}, \tag{8.1-2}$$

where a is an arbitrary real coefficient, then the field $\mathbf{E}'(z, x, y, \omega)$ is defined as the frequency-degenerate and backward PCW of the original field $\mathbf{E}(z, x, y, \omega)$, as shown schematically in Fig. 8.1(a).

8.1.2.2 Backward nondegenerate PCWs

Moreover, in a more general case, if there is a backward propagating optical field with a frequency ω' different from the frequency ω of the incident wave, which can be written as

$$\mathbf{E}'(z,x,y,\omega') = a\mathbf{A}_0(z,x,y)e^{-i[k'z+\varphi(z,x,y)]}e^{-i\omega' t}, \tag{8.1-3}$$

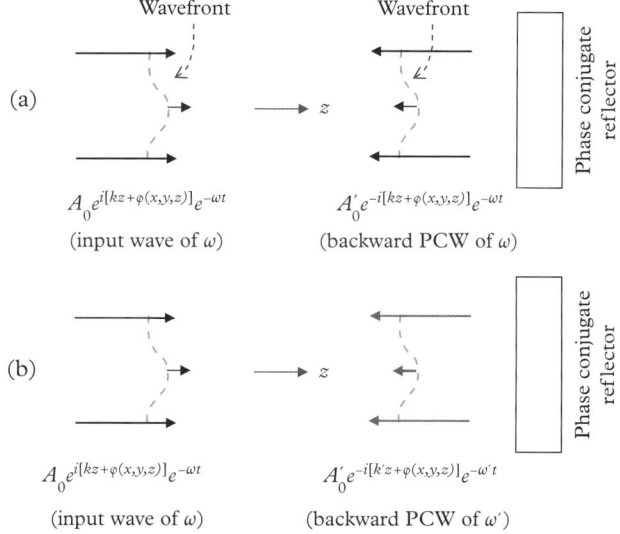

Figure 8.1 *Wavefront structures of the backward PCWs: (a) frequency-degenerate and (b) frequency-nondegenerate.*

224 *Optical Phase Conjugation*

then $\mathbf{E}'(z, x, y, \omega')$ is termed the frequency-nondegenerate backward PCW of the original field $\mathbf{E}(z, x, y, \omega)$ expressed by Eq. (8.1-1), as shown schematically in Fig. 8.1(b).

As shown in Fig. 8.1, the wavefront structures of the backward PCW is reversed with respect to the propagation direction, and where the backward PCW is generated from a phase-conjugate reflector that is a special nonlinear optical device we shall discuss in detail later.

8.1.2.3 Forward degenerate PCWs

Assume the input wave still is described by Eq. (8.1-1); its frequency-degenerate forward PCW can be defined as

$$\mathbf{E}'(z, x, y, \omega) = a\mathbf{A}_0(z, x, y)e^{i[kz-\phi(z,x,y)]}e^{-i\omega t}, \tag{8.1-4}$$

where a is an arbitrary real coefficient. In this case, two related waves with the same frequency propagate along the same direction, but have reversed wavefront structures to each other, as shown in Fig. 8.2(a).

8.1.2.4 Forward nondegenerate PCWs

Furthermore, one can also generate a forward PCW with a shifted frequency, such as

$$\mathbf{E}'(z, x, y, \omega') = a\mathbf{A}_0(z, x, y)e^{i[k'z-\phi(z,x,y)]}e^{-i\omega' t}, \tag{8.1-5}$$

where $\omega \neq \omega'$. The wave described by Eq. (8.1-5) is termed the frequency-nondegenerate forward PCW, as shown in Fig. 8.2(b).

Figure 8.2 shows schematically the relative wavefront structures of the forward PCWs with respect to that of a given input wave, where the forward PCWs are generated through a

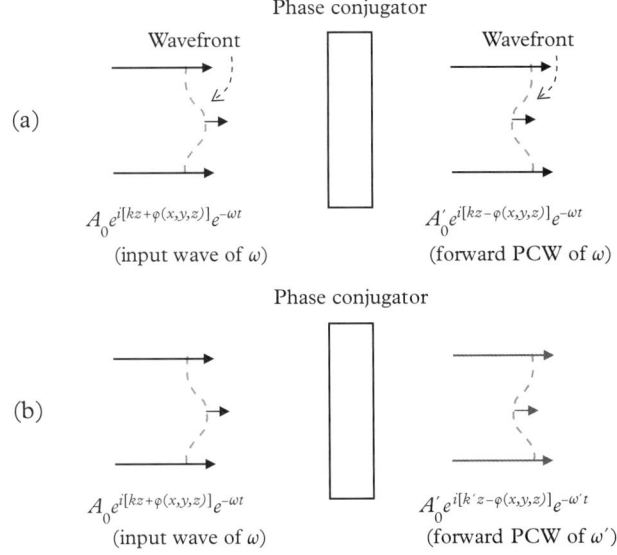

Figure 8.2 *Wavefront structures of the forward PCWs: (a) frequency-degenerate and (b) frequency-nondegenerate.*

transmission nonlinear optical device (phase-conjugator) we shall discuss later too. Once again, we can see that the forward PCWs have a reversed wavefront structures in respect to that of the input signal wave.

8.1.3   Special capability of optical PCWs

With specially designed arrangements, PCWs can manifest a unique capability to automatically remove or compensate for the harmful influences from imperfect transmission media and optical elements on a light signal beam. Actually, Yariv in 1976 first proposed the concept of an optical PCW and the use of PCWs for overcoming phase distortion influences from transmission media.[6,7]

Figure 8.3 depicts schematically the behavior of two different waves backward passing through an inhomogeneous or randomly disturbing optical medium: one is reflected from an ordinary plane mirror and the other is from a phase conjugate reflector. As shown in Fig. 8.3(a), an initial ideal plane wave becomes an aberrated wave after passing once through the medium. If one places a plane mirror perpendicularly to the propagation direction of this wave, after reflection from this mirror and backward passing through the same disturbing medium, the aberration influence from the medium on the output wave will be accumulated. In contrast, as shown in Fig. 8.3(b), if one replaces the plane mirror with a phase-conjugate reflector that generates a backward PCW, after backward passing through the same disturbed medium, the aberration influence will be removed from the final output wave. This is an example of the unusual property of the backward PCWs compared to ordinary reflected waves.

For forward PCWs, there are two ways to exploit their potential capability for overcoming aberration influences from transmission media, which are schematically shown in Fig. 8.4, where

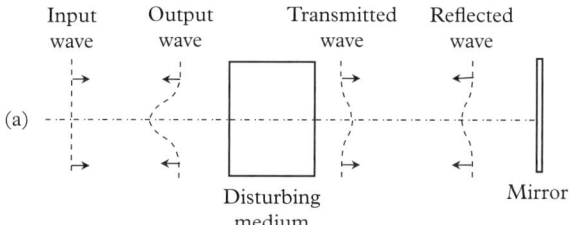

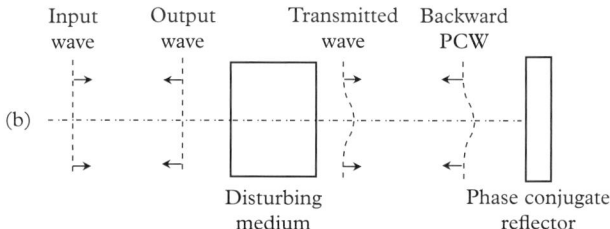

Figure 8.3 *(a) An ordinary reflected wave backward passes through a disturbing medium, the aberration influence is accumulated. (b) A PCW backward passes through the same disturbing medium; the aberration influence is removed.*

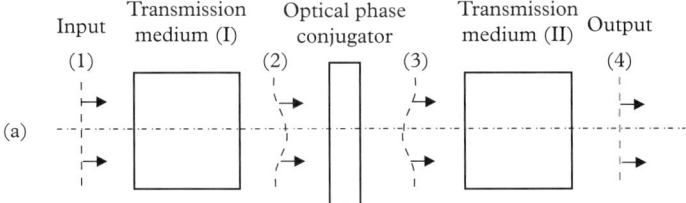

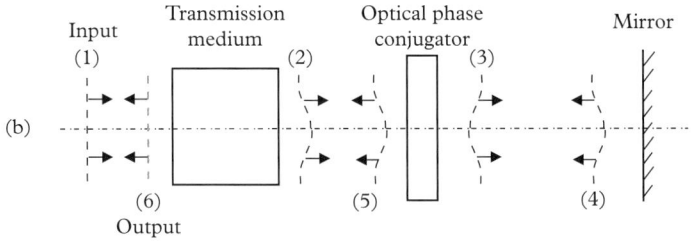

Figure 8.4 (a) After a forward PCW (3) passed through the second identical transmission medium, the aberration influence is finally removed from the output wave (4); (b) After a forward PCW (3) is reflected from a mirror and backward passed through the same transmission medium, the aberration influence is removed from the output wave (6).

the forward PCW is generated through a phase conjugator. In the case (a), the input wave (1) is an ideal plane wave, after passing through a transmission medium the transmitted wave (2) exhibits a distorted wavefront. If the phase conjugator is placed between two identical transmission media and it generates a forward PCW, after the latter passed through the second medium the aberration influence can be finally removed. Alternatively, in the case (b), the forward PCW (3) is reflected from a mirror and then backward passes through the same transmission medium, the final output wave (6) will restore an ideal plane wavefront. Here it is assumed that the wavefront of the reflected wave (4) does not change when it backward passed through the phase conjugator.

8.2 PCW generation via nonlinear wave mixing

8.2.1 Backward PCW generation via degenerate four-wave mixing (FWM)

One of the well-studied methods to generate forward PCWs is based on the degenerate four-wave mixing (FWM) method. This method was first proposed by Hellwarth in 1977.[8] In this case, as shown in Fig. 8.5, a third-order nonlinear medium is illuminated simultaneously with two counter-propagating strong plane waves and a signal beam that has an arbitrary wavefront distortion and different propagation direction. If these three incident waves have the same frequency ω, we may observe a newly generated wave of the same frequency ω, but propagating along the opposite direction of the signal beam. The following derivations can simply show that this newly generated wave is the frequency-degenerate backward PCW of the incident signal beam.

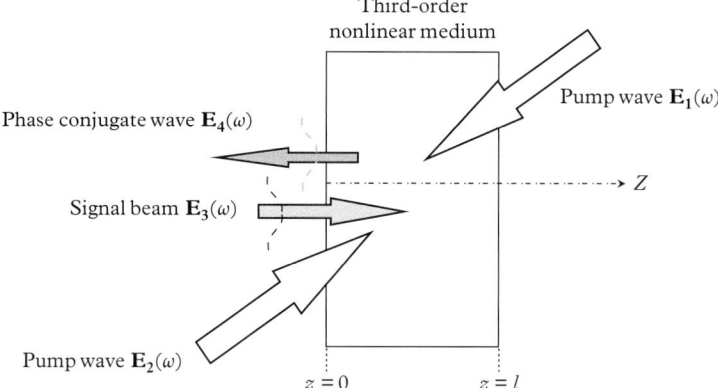

Figure 8.5 *Backward PCW generation via degenerate FWM.*

Assuming the signal beam is propagating along the z-axis, the three incident monochromatic waves can be expressed as

$$\left.\begin{aligned} E_1(\omega) &= \mathbf{a}_1 A_1(\mathbf{r}) e^{-i(\omega t - \mathbf{k}_1 \cdot \mathbf{r})} \\ E_2(\omega) &= \mathbf{a}_2 A_2(\mathbf{r}) e^{-i(\omega t - \mathbf{k}_2 \cdot \mathbf{r})} \\ E_3(\omega) &= \mathbf{a}_3 A_3(z) e^{-i(\omega t - k_3 z)} \end{aligned}\right\}, \qquad (8.2\text{-}1)$$

where $\mathbf{a}_i$ is the unit vector along the light polarization direction of the ith wave, $\mathbf{k}_1 = -\mathbf{k}_2$ is the wavevector of the one pump wave, k_3 is the absolute value of the wavevector of the signal beam, A_1 and A_2 are the real amplitude functions of the two plane pump waves, and A_3 is the complex amplitude function (involving certain information) of the signal wave.

According to the principle of FWM, the fourth coherent wave can be generated through the third-order nonlinear polarization response of the medium. This wave (E_4) and its corresponding nonlinear electric polarization field (P_4) can be written as

$$\left.\begin{aligned} \mathbf{P}_4^{(3)}(\omega) &= \varepsilon_0 \chi^{(3)}(\omega, \omega, -\omega) \mathbf{a}_1 \mathbf{a}_2 \mathbf{a}_3 A_1 A_2 A_3^* e^{-i(\omega t + k_3 z)} \\ \mathbf{E}_4(\omega) &= \mathbf{a}_4 A_4(z) e^{-i(\omega t + k_3 z)} \end{aligned}\right\}. \qquad (8.2\text{-}2)$$

Here, the newly generated wave is propagating along the $-z$ direction, and the phase-matching condition is always satisfied because $\mathbf{k}_1 + \mathbf{k}_2 = \mathbf{k}_3 + \mathbf{k}_4 \equiv 0$. One may note that this special FWM can be regarded as the annihilation of two counter-propagating pump photons and the simultaneous creation of one photon for the E_3 wave and another photon for the E_4 wave. Because of such a process, the signal wave will be amplified while the E_4 wave is created; this is a four-photon parametric interaction process.

For simplicity, it is assumed that the three incident waves are linearly polarized along the same direction (x-axis) and the third-order nonlinear medium is isotropic, therefore, the E_4 wave will be also polarized along the x-axis direction. Under this condition, we can neglect the vector property of the fields and write the third-order nonlinear polarization fields for E_3 and E_4 as

$$\left.\begin{aligned} P_3^{(3)}(\omega) &= \varepsilon_0 \chi_e^{(3)} A_1 A_2 A_4^* e^{-i(\omega t - k_3 z)} \\ P_4^{(3)}(\omega) &= \varepsilon_0 \chi_e^{(3)} A_1 A_2 A_3^* e^{-i(\omega t + k_3 z)} \end{aligned}\right\}. \qquad (8.2\text{-}3)$$

Here, $\chi_e^{(3)} = \chi_{xxxx}^{(3)}(\omega, \omega, -\omega)$ is the effective third-order nonlinear susceptibility value of the nonlinear medium, which is a real quantity for non-resonant interaction or a complex quantity for resonant or near-resonant interaction.

Substituting Eq. (8.2-3) into Eq. (2.5-13), i.e., the general expression of nonlinear coupled-wave equations, and assuming that the depletion for both strong pump waves is negligible within a short interaction length, we then obtain the equations describing the amplitude variation along the z-axis for $\mathbf{E}_3$ and $\mathbf{E}_4$ waves:

$$\left.\begin{array}{l}\dfrac{\partial A_3^*(z)}{\partial z} = i\gamma^* A_4(z) \\[6pt] \dfrac{\partial A_4(z)}{\partial z} = i\gamma A_3^*(z)\end{array}\right\}, \tag{8.2-4}$$

where γ is in general a complex coupling coefficient determined by

$$\gamma = \frac{k_3(\omega)}{2\varepsilon_r(\omega)} \chi_e^{(3)} A_1 A_2 = \frac{\omega}{2cn_0(\omega)} \chi_e^{(3)} A_1 A_2. \tag{8.2-5}$$

Here, ε_r and n_0 are the linear dielectric constant and refractive index of the nonlinear medium.

In the condition that A_1, A_2, and therefore γ are approximately constant, the solutions of Eq. (8.2-4) can be expressed as [9]

$$\left.\begin{array}{l}A_3(z) = \dfrac{\cos[|\gamma|(z-l)]}{\cos[|\gamma|l]} A_3(0) - i\dfrac{|\gamma|}{\gamma^*}\dfrac{\sin[|\gamma|z]}{\cos[|\gamma|l]} A_4^*(l) \\[8pt] A_4(z) = \dfrac{\cos[|\gamma|z]}{\cos[|\gamma|l]} A_4(l) + i\dfrac{\gamma}{|\gamma|}\dfrac{\sin[|\gamma|(z-l)]}{\cos[|\gamma|l]} A_3^*(0)\end{array}\right\}, \tag{8.2-6}$$

where $A_3(0)$ is the initial amplitude of the signal wave on the incident surface of the nonlinear medium. Considering the boundary condition of $A_4(l) = 0$, we obtain the final solutions:

$$\left.\begin{array}{l}A_3(l) = \dfrac{A_3(0)}{\cos[|\gamma|l]} \\[8pt] A_4(0) = -i\dfrac{\gamma}{|\gamma|} \tan[|\gamma|l] A_3^*(0)\end{array}\right\}. \tag{8.2-7}$$

On the one hand, from Eq. (8.2-2) we know that the wave $\mathbf{E}_4$ is counter-propagating to the wave $\mathbf{E}_3$; on the other hand, from the second expression of Eq. (8.2-7) we know that the complex amplitude $A_4(0)$ is proportional to $A_3^*(0)$ near the incident surface. Therefore, one can conclude that the waves $\mathbf{E}_4$ and $\mathbf{E}_3$ are phase-conjugated to each other and the whole system plays the role of a phase-conjugate reflector. The nonlinear intensity-reflectivity of the system can be determined from Eq. (8.2-7) as

$$R = |A_4(0)|^2 / |A_3(0)|^2 = \tan^2[|\gamma|l]. \tag{8.2-8}$$

When $|\gamma|l > \pi/4$, $R > 1$. From the first expression of Eq. (8.2-7) we know that $A_3(l) > A_3(0)$, so the signal wave $\mathbf{E}_3$ will be always amplified through the nonlinear medium.

In particular, when the following condition is fulfilled:

$$|\gamma|l = \frac{\pi}{2}, \tag{8.2-9}$$

from Eq. (8.2-7) we have

$$\frac{|A_4(0)|}{|A_3(0)|} = \infty, \quad \frac{|A_3(l)|}{|A_3(0)|} = \infty. \tag{8.2-10}$$

These expressions mean a self-oscillation of both $\mathbf{E}_3$ and $\mathbf{E}_4$ waves. In reality, this situation will not happen because the solutions given by Eq. (8.2-7) are valid only for small signal gain. When the amplitude changes for $\mathbf{E}_3$ and $\mathbf{E}_4$ waves are considerably large, one can no longer treat the amplitude functions A_1 and A_2 of the pump beams as constants, therefore the solutions expressed by Eq. (8.2-7) are not suitable to describe the gain behavior of a strong signal wave. Nevertheless, one still can use Eq. (8.2-9) as a rough estimation of the effective interaction length (l_0) for the nonlinear medium, which is

$$l_0 \approx \frac{\pi}{2|\gamma|} = \frac{cn_0(\omega)}{\omega} \frac{\pi}{\left|\chi_e^{(3)}\right|} \frac{1}{A_1 A_2}. \tag{8.2-11}$$

From Eq. (8.2-7) we can see that the greater value of $\left|\chi_e^{(3)} = \chi_{xxxx}^{(3)}(\omega,\omega,-\omega)\right|$ the nonlinear medium has, the stronger reflected wave $\mathbf{E}_4$ the medium can generate. For non-resonant interaction, $\chi_e^{(3)}$ is a real and small quantity, and consequently, the nonlinear reflectivity R is quite low. To reach a higher nonlinear reflectivity, researchers have to use resonant or near-resonant effect to increase the $\left|\chi_e^{(3)}\right|$ value of the nonlinear medium.[10]

8.2.2 Holographic model of backward PCW generation via degenerate FWM

In the preceding subsection, it was mathematically proved that a PCW can be generated through a special degenerate FWM arrangement in a third-order nonlinear medium. Such a process is essentially related to the four-photon parametric interaction, and therefore the three input waves should interact to each other in the same medium region as well as at the same time. In this sense, strictly speaking, the described mathematical treatment is valid only for the nonlinear polarization processes arising from the electronic-cloud distortion mechanism or intramolecular motion mechanism.

In this subsection, we shall consider another more general interpretation that can explain why the PCW can be generated in such a special FWM arrangement. This model is based on the principle of holography, i.e., a two-beam induced volume grating formation via intensity-dependent refractive-index changes and subsequent wavefront reconstruction with the third beam.[2,11]

As shown in Fig. 8.6, there are two counter-propagating plane pump waves (A_1 and A_2) passing through a third-order nonlinear medium; the signal wave A_3 is incident upon the medium at an angle θ with respect to the pump wave A_1. With this arrangement, the backward propagating wave A_4 can be generated through the reflection from two possible induced gratings. First, let us consider the interference between the two waves A_1 and A_3 that will produce nearly parallel interference fringes along the bisector direction of the crossing angle θ, as shown in Fig. 8.6(a). Considering that the induced refractive-index change of the medium is proportional to the local light intensity, one may realize that the interference fringes can produce an induced holographic grating within the nonlinear medium. In this case, the pump wave A_2 can be viewed as a reading beam and during its passing through the induced grating, a diffracted (or reflected) wave A_4 will be created. According to the principle of holography, this diffracted wave A_4 will restore the spatial

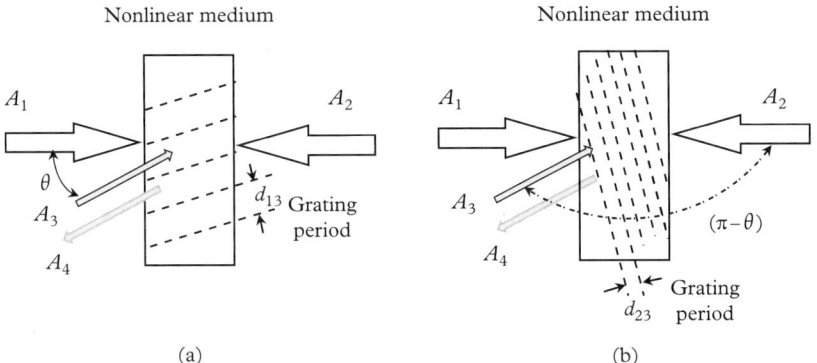

Figure 8.6 *Schematics describing the generation of PCWs via reflection from the induced holographic gratings.*

information carried by the incident signal wave A_3. In other words, the waves A_4 and A_3 are phase conjugated to each other. To further justify this conclusion, we can treat the nonlinear medium as a holographic medium whose transmission function is determined by the interference-induced refractive-index modulation and can be phenomenologically expressed as

$$T \propto (A_1 + A_3)(A_1 + A_3)^* = |A_1|^2 + |A_3|^2 + A_1^* A_3 + A_1 A_3^*. \qquad (8.2\text{-}12)$$

As assumed, the waves A_1 and A_2 are two counter-propagating plane waves and $A_2 = A_1^*$, so that the transmitted field of the reading wave A_2 will be

$$A_2' \propto TA_2 = TA_1^* = [|A_1|^2 + |A_3|^2]A_2 + A_3(A_1^*)^2 + A_1 A_2 A_3^*. \qquad (8.2\text{-}13)$$

Here, on the right-hand side of the equation, the first term proportional to A_2 represents the zero-order diffracted wave that does not involve any spatial information and therefore is not interesting for us. The contribution from the second term that actually involves a phase factor of $\exp[-i(2\mathbf{k}_1 \cdot \mathbf{r} - k_3 z)]$ can be neglected because its phase-matching condition could not be fulfilled. The third term corresponds to the diffracted wave that involves the spatial information carried by the signal wave A_3, and can be written separately as

$$A_4 \propto A_1 A_2 A_3^*. \qquad (8.2\text{-}14)$$

On the other hand, from Eq. (8.2-7) given in the previous subsection we know that under the condition of $|\gamma| l \ll 1$, we have

$$A_4(0) = -i\gamma l A_3^*(0) \propto A_1 A_2 A_3^*(0). \qquad (8.2\text{-}15)$$

Comparing the above two equations we can see that the two different physical models lead to the same conclusion.

As shown in Fig. 8.6(b), the backward wave A_4 can also be generated through the diffraction of the holographic grating induced by the waves A_2 and A_3. In that case, the pump wave A_1 plays the role of a reading plane wave, and the reflected wave A_4 is still phase conjugated to the wave A_3.

Although these two gratings can both contribute to the generation of the PCW A_4, the periods of these two grating are different and can be written as (see Fig. 8.6(a) and (b))

$$\left. \begin{array}{l} d_{13} = \lambda' \left(2 \sin \dfrac{\theta}{2} \right)^{-1} \\ d_{23} = \lambda' \left(2 \cos \dfrac{\theta}{2} \right)^{-1} \end{array} \right\}, \tag{8.2-16}$$

where λ' is the wavelength of the waves in the medium. It is obvious that these periods of the induced gratings are determined by the corresponding spacing of the interference fringes formed by two appropriate waves.

Until now all mathematical derivations are based on the assumption that all involved waves have the same frequency and the same linear polarization, thereby there are two gratings that both can contribute to the formation of the reconstructed wave with the same linear polarization, as indicated in Fig. 8.7(a). Now let us further consider that the three incident waves have the same frequency but only two of them have the same linear polarization. Two examples of combinations of polarization states among the four waves are schematically shown in Fig. 8.7(b) and (c).

In the case shown in Fig. 8.7(b), the waves A_1 and A_3 are both linearly polarized along the direction perpendicular to the plane of paper (s components), whereas the wave A_2 is linearly polarized in the plane of paper (p component). It is obvious that only the waves A_1 and A_3 can interfere with each other to produce the phase grating, while the wave A_2 plays the role of a reading wave that creates the PCW A_4 with the same polarization state (p component). Similarly, in the case shown in Fig. 8.7(c), only the waves A_2 and A_3, which have the same polarization state, can produce the phase grating, whereas the reading wave A_1 and the PCW A_4 have the other polarization state. In more general cases, the polarization states of the three incident waves and the PCW might be more complicated,

Next, we shall extend our discussions from degenerate FWM to partially degenerate FWM. In the latter case, two waves have the same frequency (ω_1), while the other two waves have another frequency (ω_2). Figure 8.8 shows two schematics describing the generation of nondegenerate backward PCW via partially degenerate FWM. In the case shown in Fig. 8.8(a), the pump wave $A_1(\omega_1)$ and the signal wave $A_3(\omega_1)$ have the same frequency and the same polarization state, and

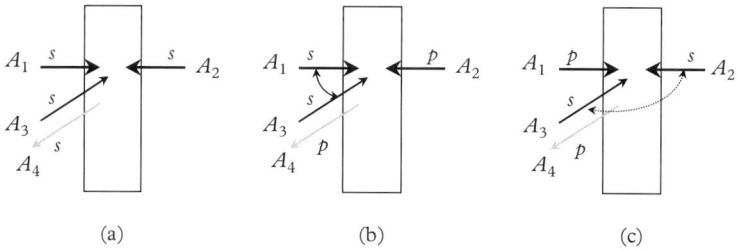

Figure 8.7 *Polarization property of the PCW under different polarization combinations of the three incident beams. (a) Three beams have the same s-polarization; (b) A_1 and A_3 beams have the same s-polarization while A_2 beam has the p-polarization; (c) A_2 and A_3 beams have the same s-polarization while A_1 beam has the p-polarization.*

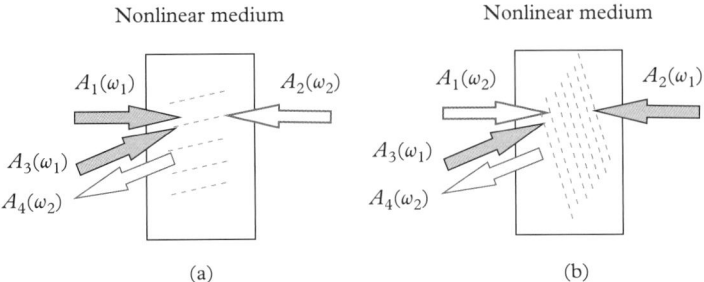

Figure 8.8 *Two schematics describing the generation of the frequency-nondegenerate backward PCW via partially degenerate FWM in a third-order nonlinear medium.*

therefore can produce an induced phase grating; whereas the reading wave $A_2(\omega_2)$ of another frequency will create the diffracted wave $A_4(\omega_2)$ through the induced grating. In this case the spatial information carried by the signal wave $A_3(\omega_1)$ can be restored in the wave $A_4(\omega_2)$; in other words, the latter is the frequency-nondegenerate PCW of the former. Similarly, in the case shown in Fig. 8.8(b), the waves $A_2(\omega_1)$ and $A_3(\omega_1)$ having the same frequency and same linear polarization can produce the grating, while the reading wave $A_1(\omega_2)$ of another frequency will create the diffracted wave $A_4(\omega_2)$.

The processes described above are essentially the same as we use two beams of the same frequency to produce a hologram, and then use another beam of different frequency to read this hologram. In that case, the diffracted beam has the same frequency as the reading beam does.

In summary, the processes of using the FWM method to generate PCW can be interpreted by two different mechanisms: one is the four-photon parametric interaction; the other is the formation of induced holographic grating(s). Strictly speaking, the first model can be applied only when the three incident waves act on the nonlinear medium at the same time, and the third-order nonlinear polarization is mainly caused by the electronic-cloud distortion or intramolecular motion. In contrast, the second model is more general and can be applied to various experimental conditions. Firstly, if we consider the fact that the phase grating is formed through the induced refractive-index change, then not only the electronic-distortion or intramolecular-motion nonlinearity but also other mechanisms, including molecular reorientation, electrostriction, population changes, and even opto-thermal effect can also contribute to the formation of the induced holographic grating. Secondly, from the viewpoint of induced grating, the simultaneous action of three incident beams is not always necessary. For instance, if two incident pulsed waves create a phase grating that can last a time period longer than the input pulse duration, one may observe the diffracted wave A_4 by delaying the incidence of the reading pulse after the two writing pulses have passed through the sample medium already. This method of using a delayed reading pulse can be employed to investigate the temporal behavior of the induced phase-gratings in a given nonlinear medium, and to justify what is the major mechanism contributing to the observed PCW generation.

Finally, it is important to indicate that the model of induced holographic grating is not only useful to explain the generation of PCW via FWM processes, but is also pertinent to explain the PCW generation through either backward *stimulated scattering* or backward *stimulated emission* (lasing) processes.

8.2.3 Forward PCW generation via FWM and three-wave mixing

8.2.3.1 Forward FWM

Forward optical PCWs can be generated in a third-order nonlinear medium via some specially designed forward FWM arrangement.

First, let us consider a special setup of utilizing forward degenerate FWM to produce the degenerate forward PCW.[12] Such an arrangement is shown schematically in Fig. 8.9, where two strong pump beams (beam 1 and beam 2) of the same frequency are passing through the nonlinear medium with a small intersection angle (θ_{12}) in a horizontal plane. The signal beam 3 of the same frequency is incident nearly in a vertical plane at a small intersection angle (θ_{32}) with respect to beam 2. By adjusting the latter angle and the spatial overlapping among the three beams in the nonlinear medium until beam 3 is exactly located at a circle that passes through beams 1 and 2 with an angular diameter of θ_{12} in an observation screen, then a newly generated coherent emission beam 4 of the same frequency can be observed at a location symmetrical to that of beam 3 along the circle. Meanwhile, after passing through the nonlinear medium, the intensity of beam 3 is increased. This observation can be well understood if we recognize it as a degenerate four-photon parametric process. This involves the annihilation of two photons from beams 1 and 2, and the simultaneous creation of two new photons contributing to beams 3 and 4, respectively. The phase-matching condition in this case can be written as

$$\mathbf{k}_1(\omega) + \mathbf{k}_2(\omega) = \mathbf{k}_3(\omega) + \mathbf{k}_4(\omega). \tag{8.2-17}$$

Neglecting the depletion of two pump beams within a thin nonlinear medium, we can readily prove that

$$A_4(\omega) \propto A_3^*(\omega), \tag{8.2-18}$$

i.e., beam 4 is a forward PCW of the signal beam 3. Here we have further extended the definition of forward PCW where two mutually conjugated forward beams may have different propagation directions.

In the arrangement described above, the generation of beam 4 can also be explained well based on the model of holographic grating. According to this model, the newly generated beam 4 is not only a result of four-photon parametric amplification process, but also can be a result of refraction from two induced gratings: one is formed by beams 2 and 3 and read by the pump

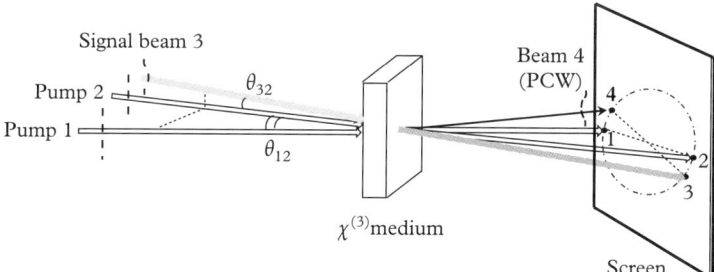

Figure 8.9 *Setup for generating forward degenerate PCW via degenerate FWM in a nonlinear $\chi^{(3)}$ medium.*

beam 1; the other is formed by beams 1 and 3 and read by the pump beam 2. However, in practice even not easy but it can be done to distinguish the grating contribution from the four-photon parametric contribution. The major difference is that the latter can take place only when three beams arrive at the nonlinear medium at exactly the same time; whereas the former can take place even when one pump beam is delayed with respect to the other pump beam within a certain time interval. In most cases, the two-beam induced gratings have a finite lifetime, depending on the specific mechanisms of the intensity-dependent refractive index changes in the nonlinear medium. Moreover, if the grating diffraction is the dominant mechanism, multi-spot structures may be observed in the screen at a high pump level, owing to a higher-order grating diffraction effect.[13]

Next, let us consider how to utilize a nondegenerate forward FWM method to create a frequency-nondegenerate forward PCW. A quite simple experimental approach is shown in Fig. 8.10(a), where a strong pump beam 1 of frequency ω_1 and a weak signal beam 2 of frequency ω_2 are passing through a third-order nonlinear medium simultaneously with an appropriate intersection angle θ. With this arrangement, a newly generated beam 3 of frequency ω_3 may be observed in a slightly different forward direction via a non-collinear four-photon parametric process, provided that the following energy conservation and phase-matching conditions are fulfilled:

$$\left. \begin{array}{l} 2\omega_1 = \omega_2 + \omega_3 \\ 2k_1 = k_2 + k_3 \end{array} \right\}. \qquad (8.2\text{-}19)$$

In this case, it can be easily proved that the beam 3 is phase-conjugated with the signal beam 2. According to Eq. (8.2-19), $\omega_3 = 2\omega_1 - \omega_2 = \omega_1 - (\omega_2 - \omega_1)$, i.e., the frequency ω_2 of the signal beam and the frequency ω_3 of the phase-conjugate beam are always located symmetrically at two sides of the pump frequency ω_1, as shown in Fig. 8.10(c). For a given combination of ω_1 and ω_2, the phase-matching angles θ and θ', shown in Fig. 8.10(b), can be calculated based on the known refractive-index dispersion of the nonlinear medium. In experiments, one may carefully adjust the intersection angle between the two incident beams until the phase conjugate beam 3 can be observed. In particular, if $\omega_2 = \omega_1 \pm \Delta\omega_r$ and $\Delta\omega_r$ is the Raman frequency shift of the medium,

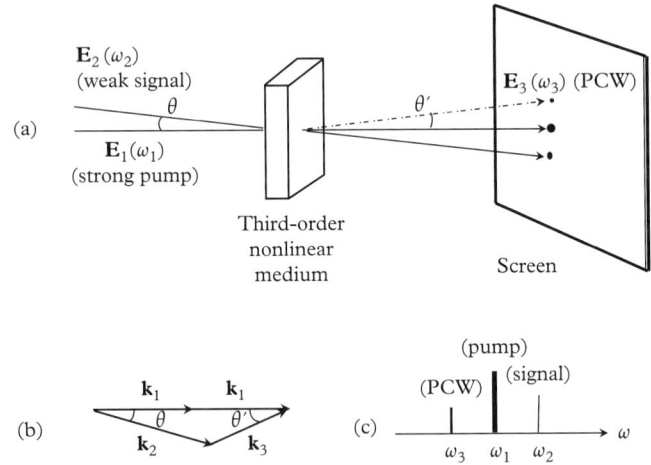

Figure 8.10 *(a) Setup for generating forward nondegenerate PCW via FWM; (b) phase-matching condition; (c) spectral allocation.*

we will see a Raman-enhanced FWM process, which is essentially similar to CARS (coherent anti-Stokes Raman spectroscopy) or CSRS (coherent Stokes Raman spectroscopy) processes.

Furthermore, if we assume that $\omega_2 \approx \omega_1 \approx \omega_3$ and the refractive-index dispersion is very small within a narrow spectral range covering these three frequencies, then $2n(\omega_1) \approx n(\omega_2) + n(\omega_3)$ and the phase-matching condition shown in Fig. 8.10(b) can be nearly fulfilled in the same forward direction. This special approach is highly useful for optical fiber communication systems where the forward PCWs can be utilized to compensate for the chromatic dispersion and nonlinear phase-distortion of a long fiber network. We shall discuss this issue in detail in Section 8.5.2.

8.2.3.2 Forward three-wave mixing in a second-order nonlinear crystal

The earliest suggestion for generating PCW through a special three-wave mixing in a second-order nonlinear crystal was given by Yariv in 1976.[6] Later on, some other special three-wave mixing arrangements were also proposed for PCW generation, and had been experimentally realized.[14–16]

For the purpose of PCW generation, one can let a strong fundamental beam of frequency ω create a second-harmonic generation (SHG) beam of 2ω through the first second-order nonlinear crystal, then let the both beams pass through another second-order nonlinear crystal as shown in Fig. 8.11(a). For the second crystal, the input fields can be written as (neglecting the polarization property of both beams):

$$\left. \begin{array}{l} E_1(\omega) = A_1(\omega)e^{-i[\omega t - k(\omega)z]} \\ E_2(2\omega) = A_2(2\omega)e^{-i[2\omega t - k(2\omega)z]} \end{array} \right\}, \qquad (8.2\text{-}20)$$

which will induce a second-order nonlinear polarization field described by

$$P_3^{(2)}(\omega) = \varepsilon_0 \chi_e^{(2)}(2\omega, -\omega) A_2(2\omega) A_1^*(\omega) e^{-i(\omega t - [k(2\omega) - k(\omega)]z)}. \qquad (8.2\text{-}21)$$

Here, $\chi_e^{(2)}$ is the effective second-order nonlinear susceptibility of the second crystal.

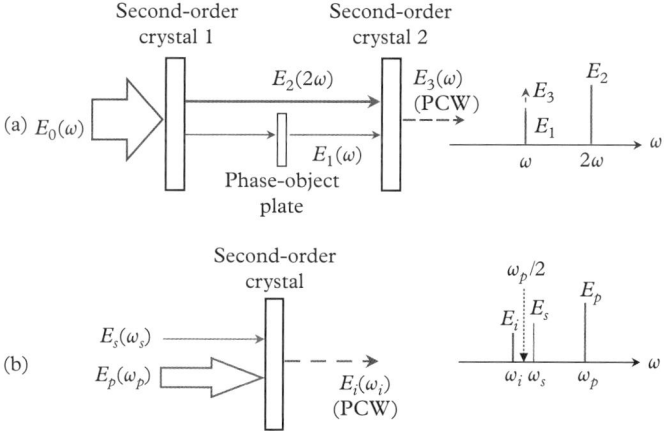

Figure 8.11 Three-wave mixing setups for generating forward PCWs in a $\chi^{(2)}$ crystal via (a) degenerate difference-frequency generation and (b) quasi-degenerate difference-frequency generation.

Assuming that the phase-matching requirement in the second crystal is satisfied, i.e., $k(2\omega) = 2k(\omega)$ or $n(2\omega) = n(\omega)$, then $P_3^{(2)}(\omega)$ will emit the third optical field of the same frequency (via the difference-frequency process):

$$E_3(\omega) \propto \chi_e^{(2)} A_2(2\omega) A_1^*(\omega) e^{-i[\omega t - k(\omega)z]}. \tag{8.2-22}$$

Here it is assumed that $E_2(2\omega)$ is a strong pump wave with a negligible depletion within the second nonlinear crystal, and $E_1(\omega)$ is a weak signal beam carrying certain amplitude/phase information. From Eq. (8.2-22) one can see that $E_3(\omega)$ is a forward PCW of the input signal wave $E_1(\omega)$.

In practice, there are two ways to separate the phase-conjugate $E_3(\omega)$ wave from the transmitted signal wave $E_1(\omega)$. The first is to distinguish one wave from the other by their polarization status. For example, under type II phase-matching condition in a negative uniaxial crystal, the pump wave $E_2(2\omega)$ and signal wave $E_1(\omega)$ can both be extraordinary rays, whereas the PCW $E_3(\omega)$ will be an ordinary ray.

The second way is to employ a near-degenerate optical parametric amplification process in a second-order nonlinear crystal, where the frequency ω_s of a signal wave is close (not equal) to the half of a strong pump frequency ω_p, i.e., $\omega_s \approx \omega_p/2$. In the similar experimental conditions as mentioned above and based on the second-order parametric interaction, a forward coherent wave can be generated in the crystal at a new (idler) frequency of ω_i, as shown in Fig. 8.11(b). The nonlinear polarization component corresponding to the newly generated (idler) wave $E_i(\omega_i)$ can be expressed as

$$P_i^{(2)}(\omega_i) = \varepsilon_0 \chi_e^{(2)}(\omega_p, -\omega_s) A_p(\omega_p) A_s^*(\omega_s) e^{-i(\omega_i t - k_i z)}. \tag{8.2-23}$$

If the phase-matching conditions of

$$\left. \begin{array}{l} \omega_i = \omega_p - \omega_s \\ k_i = k_p - k_s \end{array} \right\}$$

are fulfilled, the new emission wave $E_i(\omega_i)$ can be effectively generated,

$$E_i(\omega_i) \propto \chi_e^{(2)} A_p(\omega_p) A_s^*(\omega_s) e^{-i[\omega_i t - k(\omega_i) z]}, \tag{8.2-24}$$

which is phase conjugated with the input signal wave $E_s(\omega_s)$. In this case, the spectral positions of these two waves are mirror-symmetric in respect with the value of $\omega_p/2$, and therefore both waves can be easily separated via a spectral filter.

8.2.4 Experimental studies of backward PCW generation via FWM

8.2.4.1 System designs

The key consideration of the experimental design for using FWM to generate PCW is how to get a higher nonlinear reflectivity R. According to Eqs. (8.2-8) and (8.2-5), the value of R and the related coupling parameter γ can be expressed as

$$R = \tan^2[|\gamma| l] \text{ and } \gamma \propto \chi_e^{(3)} A_1 A_2. \tag{8.2-25}$$

Here, $\chi_e^{(3)}$ is the effective third-order nonlinear susceptibility of the medium, A_1 and A_2 are the amplitude values of the two pump waves, and l is the optical thickness (or effective interaction length) of the medium. When $|\gamma| l << \pi/4$, we have an approximate expression:

$$R \approx [|\gamma| l]^2 \propto |\chi_e^{(3)} A_1 A_2|^2 l^2 \propto |\chi_e^{(3)}|^2 I_1 I_2 l^2. \tag{8.2-26}$$

In this case, the nonlinear reflectivity R is simply proportional to $\left|\chi_e^{(3)}\right|^2$ as well as to I_1 and I_2, the intensities of two pump waves. Based on the above semiquantitative analyses, we have the following guidelines for a real experimental design: (i) to choose the media with a larger third-order nonlinearity, (ii) to provide a higher intensity for the two pump beams, and (iii) to ensure a longer effective interaction length.

Figure 8.12 shows two typical optical layouts for observing the PCW via degenerate FWM. The common feature of these two setups is that the master beam from a laser source is divided into three beams via appropriate mirrors and beam splitters: two strong pump beams pass through the nonlinear medium in the counter-propagating way, while the third weaker beam containing certain spatial information is incident upon the nonlinear medium with a small crossing angle to one of the pump beam. Based on this arrangement, the generated PCW can be measured in the opposite direction of the incident signal beam.

To ensure a longer interaction length, the angle between the signal beam and one pump beam should be relatively small. In practice, to achieve a higher local intensity the three incident beams are either weakly focused, or collimated with a reversed beam expander.

The difference between the two layouts shown in Fig. 8.12(a) and (b) is that for the former the backward pump beam is simply provided by a vertically placed mirror, whereas for the latter the two counter-propagating pump beams are provided by an optical ring-path. It is noted that the first setup is much simpler than the second setup. However, the latter is more

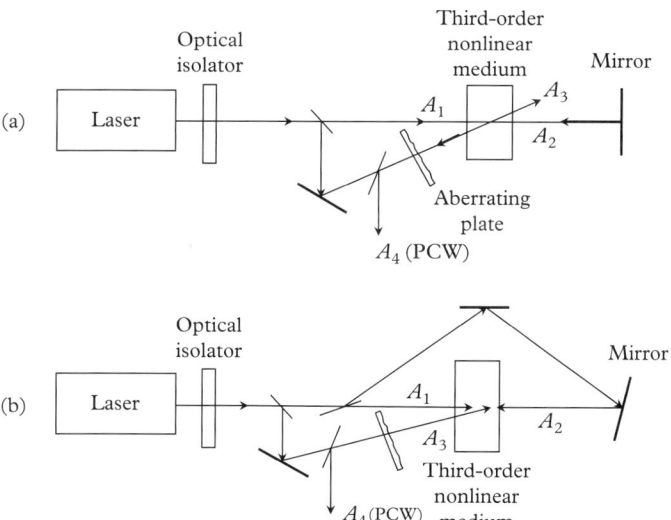

Figure 8.12 *Two experimental layouts for backward degenerate PCW generation via degenerate FWM.*

convenient when one wants to change the polarization combinations or the intensity ratio between the two pump beams. In those cases, it is easy to put additional optical elements on the ring path to control the parameters of two pump beams separately. To eliminate the possible influence of optical feedback from the backward phase-conjugate beam on the laser source, one may place an optical isolator near the output end of the laser device, or simply increase the distance between the laser source and the nonlinear medium so that the backward traveling time of the feedback light pulses is longer than the laser-pulse duration.

In experiments, the following are the major aspects to investigate:

1. *Measurement of the nonlinear reflectivity R*: It can be done by measuring the power (or energy) ratio between the incident signal beam and the 'reflected' phase-conjugate beam. Based on the measured value of R, one can determine the magnitude of the third-order nonlinearity of the studied medium.

2. *Fidelity measurement of the phase-conjugate beam*: To examine the aberration-correction capability of the phase-conjugate beam, an aberrating plate that usually is a glass slide etched by the hydrofluoric acid solution, can be placed on the incident path of the signal beam. After passing backwards through the same aberrating plate, the aberration influence on the 'reflected' phase-conjugate beam can be totally or partially removed, depending on the fidelity of the wavefront reconstruction of that beam.

3. *Polarization property of the phase-conjugate beam*: As briefly mentioned in the preceding subsection, the polarization behavior of the PCW is dependent on the polarization states of the three incident beams. Hence, from the study of influences of various polarization combinations among the three input waves on the output PCW, one may have a better understanding about the related processes.

4. *Temporal behavior of the phase-conjugate pulsed wave*: To test the dynamic response of the PCW generation via FWM arrangements, a short- or ultrashort-pulse laser source should be employed to provide the three input pulsed waves. Let two writing pulses be incident simultaneously on the sample to produce a phase-grating and delay the incidence of the third reading pulse, one may determine whether the four-photon parametric interaction or the induced phase grating is the main mechanism for the PCW generation in a given experimental condition. If the induced grating effect is the major origin, one can further determine the dominant mechanism leading to the induced refractive-index change.

In general, the nonlinear reflectivity is rather low when non-resonant media are employed. In contrast, the nonlinear reflectivity can be significantly increased when the nonlinear medium exhibits a resonant-enhancement of the intensity-dependent refractive-index change (via either one- or two-photon absorption resonance or Raman resonance) for the input laser beams with specific wavelength(s).

8.2.4.2 Backward PCW generation in non-resonant media

In early experiments on phase conjugation via FWM, CS_2 liquid was used as a non-resonant nonlinear medium and the three incident beams were the Q-switched visible laser pulses. The spatial resolution of the wavefront reconstruction for the phase-conjugate beam could be up to 30 lines/mm,[17] and the measured nonlinear reflectivity was $R \approx 10\%$.[18]

Some other early experimental studies were performed on the semiconductor crystals with 10.6 μm radiation from the pulsed CO_2 laser device. For example, using a polycrystalline Ge bulk sample as the nonlinear medium, the measured nonlinear reflectivity was $R \approx 2\%$ at the pump intensity level of ~1 MW/cm$^{2.19}$ In a similar experiment, the Hg$_{1-x}$Cd$_x$Te alloy material was used as a nonlinear medium, and the measured nonlinear reflectivity was $R \approx 9\%$; it was suggested that the nonlinearity of this medium was due to conduction-band nonparabolicity.[20]

Another example was the use of the MBBA liquid crystal as a non-resonant nonlinear medium. In this case, three input beams were from a 694.3-nm Q-switched ruby laser operated on a single longitudinal/transverse mode in the temperature range 45–55 °C. The measured maximum nonlinear reflectivity could be $R \approx 230\%$, provided that the two pump beams were circularly opposite-polarized.[21]

8.2.4.3 Backward PCW generation in resonant media

As mentioned before, in order to achieve a higher nonlinear reflectivity, the effective third-order nonlinear susceptibility value of the medium should be as large as possible. For this reason, researchers prefer to choose resonant media in which certain types of resonance enhancements of $\chi_e^{(3)}$ can be utilized.[1]

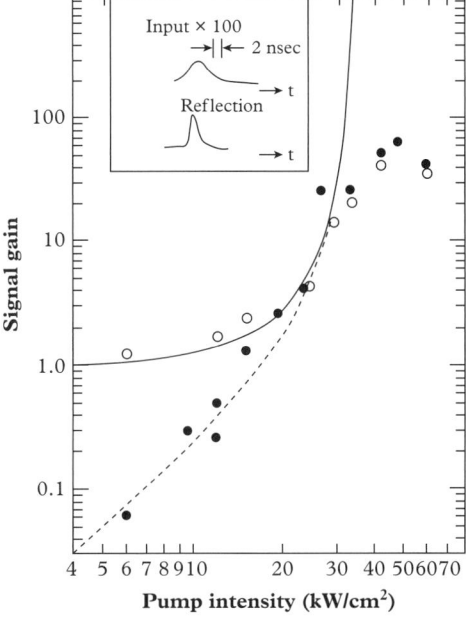

Figure 8.13 Measured intensity gain of the phase-conjugate beam (solid points) and transmitted signal beam (open circles) in Na vapor as a function of the pump intensity. The dashed and solid curves are theoretical fits for these two beams. Inset: temporal profiles of the input signal pulse and phase-conjugate pulse. (Reproduced with permission from Bloom et al.[23] Copyright 1978, Optical Society of America).

In early experiments the metal vapors were used as resonant nonlinear media.[22,23] One example was the use of Na vapor pumped by 0.5896 μm laser beams near the D-resonant line position with a detuning amount of 1.25 cm^{-1}. When the intensities for the pump beam and signal beam were 40 kW/cm^2 and 0.3 kW/cm^2, respectively, the measured nonlinear reflectivity was as high as $R \approx 10^2$, and the spatial resolution of the phase-conjugate beam was measured to be 4 lines/mm.[23] Figure 8.13 shows the measured gain data for the signal beam and the phase-conjugate beam as a function of the pump intensity; the theoretical curves shown in the same figure were given by Eq. (8.2-7). One can see in Fig. 8.13 that when the pump intensity was higher than ∼30 kW/cm^2, the obvious gain saturation for both beams occurred. In addition, the oscilloscope traces of the input pump pulse and phase-conjugate pulse are shown in the inset of Fig. 8.13, from which one can see that the pulse duration of the PCW pulse was three times shorter than that of the signal pulse.

There are many resonant media that can be utilized for degenerate FWM studies with one-photon resonant enhancement.[11,24–29] Among them, one interesting medium is SF$_6$ gas, a saturable absorbing medium for mid-IR radiation, and it can be employed for resonant degenerate FWM. In the experiment the input 10.6 μm beams were provided by a CO$_2$ laser source, the saturable absorption in SF$_6$ led to an enhanced refractive-index change, and the measured non-linear reflectivity was $R = 7\%$.[11,25] Figure 8.14 shows the photographs of far-field distributions of the input signal beam and the phase-conjugate beam from a SF$_6$ gas cell in the condition with and without an aberrator, respectively.[25] Here, it can be seen that the aberration influence can be entirely removed from the phase-conjugate beam.

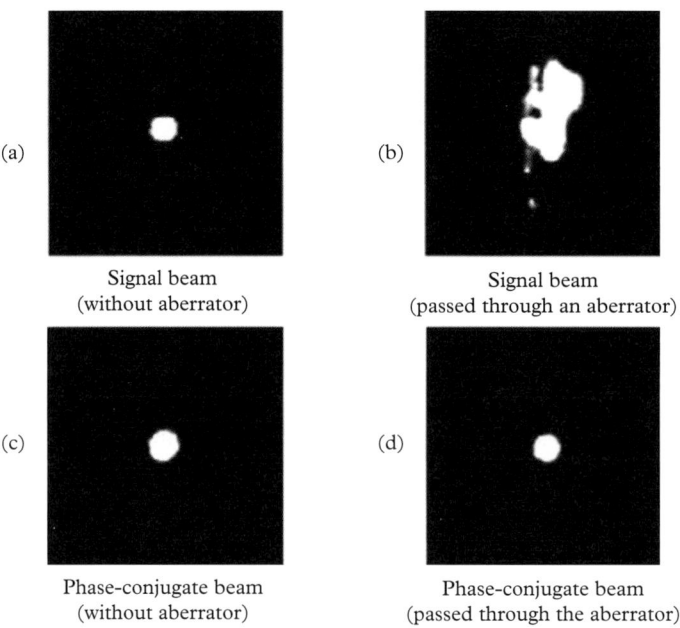

Figure 8.14 *Far-field photographs of the input signal beam and the phase-conjugate (PC) beam via FWM in SF$_6$. (Reproduced with permission from Lind et al.[25] Copyright 1979, AIP Publishing LLC).*

It is expected that the semiconductor materials can be used to generate PCW in the IR range with a certain type of resonant enhancement of the third-order nonlinearity. A commonly employed approach is the use of near-resonance between the band-gap of a semiconductor medium and the one- or two-photon energy of the input laser beams. There is an example of two-photon enhanced degenerate FWM using the Ge crystal as a nonlinear medium. The input radiation provided by a pulsed DF laser source involved multi-lines in the spectral range 3.6–4.0 μm. Since germanium exhibits two-photon absorption at 3.4 μm wavelength, a near-resonant enhancement can be achieved. Under the condition of 3.8 μm single-line excitation at an intensity level of 12 MW/cm^2, the measured nonlinear reflectivity was $R = 0.14\%$.[27] In another experiment, the nonlinear medium was a p-type Si single crystal pumped with 1.06 μm beams from a Q-switched Nd:YAG laser source. Because the band-gap energy for the Si sample at room temperature is $E = 1.11$ eV, which is close to the single pump photon energy of $h\nu = 1.116$ eV, a one-photon near-resonance enhancement can be achieved. For the sample with a residual carrier concentration of less than 10^{14} cm^{-3}, the measured nonlinear reflectivity was $R = 105\%$.[28]

It is also known that dye solutions or dye-doped solid materials have very strong linear (one-photon) or relatively weak nonlinear (two- or multi-photon) absorption bands over appropriate spectral regions, therefore they are good candidates for degenerate or partially degenerate FWM studies because of their largely enhanced third-order nonlinearity. In these cases, the observed phase-conjugate signals are generated from the diffraction of the induced phase gratings, which are most likely formed by population changes and/or opto-thermal effects. For example, the BDN dye solution is a well-known saturable absorption medium for passive Q-switching at 1.06 μm wavelength. Utilizing this dye solution as a nonlinear medium for the FWM experiment pumped with 1.06-μm laser beams, when the pump intensities were much higher than the saturation intensity value, the measured nonlinear reflectivity was $R = 600\%$.[29]

8.2.4.4 *Backward PCW generation via partially degenerate FWM*

Until now, we have only described the experimental results based on the degenerate FWM, in which all four waves had the same wavelength or frequency. However, as mentioned in the end of Section 8.2.2, the PCW can also be generated via partially degenerate FWM processes. In these cases, the two input waves having the same frequency (ω_1) are used for interference with each other and to produce an induced phase grating, while the third reading beam and the diffracted PCW have another frequency (ω_2).[30–34]

The first demonstration of using partially degenerate FWM for generating backward PCW was reported by Martin and Hellwarth.[30] Their experimental setup is shown in Fig. 8.15(a), where a Q-switched and frequency-doubled Nd:YAG laser could simultaneously provide two stronger 1.06-μm IR laser beams and one weaker 0.53-μm green laser beam. One 1.06-μm laser beam passed through a hole-pattern plate and then interacted with another 1.06-μm beam to produce a phase grating inside a nonlinear medium. In the mean time, the green laser beam was incident on the medium to read the grating and to generate a backward diffracted beam at the same wavelength of 0.53 μm. In this experiment, more than 14 liquids (including CS$_2$, water, acetone, benzene, etc.) and an IR filter glass sample were tested, and the observed diffracted beam was a visible replica of the IR signal beam. The image patterns carried by the IR signal beam and restored by the green phase-conjugate beam are shown in Fig. 8.15(b). The apparent difference of the image sizes is due to the fact that, the signal beam was slightly divergent towards the observation screen, while the backward-diffracted beam from the grating was slightly convergent. One can see that there is a good fidelity for the wavefront reconstruction of the replica beam. Moreover, by changing the time delay between the two IR writing beams and the green reading beam, the lifetime of the induced grating can be measured. For most liquid samples tested in this experiment, the

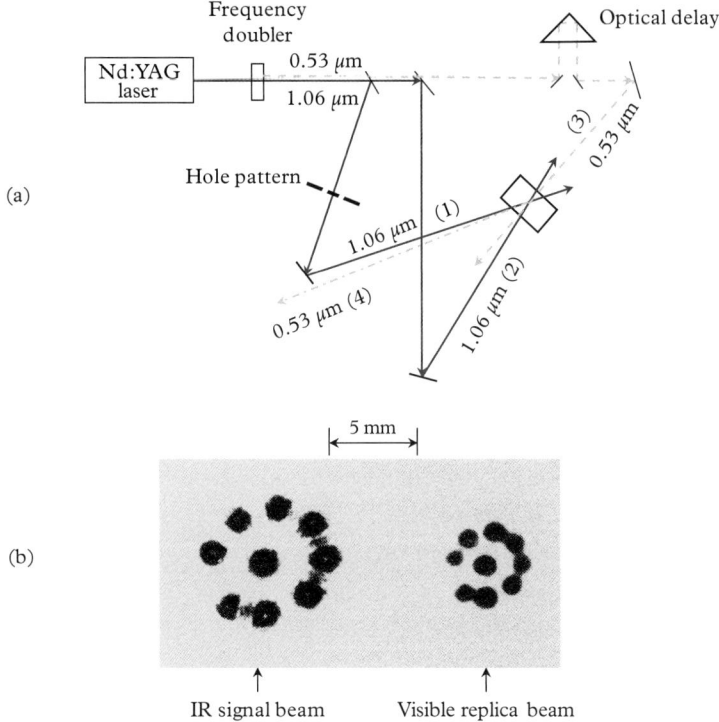

Figure 8.15 *Optical layout for generating backward PCW via a partially degenerate FWM. (Reproduced with permission from Martin and Hellwarth.[30] Copyright 1979, AIP Publishing LLC).*

results showed that the induced gratings could last even more than microseconds. This implies that the induced grating was mainly caused by opto-thermal effect. Usually the formation of thermal gratings in a resonant medium may relate to several different processes such as residual linear absorption, two- or multi-photon absorption, or stimulated scattering. During these processes, a part of pump laser energy can be converted to the thermal energy of the nonlinear medium and leads to a temperature dependent refractive-index change. The thermal grating is characterized by its longer decaying time (usually more than milliseconds).

Another type of partially degenerate FWM, the so-called Brillouin-enhanced FWM, can also be employed to generate backward nondegenerate PCW.[35-38] In this approach, the frequency difference between two beams is in resonance with the frequency shift of stimulated Brillouin scattering in the nonlinear medium.

The significance of using a partially degenerate FWM process to generate PCW is that: (i) this is a new technical approach to transfer images from one wavelength to another wavelength with high fidelity, (ii) the transferred image can be amplified or intensified because it can extract partial energy from the pump beam via the grating diffraction, and (iii) this process reveals the key insight that explains the phase-conjugate nature of the backward stimulated scattering and the backward stimulated emission (lasing).

8.3 PCW generation via backward stimulated scattering

8.3.1 Findings of phase-conjugation behavior of backward stimulated scattering

From the historical viewpoint, the earliest optical phase-conjugation phenomenon was accidently observed in an experiment of stimulated Brillouin scattering (SBS) by Zel'dovich et al. in 1972;[39] that was four years earlier than the proposition of the concept of OPC.[6,7] The experimental arrangement used for that observation is shown schematically in Fig. 8.16. Here, the pump source was a single-axial-mode and Q-switched ruby laser; the pump beam passed through an aberration plate and then was focused into the scattering medium. After passing back through the same aberration plate, the spatial structure of the backward SBS could be compared to that of the incident pump beam. In the experiment, the scattering medium was high-pressure CF_6 gas filling in a 94-cm-long cell, the divergence of the input $\lambda_0 = 694.3$ nm pump beam was 0.14×1.3 mrad, and after passing through the aberration plate it became ~ 3.5 mrad. If a plane mirror was put in the path of the pump beam, and the reflected pump beam was allowed to pass through the aberrator second time, the beam divergence was increased to ~ 6.5 mrad. Whereas, it was found that after passing through the same aberrator, the backward SBS had nearly the same small divergence as the initial pump beam had, i.e., the aberration influence was removed. The measured energy-transfer efficiency from the input pump pulse to the backward SBS pulse was $\eta \approx 25\%$.[39] In another similar early experiment, the scattering medium was CS_2 liquid in a glass cell and the aberration plate was replaced by a poor-quality ruby amplifier rod. The divergence of the initial input pump beam was 0.13 mrad; after passing through the amplifier rod it increased to 2.5 mrad. In contrast, after passing through the same amplifier rod, the backward SBS exhibited a divergence angle of only ~ 0.15 mrad, and the measured energy transfer efficiency was $\eta \approx 60\%$.[40]

These two experiments essentially showed that the aberration influence imposed on the input pump beam could be automatically canceled in the backward SBS beam. However, during the time of that observation this unexpected phenomenon could not be well explained by the known theory of SBS, and therefore did not attract enough attention in the research community until several years later.

In 1977, the concept of an optical PCW and its capability of overcoming aberration influence was proved experimentally.[8,9] Soon after, researchers began to realize that the above-mentioned observation could be well understood if we recognize that the backward SBS beam is phase conjugated with the input pump beam.

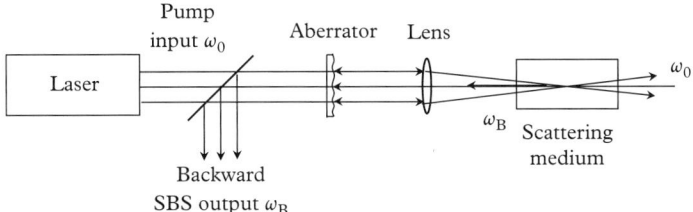

Figure 8.16 *Experimental setup for observing the phase-conjugate backward SBS.*

244 Optical Phase Conjugation

More importantly, it had been further found that the PCW can be generated not only by the backward SBS process,[41–45] but also by other known backward stimulated scattering processes, such as stimulated Raman scattering (SRS)[46–51] and stimulated Rayleigh-wing scattering (SRWS).[52–54] For all these three types of stimulated scattering processes, there is always a frequency difference between the input pump beam and the backward stimulated scattering beam; therefore, here we should deal with the frequency-nondegenerate backward PCW generation. More recently, it has been further found that the frequency-degenerate backward PCW can be generated based on the newly developed stimulated Rayleigh–Bragg scattering (SRBS)[55] and stimulated Mie scattering (SMS);[56,57] in these two cases, there is no frequency shift between the input pump beam and the backward stimulated scattering beam (see Sections 7.6 and 7.7.).

Compared to the methods of using four- or three-wave mixing to generate PCWs, the methods of utilizing backward stimulated scattering exhibit some obvious advantages, such as (i) there is no phase-matching requirement, (ii) the optical setup is very simple with no critical beam alignment requirement, and (iii) it is easy to obtain a higher nonlinear reflectivity.

8.3.2 Experimental studies on phase-conjugation properties of different types of backward stimulated scattering

Experimental studies on the phase-conjugation properties of backward stimulated scattering involve the measurements of beam divergence, restoration of the image information carried by the input pump beam, and the nonlinear reflectivity. In the following, we highlight some typical results related to different stimulated scattering mechanisms.

As an early example, Fig. 8.17 shows the measured intensity distributions and photographs of far-field patterns for the original pump beam, the aberrated pump beam, and the aberration-corrected backward SBS beam from CS_2 liquid.[41] The pump beam was a TEM_{00} single-axial-mode 694.3-nm pulsed laser beam of 17 ns duration with ∼0.57 mrad divergence angle. After

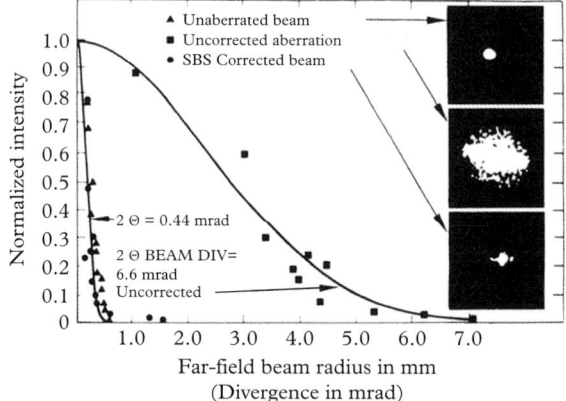

Figure 8.17 *Normalized far-field intensity distributions and photographs for the original pump beam, the aberrated pump beam, and the aberration-corrected backward SBS beam. (Reproduced with permission from Wang and Giuliano.[41] Copyright 1978, Optical Society of America).*

twice passing through an aberrator (a microscope slide etched in hydrofluoric acid), the divergence of the pump beam was increased to 6.6 mrad. In contrast, after passing through the same aberrator the backward SBS beam exhibited a much smaller divergence angle of 0.44 mrad; in other words, the aberration influence was automatically corrected.

According to the definition of OPC, the two beams conjugated to each other should have the same transverse intensity distribution and reversed wavefront structure; in other words, these two beams should have the same near-field and far-field distributions. Therefore, the near-field image reconstructed by the backward stimulated scattering beam should be identical to the original image carried by the pump beam.

The wavefront reconstruction of the backward stimulated scattering beam can also be examined by using interference techniques. One experimental setup based on that method is shown schematically in Fig. 8.18.[58] In this case the parameters of the pump beam were wavelength of 532 nm, pulse duration of 10 ns, beam size of 4 mm, and divergence angle of 1 mrad. After passing through a beam-splitter, the pump beam was divided into two beams, which were finally focused into two CS_2 liquid cells and to generate their own backward SBS separately.

Based on the setup shown in Fig. 8.18, one can observe the interferograms formed by the two incident pump laser beams and by two backward SBS beams, respectively. The measured results are shown in Fig. 8.19. From Fig. 8.19(a) one can see that without placing an aberrator, the two-pump-beam interference fringes are quite clear and regular. However, as shown in Fig. 8.19(b), after one pump beam passed through an aberrator, the interference fringes are severely deteriorated due to the aberration influence. On the other hand, with no aberrator in the optical path, the interference fringes formed by two backward SBS beams are still clear and regular, as shown in Fig. 8.19(c). This implies that the two backward SBS beams were good replicas of the two pump beams when there was no induced aberration influence. If an aberrator is put in the position B of the optical path (I), the quality of the interference fringes formed by the two

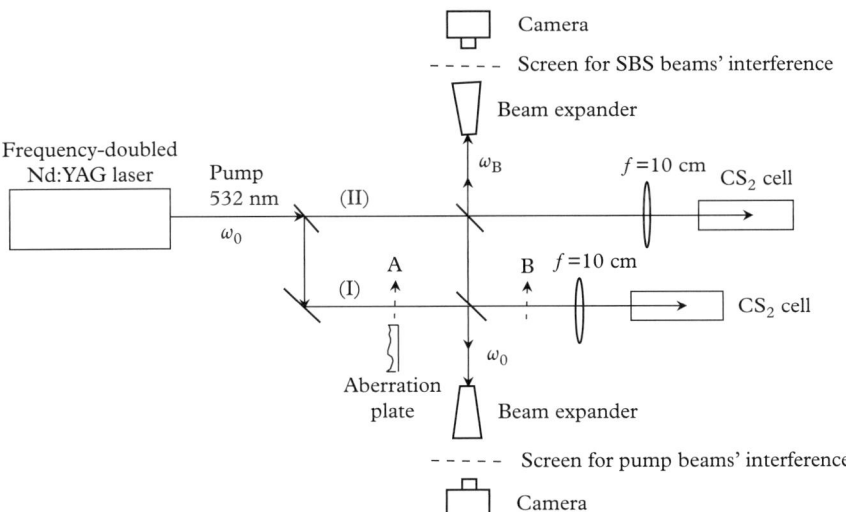

Figure 8.18 *Experimental setup for examining the wavefront reconstruction of backward SBS with two-beam interference method. The aberrator can be inserted in position A or B of the optical path (I). (After Liu and He.[58] Copyright 1999, AIP Publishing LLC).*

246 *Optical Phase Conjugation*

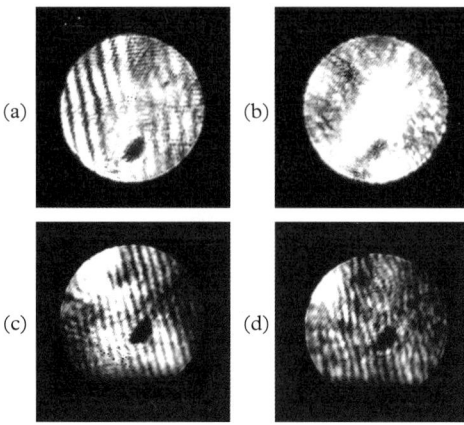

Figure 8.19 *Interferograms formed by (a) two pump beams without use of an aberrator, (b) two pump beams with the aberrator in position A of the optical path (I), (c) two backward SBS beams without use of the aberrator, and (d) two backward SBS beams with the aberrator in position B of the optical path (I). The optical setup is shown in Fig. 8.18. (After Liu and He.[58] Copyright 1999, AIP Publishing LLC).*

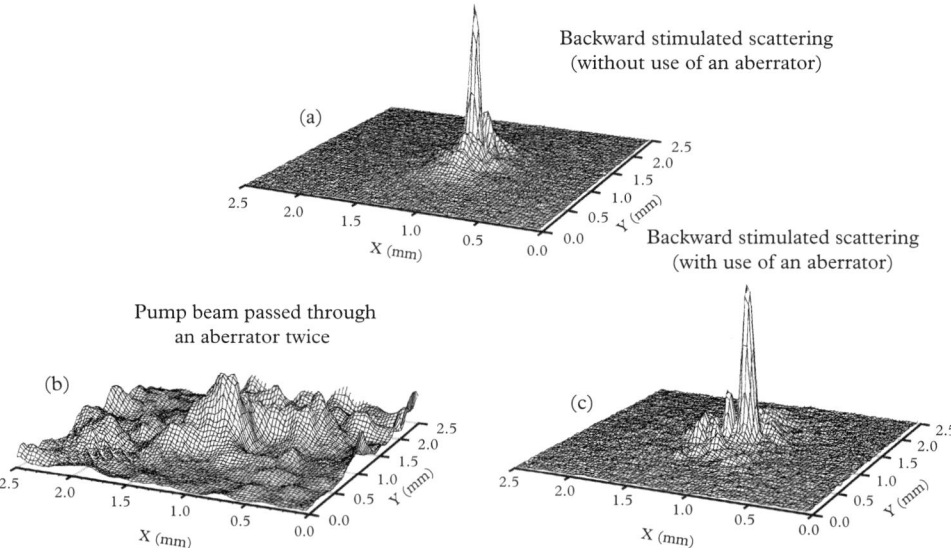

Figure 8.20 *Measured far-field intensity distributions of the backward SRBS beam from a three-photon absorbing dye solution. (a) Backward SRBS beam without the use of an aberrator; (b) pump beam twice passed though an aberrator; (c) backward SRBS beam passed through the same aberrator. Focusing length is 50 cm. (After He et al.[55] Copyright 2007, Optical Society of America).*

SBS beams is still fairly good as shown in Fig. 8.19(d). This is the further experimental evidence demonstrating the aberration-correction capability of backward stimulated scattering.

The latest progress in stimulated scattering studies is the findings of two new types of stimulated scattering mechanism, i.e., the SRBS in multi-photon absorbing liquids (solutions) and the SMS in nanoparticle suspension systems. In comparison with other stimulated scattering mechanisms (such as SBS and SRS), SRBS and SMS exhibit the advantages of low pump threshold requirement, no frequency shift, and easy control of the scattering media. In addition, both mechanisms can be utilized to produce efficient backward degenerate phase-conjugate beams.

Figure 8.20 shows three-dimensional (3D) representations of the measured far-field patterns, based on a SRBS experiment in which the scattering medium was a three-photon absorbing dye (PRL-OT04) solution in chloroform, and the pump beam was the 1064-nm and 10-ns laser pulses from a Q-switched Nd:YAG laser.[55] Figure 8.20(a) shows the far-field pattern without inserting an aberrator in the path of the pump beam, indicating a small divergence angle. Figure 8.20(b) shows the far-field pattern of the pump beam passed through an aberrator twice, indicating a very large divergence and random intensity distribution. Finally, Fig. 8.20(c) shows the pattern of the backward stimulated scattering beam passed through the same aberrator, indicating the removal of severe aberration influences.

A recent experiment has also demonstrated the good phase-conjugation property of the backward SMS.[57] In this work, CdSe/CdS/ZnS double-shell structured nanocrystals (NCs) dispersed in chloroform ($CHCl_3$) is employed as a Mie scattering medium, pumped by 816-nm and 8-ns laser pulses from a dye laser. The experimental setup is shown in Fig. 8.21(a), and the NC

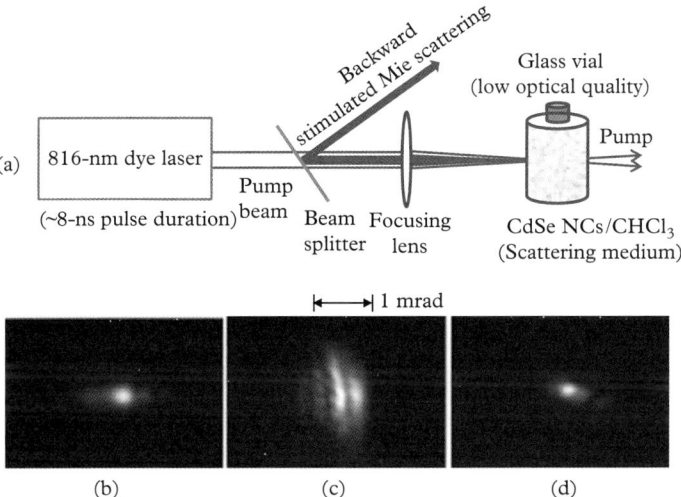

Figure 8.21 *(a) Experimental setup for measuring the far field of the backward SMS from a semiconductor nanocrystal suspension (solution) in a poor-quality glass vial. The far-field photographs of (b) the pump beam without passing through the sample vial, (c) the transmitted pump beam after passing through the sample vial, and (d) the backward SMS beam from the sample vial. Here the glass vial plays the role of an aberrator. (After He et al.[57] Copyright 2012, AIP Publishing LLC).*

suspension is in an ordinary glass vial of poor optical quality. Figure 8.21(b) shows the photograph of the far-field pattern of the pump beam without passing through the sample vial, indicating a small divergence angle of ∼0.35 mrad. Figure 8.21(c) shows the far-field pattern of the pump beam after passing through the sample vial, indicating a severe aberration influence imposed by the glass vial. In contrast, the far-field pattern of the backward SMS beam from the sample vial is shown in Fig. 8.21(d), indicating that the aberration influence from the glass vial is well removed.

Based on the existing experimental results, we can draw the following conclusions: (i) without or under small aberration influence, the backward stimulated scattering beam can be a perfect replica of the incident pump beam; (ii) under a larger aberration influence, the backward stimulated scattering beam can only be an approximate replica of the pump beam; (iii) the forward stimulated scattering beam has no phase-conjugation property. The fidelity of the reconstructed wavefront of a backward stimulated scattering beam is determined by many experimental factors, such as the pump intensity level, pump focusing geometry, gain length, sort of scattering medium, frequency shift range, and the extent of the imposed aberration.

8.3.3 Theoretical explanations: quasi-collinear FWM model

Although various types of backward stimulated scattering (BSS) can readily be utilized to efficiently generate optical PCWs, it took quite a time for researchers to understand the substance of these processes and to give a clear physical interpretation without invoking any cumbersome mathematical treatments.

A considerable number of early theoretical papers were published on this issue.[59–67] Many of these papers were based on a phenomenological assumption that there is an exponential gain discrimination (by a factor of 2?) between the phase-conjugate portion and non-phase-conjugate portion of BSS; as a result, only the former could obtain the maximum gain and be effectively amplified in the scattering medium. However, for a long time, there was a lack of a clear theoretical model or physical mechanism to support this assumption. Even ignoring the confusion and discrepancies among some of these papers, most theoretical analyses could not answer a key question: why could only the phase-conjugate portion of the BSS finally become the predominant output? For instance, it is known that a higher pump intensity is needed to have a better phase-conjugation fidelity for BSS. However, the gain saturation will take place under these circumstances; as a result, the gain difference between these two portions should become smaller, and the worse phase-conjugate fidelity would be expected according to the phenomenological assumption mentioned above. In some other theoretical analyses, both the pump field and the BSS field were represented by an infinite series function. In those cases, it was extremely difficult to obtain explicit analytical solutions for the nonlinear wave equations.

On the one hand, we know well that the principles of using degenerate or nondegenerate FWM to produce PCWs are based on the wavefront reconstruction via induced holographic gratings in the nonlinear medium. On the other hand, it is also known that the PCWs can also be generated via various types of BSS processes, no matter how different the scattering mechanisms. The latter fact implies that there should be a common origin, which determines the phase-conjugation property of any types of BSS and should not depend on the specific scattering mechanism. To find this common origin, the first insight suggested by Sokolovskaya et al. was that the phase-conjugation nature of BSS might be related with a special holographic process occurring in the scattering medium.[47,48] Later, a quasi-collinear FWM model was proposed by He et al. to explain the phase-conjugation property of the BSS, which was supported by a rigorous mathematical formulation in an unfocused-beam approximation.[68,69] According to such a model, the PCW generation via any type of BSS can be visualized as a quasi-collinear FWM process: the input pump beam contains two waves (E_1 and E_2) of frequency ω_0, while the BSS beam contains other two waves (E_3 and E_4) of frequency ω'. It can be theoretically proven that under certain conditions: $(E_3 + E_4) \propto (E_1 + E_2)^*$.[58,69]

To explain this model more clearly, it is better to invoke Gabor's original idea of holograph. In that case, for a input coherent light wave passing through a transparent object (phase object), the object is assumed to be such that a considerable part of the wave penetrates undisturbed through it, and a hologram is formed by the interference of the secondary wave arising from the presence of the object with the strong background wave, as clearly described by Born and Wolf.[70] According to this principle, after passing through a phase object the total optical field can be expressed as a superposition of two parts:[70]

$$U = U^{(i)} + U^{(s)} = A^{(i)} e^{i\phi_i} + A^{(s)} e^{i\phi_s} = e^{i\phi_i}[A^{(i)} + A^{(s)} e^{i(\phi_s - \phi_i)}]. \tag{8.3-1}$$

Here, $U^{(i)}$ is the undisturbed part of the transmitted field, $U^{(s)}$ is the disturbed part arising from the presence of the object, $A^{(i)}$ and $A^{(s)}$ are their amplitude functions, and ϕ_i and ϕ_s are the corresponding phase functions.

The Gabor principle is applicable to most phase-conjugation experiments based on BSS. In these cases, as schematically shown in Fig. 8.22, $E(\omega_0)$ is a quasi-plane pump wave. After passing through an aberration plate or a phase subject, the pump field manifests itself as a superposition of two portions: a stronger undisturbed wave $E_1(\omega_0)$ and a weaker distorted wave $E_2(\omega_0)$. These two waves interfere with each other in a scattering (gain) medium and create an induced volume holographic grating due to intensity-dependent refractive index changes of the medium. Only the undisturbed pump wave $E_1(\omega_0)$ is strong enough to fulfill the threshold requirement and to generate an initial BSS wave $E_3(\omega')$ that exhibits a regular wavefront as wave $E_1(\omega_0)$. When wave $E_3(\omega')$ passes back through the induced holographic grating region, a diffracted wave $E_4(\omega')$ can be created. Here we see a typical holographic wavefront-reconstruction process: the induced grating is formed by the regular $E_1(\omega_0)$ wave (reference beam) and the irregular

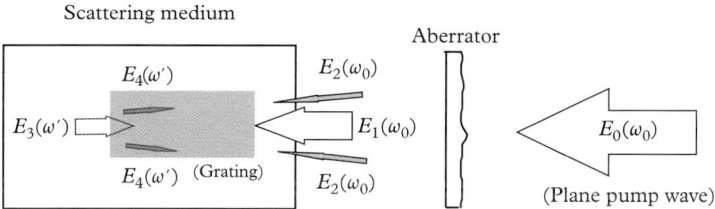

$[E_3(\omega' \leq \omega_0) + E_4(\omega' \leq \omega_0)] = a\,[E_1(\omega_0) + E_2(\omega_0)]^*$

E_1: Undistorted portion of pump wave;
E_2: Distorted portion of pump wave;
E_3: Initial backward stimulated scattering wave (reading beam);
E_4: Diffracted wave from the induced grating.

Figure 8.22 *Schematic illustration of the partially degenerate FWM model for the phase-conjugation formation of BSS (in quasi-plane wave approximation).*

$E_2(\omega_0)$ wave (signal beam), the initial backward stimulated scattering $E_3(\omega')$ is a reading beam with a regular wavefront, and the diffracted wave $E_4(\omega')$ will be the phase-conjugate replica of the $E_2(\omega_0)$ wave.

Moreover, the wave $E_4(\omega')$ will experience an amplification with the wave $E_3(\omega')$ together because both waves have the same scattering frequency. In the sense of phase-conjugation formation, for SRBS and SMS ($\omega_0 = \omega'$) it is due to a quasi-collinear and frequency-degenerate FWM process, for SBS ($\omega_0 \approx \omega'$) it is due to a nearly frequency-degenerate FWM process, and for SRS ($\omega_0 > \omega'$) it is due to a partially degenerate FWM. Based on the explanation described above, here one can see a common mechanism (i.e., pump field-induced holographic grating) plays the same role for phase-conjugation formation by using either FWM or BSS method. This common mechanism is applicable to all types of backward stimulated scattering processes, even though the specific scattering mechanisms can be different for each other.

Although there is a common ground for PCW generation via conventional FWM or BSS, there is still a remarkable difference between these two processes. For the former, only two waves (the signal wave and the backward diffracted wave) are phase conjugated to each other; however, for the latter, the sum of the two portions of the BSS beam should be phase-conjugated to the sum of the two portions of the input pump beam. In the following subsection, it can be mathematically proved that the latter situation can indeed be realized.

According to the model described here, the forward stimulated scattering beam will not have any phase-conjugation property, which is in agreement with the observations in SRS or other-type forward (SRWS or SKS) stimulated scattering experiments.

8.3.4 Theoretical treatment in unfocused-beam approximation[69]

According to the aforementioned quasi-collinear FWM model of the BSS process, after passing through an aberration plate the total input pump field can be expressed by the sum of two portions:

$$E_p(\omega, z) = [A_1(z) + A_2(z)e^{i\theta(z)}]e^{-i(\omega t + kz)}. \tag{8.3-2}$$

Here, A_1 and A_2 are the real amplitude functions of the undistorted portion and the distorted portion of the pump field, respectively, θ is a phase change function, the pump field is propagating in the -z direction, and k is the magnitude of the wavevector. For simplicity, we assume that the original input pump field is an ideal monochromatic and uniform plane wave before passing through the aberration plate. Thus, A_1 and A_2 are only the functions of z. Here, θ represents the deviation of wavefront of the distorted portion from an ideal plane wavefront.

Based on the theoretical model illustrated in Fig. 8.22, the total backward stimulated emission field can also be expressed by the sum of two portions:

$$E'(\omega', z) = [A'_1(z) + A'_2(z)e^{-i\theta'(z)}]e^{-i(\omega' t - k'z)}. \tag{8.3-3}$$

Here, A'_1 is a real amplitude function of the initial BSS wave (reading beam), A'_2 is a real amplitude function of the diffracted wave through FWM or grating diffraction, and θ' is the relative phase change function between the A'_1 and A'_2 waves. In writing Eq. (8.3-3) we have assumed that the initial backward stimulated scattering is also an ideal plane wave; this assumption is supported by experimental results without using an aberration plate. In a steady-state condition, A_1, A_2, and θ and A'_1, A'_2, and θ' are independent of time t.

In small-signal approximation, the initial BSS wave propagating along the z direction experiences an exponential gain described by

$$A'_1(z) = A'_1(0) e^{\frac{1}{2} g A_1^2 z}, \tag{8.3-4}$$

where g is the exponential gain factor due to stimulated amplification from a scattering system. Furthermore, within a short propagation distance, the depletion of the pump field can be approximately neglected, thus the wave equation for the pump field can be written as

$$\nabla^2 E_p - \frac{n_0^2(\omega)}{c^2} \frac{\partial^2 E_p}{\partial t^2} = 0, \tag{8.3-5}$$

where $n_0(\omega)$ is the linear refractive index at the pump frequency and c is the light speed in free space. Substituting Eq. (8.3-2) into Eq. (8.3-5), we can obtain the following equations obeyed by the phase change function of the pump field:

$$\left. \begin{array}{c} \dfrac{\partial \theta}{\partial z} + \dfrac{1}{2k} (\nabla \theta)^2 = 0 \\ \nabla^2 \theta = 0 \end{array} \right\}. \tag{8.3-6}$$

As a result of interaction among the three waves of A_1, A_2, and A'_1, the fourth wave of A'_2 can be generated through the mechanism of FWM or grating diffraction. The wave equation obeyed by the diffracted wave of $E'_2 = A'_2 e^{-i\theta'(z)} e^{-i(\omega' t - k' z)}$ can be written as

$$\nabla^2 E'_2 - \frac{n_0^2(\omega')}{c^2} \frac{\partial^2 E'_2}{\partial t^2} = \mu_0 \frac{\partial^2 P_2^{(3)}}{\partial t^2}, \tag{8.3-7}$$

where μ_0 is the permeability of free space and $P_2^{(3)}$ is the third-order nonlinear polarization contribution to the wave E'_2, which can be expressed as

$$P_2^{(3)} = \varepsilon_0 \chi_e'^{(3)} A_1 A'_1 A_2 e^{-i\theta(z)} e^{-i(\omega' t - k' z)}. \tag{8.3-8}$$

Here, $\chi_e'^{(3)}$ is a formally introduced effective third-order nonlinear susceptibility value. If we only consider the gain behavior of the amplitude A'_2, $\chi_e'^{(3)}$ can be viewed as a pure imaginary quantity, i.e., $\chi_e'^{(3)} = i \chi_e''^{(3)}$, where $\chi_e''^{(3)}$ is a real quantity. Substituting Eq. (8.3-8) into Eq. (8.3-7) and using slowly varying amplitude approximation lead to

$$\left. \begin{array}{c} \dfrac{\partial A'_2}{\partial z} + \dfrac{A'_2}{2k'} \nabla^2 \theta' = \dfrac{k' \chi_e''^{(3)}}{2} A_1 A_2 A'_1 \cos \delta\theta \\ \dfrac{\partial \theta'}{\partial z} + \dfrac{1}{2k'} (\nabla \theta')^2 = \dfrac{k' \chi_e''^{(3)}}{2} A_1 A_2 A'_1 \dfrac{1}{A'_2} \sin \delta\theta \end{array} \right\}, \tag{8.3-9}$$

where

$$\delta\theta = \theta - \theta'. \tag{8.3-10}$$

In the small aberration approximation, the second-order spatial derivative of $\theta'(z)$ can be neglected; thus the first equation in Eq. (8.3-9) is simplified to

$$\frac{\partial A'_2}{\partial z} = \frac{1}{2} g' A_1 A_2 A'_1 \cos \delta\theta, \qquad (8.3\text{-}11)$$

where $g' = k' \chi_e''^{(3)}$. On the other hand, subtracting the second equation of Eq. (8.3-9) from the first equation of Eq. (8.3-6) leads to

$$\frac{\partial (\delta\theta)}{\partial z} + \frac{1}{2k}(\nabla\theta)^2 - \frac{1}{2k'}(\nabla\theta')^2 = -\frac{1}{2} g' A_1 A_2 A'_1 \frac{1}{A'_2} \sin \delta\theta. \qquad (8.3\text{-}12)$$

In the small aberration approximation, the square of the first derivative of the phase function θ or θ' is relatively small, and therefore one can assume that

$$\left[\frac{1}{2k}(\nabla\theta)^2 - \frac{1}{2k'}(\nabla\theta')^2 \right] \cong 0.$$

Based on the above assumption, Eq. (8.3-12) is simplified to

$$\frac{\partial (\delta\theta)}{\partial z} = -\frac{1}{2} g' A_1 A_2 A'_1 \frac{1}{A'_2} \sin \delta\theta. \qquad (8.3\text{-}13)$$

Now we have obtained two coupled equations, Eqs. (8.3-11) and (8.3-13), to describe the changes of the amplitude and the phase function of the wave E'_2.

Dividing Eq. (8.3-11) by Eq. (8.3-13) leads to

$$\frac{\partial A'_2}{\partial (\delta\theta)} = -\frac{\cos \delta\theta}{\sin \delta\theta} A'_2. \qquad (8.3\text{-}14)$$

The solution of this equation is

$$A'_2 \sin \delta\theta = B = A'_2(0) \sin \delta\theta(0), \qquad (8.3\text{-}15)$$

where B is a constant while $A'_2(0)$ and $\sin \delta\theta(0)$ can be recognized as the initial values of the diffracted wave E'_2. From Eq. (8.3-15) we have

$$\sin \delta\theta = B/A'_2. \qquad (8.3\text{-}16)$$

Substituting this equality into Eq. (8.3-11) leads to

$$\frac{\partial A'_2}{\partial z} = \frac{1}{2} g' A_1 A_2 A'_1 \frac{1}{A'_2} \sqrt{A'^2_2 - B^2}. \qquad (8.3\text{-}17)$$

In the initial step of the diffracted wave generation, the value of $A'_2(0)$ is so small that we can assume $B \approx 0$ regardless of the specific value of $\sin \delta\theta(0)$. Then the above equation is simplified to

$$\frac{\partial A'_2(z)}{\partial z} = \frac{1}{2} g' A_1 A_2 A'_1(z). \qquad (8.3\text{-}18)$$

Here, we assume that within a short propagation length only the waves of A'_1 and A'_2 experience a rapid growth while the amplitude change of the waves of A_1 and A_2 can be neglected. A trial solution of Eq. (8.3-18) is

$$A'_2(z) = \frac{A_2}{A_1} A'_1(z) = \frac{A_2}{A_1} A'_1(0) e^{\frac{1}{2} g A_1^2 z}, \quad (8.3\text{-}19)$$

where the expression of $A'_1(z)$ is given by Eq. (8.3-4). A derivative manipulation on both sides of Eq. (8.3-19) leads to

$$\frac{\partial A'_2(z)}{\partial z} = \frac{1}{2} g A_1 A_2 A'_1(z). \quad (8.3\text{-}20)$$

Comparing Eq. (8.3-20) to Eq. (8.3-18) one can see that they are identical if we assume $g' = g$. Here g' is the exponential gain factor due to FWM or grating diffraction, while g is the exponential gain factor due to stimulated scattering of the medium. In principle the diffracted wave (as a seed signal) should get gain from both mechanisms; however, as this wave is backward propagating over a longer length, the gain contribution from stimulated amplification will be predominant. For this reason g' can be replaced by g and the expression of Eq. (8.3-19) should be the real solution of $A'_2(z)$.

In order to obtain the solution for the phase change function, we can substitute Eq. (8.3-19) into Eq. (8.3-15) and obtain

$$\sin\delta\theta(z) = \frac{A_1 A'_2(0)}{A_2 A'_1(0)} \sin\delta\theta(0) e^{-\frac{1}{2} g A_1^2 z}. \quad (8.3\text{-}21)$$

Since $A_1 \gg A_2$ and $A'_1(0) \gg A'_2(0)$, the fractional term on the right-hand side of Eq. (8.3-21) can be approximately assumed $\Rightarrow 1$, then we have

$$\sin\delta\theta(z) = \sin\delta\theta(0) e^{-\frac{1}{2} g A_1^2 z}. \quad (8.3\text{-}22)$$

If we further assume that the pump intensity (A_1^2) is strong enough, the exponential gain factor g is large enough or the effective gain length z is long enough, so that the following condition is fulfilled:

$$g A_1^2 z \gg 1, \quad (8.3\text{-}23)$$

then we obtain the result

$$\delta\theta(z) = \theta(z) - \theta'(z) \Rightarrow 0. \quad (8.3\text{-}24)$$

The physical meaning of the above expression is that while the wave $A'_2(z)$ is getting stronger through the exponential gain, its wavefront is closer and closer to the wavefront of the wave A_2.

Based on Eqs. (8.3-3), (8.3-4), (8.3-19), and (8.3-24), the final solution of the sum backward stimulated scattering field is

$$E' = \frac{A'_1(0)}{A_1} e^{\frac{1}{2} g A_1^2 z} [A_1 + A_2 e^{-i\theta(z)}] e^{-i(\omega' t - k' z)}. \quad (8.3\text{-}25)$$

Comparing this expression with the following expression for the sum pump field

$$E_p = [A_1 + A_2 e^{i\theta(z)}] e^{-i(\omega t + k z)}, \quad (8.3\text{-}26)$$

we see that the output backward stimulated scattering field can be a phase-conjugate replica of the input pump field, provided that the high gain condition of Eq. (8.3-23) is satisfied. In practice, this requirement can be achieved by increasing the gain length and/or the pump intensity.

It should be pointed out that all theoretical treatments described above are based on the small-aberration influence. Experimental results have shown that under a small aberration influence, the output backward stimulated scattering can be a good PCW of the input pump field.

So far, we have not considered the influence of the difference between ω and ω' or between k and k' on the phase-conjugation property of the BSS beam. This influence resembles such a situation when someone creates a hologram with two beams of frequency ω and then reads this hologram with another beam of frequency ω'. In that case, the size or position of the reconstructed image might be slightly changed.

Finally, the aforementioned theoretical treatment is based on the near-plane wave approximation. However, in most experiments the pump beam is focused into the scattering medium. In that case, we should invoke the Hermite-Gaussian beam approximation, and the mathematical treatment will be somewhat lengthy.[58] Nevertheless, similar conclusions can be reached for the circumstance of a focused pump field, and the high gain is still the basic requirement. In this case it can be derived that a near perfect PCW can only be obtained for the small aberration influence; under a larger aberration influence the fidelity of wavefront reconstruction of the BSS beam will deteriorate.

8.4 PCW generation via backward stimulated emission (lasing)

8.4.1 Mechanism of generating phase-conjugate backward stimulated emission

At the end of Section 7.1.3, it was pointed out that basically stimulated scattering and stimulated emission (lasing) have the same physical properties, such as threshold requirement, high brightness, high photon degeneracy, and high directionality. The only difference between these two stimulated processes is that the generation of stimulated emission requires population inversion in a lasing medium, but stimulated scattering does not require that. As it is already known that the PCWs can be generated via various stimulated scattering processes, one would naturally expect that the phase-conjugation property should also be observed for backward stimulated emission (lasing) from a gain medium with population inversion, provided that the backward single-pass gain is high enough and no cavity-enhancement is needed. Here we specifically consider such a type of stimulated emission in a high gain lasing medium, pumped by a focused laser beam with an appropriate wavelength. The population inversion of the gain medium can be created by one-, two-, or multi-photon absorption of the pump laser beam. In those lasing media such as dye-solutions, if the pump intensity and single-pass gain is high enough, the highly directional stimulated emission (*cavityless lasing*) can be generated in both forward and backward directions. In this condition, like the case of cavityless stimulated scattering, the high directionality is ensured by the focusing geometry of the input pump laser beam.

An early experiment of phase-conjugation property of backward stimulated emission in a one-photon pumped dye solution was reported in 1978 by Koptev et al.[71] In that experiment, after passing through an aberrator, only 10–20% backward stimulated emission energy could be restored within the original small divergence angle with absence of an aberrator.

The primary observation of an almost perfect phase-conjugation property of two-photon pumped backward stimulated emission was reported in 1997 by He et al.;[72] one year later they presented the detailed experimental results and theoretical descriptions of this property.[73,74] Since then the further studies have showed that the same superior phase-conjugation behavior can be observed in one-, two-, and three-photon pumped backward lasing systems, working in both nanoseconds and femtoseconds regimes.[75–77]

Since the property of OPC of backward stimulated emission does not depend on the pump mechanisms, pump pulse durations, and the sort of gain materials, one can logically infer that the origin of phase-conjugation capability of the backward stimulated emission should be exactly the same as that of BSS. When an aberrator is involved, there is an equivalent quasi-collinear FWM process, and the total backward stimulated emission fields can be phase conjugated to the total of the input pump fields.

Based on Fig. 8.23, we can briefly interpret the phase-conjugation behavior of the backward stimulated emission in the gain medium pumped by a focused laser beam. Assume that the original pump beam is a plane wave denoted by $E(\omega)$, after passing through an aberrator, it becomes a superposition of into two wave fields: a strong undistorted field $E_1(\omega)$ and a weak distorted field $E_2(\omega)$; both wave fields are then focused onto the gain medium and generate an intensity-dependent holographic grating via interference of each other. If the pump intensity is high enough and the effective gain length is long enough, the initial backward stimulated emission wave $E_3(\omega')$, that is a phase-conjugate replica of the $E_1(\omega)$ wave, will continually be amplified and create a diffractive wave $E_4(\omega')$ during passing through the induced grating region.

Quite similar to the case of BSS, the rigorous theoretical analyses can prove that under appropriate conditions, the sum backward lasing field (E_3+E_4) can be the phase-conjugate replica of the sum pump field (E_1+E_2), i.e.,[74–76]

$$[E_3(\omega') + E_4(\omega')] \propto [E_1(\omega) + E_2(\omega)]^*. \tag{8.4-1}$$

Briefly speaking, the requirements for obtaining a superior phase-conjugation performance of the backward stimulated emission are (i) a high lasing gain, (ii) a longer gain length, and (iii) a larger nonlinear refractive-index coefficient of the lasing medium. The requirements (i) and (ii) are the preconditions for generating the single-pass lasing without the need of a cavity, whereas the requirements (ii) and (iii) are necessary to ensure a higher diffraction efficiency of the induced holographic grating in the gain region of a lasing medium.

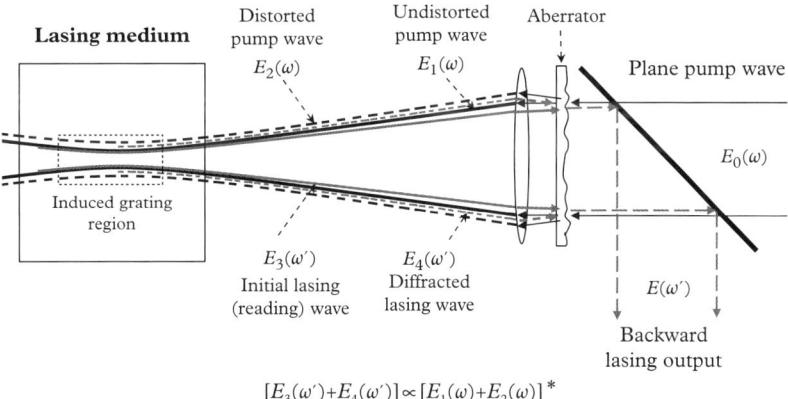

Figure 8.23 Schematic illustration of the partially degenerate FWM model for the phase-conjugation formation of backward lasing (in quasi-plane wave approximation).

256 *Optical Phase Conjugation*

8.4.2 Experimental studies on PCW generation via backward stimulated emission

In general, the phase-conjugation properties of a backward stimulated emission beam can be fully determined by comparing its far-field and near-field patterns with that of the pump beam. Figure 8.24 shows two experimental setups used to measure the far-filed pattern and the near-field image of the backward stimulated emission separately.

In the case shown in Fig. 8.24(a), a high gain lasing dye solution is in a high optical-quality quartz cuvette and can provide single-pass lasing under one-, two-, or multi-photon pump conditions. An aberration plate is placed in front of a focusing lens, while the backward stimulated emission is recollimated by the same lens, after passing through the same aberrator it is focused via a long focal-length lens to form a far-field pattern on an observation screen and to be recorded by a digital CCD camera.

In another case, shown in Fig. 8.24(b), the dye solution is in a glass vial of poor optical quality, so that the vial itself plays the role of an aberrator. To demonstrate the image-reconstruction capability of the backward stimulated emission beam, a pattern plate (such as a black letter printed in a transparent film) is placed between the beam splitter and the beam expander. If the backward stimulated emission has the phase-conjugation capability, a clear image of the pattern plate should be observed on the screen.

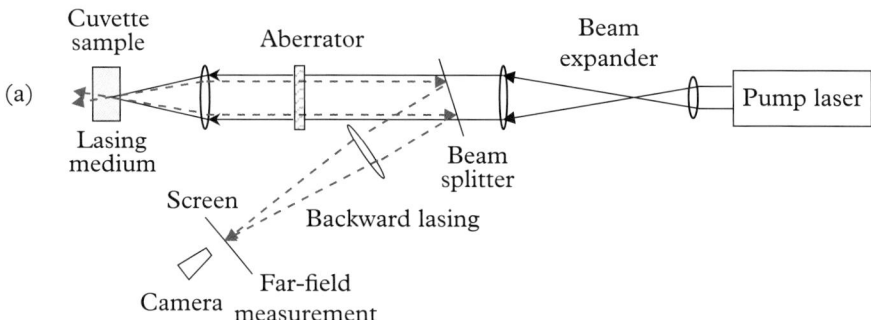

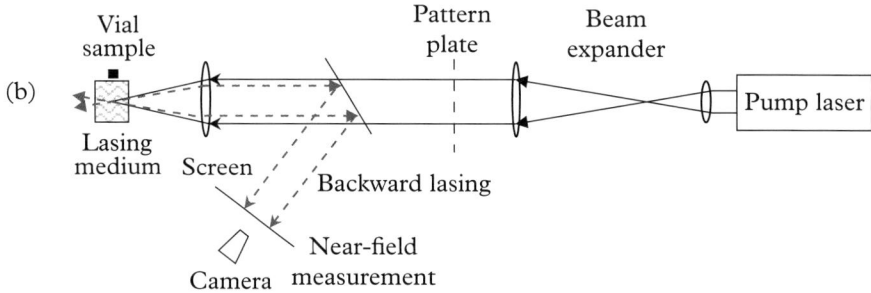

Figure 8.24 *Experimental setups for demonstrating phase-conjugation properties of the backward stimulated emission: (a) configuration for the far-field measurement and (b) configuration for the near-field image measurement.*

To avoid possible feedback influence from the cuvette or glass vial, the incident angle of the pump beam is kept larger than 5–10°.

Figure 8.25 shows typical far-field measurement results of a two-photon pumped lasing experiment with the setup of Fig. 8.24(a).[72] In this experiment, the two-photon pumped lasing medium was a 1-cm long ASPI dye solution in a quartz cuvette pumped by 1064-nm and ~11-ns laser pulses, the frequency up-conversion backward lasing wavelength was ~616 nm. Without placing an aberrator, as shown in Fig. 8.25(a) the measured divergence angles for the pump beam, the backward lasing beam, and the forward lasing beam were ~0.25, ~0.23, and ~0.23 mrad, respectively. In contrast, after placing an aberrator the measured divergence angle for the three beams were ~0.9, ~0.23, and ~0.25 mrad, respectively. In the latter case, shown in Fig. 8.25(b), the aberration influence is completely removed from the backward lasing beam, while some influence still remains on the forward lasing beam.

A similar experiment was carried out by employing the setup shown in Fig. 8.24(b), in which the same ASPI dye solution was in a glass vial of poor quality.[73] The feature of this work was to demonstrate that the original image message carried by the input pump beam could be reconstructed by the backward stimulated emission beam after passing through the aberrator or a disturbed medium. In that case, a letter-printed transparent film was placed between the focusing lens and the beam splitter. After reflection from the beam splitter, the collimated backward

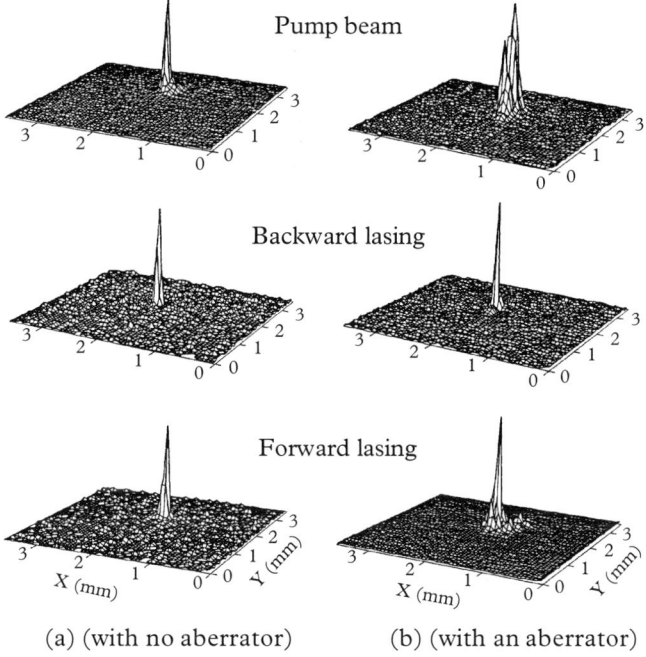

(a) (with no aberrator) (b) (with an aberrator)

Figure 8.25 *Normalized far-field intensity distributions for the input pump beam (upper), the backward stimulated emission beam (middle), and the forward stimulated emission beam (bottom): (a) without placing an aberrator and (b) with placing an aberrator. (After He et al.[72] Copyright 1997, Optical Society of America).*

lasing beam was directly projected onto a paper screen. For comparison purpose, under the same experimental condition, the lasing dye solution in the glass vial could be replaced by a SBS medium (acetone) or a SRS medium (benzene) to generate corresponding stimulated scattering, separately.

Figure 8.26(a) shows the near-field image of the 532-nm and ~6-ns pulsed pump beam for generating stimulated scattering (or the 1064-nm and ~10-ns pulsed pump beam for two-photon pumped lasing), while Fig. 8.26(b) shows the image of the backward SBS beam from an acetone-filled glass vial, indicating a quite good fidelity of the image reconstruction. Furthermore, Fig. 8.26(c) and (d) show the near-field images of the forward and backward SRS beams with 992-cm^{-1} Raman shift from a benzene-filled glass vial, indicating a poor image quality for the former and a good quality for the latter, respectively. Finally, Fig. 8.26(e) and (f) show the

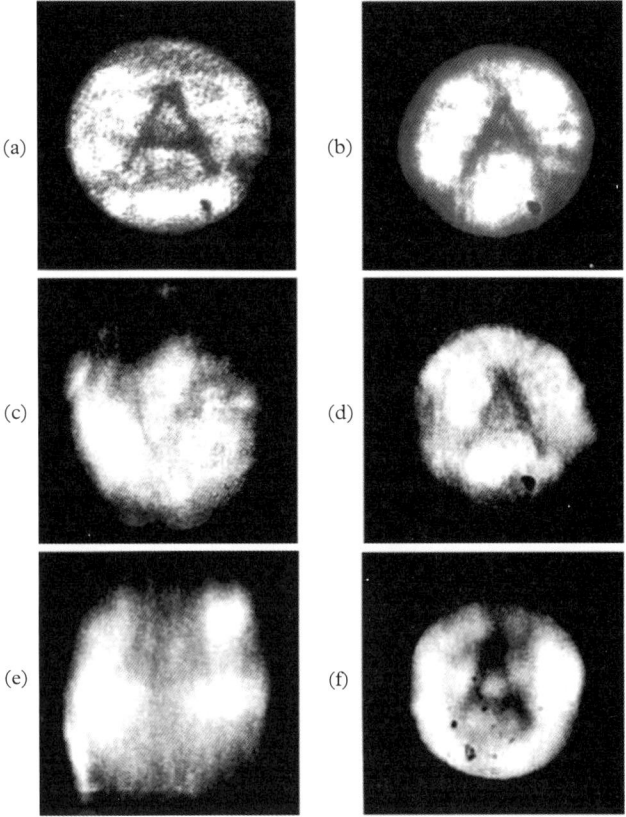

Figure 8.26 *Near-field images of the (a) the 532-nm input pump beam, (b) the backward SBS beam from an acetone-filled vial, (c, d) the forward and backward SRS beams of 562-nm wavelength from a benzene-filled vial, (e, f) 1064-nm laser pumped lasing beams of ~616 nm wavelength from a dye solution vial. (After He and Prasad[73] Copyright 1998, Optical Society of America). (See also Color fig. 12.).*

near-field images of the forward and backward two-photon pumped lasing beams from a dye solution vial, indicating that only the latter exhibits the phase-conjugation property.

In addition, the same phase-conjugation behavior is also observed in two- and three-photon pumped backward lasing systems pumped by ultra-short fs-laser pulses. In one experiment, the two-photon pumped lasing medium was a ADN dye solution, pumped by ~775-nm and ~160-fs laser pulses to generate ~470 nm forward and backward lasing output without using a cavity.[77] Figure 8.27 shows 3D displays of the far-field patterns resulting from the input pump beam without inserting an aberrator, the pump beam after passing twice through an aberrator, and the backward lasing beam after passing through the aberrator. It can be seen from Fig. 8.27(c) that the severe aberration influence on the backward lasing beam is basically removed even in the ultra-short pump pulse regime.

It is interesting to notice that the phase-conjugation nature of the backward lasing beam can also reduce the spatiotemporal fluctuations of the backward lasing pulses, as indicated by a three-photon pumped cavityless lasing experiment, in which a dye solution was pumped by ~1500-nm, ~150-fs, and 1-kHz laser pulses to generate the frequency up-conversion lasing of ~606 nm wavelength.[76] In this case, a photodiode with an effective detecting area of 0.3×0.3 mm^2 was placed in the central position of the near field of the forward and the backward lasing beam to record their intensity fluctuations separately; the output of this photodiode was connected to a gated integrator. Figure 8.28 shows the measured results under 1-kHz pump excitation at two different average rates of the integrator. From this figure, one can see that the forward lasing pulses exhibit a much greater fluctuation due to opto-thermal effect induced random fluctuation of the refractive-index distribution inside the gain medium, whereas this fluctuation influence could be significantly reduced from the backward lasing beam.

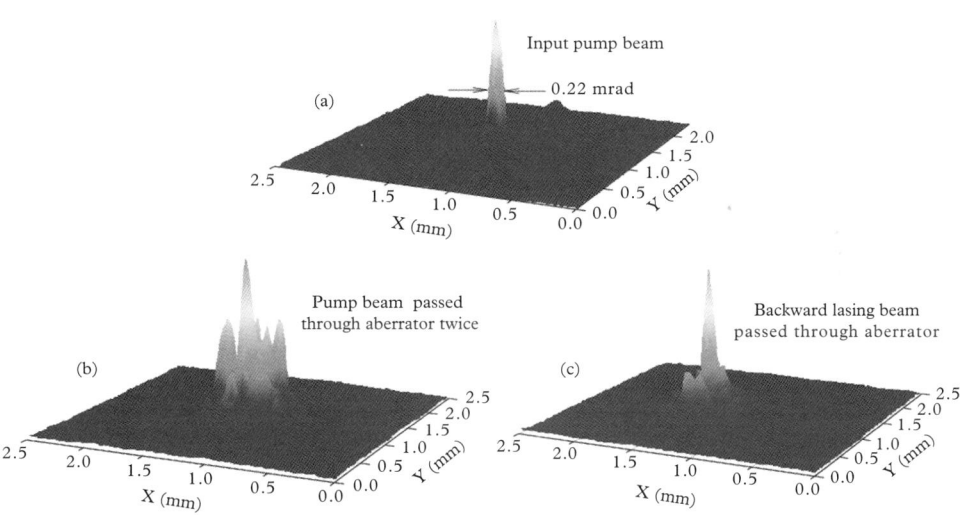

Figure 8.27 *Measured far-field distributions of (a) the ~775-nm and ~160-fs pulsed pump beam without inserting an aberrator, (b) the pump beam after passing twice through an aberrator, and (c) the two-photon pumped backward lasing beam of ~470 nm wavelength after passing through the aberrator. (After He et al.[77] Copyright 2008, American Physical Society).*

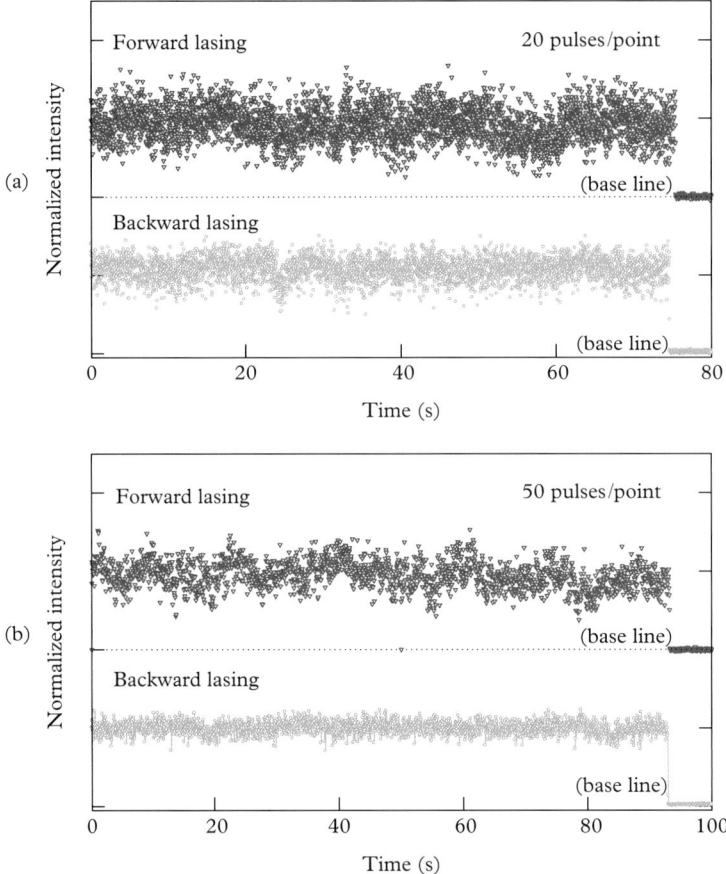

Figure 8.28 *Pulse fluctuations of the forward (up-trace) and backward (low-trace) three-photon pumped lasing beams, measured at two different average rate: (a) 20 pulses/point and (b) 50 pulses/point. (After He at al.[76] Copyright 2007, World Scientific Publishing).*

8.5 Applications of optical phase conjugation

8.5.1 Applications in special laser-device systems

From the viewpoint of fundamental research, the studies of backward PCW generation via FWM have provided a clear and straightforward picture about the nature of OPC, and also revealed an insight to interpret the phase-conjugation properties of backward stimulated scattering and stimulated emission. However, as the FWM method requires a more complex optical layout and critical optical alignment, this method sometimes is not suitable for applications. In contrast, the method of backward PCW generation via BSS has advantages of the simplicity, high nonlinear reflectivity, and capability of withstanding a high-intensity (energy) pump beam without optical damage. Among various types of backward stimulated scattering, SBS has a frequency shift that is much smaller than the spectral gain linewidth of the lasing medium, whereas SRBS and SMS have no frequency shift at all; therefore the backward phase-conjugation beam generated via these mechanisms can be further amplified by the same lasing medium.

Figure 8.29 shows some distinctive applications of the phase-conjugation technique in high-bright laser oscillator and amplifier systems. In all these cases, a suitable stimulated scattering medium is employed as a phase-conjugate reflector that can generate the backward PCW with a higher nonlinear reflectivity.

1. *Laser oscillator systems with a phase-conjugate reflector*: As shown in Fig. 8.29(a), the rear mirror of a laser cavity is replaced by a phase conjugate reflector. Compared to a conventional laser oscillator, this design is capable of producing laser output with a higher brightness and smaller beam divergence because the aberration influence of the gain medium can be automatically compensated.

2. *Laser amplifier systems with a phase-conjugate reflector*: It is known that the dimensions of the gain medium in a laser amplifier are considerably larger than that in a laser oscillator. A larger volume of the gain medium usually yields a greater aberration influence on the

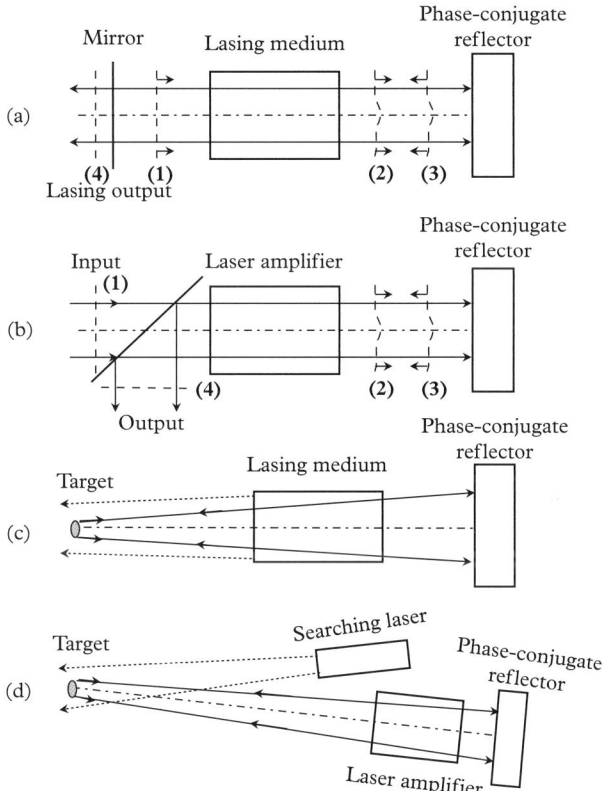

Figure 8.29 *(a) High-brightness laser oscillator with a phase conjugate reflector as the rear cavity-reflector. (b) High-brightness two-pass laser amplifier with a phase-conjugate reflector. (c) Laser target-aiming and auto-focusing system. (d) Laser target-aiming and weapon system.*

lasing beam. For this reason, the divergence and brightness of the output beam from the amplifier will be limited by the optical quality of the gain medium itself. If adopt a phase conjugate reflector in connection with the amplifier and let the reflected laser beam backward pass through the amplifying medium once again, as shown in Fig. 8.29(b), the aberration influence from the gain medium will be removed, and a high-brightness output laser beam can be obtained.

3. *Laser target-aiming and auto-focusing systems*: Another special application of optical PCWs is related to laser-based target-aiming systems. For the system shown in Fig. 8.29(c), the lasing medium emits a superradiant (stimulated) emission beam that exhibits a larger beam divergence and can easily cover the target; the partially reflected beam from the target surface can pass through the lasing medium twice with the use of a phase-conjugate reflector. After two-pass amplification, the high-power and backward output laser beam will be automatically focused on the target. This kind of technique is especially useful for those applications, such as laser fusion devices, laser trapping, laser-based rangefinder, lidar, laser target locking, and laser microfabrication.

4. *Laser weapon systems*: Laser target-aiming principle can be perfectly applied to laser weapon systems for destroying either static or high-speed flying targets. A similar but modified design is shown in Fig. 8.29(d), where an auxiliary low-power laser device is utilized to provide a broad illumination beam to catch the target, so that the partially reflected beam from the target surface will pass through a laser amplifier twice by invoking a phase-conjugate reflector. In this case, the aberration influences from the irregularity of the target surface, from the inhomogeneity of the gain medium, as well as from the disturbance in transmission medium can be automatically removed, and the full energy of the amplified laser beam can be efficiently re-focused on the target to destroy it.

The following are some specific examples of experimental studies for application purposes.

For high-energy and high-power pulsed laser oscillator/amplifier systems, the phase-conjugate reflector usually is a SBS device using a suitable liquid, high-pressure gas, or solid as the Brillouin scattering medium, the nonlinear reflectivity can be higher than 50–70%.[78–81] Multi-mode optical fibers with long gain length can be used as a SBS mirror to provide a phase-conjugate feedback in a low-energy or cw laser oscillator system.[82,83] Moreover, the phase-conjugate SBS mirrors can also be employed for laser beam combination and stabilization purposes.[84,85]

In principle, the FWM method is also applicable to provide a phase-conjugate feedback by using suitable medium of high third-order nonlinearity.[86–88] In practice, however, there are two sorts of materials have been often adopted to generate PCWs by using the FWM method: (i) the dye-solution or dye-doped solids or liquid crystals,[89–91] and (ii) the photorefractive crystals.[92–94] The drawback for the former is that the opto-thermal effect often dominates the induced refractive-index changes, whereas for the latter the response time of the induced refractive-index change is too slow (in the order of millisecond to seconds).

More recently, the modified OPC technique has been utilized for high-resolution imaging in biological tissues[95–97], as well as imaging of nano-objects through turbid media.[98,99]

8.5.2 Applications in high-speed and long-distance optical fiber communication systems

8.5.2.1 Limitations imposed on fiber communication systems

It is known that to achieve high-data-rate transmission through a super-long optical fiber network, the following two requirements are essentially important. First, suitable optical wavelengths should be chosen at which the transmission loss in the fiber network is minimized. Second, an appropriate multichannel mechanism should be employed, either in the wavelength-division multiplexing (WDM) regime or in the time-division multiplexing (TDM) regime. To meet the first requirement, the wavelengths of the optical carrier wave are usually chosen either in the ~1.3 μm or in ~1.5 μm region where two windows of transmission loss of fibers are located. Concerning the second requirement, however, there are two major limitations imposed on the transmission bit-rate as well as on the transmission distance via an optical fiber network. The first limitation comes from the so-called chromatic dispersion or group-velocity dispersion (GVD) that means the optical pulses having different spectral components exhibit different propagation velocities along a fiber transmission system. Because of this effect, an ultra-short optical pulse, which contains multi-spectral-components within a considerable spectral width, will become broader and broader in a time-scale during their propagation within a long fiber system. For this reason, the temporal spacing between two adjacent optical pulses cannot be less than a certain value for a given fiber length, therefore the transmission bit-rate is limited. The chromatic dispersion of an optical fiber system is defined as[100]

$$D(\lambda) = \frac{1}{L}\frac{d\tau}{d\lambda} = \frac{1}{L}\frac{d}{d\lambda}\left[-\frac{L\lambda_s^2}{2\pi c}\left(\frac{d\beta}{d\lambda}\right)_{\lambda_s}\right] = -\frac{\lambda_s^2}{2\pi c}\left(\frac{d^2\beta}{d\lambda^2}\right)_{\lambda_s}, \qquad (8.5\text{-}1)$$

where τ is the time delay for an optical pulse propagating in a single-mode fiber of length L, β is the propagation constant of the fiber, λ_s is the central wavelength of the optical signal pulse, and c is the speed of light. For most standard single-mode fibers the zero-dispersion wavelength is $\lambda_0 \approx 1.27$ μm at which $D(\lambda_0) \approx 0$; in a signal wavelength region of $\lambda_s \approx 1.5$ μm, the chromatic dispersion $D(\lambda_s) \approx 15\text{--}18$ ps/km/nm. The chromatic-dispersion limited transmission distance over this type of fiber (for a 1-dB penalty) is given by[101]

$$B^2L \cong 6500 \text{ (Gb/s)}^2\text{km}. \qquad (8.5\text{-}2)$$

Here, B is the bit-rate and L is the fiber length. The limit, for example, is ~65 km at a bit-rate of 10 Gb/s.

In addition to the limitation from chromatic dispersion, there is another limitation that is related to the spectral broadening of optical pulses propagating in a fiber system. It is known that an optical fiber is a good third-order nonlinear medium, in which many nonlinear optical effects can easily take place due to its extremely long interaction length. Among these effects, the light-intensity dependent refractive-index change may produce the self-phase-modulation of an intense short light pulse during its propagation in the fiber system. Sequentially, this self-modulation effect leads to a spectral self-broadening of the given light pulse. Some other nonlinear process, such as FWM or stimulated scattering may also produce new spectral components to cause further spectral broadening. This kind of spectral broadening may degrade the quality of signal transmission and particularly impose a limitation on the minimum spectral spacing between two adjacent channels in a fiber communication system adopting the WDM regime.

8.5.2.2 Functions of a midway optical phase conjugator

The development of optical fiber communication technology in recent decades has shown that the OPC technique can be successfully employed to overcome the limitations from both the chromatic dispersion and the self-modulation (spectral self-broadening). For this purpose, a special optical phase-conjugator, which can generate forward PCWs of the input signal waves with an inverse spectral structure (see Section 8.2.3), should be placed in the middle of two identical single-mode

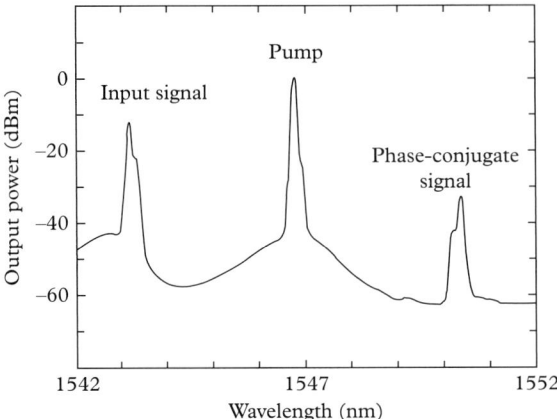

Figure 8.30 *Frequency-structure reversal of the forward PCW via FWM in a 21-km long dispersion shifted fiber. (Reproduced with permission from Jopson et al.[105] Copyright 1993, IET).*

fiber transmission systems. Such a special technique adopted for fiber communications is termed "midway optical phase-conjugation (MOPC)" or "mid-span spectral inversion (MSSI)". The principle of this technique is based on the theoretical proposal suggested by Yariv et al. in 1979.[102] In this proposal, a midway phase conjugator is placed between two identical fiber systems and to generate a PCW at frequency $\omega_{pc} = (\omega_p - \Omega)$ with respect to the input signal wave of frequency $\omega_s = (\omega_p + \Omega)$, where ω_p is the pump frequency for FWM satisfying $2\omega_p = \omega_s + \omega_{pc}$, and Ω is an arbitrary small frequency shift value. It was theoretically proved that in this specific case, the chromatic dispersion and self-modulation influence from the first fiber system on the optical signal could be cancelled by that from the second fiber system. In 1992, Inoue reinvestigated this issue theoretically and suggested the use of forward nondegenerate FWM in an optical fiber as a new approach to achieve midway phase-conjugation.[103] In order to meet the phase-matching requirement of FWM, it was also suggested that ω_0, ω_s and ω_{pc} should be chosen working in the spectral region near to zero-dispersion wavelength. The first experimental demonstration of MSSI in a dispersion-shifted fiber section via forward FWM was reported in 1993 by Watanabe et al.,[104] and subsequently by Jopson et al.[105]

As an example, Fig. 8.30 shows the measured spectral structures of the input signal pulse, the pump laser beam for FWM, and the forward phase-conjugate pulse generated in a 21-km long dispersion shifted single-mode fiber as the phase-conjugator.[105] The central wavelength for the signal pulse was 1543.1 nm, the pump wavelength was 1546.7 nm (just located at zero-GVD position of the fiber), and the central wavelength of the phase-conjugate pulse was 1550.3 nm, respectively. It can be seen that the spectral position and shape of the latter are inverted in respect to the input signal pulse. In the same experiment, without adopting such a midway phase-conjugator, the effective transmission distance of the signals was limited to ∼80 km due to the influences from both DVD and self-modulation effects, whereas it was increased to more than 400 km after adopting the phase-conjugator.

Figure 8.31 shows schematically how a midway phase conjugator between two identical optical fiber spans can balance chromatic dispersion influences from the two fiber spans. For simplicity,

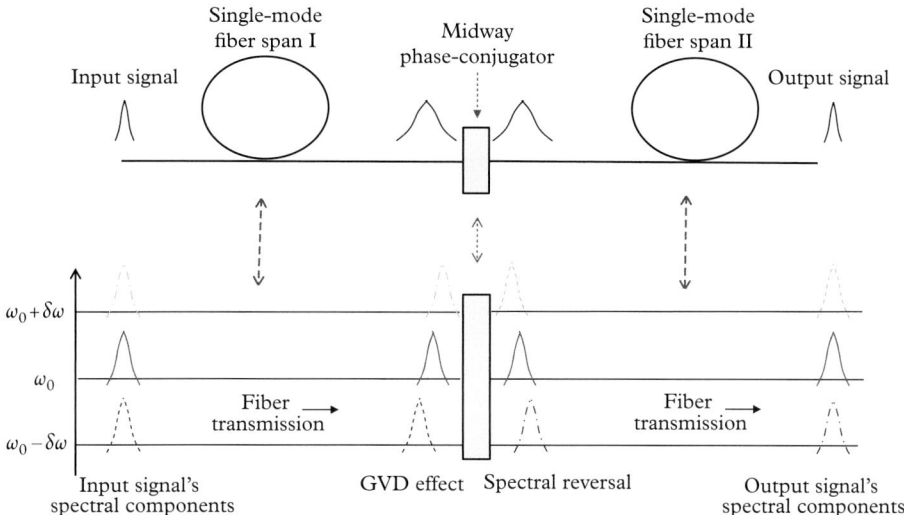

Figure 8.31 *Schematic diagram of using a midway optical phase conjugator (between two identical fiber spans) to compensate for the GVD effect. Here ω_0 denotes the signal central frequency (before entering the conjugator) and the central frequency of the output PCW pulse from the conjugator. The bottom of the figure shows the pulse propagation of three different spectral components within the original spectral width.*

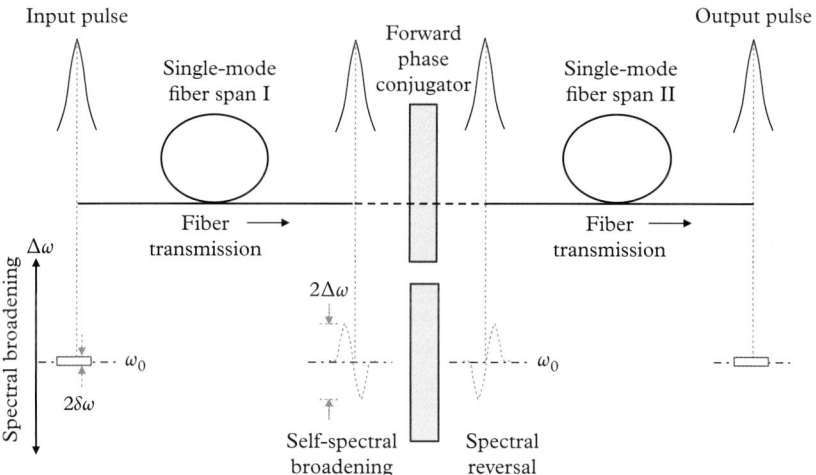

Figure 8.32 *Schematic diagram of using a midway optical phase conjugator (between two identical fiber spans) to compensate for the influence of spectral self-broadening. Here, $2\delta\omega$ is the initial spectral width of the input signal pulse, $2\Delta\omega$ is the maximum self-broadening after passing through the first fiber span, and ω_0 is as defined in Fig. 8.31.*

it is assumed that the input light pulse has a central frequency of ω_0 and spectral width of $2\delta\omega$. During passing through the first fiber span, the pulse envelope gradually broadens due to GVD dispersion effect, because different spectral components of the pulse have different group velocities, as schematically shown in the bottom of the figure. Here we have assumed that both fiber spans have a negative GVD effect, so that the blue-side spectral components of the signal pulse propagate faster in the fiber than the red-side components. After passing through a midway phase-conjugator, the red-side components of the signal pulse are converted into the blue-side components of the phase-conjugate pulse and, therefore, will propagate faster in the second fiber span. For that reason, the chromatic dispersion influences from two fiber spans can be offset by each other, and the final output pulse from the second fiber span will restore the original pulse envelop.

Another function of the midway phase conjugator is that the self-phase-modulation induced spectral broadening influence on the signal pulse can also be finally removed via the spectral structure inversion. Using the illustration in Fig. 8.32, we can provide a brief interpretation of such a function. It is still assumed that the signal pulse input on the first fiber span has a central

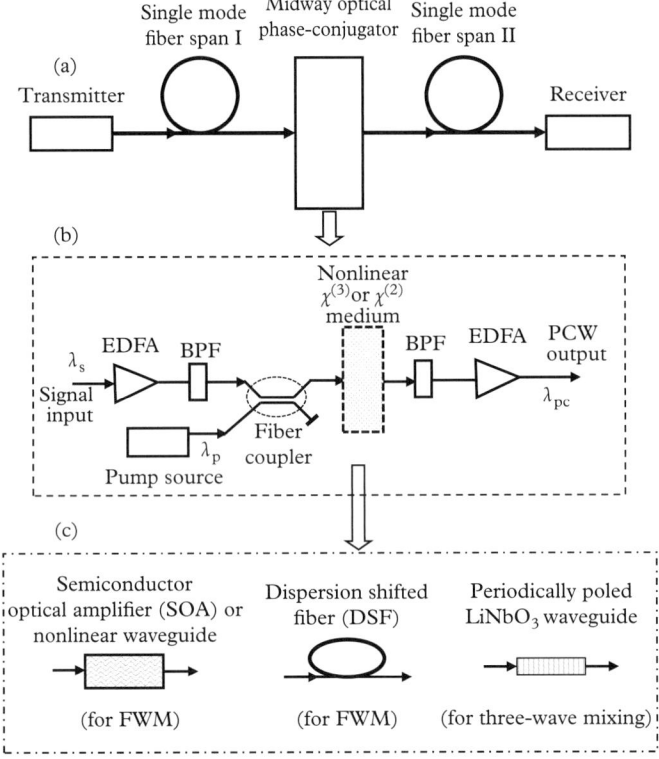

Figure 8.33 (a) Midway optical phase conjugator employed in a long-distance and high bit-rate optical fiber communication system. (b) Basic components of a midway optical phase-conjugator. (c) Three types of nonlinear media can be adopted for phase conjugator. BPF: band-pass filter; EDFA: erbium-doped fiber amplifier.

frequency of ω_0 and spectral width of $2\delta\omega$. Owing to the influence of self-phase-modulation (see Section 6.3.1), after passing through the first fiber span the spectral width of the signal pulse increases to $2\Delta\omega$: the leading half of the pulse undergoes a frequency redshift while the lagging half experiences a blueshift. The midway forward phase-conjugator can generate a pulse output with inverse spectral structure, so that the leading half of the phase-conjugate pulse has the higher frequency components, while the lagging half has the lower frequency components. If the second fiber span exhibits the same nonlinear optical property and propagation length as the first fiber span, then after finishing the propagation in the second span, the influences of self-modulation and self-spectral-broadening on the signal transmission in a fiber network can be effectively removed.

As described in Section 8.2.3, two technical approaches can be used to generate forward PCW with spectral inversion: one is the frequency nondegenerate FWM in a third-order ($\chi^{(3)}$) nonlinear medium, and another is the three-wave mixing in a second-order ($\chi^{(2)}$) nonlinear medium. In the first case, the pump frequency is chosen to be $2\omega_p = \omega_s + \omega_{pc}$; whereas in the second case, the pump frequency should be $\omega_p = \omega_s + \omega_{pc}$. In practice, the third-order nonlinear media for phase-conjugation purpose in fiber communication systems can be (i) a dispersion shifted single-mode fiber of a length from 10^{-1} to 10^1 km,[104–111] (ii) a semiconductor optical amplifier (SOA),[112–119] and (iii) a nonlinear optical waveguide with high third-order nonlinearity.[120–123] The second-order nonlinear medium adopted for a phase-conjugator in the fiber communication system usually is a periodically poled LiNbO$_3$ (PPLN) crystal.[124-126]

Figure 8.33 shows schematically the basic components of a generalized midway optical phase conjugator in a modern optical fiber communication system. As shown in Fig. 8.33(a), after propagating over the first transmission fiber span, the distorted signal pulses pass through an erbium-doped fiber amplifier (EDFA) and a band-pass filter (BPF) then enter the nonlinear medium for

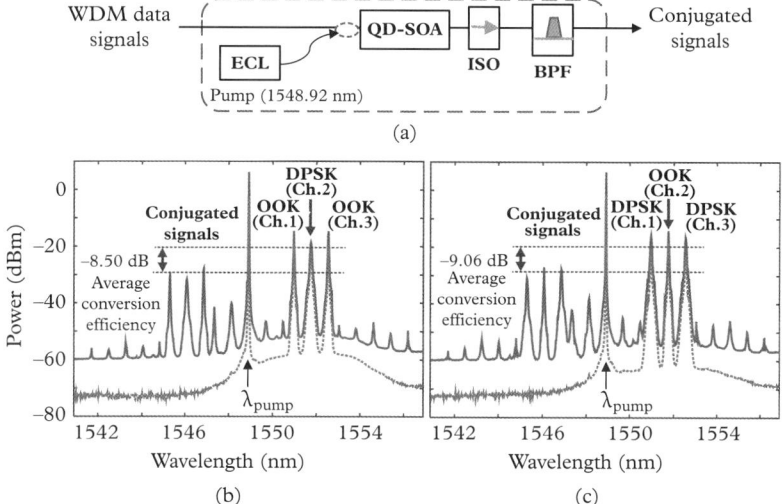

Figure 8.34 *(a) Midway optical phase conjugator employed a quantum-dot semiconductor optical amplifier (QD-SOA) as the FWM medium. (b, c) Input and output signals of three-channel combinations at the conjugator. ECL: external-cavity laser diode; ISO: isolator; BPF: band-pass filter. (Reproduced with permission from Matsuura and Kishi.[119] Copyright 2013, Optical Society of America).*

nonlinear wave mixing. In the mean time, the pump light usually from a semiconductor laser or fiber distributed feedback laser is launched into the same nonlinear medium through a fiber coupler. After completion of four-wave or three-wave mixing, the newly generated phase-conjugate output passes through another BPF and EDFA and then enters the second transmission fiber span.

8.5.2.3 Some examples of system performances

Figure 8.34(a) shows the structure of a real mid-span optical phase conjugator employed a quantum-dot SOA as the third-order nonlinear medium for FWM.[119] The input is a group of three-channel signals of wavelength 1550.92, 1551.72, and 1552.52 nm, respectively, with a channel spacing of 100 GHz. For each channel, the signal can be modulated as either on-off-keying (OOK) signal or differential-phase-shift keying (DPSK) signal. The pump source of 1548.92 nm wavelength is an external-cavity laser diode, and the whole conjugator system

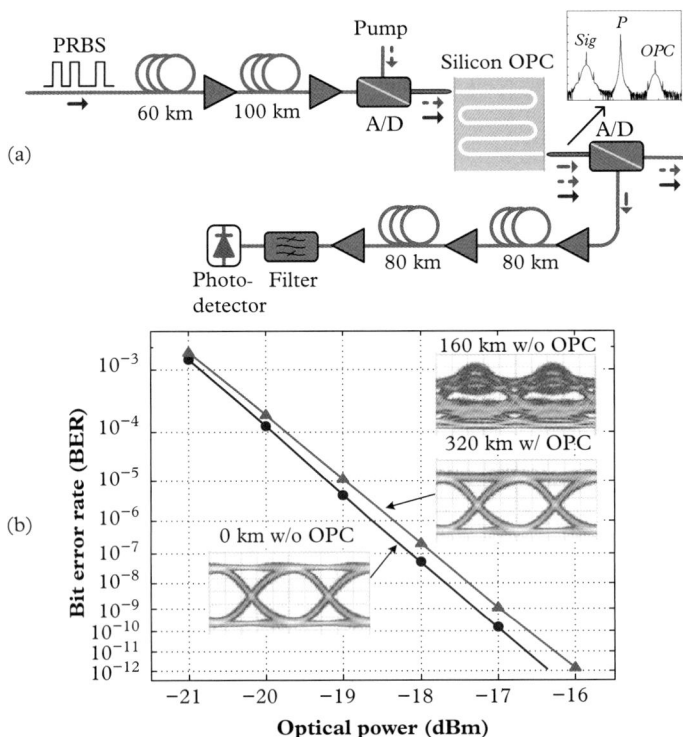

Figure 8.35 *(a) Experimental setup of utilizing a silicon-waveguide as the optical phase-conjugator. PRBS: pseudorandom-bit-sequence; A/D: add/drop multiplexer; big-triangles represent EDFAs; inset: optical spectrum of conjugator output. (b) Measured BER of 10 Gbit/s PRBS signal against optical power before (circles) and after (triangles) transmission through a 320-km standard fiber network with the silicon conjugator at mid-span. The corresponding eye diagrams are shown as insets. (Reproduced with permission from Ayotte et al.[120] Copyright 2007, IET).*

is placed between two 50-km long standard single-mode fiber spans. The measured spectra of the input (dotted lines) and output (solid lines) signals at the conjugator for two different signal combinations are shown in Fig. 8.34(b) and (c), respectively. The FWM conversion efficiency is about −0.9 dB, and the signal/noise ratio of the output conjugation signals is up to ∼30 dB. This conjugator system has been used to achieve multichannel transmission of 10 Gbit/s signals over 100 km with low power penalties.

Figure 8.35(a) shows an example of a fiber communication system that uses a rib silicon waveguide as the phase conjugator for FWM.120 The waveguide's width, height, and etch depth are 1.5, 1.55, and 0.7 μm, respectively, with a double S-bend shape of total length of 8 cm. After passing through the first fiber span of 160-km length, the 1555.8-nm signal beam carrying 40 Gbit/s pseudorandom-bit-sequence (PRBS) data in a single channel is combined with a pump beam of 1557.3 nm wavelength and 600 mW power, and the 1558.8-nm phase-conjugate output is generated in the waveguide via FWM. The measured BER (bit error rate) values of a 10 Gbit/s optical signal against the signal power before (circles) and after (triangles) transmission through a 320-km fiber network are shown in Fig. 8.35(b), in which the corresponding eye diagrams are also shown as insets. As can be seen, the power penalty between the two curves at BER of 10^{-9} is less than 0.3 dB.

Figure 8.36(a) shows the structure of a mid-span phase conjugator adopting a Ti:PPLN crystal waveguide for three-wave-mixing and generating a forward PCW.[126] This configuration is polarization-insensitive and can provide a higher conversion efficiency of second-order nonlinear processes. The input of the conjugator contains a pulsed signal beam of wavelength $\lambda_s = 1551$ nm

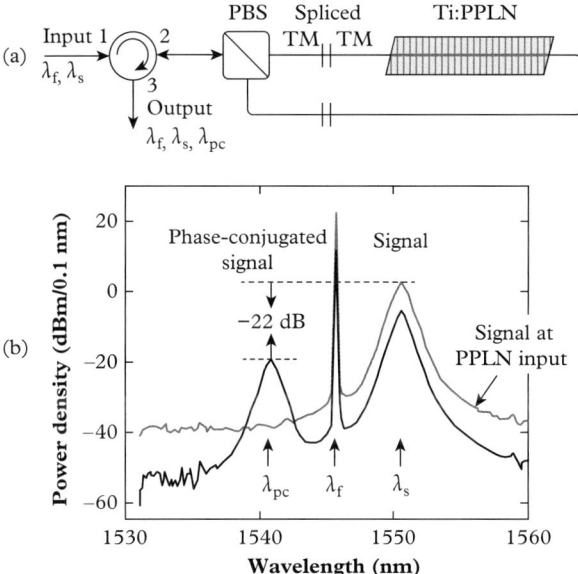

Figure 8.36 *(a) Polarization-insensitive PPLN conjugator system; PBS is a polarization beam splitter. (b) Optical spectrum at the input and output of the conjugator system. (Reproduced with permission from Hu et al.[126] Copyright 2010, Optical Society of America).*

and a cw fundamental pump beam of $\lambda_f = 1546.1$ nm. During its transmission in the Ti:PPLN crystal, the fundamental beam creates a second-harmonic generation (SHG) beam of wavelength $\lambda_f/2 = 773.05$ nm; the SHG beam then interacts with the signal beam to generate a pulsed phase-conjugate beam of wavelength $\lambda_{pc} = 1541$ nm. When this phase conjugator is placed between two fiber spans of length 52.8 and 57.6 km, the measured spectra of the input and output at the conjugator are as shown in Fig. 8.36(b), indicating a conversion efficiency of −22 dB for the phase conjugated 160 Gbit/s signal. The input signal pulse width was ∼2.7 ps, while the phase-conjugated pulse width measured at the end of the second fiber span was ∼3.5 ps; this slight broadening is mainly due to the third-order dispersion, which cannot be compensated by the OPC.

Another recent experiment, using a highly nonlinear As_2S_3 glass waveguide as the mid-span phase conjugator, has demonstrated the capability of dispersion-free transmission over 225 km for a 3×40 Gbit/s WDM signal.[123]

PROBLEMS

1. If the input laser beam is expressed by $E(x, y, z) = A_0 \exp\{i[kz + \varphi(x, y, z)]\}$, please write the expressions of the corresponding frequency-degenerate backward and forward phase conjugate waves, and sketch the ways in which the distortion influence from a propagation medium can be removed by using these two waves.
2. For the degenerate four-wave mixing arrangement shown in Fig. 8.6, if the incident light wavelength is 800 nm, the refractive index of the nonlinear medium is $n_o = 1.4$, the crossing-angle between the pump beam 1 and the signal beam 3 is 2.5°, please use Eq. (8.2-16) to calculate the periods of the two induced holographic gratings.
3. In an experiment of using degenerate four-wave mixing to generate the backward phase conjugate wave, how would you distinguish the contributions from two different mechanisms (four-photon parametric interaction and induced grating's reflection)?
4. How would you explain the phase-conjugate property of backward stimulated scattering?
5. In a fiber telecommunication system, how would you generate a spectral reversed forward phase-conjugate signal? Why could this signal, after propagation in the second fiber span, finally restore the original signal input to the first fiber span?

REFERENCES

1. G. S. He, *Prog. Quantum Electron.* **26**, 131 (2002).
2. A. Yariv, *IEEE J. Quantum Electron.* QE-14 (1978), 650.
3. R. W. Hellwarth, *Opt. Eng.* **21**, 257 (1982).
4. R. A. Fisher, Ed., *Optical Phase Conjugation* (Academic, New York, 1983); A. Brignon and J.-P. Huignard, Eds., *Phase Conjugate Optics* (Wiley, New York, 2003).
5. B.Ya. Zel'dovich, N. F. Pilipetsky, and V. V. Shkunov, *Principles of Phase Conjugation*, (Springer-Verlag, Berlin, 1985).
6. A. Yariv, *Appl. Phys. Lett.* **28**, 88 (1976).
7. A. Yariv, *J. Opt. Soc. Am.* **66**, 301 (1976).

8. R. W. Hellwarth, *J. Opt. Soc. Am. A* **67**, 1 (1977).
9. A. Yariv and D. M. Pepper, *Opt. Lett.* **1**, 16 (1977).
10. R. L. Abrams and R. C. Lind, *Opt. Lett.* **2**, 94 (1978); erratum, *ibid*, **3**, 205 (1978).
11. D. G. Steel, R. C. Lind, J. F. Lam, and C. R. Giuliano, *Appl Phys. Lett.* **35**, 376 (1979).
12. A. Khyzniak, V. Kondilenko, Yu. Kucherov, S. Lesnik, S. Odoulov, and M. Soskin, *J. Opt. Soc. Am. A* **1**, 169 (1984).
13. J. Takacs, H. C. Ellin, and L. Solymar, *Opt. Commun.* **93**, 223 (1992).
14. P. V. Arizonis, F. A. Hopf, W. D. Bamberger, S. F. Jacobs, A. Tomita, and K. H. Womack, *Appl. Phys. Lett.* **31**, 435 (1977).
15. L. Lefort and A. Barthelemy, *Opt. Lett.* **21**, 848 (1996).
16. F. Devaux, E. Guiot, and E. Lantz, *Opt. Lett.* **23**, 1597 (1998).
17. D. M. Bloom and G. C. Bjorklund, *Appl. Phys. Lett.* **31**, 592 (1977).
18. S. M. Jensen and R. W. Hellwarth, *Appl. Phys. Lett.* **32**, 166 (1978).
19. E. E. Bergmann, I. J. Bigio, B. J. Feldman, and R. A. Fisher, *Opt. Lett.* **3**, 82 (1978).
20. M. A. Khan, P. W. Kruse, and J. F. Ready, *Opt. Lett.* **5**, 261 (1980).
21. D. Fekete, J. AuYeung, and A. Yariv, *Opt. Lett.* **5**, 51 (1980).
22. D. Grischkowsky, N. S. Shiren, and R. J. Bennett, *Appl. Phys. Lett.* **33**, 805 (1978).
23. D. M. Bloom, P. F. Liao, and N. P. Ecnomou, *Opt. Lett.* **2**, 58 (1978).
24. P. F. Liao and D. M. Bloom, *Opt. Lett.* **3**, 4 (1978).
25. R. C. Lind, D. G. Steel, M. B. Klein, R. L. Abrams, C. R. Giuliano, and R. K. Jain, *Appl. Phys. Lett.* **34**, 457 (1979).
26. A. Tomita, *Appl. Phys. Lett.* **34**, 463 (1979).
27. D. Depatie and D. Haueisen, *Opt. Lett.* **5**, 252 (1980).
28. R. K. Jain and M. B. Klein, *Appl. Phys. Lett.* **35**, 454 (1979).
29. E. I. Moses and F. Y. Wu, *Opt. Lett.* **5**, 64 (1980).
30. G. Martin and R. W. Hellwarth, *Appl. Phys. Lett.* **34**, 371 (1979).
31. J. Nilsen and A. Yariv, *Opt. Commun.* **39**, 199 (1981).
32. Y. L. Guo, M.-Y. Yao, and S.-M. Pang, *IEEE J. Quantum Electron.* **20**, 328 (1984).
33. H. A. Mac Kenzie, D. J. Hagan, and H. A. Al-Attar, *IEEE J. Quantum Electron.* **22**, 1328 (1986).
34. M. T. De Araujo, S. S. Vianna, and G. Grynberg, *Opt. Commun.* **80**, 79 (1990).
35. N. A. Andreev, V.-I. Bespalov, A. M. Kiselev, A. Z. Matveev, G. A. Pasmanik, and A. A. Shilov, *JETP Lett.* **32**, 625 (1980).
36. M. D. Skeldon, P. Narum, and R. W. Boyd, *Opt. Lett.* **12**, 343 (1987).
37. M. W. Bowers and R. W. Boyd, *IEEE J. Quantum Electron.* **34**, 634 (1998).
38. Y.-S. Kuo, Y.-C. Fong, C. Wu, J. G. McInerney, and K. McIver, *Opt. Commun.* **97**, 228 (1993).
39. B. Ya. Zel'dovich, V. I. Popovichev, V. V. Ragul'skii, and F. S.Faizullov, *JETP Lett.* **15**, 109 (1972).
40. O.Yu. Nosach, V. I. Popovichev, V. V. Ragul'skii, and F. S. Faizullov, *JETP Lett.* **16**, 435 (1972).
41. V. Wang and C. R. Giuliano, *Opt. Lett.* **2**, 4 (1978).
42. M. Slatkine, I. J. Bigio, B. J. Feldman, and R. A. Fisher, *Opt. Lett.* **7**, 108 (1982).
43. E. Armandillo and D. Proch, *Opt. Lett.* **8**, 523 (1983).
44. I. V. Tomov, R. Fedosejevs, and D. C. D. McKen. *Opt. Lett.* **9**, 405 (1984).
45. M. T. Duignan, B. J. Feldman, and W. T. Whitney, *Opt. Lett.* **12**, 111 (1987).
46. B. Ya. Zel'dovich, N. A. Mel'nikov, N. F. Pilipetskii, and V. V. Ragul'skii, *JETP Lett.* **25**, 36 (1977).
47. A. I. Sokolovskaya, G. L. Brekhovskikh, and A. D. Kudryavtseva, *Sov. Phys. Dokl.* **22**, 156 (1977).
48. A. I. Sokolovskaya, G. L. Brekhovskikh, and A. D. Kudryavtseva, *Opt. Commun.* **24**, 74 (1978).
49. R. Mays and R. J. Lysiak, *Opt. Commun.* **31**, 89 (1979).
50. I. V. Tomov, R. Fedosejevs, D. C. D. McKen, C. Domier, and A. A. Offenberger, *Opt. Lett.* **8**, 9 (1983).
51. A. I. Sokolovskaya, G. L. Brekhovskikh, and A. D. Kudryavtseva, *IEEE J. Quantum Electron.* **23**, 1332 (1987).

52. A. D. Kudryavtseva, A. I. Sokolovskaya, J. Gazengel, N. Phu Xuan, and G. Rivoire, *Opt. Commun.* **26**, 446 (1978).
53. J. L. Ferrier, Z. Wu, J. Gazengel, N. Phu Xuan, and G. Rivoire, *Opt. Commun.* **41**, 135 (1982).
54. E. J. Miller, M. S. Malcuit, and R. W. Boyd, *Opt. Lett.* **15**, 1188 (1990).
55. G. S. He, Q. Zheng, and P. N. Prasad, *J. Opt. Soc. Am. B* **24**, 1166 (2007).
56. G. S. He, K.-T. Yong, H.-Y. Qin, Q. Zheng, P. N. Prasad, S. He, and H. Ågren, *IEEE Quantum Electron.* **44**, 894 (2008).
57. G. S. He, W.-C. Law, L. Liu, X. Zhang, and P. N. Prasad, *Appl. Phys. Lett.* **101**, 011110 (2012).
58. D. Liu and G. S. He, *Sov. Phys. JETP* **88**, 235 (1999); erratum, *ibid*, **88**, 1241 (1999).
59. R. W. Hellwarth, *J. Opt. Soc. Am.* **68**, 1050 (1978).
60. H. Hsu and S. S. Bor, *IEEE J. Quantum Electron.* **25**, 430 (1989).
61. G. G. Kochemasov and V. D. Nikolaev, *Sov. J. Quantum Electron.* **7**, 60 (1977).
62. I. M. Bel'dyugin, M. G. Galushkin, E. M. Zemskov, and V. I. Mandrosov, *Sov. J. Quantum Electron.* **6**, 1349 (1976).
63. B. Ya. Zel'dovich and V. V. Shkunov, *Sov. J. Quantum Electron.* **7**, 610 (1977).
64. N. B. Baranova, B.Ya. Zel'dovich, and V. V. Shkunov, *Sov. J. Quantum Electron.* **8**, 559 (1978).
65. N. B. Baranova and B.Ya. Zel'dovich, *Sov. J. Quantum Electron.* **10**, 555 (1980).
66. V. I. Bespalov, V. G. Manishin, and G. A. Pasmanik, *Sov. Phys. JETP* **50**, 879 (1979).
67. V. G. Sidorovich and V. V. Shkunov, *Sov. Phys. Tech. Phys.* **24**, 472 (1979).
68. G. S. He, D. Liu, and S. H. Liu, *Bull. Am. Phys. Soc.* **30**, 1800 (1985).
69. G. S. He, D. Liu, and S. H. Liu, *Chin. Phys. Lasers* **13**, 713 (1986).
70. M. Born and E. Wolf, *Principles of Optics*, 6th ed. (Pergamon, London, 1983), p. 453.
71. V. G. Koptev, A. M. Lazaruk, I. P. Petrovich, and A. S. Rubanov, *JETP Lett.* **28**, 434 (1978).
72. G. S. He, Y. Cui, M. Yoshida, and P. N. Prasad, *Opt. Lett.* **22**, 10 (1997).
73. G. S. He and P. N. Prasad, *J. Opt. Soc. Am. B* **15**, 1078 (1998).
74. G. S. He, N. Cheng, P. N. Prasad, D. Liu, and S. H. Liu, *J. Opt. Soc. Am. B* **15**, 1086 (1998).
75. G. S. He and P. N. Prasad, *IEEE J. Quantum Electron.* **34**, 473 (1998).
76. G. S. He, Q. Zheng, N. Cheng, F. Xu, and P. N. Prasad, *J. Nonlinear Opt. Phys. Mater.* **16**, 137 (2007).
77. G. S. He, H.-Y. Qin, Q. Zheng, P. N. Prasad, S. Jockusch, N. J. Turro, M. Halim, D. Sames, H. Ågren, and S. He, *Phys. Rev. A* **77**, 013824 (2008).
78. M. Sugii, M. Okabe, A. Watanabe, and K. Sasaki, *IEEE J. Quantum Electron.* **24**, 2264 (1988).
79. N. A. Kurnit and S. J. Thamas, *IEEE J. Quantum Electron.* **25**, 421 (1989).
80. R. J. St. Pierre, D. W. Mordaunt, H. Injeyan, J. G. Berg, R. C. Hillard, M. E. Weber, M. G. Wickham, G. M. Harpole, and R. Senn, *IEEE J. Sel. Top. Quantum Electron.* **3**, 53 (1997).
81. B. Kralikova, J. Skala, P. Straka, and H. Turcicova, *Appl. Phys. Lett.* **77**, 627 (2000).
82. T. Riesbeck, E. Risse, and H. J. Eichler, *Appl. Phys. B* **73**, 847 (2001).
83. V. I. Kovalev and R. G. Harrison, *Opt. Lett.* **30**, 1375 (2005).
84. H. Yoshida, M. Nakatsuka, T. Hatae, S. Kitamura, T. Sakuma, and T. Hamano, *Jpn. J. Appl. Phys.* Part 2, **43**, L1038 (2004).
85. H. J. Kong, J. W. Yoon, J. S. Shin, D. H. Beak, and B. J. Lee, *Laser Part. Beams* **24**, 519 (2006).
86. X.-W. Xia, D. Hsiung, P. S. Bhatia, M. S. Shahriar, T. T. Grove, and P. R. Hemmer, *Opt. Commun.* **191**, 347 (2001).
87. S.-Y. Zhou, T. Xia, Z. Xu, and Y.-Z. Wang, *Chin. Phys. Lett.* **27**, 014211 (2010).
88. Z. Zhai, Y. Dou, J. Xu, and G. Zhang, *Phys. Rev. A* **83**, 043825 (2011).
89. S. Bian, W. Zhang, S. Il. Kim, N. B. Embaye, G. J. Hanna, J. J. Park, B. K. Canfield, and M. G. Kuzyk, *J. Appl. Phys.* **92**, 4186 (2002).
90. R. C. Sharma, T. A. Waigh, and J. P. Singh, *Appl. Phys. Lett.* **92**, 101125 (2008).
91. P. Karpinski and A. Miniewicz, *Appl. Phys. Lett.* **101**, 161108 (2012).
92. T. Bach, K. Nawata, M. Jazbinsek, T. Omatsu, and P. Gunter, *Opt. Express* **18**, 87 (2010).

93. M. Jazbinsek, D. Haertle, G. Montemezzani, P. Gunter, A. A. Grabar, I. M. Stoika, and Y. M. Vysochanskii, *J. Opt. Soc. Am. B* **22**, 2459 (2005).
94. M. Woerdemann, K. Berghoff, and C. Denz, *Opt. Express* **18**, 22348 (2010).
95. Z. Yaqoob, D, Psaltis, Demetri, M. S. Feld, and C. Yang, *Nat. Photonics* **2**, 110 (2008).
96. I. M. Vellekoop, A. Lagendijk, and A. P. Mosk, *Nat. Photonics* **4**, 320 (2010).
97. I. M. Vellekoop, M. Cui, and C. Yang, *Appl. Phys. Lett.* **101**, 081108 (2012).
98. C.-L. Hsieh, Y. Pu, R. Grange, and D. Psaltis, *Opt. Express* **18**, 12283 (2010).
99. P.-Y. Chen and A. Alu, *Nano Lett.* **11**, 5514 (2011).
100. C. Lin, H. Kogelnik, and L. G. Cohen, *Opt. Lett.* **5**, 476 (1980).
101. A. H. Gnauck, R. M. Jopson, and R. M. Derosier, *IEEE Photonics Technol. Lett.* **5** 663 (1993).
102. A. Yariv, D. Fekete, and D. M. Pepper, *Opt. Lett.* **4**, 52 (1979).
103. K. Inoue, *J. Lightwave Technol.* **10**, 1553 (1992).
104. S. S. Watanabe, T. Naito, and T. Chikama, *IEEE Photonics Technol. Lett.* **5**, 92 (1993).
105. R. M. Jopson, A. H. Gnauck, and R. M. Derosier, *Electron. Lett.* **29**, 576 (1993).
106. W. Pieper, C. Kurtzke, R. Schnabel, D. Breuer, R. Lundwig, K. Petermann, and H. G. Weber, *Electron. Lett.* **30**, 724 (1994).
107. A. H. Gnauck, R. M. Jopson, and R. M. Derosier, *IEEE Photonics Technol. Lett.* **7**, 582 (1995).
108. A. Royset, S. Y. Set, I. A. Goncharenko, and R. I. Laming, *IEEE Photonics Technol. Lett.* **8**, 449 (1996).
109. K. Inoue, *Opt. Lett.* **22**, 1772 (1997).
110. M. D. Pelusi and B. J. Eggleton, *Opt. Express* **20**, 8015 (2012).
111. L. B. Du, M. M. Morshed, and A. J. Lowery, *Opt. Express* **20**, 19921 (2012).
112. M. C. Tatham, G. Sherlock, and L. D. Westbrook, *Electron. Lett.* **29**, 1851 (1993).
113. R. M. Jopson and R. E. Tench, *Electron. Lett.* **29**, 2216 (1993).
114. J. Inoue and H. Kawaguchi, *IEEE Photonics Technol. Lett.* **10**, 349 (1998).
115. H. C. Lim, F. Futami, and K. Kikuchi, *IEEE Photonics Technol. Lett.* **11**, 578 (1999).
116. H. Sotobayashi and K. Kitayama, *J. Lightwave Technol.* **17**, 2488 (1999).
117. S. Watanabe, H. Kuwatsuka, S. Takeda, and H. Isshikawa, *Electron. Lett.* **33**, 316 (1997).
118. T. Merker, P. Meissner, and U. Feiste, *IEEE J. Sel. Top. Quantum Electron.* **6**, 258 (2000).
119. M. Matsuura and N. Kishi, *Opt. Lett.* **38**, 1700 (2013).
120. S. Ayotte, S. Xu, H. Rong, O. Cohen, and M. J. Paniccia, *Electron. Lett.* **43**, 1037 (2007).
121. Y. Dai, X. Chen, Y. Okawachi, A. C. Turner-Foster, M. A. Foster, M. Lipson, A. L. Gaeta, and C. Xu, *Opt. Express* **17**, 7004 (2009).
122. O. Kuzucu, Y. Okawachi, R. Salem, M. A. Foster, A. C. Turner-Foster, M. Lipson, and A. L. Gaeta, *Opt. Express* **17**, 20605 (2009).
123. M. D. Pelusi, F. Luan, D.-Y. Choi, S. J. Madden, D. A. P. Bulla, B. Luther-Davies, and B. J. Eggleton, *Opt. Express* **18**, 26686 (2010).
124. I. Brener, B. Mikkelsen, G. Raybon, R. Harel, K. Parameswaran, J. R. Kurz, and M. M. Fejer, *Electron. Lett.* **36**, 1788 (2000).
125. D. Kunimatsu, C. Q. Xu, M. D. Pelusi, X. Wang, K. Kikuchi, H. Ito, and A. Suzuki, *IEEE Photonics Technol. Lett.* **12**, 1621 (2000).
126. H. Hu, R. Nouroozi, R. Ludwig, C. Schmidt-Langhorst, H. Suche, W. Sohler, and C. Schubert, *Opt. Lett.* **35**, 2867 (2010).

9
Nonlinear and Ultrahigh Resolution Laser Spectroscopy

In this chapter, we describe the principles of major nonlinear and ultrahigh resolution laser spectroscopic techniques, including saturation spectroscopy, two-photon spectroscopy, coherent Raman spectroscopy, nonlinear polarization spectroscopy, and laser cooling and trapping spectroscopy.[1–4] All these spectral techniques are based on the use of one or several laser beams (at least one of them is tunable), and there is no need of any ordinary spectrometers or dispersion elements (such as prisms, gratings, or Fabry–Perot etalon). The most remarkable advantages of these novel spectroscopic techniques are their Doppler-free capability and ultrahigh spectral resolution, which render researchers capable to measure the hyperfine structures, isotope shifts, Stark and Zeeman splitting, and to establish new optical frequency standards (atomic clocks) as well.

9.1 Major mechanisms of spectral broadening

The spectral resolution is one of the most essential criteria for evaluating a given spectroscopic method, which is defined as the minimum interval ($\delta\nu$ or $\delta\lambda$) between two spectral lines which can just be resolved by this method. An alternative parameter is the spectral resolving power that is defined as $\nu_0/\delta\nu$ or $\lambda_0/\delta\lambda$, where ν_0 and λ_0 are the central frequency and wavelength of these two resolvable lines. If there is a single spectral line with an infinitely narrow width, the measured linewidth cannot be smaller than the spectral resolution ($\delta\nu$ or $\delta\lambda$) of the method used. In order to determine the real spectral line shapes and possible hyperfine structures of the concerned atomic transitions, researchers must choose an appropriate spectroscopic method that has a high enough spectral resolution.

The spectral resolution for any type of conventional spectroscopic apparatus is limited by the so-called instrumental broadening, which is determined by the slit-width, the angular dispersion of the spectral-dispersion elements (such as prisms, gratings, or Fabry–Perot etalons), and the diffraction effect of a limited light-beam aperture. For these reasons, the spectral resolving power of the conventional spectroscopic apparatuses using prisms, gratings, or Fabry–Perot etalons as spectral dispersion elements, is usually in the order of 10^4, 10^5, and 10^6, respectively. It is obvious that these traditional apparatuses are not suitable for spectroscopic studies requiring a higher spectral resolution. The frequency-tunable lasers have provided many new opportunities, so that researchers can develop various novel spectroscopic techniques, which do not need any conventional spectral dispersion elements or apparatuses, and the limitation from those apparatuses can be entirely eliminated.

Nonlinear Optics and Photonics. First Edition. Guang S. He. © Guang S. He 2015.
Published in 2015 by Oxford University Press.

Although there are a number of new laser spectroscopic methods having a much higher spectral resolution than the conventional methods, the spectral resolution of these novel spectral approaches is still limited by some other effects, which are related to the investigated medium and its interaction with light. In the following part of this section, we shall briefly describe those effects that may cause spectral broadening of a narrow spectral line and will impose different limitations on the final spectral resolution.

9.1.1 Doppler broadening of the gaseous medium

The molecules (or atoms) of a gaseous medium are always in a state of random thermal motion. Assume that v_z is the thermal velocity in the observation direction (z-axis) of a given molecule that makes a resonant transition, owing to the Doppler effect the observed frequency ω of that molecular transition will be

$$\omega = \omega_0(1 + v_z/c), \tag{9.1-1}$$

where ω_0 is the transition frequency of a molecule at rest and c is the light speed in the medium. According to the Maxwell distribution law of molecular velocities, the probability of molecules having a thermal velocity v_z can be expressed as

$$f(v_z) = \sqrt{\frac{m}{2\pi k_b T}} e^{-\frac{mv_z^2}{2k_b T}}, \tag{9.1-2}$$

where m is the mass of the molecule, k_b is the Boltzmann constant, T is the temperature in degrees kelvin. Based on Eq. (9.1-1), the above expression can be rewritten as the probability of observing the transition frequency ω of molecules, i.e.,

$$f(\omega) \propto \sqrt{\frac{m}{2\pi k_b T}} e^{-(\omega-\omega_0)^2 \frac{mc^2}{\omega_0^2 2k_b T}}. \tag{9.1-3}$$

The physical meaning of Eq. (9.1-3) is that the observed spectral profile of contributions from all molecules manifests a Gaussian shape with a spectral width determined by

$$(\Delta\omega)_D = 2\omega_0 \sqrt{\frac{2k_b T}{mc^2} \ln 2}. \tag{9.1-4}$$

As is known, the Doppler effect causes typical inhomogeneous broadening applicable for any gaseous medium. For common gas samples in room temperature, the ratio of $\omega_0/(\Delta\omega)_D$ usually ranges from 10^5 to 10^6. Consequently, the spectral resolving power of an ordinary spectroscopic method cannot be higher than the above values unless a special Doppler-free technique is employed. The saturation spectroscopy and two-photon spectroscopy are among the most successful Doppler-free nonlinear spectroscopic techniques.

9.1.2 Collision broadening of the gaseous medium

In a gaseous medium, the random molecular thermal motion not only causes the Doppler broadening, but also leads to molecular collisions that may introduce a sudden interruption among the phase relation of molecular wave functions. According to the principle of uncertainty, an interrupting change of phase relation of molecular wave functions implies an additional frequency uncertainty, which is the origin of the so-called collision broadening. In this case, the spectral broadening is determined by the average time interval τ' between two collision events for a given molecule. The collision-induced broadening belongs to the category of homogeneous broadening and its spectral profile can be expressed as

$$f(\omega) = f_{\max} \frac{(1/\tau')^2}{(\omega - \omega_0)^2 + (1/\tau')^2}. \tag{9.1-5}$$

This implies a Lorentzian line shape with a spectral linewidth

$$(\Delta\omega)_{\text{collision}} = \frac{2}{\tau'}. \tag{9.1-6}$$

Assuming the gaseous medium consists of the identical molecules and the total molecular number is N involved in a volume V, then the average collision rate of an arbitrary molecule with others is

$$\frac{1}{\tau'} = \frac{N}{V}\sigma_c \bar{v}, \tag{9.1-7}$$

where $\bar{v} = 4(k_b T/\pi m)^{1/2}$ is the average relative molecular velocity and σ_c is the molecular collision cross-section. Substituting the expression of $\bar{v}$ into Eq. (9.1-7) leads to

$$\frac{1}{\tau'} = 4\sigma_c \sqrt{\frac{k_b T}{\pi m} \frac{N}{V}} = 4\sigma_c \sqrt{\frac{1}{\pi m k_b T}} P, \tag{9.1-8}$$

where P is the pressure of the gaseous medium. Based on this expression, Eq. (9.1-6) becomes

$$(\Delta\omega)_{\text{collision}} = 8\sigma_c \sqrt{\frac{1}{\pi m k_b T}} P. \tag{9.1-9}$$

One can see from this expression that the collision broadening is proportional to the gas pressure; therefore, it is also termed pressure broadening. At room temperature and normal pressure, the influence from pressure broadening is much smaller than that from Doppler broadening. Nevertheless, once the latter influence has been removed, pressure broadening may become the next major influence on the spectral resolution. In such a case, the pressure of the gaseous medium should be kept at a low level. In order to obtain a very high spectral resolution, sometimes the gas pressure should remain in a range less than 10^{-2}–10^{-3} Torr.

9.1.3 Transit-time broadening

This is a spectral broadening effect that takes place when a laser beam interacts with a gaseous medium or a molecular (or atomic) beam. In this case, the interaction time between a moving molecule and the laser beam is limited; therefore, according to the principle of uncertainty, the measured spectral line undergoes an additional broadening influence. The maximum broadening range is determined by the minimum transit time (flight time) of molecules traveling across the laser beam,

$$(\Delta\omega)_{\text{transit}} \approx \frac{\bar{v}}{a_0}. \tag{9.1-10}$$

Here, a_0 is the transverse size of the laser beam used for the spectral measurement, and $\bar{v}$ is the average thermal velocity of the molecules. In common experimental conditions, the transit-time broadening is much smaller than Doppler broadening and pressure broadening. For example, the transit-time broadening of gaseous molecules traveling across a 1-mm laser beam is in the range 10^5–10^6 Hz. In cases requiring very high spectral resolution, the transit-time broadening can be further reduced by expanding the effective laser beam aperture or greatly reducing the velocity of molecules transversely passing through the laser beam.

9.1.4 Second-order (transverse) Doppler broadening

If we consider a directional molecular beam moving along a certain direction and examine the molecular spectroscopic behavior from the perpendicular direction, then the Doppler broadening can be eliminated because the component of molecular velocity along the observation direction is zero. This conclusion is valid only in the first-order approximation. According to special relativity, the apparent frequency shift due to the Doppler effect can be generally expressed as

$$\omega = \omega_0 \sqrt{1 - (v/c)^2}/[1 - (v/c)\cos\theta], \qquad (9.1\text{-}11)$$

where v is the absolute value of the molecular velocity and θ is the angle between the molecular moving direction and the observation direction. If $\theta = 0$, $\cos\theta = 1$, and the term $(v/c)^2$ can be neglected in the first-order approximation, Eq. (9.1-11) becomes

$$\omega = \omega_0 \frac{\sqrt{1-(v/c)^2}}{1-(v/c)} \approx \omega_0 \frac{1}{1-(v/c)} \approx \omega_0[1+(v/c)]. \qquad (9.1\text{-}12)$$

This is the linear Doppler effect we presented earlier in Eq. (9.1-1). In the case of $\theta = 90°$ and $\cos\theta = 0$, Eq. (9.1-11) simply becomes

$$\omega = \omega_0 \sqrt{1-(v/c)^2} \approx \omega_0 \left[1 - \frac{1}{2}(v/c)^2\right]. \qquad (9.1\text{-}13)$$

This is the second-order (or transverse) Doppler effect, and the frequency shift is simply proportional to the square of the absolute value of molecular velocity, i.e.,

$$(\Delta\omega)_{D'} = \omega_0 - \omega \approx \frac{\omega_0}{2}(v/c)^2. \qquad (9.1\text{-}14)$$

At normal temperatures, the ratio $\omega_0/(\Delta\omega)_{D'}$ is in the range 10^{11}–10^{12}; therefore the influence from the second-order Doppler effect on the spectral resolution can usually be neglected. If it is necessary to further eliminate the second-order Doppler broadening, the absolute velocity of molecules must be significantly reduced. For this purpose, researchers have developed the so-called laser cooling and trapping technique, in which the thermal velocity of the atoms can be reduced to an extremely low value.

9.1.5 Recoil broadening

When the atom of a gaseous medium emits or absorbs a photon of frequency ω_0, momentum conservation requires that the molecular velocity along the light-traveling direction will change with the amount of

$$\Delta v = \frac{\hbar \omega_0}{2\pi mc}, \qquad (9.1\text{-}15)$$

where m is the mass of the atom and $(\hbar\omega_0/2\pi c)$ is the momentum of the absorbed or emitted photon. Owing to an influence from the linear Doppler effect, the emission or absorption spectral line will undergo a broadening influence expressed by

$$(\Delta\omega)_{\text{recoil}} = \frac{\hbar \omega_0^2}{2\pi mc^2}. \qquad (9.1\text{-}16)$$

This is the recoil broadening, which is generally smaller than other broadening mechanisms described above.

9.1.6 Influence from laser spectral linewidth

For all nonlinear laser spectroscopic techniques described in this chapter, one or two (or even more) laser beams are necessary, and at least one of them must be tunable. It is obvious that the final spectral resolution of any laser spectroscopic method cannot be better than the linewidth or frequency stability of the laser beam(s) used. Suppose all major broadening effects mentioned above can be basically eliminated; the remaining issue is to improve the lasing frequency stability and reduce the lasing spectral linewidth. If the lasing frequency is ω and the linewidth is $\delta\omega$, the advanced tunable laser devices could provide a $(\omega/\delta\omega)$ range greater than 10^{13}–10^{15}.

9.2 Saturation spectroscopy

9.2.1 General description of the saturated-absorption effect

It is well known that for a gaseous medium under normal conditions of temperature and pressure, the absorptive spectral linewidth is mainly limited by linear Doppler broadening, which is much larger than the natural linewidth. Before the advent of lasers, researchers employed directional atomic or molecular beams to reduce the Doppler broadening. However, that improvement was rather limited. The first breakthrough in high-resolution laser spectroscopy was the invention of so-called Doppler-free saturation spectroscopy.

In the early 1960s, when researchers studied the axial-modes lasing behavior of the He–Ne laser devices, it was found that the spectral gain curve of Ne lasing transition was basically determined by the linear Doppler broadening profile.[5] Because of the inhomogeneous broadening nature, the excited atoms that correspond to different frequency positions within the gain curve, cannot exchange their excitation energies with each other. As a result, the lasing of several axial modes leads to a drop of population inversion (and hence gain) at several discrete frequency positions; or in other words, there will be several narrow dips in the continuous Doppler-broadening gain profile. This is the so-called "hole-burning" or "Lamb dip" effect.

Conversely, if the considered medium is an absorbing gas sample excited by a monochromatic laser beam at frequency ω' that is within a Doppler-broadened absorption line of the medium, then only a certain group of molecules can make the real absorptive transitions. The thermal velocity v' of this group of molecules should satisfy the following relation:

$$kv' = \omega' - \omega_0,$$

where k is the magnitude of wavevector of the incident laser beam and ω_0 is the central frequency of the Doppler line. Suppose the intensity of the laser beam is high enough, so that a considerable amount of molecules satisfying the above requirement are excited to an upper level while the population in the lower (ground) level decreases accordingly. Since the linear absorption coefficient of the medium is proportional to the population difference between the lower and the upper level (see Eq. (9.2-5) below), as a result of such strong selective excitation of certain groups of molecules, there will be a narrow dip at the position of laser frequency ω' within the broad Doppler profile of the absorption line.

This elementary process of saturated absorption in a resonant medium with inhomogeneous broadening can be schematically illustrated in Fig. 9.1. In Fig. 9.1(a), we present the population distribution in the lower and upper level as a function of molecular velocity, while in Fig. 9.1(b) we demonstrate the absorption coefficient $\alpha(\omega)$ as a function of the frequency. For strong laser excitation, we expect a narrow dip of absorption at the excitation frequency ω', as shown in Fig. 9.1(b). The width of the absorption dip at a given frequency position may be determined by other spectral broadening effects that have much smaller broadening influences, and in the best situation the dip width can be as narrow as the natural linewidth of a molecular transition.

If we let only one tunable laser beam pass through the gas sample and scan the lasing frequency across the overall absorption spectral linewidth of the medium, we still will observe a Doppler line shape even though the saturation effect occurs. However, if we use two laser beams, one a strong beam with a fixed frequency ω' and the other a weak beam with a tunable frequency ω'', then the

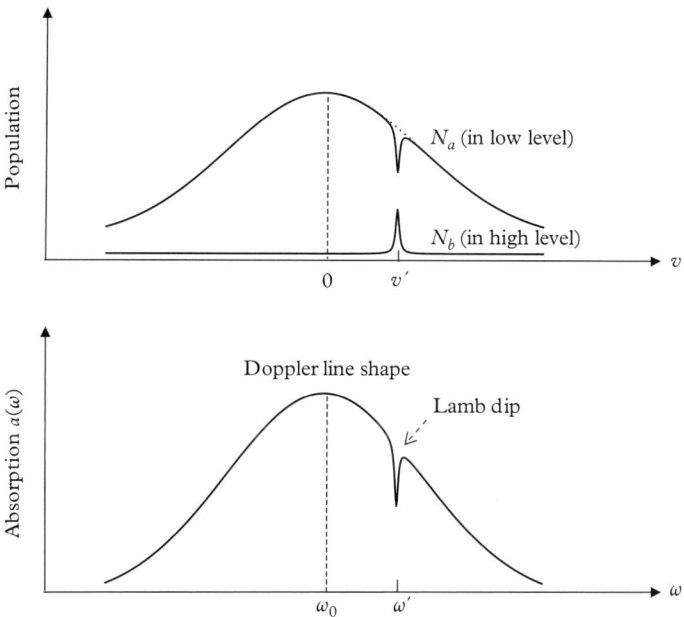

Figure 9.1 *A two-level gaseous absorbing medium interacting with an intense monochromatic laser beam of frequency ω': (a) the population distribution as a function of the thermal velocity of molecules; (b) the absorption coefficient as a function of optical frequency.*

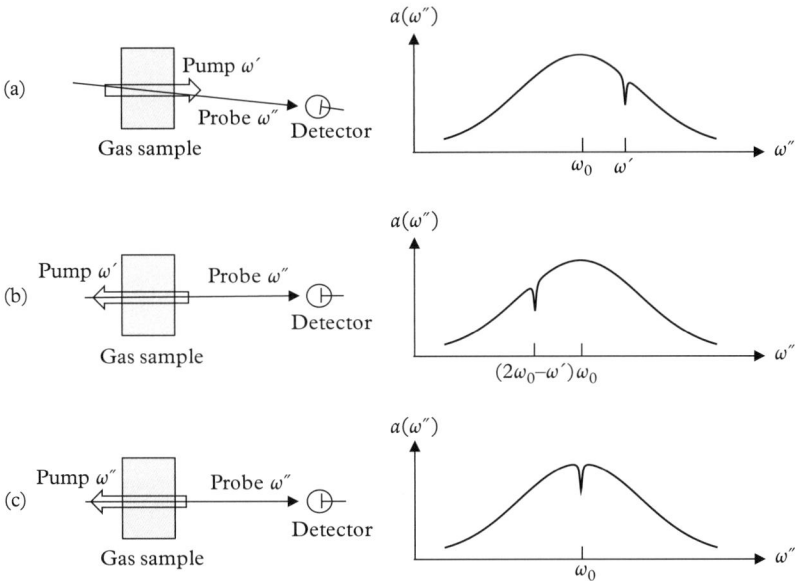

Figure 9.2 *Schematic diagrams of three approaches for Doppler-free saturation spectroscopy. (a) The strong saturating beam with a fixed frequency ω' and the weak probe beam with a tunable frequency ω'' pass through the sample nearly along the same direction, (b) these two beams pass through the sample in opposite directions, and (c) these two beams have the same tunable frequency ω'' and pass through the sample in opposite directions.*

narrow absorption-dip induced by the first strong laser beam at position ω' can be detected by the second weak probe beam. Three different approaches can be employed to do this, which are shown schematically in Fig. 9.2.

In the case shown in Fig. 9.2(a), the two laser beams are forward passing through the gas sample with a very small crossing angle; the strong (saturating) beam caused an narrow absorption dip at frequency ω', while the second beam is so weak that its saturation influence can be neglected. Thus scanning the frequency of the second (probe) beam, the narrow saturation dip induced by the first beam can be detected. A small crossing angle between these two beams is necessary to eliminate the possible background influence from the strong saturating beam at the detector position.

In the case shown in Fig. 9.2(b), the two laser beams pass through the gas sample in opposite directions, and the absorption dip induced by the strong saturating beam can still be detected by tuning the weak probe beam. In this case, however, the detected dip will be located at the frequency position of $\omega'' = \omega_0 - (\omega' - \omega_0) = 2\omega_0 - \omega'$, due to the fact that the sign of Doppler frequency shift depends on the moving direction of molecules.

Finally, in the case shown in Fig. 9.2(c), the saturating beam and the probe beam have the same tunable frequency ω'' and are counter-propagating through the gas sample. By continuously tuning the lasing frequency across the Doppler profile and measuring the transmitted intensity of the probe beam, one can detect a narrow absorption dip only at the central frequency ω_0, because in this case only those molecules, the velocity components of which along the beam direction are nearly zero, can interact in resonance with the two beams together.

For the second and third approaches shown in Fig. 9.2(b) and (c), the two laser beams can be parallel to each other; therefore the additional broadening influence from the non-parallelism of these two beams can be eliminated. On the other hand, the first and the second approaches shown in Fig. 9.2(a) and (b) can be used to study the narrow absorption spectra of molecular groups with any values of velocity, whereas the third approach shown in Fig. 9.2(c) can only study the narrow spectrum of the molecules with a near-zero velocity along the laser beam direction. In practice, the configuration shown in Fig. 9.2(c) is often used because it needs only one (tunable) laser source.

9.2.2 Basic theoretical considerations

For a monochromatic light beam passing through a resonant absorbing medium the intensity attenuation can be written as

$$I(\omega, z) = I_0(\omega, 0) \exp[-\alpha(\omega)z], \qquad (9.2\text{-}1)$$

where $I_0(\omega, z)$ is the input intensity, z is the propagation length in the medium, and $\alpha(\omega)$ is the linear absorption coefficient of the medium. According to the semiclassical theory of radiation, $\alpha(\omega)$ is determined by the imaginary part of the linear resonant susceptibility $\chi_R^{(1)}(\omega)$, i.e.,

$$\alpha(\omega) = \frac{\omega}{n_0 c} \text{Im} \chi_R^{(1)}(\omega), \qquad (9.2\text{-}2)$$

where n_0 is the linear refractive index of the medium and c is the light speed in a vacuum. The linear susceptibility for a resonant gas medium can be expressed as (see Eq. (18.4-2))

$$\chi_R^{(1)}(\omega) = \frac{N_a - N_b}{\varepsilon_0 \hbar} \frac{(p_0)^2}{\omega_{ab} - \omega - i\Gamma}. \qquad (9.2\text{-}3)$$

Here, $N_a - N_b$ is the population difference between the lower level a and the upper level b, p_0 is the matrix element of the dipole moment for the transition between levels a and b with the resonant frequency ω_{ab}, and Γ is the linewidth parameter determined by

$$\Gamma = \frac{1}{2}(\gamma_a + \gamma_b), \qquad (9.2\text{-}4)$$

where γ_a and γ_b are the radiation relaxation ratios of the population in levels a and b, respectively. In principle, the natural linewidth is the highest spectral resolution that could be achieved from absorption spectroscopy when all other spectral broadening influences can be eliminated.

Substituting Eq. (9.2-3) into Eq. (9.2-2) leads to

$$\alpha(\omega) = \frac{\omega}{n_0 c} \frac{(N_a - N_b)(p_0)^2}{\varepsilon_0 \hbar} \frac{\Gamma}{(\omega_{ab} - \omega)^2 + \Gamma^2}. \qquad (9.2\text{-}5)$$

It is known that the molecular absorption transition has a Lorentzian shape with a natural linewidth of

$$(\Delta \omega)_{\text{natural}} = 2\Gamma, \qquad (9.2\text{-}6)$$

provided that Doppler and other broadening effects can be eliminated. From Eq. (9.2-5) we can see that for a weak input light beam with a frequency close to ω_{ab}, the change of $(N_a - N_b)$ is negligible and the absorption coefficient is nearly independent of the intensity of the input beam. However, once the input light intensity is high enough that the change of $(N_a - N_b)$ is no longer negligible, the absorption coefficient α will be dependent not only on the frequency but also on the input intensity. In the latter case, the absorption is getting smaller as the input intensity is getting higher. This is the saturable absorption phenomenon known before the advent of lasers.

As an example, let us consider the simple case shown in Fig. 9.2(a). If the absorption change induced by the saturating beam is $\Delta\alpha(\omega'')$, and the unsaturated absorption coefficient is α_0 that is assumed constant within a small spectral range comparable to the width of 2Γ, the ratio between $\Delta\alpha(\omega'')$ and α_0 can be expressed as[1]

$$\frac{\Delta\alpha(\omega'')}{\alpha_0} = -\frac{I_1}{2I_s}\frac{\Gamma^2}{(\omega'' - \omega')^2 + \Gamma^2}, \qquad (9.2\text{-}7)$$

where I_1 is the intensity of the strong saturating beam, ω' is the frequency of the same beam and equal to the resonant transition frequency ω_{ab} of a group of molecules with certain velocity, and I_s is the saturation intensity parameter of the molecular system and can be written as

$$I_s = \frac{\varepsilon_0 c n_0}{8\pi^2}\left(\frac{h}{p_0}\right)^2 \gamma_a \gamma_b. \qquad (9.2\text{-}8)$$

If the lower level is the ground state, apparently we have $\gamma_a \to 0$ and hence $I_s \to 0$. Actually, there will be an equivalent $\gamma_a \neq 0$, owing to the finite flight time of moving molecules across the saturating laser beam.

Equation (9.2-7) shows that the saturation induced absorption dip exhibits a Lorentzian shape, the line width of which is equal to the natural linewidth of molecular transition. Although saturation spectroscopic methods can effectively eliminate the Doppler broadening of the gaseous medium, in practice the dip shape might be affected by other broadening mechanisms as mentioned early, such as collision broadening, transit-time broadening, and the second-order Doppler broadening. Furthermore, for saturation spectroscopy there exist some additional broadening influences. For example, if the saturating beam and the probe beam are not parallel perfectly but have a small crossing angle θ, as shown in Fig. 9.2(a), there will be a residual influence from the Doppler effect, which is termed the angular broadening or geometric broadening and can be expressed as

$$(\Delta\omega)_{angular} = 2(\Delta\omega)_D \cdot \tan\theta \sin\theta \approx 2(\Delta\omega)_D \cdot \theta^2. \qquad (9.2\text{-}9)$$

Even if one can make the two laser beams parallel perfectly ($\theta = 0$) as in the cases shown in Fig. 9.2(b) and (c), owing to the finite divergence angle ($\delta\theta \neq 0$) of the laser beam(s), there is still an influence from residual geometric broadening, which can be expressed by the same formula, as shown in Eq. (9.2-9) by replacing θ by $\delta\theta$.

As mentioned earlier, the saturating beam is usually much stronger than the probe. However, if the intensity of the saturating beam is too high, the effective saturation not only occurs at the central part of the dip but may also extend to the side areas of this dip. This causes the so-called power broadening in saturation spectroscopy, which is determined by[1]

$$(\Delta\omega)_{power} = \Gamma\left[1 + \sqrt{1 + (I_1/I_s)}\right]\bigg/2. \qquad (9.2\text{-}10)$$

This expression implies that, to eliminate the influence from the power broadening, the intensity I_1 of the saturating beam should not be considerably greater than the saturation intensity parameter I_s of the absorbing medium.

9.2.3 Experimental setups and results

Doppler-free saturation spectroscopy is mainly used to study the visible and IR absorption spectra of gas atoms and molecules. In practice, a typical experimental setup of saturation spectroscopy is usually composed of (i) the tunable laser source(s), (ii) two laser beams (at least one should be tunable), (iii) a low-pressure gas sample, and (iv) opto-electric detector(s) connected to a boxcar integrator or a lock-in amplifier. The laser source(s) should have very narrow linewidth and high lasing frequency stability. In order to increase the signal-to-noise ratio of the detected laser signal, the saturating beam can be modulated with a chopper or other modulator.

Figure 9.3 shows three typical and often used experimental setups for saturation spectroscopy studies. In the first setup shown in Fig. 9.3(a), the absorbing gas sample is placed simply inside the cavity of a tunable laser device. As the lasing frequency ω scans into the Doppler absorption

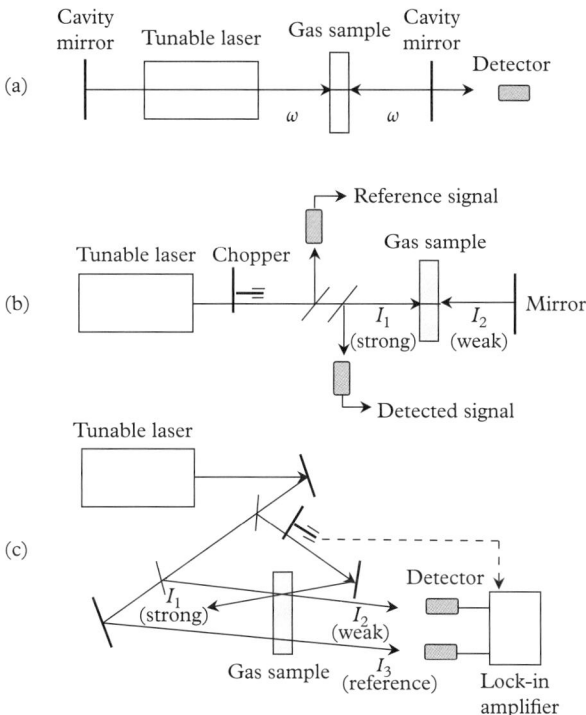

Figure 9.3 *Typical experimental setups for saturation spectroscopy measurements. (a) The gas sample is placed inside the cavity of a tunable laser device, (b) the backward probe beam is provided by a vertically placed external mirror, and (c) two counter-propagating beams have a small crossing angle.*

line of the gas sample, the transmission of that sample becomes lower and the overall lasing output decreases. However, when the lasing frequency ω is tuning to equal the central frequency ω_0 of the Doppler line of the absorbing sample, a maximum absorption dip will appear at the position of $\omega = \omega_0$ and, consequently, a narrow peak on the output intensity curve as a function of lasing frequency will be observed.[6,7] Under the best conditions, the spectral width of this peak (the so-called inverted Lamb dip) can be limited only by the natural linewidth of the molecular transition of the gas sample. The earliest successful example of this type of experimental studies was the Doppler-free measurement of narrow absorption spectrum of a methane (CH_4) sample that was placed inside the cavity of a He–Ne laser system operating at 3.39 μm.[7] The spectral resolving power was up to 10^{11}, and the hyperfine splitting of molecular transitions could be measured. This method is also highly useful for high-precision lasing frequency locking control.

In more common arrangements, the absorbing gas sample is placed outside a tunable laser source which provides both saturating and probe beams via the beam splitter or reflector, as shown in Fig. 9.3(b) and (c).[8–10] In the case shown in Fig. 9.3(b), the strong laser beam passes through the absorbing sample and creates a Lamb dip within the Doppler line profile; the partial transmitted beam is reflected by a vertically placed reflector and reenters the sample as a weak probe beam. When the laser frequency is tuned to the center of a Doppler profile of the absorbing sample, a narrow absorption dip for the probe beam can be detected. To eliminate the influence from the intensity fluctuation of the laser beam, the partial incident laser beam is reflected as a reference signal and directed to a detector. The feature of the setup shown in Fig. 9.3(c) is that the reference beam also passes through the gas sample; thus the additional errors from the existence of the sample can also be canceled. Another advantage of this last setup is the convenience to control the saturating and probe beams separately, although the angle broadening influence exists due to the non-parallelism of these two beams.

In the early stages of the development of tunable laser devices, tunable dye lasers were mostly adopted for high-resolution laser spectroscopic studies. Now tunable semiconductor diode lasers with an external cavity are more commonly employed for this purpose.

As an early experimental example of saturation spectroscopy, Fig. 9.4 shows the absorption spectral curve of water vapor under low pressure at 1889.58 cm^{-1}, measured with a setup similar to that shown in Fig. 9.3(b). The laser source was a tunable InSb spin-flip Raman laser pumped with a cw CO laser; the pressure of the water vapor sample was 30 mTorr and the absorbing gas cell was 40 cm long; the size and the spectral linewidth of the tunable laser beam were ∼2 cm and 50–100 kHz, respectively.[11] The central frequency of the absorption transition $\nu_2(5_{3,2}\to 6_{4,3})$ for H_2O molecules is located at 1889.58 cm^{-1} and the result shown in Fig. 9.4 is obtained by precisely tuning the laser frequency around the above frequency position and measuring the ratio between the input and output intensities. One can see from Fig. 9.4 that the whole absorption profile is Doppler broadened with a linewidth of ∼165 MHz, but the width of the central Lamb dip is only about 200 kHz. In the present experimental conditions, the natural linewidth of this transition is ∼10 kHz, pressure broadening ∼50 kHz, transit-time broadening 20–30 kHz, and the power broadening is negligible for the 10 mW laser beam. Considering all these spectral broadening factors, the expected dip width should be 100 kHz, which is very close to the measured value.

Doppler-free saturation spectroscopy can be employed to investigate the fine and hyperfine structures of optical transitions for atoms or molecules. As another example of the early studies, Fig. 9.5(b) shows the fine structure of the Balmer α-line (D_α line of the $n = 2 \to 3$ transition) of atomic deuterium, obtained by using the method of saturation spectroscopy and a setup similar to that shown in Fig. 9.3(c).[1,12] The deuterium atoms were generated in a Wood-type dc discharge tube at low pressure, where they were excited to the atomic $n = 2$ state. The laser source was a

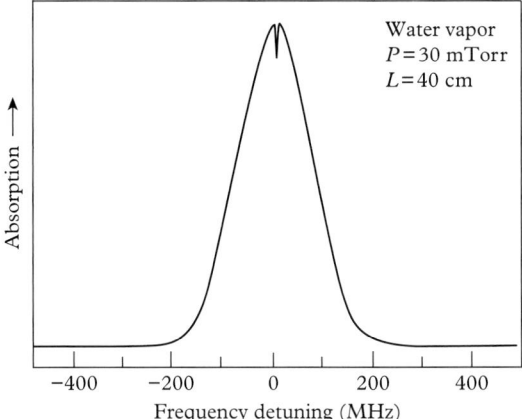

Figure 9.4 *Saturation spectroscopic curve at 1889.58 cm^{-1} of water vapor in pressure of 30 mTorr. (Reproduced with permission from Patel.[11] Copyright 1974, AIP Publishing LLC).*

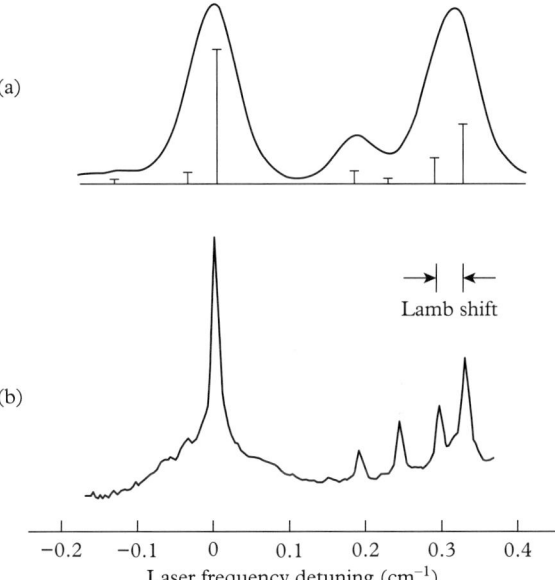

Figure 9.5 *Balmer α-line of atomic deuterium (D): (a) emission line profile of a cooled deuterium gas discharge (T = 50 K) and theoretical fine structure with relative transition probabilities (without Doppler effect); (b) saturated absorption spectrum with optically resolved Lamb shift. (Reproduced with permission from Hänsch et al.[12] Copyright 1974, American Physical Society).*

pulsed dye laser with a 30 MHz linewidth, 8 ns pulse duration, 1 mm beam diameter, and tunable around 656.3 nm, the central wavelength of the D_α line. This line consists of seven closely spaced fine-structure components, which are hardly resolved with conventional spectroscopic methods due to the Doppler broadening even operating at cryogenic temperature, as shown in Fig. 9.5(a). In contrast, from the result of saturation spectroscopic measurements shown in Fig. 9.5(b), the four strongest fine-structure components can be clearly resolved, and the Lamb shift could be readily measured with a narrowest linewidth about 250 MHz. Using a frequency-stabilized He–Ne laser in conjunction with a Fabry–Perot interferometer, the absolute wavelength values of the Balmer α-line for atomic hydrogen and deuterium could be measured. Consequently, a new Rydberg value, $R_\infty = 10973731.43(10)$ m^{-1}, was obtained with an almost tenfold improvement in the accuracy over the best conventional measurements.[1,12]

To indicate the development of the saturated absorption spectroscopic technique, the following are some examples of the use of this technique to investigate the hyperfine spectral structures of a variety of atoms and molecules in the gaseous state:

i. Metal atoms: Hg,[13] V,[14] Na,[15] Ca,[16] Zr,[17] Mg,[18] Li,[19] F,[20] Cs,[21] Sm,[22] Rb,[23] K,[24] and so on.

ii. Nonmetal atoms: O,[25] Cl,[26] N,[27] Ar,[28] Ps (positronium),[29] and others.

iii. Molecules: MeBr,[30] BCl$_3$,[31] CF$_3$Br,[32] C$_2$H$_2$,[33] C$_2$H$_4$,[34] CO$_2$,[35] YbF,[36] NO$_2$,[37] and so on.

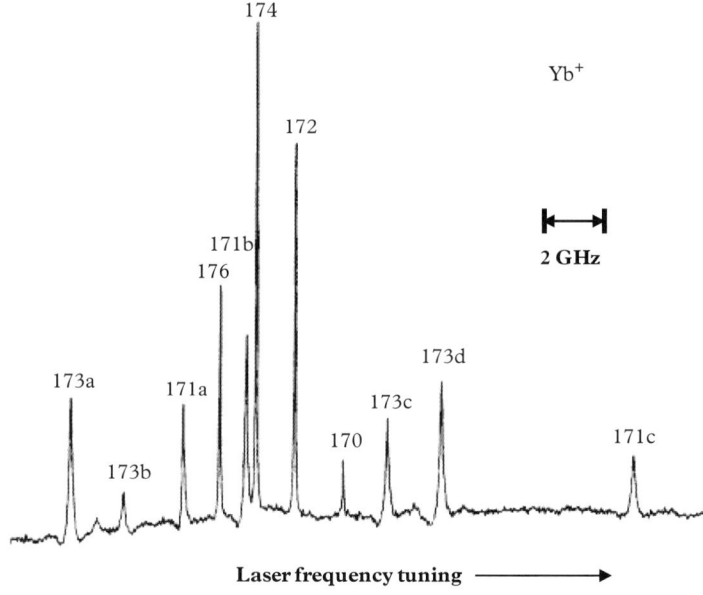

Figure 9.6 *Saturated absorption spectrum for the Yb$^+$ 369.4-nm 6s–6p$_{1/2}$ transition, indicating the isotope shifts and hyperfine structure. (Reproduced with permission from Maartensson-Pendrill et al.[38b] Copyright 1994, American Physical Society).*

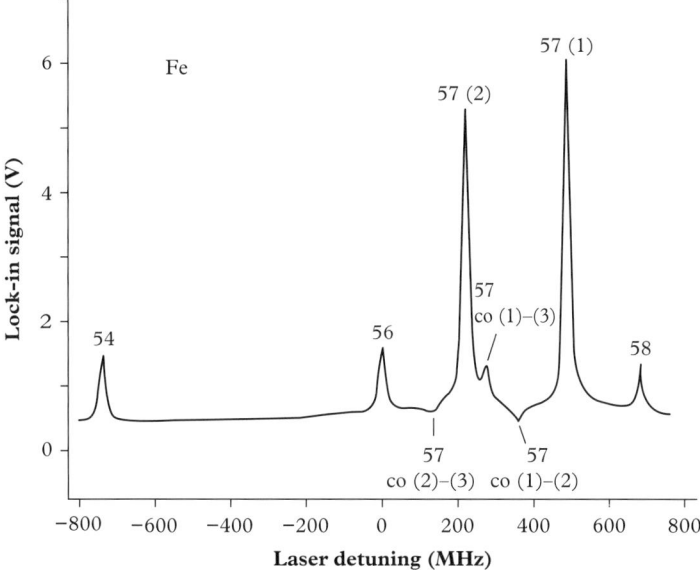

Figure 9.7 *Doppler-free saturated absorption spectrum, recorded with an enriched sample of iron (~10% ^{54}Fe, ~10% ^{56}Fe, ~70% ^{57}Fe, and ~10% ^{58}Fe). Small peaks from the crossover (co) resonance are also indicated. (Reproduced with permission from Krins et al.[39] Copyright 2009, American Physical Society).*

The saturated absorption technique can also be used to study the spectral structure of the ionized atoms with high spectral resolution. As an example, Fig. 9.6 shows the measured Doppler-free saturated absorption spectrum for the Yb$^+$ 369.4-nm $6s$–$6p_{1/2}$ transition, measured in a vapor sample of Yb$^+$ ions produced by cathode sputtering in a low-pressure rare-gas discharge.[38b] The isotope shift and hyperfine structure can be readily determined with very high precision.

Moreover, Fig. 9.7 shows the saturated absorption spectrum of different isotopes of Fe atoms, measured in an enriched sample (involving ~10% ^{54}Fe, ~10% ^{56}Fe, ~70% ^{57}Fe, and ~10% ^{58}Fe) by using a ~372-nm tunable diode laser in resonance with the $3d^6 4s^2$ a $^5D_4$–$3d^6 4s4p$ z $^5F_5^0$ transition.[39] The laser spectral linewidth was less than 1 MHz, and the sample was produced in a Fe–Ar hollow cathode discharge tube. From Fig. 9.7, it can be seen that the three lines of the even isotopes have nearly the same intensity, while the ^{57}Fe line is much stronger due to its high isotope abundance. In addition to these lines, one also can see the small peaks due to crossover resonances between the hyperfine components of ^{57}Fe. The physical origin of crossover resonance will be described in the next subsection.

9.2.4 Crossover resonances in saturation spectroscopy

In saturation spectroscopy experiments using counter-propagating beams with the same tunable frequency, the recorded absorption dips of the probe beam as a function of the scanning frequency should indicate simply the exact positions of different molecular transitions, provided that all of these transitions are independent of each other. In some cases, however, two (or even more) of these transitions may be coupled to each other if they have a common initial level or a common terminal level. Under these circumstances, one may observe an additional dip signal positioned in the middle between two principal dips that correspond to two coupled transitions. This additional dip signal originates from the so-called crossover resonance mechanism.

For simplicity, let us consider a three-level system in which two absorptive transitions ω_0^{ab} and ω_0^{ac} are initiated from the same low level a, as shown in Fig. 9.8(a). In this case, as the frequency of the two counter-propagating laser beams is tuned to equal ω_0^{ab} and ω_0^{ac} sequentially, one could observe two separate saturation dips with the weaker probe beam. Moreover, when the laser frequency ω is tuned to the position of

$$\omega = \omega' = (\omega_0^{ab} + \omega_0^{ac})/2, \tag{9.2-11}$$

the stronger saturating beam will create two narrow holes in the Doppler profile of the population curve of level a via transitions of ω^{ac} and ω^{ab} that are attributed to the molecular groups having the particular axial velocities of

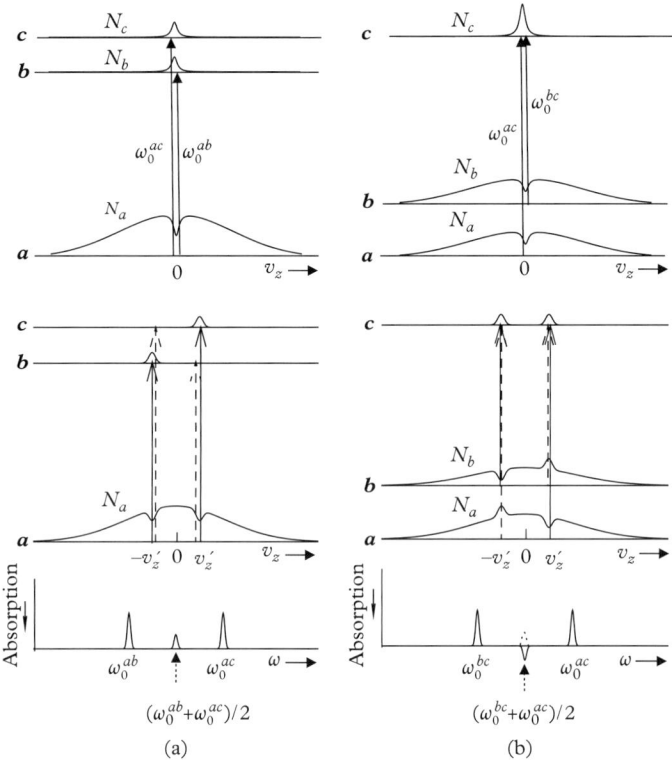

Figure 9.8 *Crossover resonances for two coupled transitions with (a) a common lower level a or (b) a common upper level c. Transitions induced by the saturating beam are indicated by solid vertical lines, whereas the transitions induced by the probe beam are represented by dashed vertical lines.*

$$v'_z = \pm \left| c \frac{\omega_0^{ac} - \omega_0^{ab}}{\omega_0^{ac} + \omega_0^{ab}} \right|. \tag{9.2-12}$$

Under this particular excitation condition, the counter-propagating probe beam will interact with the same groups of molecules restricted by Eq. (9.2-12) and, therefore, encounter a dip of saturated absorption. This is the crossover resonance process related to the principal transitions originating from a common lower level.

The other crossover resonance process, which is related to the transitions sharing a common upper level, is shown schematically in Fig. 9.8(b). In this case, the two absorptive transitions ω^{ac} and ω^{bc} are sharing the same final level c. When the laser frequency is tuned to the frequency of

$$\omega = \omega' = (\omega_0^{ac} + \omega_0^{bc})/2, \tag{9.2-13}$$

the strong saturating beam will create two narrow holes in curves N_a and N_b and two population peaks in curve N_c through the transitions ω^{ac} and ω^{bc}, respectively, which are attributed to the molecular groups having the specific velocities of

$$v'_z = \pm \left| c \frac{\omega_0^{ac} - \omega_0^{bc}}{\omega_0^{ac} + \omega_0^{bc}} \right|. \tag{9.2-14}$$

Partial molecules excited into the common level c may decay into the lower level a or b. Considering that the influence of population relaxation backing to its initial levels is negligible because of the continuous pump action of the saturating beam, but part of the excited molecules created by the strong transition ω^{ac} may relax to the level b while part of excited molecules created by the transition ω^{bc} may decay to the level a. The net result is that the backward propagating probe beam will see the formation of two narrow population peaks on the curves N_a and N_b at the positions of $-v'_z$ and v'_z, respectively. Under this particular condition, the probe beam interacts with the same groups of molecules restricted by Eq. (9.2-14) and experiences a certain change of absorption. The observed absorption change can be positive or negative depending on the relaxation behavior of the molecular system. For example, more absorption will be observed if the population peaks on the curves N_a and N_b are greater than those on the curve N_c, or otherwise a reduced absorption could be seen in the opposite situation.

A typical experimental result demonstrating the crossover contributions on the observed saturation spectrum will be presented in Section 9.5.1.

9.3 Two-photon absorption spectroscopy

9.3.1 General description

The first theoretical suggestion of using two-photon absorption (2PA) to eliminate the Doppler effect was reported in 1970.[40] In 1974, several research groups independently reported their experimental results of the Doppler-free two-photon spectroscopy in a sample of Na vapor.[41–44]

The principle of the Doppler-free 2PA spectroscopy is rather straightforward. Assume that two counter-propagating laser beams with the same frequency ω are traveling through an absorbing gas sample in the same time and overlapping each other. For a group of molecules that have a velocity v in the axial direction of the two light beams, the apparent 2PA frequency will be $\omega' = 2\omega(1 - v/c)$ for one beam and $\omega'' = 2\omega(1 + v/c)$ for the other beam; thereby, the Doppler effect cannot be eliminated by 2PA from any single light beam. However, if the molecule absorbs one photon from one beam and one photon from the other beam simultaneously, the net 2PA frequency will be

$$\omega\left(1-\frac{v}{c}\right) + \omega\left(1+\frac{v}{c}\right) = 2\omega. \tag{9.3-1}$$

For this process, the 2PA frequency is always the same for all molecules having different velocity values, and a very narrow resonance occurs if the light frequency is tuned to

$$\omega = \omega_0/2, \tag{9.3-2}$$

where ω_0 is the resonant 2PA transition frequency of the molecules at rest. In this way, the first-order Doppler effect can be simply eliminated.

Figure 9.9 shows a schematic diagram of the elementary Doppler-free 2PA process. Upon the excitation of two counter-propagating laser beams with the same tunable circular frequency ω, all gas molecules may absorb one photon from one beam and one photon from the other beam simultaneously. This process can take place effectively only when the 2ω value of the laser beams is equal or very close to the 2PA transition frequency ω_0 value, as required by Eq. (9.3-2). In the general case, after being excited to an upper state (t) the molecules may relax to a lower level (c) with fluorescence emission. If one tunes the frequency ω of the two input laser beams and simultaneously measures the fluorescence intensity as a function of ω, the Doppler-free 2PA spectrum can be obtained.

It can be proved that the narrowest spectral width of the 2PA resonance curve for counter-propagating beams can be limited only by the natural linewidth. In practice, however, there exist some other spectral broadening influences as we have discussed in the above two sections. In addition, the 2PA created by any single beam is also possible, which will lead to a Doppler-broadening background due to single-beam TPA processes. In this case, a proper measure should be taken to eliminate or reduce this residual influence.

In comparison with saturation spectroscopy, the Doppler-free two-photon spectroscopy may offer a much higher sensitivity of spectral measurements. The reason is that for the former the laser beams interact only with a group of molecules of certain velocity value, whereas for the latter the laser beams interact with all molecules having different velocity values.

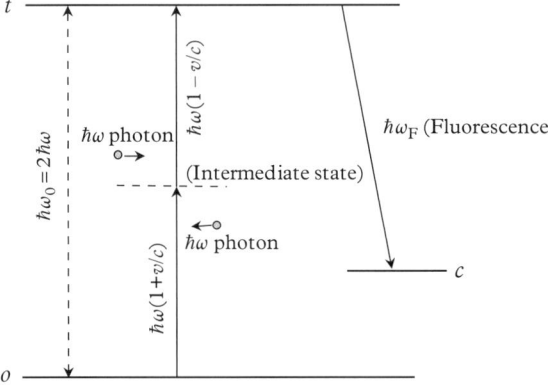

Figure 9.9 Schematic diagram of the elementary process of Doppler-free 2PA spectroscopy.

9.3.2 Theoretical considerations of 2PA

We consider two coherent monochromatic beams, which are of slightly different frequencies ω_1 and ω_2 and both nearly in resonance with an isotropic two-photon absorbing medium. According to Eqs. (5.4-1)–(5.4-3), the 2PA-enhanced complex refractive-index changes for these two beams are

$$\left.\begin{array}{l}\Delta n'(\omega_1) = n'_2 |E_0(\omega_2)|^2 \\ \Delta n'(\omega_2) = n'_2 |E_0(\omega_1)|^2\end{array}\right\}. \tag{9.3-3}$$

Here, $E_0(\omega_1)$ and $E_0(\omega_2)$ are the field functions of the two incident beams, and here n'_2 is a complex nonlinear refractive-index coefficient,

$$n'_2 = \frac{K'}{2n_0} \cdot \frac{1}{[\omega_{to} - (\omega_1 + \omega_2)] - i\Gamma}, \tag{9.3-4}$$

where Γ is the linewidth factor of the molecular 2PA transition and K' is a material coefficient determined by

$$K' = \frac{N_o - N_t}{\varepsilon_0 6\hbar^3} \left| \sum_b \left[\frac{(p_1)_{ob}(p_2)_{bt}}{\omega_{bo} - \omega_1} + \frac{(p_2)_{ob}(p_1)_{bt}}{\omega_{bo} - \omega_2} \right] \right|^2. \tag{9.3-5}$$

Here, N_o and N_t are the molecular population densities (in units of cm^{-3}) in the ground state (o) and an upper excited state (t) shown in Fig. 9.9, p_1 and p_2 are the components of the molecular dipole moment operator along the polarization directions of the two light beams, and the summation is over all possible states (b).

For the beam of frequency ω_1, the explicit expression of the resonance-induced complex refractive-index change is

$$\Delta n'(\omega_1) = \frac{K'}{2n_0(\omega_1)} \frac{[\omega_{to} - (\omega_1 + \omega_2)] + i\Gamma}{[\omega_{to} - (\omega_1 + \omega_2)]^2 + \Gamma^2} |E_0(\omega_2)|^2 \tag{9.3-6}$$

$$= \text{Re}\{\Delta n'(\omega_1)\} + i\text{Im}\{\Delta n'(\omega_1)\},$$

where

$$\left.\begin{array}{l}\text{Re}\{\Delta n'(\omega_1)\} = \dfrac{K'}{2n_0(\omega_1)} \dfrac{\omega_{to} - (\omega_1 + \omega_2)}{[\omega_{to} - (\omega_1 + \omega_2)]^2 + \Gamma^2} |E_0(\omega_2)|^2 \\ \text{Im}\{\Delta n'(\omega_1)\} = \dfrac{K'}{2n_0(\omega_1)} \dfrac{\Gamma}{[\omega_{to} - (\omega_1 + \omega_2)]^2 + \Gamma^2} |E_0(\omega_2)|^2\end{array}\right\}. \tag{9.3-7}$$

In the absorbing medium, the electric-field change of the beam of frequency ω_1 along the propagation direction z can be written as

$$E(\omega_1, z) = E_0(\omega_1) e^{-i(\omega_1 t - k_1 z)}$$

$$= E_0(\omega_1) e^{-i\{\omega_1 t - \frac{2\pi}{\lambda_1}[n_0(\omega_1) + \text{Re}\Delta n'(\omega_1) + i\text{Im}\Delta n'(\omega_1)]z\}} \tag{9.3-8}$$

$$= E_0(\omega_1) e^{-i\{\omega_1 t - \frac{2\pi}{\lambda_1}[n_0(\omega_1) + \text{Re}\Delta n'(\omega_1)]z\}} e^{-\frac{2\pi}{\lambda_1}\text{Im}\Delta n'(\omega_1)z},$$

where λ_1 is the wavelength in a vacuum of this light beam. We then obtain

$$|E(\omega_1,z)|^2 = |E_0(\omega_1)|^2 e^{-\frac{4\pi}{\lambda_1}\text{Im}\Delta n'(\omega_1)z}. \tag{9.3-9}$$

Since $I(\omega_1,z) \propto |E(\omega_1,z)|^2$, from Eq. (9.3-9) we obtain

$$I(\omega_1,z) = I_0(\omega_1) e^{-\frac{4\pi}{\lambda_1}\text{Im}\Delta n'(\omega_1)z} = I_0(\omega_1) e^{-B_1 z}, \tag{9.3-10}$$

where the nonlinear attenuation coefficient B_1 (in units of cm^{-1}) due to 2PA is given by

$$B_1 = \frac{4\pi}{\lambda_1}\text{Im}\{\Delta n'(\omega_1)\} = \frac{2\pi K'}{\lambda_1 n_0(\omega_1)}\frac{\Gamma}{[\omega_{to}-(\omega_1+\omega_2)]^2+\Gamma^2}|E_0(\omega_2)|^2. \tag{9.3-11}$$

Using the following relation (cf. Eq. (1.4-1)):

$$I_0(\omega_2) = \frac{1}{2}\varepsilon_0 c n_0(\omega_2)|E_0(\omega_2)|^2,$$

and substituting Eq. (9.3-5) into Eq. (9.3-11) leads to

$$B_1 = (N_o - N_t)\sigma_{2PA} I_0(\omega_2), \tag{9.3-12}$$

where σ_{2PA} is the molecular 2PA cross-section (in units of cm^4/W) determined by

$$\sigma_{2PA} = \frac{(2\pi)^3 \omega_1}{3c^2 \varepsilon_0^2 n_0^2 h^3}\left|\sum_b \left[\frac{(p_1)_{ob}(p_2)_{bt}}{\omega_{bo}-\omega_1} + \frac{(p_2)_{ob}(p_1)_{bt}}{\omega_{bo}-\omega_2}\right]\right|^2 \cdot \frac{\Gamma}{[\omega_{to}-(\omega_1+\omega_2)]^2+\Gamma^2}. \tag{9.3-13}$$

For a gas sample, $n_0 \approx 1$ and the 2PA effect is quite weak; the term $B_1 z$ is usually much smaller than 1 and Eq. (9.3-10) can be simplified as

$$I(\omega_1,z) = I_0(\omega_1)e^{-B_1 z} \approx I_0(\omega_1)[1-B_1 z]; \tag{9.3-14}$$

then the relative absorption of this beam as a function of $(\omega_1+\omega_2)$ should be

$$\Delta I(\omega_1)/I_0(\omega_1) \approx B_1 z \propto \frac{\Gamma}{[\omega_{to}-(\omega_1+\omega_2)]^2+\Gamma^2}I_0(\omega_2)z. \tag{9.3-15}$$

Following the same procedure, we could obtain a similar expression for another beam of ω_2:

$$\Delta I(\omega_2)/I_0(\omega_2) \approx B_2 z \propto \frac{\Gamma}{[\omega_{to}-(\omega_1+\omega_2)]^2+\Gamma^2}I_0(\omega_1)z. \tag{9.3-16}$$

From the above two expressions we can see that the two-beam induced 2PA exhibits a natural spectral lineshape without considering the Doppler effect and other broadening mechanisms.

Considering the Doppler effect in a gas sample, the probability for a molecule to make a transition from the lower state o to the upper real state t via 2PA can be expressed as

$$W_{ot} \propto I_0(\omega_1)I_0(\omega_2)\frac{\Gamma}{[\omega_{to}-(\omega_1+\omega_2)+(\mathbf{k}_1+\mathbf{k}_2)\cdot\mathbf{v}]^2+\Gamma^2}, \tag{9.3-17}$$

where $\mathbf{k}_1$ and $\mathbf{k}_2$ are the wavevectors of the two beams and $\mathbf{v}$ is the velocity vector of the molecule. If these two beams are counter-propagating through the gas sample and their frequency difference is small, we have $\mathbf{k}_1 \approx -\mathbf{k}_2$ and Eq. (9.3-17) becomes

$$W_{ot} \propto I_0(\omega_1) I_0(\omega_2) \frac{\Gamma}{[\omega_{to} - (\omega_1 + \omega_2)]^2 + \Gamma^2}. \tag{9.3-18}$$

This means that under the excitation of two counter-propagating beams the 2PA transitions of different molecules are independent of their velocity distribution; therefore, the Doppler broadening can be eliminated.

To reduce the Doppler-broadening background originating from 2PA of one beam by itself, researchers may properly choose frequencies for these two beams to ensure $(\omega_1 + \omega_2) = \omega_{to}$, but $2\omega_1 \neq 2\omega_2 \neq \omega_{to}$.

The selection rules of 2PA transitions require that in the electric-dipole approximation, the initial and final states of a 2PA transition must be of the same parity. In particular, for the transitions between two atomic S-states, the atomic angular momentum should remain unchanged. In this case, even the two counter-propagating laser beams are of the same frequency but have two opposing circular polarization senses, only the 2PA from the two beams is allowed, and the 2PA from any one beam is forbidden. This is one of the methods to remove or reduce the background influence from one-beam 2PA.[1,2]

For Doppler-free two-photon spectroscopy, the two counter-propagating laser beams can be perfectly parallel; therefore the angular or geometric broadening is negligible. By increasing the size of the light beams or by using a directional atom (or molecule) beam running along the axial direction of the light beams, the transit-time broadening can also be decreased. Nevertheless, other influences such as collision broadening, recoil broadening, and second-order Doppler broadening still exist.

9.3.3 Experimental studies

Doppler-free two-photon spectroscopy can be used to study the hyperfine structure as well as the Stark or Zeeman spectral splitting of atomic or molecular gas samples. In comparison with saturation spectroscopy, the two-photon spectroscopy has some unique features: (1) it can be employed to investigate the high excited states including those prohibited by the one-photon transition; (2) the two counter-propagating laser beams interact with all atoms having different velocities; and (3) the 2PA resonance can be determined conveniently by detecting the fluorescence emission rather than directly measuring the transmission of a probe laser beam.

Figure 9.10 shows the schematic setup for Doppler-free two-photon spectroscopic measurements. It is composed of the following basic parts: (i) a tunable laser source, (ii) a beam-reflecting system to provide the backward running beam, (iii) a gas sample cell with optical windows, and (iv) a detection system for the two-photon-induced fluorescence.

As an example of early experimental results, Fig. 9.11 shows the measured hyperfine-structure spectrum of the 3S–4D transition of Na atoms by using Doppler-free two-photon spectroscopy.[43] The laser source was a single-mode cw dye laser working at 578.73 nm, and the two counter-propagating laser beams were focused into a sample cell filled with Na vapor of 3×10^{-7} Torr. The 330.2-nm ultraviolet fluorescence emission from the excited atoms through the 4P–3S transition was detected from the side window of the cell. From Fig. 9.11 one can clearly see the hyperfine splitting of the 3S-state ($F = 2, 1$) corresponding to the separation of a–c lines and b–d lines, and the fine-structure splitting of the 4D-state ($J = 5/2, 3/2$) corresponding to the separation of

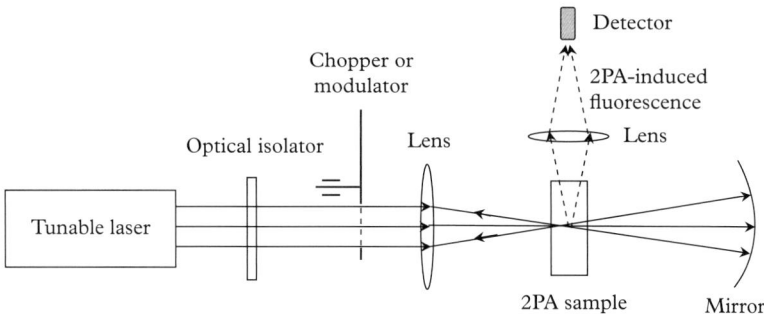

Figure 9.10 *Experimental setup for Doppler-free 2PA spectroscopy.*

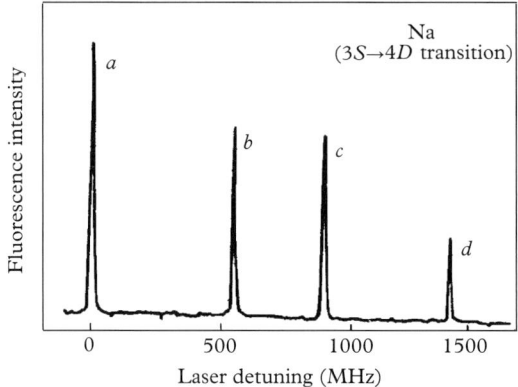

Figure 9.11 *Doppler-free 2PA spectrum of the 3S–4D transition of Na atoms. The hyperfine splitting of the 3S ground state and the fine structure of the 4D-state are resolved. (Reproduced with permission from Hänsch et al.[43] Copyright 1974, Elsevier).*

a–b lines and c–d lines. The minimum linewidth was measured as ∼15 MHz (less than 1% of the Doppler width).

Another experimental example is the Doppler-free two-photon spectroscopic study of the 1S–2S transition of atomic hydrogen and deuterium using the pulsed dye laser beams.[45,46] In addition, the same two-photon spectroscopic method was used to study the Zeeman splitting and Stark splitting of the 3S–5S transition and 3S–4D transition of Na atoms.[44,47,48]

After the initial stage, this Doppler-free spectroscopic technique has been continuously employed to study the gas samples of atoms or molecules including (but not limited) H and D,[49,50] alkali-metal atoms of Na,[51] Rb,[52–54] and Cs,[52] as well as polyatomic molecules.[55,56]

As a recent example of the 2PA spectroscopic study of Rb vapor, the Doppler-free spectrum of $5S_{1/2} \rightarrow 7S_{1/2}$ two-photon transitions are shown in Fig. 9.12, as measured by using the ∼760-nm beam from a tunable external-cavity diode laser. The vapor temperature and pressure were ∼110 °C and ∼0.5 mTorr, respectively, providing an atom density of ∼5×10^{21} cm^{-3}.[53]

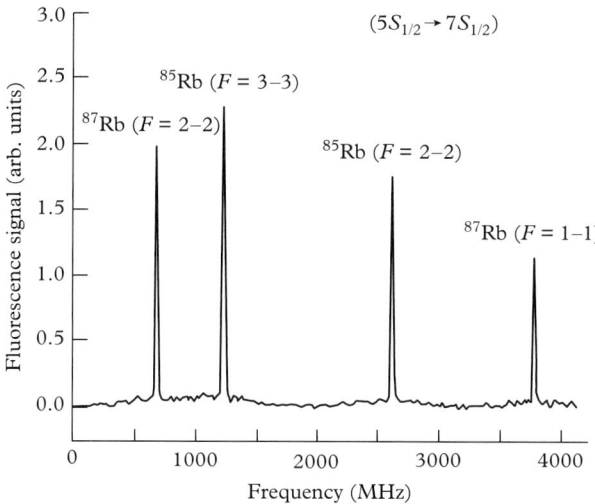

Figure 9.12 Doppler-free 2PA spectrum of $5S_{1/2}$–$7S_{1/2}$ transitions for two Rb isotopes. (Reproduced with permission from Ko and Liu.[53] Copyright 2004, Optical Society of America).

It is shown in Fig. 9.12 that the hyperfine splitting values for the two isotopes of ^{85}Rb and ^{87}Rb are 2754 and 6202 MHz, respectively. The measured linewidth of the ^{85}Rb $5S_{1/2}(F=3) \rightarrow 7S_{1/2}(F=3)$ transition was ∼3 MHz.

9.4 Coherent Raman spectroscopy

9.4.1 General description

The term "coherent Raman spectroscopy," in general, covers the following nonlinear spectroscopic techniques: (1) coherent anti-Stokes Raman spectroscopy, (2) Raman-induced Kerr effect spectroscopy, (3) Raman gain spectroscopy, and (4) inverse Raman spectroscopy. Techniques (1) and (2) are based on the Raman resonance-enhanced four-wave frequency mixing (FWFM) effects, and techniques (3) and (4) are based on the stimulated Raman scattering (SRS) mechanism.

With respect to experiments, these techniques have some common features. First, there must be two laser beams with different frequency, and at least one of them should be tunable. Second, the frequency difference of these two beams can be tuned to one or several different Raman resonance frequencies of a given sample medium. Finally, the spectral information of the sample is obtained either by measuring one transmitted laser beam or to detect a newly generated coherent emission beam. Since all interacting light beams are coherent and the Raman resonance is involved for all these techniques, they can be generally termed as coherent Raman spectroscopy that differs from conventional Raman spectroscopy.

The coherent Raman spectroscopic techniques are highly useful for studying the Raman spectra of various media (including gases, solids, liquids, plasmas, flames, surfaces, waveguides, and films). Using these new spectral techniques, researchers can acquire valuable data and a great deal of information about the structures of energy levels, transition properties, lifetime of levels, and relaxation behavior.

9.4.2 Coherent anti-Stokes Raman spectroscopy (CARS)

Coherent anti-Stokes Raman spectroscopy (CARS) is a special Raman enhanced FWFM effect. To observe this effect, two laser beams having different frequency are incident upon a Raman medium with a certain crossing angle; by scanning the frequency difference of these two beams across the Raman resonant range of the medium, a Raman-enhanced coherent anti-Stokes emission could be observed in the phase-matched direction.

During the early study of SRS, it was found that in addition to the transmitted pump beam of frequency ω_p and the Stokes-shifted SRS beam of frequency ω_s, one could also observe a ring-shape coherent emission at the anti-Stoles shifted frequency of ω_{as}.[57] The angle diameter of the ring emission is determined by the phase-matching requirement of this special FWFM process (see Section 7.3.3). It was further found that using this effect one may measure the Raman spectral structure and the corresponding Raman-enhanced third-order nonlinear susceptibility of a sample medium.[58-62] This type of nonlinear spectroscopic study has been termed "coherent anti-Stokes Raman spectroscopy," abbreviated as CARS,[63] and is commonly accepted by the research community.[4,64-67] However, subsequently, in some of the literature, the term of CARS is also interpreted as "coherent anti-Stokes scattering." The same abbreviation with two different interpretations may actually cause certain ambiguity. According to its original definition, the CARS effect is a Raman resonance-enhanced FWFM process, or a Raman resonance-enhanced four-photon parametric interaction process, which should meet the phase-matching requirement. At the beginning and the end of this process, the molecules of a Raman medium always stay in the same ground state and no real Raman scattering process is involved in any step. In contrast, the real Raman scattering may occur along any direction without phase-matching requirement, and it is always accompanied by the molecular transition from the initiate state to a different final state in which the excited molecules will stay for certain period of time (see Section 7.2.1). For a rigorous CARS process, the four waves involved interact with each other at the same time, i.e., none of them should be delayed.

In the case of CARS, it is assumed that there is a laser beam of frequency ω_1 and a weaker laser beam of frequency ω_2, with the frequency relation $\omega_1 \cong \omega_2 + \Delta\omega_r$, where $\Delta\omega_r$ is a Raman-mode frequency of the sample medium. When these two beams are incident upon the Raman medium with an appropriate small crossing angle θ_0, a coherent emission at anti-Stokes frequency ω_{as} can be generated through the Raman resonance-enhanced third-order nonlinear polarization process. The schematic quantum-transition diagram and the phase-matching geometry of this process are shown in Fig. 9.13(a). The frequency ω_{as} and wavevector $\mathbf{k}_{as}$ of the coherent anti-Stokes Raman emission are determined by

$$\left.\begin{aligned}\omega_{as} &= 2\omega_1 - \omega_2 = \omega_1 + \Delta\omega_r \\ \mathbf{k}_{as} &= 2\mathbf{k}_1 - \mathbf{k}_2\end{aligned}\right\}, \tag{9.4-1}$$

where $\mathbf{k}_1$ and $\mathbf{k}_2$ are the wavevectors of the two incident beams. The physical essence of an elementary CARS process involves the annihilation of two photons of ω_1 and the simultaneous creation of a pair of photons with different frequencies of ω_2 and ω_{as}.

In another case, if the beam of ω_1 is weaker than the beam of ω_2, a coherent Stokes Raman emission can be generated in a new direction. This is the so-called coherent Stokes Raman

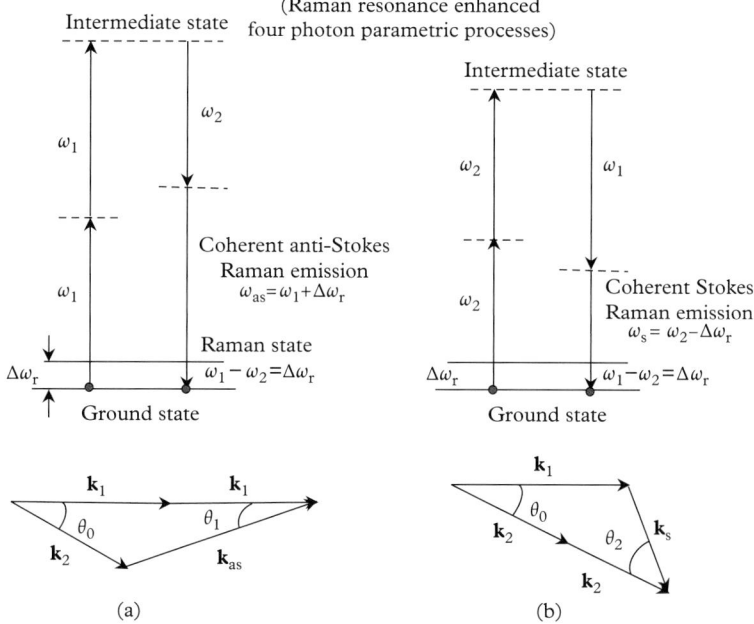

Figure 9.13 *The quantum-transition diagram and phase-matching condition for (a) CARS and (b) CSRS processes.*

spectroscopy (CSRS) process; its quantum-transition diagram and the phase-matching geometry are shown in Fig. 9.13(b). The frequency ω_s and wavevector $\mathbf{k}_s$ of the coherent Stokes Raman emission are determined by

$$\left.\begin{array}{l}\omega_s = 2\omega_2 - \omega_1 = \omega_2 - \Delta\omega_r \\ \mathbf{k}_s = 2\mathbf{k}_2 - \mathbf{k}_1\end{array}\right\}. \tag{9.4-2}$$

The physical essence of an elementary CSRS process involves the annihilation of two photons of ω_2 and the simultaneous creation of a pair of photons with different frequencies of ω_1 and ω_s.

If the value of $\Delta\omega_r$ is not too large, the three angles θ_0, θ_1, and θ_2 shown in Fig. 9.13 are nearly the same, and are determined by the $\Delta\omega_r$ value and the refractive-index dispersion of the Raman medium. For common transparent liquid and solid media, $\theta_0 \approx 1\text{--}3°$, whereas for gas samples, $\theta_0 \approx 0$.

If the two input laser beams have nearly the same intensity level, both CARS and CSRS signals can be observed along different phase-matched directions. This feature helps researchers to spatially separate the CARS or CSRS signal from the strong incident laser beams. On the other hand, under some circumstances the intense input laser beams may induce additional fluorescence emission in the Raman samples, which may bring an undesirable strong background on spectral measurements. However, considering that this fluorescence emission is usually located on the low-frequency side of the laser excitation frequency, a coherent anti-Stokes signal can be easily detected upon a higher signal-to-noise ratio. This is the reason why the CARS method is more interesting and useful than the CSRS method.

According to the nonlinear polarization theory, the third-order polarization component contributing to CARS processes in a Raman medium is

$$P^{(3)}(\omega_{as}) = \varepsilon_0 \chi_{as}^{(3)} E(\omega_1) E(\omega_1) E^*(\omega_2), \tag{9.4-3}$$

where $\chi_{as}^{(3)}$ is the CARS related third-order susceptibility of the medium. If the medium is isotropic and the two incident laser beams are linearly polarized along the same x-axis direction, following the same procedure as described in Section 4.2 for third-harmonic generation, we could then obtain the expression for the intensity of the CARS signal:

$$I_{as}(z_0) \propto \left|\chi_{e,as}^{(3)}\right|^2 I_1^2(0) I_2(0) z_0^2. \tag{9.4-4}$$

Here, $I_1(0)$ and $I_2(0)$ are the intensities of the two input beams, z_0 is the interaction length between the two beams inside the medium, and $\chi_{e,as}^{(3)}$ is the effective third-order susceptibility value for the CARS process. In obtaining Eq. (9.4-4), we have assumed that the depletion of the two input laser beams can be neglected and that the phase-matching requirement is met. Furthermore, $\chi_{e,as}^{(3)}$ can be written as

$$\chi_{e,as}^{(3)} = \chi_{e,NR}^{(3)} + \chi_{e,R}^{(3)}, \tag{9.4-5}$$

where the first term represents the non-resonant contribution from the electron-cloud distortion mechanism and is a real quantity independent of frequency; the second term represents the Raman-enhanced contribution, which is a complex quantity and is a sensitive function of the frequency difference of the two input beams. The explicit expression for the second term is given in Section 18.4 by Eq. (18.4-12) and can be rewritten here as

$$\chi_{e,R}^{(3)}(\omega_1, \omega_1, -\omega_2) = \frac{N_0}{\varepsilon_0 6 \hbar^3} \left| \sum_b \left[\frac{(p_x)_{ab}(p_x)_{bc}}{\omega_{ba} - \omega_1} + \frac{(p_x)_{ab}(p_x)_{bc}}{\omega_{ba} + \omega_2} \right] \right|^2$$
$$\times \frac{1}{\Delta\omega_r - (\omega_1 - \omega_2) - i\Gamma_r}, \tag{9.4-6}$$

where N_0 is the molecular density in the ground state and Γ_r is the linewidth factor of the Raman transition. Using Eq. (7.2-21) for the expression of Raman cross-section, σ_r, Eq. (9.4-6) is simplified to

$$\chi_{e,R}^{(3)}(\omega_1, \omega_1, -\omega_2) = N_0 \sigma_r \frac{\varepsilon_0 (2\pi)^3 c^4}{\hbar \omega_2^4} \frac{1}{\Delta\omega_r - (\omega_1 - \omega_2) - i\Gamma_r}. \tag{9.4-7}$$

From Eqs. (9.4-4) and (9.4-5) we can see that if the non-resonant term that gives a continuous spectral background contribution is negligible, the intensity of the CARS signal is simply proportional to the square of N_0 and σ_r, respectively. In addition, the spectral intensity distribution of the CARS signal will be

$$I_{as}(\omega_1 - \omega_2) \propto \left|\chi_{e,NR}^{(3)} + \text{Re}\{\chi_{e,R}^{(3)}\}\right|^2 + \left|\text{Im}\{\chi_{e,R}^{(3)}\}\right|^2, \tag{9.4-8}$$

where

$$\begin{aligned}\operatorname{Re}\{\chi_{e,R}^{(3)}(\omega_1,\omega_1,-\omega_2)\} &= N_0\sigma'\frac{\varepsilon_0(2\pi)^3 c^4}{\hbar\omega_2^4}\frac{\Delta\omega_r-(\omega_1-\omega_2)}{[\Delta\omega_r-(\omega_1-\omega_2)]^2+\Gamma_r^2}\\ \operatorname{Im}\{\chi_{e,R}^{(3)}(\omega_1,\omega_1,-\omega_2)\} &= N_0\sigma'\frac{\varepsilon_0(2\pi)^3 c^4}{\hbar\omega_2^4}\frac{\Gamma_r}{[\Delta\omega_r-(\omega_1-\omega_2)]^2+\Gamma_r^2}\end{aligned}\right\}. \quad (9.4\text{-}9)$$

The term $\chi_{e,NR}^{(3)}$ is usually a positive constant over a small frequency range, but the term $\operatorname{Re}\{\chi_{e,R}^{(3)}\}$ changes its amplitude and sign across the Raman resonance tuning range; hence, the CARS spectra for some samples may manifest a spectral profile like an unsymmetrical dispersion curve.[59–63]

Figure 9.14 shows the typical experimental setup for CARS, as well as for CSRS or some other coherent Raman spectroscopic measurements described later. There must be two laser sources, and one of them is tunable. To ensure a higher local intensity two laser beams of ω_1 and ω_2 are usually focused into the sample medium with a small phase-matched crossing angle θ_0. The spatially separated CARS (or CSRS) beam can be detected directly, or passes through a spectral filtering system prior entering the opto-electronic detectors. After the electronic amplifier system, the detected CARS (or CSRS) signal can be recorded as a function of the frequency-difference tuning of these two input laser beams.

As an example of CARS experimental results, Fig. 9.15(a) shows the measured spectra of the CARS signal in the 1:1 liquid mixture of benzene and cyclohexane, excited with two laser beams (one is tunable) from two dye laser sources.[62] Both benzene and cyclohexane have a very strong Raman mode at 992 cm^{-1} and 802 cm^{-1}, respectively, and are shown in Fig. 9.15(b) by the spontaneous Raman scattering spectra (the solid-line curve), obtained with a 514.5-nm Ar$^+$ laser line excitation. It is known that the spectral profile of spontaneous Raman scattering is proportional to the imaginary part of the Raman-enhanced third-order nonlinear susceptibility, and the real part of the latter could be theoretically determined by virtue of Kramers–Kronig relations and is shown in Fig. 9.15(b) by the dotted curve. The non-resonant contributions of the third-order nonlinear susceptibility for these two liquids can be determined by the best fits of the measured CARS curves, and are shown by the two horizontal lines in Fig. 9.15(b).

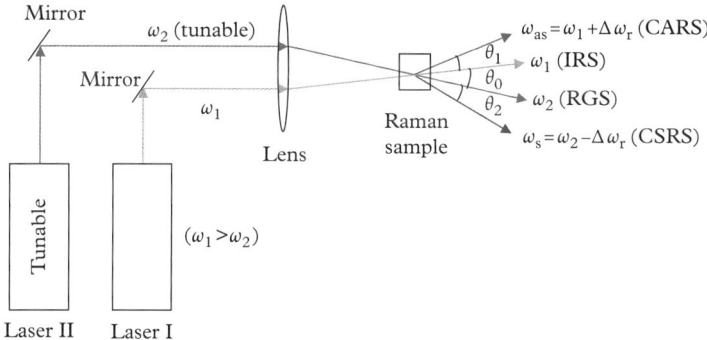

Figure 9.14 *Experimental setup for observing various coherent Raman spectroscopic effects.*

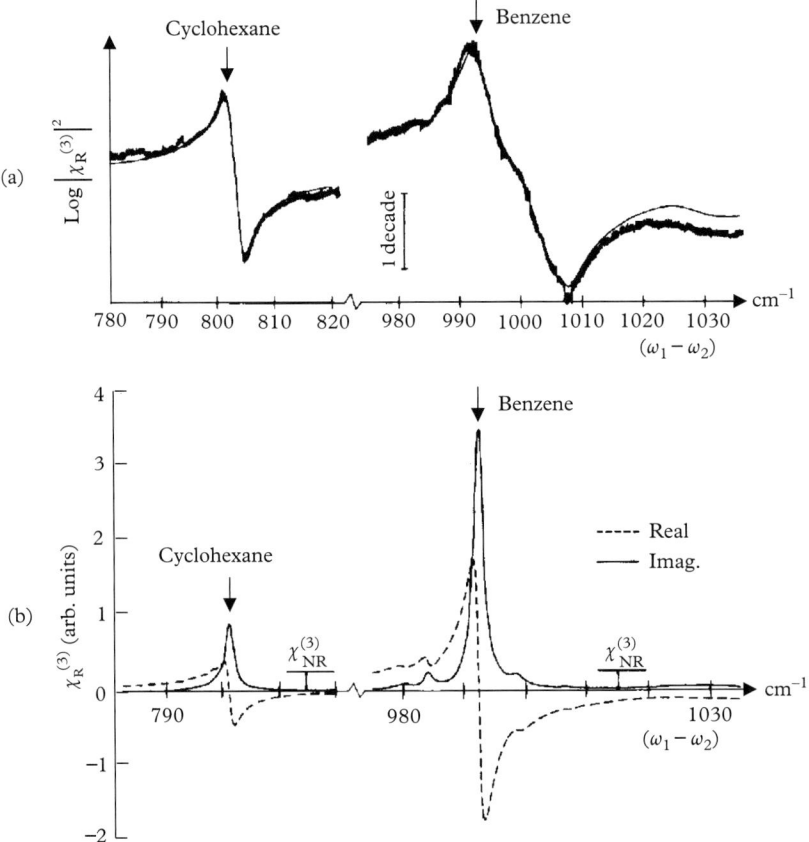

Figure 9.15 (a) Experimental spectral curves (thick solid line) and theoretical fits (thin solid line) of CARS signal for a 1:1 benzene–cyclohexane mixture; (b) spontaneous Raman scattering profile (solid line), corresponding real part (dashed line) of Raman susceptibility, and the estimated value of non-resonant contribution $\chi_{NR}^{(3)}$ by the best fits. (Reproduced with permission from Lotem et al.[62] Copyright 1976, American Physical Society).

From Eq. (9.4-8) and the experimental example shown in Fig. 9.15, we can see that for liquid and solid samples when the non-resonant contribution is comparable with the real part of the resonant Raman susceptibility, the observed CARS spectra may differ remarkably from those obtained by ordinary Raman spectroscopy. This is the major drawback of CARS technique. However, for gas media the non-resonant contribution is often negligible, and the peak positions of the CARS spectra can directly indicate the Raman-mode positions of the tested samples with a higher spectral resolution that is mostly limited by the laser spectral linewidth, because in this case the Doppler effect has no influence on the Raman resonance behavior of the sample medium.

For gas media, the molecular density N_0 is much lower than that of liquids and solids. However, this reduction can be compensated by great increase of the interaction length. Since

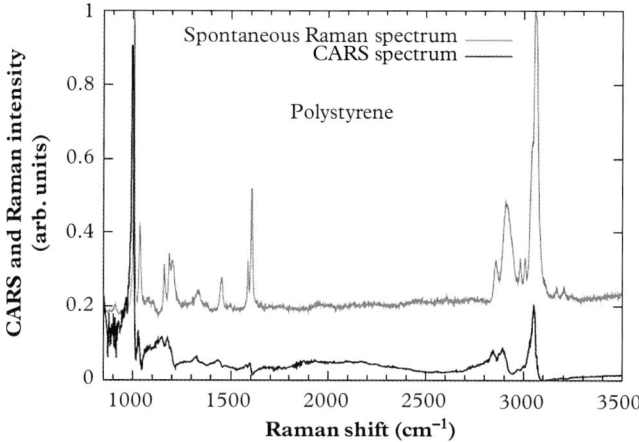

Figure 9.16 *The Raman spectral curves of polystyrene recorded by ordinary spontaneous Raman scattering method and CARS method separately. The ordinary Raman spectrum is arbitrarily offset for clarity. (Reproduced with permission from Chimento et al.[68] Copyright 2009, IOP Publishing).*

the refractive-index dispersion in a small frequency range is negligible in a gas medium and the phase-matching angle is $\theta_0 \approx 0$, the collinear interaction between the two input laser beams is available. Although most CARS measurements in gases are done with the vibrational or vibration-rotational Raman transitions, the same technique can also apply to measure the pure rotational Raman transitions of gas samples.[64]

To demonstrate the different Raman spectral appearances of a solid sample measured by the ordinary Raman method and the CARS method, Fig. 9.16 shows the Raman spectra over the Raman shift range 850–3500 cm^{-1} for a polystyrene polymer sample measured by these two methods separately.[68] It can be seen from this figure that some spontaneous Raman peaks are missed in the CARS spectrum, and there is a shift of some other weaker peaks of the latter compared to the former. In the sense of spectral details and fidelity of Raman-mode peak positions, the CARS method seems not beneficial very much in comparing with the ordinary Raman spectroscopy.

Nevertheless, as a typical Raman-enhanced FWFM process, the study of CARS processes had some significance in the early fundamental research of nonlinear optics. From the application viewpoint, CARS has the advantages of high sensitivity, high spectral resolution, high spatial resolution (with focused laser beams), and high temporal resolution (with narrow laser pulses). It can be effectively utilized to investigate flames, combustion materials, plasmas, gas discharges, chemical-reaction products, and various fluorescent samples. This technique can also be used to determine the Raman cross-section, the real and imaginary parts of the Raman susceptibility, as well as the concentration and thermal (temperature) distribution of the analyzed species.[3,4,65–67]

9.4.3 Raman-induced Kerr effect spectroscopy (RIKES)

In Section 5.5 we have described the Raman resonance-enhanced refractive-index change due to two-beam coupling under the conditions that they are linearly polarized along the same direction and have a frequency difference near to a Raman-mode frequency of the medium. In a more general case, if one of these two beams is much stronger than the other and they have different initial polarization states, then the resonant refractive-index change induced by the stronger pump beam exhibits birefringence, which can be detected by measuring the change of polarization state of the weaker probe beam. By scanning the frequency difference of these two beams and recording simultaneously the polarization change of the probe beam, the Raman spectral structure of the sample medium can be obtained. This is the principle of the so-called Raman-induced Kerr effect spectroscopy (RIKES).[69,70]

To performance a RIKES experiment, we can still utilize the setup shown in Fig. 9.14, which is based on two laser sources and one laser is tunable. In the present case, however, the angle between the two laser beams in the medium can be arbitrarily small, because no phase matching is necessary here for measuring the induced refractive-index change.

We assume the ω_1 beam is a weaker probe beam than the tunable ω_2 pump beam, i.e., $\omega_1 - \omega_2 \approx \Delta\omega_r$. Both beams are propagating along the z-direction, and the probe beam is linearly polarized along the x-direction, whereas the pump beam is either linearly polarized at a 45° angle to the x-axis or circularly polarized. After passing through an isotropic Raman medium, the pump beam-induced polarization change of the probe beam can be detected by letting the transmitted probe beam further pass through a crossed linear polarizer. Under this special arrangement, there exist only two independent and nonzero elements of the third-order susceptibility (see Appendix 6), and the detected signal intensity of the y-component of the probe beam will be determined by[70,71]

$$I_{1y}(\omega_1) \propto \left| P_y^{(3)}(\omega_1) \right|^2$$
$$\propto \left| \chi_{yxyx}^{(3)}(\omega_2, -\omega_2, \omega_1) E_{2x} E_{2y}^* E_{1x} + \chi_{yyxx}^{(3)}(\omega_2, -\omega_2, \omega_1) E_{2y} E_{2x}^* E_{1x} \right|^2 \quad (9.4\text{-}10)$$
$$\propto \left| \chi_{yxyx}^{(3)} E_{2x} E_{2y}^* + \chi_{yyxx}^{(3)} E_{2y} E_{2x}^* \right|^2 |E_{1x}|^2.$$

where E_{2x} and E_{2y} are the Cartesian components of the pump field of frequency ω_2 and E_{1x} is the x-component of the probe field of frequency ω_1.

If the input pump wave is linearly polarized at a 45° angle to the x-axis, we have $E_{2x} = E_{2y}$ and Eq. (9.4-10) becomes

$$I_{1y} \propto \left| \chi_{yxyx}^{(3)} + \chi_{yyxx}^{(3)} \right|^2 |E_{2y}|^4 |E_{1x}|^2. \quad (9.4\text{-}11)$$

This expression reminds us of Eq. (9.4-4) for the CARS process; it means that the observed spectrum will be a Raman resonance structure superimposed on the non-resonant background.

For a circularly polarized pump beam, we have $E_{2x} = \pm i E_{2y}$ and Eq. (9.4-10) will be[71]

$$I_{1y} \propto \left| \chi_{yxyx}^{(3)} - \chi_{yyxx}^{(3)} \right|^2 |E_{2y}|^4 |E_{1x}|^2. \quad (9.4\text{-}12)$$

Note that both $\chi_{yxyx}^{(3)}$ and $\chi_{yxyx}^{(3)}$ have real non-resonant parts and complex resonant parts (cf. Eq. (9.4-5)); the non-resonant parts are independent of their Cartesian indices and thereby cancel each other. In this case, one could observe the Raman resonance structure without the influence from non-resonant background. Moreover, if the pump beam has an arbitrary elliptical polarization state, the shape of the observed spectral curve will be between these two extreme cases, as discussed above.

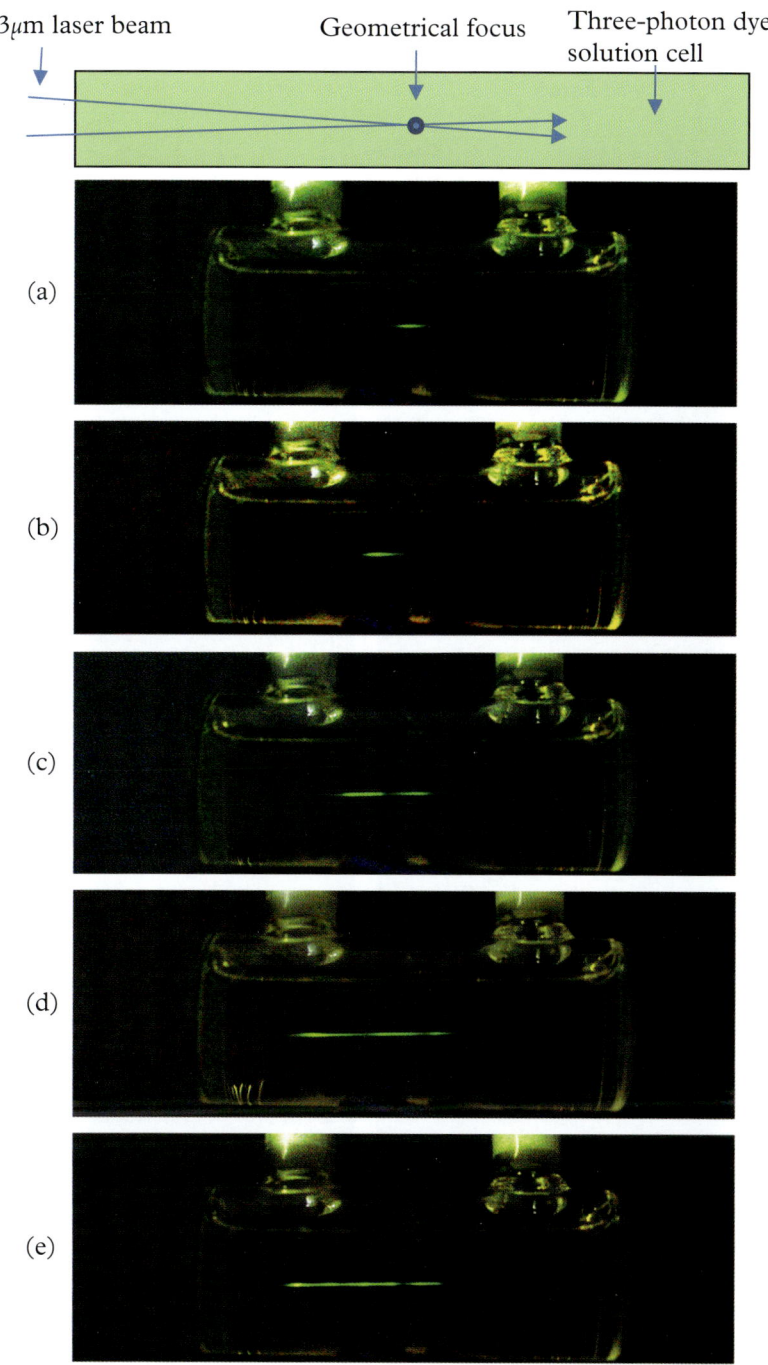

Color fig. 1. *Self-focusing behavior of ultrashort laser pulses at different intensity levels and in a three-photon dye solution, showing the focus moving (b), splitting (c), and multi-foci formation (d,e). (See the text in Section 6.2).*

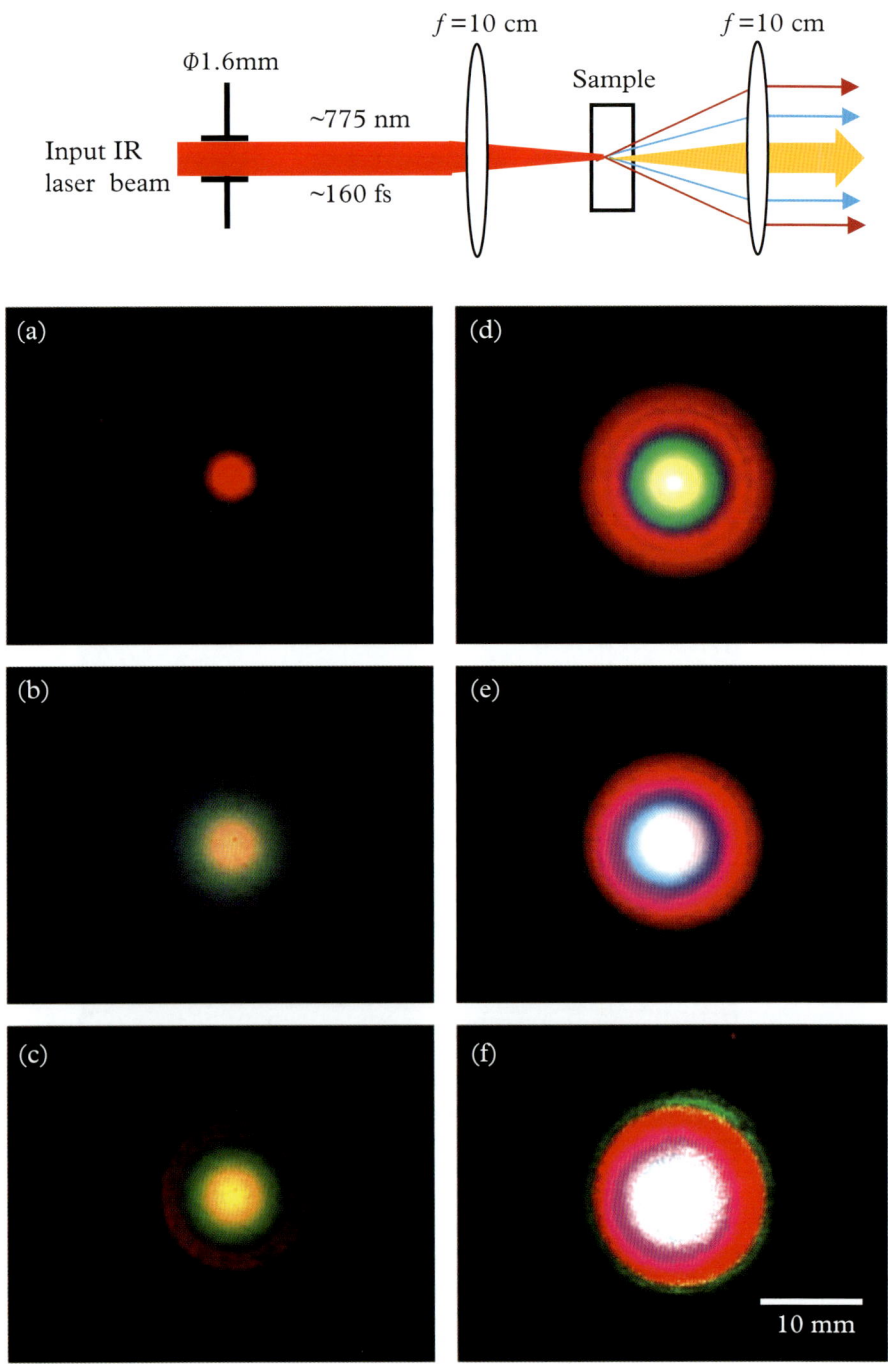

Color fig. 2. *Near-field patterns of supercontinuum generation in heavy water pumped by 775-nm and 160-fs laser pulses at different intensity levels. (See the text in Section 6.4).*

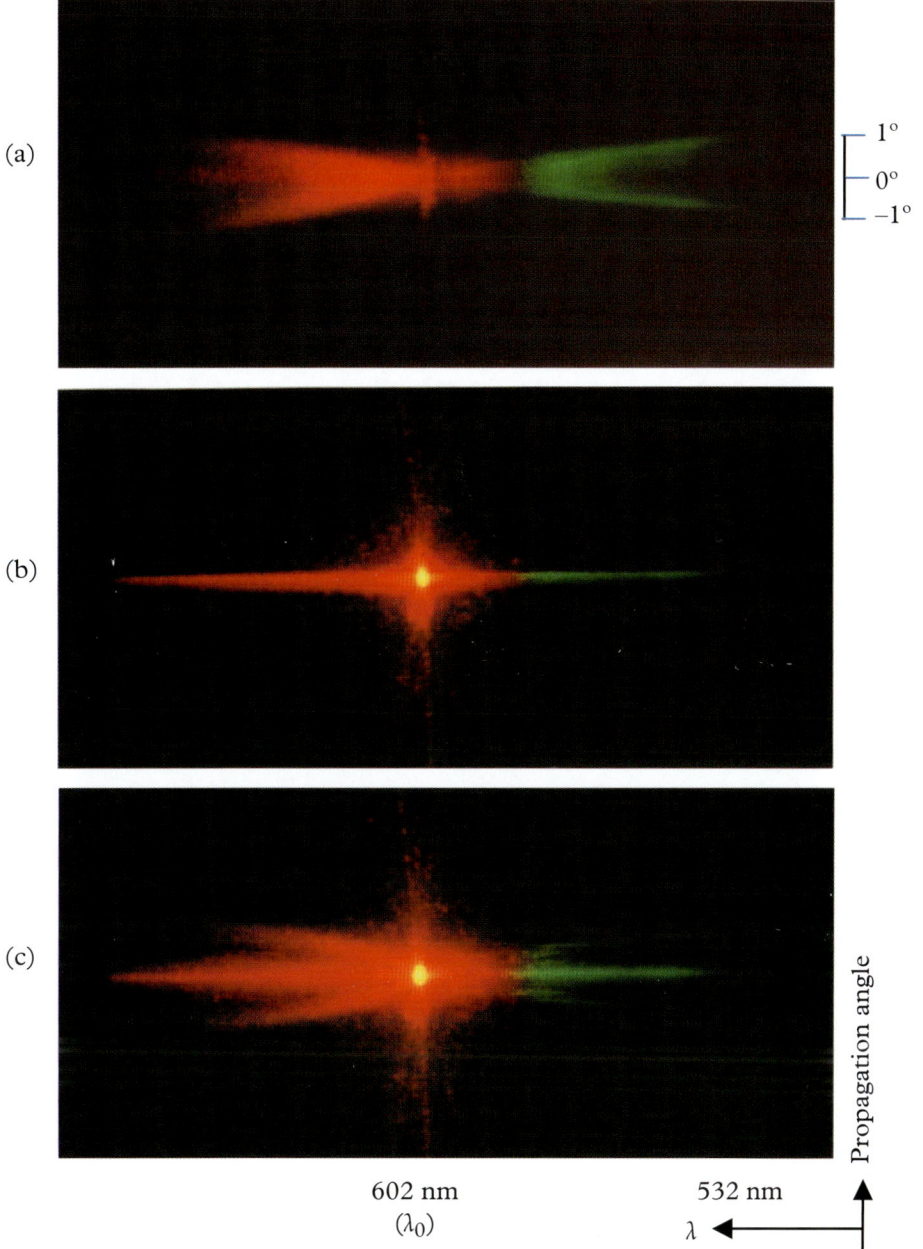

Color fig. 3. *Angle-dependent spectral structure of continuum generation in heavy water pumped by 602-nm and 0.5-ps laser pulses: (a) non-axial coherent emission due to four-wave mixing, (b) axial coherent emission due to self-broadening, and (c) overall coherent output emission. (See the text in Section 6.4.2).*

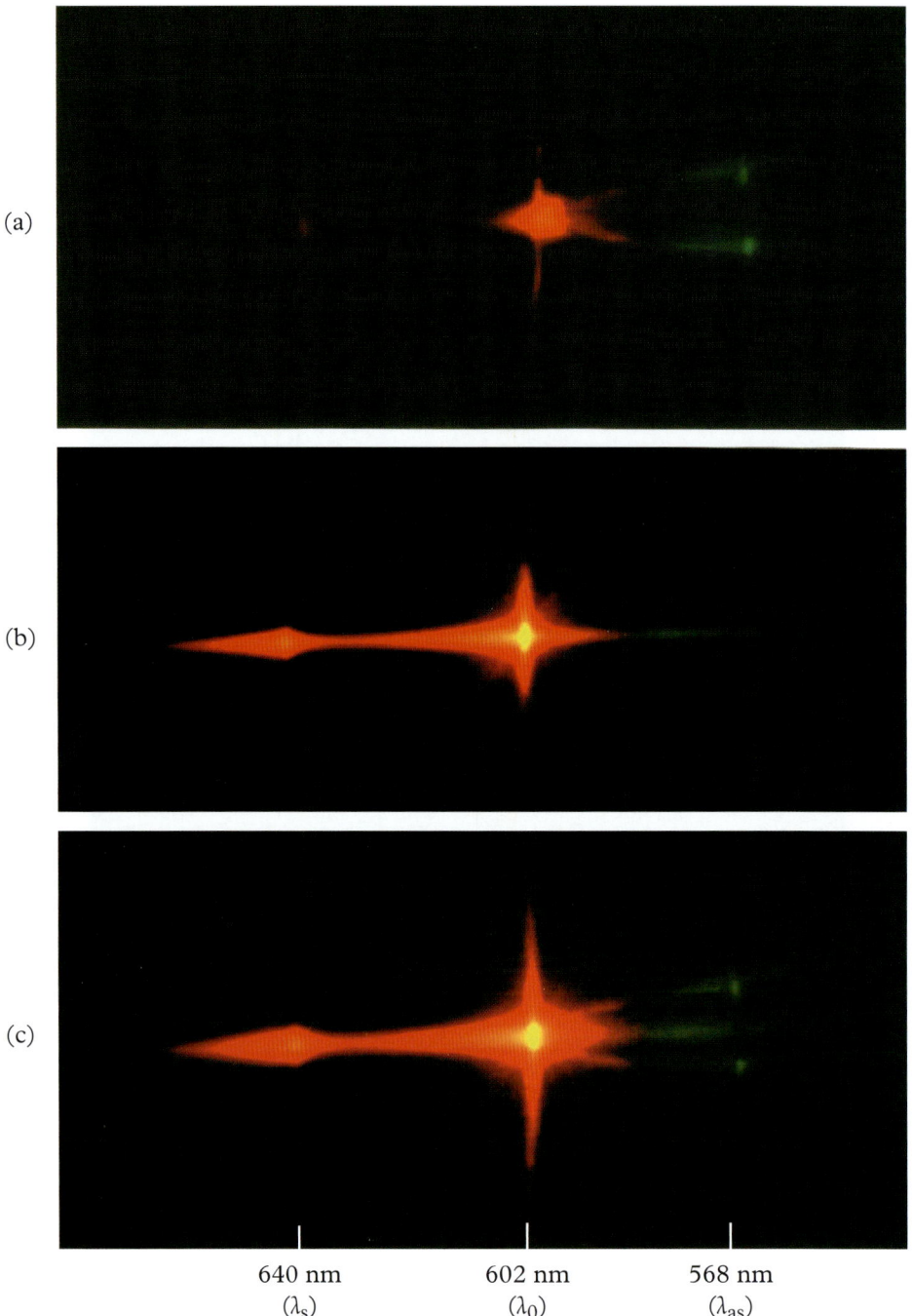

Color fig. 4. *Angle-dependent spectral structure of continuum generation in benzene. The others are the same as Color fig.3. (See the text in Section 6.4.2).*

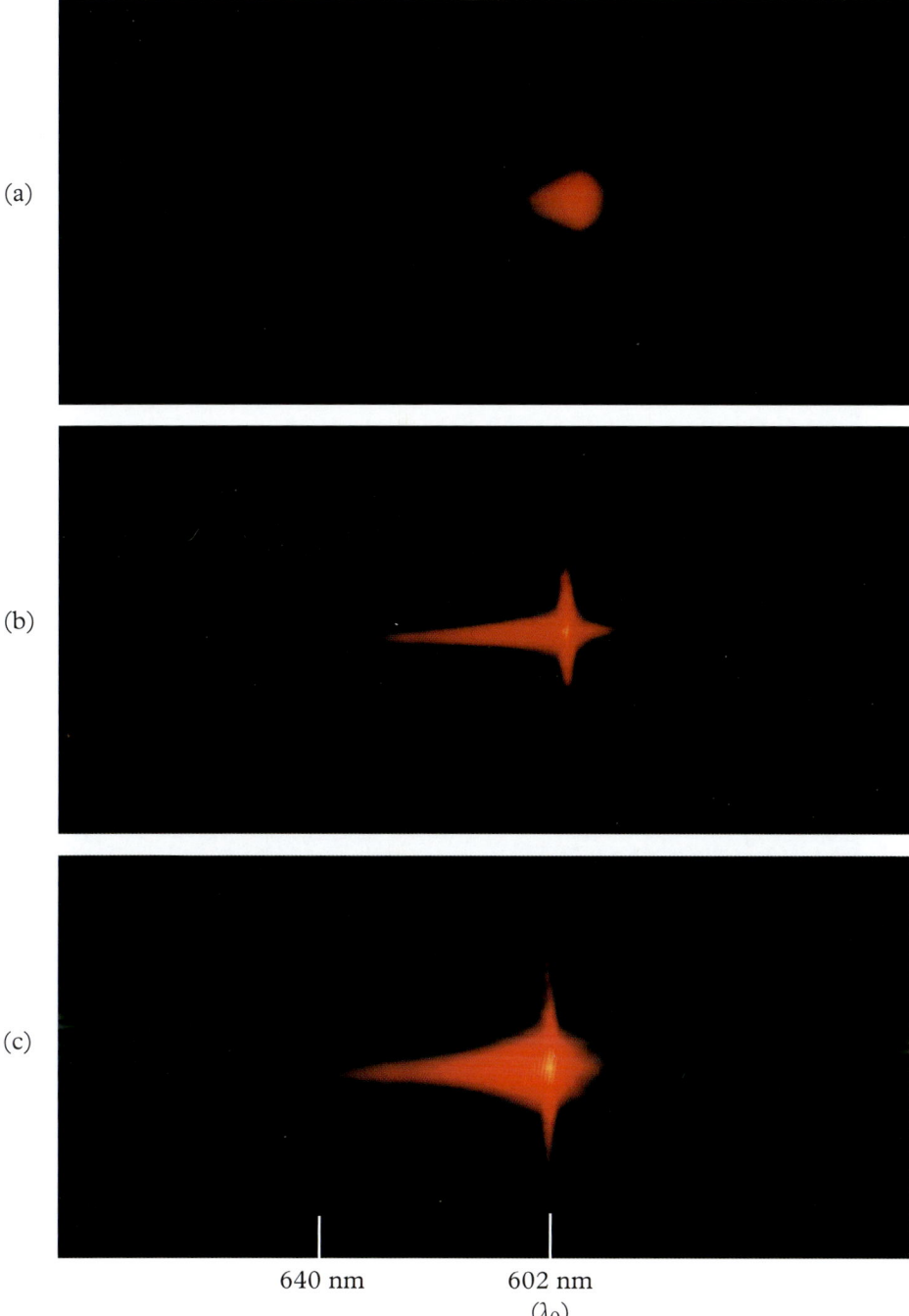

Color fig. 5. *Spectral structure of coherent emission in carbon disulfide. The others are the same as Color fig.3. (See the text in Section 6.4.2).*

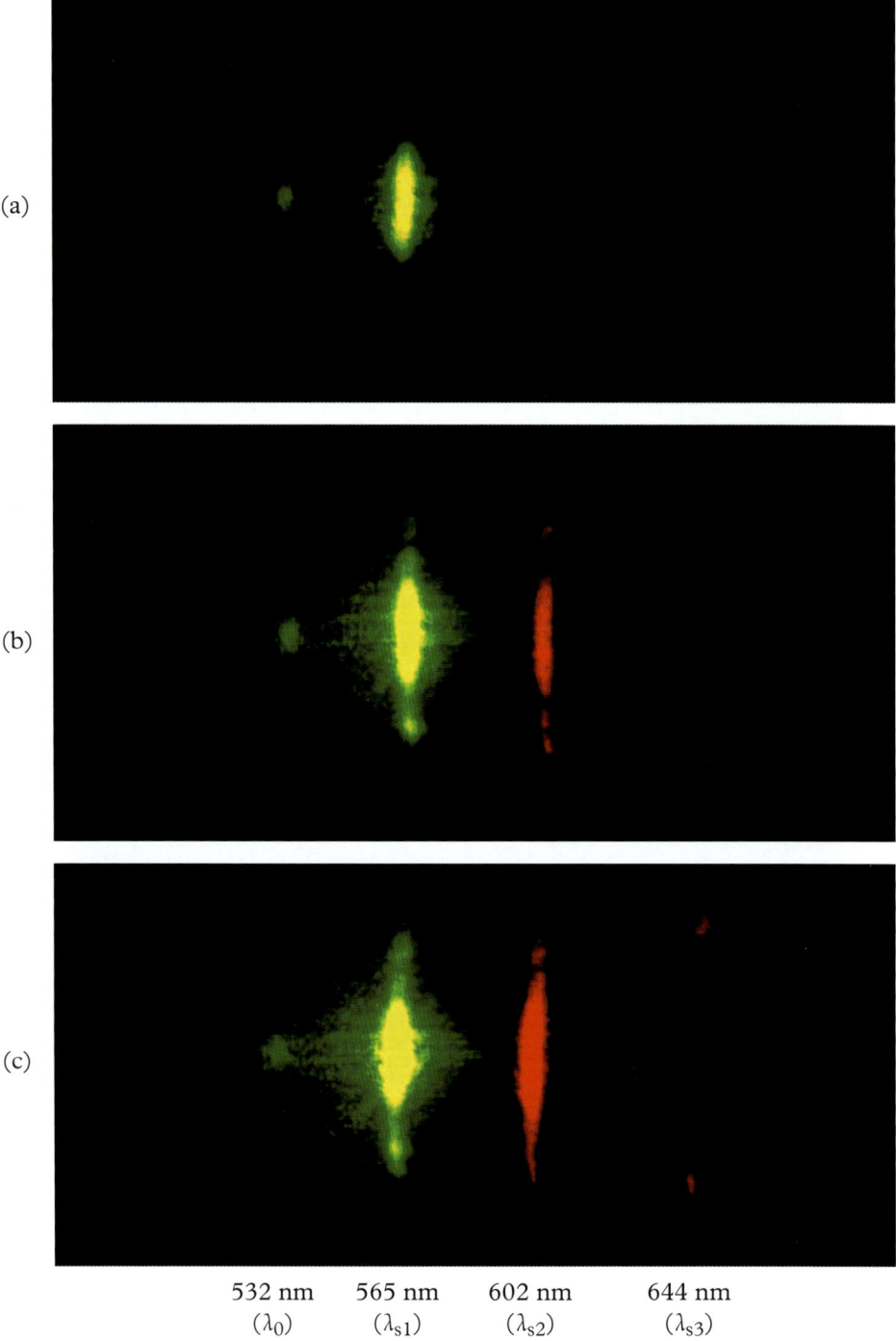

Color fig. 6. *Spectral structures of Stokes components of the axial stimulated Raman scattering (SRS) and the non-axial Raman-enhanced four-wave frequency mixing (FWFM) in calcite pumped by 532-nm and 8-ns laser pulses at three different intensity levels. (See the text in Section 7.3.3).*

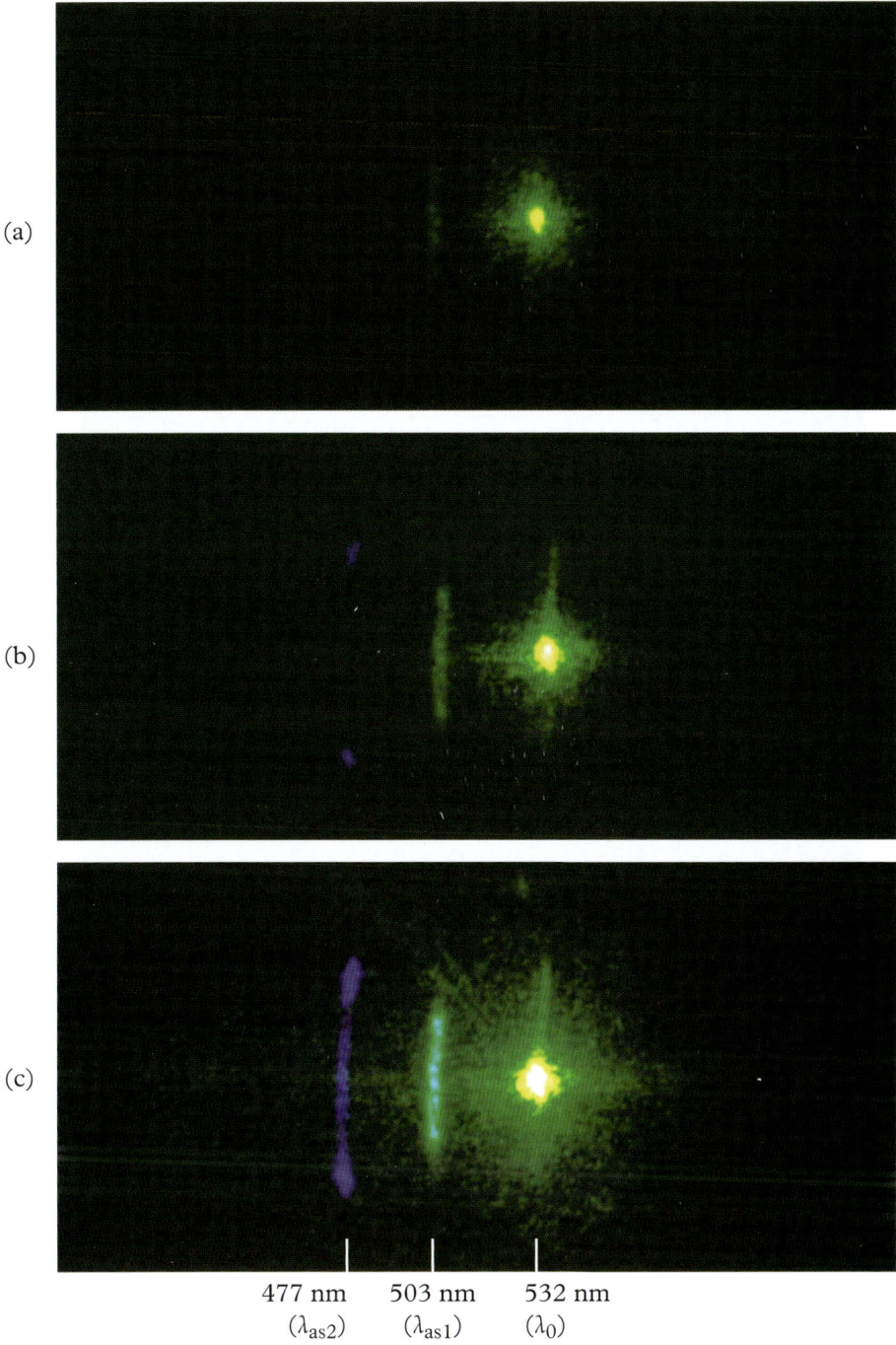

Color fig. 7. *Spectral structures of anti-Stokes components of SRS and FWFM in calcite. The others are the same as Color fig. 6. (See the text in Section 7.3.3).*

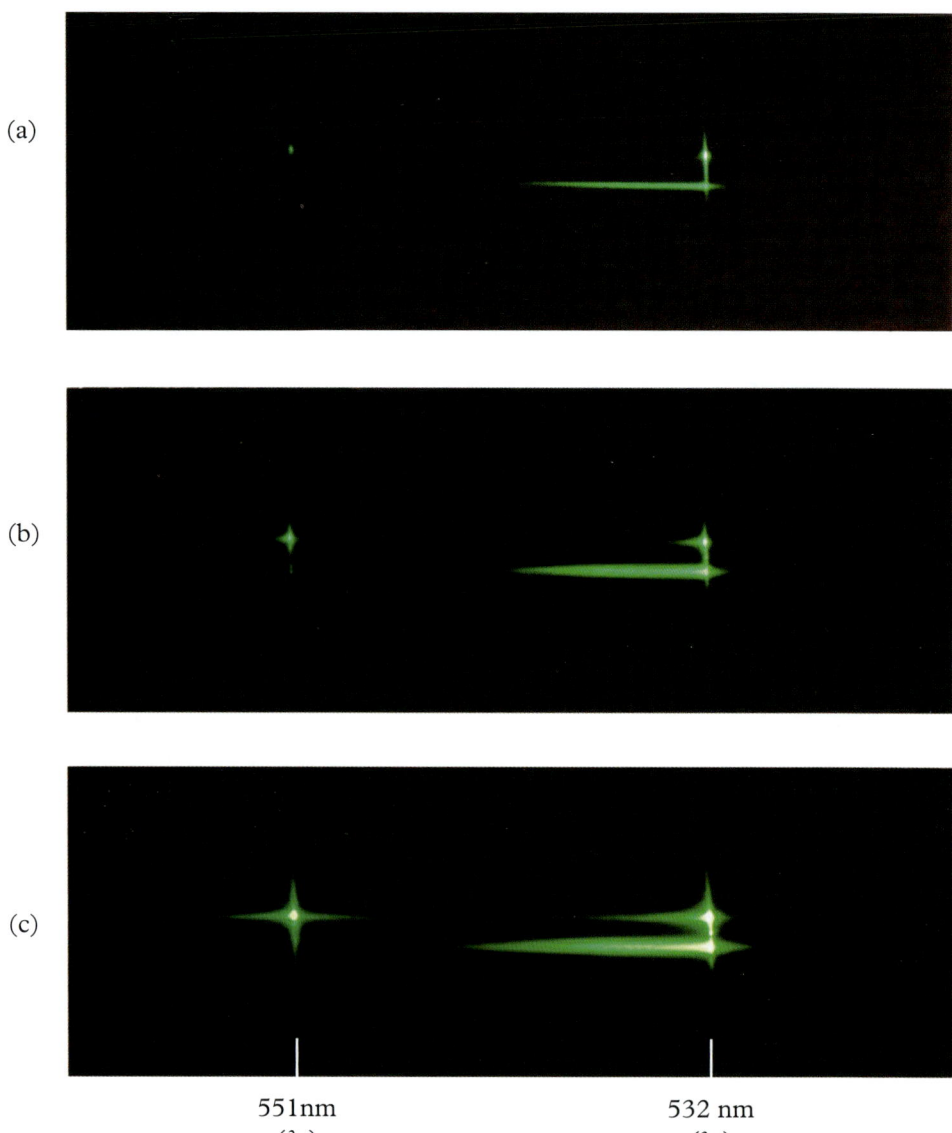

Color fig. 8. *Spectra of the forward SRS (upper trace) and the backward stimulated Rayleigh–Kerr scattering (SRKS) (lower trace) in 10-cm long CS_2 liquid cell, pumped by 532-nm and 10-ns laser pulses at three different intensity levels. (See the text in Section 7.5.7).*

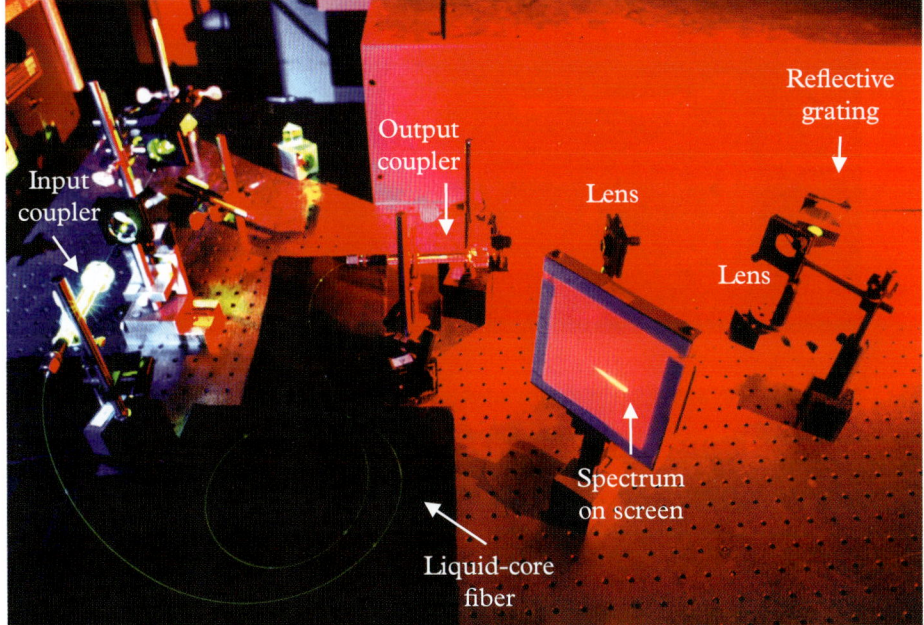

Color fig. 9. *Experimental setup (a) showing the cascaded stimulated Kerr scattering with a super-broad band spectrum output (b) from a CS_2-filled hollow fiber system, pumped by 532-nm and ~10-ns laser pulses. (See the text in Section 7.5.2).*

(a)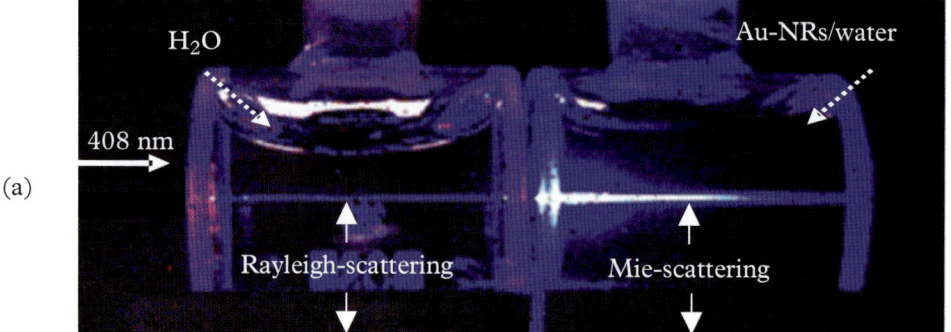

(b)

(c)

Color fig. 10. *Laser beam induced Rayleigh scattering in water (left cell) and Mie scattering in Au-nanorods suspended water (right cell) with three different incident laser wavelengths. The Rayleigh scattering sensitively depends on the incident wavelength, whereas the Mie scattering does not. (See the text in Section 7.1).*

(a)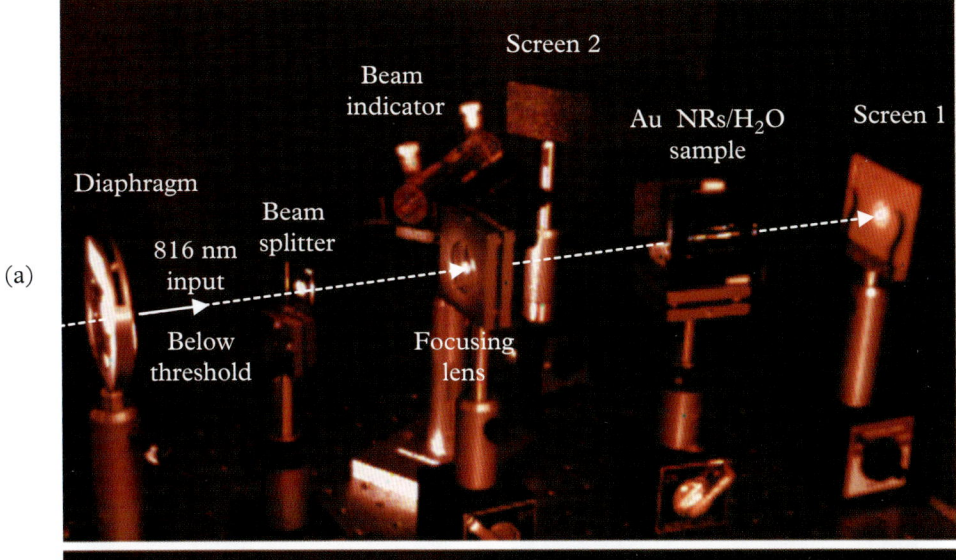

(b)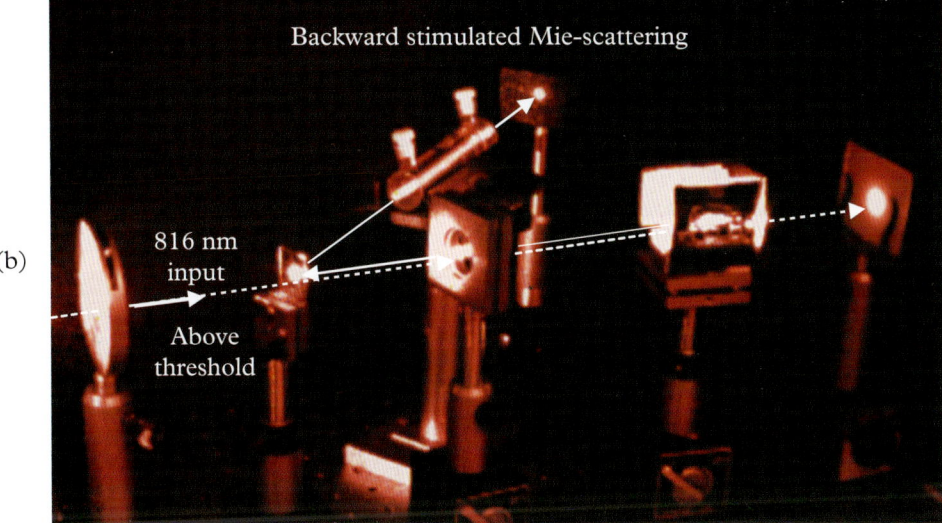

Color fig. 11. *Experimental demonstration for (a) spontaneous Mie scattering and (b) stimulated Mie scattering in Au-nanorods/water, pumped by ~816-nm and ~10-ns laser pulses. (See the text in Section 7.7).*

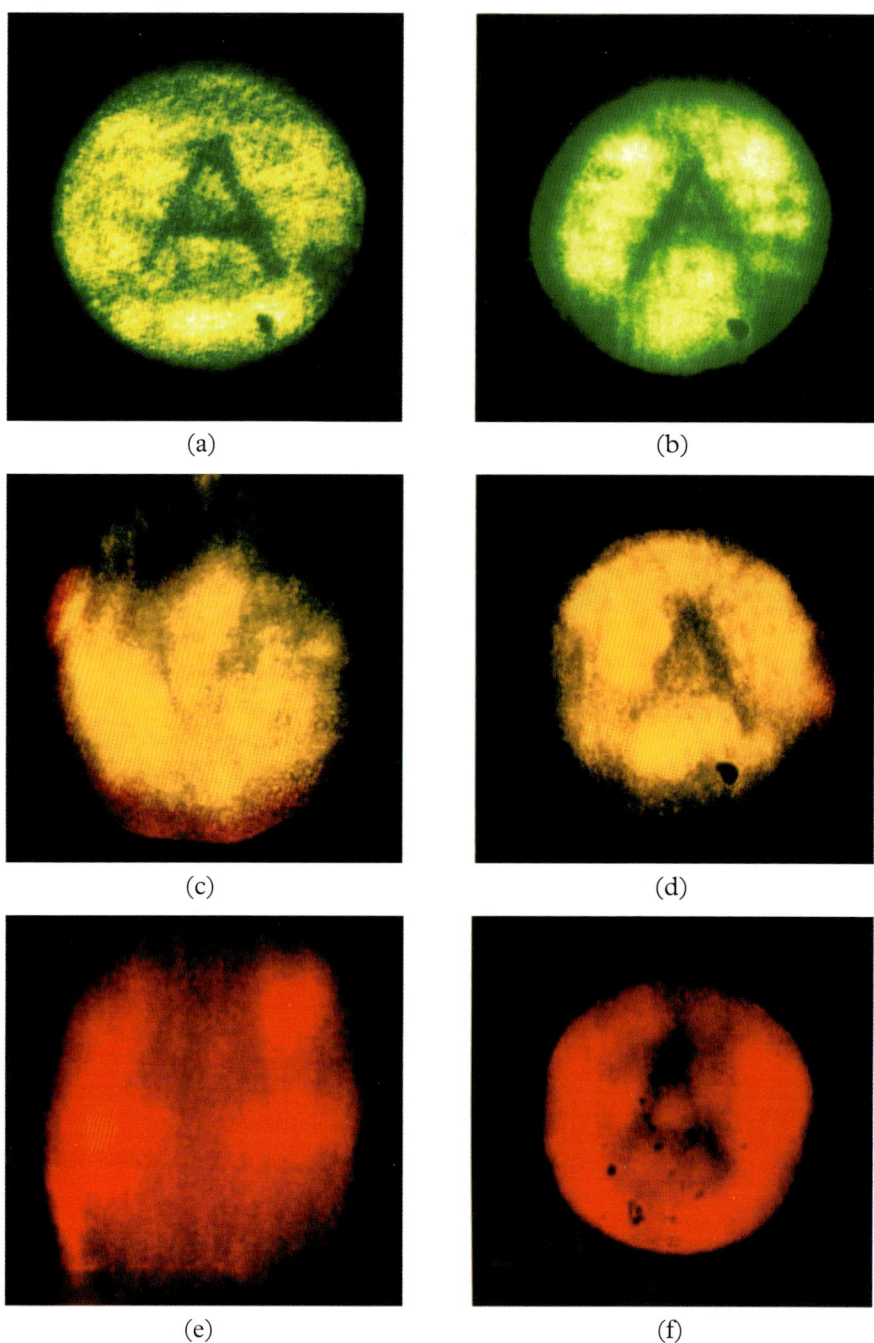

Color fig. 12. *Experimental demonstration of optical phase-conjugation for backward stimulated scattering and lasing. Near-field images of (a) the 532-nm input pump beam, (b) the backward SBS beam from an acetone-filled vial, (c, d) the forward and backward SRS beams of 562-nm wavelength from a benzene-filled vial, (e, f) 1064-nm laser pumped forward and backward lasing beams of ~616 nm wavelength from a dye solution vial. (See the text in Section 8.4.2).*

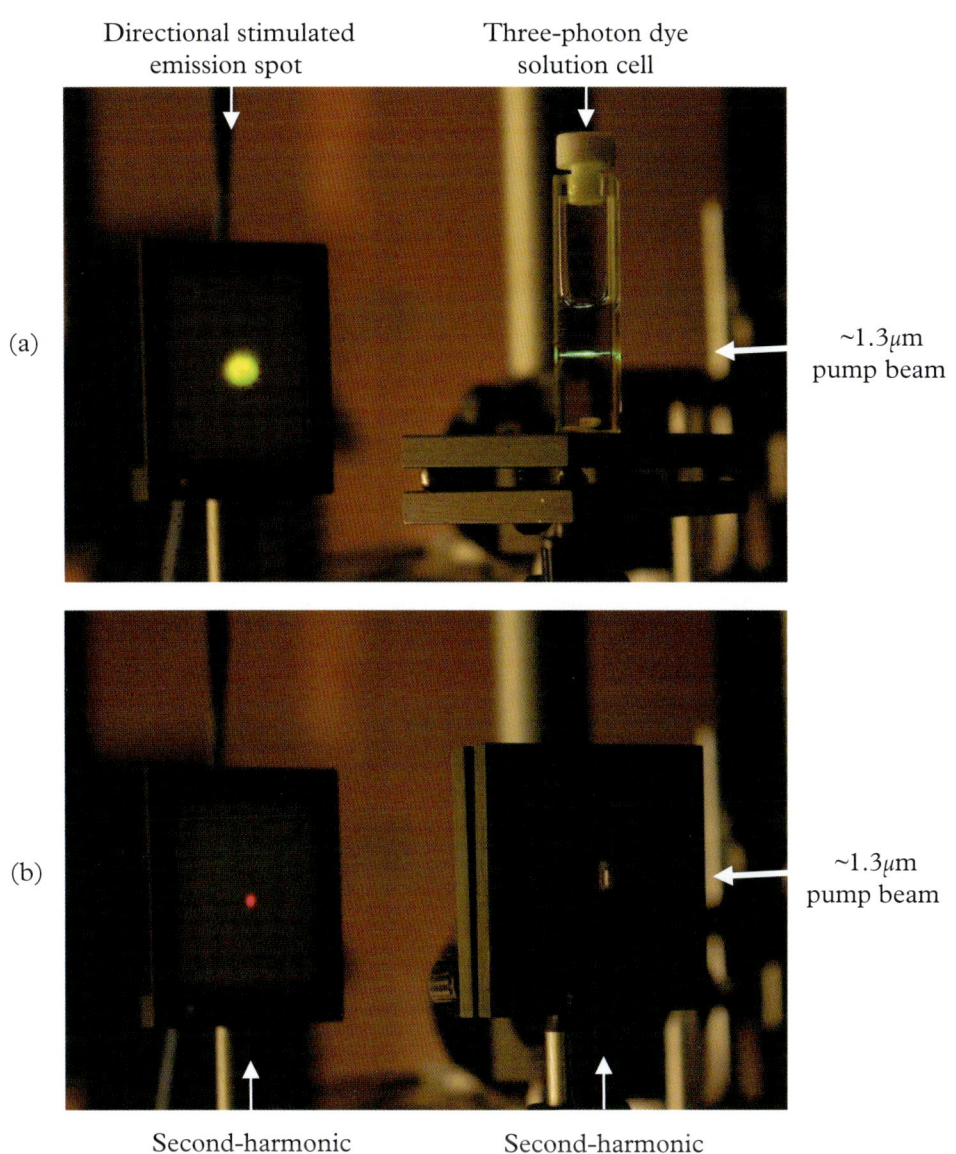

Color fig. 13. *Experimental demonstration of (a) three-photon pumped frequency upconversion lasing beam from a dye-solution filled cuvette and (b) second-harmonic generation from a BBO crystal. (See the text in Section 14.4.3).*

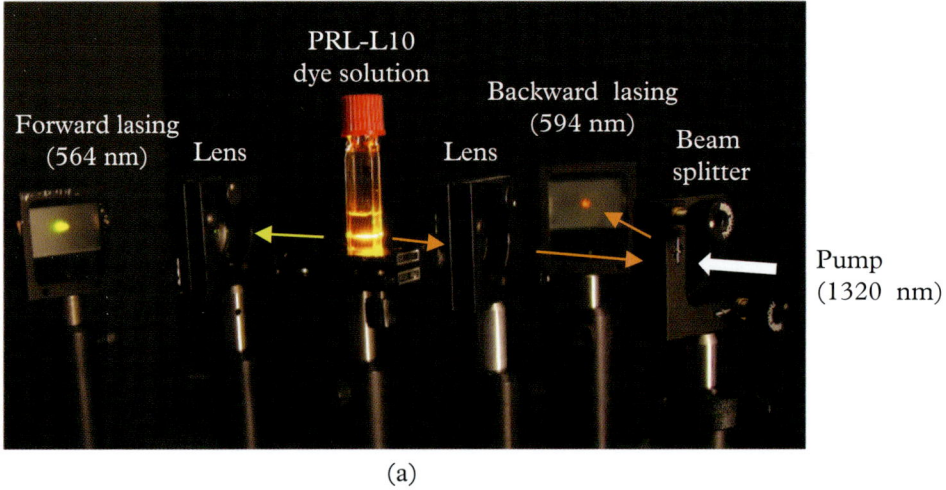

(a)

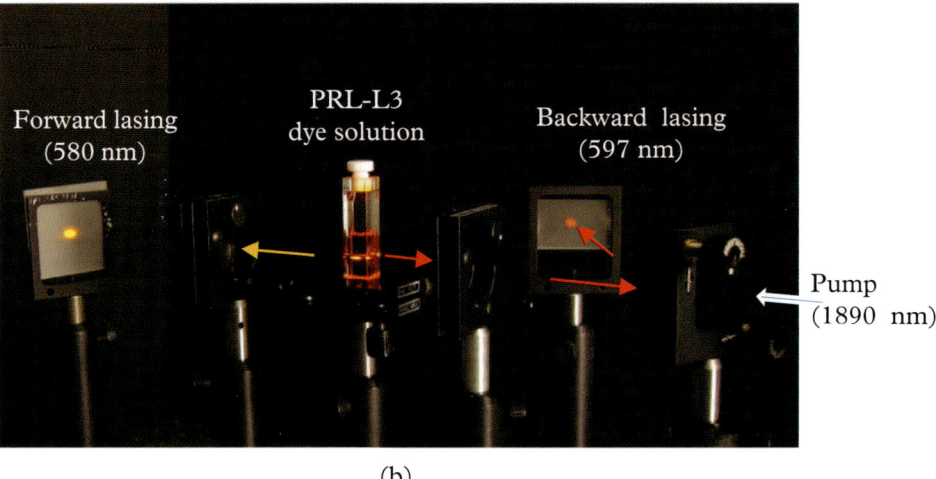

(b)

Color fig. 14. *Spectral asymmetry of the forward and backward three-photon pumped lasing (a) from PRL-L10 dye solution showing ∼30-nm lasing wavelength difference and four-photon pumped lasing (b) from PRL-L3 dye solution showing ∼17-nm lasing wavelength difference. (See the text in Section 14.4.3).*

Color fig. 15. *Three-dimensional (3D) data storage in a two-photon active dye-doped polymer block of 3 × 3 × 3 mm size. The spacing between adjacent layers is 5 μm, and the transverse scale bar is 50 μm. (See the text in Section 14.6.2).*

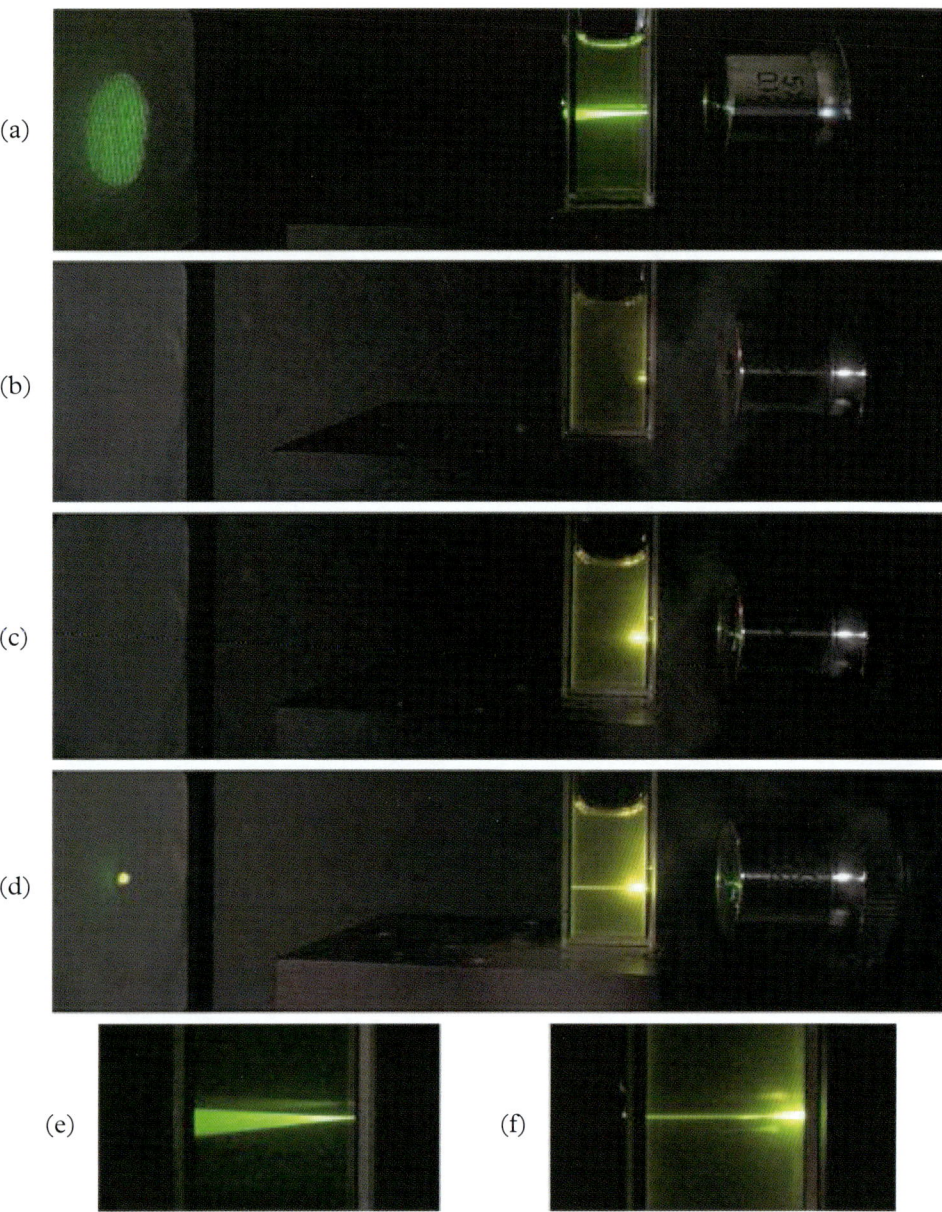

Color fig. 16. *Saturation effect of linear absorption and spatial-soliton lasing in Rhodamine-6G dye solution pumped by 532-nm and 8-ns laser beam. (a) The pump beam was focused on the front surface of the cuvette filled with low-concentration dye solution; (b) The pump beam of low intensity was focused onto the high-concentration dye solution, and could penetrate only a very short distance owing to high linear absorption; (c) The pump beam of moderate intensity could penetrate a longer distance owing to the saturation effect of absorption; (d) The pump beam of high intensity could produce a spatial-soliton lasing at the wavelength of ∼570 nm with an extremely small divergent angle; (e) & (f) The details of (a) & (d). For saturated absorption cf. Section 9.2.1, for spatial soliton cf. Section 13.1.*

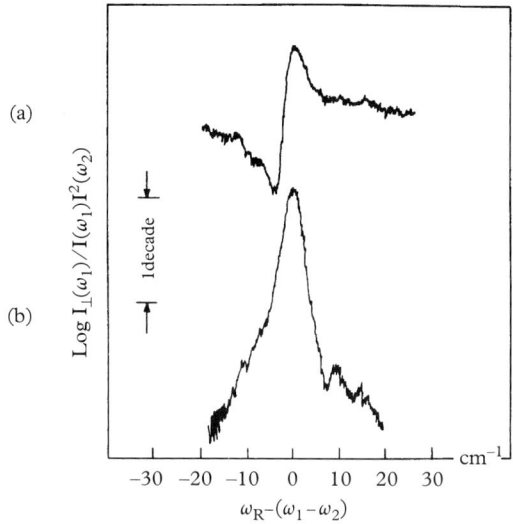

Figure 9.17 *RIKES spectra of benzene near the 992 cm^{-1} mode: (a) the pump beam is linearly polarized at a 45° angle to the polarization direction of the probe beam; (b) the pump beam is circularly polarized. (Reproduced with permission from Levenson and Song.[70] Copyright 1976, Optical Society of America).*

Figure 9.17 shows the RIKES spectra of liquid benzene around the 992 cm^{-1} mode, as excited with two different polarization states of a pump beam.[70] These results are obtained by utilizing a linearly polarized 555-nm laser beam of narrow bandwidth (∼0.01 cm^{-1}) as the probe and a tunable ∼587-nm laser beam as the pump; the 0.65 cm^{-1} linewidth of the latter limits the experimental resolution. In obtaining the curve (a) in Fig. 9.17, the pump beam is linearly polarized at a 45° angle to the probe beam, showing a strong interference between the non-resonant part and the real part of the Raman susceptibility, whereas in obtaining curve (b) the pump beam is circularly polarized, indicating a minimum contribution from the non-resonant background.

In summary, it is more significant that the RIKES suggested a new technical approach to study the nonlinear spectra of the medium based on a stronger pump beam induced polarization changes of a weaker probe beam. This is the common principle of various types of nonlinear polarization spectroscopy that will be addresses later.

9.4.4 Raman gain spectroscopy (RGS) and inverse Raman spectroscopy (IRS)

Both of these spectroscopic techniques are based on the SRS mechanism. Let us first consider the principle of Raman gain spectroscopy (RGS). When a stronger pump beam of frequency ω_1 passes through a Raman medium this beam will provide an exponential gain at the Stokes frequency $\omega_s \approx \omega_1 - \Delta\omega_r$. The gain coefficient is determined by (cf. Eq. (7.2-31))

$$G(\omega_s) = N_a \sigma_r \frac{\lambda_s^2}{\hbar \omega_s} I_0(\omega_1) \cdot S(\omega_1 - \omega_s), \qquad (9.4\text{-}13)$$

where N_a is the density of scattering molecules, σ_r is the Raman cross-section for a given Raman mode, $I_0(\omega_1)$ is the intensity of the pump beam, and $S(\omega_s)$ is a normalized spectral profile function that reaches its maximum when $\omega_s = \omega_1 - \Delta\omega_r$. The SRS gain described by Eq. (9.4-13) can be readily measured using a weaker probe beam with a frequency $\omega_2 = \omega_s$. Furthermore, if one beam is tunable and the frequency difference between these two beams can be scanned over a larger range, the complete Raman spectral structures involving more Raman modes with different cross-sections and linewidths can be revealed. This is the so-called RGS technique.[72-74]

In practice, the experimental setup shown in Fig. 9.14 is also suitable for performing a RGS measurement. In this case, a stronger pump beam of frequency ω_1 and a weaker probe beam with a tunable frequency ω_2 are sent to the Raman medium with an arbitrary small crossing angle, because no phase-matching requirement is needed for the SRS process. A smaller crossing angle is preferable for achieving a longer interaction (gain) length, but this angle cannot be too small, otherwise it will be more difficult to separate the transmitted probe beam from the pump beam at the position of the detector.

As an example, Fig. 9.18 shows the Raman spectrum of benzene near the 992 cm^{-1} mode frequency measured by the RGS method.[75] This curve is obtained by using a $\lambda_1 \sim 577$ nm tunable dye laser as the pump beam and another $\lambda_2 = 612$ nm dye laser as the probe beam, focused into a 16 μm spot in the sample with an interaction length of 130 μm. The measured linewidth of the

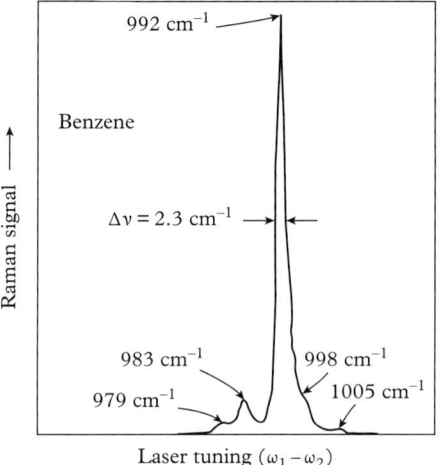

Figure 9.18 *Stimulated Raman gain spectrum of benzene near the 992 cm^{-1} mode frequency. The isotope vibration at 983 cm^{-1} and other weaker modes at 979, 998, and 1005 cm^{-1} are clearly visible. (Reproduced with permission from Levine and Bethea.[75] Copyright 1980, IEEE).*

992 cm^{-1} vibrational mode is 2.3 cm^{-1} with an experimental resolution better than 1 cm^{-1}. In Fig. 9.18, the isotope mode at 983 cm^{-1} as well as other weaker modes at 979, 998, and 1005 cm^{-1} can be clearly identified.

The major advantage of RGS is its high sensitivity. The detectable gain level could be as low as 10^{-5}–10^{-8} cm^{-1}; thereby it is possible to measure very weak Raman modes and spectral structures. In conjunction with its high spatial resolution, this technique can be utilized to study monomolecular-layer samples.

Considering the conservation of energy, the gain of the probe beam is based on the attenuation of the pump beam. Therefore, in principle, the same Raman resonance structure of a given medium can also be detected by measuring the attenuation of the pump beam as a function of the scanned frequency difference between these two beams. This is the basic idea of the so-called inverse Raman spectroscopy (IRS).[76–80] In practice, the intensity of the beam of frequency $\omega_1 > \omega_2$ is comparable with the beam of frequency ω_2, and the attenuation of the former can be written as

$$I_1(\omega_1, z) = I_1(0)e^{-G(\omega_1-\omega_2)z} \approx I_1(0)[1 - G(\omega_1 - \omega_2)z]. \tag{9.4-14}$$

Here, $I_1(0)$ is the initial intensity of beam ω_1, z is the interaction length between the two beams, and the exponential attenuation coefficient G is a function of $(\omega_1 - \omega_2)$ and determined by the same expression, as shown in Eq. (9.4-13).

The experimental setup shown in Fig. 9.14 can also be adopted for IRS. The Raman spectrum of a sample can be obtained by recording the intensity change of the transmitted beam of frequency ω_1 while scanning the frequency difference $\omega_1 - \omega_2$, and no conventional spectrometer device is needed.

On the other hand, as shown in Fig. 9.19, an alternative approach is to use a broadband coherent emission beam in conjunction with a narrow-line laser beam of lower frequency ω_2.

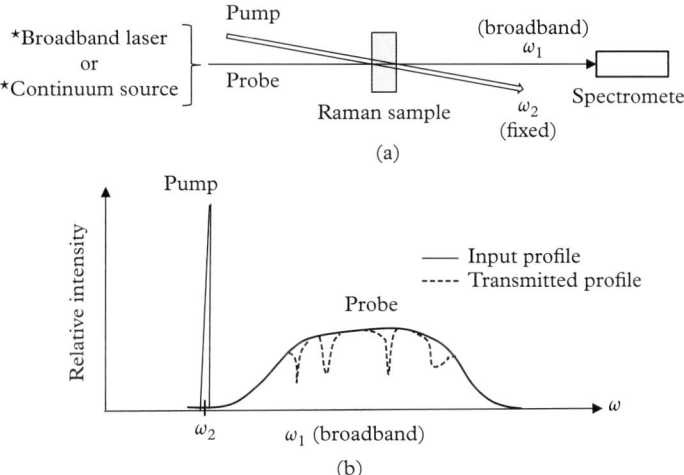

Figure 9.19 *(a) Experimental setup for IRS measurements using a broadband laser beam and a narrow-line laser beam of lower frequency. (b) Schematic illustration of IRS spectrum obtained via the broadband beam interacting with the narrow-line beam.*

Simultaneously measuring the input and the output spectral distribution of the broadband probe beam via a conventional spectrometer device, the Raman attenuation spectrum of the sample medium can be revealed. In that case, no frequency tuning is needed.

9.5 Laser polarization spectroscopy

Laser polarization spectroscopy has been a novel spectroscopic technique, established since the middle of the 1970s. The unique feature of this technique is the detection of change of polarization status of a probe beam; this change is induced by a stronger pump laser beam through resonant interaction with the molecular system. Assuming the sample is an isotropic medium and the frequency of the pump beam is tuned in resonance with one of its absorptive transitions, there will be an induced electric polarization change of the medium. This change will lead to two effects: (i) the induced anisotropic absorption change (dichroism), and (ii) the induced anisotropic refractive-index change (birefringence). These two effects may further cause the change of the polarization status of a probe beam. In practice, even a small polarization change of the probe beam can easily be detected using the crossed-polarizer method. Strictly speaking, the RIKES described in Section 9.4.3 is the first successful demonstration of this novel spectroscopic technique. In this section we shall further discuss some other spectroscopic applications based on the principle of laser polarization spectroscopy.[81–87]

9.5.1 Doppler-free saturated-absorption polarization spectroscopy

For simplicity, we assume that the sample is an atomic gas medium and the saturating beam is a circularly polarized and monochromatic laser beam of frequency $\omega \approx \omega_0$, where ω_0 is the central frequency of an absorptive transition of the medium. A weaker counter-propagating probe beam is linearly polarized and has the same frequency as the stronger saturating beam. In this case we can decompose the probe into two circularly polarized components, one rotating in the same (+) sense and the other in the opposite (−) sense of the saturating beam. The saturating beam induced anisotropic absorption coefficient change (dichroism) is described by $\Delta\alpha^+$ and $\Delta\alpha^-$ for the two polarization components of the probe beam; and the induced anisotropic refractive-index change (birefringence) for these two components can be described by Δn^+ and Δn^-, respectively. The difference $\Delta\alpha^+ - \Delta\alpha^-$ describes a circular dichroism that makes the probe beam elliptically polarized, whereas the difference $\Delta n^+ - \Delta n^-$ implies a gyrotropic birefringence that will further rotate the axes of the elliptical polarization of the probe light.

After the probe beam passes through the medium and a nearly crossed linear polarizer, the complex amplitude of the probe field can be written as[81]

$$E_\perp = E_0 \left[\theta + (\Delta n^+ - \Delta n^-)\frac{\omega z}{2c} - i(\Delta\alpha^+ - \Delta\alpha^-)\frac{z}{4} \right], \quad (9.5\text{-}1)$$

where E_0 is the initial amplitude of the probe light, θ is a small angle by which the polarizer is rotated from the perfect crossed position, and z is the interaction length of these two beams in the medium. As in the typical saturation spectroscopy, the saturating beam induced absorption change as a function of the laser frequency is expressed by

$$\Delta\alpha^+ = \Delta\alpha^-/d = -\frac{1}{2}\alpha_0 \frac{I_1}{I_s(1+\delta^2)}. \quad (9.5\text{-}2)$$

Here, α_0 is the unsaturated absorption coefficient, I_1 is the intensity of the saturating beam, I_s is the saturation parameter, d is a parameter dependent on the related level structure and transition property of the medium, and $\delta = (\omega - \omega_0)/\Gamma$, where Γ is the natural linewidth factor of the considered transition.

It is known that the imaginary part of the third-order nonlinear polarization of a medium determines the absorption change, and the real part determines the induced refractive-index change. According to the Kramers–Kronig relations we then have

$$\Delta n^{\pm} = -\frac{1}{2\omega}\Delta\alpha^{\pm}\delta c. \quad (9.5\text{-}3)$$

Substituting Eqs. (9.5-3) and (9.5-2) into Eq. (9.5-1) one can finally obtain the expression for the intensity of the transmitted probe beam passing through the blocking polarizer:

$$I_{\perp} = I_0\left[\theta^2 + \frac{\theta\delta s}{2(1+\delta^2)} + \frac{s^2}{16(1+\delta^2)}\right], \quad (9.5\text{-}4)$$

where I_0 is the initial intensity of the probe and

$$s = -(1-d)\frac{\alpha_0 z}{2}\frac{I_1}{I_s} \quad (9.5\text{-}5)$$

represents the maximum relative intensity difference between the two counter-rotating probe components.

One can see from Eq. (9.5-4) that at the perfect crossed polarizer position, $\theta = 0$, the resonant signal is proportional to s^2 and has a Lorentzian shape. If the s value is rather small, we can make $\theta \gg s$ and the third term in Eq. (9.5-4) negligible; then we will see a dispersion-shaped signal on a constant background with a higher sensitivity.

An experimental setup for saturated-absorption polarization spectroscopy is shown in Fig. 9.20(a).[81] Two beams are from a tunable dye laser source: one is circularly polarized through a quarter-wave plate as the saturating beam, and the other is linearly polarized through a polarizer as the probe beam. After sequentially passing through the sample medium, a nearly crossed polarizer, and a spatial filter the transmitted probe beam is measured by a photodetector. Based on this setup, the recorded polarization spectral signal of a portion of the deuterium Balmer-β line spectrum as a function of the frequency scanning is shown in Fig. 9.20(c). To obtain this result, a cw dye laser was in single-frequency operation and tunable around 486-nm with a linewidth of 20 MHz. The power level was 1mW for the saturating beam and 0.1 mW for the probe. The deuterium atoms were excited to the absorbing $n = 2$ state in a Wood-type discharge tube (1 m long, 8 mm in diameter, 0.2 Torr, and 3 mA dc current). For comparison, the theoretically expected Stark pattern due to an axial dc electric field of 10 V/cm, as well as the contributions from crossover resonances are shown in Fig. 9.20(b). From Fig. 9.20 we can see that the observed Doppler-free spectral structures are in good agreement with the theoretical calculation.[81]

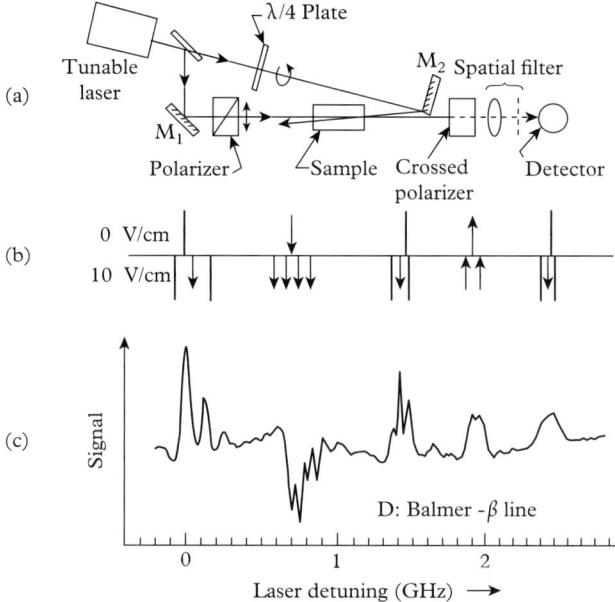

Figure 9.20 *(a) Experimental setup for saturated-absorption polarization spectroscopy measurements. (b) Theoretically expected resonance lines for two axial electric field values, the crossover lines due to a common upper (↑) or lower (↓) level are indicated by arrows. (c) Polarization spectrum of portion of the deuterium Balmer-β line. (Reproduced with permission from Wieman and Hänsch.[81] Copyright 1976, American Physical Society).*

9.5.2 CARS polarization spectroscopy

It has been mentioned in Section 9.4.2 that for a CARS measurement there is a background influence due to the non-resonant part ($\chi^{(3)}_{NR}$) of the third-order nonlinear susceptibility. In some cases, this background may reduce signal-to-noise ratio and limit the sensitivity of measurements. It is found that this non-resonant background could be eliminated using a polarization detection technique.

Assume that two monochromatic laser beams pass through a Raman medium along the z-axis: one is stronger (with frequency ω_1) and linearly polarized along the x-axis; the other is weaker (with frequency ω_2) and linearly polarized along a direction at an angle ϕ with the x-axis. The frequency difference between these two beams satisfies the Raman resonance condition, i.e., $\omega_1 - \omega_2 \approx \Delta\omega_r$. With such an arrangement of light polarization of these two beams, the third-order nonlinear polarization component of the medium at frequency $\omega_{as} = 2\omega_1 - \omega_2$ is given by[82]

$$\left. \begin{array}{l} P_x^{(3)}(\omega_{as}) = 3\varepsilon_0 \cos\phi (\chi^{(3)}_{xxxx.NR} + \chi^{(3)}_{xxxx.R}) E_1^2 E_2^* = P^{(3)}_{x.NR} + P^{(3)}_{x.R} \\ P_y^{(3)}(\omega_{as}) = 3\varepsilon_0 \sin\phi (\chi^{(3)}_{yxxy.NR} + \chi^{(3)}_{yxxy.R}) E_1^2 E_2^* = P^{(3)}_{y.NR} + P^{(3)}_{y.R} \end{array} \right\}, \quad (9.5\text{-}6)$$

where E_1 and E_2 are the field amplitude functions of the two beams. The angle between the directions of the non-resonant polarization component $P_{NR}^{(3)}$ and the x-axis is determined by

$$\beta = \tan^{-1}\left[\left(\chi_{yxxy.NR}^{(3)}/\chi_{xxxx.NR}^{(3)}\right)\tan\phi\right]. \quad (9.5\text{-}7)$$

Let the new axes x' and y' denote the directions parallel and perpendicular to $P_{NR}^{(3)}$, respectively; then the third-order polarization components in the new axes system will be

$$\left.\begin{array}{l}P_{x'}^{(3)}(\omega_{as}) = \varepsilon_0[3\chi_{xxxx.NR}^{(3)}\cos\phi/\cos\beta + 3\chi_{xxxx.R}^{(3)}\cos\phi\cos\beta(1+\rho\tan\phi\tan\beta)]E_1^2 E_2^* \\ P_{y'}^{(3)}(\omega_{as}) = -\varepsilon_0\{3\chi_{xxxx.R}^{(3)}\cos\phi\sin\beta[1-\rho(\tan\phi/\tan\beta)]\}E_1^2 E_2^*\end{array}\right\}, \quad (9.5\text{-}8)$$

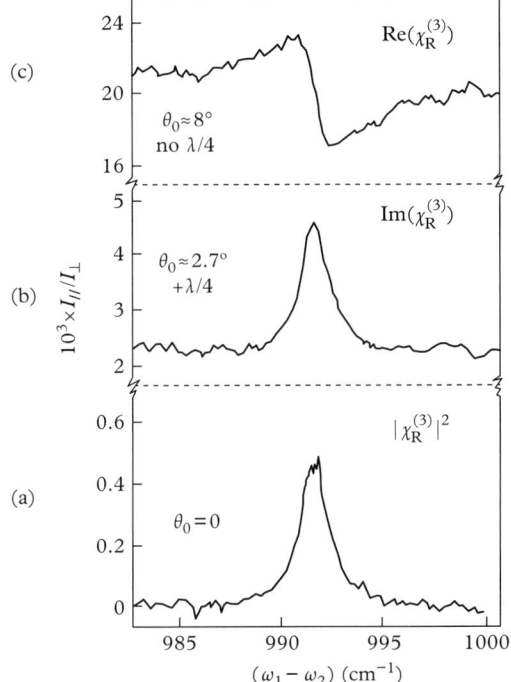

Figure 9.21 CARS polarization spectra near 992 cm^{-1} of 0.011M benzene in carbon tetrachloride: (a) $|\chi_R^{(3)}|^2$ with suppression of non-resonant background, (b) $\text{Im}\{\chi_R^{(3)}\}$ superimposed on a weak non-resonant background, and (c) $\text{Re}\{\chi_R^{(3)}\}$ superimposed on a weak non-resonant background. (Reproduced with permission from Oudar et al.[82] Copyright 1979, AIP Publishing LLC).

where

$$\rho = \chi^{(3)}_{yxxy.R} / \chi^{(3)}_{xxxx.R} . \tag{9.5-9}$$

From Eq. (9.5-8) one can see that in general the CARS signal is elliptically polarized because the resonant third-order susceptibility elements are complex. Nevertheless, if we put a linear polarizer (analyzer) along the y'-axis direction, the non-resonant contribution can be eliminated effectively, so that the measurement sensitivity can be improved considerably. Furthermore, if we set the analyzer at a small angle θ_0 with the y'-axis, the transmitted CARS intensity will be

$$I(\omega_{as}) \propto \left| P^{(3)}_{x'} \sin\theta_0 - P^{(3)}_{y'} \cos\theta_0 \right|^2. \tag{9.5-10}$$

Adjusting the angle θ_0 properly and putting a quarter-wave plate in front of the analyzer, one may also measure the real and imaginary part of $\chi^{(3)}_R$ separately.

As an example, Fig. 9.21 shows the measured CARS polarization spectra of the 992 cm^{-1} mode of benzene under different conditions of polarization detection.[82] The sample was a dilute mixture of benzene (0.011M) in carbon tetrachloride. The two beams were from two separate pulsed dye lasers. The wavelength and peak power of the ω_1 beam were $\lambda_1 = 595$ nm and 7 kW; the wavelength of the ω_2 beam was tunable and the peak power was 3 kW. The linewidth for both beams was 0.2 cm^{-1}. The curves shown in Fig. 9.21(a), (b), and (c) indicate the contributions from the overall, the imaginary part, and the real part of the resonant Raman susceptibility, respectively.

9.5.3 Polarization labeling molecular spectroscopy

The principle of polarization spectroscopy is also applicable to the study of molecular spectra. It is well known that when the investigated sample is a molecular gas system, the spectral structure might be rather complicated and numerous lines may crowd together, because there are too many allowed transitions between a great number of levels. If in some way we could selectively excite or detect part of these transitions, the observed spectral structures and line numbers can be greatly simplified and reduced, respectively. Polarization spectroscopic technique is one of the solutions. The basic idea is that a circularly polarized and frequency-tunable pump laser beam can be employed to selectively excite a group of designated transitions, which are initiated from a specific low level among a lot of others (various degenerate angular momentum sublevels of molecules). This type of selective excitation produces a light-polarization dependent depopulation of the specific low level, which leads to an induced circular dichroism and gyrotropic birefringence related to the change of the physical property of the depopulated low level. This induced dichroism and birefringence may alter the polarization status of a weak probe beam that has the same frequency as the pump beam and is initially linearly polarized, as we described in Section 9.5.1. If the probe frequency is not the same and away from any relevant resonant frequencies, then the polarization status of the probe remains unchanged. However, when the probe frequency is tuned to another absorption frequency that corresponds to a transition from the same depopulated low level but to another upper level, the polarization status of the probe beam will also be disturbed, and this polarization variation can readily be detected with a crossed analyzer. Based on these considerations, if we fix the pump frequency in resonance with a designated absorption transition, and continuously scan the frequency of the probe beam, a series of absorption lines can be recorded through the crossed analyzer, all originating from the same selectively depopulated low level. This is the principle of polarization labeling molecular spectroscopy.[83-87]

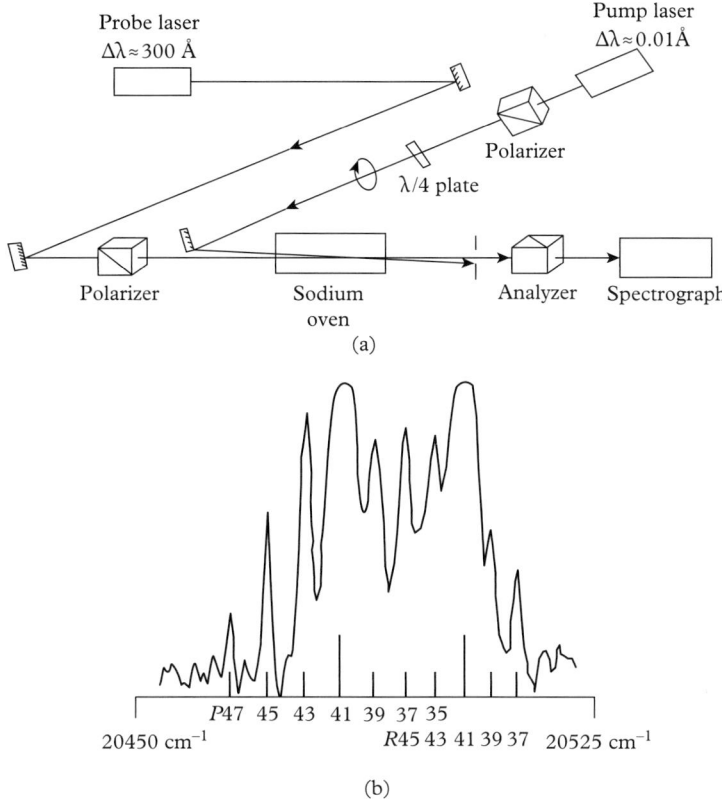

Figure 9.22 *(a) Experimental setup for polarization labeling spectroscopic measurements. (b) Details of polarization spectrum of Na_2 molecules for the $v' = 2$ transition of $B^1\Pi_u$ band. The circularly polarized pump laser was tuned to the $X(0, 41) \to B(4, 40)$ transition. (Reproduced with permission from Teets et al.[84] Copyright 1976, American Physical Society).*

Figure 9.22(a) shows the experimental setup used for the early performance of polarization labeling spectroscopic measurements.[84] The sample medium was sodium molecular vapor at a temperature near 300 °C corresponding to a density of about 5×10^{12} molecules/cm^3. The circularly polarized pump laser was tunable near 482.5 nm with a line width of 1 GHz, a pulsewidth of ~5 ns, and a peak power of several kilowatts, while the linearly polarized probe beam was working in the same wavelength range with a peak power of 20 kW and a broad spectral band of ~30 nm. After forward passing through the vapor cell and a crossed analyzer, the polarization spectra of the transmitted probe beam could be recorded by a conventional spectrometer. With this arrangement, for the selected single vibration-rotational transitions in the blue-green band ($X^1\Sigma_g^+ \to B^1\Pi_u$), a series of other vibration–rotational transition lines, originating from the same pump-depopulated low level, could be recorded within the broad spectral band of the probe beam without the need of tuning the latter. Furthermore, when the pump frequency was tuned to another absorption transition or to depopulate another selected low level, a new series of transition lines from the common low level could be identified. In this case, the recorded spectra were much simpler and better characterized than the ordinary molecular absorption spectra.

It should be noted that as the gas pressure is quite high, the molecular collision may impose its influence on the structure of the observed polarization spectrum. In fact, collisions may change the rotational angular moment of the molecules without randomizing the molecular orientation. Na_2 is a homonuclear diatomic molecule and exists in the ortho and para forms which are not mixed by collisions. For this reason, the polarized pumping induced anisotropic orientation change of molecules in the specific low level may transfer to other neighboring low levels through collisions, and the resulting polarization spectrum might be a superposition of those lines originating from these neighboring low levels. Such a type of polarization spectrum of $v' = 2$ transition within the $(X^1\Sigma_g^+ \rightarrow B^1\Pi_u)$ band of the sodium molecule is shown in Fig. 9.22(b) with the identified transitions from the relevant neighboring rotational levels.[84] The pump frequency was tuned to excite the $X(v = 0, \mathcal{J} = 41) \rightarrow B(v' = 4, \mathcal{J}' = 40)$ transition only. To obtain this result, a buffer gas of nitrogen at several hundred Torr was introduced into the vapor cell to increase the collision effect. From the spectrum shown in Fig. 9.22(b), one could further calculate the angular momentum and the rotational constants of both the upper and lower states.

9.6 Laser cooling and trapping spectroscopy

9.6.1 Principles of laser cooling and trapping

So far, in this chapter we have described several different nonlinear spectroscopic techniques that can be adopted for Doppler-free spectroscopic studies in gas samples with a resolution only limited by second-order Doppler broadening and transient-time broadening. To overcome these two limitations, the best solution is to cool the atoms to very low temperature by significantly reducing their absolute velocity and finally to trap them into a certain spatial region, so that the interaction time between the laser beam(s) and these trapped atoms can be much longer.

The principles of optical cooling and trapping of neutral atoms are based on either the laser driven force in a gradient light field, or the resonant radiation pressure imposed on the atoms.

When a non-uniform monochromatic light field interacts with an atomic gas medium, each atom undergoes a ponderomotive force expressed by

$$\mathbf{f}(\omega) = \frac{1}{8\pi}\alpha_{real}^{(1)}(\omega)\mathrm{grad}\,|E(\omega)|^2, \quad (9.6\text{-}1)$$

where $\alpha_{real}^{(1)}$ is the real part of the atomic linear polarizability and $E(\omega)$ is the amplitude function of the optical field. This expression means that the gas atoms will experience a force proportional to the gradient of the square of the optical electric field. Letokhov proposed in 1968 that upon the action of a strong monochromatic standing-wave field, the atoms may be trapped around the antinode positions if the real part of the atomic polarizability is chosen to be < 0, as shown schematically in Fig. 9.23(a), provided that the atoms are pre-cooled to low enough temperature, i.e., the atomic velocity along the z-axis should be slower than a critical value of [88]

$$v_{cr.} = E_0\sqrt{\left|\alpha_{real}^{(1)}\right|/2m}. \quad (9.6\text{-}2)$$

Here, E_0 is the amplitude of the standing wave and m is the mass of the atom.

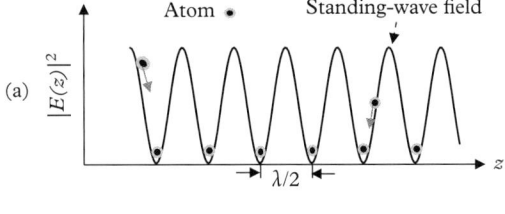
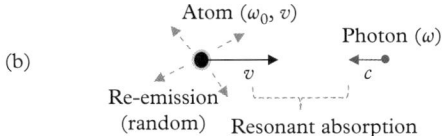
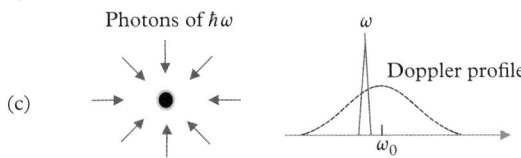

Figure 9.23 (a) Schematic diagram showing that the atoms may be trapped around the antinode positions of a strong monochromatic standing-wave field. (b) Schematic indicating that the resonant absorption of a counter-propagating photon could reduce the atomic velocity. (c) The atoms could be cooled by the photons of frequencies located in the red-side of the Doppler profile and incident from all directions.

As another different approach, in 1970 Ashkin suggested that the atomic velocity in a gaseous medium could be controlled by the resonant radiation pressure from a laser beam.[89] As shown schematically in Fig. 9.23(b), when a forward moving atom absorbs a counter-propagating photon, according to momentum conservation this atom acquires a recoil momentum of $-\hbar\omega/c$, and the corresponding atomic velocity will be reduced by

$$\Delta v = -\frac{\hbar\omega}{mc} = -\frac{h}{m\lambda}. \tag{9.6-3}$$

Through sequential and repeated actions of such kind of resonant absorption process, the forward moving atom could be gradually cooling down. Although the photon energy absorbed by the atom can be released via fluorescence emission and therefore there is a recoil influence on the atom with each event of the re-emission, owing to the fact that the re-emission goes in different directions randomly, the averaged result of the re-emission leaves no influence on the atom.

Moreover, considering the Doppler effect in the case shown in Fig. 9.23(b), one could find that the atom to be cooled with a velocity of v can only absorb a counter-propagating photon with a frequency satisfying the following resonance condition:

$$\omega = \omega_0(1 - v/c). \tag{9.6-4}$$

where ω_0 is the absorption frequency for the atom at rest. With this consideration, Hänsch and Schawlow proposed in 1975 that the more effective cooling could be achieved by irradiating the atoms from all directions with the red-detuned laser radiation in respect to the center of the Doppler broadening profile, as shown schematically in Fig. 9.23(c).[90]

However, in the case shown in the right side of Fig. 9.23(c), if the laser beam exhibits quite narrow spectral linewidth compared to the Doppler linewidth, so once the atoms have been slowed down, they cannot absorb the incident photon any more as the requirement of Eq. (9.6-4) can no longer be met. In order to make the resonant absorption and deceleration of the atoms continuously take place, two measures can be adopted. One method is to timely sweep the tunable laser frequency across the low-frequency side of the Doppler profile during optical cooling processes, to ensure all atoms of different velocities can be continuously decelerated. The other method is suitable for the case of using a laser beam with fixed frequency to cool a quasi-directional atom beam. In the latter case, we could let the atom beam pass though a spatially varying magnetic field (produced by a tapered solenoid) to experience a gradually varying Zeeman shift of the resonance frequency ω_0; thus even for a fixed laser of frequency ω and a continuously reduced velocity v the resonant interaction condition of Eq. (9.6-4) can remain satisfied.

9.6.2 Techniques for laser cooling and trapping

One of the earliest laser cooling experiments was accomplished by Wineland et al. in 1978,[91] based on the mechanism of resonant radiation pressure applied by a red-detuned laser beam. In their experiment, the sample medium was a gas of Mg^+ ions with $\sim 2 \times 10^7$ cm^{-3} density constrained in a copper Penning trap, irradiated by a frequency-doubled dye laser beam of ~ 280 nm wavelength tuned in resonance with the $^2S_{1/2} \leftrightarrow ^2P_{3/2}$ transition ($M_J = +1/2 \leftrightarrow +3/2$ or $M_J = -1/2 \leftrightarrow -3/2$). When the laser frequency was tuned to the red-side of the Doppler profile of this transition, the velocity of the Mg^+ ions along the opposite direction of the input laser beam could be greatly reduced, and the corresponding kinetic temperature of the ions could be cooled from ~ 700 K to lower than 40 K.

Later on, in 1985 Prodan et al. reported their experimental result of optical cooling on a Na atom beam by using a single counter-propagating laser beam and a Zeeman shifter provided by tapered solenoid. The atoms' mean velocity could be cooled from ~ 1000 m/s to ~ 15 m/s and the latter corresponds to a kinetic temperature of ~ 100 mK.[92] In the same work, the deeply cooled Na atoms could be further trapped in a three-dimensional (3D) inhomogeneous magnetic field that exerts forces on atoms with a magnetic dipole moment.

In the same year, Chu et al. reported their much improved laser cooling result on the Na vapor.[93] In their work, the Na beam was first pre-cooled by using a tunable dye laser beam with frequency sweep over the low-frequency side of the Doppler line. The atom velocity was reduced from an initial ~ 200 m/s to ~ 20 m/s; such pre-cooled atoms then could be further confined in a volume of ~ 0.2 cm^3 by applying three pairs of counter-propagating laser beams with their frequency located in the red-side of the Doppler profile. By this way, an atom cloud of $\sim 10^6$ cm^{-3} density was formed within which the atom velocity was established to be as low as ~ 60 cm/s corresponding to a kinetic temperature of ~ 240 μK.

The next progress was made by Raab et al. in 1987 by virtue of the use of a magneto-optical trap (MOT) for neutral atoms.[94] Using this MOT, they were able to trap 10^7 pre-cooled Na atoms in a volume of 0.5 mm diameter for over 120 s; the estimated temperature of the trapped atoms was ~ 600 μK.

To explain the idea of the MOT design, we can rely on the schematic diagram shown in Fig. 9.24. An assumed atom system is in a magnetic field with a gradient along the z-axis; the

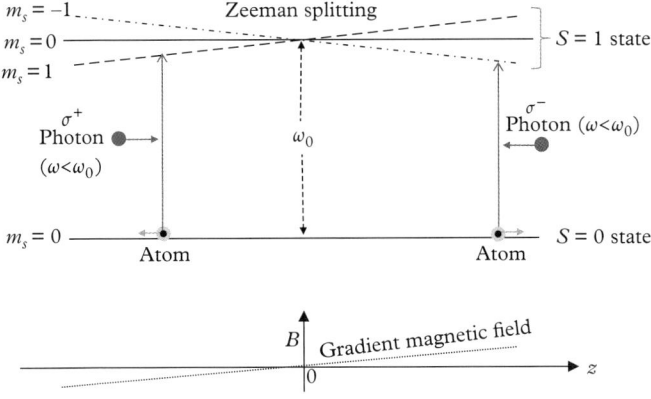

Figure 9.24 *Energy-level diagram showing that a gradient magnetic field in conjunction with a pair of counter-propagating laser beams circularly polarized with opposite signs can more effectively reduce the atomic velocity in both moving directions along the z-axis.*

absorption transition is assumed to take place between the $S = 0$ ground state and the $S = 1$ excited state. Without an externally applied magnetic field, the resonance frequency is ω_0 for the atom at rest, whereas upon the action of a gradient magnetic field, the excited state splits into three sublevels due to the Zeeman effect, as shown at the top of Fig. 9.24. If there is a pair of counter-propagating laser beams along the z-axis with a fixed frequency of ω (quite close to but lower than ω_0) and both beams are circularly polarized with opposite signs, the atoms of different velocity components along the z-axis can be more effectively cooled by the laser beams from both $\pm z$ sides. For instance, let us examine those atoms moving along the $+z$ direction, which more easily absorb those photons coming from the right-side laser beam and are excited to the sublevel of $m_s = -1$. By considering both the Doppler effect of the moving atoms and the resonance requirement of Eq. (9.6-4), all atoms moving along the $+z$ direction with different velocity values can remain in resonance with the laser beam from the right side within an appropriate z-range. For the same reason, all atoms moving along the $-z$ direction can also be effectively cooled by the laser beam from the left side.

Furthermore, if these pre-cooled atoms are situated in a 3D gradient magnetic field and irradiated by three pairs of laser beams along the x, y, and z directions, these atoms can be finally trapped in the intersecting region of the six laser beams. This is the physical basis of a MOT design.

In practice, as shown schematically in Fig. 9.25, a typical experimental setup for laser cooling and trapping usually involve two major parts: the first is for pre-cooling of an investigated atoms beam and the second is for magneto-optical trapping. These two parts are both enclosed in an ultra-high vacuum environment (e.g., $\leq 10^{-9}$–10^{-10} Torr). In the first stage, the atom beam passes through a one-dimensional gradient magnetic field and is pre-cooled by a single laser beam. The pre-cooled atoms then migrate to the MOT region and to be trapped by six laser beams. The 3D gradient (spherical quadrupole) magnetic field is usually generated by two coils of opposing current working in an anti-Helmholtz configuration.

Later on, in 1990 Monroe et al. reported their optical cooling results of low pressure cesium vapor by utilizing a compact MOT design in a glass cell.[95] In their experiment, the three pairs

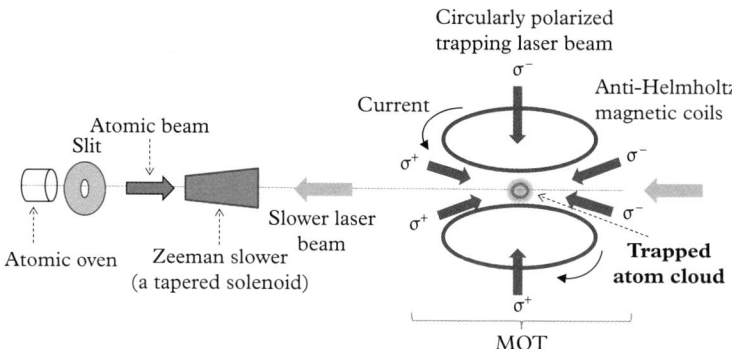

Figure 9.25 *A typical experimental setup for laser pre-cooling by using a single laser beam (left) and for magneto-optical trapping by using three pairs of counter-propagating laser beams circularly polarized with opposite signs (right). The 3D gradient magnetic field is produced by a pair of coils working in an anti-Helmholtz configuration.*

of ∼852-nm diode laser beams were tuned to the red-side of the $6S_{1/2}(F=4) \to 6P_{3/2}(F=5)$ transition of Cs atoms. There were ∼1.8×10^7 atoms contained in a trapped atom cloud of density ∼5×10^{10} cm^{-3} with a kinetic temperature ∼30 μK. The same experiment also showed that once the magneto-optical trapping procedure was done, the very cold atom could be loaded into another pure magnetic trap after turning off all laser beams; thus the kinetic temperature of the magnetically trapped atoms could be as low as ∼1 μk.

9.6.3 Ultrahigh resolution spectroscopy using laser cooling and trapping techniques

The laser cooling and trapping techniques can substantially eliminate the spectral broadening influences from both the first- and second-order Doppler effects as well as from the transient-time effect, and have enabled researchers to implement a new type of spectroscopic study with ultrahigh resolution.

By using these new techniques, the natural linewidth, hyperfine structures, isotope shifts, and transition dynamics of a series of atomic gas systems have been investigated, including the alkali metal atoms Cs,[96] Rb,[97–100] and Fr,[101,102] alkali-earth metal atoms Ca and Sr,[103–105] other metal atoms of Yb,[106,107] Er,[108,109] Dy,[110] Hg,[111,112] as well as other nonmetal atoms.[113–115] In most of these studies, the sample atoms could be optically cooled and then trapped by a MOT, reaching a temperature level lower than 10^{-4}–10^{-6} K. In the following, we shall mention several successful examples of these studies.

First, Fig. 9.26 shows the spectral profiles of the 2^3S_1–3^3P_2 transition of atoms of ^{4}He and ^{6}He, measured by using laser cooling and trapping techniques.[114] In this work reported by Wang et al., the pre-cooling and magneto-optical trapping were done by utilizing 1083-nm laser beams in resonance with the 2^3S_1–2^3P_2 transition; then tunable 389-nm diode laser light was employed to excite the 2^3S_1–3^3P_2 transition; the re-emitted resonance fluorescence signal could be detected by a photomultiplier. Based on the results shown in Fig. 9.26, the spectral linewidth (FWHM) is 6.8±0.1 MHz for ^{4}He and 6.2±0.4 MHz for ^{6}He, respectively, while the natural linewidth of this

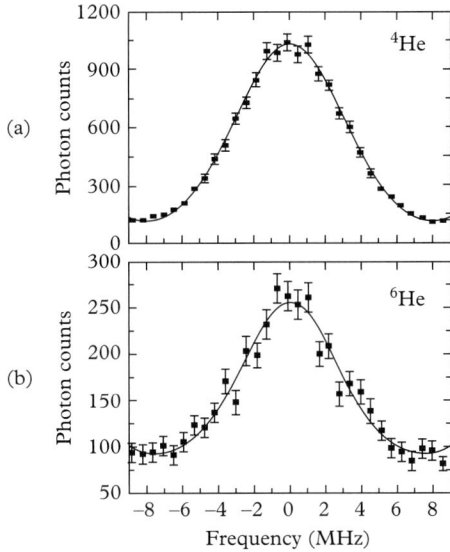

Figure 9.26 Measured spectral profiles of the 2^3S_1–3^3P_2 transition for the ^{4}He atoms (a) and the ^{6}He atoms (b) by using laser cooling and trapping techniques. The solid curves are the best Gaussian fits. (Reproduced with permission from Wang et al.[114] Copyright 2004, American Physical Society).

transition is ∼5 MHz, which means that the measured spectral linewidth value is very close to the latter. In addition, the measured isotope shift for this particular transition between ^{4}He and ^{6}He is 43 194.772±0.056 MHz.

In the effort to realize new optical clocks and frequency standards, searching for a highly stable clock transition with extremely narrow spectral linewidth from a suitable atom or ion system is essentially important. It has been recognized that the ytterbium $1S_0 \leftrightarrow 3P_0$ transition is an excellent candidate for that purpose. Actually, the optical transition $(6s^2)^1S_0 \leftrightarrow (6s6p)^3P_0$ of the neutral Yb atom is dipole-forbidden from spin and orbital angular momentum considerations, but still exhibits certain transition probability through hyperfine mixing of the $^3P_0$ level with nearby states, and the calculated natural linewidth of this transition is extremely narrow (∼10 mHz).[107] In the experiment performed by Hoyt et al., the spectral profiles of this transition for two isotopes of ^{171}Yb and ^{173}Yb were measured by using laser cooling and trapping techniques and the results are shown in Fig. 9.27.[107] To obtain these results, the atoms of these two isotopes were cooled and trapped via two stages, then the trapped atom cloud was excited by 578.4 nm laser radiation of ∼5 kHz linewidth in resonance to the $(6s^2)^1S_0 \leftrightarrow (6s6p)^3P_0$ transition. The ordinate in Fig. 9.27 represents the relative depletion of the population in the lower $(6s^2)^1S_0$ state, which is the function of the 578.4-nm laser frequency detuning across the center frequency of the clock transition. The solid curves shown in Fig. 9.27 exhibit a Gaussian shape due to the residual Doppler effect; from these experimental curves it is determined that the temperature for these two tested isotopes is ∼84 µK and ∼48 µK, respectively.

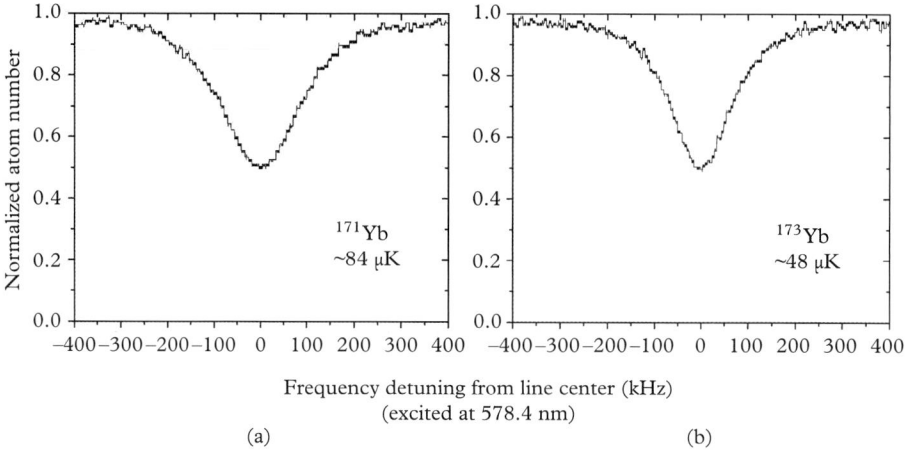

Figure 9.27 Spectral profiles of the $(6s^2)^1 S_0 \leftrightarrow (6s6p)^3 P_0$ transition at 578.4 nm wavelength for ^{171}Yb (a) and ^{173}Yb (b) measured by using laser cooling and trapping techniques. The ordinate indicates the relative depletion of the atom population in the lower state due to resonance absorption. (Reproduced with permission from Hoyt et al.[107] Copyright 2005, American Physical Society).

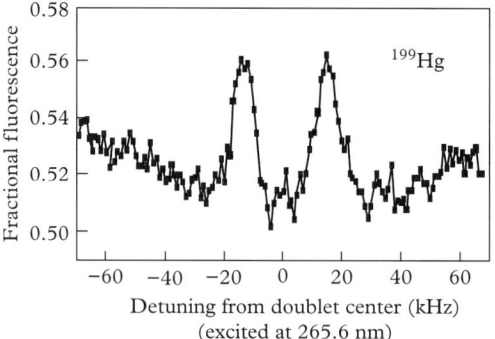

Figure 9.28 Spectral profiles of the $(6s^2)^1 S_0 \leftrightarrow (6s6p)^3 P_0$ clock transition of ^{199}Hg atoms probed by a pair of counter-propagating ~265.6-nm laser beams. The ordinate reflects the depletion of the population in the low state of the clock transition due to resonance absorption of the probe laser beams. (Reproduced with permission from Petersen et al.[112] Copyright 2008, American Physical Society).

On the other hand, mercury is also an interesting candidate to provide an optical clock transition with high stability and extremely narrow linewidth. Using the laser MOT method, Petersen et al. have measured the spectral profiles of the $(6s^2)^1 S_0 \leftrightarrow (6s6p)^3 P_0$ clock transition of the two isotopes ^{109}Hg and ^{201}Hg.[112] In their experiment, the Hg atoms were first cooled and trapped in a MOT by using 253.7-nm (cooling) laser beams red-detuned in resonance to the $^1S_0 \rightarrow {}^3P_1$

allowed transition, then a pair of counter-propagating (probe) laser beams tunable around the 265.6 nm wavelength were employed to do the saturation absorption spectroscopic measurement. In the latter case, the population depletion in the low state of the clock transition can be measured as a function of the probe laser frequency through the detection of the fluorescence signal excited by other laser beam with a different wavelength. One result, the measured spectral profile of the $(6s^2)^1S_0 \leftrightarrow (6s6p)^3P_0$ clock transition of the ^{199}Hg atoms, is shown in Fig. 9.28, in which two saturation peaks can be more clearly resolved. This observed doublet structure is due to the recoil effect of photon absorption with respect to the two counter-propagating probe laser beams. It should be noted that the physical meaning of the ordinate for Figs. 9.27 and 9.28 is the same.

PROBLEMS

1. For a Na vapor sample at $T = 500$ K temperature, the linear absorption wavelength is 589 nm. Calculate the (first-order) Doppler broadening and the second-order Doppler broadening (in units of MHz).
2. For the saturation spectroscopic experiment shown Fig. 9.2(a) assume that $\omega' = \omega_0$ and $(\Delta\omega)_D \approx 40\Gamma$, plot the Doppler-broadening absorption curve of $(\alpha + \Delta\alpha)/\alpha(\omega_0)$ as a function of $(\omega'' - \omega')/\Gamma$ at the intensity level of $I_1 = 0.2I_s$ and determine the full width of the Lamb dip (in units of Γ).
3. Describe the mechanism difference between the stimulated Raman scattering effect and the CARS effect.
4. Explain the physical meanings of the three terms in Eq. (9.4-8), i.e., in what effects they will play a role.
5. In order to reduce the speed of a Na atom from 1000 m/s to 50 m/s by using the resonant radiation pressure method shown schematically in Fig. 9.23(b), how many photons from a 589-nm laser beam have to be absorbed by this atom?

REFERENCES

1. T. W. Hänsch, in *Nonlinear Spectroscopy*, edited by N. Bloembergen (North-Holland, Amsterdam, 1977), pp. 17–86.
2. V. K. Letokhov and V. P. Chebotayev, *Nonlinear Laser Spectroscopy* (Springer-Verlag, Berlin, 1977).
3. M. D. Levenson and S. S. Kano, *Introduction to Nonlinear Laser Spectroscopy* (Academic Press, New York, 1988).
4. W. Demtröder, *Laser Spectroscopy: Basic Concepts and Instrumentation*, 2nd ed. (Springer-Verlag, Berlin, 1996).
5. W. R. Bennett, Jr., *Phys. Rev.* **126**, 580 (1962).
6. P. H. Lee and M. L. Skolnick, *Appl. Phys. Lett.* **10**, 303 (1967).
7. R. L. Barger and J. L. Hall, *Phys. Rev. Lett.* **22**, 4 (1969).
8. R. G. Brewer, M. J. Kelley, and A. Javan, *Phys. Rev. Lett.* **23**, 559 (1969).
9. T. W. Hänsch, M. D. Levenson, and A. L. Schawlow, *Phys. Rev. Lett.* **26**, 946 (1971).
10. T. W. Hänsch, I. S. Shahin, and A. L. Schawlow, *Phys. Rev. Lett.* **27**, 707 (1971).

11. C. K. N. Patel, *Appl. Phys. Lett.* **25**, 112 (1974).
12. T. W. Hänsch, M. H. Nayfeh, S. A. Lee, S. M. Curry, and I. S. Shahin, *Phys. Rev. Lett.* **32**, 1336 (1974).
13. B. Couillaud, L. A. Bloomfield, J. E. Lawler, A. Siegel, T. W. Hansch, *Opt. Commun.* **35**, 359 (1980); C. J. Sansonetti and D. Veza, *J. Phys. B* **43**, 205003 (2010).
14. S. Kroell and A. Persson, *Opt. Commun.* **54**, 277 (1985).
15. S. Svanberg, G. Y. Yan, T. P. Duffey, and A. L. Schawlow, *Opt. Lett.* **11**, 138 (1986).
16. F. Riehle, J. Ishikawa, and J. Helmcke, *Phys. Rev. Lett.* **61**, 2092 (1988).
17. D. S. Gough and P. Hannaford, *Opt. Commun.* **67**, 209 (1988); S. Bouazza, D. S. Gough, P. Hannaford, M. Wilson, and C. Lim, *J. Phys. B* **35**, 651 (2002).
18. A. Amy-Klein, O. Gorceix, S. Le Boiteux, J. R. Rios Leite, and M. Ducloy, *Opt. Commun.* **90**, 265 (1992).
19. C. J. Sansonetti, B. Richou, R. Englenman, Jr., and L. J. Radziemski, *Phys. Rev. A* **52**, 2682 (1995).
20. D. A. Tate and D. N. Aturaliye, *Phys. Rev. A* **56**, 1844 (1997).
21. O. Schmidt, K.-M. Knaak, R. Wynands, and D. Meschede, *Appl. Phys. B* **59**, 167 (1994); S. Briaudeau, D. Bloch, and M. Ducloy, *Phys. Rev. A* **59**, 3723 (1999).
22. D.-H. Kwon and Y. Rhee, *J. Opt. Soc. Am. B* **21**, 1250 (2004).
23. D. C. Heinecke, A. Bartels, T. M. Fortier, D. A. Braje, L. Hollberg, and S. A. Diddams, *Phys. Rev. A* **80**, 053806 (2009); G. Skolnik, N. Vujicic, and T. Ban, *Opt. Commun.* **282**, 1326 (2009).
24. L. Mudarikwa, K. Pahwa, and J. Goldwin, *J. Phys. B* **45**, 065002 (2012).
25. G. M. Tino, L. Hollberg, A. Sasso, M. Inguscio, and M. Barsanti, *Phys. Rev. Lett.* **64**, 2999 (1990); R. M. Jennerich, and D. A. Tate, *Phys. Rev. A* **62**, 042506 (2000).
26. D. A. Tate and J. P. Walton, *Phys. Rev. A* **59**, 1170 (1999).
27. R. M. Jennerich, A. N. Keiser, and D. A. Tate, *Eur. Phys. J. D* **40**, 81 (2006).
28. W. Williams, Z.-T. Lu, K. Rudinger, C.-Y. Xu, R. Yokochi, and P. Mueller, *Phys. Rev. A* **83**, 012512 (2011).
29. D. B. Cassidy, T. H. Hisakado, H. W. K. Tom, and A. P. Mills, Jr., *Phys. Rev. Lett.* **109**, 073401 (2012).
30. F. Herlemont, J. Fleury, J. Lemaire, and J. Demaison, *J. Chem. Phys.* **76**, 4705 (1982).
31. D. P. Godwin and R. J. Butcher, *Infrared. Phys.* **34**, 213 (1993).
32. D. P. Godwin, A. J. Gray, and R. J. Butcher, *J. Mol. Spectrosc.* **158**, 147 (1993).
33. M. de Labachelerie, K. Nakagawa, and M. Ohtsu, *Opt. Lett.* **19**, 840 (1994); C. S. Edwards, G. P. Barwood, H. S. Margolis, P. Gill, and W. R. C. Rowley, *J. Mol. Spectrosc.* **234**, 143 (2005).
34. C. R. Bucher, K. K. Lehmann, D. F. Plusquellic, and G. T. Fraser, *Appl. Opt.* **39**, 3154 (2000).
35. D. Mazzotti, P. De Natale, G. Giusfredi, C. Fort, J. A. Mitchell, and L. Hollberg, *Opt. Lett.* **25**, 350 (2000); A. Castrillo, E. De Tommasi, L. Gianfrani, L. Sirigu, and J. Faist, *Opt. Lett.* **31**, 3040 (2006).
36. S. M. Skoff, R. J. Hendricks, C. D. J. Sinclair, M. R. Tarbutt, J. J. Hudson, D. M. Segal, B. E. Sauer, and E. A. Hinds, *New. J. Phys.* **11**, 123026 (2009).
37. P. Dupre, *Phys. Rev. A* **85**, 042503 (2012).
38. A. M. Maartensson-Pendrill, D. S. Gough, and P. Hannaford, *Phys. Rev. A* **49**, 3351 (1994).
39. S. Krins, S. Oppel, N. Huet, J. von Zanthier, and T. Bastin, *Phys. Rev. A* **80**, 062508 (2009).
40. L. S. Vasilenko, V. P. Chebotayev, and A. V. Shishaev, *JETP Lett.* **12**, 113 (1970).
41. F. Biraben, B. Cagnac, and G. Grynberg, *Phys. Rev. Lett.* **32**, 643 (1974).
42. M. D. Levenson and N. Bloembergen, *Phys. Rev. Lett.* **32**, 645 (1974).
43. T. W. Hänsch, K. C. Harvey, G. Meisel, and A. L. Schawlow, *Opt. Commun.* **11**, 50 (1974).
44. N. Bloembergen, M. D. Levenson, and M. M. Salour, *Phys. Rev. Lett.* **32**, 867 (1974).
45. T. W. Hänsch, S. A. Lee, R. Wallenstein, and L. Wieman, *Phys. Rev. Lett.* **34**, 307 (1975).
46. S. A. Lee, R. Wallenstein, and T. W. Hänsch, *Phys. Rev. Lett.* **35**, 1262 (1975).
47. F. Biraben, B. Cagnac, and G. Grynberg, *Phys. Lett.* **48A**, 469 (1974).
48. K. C. Harvey, R. T. Hawkins, G. Meisel, and A. L. Schawlow, *Phys. Rev. Lett.* **34**, 1073 (1975).

49. E. A. Hildum, U. Boesl, D. H. McIntyre, R. G. Beausoleil, and T. W. Hänsch, *Phys. Rev. Lett.* **56**, 576 (1986).
50. C. Schwob, L. Jozefowski, B. de Beauvoir, L. Hilico, F. Nez, L. Julien, F. Biraben, O. Acef, and A. Clairon, *Phys. Rev. Lett.* **82**, 4960 (1999).
51. S. R. Bramwell, A. I. Ferguson, and D. M. Kane, *Opt. Lett.* **12**, 666 (1987).
52. F. Arqueros, P. E. LaRocque, M. S. O'Sullivan, and B. P. Stoicheff, *Opt. Lett.* **9**, 82 (1984).
53. M.-S. Ko and Y.-W. Liu, *Opt. Lett.* **29**, 1799 (2004).
54. V. Jacques, B. Hingant, A. Allafort, M. Pigeard, and J. F. Roch, *Eur. J. Phys.* **30**, 921 (2009).
55. E. Riedle, H. J. Neusser, and E. W. Schlag, *J. Chem. Phys.* **75**, 4231 (1981).
56. M. Misono, J. Wang, M. Okubo, S. Kasahara, H. Kato, and M. Baba, *Mol. Phys.* **100**, 1147 (2002).
57. P. D. Maker and R. W. Terhune, *Phys. Rev.* **137**, A801 (1965).
58. J. J. Wynne, *Phys. Rev.* **178**, 1295 (1969).
59. M. D. Levenson, C. Flytzanis, and N. Bloembergen, *Phys. Rev. B* **6**, 3962 (1972).
60. J. J. Wynne, *Phys. Rev. Lett.* **29**, 650 (1972).
61. M. D. Levenson and N. Bloembergen, *Phys. Rev. B* **10**, 4447 (1974).
62. H. Lotem, R. T. Lynch, Jr., and N. Bloembergen, *Phys. Rev. A* **14**, 1748 (1976).
63. R. F. Begley, A. B. Harvey, and R. L. Byer, *Appl. Phys. Lett.* **25**, 387 (1974).
64. L. P. Goss, J. W. Fleming, and A. B. Harvey, *Opt. Lett.* **5**, 345 (1980).
65. W. M. Tolles, J. W. Nibler, J. R. McDonald, and A. B. Harvey, *Appl. Spectrosc.* **31**, 253 (1977).
66. A. B. Harvey and J. W. Nibler, *Appl. Spectrosc. Rev.* **14**, 101 (1978).
67. G. L. Eesley, *Coherent Raman Spectroscopy* (Pergamon, Oxford, 1981).
68. P. F. Chimento, M. Jurna, H. S. P. Bouwmans, E. T. Garbacik, L. Hartsuiker, C. Otto, J. L. Herek, and H. L. Offerhaus, *J. Raman Spectrosc.* **40**, 1229 (2009).
69. D. Heiman, R. W. Hellwarth, M. D. Levenson, and G. Martin, *Phys. Rev. Lett.* **36**, 189 (1976).
70. M. D. Levenson and J. J. Song, *J. Opt. Soc. Am.* **66**, 641 (1976).
71. Y. R. Shen, *The Principle of Nonlinear Optics* (Wiley, New York, 1984), p. 276.
72. I. Reinhold and M Maier, *Opt. Commun.* **5**, 31 (1972).
73. A. Owyoung and E. D. Jones, *Opt. Lett.* **1**, 152 (1977).
74. B. F. Levine, C. V. Shank, and J. P. Heritage, *IEEE J. Quantum Electron.* **15**, 1418 (1979).
75. B. F. Levine and C. G. Bethea, *IEEE J. Quantum Electron.* **16**, 85 (1980).
76. W. J. Jones and B. P. Stoicheff, *Phys. Rev. Lett.* **13**, 657 (1964).
77. R. A. McLaren and B. P. Stoicheff, *Appl. Phys. Lett.* **16**, 140 (1970).
78. R. R. Alfano and S. L. Shapiro, *Chem. Phys. Lett.* **8**, 631 (1971).
79. W. Werncke, J. Klein, A. Lau, K. Lenz, and G. Hunsalz, *Opt. Commun.* **11**, 159 (1974).
80. E. S. Yeung, *J. Mol. Spectrosc.* **53**, 379 (1974).
81. C. Wieman and T. W. Hänsch, *Phys. Rev. Lett.* **36**, 1170 (1976).
82. J-L. Oudar, R. W. Smith, and Y. R. Shen, *Appl. Phys. Lett.* **34**, 758 (1979).
83. M. E. Kaminsky, R. T. Hawkins, F. V. Kowalski, and A. L. Schawlow, *Phys. Rev. Lett.* **36**, 671 (1976).
84. R. Teets, R. Feiberg, T. W. Hänsch, and A. L. Schawlow, *Phys. Rev. Lett.* **37**, 683 (1976).
85. N. W. Carlson, A. J. Taylor, K. M. Jones, and A. L. Schawlow, *Phys. Rev. A* **24**, 822 (1981).
86. P. F. Liao and G. C. Bjorklund, *Phys. Rev. Lett.* **36**, 584 (1976).
87. M. H. Kabir, S. Kasahara, W. Demtroder, Y. Tatamitani, M. Okubo, M. Misono, J. Wang, M. Baba, D. L. Joo, J. O'Reilly, A. Doi, Y. Kimura, and H. Kato, *Chem. Phys.* **283**, 237 (2002).
88. V. S. Letokhov, *JETP Lett.* **7**, 272 (1968).
89. A. Ashkin, *Phys. Rev. Lett.* **25**, 1321 (1970).
90. T. W. Hänsch and A. L. Schawlow, *Opt. Commun.* **13**, 68 (1975).
91. D. J. Wineland, R. E. Drullinger, and F. L. Walls, *Phys. Rev. Lett.* **40**, 1639 (1978).

92. J. Prodan, A. Migdall. W. D. Phillips, I. So, H. Metcalf, and J. Dalibard, *Phys. Rev. Lett.* **54**, 992 (1985); A. L. Migdall, J. V. Prodan, W. D. Philips, T. H. Bergeman, and H. J. Metcalf, *Phys. Rev. Lett.* **54**, 2596 (1985).
93. S. Chu, L. Hollberg, J. E. Bjorkholm, A. Cable, and A. Ashkin, *Phys. Rev. Lett.* **55**, 48 (1985).
94. E. L. Raab, M. Prentiss, A. Cable, S. Chu, and D. E. Fritchard, *Phys. Rev. Lett.* **59**, 2631 (1987).
95. C. Monroe, W. Swann, H. Robinson, and C. Wieman, *Phys. Rev. Lett.* **65**, 1571 (1990).
96. N. P. Georgiades, E. S. Polzik, and H. J. Kimble, *Opt. Lett.* **19**, 1474 (1994).
97. P. S. Jessen, C. Gerz, P. D. Lett, W. D. Phillips, S. L. Rolston, R. J. C. Spreeuw, and C. I. Westbrook, *Phys. Rev. Lett.* **69**, 49 (1992).
98. M. J. Snadden, A. S. Bell, E. Riis, and A. I. Ferguson, *Opt. Commun.* **125**, 70 (1996).
99. A. Marian, M. C. Stowe, D. Felinto, and J. Ye, *Phys. Rev. Lett.* **95**, 023001 (2005).
100. M. Maric, J. J. McFerran, and A. N. Luiten, *Phys. Rev. A* **77**, 032502 (2008).
101. J. E. Simsarian, W. Shi, L. A. Orozco, G. D. Sprouse, and W. Z. Zhao, *Opt. Lett.* **21**, 1939 (1996).
102. J. M. Grossman, R. P. Fliller III, T. E. Mehlstaubler, L. A. Orozco, M. R. Pearson, G. D. Sprouse, and W. Z. Zhao, *Phys. Rev. A* **62**, 052507 (2000).
103. E. A. Curtis, C. W. Oates, and L. Hollberg, *J. Opt. Soc. Am. B* **20**, 977 (2003).
104. H. Katori, T. Ido, Y. Isoya, and M. Kuwata-Gonokami, *Phys. Rev. Lett.* **82**, 1116 (1999).
105. I. Courtillot, A. Quessada, R. P. Kovacich, J.-J. Zondy, A. Landragin, A. Clairon, and P. Lemonde, *Opt. Lett.* **28**, 468 (2003).
106. T. Kuwamoto, K. Honda, Y. Takahashi, and T. Yabuzaki, *Phys. Rev. A* **60**, R745 (1999).
107. C. W. Hoyt, Z. W. Barber, C. W. Oates, T. M. Fortier, S. A. Diddams, and L. Hollberg, *Phys. Rev. Lett.* **95**, 083003 (2005).
108. J. J. McClelland and J. L. Hanssen, *Phys. Rev. Lett.* **96**, 143005 (2006).
109. A. J. Berglund, J. L. Hanssen, and J. J. McClelland, *Phys. Rev. Lett.* **100**, 113002 (2008).
110. M. Lu, S. H. Youn, and B. L. Lev, *Phys. Rev. A* **83**, 012510 (2011).
111. H. Hachisu, K. Miyagishi, S. G. Porsev, A. Derevianko, V. D. Ovsiannikov, V. G. Pal'chikov, M. Takamoto, and H. Katori, *Phys. Rev. Lett.* **100**, 053001 (2008).
112. M. Petersen, R. Chicireanu, S. T. Dawkins, D. V. Magalhaes, C. Mandache, Y. Le Coq, A. Clairon, and S. Bize, *Phys. Rev. Lett.* **101**, 183004 (2008).
113. H. J. Metcalf, and P. van der Straten, *Laser Cooling and Trapping* (Springer, New York, 1999).
114. L.-B. Wang, P. Mueller, K. Bailey, G. W. F. Drake, J. P. Greene, D. Henderson, R. J. Holt, R. V. F. Janssens, C. L. Jiang, Z.-T. Lu, T. P. O'Connor, R. C. Pardo, K. E. Rehm, J. P. Schiffer, and X. D. Tang, *Phys. Rev. Lett.* **93**, 142501 (2004).
115. P. Mueller, I. A. Sulai, A. C. C. Villari, J. A. Alcántara-Núñez, R. Alves-Condé, K. Bailey, G. W. F. Drake, M. Dubois, C. Eléon, G. Gaubert, R. J. Holt, R. V. F. Janssens, N. Lecesne, Z.-T. Lu, T. P. O'Connor, M.-G. Saint-Laurent, J.-C. Thomas, and L.-B. Wang, *Phys. Rev. Lett.* **99**, 252501 (2007).

10
Optical Coherent Transient Effects

When an intense and short laser pulse or a fast-switching laser field interacts with a resonant absorbing medium, several optical coherent transient effects can be observed, including the self-induced transparency effect, photon echo effect, optical nutation effect, and the optical free-induction decay effect. To observe these optical coherent transient effects, the following conditions should be satisfied: (i) the frequency of the incident light field must be equal or very close to the resonant absorption frequency of the medium, (ii) the pulse duration or the switching time of an applied optical field is much shorter than the transverse relaxation (dephasing) time T_2 of the absorbing molecular system, and (iii) the applied optical field is strong enough so that the Rabi frequency Ω is much greater than $1/T_2$. The study of these effects enables us to obtain a deeper understanding of the coherent transient interaction between the intense pulsed or rapidly switched optical field and the resonant medium, and to measure the magnitude of the dipole-moment matrix elements, relaxation parameters, and spectral broadening mechanisms of the investigated materials. On the other hand, these optical coherent transient effects can also be applied to develop some special techniques for optical pulse delay, pulse reshaping, and optical pulse manipulation.

10.1 Coherent transient interaction of intense light with a resonant medium

So far in this book, we have mainly considered the steady-state (or quasi-steady-state) interaction between the laser fields and the transparent (non-absorbing) materials. On these conditions, we have assumed that (i) the electric polarization amplitude function $P_0(t)$ of a given medium is dependent only on the instantaneous value of the optical field amplitude function $E_0(t)$, and (ii) the number of molecules (or atoms) making an absorptive transition from the ground state to a higher real exited state is negligible.

Now let us consider another kind of interaction wherein an intense short pulsed or fast-switching optical field is incident upon a resonant absorbing medium. In such a case, the temporal change of population distribution between two optically correlated states and the phase relaxation of the excited-state population of the medium play a vital role. Meanwhile, the response of the medium may depend not only on the instantaneous value of the applied field function $E_0(t)$, but also on the integration value of $E_0(t)$ over a certain time interval. This kind of interaction is termed the coherent transient interaction.

For simplicity, we suppose that the resonant medium possesses only two energy levels with a characteristic transition frequency ω_0, which is equal to the frequency of the applied laser field.

In this case, we usually use the following three parameters to describe the relaxation property and spectral broadening of the transition of molecules: (i) the longitudinal relaxation time T_1, defined by $T_1 = 1/\delta\nu_{nat}$ where $\delta\nu_{nat}$ is the natural linewidth of the transition, and T_1 actually is the lifetime of the excited molecules staying in the high energy level; (ii) the transverse relaxation time T_2, defined by $T_2 \approx 1/(\delta\nu)'$, where $(\delta\nu)'$ is the homogeneous broadening linewidth of the related transition; and (iii) the inhomogeneous relaxation (dephasing) time T_2^*, defined by $T_2^* \approx 1/(\delta\nu)''$, where $(\delta\nu)''$ is the inhomogeneous broadening linewidth of the same resonant transition. For solid-state media in very low temperature, it usually holds that $T_2^* < T_2 < T_1$; for gaseous media in normal temperature, we usually have $T_2^* << T_2 < T_1$.

Suppose that the pulse duration or the rapidly switching time of an incident coherent optical field is Δt; the condition for coherent transient interaction can be described as

$$\Delta t << T_2, T_1. \qquad (10.1\text{-}1)$$

We say, in this case, the process is transient because the considered time duration of interaction is shorter than the characteristic times T_1 and T_2 of the molecular system. We also say this interaction is coherent, because within such a short period of time all molecules can react synchronously to the optical field, since the relaxation influences from spontaneous emission and homogeneous broadening can be negligible. On the other hand, for such a short input laser pulse, the pulse duration is actually equal to the coherent time of the optical field, and therefore the applied field can be recognized as a nearly perfect coherent electromagnetic field.

From the above description, one realizes that we are now dealing with the fast interaction between an intense coherent optical field and a resonant molecular system, and the relatively slow relaxation processes characterized by T_1 and sometime also by T_2 can be neglected within the interaction period. One unique feature of this kind of interaction is that the response of the resonant medium may depends not only on the optical electric field $E_0(t)$, but also on the pulse "area," i.e., the integration of $E_0(t)$ over time t.

In the following section we will find that when the area of the integral of $E(t)$ over t is equal to the value of 2π, the resonant medium will manifest no absorption for the incident optical pulse, i.e., after passing through this medium the incident light pulse will retain its pulse shape and energy unchanged. This is the so-called self-induced transparency effect, and the light pulse satisfying the above requirement is termed "2π-pulse."

10.2 Self-induced transparency

10.2.1 Definition of 2π-pulse and self-induced transparency

Self-induced transparency means that under the condition of transient coherent interaction, the transmissivity of a resonant absorbing medium is dependent sensitively on the integral area of amplitude function $E_0(t)$ with respect to time t of an input pulsed optical field. In particular, if the value of this integral area (multiplying by $2\pi p_0/h$, where p_0 is the expectation value of the electric dipole moment operator of a molecular transition) is equal to 2π, the medium will show an entirely transparent behavior, i.e., the shape and the energy of the transmitted optical pulse remain unchanged after passing through this absorptive medium. This is a 2π-pulse-induced self-transparency process.[1,2] This process can be explained by the following physical picture: the absorbed energy from the first half of the input light pulse is re-emitted and added to the second half of the transmitted light pulse. Because of this special energy exchange between the optical field and resonant medium, the measured propagation speed of the pulse signal will be slower than that of an ordinary light pulse traveling in the same medium without satisfying the 2π-pulse condition.

For simplicity, we assume that the absorptive molecules are doped in a transparent and non-resonant host medium. In this case, the overall electric polarization of the medium includes two parts:

$$\overline{P} = P_0 + P. \tag{10.2-1}$$

Here, P_0 is the non-resonant polarization contribution from the host medium, which determines the linear refractive index n_0 of this medium, and P is the resonant polarization contribution from the dopant molecules, which is the key subject we shall discuss here.

We further assume that the dopant molecule has two energy levels with a resonant transition frequency ω_0, the density of these molecules is N_0, the difference of population densities in the two levels is $N = N_1 - N_2$, and before applying an external optical field all dopant molecules are in the lower level, i.e., $N_1(t=0) = N_0$. If the incident optical field is a linearly polarized monochromatic plane wave propagating along the z-axis, we can neglect the vector property of the field and write the wave equation as (cf. Eq. (6.1-5) and Eq. (5.1-1)):

$$\frac{\partial^2 E(t)}{\partial z^2} - \frac{n_0}{c^2}\frac{\partial^2 E(t)}{\partial t^2} = \mu_0 \frac{\partial^2 P(t)}{\partial t^2}. \tag{10.2-2}$$

Here, c is the speed of light in a vacuum, n_0 is the linear refractive index of the medium, μ_0 is the permeability of vacuum, and $P(t)$ is the resonant part of electric polarization attributed to the absorptive molecules and determined by the following material equations[3,4]:

$$\left.\begin{array}{l}\dfrac{\partial^2 P}{\partial t^2} + \dfrac{2}{T_2}\dfrac{\partial P}{\partial t} + \omega_0^2 P = \dfrac{4\pi \omega_0 p_0^2}{h} NE \\[2mm] \dfrac{\partial N}{\partial t} + \dfrac{1}{T_1}[N - N_0] = -\dfrac{4\pi}{h\omega_0} E \dfrac{\partial P}{\partial t}\end{array}\right\}, \tag{10.2-3}$$

where p_0 is the quantum expectation value of the molecular dipole moment operator and h is Planck's constant.

In the semi-classical approach, the optical field E and the resonant electric polarization P of the medium can be expressed in the form of real functions

$$\left.\begin{array}{l}E(z,t) = E_0(z,t)\cos[\omega t - kz + \varphi]\\ P(z,t) = P_0(z,t)\cos[\omega t - kz + \phi]\end{array}\right\}, \tag{10.2-4}$$

where k is the magnitude of the wavevector of the optical field, ϕ and φ are the phase factors, and $E_0(z,t)$ and $P_0(z,t)$ are slowly-varying amplitude functions that satisfy the following conditions:

$$\left|\frac{\partial E_0}{\partial z}\right| \ll k|E_0|\,;\; \left|\frac{\partial P_0}{\partial t}\right| \ll \omega |P_0|.$$

Substituting Eq. (10.2-4) into Eqs. (10.2-2) and (10.2-3) leads to the following equations:

$$\left.\begin{aligned} \frac{\partial E_0}{\partial t} + \frac{c}{n_0}\frac{\partial E_0}{\partial z} &= \frac{\omega P_0}{2n_0^2\varepsilon_0}\sin(\phi-\varphi) \\ E_0\left(\frac{\partial \varphi}{\partial t} + \frac{c}{n_0}\frac{\partial \varphi}{\partial z}\right) &= -\frac{\omega P_0}{2n_0^2\varepsilon_0}\cos(\phi-\varphi) \\ \frac{\partial P_0}{\partial t} + \frac{P_0}{T_2} &= -\frac{2\pi p_0^2}{h}NE_0\sin(\phi-\varphi) \\ \left[\frac{\partial \phi}{\partial t}+(\omega-\omega_0)\right]P_0 &= -\frac{2\pi p_0^2}{h}NE_0\cos(\phi-\varphi) \\ \frac{\partial N}{\partial t} + \frac{N-N_0}{T_1} &= \frac{2\pi}{h}P_0 E_0\sin(\phi-\varphi) \end{aligned}\right\}, \quad (10.2\text{-}5)$$

where ε_0 is the vacuum permittivity. If the optical field is in exact resonance with the absorbing molecular system, i.e., $\omega = \omega_0$, we then have $\phi - \varphi = \pi/2$ and

$$\left.\begin{aligned} \frac{\partial E_0}{\partial t} + \frac{c}{n_0}\frac{\partial E_0}{\partial z} &= \frac{1}{2n_0^2\varepsilon_0}\omega_0 P_0 \\ \frac{\partial P_0}{\partial t} + \frac{1}{T_2}P_0 &= -\frac{2\pi p_0^2}{h}NE_0 \\ \frac{\partial N}{\partial t} + \frac{1}{T_1}(N-N_0) &= \frac{2\pi}{h}P_0 E_0 \end{aligned}\right\}. \quad (10.2\text{-}6)$$

For coherent transient interaction, as indicated in the preceding section, the pulse duration is much shorter than T_1 and T_2; thus in Eq. (10.2-6) we can let $T_1, T_2 \Rightarrow \infty$, and Eq. (10.2-6) is simplified to

$$\left.\begin{aligned} \frac{\partial E_0}{\partial t} + \frac{c}{n_0}\frac{\partial E_0}{\partial z} &= \frac{1}{2n_0^2\varepsilon_0}\omega_0 P_0 \\ \frac{\partial P_0}{\partial t} &= -\frac{2\pi p_0^2}{h}NE_0 \\ \frac{\partial N}{\partial t} &= \frac{2\pi}{h}P_0 E_0 \end{aligned}\right\}, \quad (10.2\text{-}7)$$

The solutions of the last two equations in Eq. (10.2-7) are

$$\left.\begin{aligned} N &= N_0\cos\theta \\ P_0 &= -p_0 N_0 \sin\theta \end{aligned}\right\}, \quad (10.2\text{-}8)$$

where

$$\theta(z,t) = \frac{2\pi p_0}{h}\int_{-\infty}^{t} E_0(z,t')dt'. \quad (10.2\text{-}9)$$

In the following analyses, the key parameter is the θ value determined by the integral area of the input light pulse, i.e.,

$$\theta(z) = \frac{2\pi p_0}{h} \int_{-\infty}^{\infty} E_0(z, t') dt'. \tag{10.2-10}$$

It can be seen from the first part of Eq. (10.2-8) that when $\theta(z) = \pi/2, N = 0$ and $N_1 = N_2$, so the population densities in the two energy levels are equal to each other, and this is the so-called $\pi/2$-pulse interaction. When $\theta(z) = \pi, N = -N_0$, so all molecules are excited to the high energy level, and this is the so-called π-pulse interaction. Finally, if $\theta(z) = 2\pi$ and $N = N_0$, all molecules will still remain in the low energy level; this corresponds to the so-called 2π-pulse interaction. In the last case, the coherent transient interaction does not produce any net change of the molecule distribution between these two energy levels, and therefore the energy of the light pulse remains unchanged. To justify this conclusion, we must further consider the energy flux density S of the input light pulse as a function of the propagation distance z:

$$S(z) = \frac{1}{2} \varepsilon_0 c n_0 \int_{-\infty}^{\infty} E_0^2(z, t) dt.$$

The variation of $S(z)$ following the propagation distance in the resonant medium can be derived as (see Appendix 8.1)

$$\frac{dS(z)}{dz} = \frac{\hbar \omega_0}{4\pi} N_0 [\cos \theta(z) - 1]. \tag{10.2-11}$$

When $\vartheta(z) = 2\pi$ and $dS(z)/dz = 0$, we indeed see the self-induced transparency of a 2π-pulse. On the other hand, based on the numerical solution of Eq. (10.2-7), for several different input pulse shapes and on the condition of $\theta(z) < 3\pi$, the following formula is approximately valid:[2,5]

$$\frac{dS(z)}{dz} = -\alpha S(z) \frac{2[1 - \cos \theta(z)]}{\theta^2(z)}, \tag{10.2-12}$$

where α is the ordinary linear absorption coefficient of the absorbing medium. If the incident pulse is so weak that $\theta(z) \ll \pi$ and $\cos \theta \approx 1 - (\theta^2/2)$, then the above equation becomes much simpler and leads to the following solution:

$$S(z) = S(0) e^{-\alpha z}. \tag{10.2-13}$$

This is the ordinary exponential decay of a weak light pulse. Once again, when $\theta(z) = 2\pi$, $\cos \theta(z) = 1$ and $dS(z)/dz = 0$, so we obtain the same result for the 2π-pulse.

10.2.2 Shape and speed of the 2π-pulse

Now we shall determine the shape and propagation speed of a stable 2π-pulse passing through the resonant medium. For this purpose, we substitute the second equation in Eq. (10.2-8) into the first equation in Eq. (10.2-7) and obtain

$$\frac{\partial E_0}{\partial t} + \frac{c}{n_0} \frac{\partial E_0}{\partial z} = -\frac{\omega_0 p_0 N_0}{2 n_0^2 \varepsilon_0} \sin \theta(z, t). \tag{10.2-14}$$

Making a substitution of $\tau = t - (z/\xi)$ and noticing that

$$\frac{\partial E_0}{\partial t} = \frac{\partial E_0}{\partial \tau}, \quad \frac{\partial E_0}{\partial z} = -\frac{1}{\xi}\frac{\partial E_0}{\partial \tau},$$

where ξ represents the propagation speed of the pulse in the resonant medium, one could rewrite Eq. (10.2-14) as

$$\frac{\partial E_0(z,\tau)}{\partial \tau} = \frac{1}{\tau_0^2}\frac{h}{2\pi p_0}\sin\theta(z,\tau), \tag{10.2-15}$$

where

$$\frac{1}{\tau_0^2} = \left[\frac{1}{\xi} - \frac{n_0}{c}\right]^{-1}\frac{\pi p_0^2 \omega_0 N_0}{n_0 c \varepsilon_0 h}. \tag{10.2-16}$$

On can readily find that Eq. (10.2-15) leads to another relation:

$$E_0(z,\tau) = \frac{h}{\pi \tau_0 p_0}\sin\frac{1}{2}\theta(z,\tau). \tag{10.2-17}$$

Based on Eqs. (10.2-15) and (10.2-17), a particular solution describing the stable shape of the 2π-pulse traveling in a resonant absorbing medium can be obtained in the form of a hyperbolic secant function (see Appendix 8.2):[2]

$$E_0^o\left(t - \frac{z}{\xi}\right) = \frac{h}{\pi \tau_0 p_0}\operatorname{sech}\left[\left(t - \frac{z}{\xi}\right)/\tau_0\right]. \tag{10.2-18}$$

Here the parameter τ_0 is determined by Eq. (10.2-16) and it represents the temporal width of the 2π-pulse. Integrating the above expression with respect to τ results in

$$\frac{2\pi p_0}{h}\int_{-\infty}^{\infty} E_0^o(\tau)d\tau = 2\pi. \tag{10.2-19}$$

This result verifies that the hyperbolic secant function indicated by Eq. (10.2-18) is indeed the appropriate expression of the shape of a stable 2π-pulse.

From Eq. (10.2-16) we can further determine the propagation speed ξ of the 2π-pulse in the resonant absorbing medium by

$$\frac{1}{\xi} - \frac{n_0}{c} = \frac{1}{2}\alpha\frac{\tau_0^2}{T_2}, \tag{10.2-20}$$

where α is the linear absorption coefficient of the medium and can be defined as

$$\alpha = \frac{2\pi p_0^2 \omega_0 N_0 T_2}{n_0 c \varepsilon_0 h}. \tag{10.2-21}$$

Here, we assume that the influence from the homogeneous broadening mechanism is the dominant factor. From Eq. (10.2-20) one can see that the speed of the 2π-pulse is always lower than that of ordinary weak light pulses.

The foregoing are the properties of a precise 2π-pulse signal. In a more general case, the initial area of an input light pulse, $\theta(z=0) = \theta_0$, may have an arbitrary value. In this case the variation of the pulse area θ following the propagation distance z can be expressed as[1,2]

$$\frac{d\theta(z)}{dz} = -\frac{\alpha}{2}\sin\theta(z). \qquad (10.2\text{-}22)$$

The general solution of this equation is

$$\tan\frac{\theta(z)}{2} = \tan\frac{\theta_0}{2}e^{-\alpha z/2}. \qquad (10.2\text{-}23)$$

Based on this solution one can examine the propagation behavior of the input pulse with various initial θ_0 values using numerical computation methods. As an example, Fig. 10.1(a) shows the calculated change of the pulse area $\theta(z)$ as a function of the propagation distance z. Figure 10.1(b) shows the variation of the pulse shape along the propagation distance z when the initial pulse area is chosen to be $\theta_0 = 0.9\pi$ and 1.1π, respectively; the shapes of the initial pulses are arbitrarily chosen to be Gaussian.

From Fig. 10.1 we can see that in case of $\theta_0 < \pi$, the input pulse decays rapidly and vanishes entirely within a distance of several multiples of $(\pi\alpha^{-1})$, whereas if $\pi < \theta_0 < 2\pi$, the pulse area increases following the propagation distance z until it reaches a stable 2π-pulse shape. Moreover, in a more general case, for an input pulse with the initial condition of

$$2\pi\left(m - \frac{1}{2}\right) < \theta_0 < 2\pi\left(m + \frac{1}{2}\right), \qquad (10.2\text{-}24)$$

the light pulse will split into m separate stable 2π-pulses during the propagation inside the resonant medium, and this pulse splitting process will be completed within a distance of several multiples of $(\pi\alpha^{-1})$.

So far, in this section we considered only the resonant absorptive medium, i.e., at the beginning all molecules are in the low energy level. On the contrary, if we consider a resonant gain medium, i.e., at the beginning all molecules are in the upper energy level and $N(t=0) = -N_0$, then the following conclusions can be obtained based on analogous theoretical analyses:[3,4]

1. There is a stable 2π-pulse and, after passing through the gain medium, its energy and pulse shape remain unchanged;

2. For an input pulse with $\theta_0 \geq m2\pi$, it may split into m stable 2π-pulses during the propagation inside the gain medium;

3. The propagation speed of the 2π-pulse in a resonant gain medium could be considerably greater than the ordinary speed of light (c/n_0). This is due to the fact that the leading part of the pulse is amplified by the resonant medium and the trailing part of the pulse is absorbed by the same medium; therefore, we may even observe a pulse energy propagation speed faster than the speed of an ordinary light pulse in the same medium without resonance.

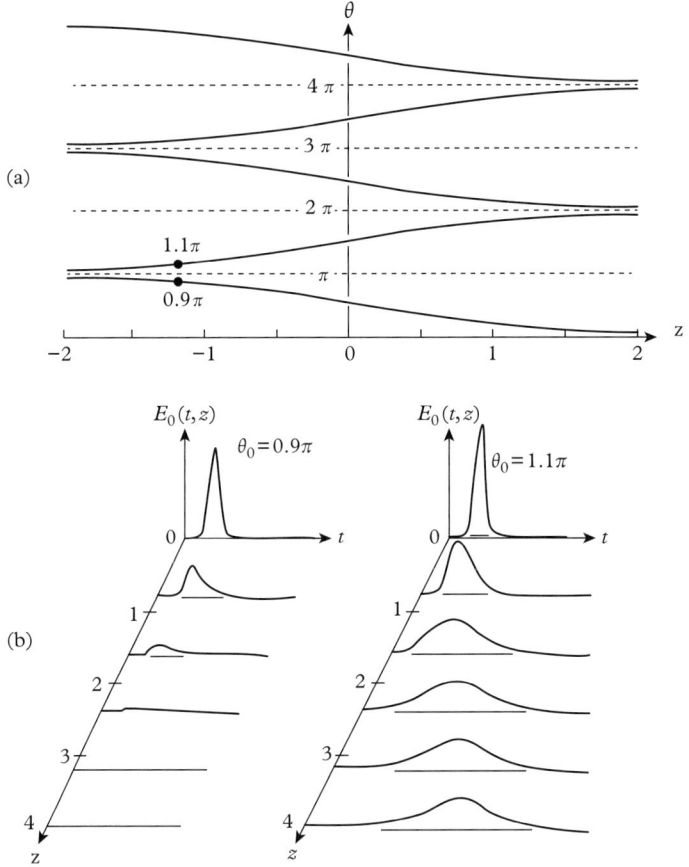

Figure 10.1 Self-induced transparency of short intense light pulses in a resonant absorbing medium: (a) the pulse area $\theta(z)$ as a function of the propagation distance z with an arbitrarily chosen $z = 0$ point; (b) the variation of pulse shape along the propagation distance z for two different initial values of θ_0. Units of z are multiples of $\pi\alpha^{-1}$. (Reproduced with permission from McCall and Hahn.[1] Copyright 1967, American Physical Society).

10.2.3 Experimental studies of self-induced transparency

Since the first observation of the self-induced transparency phenomenon in a ruby crystal in 1967, this effect in many other resonant absorbing media has been investigated. In particular, using the tunability of high-peak power pulsed lasers and optical parametric generators, researchers may considerably extend the sorts of resonant media as well as the scope of the studies.[6,7] The following are some of the resonant absorbing media which have been studied by using the self-induced transparency technique:

A. molecular gases such as SF_6,[8,9] BCl_3,[10] NH_3,[11] and HF;[12]
B. atomic gases such as Ne[13] and I;[14]
C. metal vapors such as Rb,[15,16] K,[17] Na,[18,19] Sm,[20] and Yb;[21]
D. metal ion-doped crystals or glasses such as $Cr^{3+}:Al_2O_3$,[1,2] $Pr^{3+}:LaF_3$,[22] $Cr^{3+}:LiAl_5O_8+Al_2O_3$,[23] and Er^{3+}-doped glass fiber;[24]
E. semiconductor crystals such as $CdS_{0.75}Se_{0.25}$[25] and InGaAs quantum-dot wavequides.[26]

One of the well-known experiments for the early studies of self-induced transparency was accomplished with the Rb vapor system working at the D_1 line nondegenerate transition of Rb atoms. The vapor cell was placed in an 74.5-kG magnetic field, so that its D_1 absorption line could be shifted from 794.76 nm to the wavelength of the 794.46-nm Hg^+-laser pulses of ∼7 ns duration.[15] The density of the Rb vapor was about $10^{12}/cm^3$, and the maximum intensity of the laser pulses was about 25 W/cm^2. The measured shapes of the output and input laser pulses at various input intensity levels are shown in Fig. 10.2(a), and the corresponding computer generated

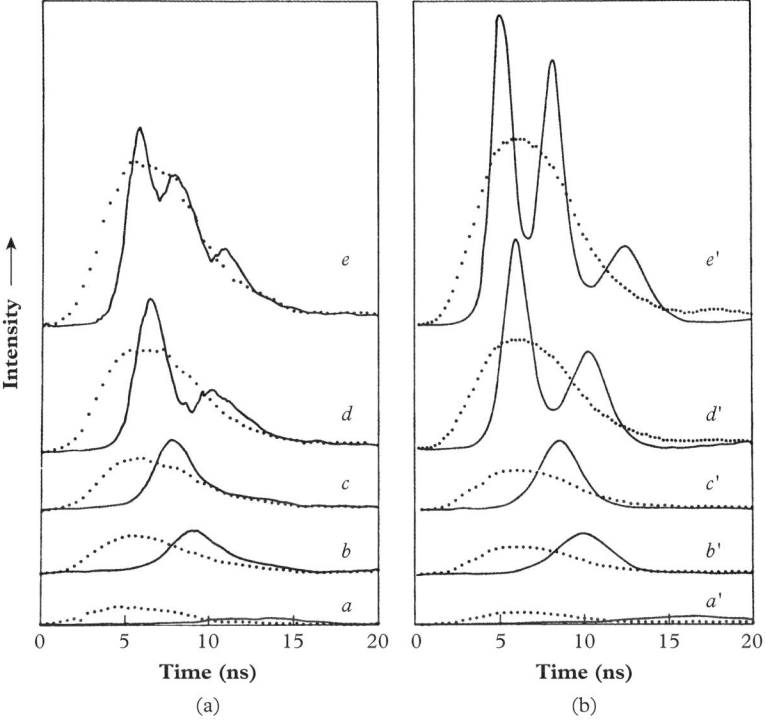

Figure 10.2 *Input (dotted-line) and output (solid-line) pulse shapes of the 794.46-nm laser beam passing through a 1-mm-thick Rb vapor cell: (a) measured curves at various input pulse area values and (b) computer simulated curves. (Reproduced with permission from Gibbs and Slusher.[15] Copyright 1970, American Physical Society).*

output pulse shapes under the same input conditions are shown in Fig. 10.2(b). In Fig. 10.2(a), the physical meaning of measured results at various input levels can be described as follows: (a) the input pulse area (θ_0) is just under $\sim\pi$ value and there is a maximum observed output pulse delay of 15–20 ns; (b) the θ_0 value is about 2π and the energy loss is nearly negligible; (c) the θ_0 value is about 3π and the peak value of the reshaped output pulse is higher than that of the input pulse; (d) $4\pi \leq \theta_0 \leq 5\pi$ and there is a double output pulse; and (e) $\theta_0 \approx 6\pi$ and the pulse broke into three 2π pulses with the peaks higher than the input pulse. These experimental results are basically in good agreement with the theoretical predictions, as shown in Fig. 10.2(b).

In most self-induced transparency studies, the absorptive transitions are from the ground state to the upper excited states. Nevertheless, the same effect can also be observed at a transition from a lower excited state to a higher excited state. As such an example, one experiment of self-induced transparency was implemented using a low-pressure ($\sim$0.75 Torr) neon dc-discharge tube as the absorber cell and the 1.15-μm He–Ne laser pulses as the excitation source. In this case, the degenerate transition from $2P_4$ level to $2S_2$ level of the atom neon is in resonance with the wavelength of the excitation laser pulses of 3-ns pulsewidth. The measured data of pulse energy transmission as a function of the input peak intensity are shown in Fig. 10.3 under conditions of two different values of the absorption αL.[13] From this figure one can see an undulating dependence of the

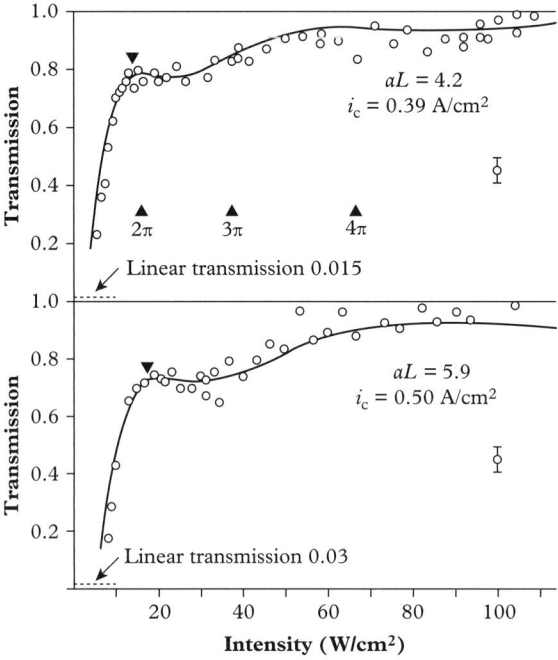

Figure 10.3 *Transmission of the 1.15-μm laser pulses passing through a Ne absorber tube as a function of the input peak intensity at two absorption values. Here i_c is the current density of the dc-discharge. (Reproduced with permission from Krieger and Toschek.[13] Copyright 1975, American Physical Society).*

transmission on the input peak intensity with the first maximum (for $\alpha L = 4.2$) at the position near 15 W/cm^2 and the second maximum at the position near 60 W/cm^2, respectively. These two maxima approximately correspond to the calculated input 2π-pulse and 4π-pulse, respectively. Here no 100% transmission could be achieved because of (i) the deviation of the input pulse shape from an ideal hyperbolic secant profile, (ii) the degeneracy of the transition, and (iii) the relaxation influence on the transition. In Fig. 10.3, the solid-line curves are obtained by using a phenomenological calculation model and fitted with the experimental value indicated by the downward arrow.

Another later experiment in ytterbium (Yb) vapor has shown a similar dependence of the nonlinear transmission on the input laser pulse area, and the reshaping effect of an input pulse with $\sim 2\pi$ area and irregular pulse shape.[21] In this experiment, the input laser pulse of ~ 5 ns duration was from a single-mode tunable pulsed dye laser, and the resonant absorbing medium was Yb vapor with a density of $\sim 4 \times 10^9$ atoms/cm^3 and propagation length of 1 m. The laser wavelength (555.648 nm) was tuned in precise resonance with the $6s^2\,^1S_0$–$6s6p\,^3P_1$ transition of the ^{174}Yb atom. Figure 10.4(a) shows the measured relative waveforms of the input pulse and the transmitted pulses corresponding to several different initial pulse areas smaller than 2π, indicating

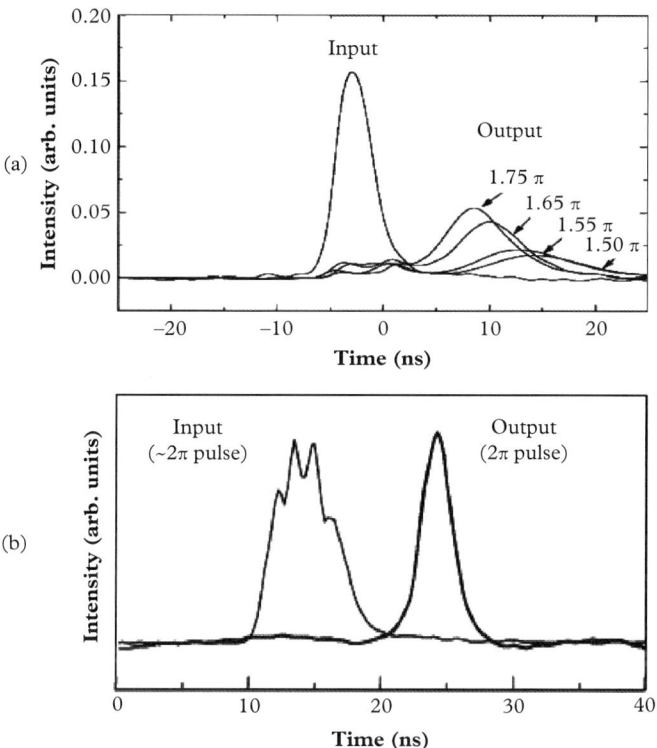

Figure 10.4 *Self-induced transparency in the Yb vapor: (a) relative waveforms of the input pulse and the transmitted pulses with different initial pulse areas; (b) waveforms of an irregular input pulse of $\sim 2\pi$ area and the transmitted pulse of 2π area. (Reproduced with permission from Yi et al.[21] Copyright 2000, Japanese Physical Society).*

that when the initial input pulse area increases from ~1.5π to 2π, the transmission of pulse energy is increased rapidly. Under the same experimental condition, another interesting result is shown in Fig. 10.4(b), where the input pulse has a ~2π area and arbitrary irregular shape while the output pulse is a stable 2π-pulse with a sech-function shape. The latter result demonstrates the pulse reshaping function based on the self-induced transparency process.

10.3 Photon echo effect

10.3.1 Concept of photon echo

The photon echo effect is an optical analog of the spin echo effect in nuclear magnetic resonance studies.[27-29] On the coherent transient interaction condition, if two intense coherent light pulses are applied to a resonant absorbing medium, the first is a $\pi/2$-pulse satisfying the initial condition of $\theta_0 = \pi/2$, the second is a π-pulse satisfying $\theta_0 = \pi$, and the temporal separation between these two pulses is τ_s satisfying the condition of $\tau_s < T_2 < T_1$, one may observe the third coherent emission pulse propagating along the transmission direction of the two excitation pulses. The third optical pulse is the so-called photon echo pulse, and the temporal interval between the echo pulse and the second pulse is equal to τ_s, as shown schematically in Fig. 10.5.

From Eq. (10.2-8), one can see that the difference of population densities of the resonant molecules in their two energy levels is determined by

$$N = N_1 - N_2 = N_0 \cos\theta, \qquad (10.3\text{-}1)$$

where N_1 and N_2 are the population densities in the lower and the upper level, N_0 is the overall density of molecules, and $\theta(z)$ is the pulse-area function determined by Eq. (10.2-10). When the first $\pi/2$-pulse is applied we have $\theta_1 = \pi/2$ and $N = 0$, meaning that half of the molecules are excited to the upper level, and the medium responds with its induced resonant electric polarization that can generate secondary optical emission. When the first excitation is terminated, the induced resonant polarization and the secondary coherent emission start to decay, because the synchronous phase relationship among different excited molecules is lost gradually owing to dephasing influences from various inhomogeneous broadening mechanisms. After a time interval τ_s (considerably shorter than T_2), if the second π-pulse is applied, we have $\theta_2 = \pi$ and $N' = -N \approx 0$. At this

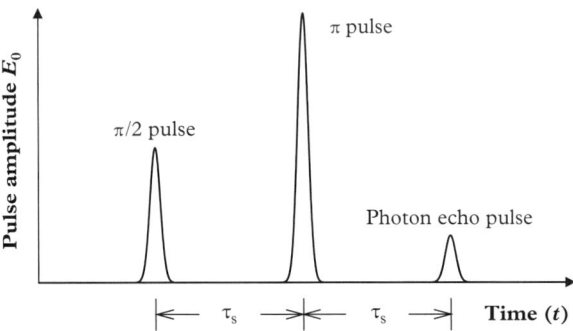

Figure 10.5 *Temporal sequence of the two excitation pulses and the photon echo pulse.*

moment, the molecule populations are still equal at the two levels, and the net energy exchange between the medium and the second pulse is negligible. However, the function of the applied second π-pulse is to reverse the dephasing processes among the excited molecules. For this reason after another time interval of τ_s, the induced resonant polarization contributions from all excited molecules will be in equal-phase again, as a result, we shall see a maximum coherent emission that is the photon echo.

10.3.2 Theoretical description of photon echo

The early-developed theory of the photon echo effect was based on the theory of spin echo effect, and its main theoretical conclusions were in good agreement with the experimental results. Here we shall briefly describe this theoretical approach and the related conclusions.[29–31]

We still assume the medium has only two levels, and the rapid dephasing process is mainly determined by the inhomogeneous broadening mechanisms. On this assumption, different absorbing molecules may have different resonant absorption frequencies ω_{0j}, where the subscript j refers to a single molecule (or a group of molecules that are identical), and the interaction Hamiltonian of an ensemble of great number of molecules can be expressed as[29,30]

$$H = H_0 + H' = \sum_j \left[\frac{h}{2\pi} \omega_{0j} \sigma_{zj} - p_0 (\sigma_{xj} E_x + \sigma_{yj} E_y) \right]. \tag{10.3-2}$$

Here, H_0 is the Hamiltonian of the molecular ensemble with no light field, H' is the interaction Hamiltonian between the molecules and the applied optical field, σ_x, σ_y, and σ_z are the three Pauli spin matrices divided by two, p_0 is the matrix element of electric dipole-moment operator, and E_x and E_y represent two electric field components of a plane optical wave traveling along the z-axis.

In order to have a straightforward picture of the related process, here we introduce a pseudo-electric polarization operator for the ensemble of a great number of molecules

$$\mathbf{P}' = \sum_j \mathbf{p}'_j = p_0 \sum_j (\sigma_{xj} \mathbf{x}_0 + \sigma_{yj} \mathbf{y}_0 + \sigma_{zj} \mathbf{z}_0), \tag{10.3-3}$$

where $\mathbf{p}'_j$ is the pseudo-electric dipole moment operator of the jth molecule and $\mathbf{x}_0$, $\mathbf{y}_0$, and $\mathbf{z}_0$ are the unit vectors along the directions of x, y, and z. If the applied field is a circularly polarized and monochromatic plane wave expressed as

$$\mathbf{E} = E_x \mathbf{x}_0 + E_y \mathbf{y}_0 = E_0(z, t)[\cos(\omega t - kz) \cdot \mathbf{x}_0 + \sin(\omega t - kz) \cdot \mathbf{y}_0],$$

where $E_0(z, t)$ is a slow-varying amplitude function and k is the magnitude of wavevector, then we can introduce the so-called pseudo-electric-field defined for the jth molecule as

$$\mathbf{E}'_j = E_x \mathbf{x}_0 + E_y \mathbf{y}_0 - \frac{h \omega_{0j}}{2\pi p_0} \mathbf{z}_0. \tag{10.3-4}$$

Thus, based on Eqs. (10.3-3) and (10.3-4), we can rewrite Eq. (10.3-2) as

$$H = -\sum_j \mathbf{p}'_j \cdot \mathbf{E}'_j. \tag{10.3-5}$$

The statistical average of **P'** is determined by (cf. Eq. (18.1-7))

$$\langle \mathbf{P'} \rangle = \mathrm{Tr}(\mathbf{P'}\rho),$$

where ρ is the density matrix operator of the molecular assembly, whose definition and physical meaning are described in Section 18.1. The temporal variation of ρ is governed by the following equation of motion:

$$\frac{\partial \rho}{\partial t} = -\frac{2\pi i}{h}[H, \rho].$$

Therefore, the temporal variation of quantum expectation of the pseudo-polarization component for the jth molecule satisfies the following equation:

$$\frac{d\langle \mathbf{p'}_j \rangle}{dt} = \frac{2\pi p_0}{h} \langle \mathbf{p'}_j \rangle \times \mathbf{E'}_j. \tag{10.3-6}$$

The mechanical picture of the above equation is that the vector $\langle \mathbf{p'}_j \rangle$ undergoes a precession movement around the vector $\mathbf{E'}_j$ in a Cartesian system that is rotating around the z-axis with the optical frequency ω. To make this picture simple, we may examine the precession behavior of the vector $\langle \mathbf{p'}_j \rangle$ in a Cartesian system that is stationary with respect to the z-axis. For this purpose, we could make the following representation transformation on the density matrix:

$$\rho = \exp\left[-i\sum_j (\omega t - kz_j)\sigma_{zj}\right] \rho^* \exp\left[i\sum_j (\omega t - kz_j)\sigma_{zj}\right].$$

In the new representation system all parameters are starred, and Eq. (10.3-6) becomes

$$\frac{d\langle \mathbf{p'}_j \rangle^*}{dt} = \frac{2\pi p_0}{h} \langle \mathbf{p'}_j \rangle^* \times \mathbf{E'}_j{}^*.$$
$$= \frac{2\pi p_0}{h} \langle \mathbf{p'}_j \rangle^* \times (\mathbf{x'}_0 E_0 + \mathbf{z}_0 \delta E_j), \tag{10.3-7}$$

where E_0 is the amplitude of the applied optical field, $\mathbf{x'}_0$ is the unit vector of the x'-axis in the new Cartesian system that is stationary with respect to the original z-axis, and

$$\delta E_j = -\frac{h}{2\pi p_0}(\omega_{0j} - \omega). \tag{10.3-8}$$

Equation (10.3-7) implies that in this new coordinate system the vector $\langle \mathbf{p'}_j \rangle^*$ undergoes a precession movement around the vector $\mathbf{E'}_j^*$ with an angular frequency determined by

$$\Omega_j = \frac{2\pi p_0}{h}|\mathbf{E'}_j^*| = \frac{2\pi p_0}{h}\sqrt{E_0^2 + (\delta E_j)^2}. \tag{10.3-9}$$

For resonant interaction, we have $\omega_{0j} - \omega \approx 0$ and $\delta E_j \ll E_0$; thus the angular frequency of precession becomes

$$\Omega = \frac{2\pi p_0}{h} E_0. \tag{10.3-10}$$

This is the so-called *Rabi frequency*, which is determined by the product of molecular parameter p_0 and the optical field amplitude E_0.

In the new representation system, the macroscopic electric polarization vector of the medium is determined by

$$<\mathbf{P'}>^* = \sum_j <\mathbf{p'}_j>^*, \qquad (10.3\text{-}11)$$

and the quantities $<\mathbf{p'}_j>$ and $<\mathbf{p'}_j>^*$ are related by[30]

$$\left.\begin{array}{l} <p'_{jx}> = <p'_{jx}>^* \cos(\omega t - kz_j) - <p'_{jy}>^* \sin(\omega t - kz_j) \\ <p'_{jy}> = <p'_{jy}>^* \cos(\omega t - kz_j) + <p'_{jx}>^* \sin(\omega t - kz_j) \\ <p'_{jz}> = <p'_{jz}>^* \end{array}\right\}. \qquad (10.3\text{-}12)$$

The physical meaning of the vectors $<\mathbf{p'}_j>^*$ can be interpreted as follows: the sum of their components along the z-axis represent the population distribution of the medium, while their components perpendicular to the z-axis represent the molecular contribution to the macroscopic polarization of the resonant medium, which is the source of the secondary light-wave emission. Based on this interpretation we can further explain the fundamental process of the photon echo generation.

Keeping in mind that the macroscopic polarization of the medium is always determined by the summation of all of the vectors $<\mathbf{p'}_j>^*$ with different subscript j, one may realize that the relative phase relations between different vectors $<\mathbf{p'}_j>^*$ are essentially important. For the sake of clearness, in Fig. 10.6 we only examine the behavior of one of a larger number of vectors $<\mathbf{p'}_j>^*$ that correspond to the molecules of different ω_{0j} values. Figure 10.6(a) shows the status of molecular systems before applying the first pulse. In this case, all vectors of $<\mathbf{p'}_j>^*$ point to the -z direction. This means that all molecules are in the lower level, and there is no macroscopic polarization at all. When the first $\pi/2$-pulse is applied, the optical field $(\mathbf{x'}_0 E_0)$ points to the x'-axis, and all vectors $<\mathbf{p'}_j>^*$ will rotate around the vector $\mathbf{E'}_j^*$ (and hence $\mathbf{x'}_0 E_0$ because of $\delta E_j \ll E_0$, see Eq. (10.3-7)) by an angle of

$$\theta_1 = \Omega \tau_1 = \frac{2\pi p_0}{h} E_0 \tau_1 = \frac{\pi}{2},$$

where τ_1 is the duration of the first pulse. In this case, as shown in Fig. 10.6(b), all vectors $<\mathbf{p'}_j>^*$ rotate to the x'–y' plane and give a maximum contribution to the macroscopic polarization that produces a maximum coherent emission although it cannot be distinguished from the input pulse signal. When the first pulse is ended, we have $E_0 \to 0$ and according to Eqs. (10.3-7) and (10.3-9), all vectors $<\mathbf{p'}_j>^*$ will continuously rotate around $\mathbf{z}_0 \delta E_j$ with an angular frequency determined by

$$\Omega'_j = \frac{2\pi p_0}{h} \delta E_j = \pm |\omega_{0j} - \omega|, \qquad (10.3\text{-}13)$$

where ω is the frequency of the input light field while ω_{0j} is the central resonant frequency of the jth molecule(s).

Owing to the inhomogeneous broadening nature, the excited molecules start to lose their synchronous phase relationships. In other words, for molecules with different ω_{0j} values, the rotating speeds of $<\mathbf{p'}_j>^*$ will be different; as a result, the net macroscopic polarization is nearly negligible because of the molecular dephasing process. Moreover, according to Eq. (10.3-13) we can divide

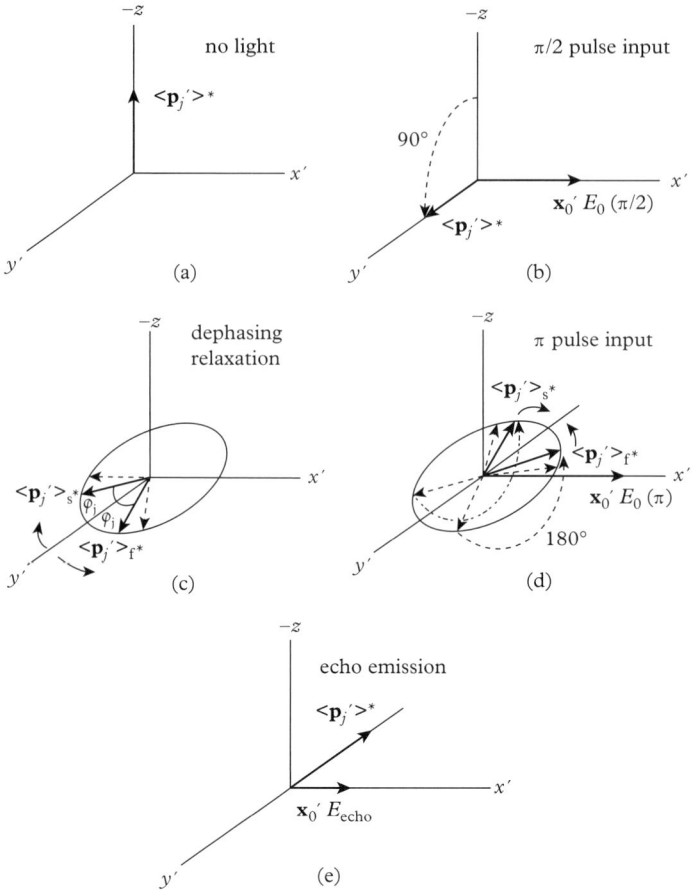

Figure 10.6 *Schematic illustration of photon echo generation: (a) before applying an optical field, (b) applying the first π/2-pulse, (c) after the first pulse is ended, (d) applying the second π-pulse, and (e) τ_s later after the second pulse.*

all dephasing molecules into two groups: one group (*f* group) of molecules have positive values of δE_j and their $<\mathbf{p}'_j>^*$ rotate along the counter-clockwise direction, while the other group (*s* group) of molecules have negative values of δE_j and their vectors $<\mathbf{p}'_j>^*$ rotate along the clockwise direction, as shown in Fig. 10.6(c). Consequently, after an arbitrary time interval τ_s the different groups of molecules will rotate over different angles in the $x'-y'$ plane determined by

$$\varphi_j = \pm \left|\Omega'_j\right| \tau_s. \tag{10.3-14}$$

For example, in Fig. 10.6(c) the dephasing behavior of another molecule group (corresponding to a different *j* value) is shown by two dotted-line arrows. At this moment if we apply the second π-pulse on the medium, there is an optical field vector $x'_0 E_0$ applying along the x'-axis again, and, as shown in Fig. 10.6(d), all vectors of $<\mathbf{p}'_j>_f^*$ and $<\mathbf{p}'_j>_s^*$ will rotate around the x'-axis by an angle of

$$\theta_2 = \Omega\tau_2 = \frac{2\pi p_0}{h} E_0 \tau_2 = \pi,$$

where τ_2 is the duration of the second pulse. After the second pulse is ended, these two groups of vectors of $<\mathbf{p}'_j>_f^*$ and $<\mathbf{p}'_j>_s^*$ continuously rotate toward the $-y$-axis direction. It is expected that after another time interval of τ_s, all vectors of $<\mathbf{p}'_j>_f^*$ and $<\mathbf{p}'_j>_s^*$ will recombine in the $-y$-axis at the same time, as shown in Fig. 10.6(e). At this moment, the synchronous phase relationship among all excited molecules is restored, and the maximum macroscopic polarization is rebuilt. As a result, a new coherent optical pulse emission (photon echo) can be observed.

Since the photon echo can be thought as being emitted by a great number of induced classical electric dipoles with a synchronous phase relationship, it should be highly directional. The theoretical analyses show that the intensity of the echo emission can be expressed as[29]

$$I_e(\mathbf{k}_e) \propto N_0^2 I_0^s \left| \{\exp[i(\mathbf{k}_e + \mathbf{k}_1 - 2\mathbf{k}_2) \cdot \mathbf{r}]\}_{av} \right|^2, \qquad (10.3\text{-}15)$$

where N_0 is the molecular density, I_0^s is the emission contribution from a single molecule, $\mathbf{k}_e$ is the wavevector of the echo wave, and $\mathbf{k}_1$ and $\mathbf{k}_2$ are the wavevectors of the first and second pulses. The average is over the different molecule positions within the interaction volume.

From Eq. (10.3-15), one can see that the emission maximum should be observed in the direction determined by

$$\mathbf{k}_e = 2\mathbf{k}_2 - \mathbf{k}_1. \qquad (10.3\text{-}16)$$

According to this requirement, when $\mathbf{k}_1$ and $\mathbf{k}_2$ are collinear, the $\mathbf{k}_e$ should be collinear with them too. However, if the first and the second pulse have slightly different propagation directions with a small crossing angle φ_0, the angle between $\mathbf{k}_e$ and $\mathbf{k}_2$ should equal φ_0. In experiments, researchers may use this property to separate the relatively weak echo signal from the transmitted excitation pulse signals.

10.3.3 Experimental studies of photon echo

Most experimental results have shown that the photon echo intensity as a function of the time separation (τ_s) between the two excitation-pulses can be expressed as[32]

$$I_e(\tau_s) = I_e(0) \exp[-4\tau_s/T_2]. \qquad (10.3\text{-}17)$$

Here, T_2 is the transverse relaxation time of the sample medium, which reflects the influences from homogeneous broadening mechanisms. However, the real T_2 value under a given sample condition (temperature, pressure, external fields, etc.) can be simply determined by varying the time delay τ_s and measuring the corresponding echo signal intensity change. On the other hand, the transverse relaxation time T_2 can also be expressed as[33–35]

$$\frac{1}{T_2} = \frac{1}{2T_1} + \frac{1}{T_2''}, \qquad (10.3\text{-}18)$$

where T_1 is the longitudinal relaxation time that is the lifetime of the population in the upper state of the transition, $1/T_2'$ is the pure dephasing rate of the excited molecule due to various homogeneous broadening mechanisms. Measuring the values of T_2 and T_1 for an investigated medium under given conditions, one may further determine the homogeneous relaxation time T_2'. Based on these two facts mentioned above, one can realize that the photon echo study is a special approach to investigate the spectral broadening and relaxation properties.

The experimental setup for observing photon echo is depicted in Fig. 10.7, where the output laser from a pulsed laser source is split into two beams: one is directly to pass through the resonant absorbing sample medium, whereas the other is delayed and then arrives at the sample with a small crossing-angle of φ_0. The intensity of each pulsed beam can be adjusted by a variable optical attenuator to meet the pulse area requirement, and the photon echo signal is detected along the $2\varphi_0$ direction.

Since the first observation of the photon echo effect in ruby crystal reported by Kurnit et al. in 1964,[28] a great number of resonant media have been studied using the photon echo technique. Some of these are summarized as follows:

A. molecular gases such as SF_6,[32] CH_3F,[36] SiF_4,[37] BCl_3,[38] and I_2;[39]

B. metal vapors such as Cs,[40] Na,[41] Li,[42] Rb,[43] and Yb;[44]

C. atomic gases and their binary mixtures;[45,46]

D. metal ion-doped crystals including $Cr^{3+}:Al_2O_3$,[28,29] $Pr^{3+}:LaF_3$,[47] $Pr^{3+}:YAlO_3$,[48] $Pr^{3+}:Y_2SiO_5$,[49] $Nd:CaWO_4$,[50] $Nd^{3+}:LaF_3$,[50] $Nd^{3+}:YAG$,[51] $Nd^{3+}:CaF_2$,[52] $Eu^{3+}:YAlO_3$,[53] $Eu^{3+}:Y_2O_3$,[54] $Er^{3+}:Y_2SiO_5$,[55] $Tb^{3+}:LiYF_4$,[56] $Tm^{3+}:Y_2Si_2O_7$,[57] $Tm^{3+}:YAG$,[58] $Pr^{3+}:KY(WO_4)_2$,[59] and $Er^{3+}:LuLiF_4$;[60]

E. metal ion-doped glasses[61-63] and semiconductor crystals;[34,64]

F. organic crystals[33,65,66] and glass-like organic systems.[67-71]

As an example of the early experimental studies of optical echoes, Fig. 10.8 shows the observed photon echo signals in a low-temperature ruby sample excited by two optical pulses from a Q-switched ruby laser operating at 77 K.[29] To ensure that the lasing spectral line of the excitation pulses could coincide with the R_1 absorption line of the ruby sample with a relatively larger value of T_2, the measured ruby sample was kept at the temperature of 4.2 K, and a dc magnetic

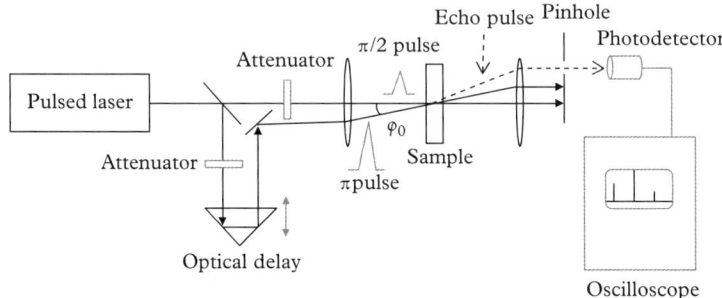

Figure 10.7 *Experimental setup for producing and measuring the photon echo effect.*

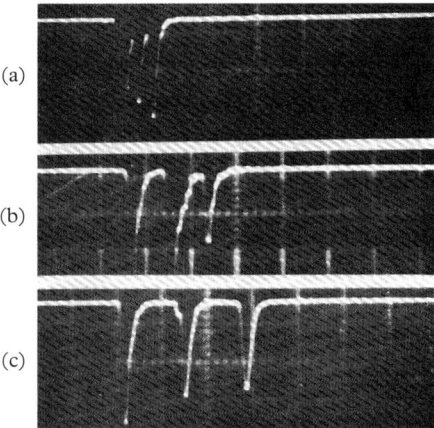

Figure 10.8 *Photon echo behavior in a low-temperature ruby sample excited by two optical pulses from a ruby laser source with time delays of τ_s = 36, 83, and 137 ns, respectively. The third pulse is the photon echo while the two excitation pulses are attenuated prior to detection. The time scale is 100 ns/div. (Reproduced with permission from Abella et al.[29] Copyright 1966, American Physical Society).*

field of a hundred Gauss was applied along the optical axis of the ruby crystal. It is seen that the photon echo always appears at the time of τ_s following the second pulse.

Figure 10.9 shows the curves of photon echo intensity (on the natural logarithm scale) plotted as a function of delay between two excitation pulses, measured in a pentacene-doped polymethyl methacrylate (PMMA) film at 1.2 K and various pressures.[35] The input pulses of ~35 ps duration were from a dye laser with a tunable wavelength around 580 nm in resonance with the S_1–S_0 electronic transition of pentacene. The central frequency and linewidth of this transition is a function of temperature and pressure of the sample. The measurement results indicate that the T_2 value is sensitively dependent on the pressure within a range from 0 to 4 kbar, whereas this dependence is dramatically reduced at pressures higher than 4 kbar.

The photon echo technique can also be useful to study the ultrafast relaxation processes of molecular vibrational transitions. Figure 10.10 shows the results of photon echo measurement implemented in a porous silicon (*p*-Si) film at 10 K.[72] This sample manifests three absorption peaks around the spectral range 4.4–4.8 μm, which correspond to three stretching modes from the defects of SiH, SiH_2, and O_3SiH, respectively. To obtain the results shown in Fig. 10.10, the input pulses were provided by a free electron laser source, and the excitation wavelengths for these three vibrational modes were 4.73, 4.82, and 4.42 μm, respectively. Based on the measured data of photon echo intensity as a function of the delay between the two input pulses, the T_2 values of these three vibrational transitions are determined and listed in the bottom of each experimental curve.

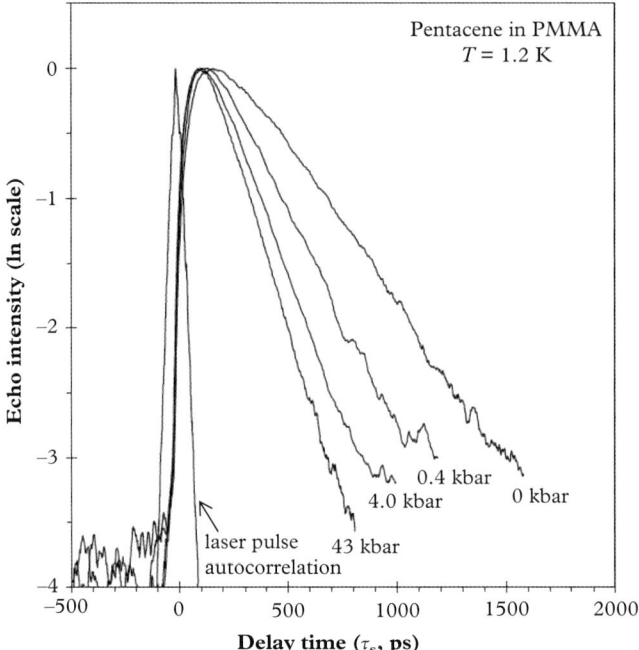

Figure 10.9 *Dependence of echo intensity on τ_s of the pentacene-doped PMMA film sample, measured at 1.2 K and different pressures. The input laser pulses are around 580 nm wavelength and ~35 ps pulse width. (Reproduced with permission from Berg and Chronister.[35] Copyright 1997, AIP Publishing LLC).*

10.4 Optical nutation effect

10.4.1 Conceptual description

The optical nutation effect is an optical analog of the spin nutation in nuclear magnetic resonance.[73–75] This effect occurs when an intense step-function light field is suddenly applied to a resonant absorbing medium and the transmissivity of the medium exhibits a quasi-periodic and damped oscillation response. In this case, one may observe a quasi-periodic oscillation in the leading part of the transmitted light signal, the oscillation period depends on the intensity of the input light field, and the decay of this oscillation is dependent on T_2 of the medium. This is the so-called optical nutation effect, as illustrated schematically in Fig. 10.11.

The optical nutation effect is referred to the transient oscillating behavior of energy exchange between the intense optical field and the resonant absorbing medium under coherent transient interaction conditions. At the beginning of applying an intense step-function light field, as schematically shown in Fig. 10.11, a great number of molecules will be excited to the upper level accompanying a remarkable energy loss of the input optical field. Shortly after that, some of the excited molecules return to the lower level and reemit the partial absorbed-energy to the optical field in form of coherent emission that enhances the intensity of the transmitted light signal.

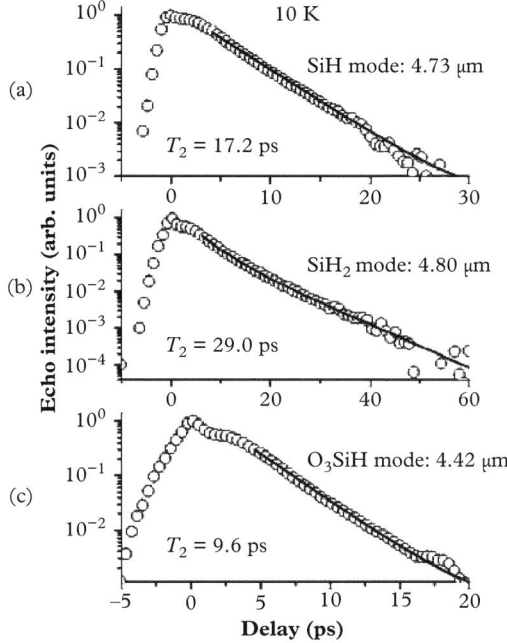

Figure 10.10 *Echo intensity decay of vibrational transitions corresponding to three stretch modes of a porous silicon sample at a temperature of 10 K. (Reproduced with permission from Jobson et al.[72] Copyright 2006, American Physical Society).*

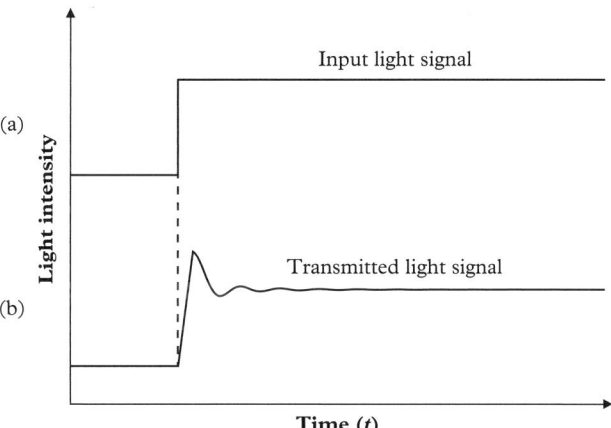

Figure 10.11 *Schematic illustration of the optical nutation effect: (a) the waveform of the incident optical signal and (b) the waveform of the transmitted optical signal.*

344 *Optical Coherent Transient Effects*

The alternate actions of these two processes (absorption and reemission) result in the oscillating behavior of the transmitted signal intensity. On the other hand, these processes can take place effectively only on the condition that all molecules interact with the optical field synchronously. Since the equal-phase relationship among different molecules is lost gradually within the period of T_2, one would expect a damped feature for the oscillating behavior of the transmitted light signal within the same period. This is the physical picture of the optical nutation effect.

In order to give a rigorous and quantitative description of the optical nutation effect, we can rely on a theoretical approach based on the Bloch equation that had been commonly employed in the nuclear magnetic resonance study. This theoretical approach can also be adopted to interpret other optical coherent transient effects, including photon echo effect discussed in the preceding section as well as optical free induction decay effect that will be discussed in the next section.

10.4.2 Optical Bloch equation

Assume that the input optical field is a monochromatic plane wave linearly polarized along the *x*-axis and propagating in the *z*-axis direction:

$$\mathbf{E}(z, t) = \mathbf{x}_0 E_0 \cos(\omega t - kz).$$

The density matrix operator of a two-level molecular system fulfills the following equation of motion:

$$\frac{i\hbar}{2\pi} \dot{\rho} = [H, \rho] + \text{relaxation term}. \tag{10.4-1}$$

Here, $H = H_0 + H'$; $H' = -\mathbf{p} \cdot \mathbf{E}(z, t)$ is the interaction Hamiltonian and $\mathbf{p}$ is the molecular dipole-moment operator. For a two-level molecular system, the above equation leads to the following equations for matrix elements:[76]

$$\left. \begin{array}{l} \dot{\rho}_{11} = i\Omega(\rho_{21} - \rho_{12}) \cos(\omega t - kz) - (\rho_{11} - \rho_{11}^0)/T_1 \\ \dot{\rho}_{22} = i\Omega(\rho_{12} - \rho_{21}) \cos(\omega t - kz) - (\rho_{22} - \rho_{22}^0)/T_1 \\ \dot{\rho}_{12} = i\Omega(\rho_{22} - \rho_{11}) \cos(\omega t - kz) + \left(i\omega_0 - \dfrac{1}{T_2}\right) \rho_{12} \\ \dot{\rho}_{21} = -i\Omega(\rho_{22} - \rho_{11}) \cos(\omega t - kz) - \left(i\omega_0 + \dfrac{1}{T_2}\right) \rho_{21} \end{array} \right\}, \tag{10.4-2}$$

where ω_0 is the frequency of molecular resonant transition, Ω is the Rabi frequency defined by Eq. (10.3-10), ρ_{11}^0 and ρ_{22}^0 are the population probabilities of levels 1 and 2 in the absence of an external optical field, and T_1 and T_2 are phenomenologically introduced relaxation factors related to the diagonal and off-diagonal density matrix elements. The physical meanings of T_1 and T_2 are explained in the first section of this chapter.

In order to eliminate the term varying with time at optical frequency, we can make the following mathematical transformation:

$$\rho_{12} = \tilde{\rho}_{12} \exp[i(\omega t - kz)], \tag{10.4-3}$$

and neglect the contribution from non-resonant terms (such as those varying with time at $2\omega t$). Equation (10.4-2) then becomes

$$\left.\begin{array}{l} \dot{\rho}_{11} = i\Omega(\tilde{\rho}_{21} - \tilde{\rho}_{12})/2 - (\rho_{11} - \rho_{11}^0)/T_1 \\ \dot{\rho}_{22} = i\Omega(\tilde{\rho}_{12} - \tilde{\rho}_{21})/2 - (\rho_{22} - \rho_{22}^0)/T_1 \\ \left(\dfrac{d}{dt} - i\Delta + \dfrac{1}{T_2}\right)\tilde{\rho}_{12} = i\Omega(\rho_{22} - \rho_{11})/2 \\ \left(\dfrac{d}{dt} + i\Delta + \dfrac{1}{T_2}\right)\tilde{\rho}_{21} = -i\Omega(\rho_{22} - \rho_{11})/2 \end{array}\right\}. \qquad (10.4\text{-}4)$$

Here, Δ represents the detuning between frequency ω of the applied optical field and the resonant transition frequency ω_0 of the molecules; for gaseous resonant medium it is expressed by

$$\Delta = -\omega + \omega_0 + kv_z, \qquad (10.4\text{-}5)$$

where v_z is the molecular thermal velocity component in the z direction. Based on Eq. (10.4-4) and combining $\dot{\rho}_{11}$ with $\dot{\rho}_{22}$ and $\dot{\tilde{\rho}}_{12}$ with $\dot{\tilde{\rho}}_{21}$, we obtain the following set of equations:

$$\left.\begin{array}{l} \dot{\tilde{\rho}}_{12} + \dot{\tilde{\rho}}_{21} = -i\Delta(\tilde{\rho}_{21} - \tilde{\rho}_{12}) - (\tilde{\rho}_{12} + \tilde{\rho}_{21})/T_2 \\ i(\dot{\tilde{\rho}}_{21} - \dot{\tilde{\rho}}_{12}) = \Delta(\tilde{\rho}_{12} + \tilde{\rho}_{21}) + \Omega(\rho_{22} - \rho_{11}) - i(\tilde{\rho}_{21} - \tilde{\rho}_{12})/T_2 \\ \dot{\rho}_{22} - \dot{\rho}_{11} = -i\Omega(\tilde{\rho}_{21} - \tilde{\rho}_{12}) - (\rho_{22} - \rho_{11})/T_1 + (\rho_{22}^0 - \rho_{11}^0)/T_1 \end{array}\right\}. \qquad (10.4\text{-}6)$$

After introducing a set of new variables:

$$\left.\begin{array}{l} u = \tilde{\rho}_{12} + \tilde{\rho}_{21} \\ v = i(\tilde{\rho}_{21} - \tilde{\rho}_{12}) \\ w = \rho_{22} - \rho_{11} \end{array}\right\}, \qquad (10.4\text{-}7)$$

Eq. (10.4-6) can be simplified as

$$\left.\begin{array}{l} \dot{u} + \Delta v + u/T_2 = 0 \\ \dot{v} - \Delta u - \Omega w + v/T_2 = 0 \\ \dot{w} + \Omega v + (w - w^0)/T_1 = 0 \end{array}\right\}. \qquad (10.4\text{-}8)$$

Here, $w^0 = \rho_{22}^0 - \rho_{11}^0$ denotes the population difference between the two levels in the absence of an external optical field. Equation (10.4-8) is known as the *optical Bloch equation*. Moreover, if all relaxation terms can be neglected, Eq. (10.4-8) can be rewritten in a more compact form:

$$\frac{d\mathbf{B}}{dt} = \mathbf{B} \times \boldsymbol{\beta}. \qquad (10.4\text{-}9)$$

Here, $\mathbf{B}$ is customarily termed the Bloch vector, and the vectors $\mathbf{B}$ and $\boldsymbol{\beta}$ are defined by their Cartesian components:

$$\left.\begin{array}{l} \mathbf{B} = \mathbf{x}'_0 u + \mathbf{y}'_0 v + \mathbf{z}_0 w = [u, v, w] \\ \boldsymbol{\beta} = \mathbf{x}'_0 \Omega - \mathbf{z}_0 \Delta = [\Omega, 0, -\Delta] \end{array}\right\}, \qquad (10.4\text{-}10)$$

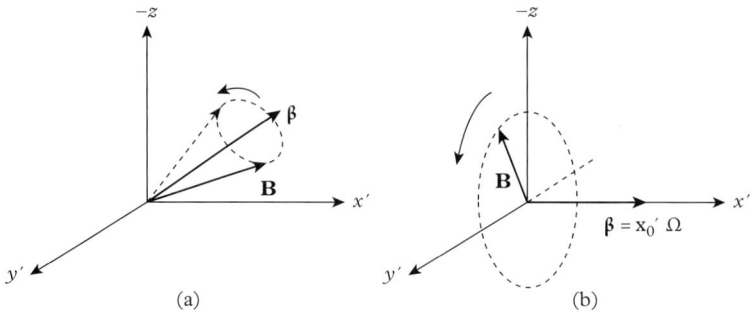

Figure 10.12 *Schematic illustration of the precession of Bloch vector* **B** *around vector* **β** *in (a) a general situation and (b) the exact-resonance situation.*

where $\mathbf{x}'_0$, $\mathbf{y}'_0$, and $\mathbf{z}_0$ are the unit vectors along the x', y', and z axes in a frame stationary with respect to the z-axis. In the above expressions, u and v describe the resonant polarization behavior of the medium; w describes the population difference between the two levels, Δ describes the detuning between the frequency of the applied field and the molecular transition frequency, and Ω is the Rabi frequency determined by the amplitude of the optical field and the dipole-moment matrix element of the molecule.

As shown in Fig. 10.12, the vector form of the Bloch equation, Eq. (10.4-9), represents the precession of the Bloch vector **B** around the vector **β**. The frequency of precession is equal to the magnitude of $|\boldsymbol{\beta}|$, i.e.,

$$\beta = \sqrt{\Omega^2 + \Delta^2}. \tag{10.4-11}$$

Under the condition of exact resonance, we have $\Delta = 0$ and $\beta = \Omega$.

Comparing Eq. (10.4-9) to Eq. (10.3-7), one can see that these two equations are essentially the same. Here, the vector **B** determines the macroscopic polarization contribution (by u and v) and the molecular population distribution (by w), while the vector **β** represents the properties (amplitude and frequency detuning) of the incident optical Field.

In order to obtain an analytical solution of the optical Bloch equation, we first consider the simplest situation where the influences from both relaxation times T_1 and T_2 can be neglected. In Eq. (10.4-8), letting $T_1, T_2 \to \infty$ leads to

$$\left. \begin{array}{l} \dot{u} + \Delta v = 0 \\ \dot{v} - \Delta u - \Omega w = 0 \\ \dot{w} + \Omega v = 0 \end{array} \right\}. \tag{10.4-12}$$

Assuming the optical field is applied to the medium at the moment of $t = 0$, then in the time period of $0 < t \ll T_1, T_2$, Eq. (10.4-12) possesses the following solution:

$$\left. \begin{array}{l} u(t) = \dfrac{\Delta w(0)}{\beta^2}(\cos \beta t - 1) \\[2mm] v(t) = \dfrac{\Omega w(0)}{\beta} \sin \beta t \\[2mm] w(t) = w(0)\left[1 + \dfrac{\Omega^2}{\beta^2}(\cos \beta t - 1)\right] \end{array} \right\}, \tag{10.4-13}$$

where $w(0) = \rho_{22}(0) - \rho_{11}(0) = \rho_{22}^0 - \rho_{11}^0 = w^0$ represents the difference of the population in the two level at $t = 0$. The above solution satisfies the following initial condition:

$$\mathbf{B}(0) = [0, 0, w(0)], \tag{10.4-14}$$

and at any other time $B(t)$ remains constant, i.e.,

$$B(t) = |\mathbf{B}(t)| = [u^2(t) + v^2(t) + w^2(t)]^{1/2} = w(0). \tag{10.4-15}$$

In exact resonance we have $\Delta = 0$, $\beta = \Omega$, and $u(t) = 0$. In this case, as shown in Fig. 10.12(b), the vector $\boldsymbol{\beta}$ is along the x'-axis, while vector $\mathbf{B}$ is rotating in the $y'-z$ plane around the x'-axis with the Rabi frequency Ω (see also Fig. 10.6(b)). Regarding the vector $\mathbf{B}$, its z-axis component represents the population difference between the two levels, and its y'-axis component represents the contribution to the macroscopic polarization; during the rotation of the vector $\mathbf{B}$ in the $y'-z$ plane, the population distribution and the secondary coherent emission will vary periodically with frequency Ω, provided that the incident optical field is strong enough and the considered time interval is shorter than T_1 and T_2. This is the geometrical picture of the optical nutation effect.

10.4.3 Solution for optical nutation effect

Assuming N_0 is the molecule number per unit volume, the macroscopic resonant electric polarization of the medium can be expressed as (ignoring its vector property):[76]

$$P(z, t) = P_0(z, t) \exp[-i(\omega t - kz)] = N_0 p_0 < \tilde{\rho}_{12} > \exp[i(\omega t - kz)], \tag{10.4-16}$$

where $P_0(z, t)$ is the amplitude function, $<\tilde{\rho}_{12}>$ represents the average value of $\tilde{\rho}_{12}$ over the thermal velocity of the molecules for a gas medium (with inhomogeneous Doppler broadening), i.e.,

$$< \tilde{\rho}_{12} > = \frac{1}{k\bar{v}\sqrt{\pi}} \int_{-\infty}^{\infty} \tilde{\rho}_{12} \exp[-(\Delta/k\bar{v})^2] d\Delta. \tag{10.4-17}$$

Here, Δ is the detuning determined by Eq. (10.4-5) and $\bar{v}$ is the root-mean-square velocity of molecules. From Eq. (10.4-7) we know $\tilde{\rho}_{12} = (u + iv)/2$, then Eq. (10.4-17) can be rewritten as

$$< \tilde{\rho}_{12} > = \frac{1}{k\bar{v}\sqrt{\pi}} \int_{-\infty}^{\infty} \frac{1}{2}(u + iv) \exp[-(\Delta/k\bar{v})^2] d\Delta. \tag{10.4-18}$$

The polarization (Pz, t), which is determined by $<\tilde{\rho}_{12}>$ according to Eq. (10.4-16), is the physical source of the transient secondary coherent emission that can be expressed as

$$E'(z, t) = E'_0(z, t) e^{-i(\omega t - kz)}, \tag{10.4-19}$$

where $E'_0(z, t)$ is the amplitude function of the emission field. Moreover, in the slowly-varying-amplitude approximation, $E'_0(z, t)$ and $P_0(z, t)$ are related to each other through the following equation:

$$\frac{\partial E'_0(z, t)}{\partial z} = -\frac{ik}{2n_0^2 \varepsilon_0} N_0 p_0 < \tilde{\rho}_{12} >. \tag{10.4-20}$$

If the effective interaction length is L, after completing the time integration over the above equation we obtain

$$E'_0(z,t) = -\frac{ik}{2n_0^2\varepsilon_0} N_0 p_0 L <\tilde{\rho}_{12}>. \qquad (10.4\text{-}21)$$

Substituting the expressions of $u(t)$ and $v(t)$ in Eq. (10.4-13) into Eq. (10.4-18) results in

$$\begin{aligned}<\tilde{\rho}_{12}> &= \frac{1}{k\bar{v}\sqrt{\pi}} \int_{-\infty}^{\infty} \frac{1}{2}[u(t) + iv(t)] e^{-(\Delta/k\bar{v})^2} d\Delta \\ &\approx \frac{i\Omega w(0)}{k\bar{v}\sqrt{\pi}} e^{-(\bar{\Delta}/k\bar{v})^2} \int_0^{\infty} \frac{\sin(\sqrt{\Delta^2+\Omega^2}\cdot t)}{\sqrt{\Delta^2+\Omega^2}} d\Delta \\ &\approx \frac{i\sqrt{\pi}}{2k\bar{v}} \Omega w(0) e^{-(\bar{\Delta}/k\bar{v})^2} \mathcal{J}_0(\Omega t), \end{aligned} \qquad (10.4\text{-}22)$$

where $\bar{\Delta}$ is the average detuning and $\mathcal{J}_0$ is the zero-order Bessel function. To obtain this result of integration, it is assumed that the frequency of the incident optical field is very close to the central frequency of the Doppler-broadening absorptive line of the medium; thus we have $<u(t)> \approx 0$ and the Gaussian distribution factor can be moved out of the integral.

Substituting Eq. (10.4-22) into Eq. (10.4-21) leads to

$$E'_0(L,t) = \frac{\sqrt{\pi}}{4n_0^2\varepsilon_0\bar{v}} N_0 p_0 L \Omega w(0) e^{-(\bar{\Delta}/k\bar{v})^2} \mathcal{J}_0(\Omega t). \qquad (10.4\text{-}23)$$

The physical meaning of this result is that the coherent transient emission manifests a quasi-periodic oscillation with the Rabi frequency Ω. This conclusion is obtained under the condition that all relaxation influences can be neglected.

If we further consider the relaxation influences and assume approximately $T_2 \approx T_1$, then based on the original Bloch equation, Eq. (10.4-8), one could finally obtain the following result:[76]

$$\begin{aligned}E'_0(L,t) &= \frac{\sqrt{\pi}}{4n_0^2\varepsilon_0\bar{v}} N_0 p_0 L \Omega w(0) e^{-(\bar{\Delta}/k\bar{v})^2} e^{-t/T_2} \mathcal{J}_0(\Omega t) \\ &= \frac{\pi^{3/2}}{2n_0^2\varepsilon_0\hbar\bar{v}} E_0 N_0 p_0^2 L w(0) e^{-(\bar{\Delta}/k\bar{v})^2} e^{-t/T_2} \mathcal{J}_0(\Omega t). \end{aligned} \qquad (10.4\text{-}24)$$

We have so far obtained the expression for the coherent transient emission field. However, the real observed signal is composed of two parts: the transmitted incident field and the coherent transient emission field. Thus the observed overall optical intensity should be

$$I_t \propto E_t^2 \propto [E_0 + E'_0(L,t)]^2, \qquad (10.4\text{-}25)$$

where E_0 is the amplitude of the incident optical field. Since $E'_0 \ll E_0$, the above expression can be simplified as

$$I_t \propto [E_0^2 + 2E_0 E'_0(L,t)]. \qquad (10.4\text{-}26)$$

Here, the first term represents the intensity of the incident field, which is assumed constant when $t > 0$, whereas the second term represents the time-varying intensity component, i.e.,

$$I'_t \propto E_0 E'_0(L, t) \propto E_0^2 e^{-t/T_2} \mathcal{J}_0(\Omega t). \tag{10.4-27}$$

The physical meaning of this result is that, at the leading edge of the transmitted optical signal, one will see an exponentially damped periodical modulation with the frequency Ω.

Equation (10.4-24) is valid for the resonant medium with a larger inhomogeneously broadened linewidth. Similarly, for the resonant medium with a homogeneously broadened spectral line, the time-varying part of the transmitted signal can be expressed as[73]

$$I'_t(t) \propto e^{-t/(2T_2)} \sin(\Omega t). \tag{10.4-28}$$

To obtain this result it is assumed that $T_2 \ll T_1$ and $1/\Omega \ll T_1$. On the other hand, in the time range $t \gg \Omega^{-1}$, we have $\mathcal{J}_0(\Omega t) \approx [\frac{2}{\pi \Omega t}]^{1/2} \cos(\Omega t - \frac{\pi}{4})$. Therefore, Eqs. (10.4-27) and (10.4-28) show us the similar intensity modulation behavior for both types of major mechanisms of spectral broadening.

10.4.4 Experimental studies of optical nutation

The first observation of optical nutation effect was accomplished in a SF_6 gas sample by Hocker and Tang in 1968.[75] Since then, optical nutation measurements have been employed to investigate various resonant absorbing media. The following are some of the materials used for optical nutation studies:

A. molecular gases such as SF_6,[77] CH_3F,[36] NH_2D,[36,78] NH_3,[79,80] I_2,[81] CO_2,[82] and N_2O;[83]
B. metal vapors such as Na,[84,85] Yb,[86] and Rb;[87,88]
C. others including atomic O,[89] Ba^+ ion beam,[90] organic crystals,[91] and Tm^{3+}:YAG.[92,93]

In experiments, researchers can use either a pulsed laser source to provide a long rectangular pulse shape or a cw laser source. In the latter case, the sudden start of the interaction between the applied optical field and the resonant medium can be achieved by using two different methods: one is based on the laser-frequency shifting, and the other is based on the resonant-frequency shifting of the medium. By using the first method, the frequency of the laser beam can be suddenly shifted to an exact resonance with a given absorptive transition of the medium. With the second method, the input laser frequency is fixed and in off-resonance with the medium, but a rectangle-shape electric pulse is applied to the medium; thus the medium can be abruptly in resonance with the laser field via the Stark effect. In both cases, the optical nutation effect can be observed.

As a typical experimental example, Fig. 10.13 shows the early observed optical nutation phenomenon in a low pressure (4.8 mTorr) $^{13}CH_3F$ gas sample, using the P(32)-branch output line (1035.474 cm^{-1}) of a cw CO_2-laser in resonance with one of the ν_3 band transitions ($\mathcal{J}, K = 4, 3 \rightarrow 5, 3$) of the sample gas.[36] In this figure, the bottom curve is the waveform of the applied Stark electric pulse on the gas sample, while the upper curve is the waveform of the transmitted optical intensity. During the period of the first half of the applied rectangle electric pulse, we see a damped periodic oscillation of the transmitted optical-field intensity. This is the optical nutation caused by one group of gas molecules, which have a certain thermal velocity and are in exact resonance with the laser field when the electric pulse is turned on. Moreover, when the electric pulse is suddenly switched off, another group of molecules having another certain velocity can

350 *Optical Coherent Transient Effects*

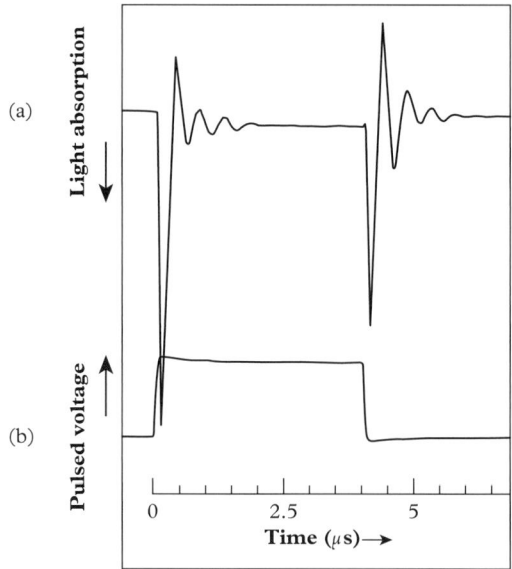

Figure 10.13 *Optical nutation phenomenon in the $C^{13}H_3F$ gas at 4.8 mTorr pressure: (a) the waveform of the transmitted optical signal and (b) the electric pulse for generating Stark effect with an amplitude of 35 V/cm. (Reproduced with permission from Brewer and Shoemaker.[36] Copyright 1971, American Physical Society).*

be in exact resonance with the same laser line, because the Stark frequency shift of the medium in the present condition is smaller than the Doppler line width of 66 MHz of the gas sample. For this reason, we can see the similar optical nutation behavior at the end of the applied electric pulse.

There is another experimental example showing the typical appearance of optical nutation in Na vapor system working at its D_2 line.[85] The measured leading edges of the tuning-in-resonance laser pulse and the transmitted pulses from a Na vapor cell heated to about 130 °C are shown in Fig. 10.14(a). To obtain these results, an electro-optical gating system was used to form the excitation light pulse from a cw ring dye laser beam. From the oscillating behavior of the optical nutation signal one may determine the Rabi frequency at various input laser intensities or power levels, and the measured results are shown in Fig. 10.14(b), indicating that the square of measured Rabi frequency is indeed proportional to the input laser intensity (or power) as predicted by the theory and expressed by Eq. (10.3-10). Knowing the values of the Rabi frequency and the input laser intensity, one can further determine the value of molecular dipole moment p_0.

Finally, Fig. 10.15(a) shows the optical nutation appearance measured in a 0.1% Tm:YAG crystal sample, working at 1.4 K with its $^3H_6(1) \to {}^3H_4(1)$ electronic transition and excited by a square pulse of ~20 μs duration and ~793 nm wavelength.[93] The measurement results shown in Fig. 10.15(b) confirm that the nutation signal intensity is proportional to the input light intensity as predicted by Eq. (10.4-27), and the logarithm of nutation modulation frequency Ω is linearly proportional to the logarithm of the input light intensity according to Eq. (10.3-10).

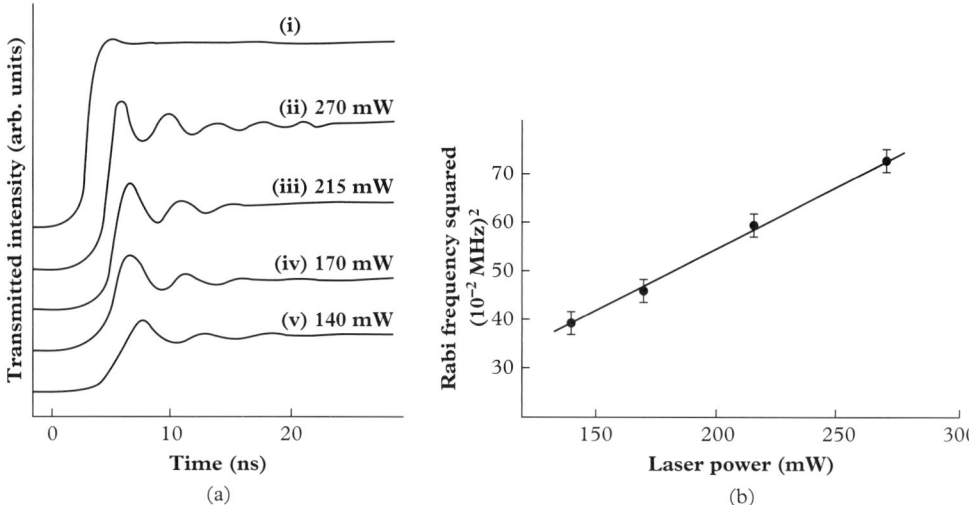

Fig 10.14 *(a) Leading edges of the input dye laser pulse (i) and the transmitted laser pulses (ii–v) at various input intensity levels. (b) Square of measured Rabi frequency versus laser intensity. The absorbing medium is Na vapor working at 130 °C with its D_2 transition line. (Reproduced with permission from Farrell et al.[85] Copyright 1985, Elsevier).*

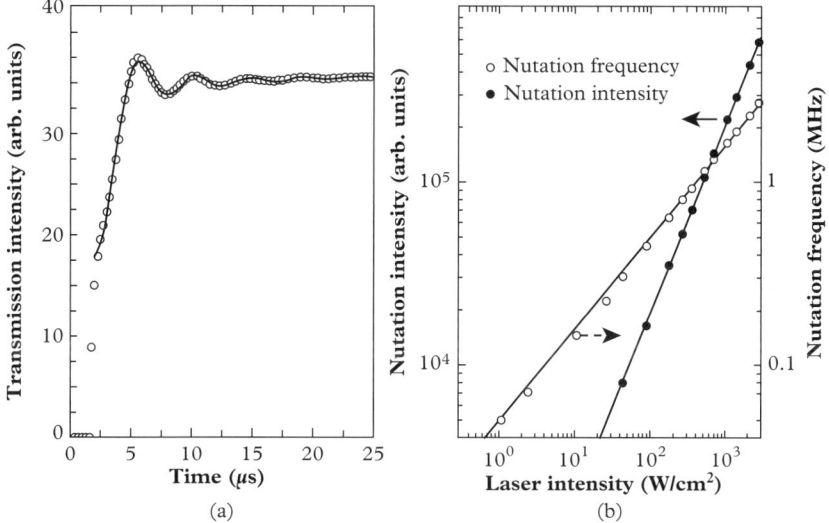

Figure 10.15 *(a) Optical nutation behavior of the Tm^{3+}:YAG crystal sample, working at 1.4 K with its ~793-nm absorption line and at the input laser intensity of 26.3 W/cm²; (b) measured nutation intensity and modulation frequency versus the input laser intensity. (Reproduced with permission from Sun et al.[93] Copyright 2000, American Physical Society).*

10.5 Optical free induction decay effect

In the preceding section, we have considered the optical nutation effect that occurs when a stationary intense optical field is suddenly imposed upon the resonant medium. Now let us consider another transient behavior of the same medium when the applied stationary resonant optical field is rapidly turned off. In this case, one may realize that the transmitted optical signal will not disappear immediately because the established macroscopic electric polarization field of the medium and the associated secondary coherent emission field will not vanish instantly. Actually, these two associated fields decay gradually following the dephasing process of the molecules during a time range $\Delta t \approx T_2$. This is the so-called optical free induction decay effect,[94] and it can be visualized as an optical analog of the corresponding free-induction decay in nuclear magnetic resonance.[95] In particular, if the offset of the resonant interaction between the laser field and a gas sample is realized by switching off the applied Stark electric field, the frequencies of the secondary coherent emission field and the ongoing incident laser field are no longer the same; therefore, one could observe a decaying optical signal modulated by the beat-frequency that equals the detuning between the medium resonance frequency and the ongoing optical field frequency, while the depletion of the beat-frequency signal reflects the dephasing property of the medium.

For a gas medium with an inhomogeneously broadened spectral line, it can be assumed that when $t \leq 0$ there is a stationary resonant interaction between the optical field and the medium, then we let all time derivatives be zero in Eq. (10.4-8) and have the following initial conditions:[76]

$$\left. \begin{array}{l} u(0) = -\Delta \Omega w^0 / (\Omega^2 T_1/T_2 + \Delta^2 + 1/T_2^2) \\ v(0) = (\Omega w^0/T_2)/(\Omega^2 T_1/T_2 + \Delta^2 + 1/T_2^2) \\ w(0) = w^0 \left[1 - (\Omega^2 T_1/T_2)/(\Omega^2 T_1/T_2 + \Delta^2 + 1/T_2^2)\right] \end{array} \right\}. \qquad (10.5\text{-}1)$$

Here, w^0 is the initial population difference between the two levels of the medium in the absence of an applied optical field. It is further assumed when $t \geq 0$ the new detuning is

$$\Delta' = \Delta + \delta\omega_0,$$

where $\delta\omega_0$ is the Stark frequency shift that is assumed here to be greater than the Doppler spectral width of the medium. Therefore, no resonant interaction now exists and we can let $\Omega \Rightarrow 0$; then Eq. (10.4-8) becomes

$$\left. \begin{array}{l} \dot{u} + \Delta'v + u/T_2 = 0 \\ \dot{v} - \Delta'u + v/T_2 = 0 \\ \dot{w} + (w - w^0)/T_1 = 0 \end{array} \right\}. \qquad (10.5\text{-}2)$$

The solution of the above equation for $t \geq 0$ can be obtained as follows

$$\left. \begin{array}{l} u(t) = [u(0)\cos\Delta't - v(0)\sin\Delta't]e^{-t/T_2} \\ v(t) = [u(0)\sin\Delta't + v(0)\cos\Delta't]e^{-t/T_2} \\ w(t) = w^0 + [w(0) - w^0]e^{-t/T_1} \end{array} \right\}. \qquad (10.5\text{-}3)$$

Here, $u(0)$, $v(0)$, and $w(0)$ are determined by Eq. (10.5-1), and $w(0)$ represents the population difference at the moment when the resonance optical field is turned off. According to Eq. (10.4-18), we have[76]

$$<\tilde{\rho}_{12}> = \frac{i}{2k\bar{v}\sqrt{\pi}} \int_{-\infty}^{\infty} e^{-(\Delta/k\bar{v})^2} \left[u(0) \sin \Delta' t + v(0) \cos \Delta' t \right] e^{-t/T_2} d\Delta$$

$$\approx \frac{i}{2k\bar{v}\sqrt{\pi}} e^{-(\bar{\Delta}/k\bar{v})^2} \cdot \Omega w^0 e^{-t/T_2} \cdot \cos \delta\omega_0 t \int_{-\infty}^{\infty} \frac{-\Delta \sin \Delta t + (1/T_2) \cos \Delta t}{\Omega^2 T_1/T_2 + \Delta^2 + 1/T_2^2} d\Delta \quad (10.5\text{-}4)$$

$$\approx \frac{i\sqrt{\pi}}{2k\bar{v}} e^{-(\bar{\Delta}/k\bar{v})^2} \cdot \Omega w^0 \cdot e^{-(t/T_2)\left[1+\sqrt{\Omega^2 T_1 T_2 + 1}\right]} \cdot \left[\frac{1}{\sqrt{\Omega^2 T_1 T_2 + 1}} - 1 \right] \cos \delta\omega_0 t.$$

In obtaining this result it is assumed that $<u(t)> = 0$ and the Gaussian distribution factor can be moved out from the integral. Following the same procedure used for deriving Eq. (10.4-27), one can finally obtain the expression for the oscillating component with a beat-frequency of $\delta\omega_0$ in the transmitted optical field

$$I'_t(t) \propto 2E_0 E'_0(L,t) = E_0^2 Q(t) \cos \delta\omega_0 t, \quad (10.5\text{-}5)$$

where

$$Q(t) = \frac{\pi^{3/2}}{n_0^2 \varepsilon_0 \hbar \bar{v}} N_0 L p_0^2 w^0 \left(\frac{1}{\sqrt{\Omega^2 T_1 T_2 + 1}} - 1 \right) e^{-(\bar{\Delta}/k\bar{v})^2} e^{-(t/T_2)\left[1+\sqrt{\Omega^2 T_1 T_2 + 1}\right]}. \quad (10.5\text{-}6)$$

The physical meaning of Eqs. (10.5-5) and (10.5-6) is that in the condition of using a cw laser beam and Stark electric field switching to suddenly terminate a stationary resonant interaction at $t = 0$, the optical free induction decay (OFID) can occur in a damped and periodically oscillating form of the transmitted optical signal in the range $t \geq 0$. The intensity modulation frequency equals the Stark frequency shift $\delta\omega_0$, while the temporary damping of this intensity oscillation is determined by the factor $Q(t)$ given by Eq. (10.5-6). In general, the intensity modulation frequency $\delta\omega_0$ for OFID is greater than the modulation frequency Ω for optical nutation, whereas the damping rate of the intensity oscillation for the OFID signal is faster than that for the optical nutation signal.

Since the first observation of OFID in a NH_2D gas sample reported by Brewer and Shoemaker in 1972,[94] a number of resonant media have been employed for OFID studies. Some of them are summarized as follows:

A. molecular gases including NH_2D,[94] I_2,[81,96,97] NH_3,[80] CH_4,[98] CHF_3,[98] CO_2,[99] HCN,[100] and HBr;[101]
B. metal vapors such as Na[102] and Rb;[103]
C. metal ion-doped crystals including $Cr^{3+}:Al_2O_3$,[104–106] $Pr^{3+}:LaF_3$,[107,108] $Er^{3+}:LaF_3$,[109] $Sm^{3+}:LaF_3$,[110] $Tm^{3+}:LaF_3$,[111] $Pr^{3+}:YAG$,[112] and $Er^{3+}:Y_2SiO_5$;[113]
D. other solid-state samples such as the F_3^+ center in NaF,[114] $Si:GaAs$,[115] and pentacene in p-terphenyl.[116]

As an example of OFID studies in metal ion-doped crystals, Fig. 10.16 shows the experimental results of the free induction decay behavior in Pr^{3+}-doped LaF_3 crystal at 2 K.[107] The sample was coherently prepared by a cw dye laser beam of 592.5 nm, whose frequency could be suddenly shifted by pre-passing through an extra-cavity acousto-optic modulator. In Fig. 10.16(a), the OFID signal intensity is recorded from the same crystal sample at three different external magnetic field levels, with a beat frequency of 5 MHz determined by the frequency shift of the input

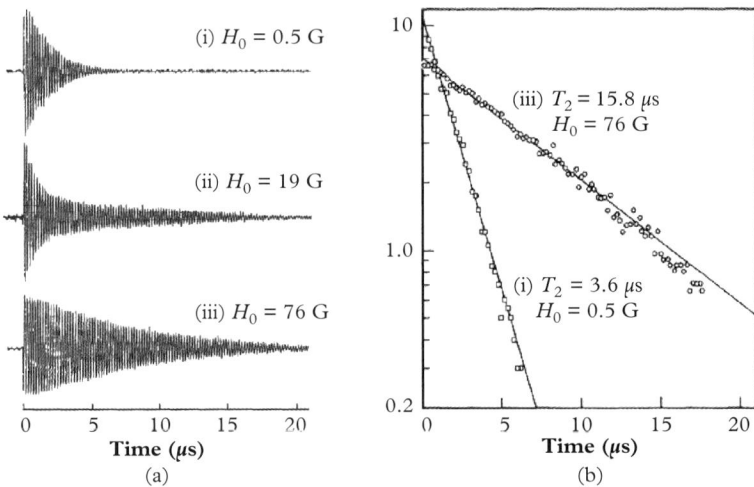

Figure 10.16 *(a) Optical free-induction decay (OFID) of 0.1 at. % Pr^{3+} in LaF_3 at 2 K in the presence of an external magnetic field of (a) $H_0 \approx 0.5$ G (Earth's field), (b) $H_0 = 19$ G, and (c) $H_0 = 76$ G with the beat frequency of 5 MHz. (b) Semilog plots of the data shown in plots (i) and (iii) of (a). (Reproduced with permission from DeVoe et al.[107] Copyright 1979, American Physical Society).*

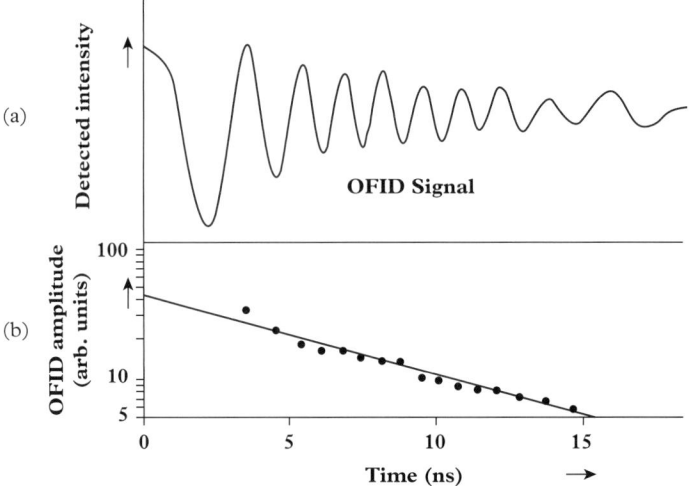

Figure 10.17 *(a) OFID signal of the O_2 site of pentacene in p-terphenyl at 1.5 K excited with the 592.01-nm laser beam of intensity 2.8 mW/mm^2. (b) Semilog plot of experimental beat amplitudes against time. (Reproduced with permission from de Bree and Wiersma.[116] Copyright 1978, Elsevier).*

laser beam. In cases (i) and (iii) of Fig. 10.16(a), the OFID signals decrease exponentially with respect to the time, proved by the semilog plots shown in Fig. 10.16(b) with two measured values of T_2. On the other hand, in the case (ii) of Fig. 10.16(a), the decay of the OFID signal exhibits a biexponential feature showing the mixed influence from two different states of Pr^{3+} energy-level structures.

Another example is related to the OFID studies in organic mixed crystals.[116] Figure 10.17(a) shows the OFID signal of the O_2 site of pentacene in p-terphenyl host crystal at 1.5 K excited with a 592.01-nm dye laser beam, where the observed $\sim$1 GHz beat signal is a result of the frequency switching of a single-mode cw dye laser passing through an extra-cavity electro-optic phase modulator. The semilog plot of the beat amplitude as a function of the time is shown in Fig. 10.17(b), which shows an exponential decay of the OFID signal, as predicted by the last term in Eq. (10.5-6).

....

PROBLEMS

1. For a linearly absorbing medium of $n_0 = 1.75$, if the absorption coefficient at 560 nm wavelength is $\alpha = 0.5 cm^{-1}$, the density of absorbing atoms is $N_0 = 1.5 \times 10^{19} cm^{-3}$, and the transverse relaxation time $T_2 = 10^{-7}$s, please use Eq. (10.2-21) to calculate the p_0 value, and then use Eq. (10.2-19) to determine the optical field E_0 value of a 2π-pulse with a pulsewidth of $\tau_0 = 10$ ns.
2. Under the same condition as mentioned above, please use Eq. (10.2-20) to calculate the propagation speed of the 2π-pulse.
3. Based on the calculated p_0 value and E_0 value from problem 1, please use Eq. (10.3-10) to determine the Rabi frequency Ω.
4. Based on the results of optical nutation measurement shown in Fig. 10.15(a), estimate the value of the Rabi frequency at the intensity level of 26.3 W/cm^2, and further calculate the corresponding p_0 value with assumed $n_0 = 1.82$.

....

REFERENCES

1. S. L. McCall and E. L. Hahn, *Phys. Rev. Lett.* **18**, 908 (1967).
2. S. L. McCall and E. L. Hahn, *Phys. Rev.* **183**, 457 (1969).
3. V. S. Letokhov, *Sov. Phys. JETP* **29**, 221 (1969).
4. P. G. Kryukov and V. S. Letokhov, *Sov. Phys. Usp.* **12**, 641 (1070).
5. S. L. MacCall and E. L. Hahn, *Phys. Rev. A* **2**, 861 (1970).
6. R. E. Slusher, in *Progress in Optics*, Vol. 12, edited by E. Wolf (North-Holland, Amsterdam, 1974), pp. 52–100.
7. I. A. Poluektov, Yu. M. Popov, and V. S. Roitberg, *Sov. Phys. Usp.* **17**, 673 (1975).
8. C. K. N. Patel and R. E. Slusher, *Phys. Rev. Lett.* **19**, 1019 (1967).
9. A. Zembrod and Th. Gruhl, *Phys. Rev. Lett.* **27**, 287 (1971).
10. S. S. Alimpiev and N. V. Karlov, *Sov. Phys. JETP* **34**, 947 (1972).

11. C. D. David, Jr., *Appl. Phys. Lett.* **23**, 306 (1973).
12. L. M. Peterson, *Appl. Phys. Lett.* **31**, 86 (1977).
13. W. Krieger and P. E. Toschek, *Phys. Rev. A* **11**, 276 (1975).
14. J. J. Bannister, H. J. Baker, T. A. King, and W. G. McNaught, *Phys. Rev. Lett.* **44**, 1062 (1980).
15. H. M. Gibbs and R. E. Slusher, *Phys. Rev. Lett.* **24**, 638 (1970).
16. D. Grischkowsky, *Phys. Rev. A* **7**, 2096 (1973).
17. D. J. Bradley, G. M. Gale, and P. D. Smith, *Nature* **225**, 719 (1970).
18. G. J. Salamo, H. M. Gibbs, and G. G. Churchill, *Phys. Rev. Lett.* **33**, 273 (1974).
19. D. W. Dolfi and E. L. Hahn, *Phys. Rev. A* **21**, 1272 (1980).
20. W.-K. Lee, M.-K. Oh, W.-S. Choi, J.-H. Jeon, J.-H. Lee, and J.-S. Chang, *Jpn. J. Appl. Phys. Part 1*, **41**, 5170 (2002).
21. J. Yi, H. Park, Y. Rhee, and J. Lee, *Jpn. J. Appl Phys. Part 1*, **39**, 1128 (2000).
22. A. Szabo and N. Takfuchi, *Opt. Commun.* **15**, 250 (1975).
23. V. V. Samartsev and R. G. Usmanov, *Sov. Phys. JETP* **45**, 892 (1977).
24. M. Nakazawa, Y. Kimura, K. Kurokawa, and K. Suzuki, *Phys. Rev. A* **45**, R23 (1992).
25. F. Bruckner, V. S. Dneprovskii, S. M. Zakharov, D. G. Koshchug, E. A. Manykin, and V. U. Khattatov, *JETP Lett.* **20**, 51 (1974).
26. S. Schneider, P. Borri, W. Langbein, U. Woggon, J. Foerstner, A. Knorr, R. L. Sellin, D. Ouyang, and D. Bimberg, *Appl. Phys. Lett.* **83**, 3668 (2003).
27. E. L. Hahn, *Phys. Rev.* **80**, 580 (1950).
28. N. A. Kurnit, I. D. Abella, and S. R. Hartmann, *Phys. Rev. Lett.* **13**, 567 (1964).
29. I. D. Abella, N. A. Kurnit, and S. R. Hartmann, *Phys. Rev.* **141**, 391 (1966).
30. S. R. Hartmann, in *Quantum Optics*, edited by R. J. Glauber (Academic Press, New York, 1969), pp. 532–550.
31. I. D. Abella, in *Progress in Optics*, Vol. 7, edited by E. Wolf (North-Holland, Amsterdam, 1969), pp. 139–168.
32. C. K. N. Patel and R. E. Slusher, *Phys. Rev. Lett.* **20**, 1087 (1968).
33. D. E. Cooper, R. W. Olson, and M. D. Fayer, *J. Chem. Phys.* **72**, 2332 (1980).
34. L. W. Molenkamp and D. A. Wiersma, *Phys. Rev. B* **32**, 8108 (1985).
35. O. Berg and E. L. Chronister, *J. Chem. Phys.* **106**, 4401 (1997).
36. R. G. Brewer and R. L. Shoemaker, *Phys. Rev. Lett.* **27**, 631 (1971).
37. R. J. Nordstrom, W. M. Gutman, and C. V. Heer, *Phys. Lett.* A**50**, 25 (1974).
38. S. S. Alimpiev and N. V. Karlov, *Sov. Phys. JETP* **36**, 255 (1973).
39. H. Nakatsuka, S. Asaka, M. Tomita, and M. Matsuoka, *Opt. Commun.* **47**, 65 (1983).
40. .40 B. Bölger and J. C. Diels, *Phys. Lett.* **28A**, 401 (1968).
41. A. Flusberg, T. Mossberg, R. Kachru, and S. R. Hartmann, *Phys. Rev. Lett.* **41**, 305 (1978).
42. N. W. Carlson, A. G. Yodh, and T. W. Mossberg, *Phys. Rev. Lett.* **51**, 35 (1983).
43. E. A. Rotberg, B. Barrett, S. Beattie, S. Chudasama, M. Weel, I. Chan, and A. Kumarakrishnan, *J. Opt. Soc. Am. B* **24**, 671 (2007).
44. N. N. Rubtsova, V. G. Gol'dort, V. N. Ishchenko, S. A. Kochubei, E. B. Khvorostov, V. A. Reshetov, and I. V. Yevseyev, *Laser Phys.* **22**, 1489 (2012).
45. T. Baer and I. D. Abella, *Opt. Lett.* **3**, 170 (1978).
46. M. R. Woodworth, *Opt. Lett.* **8**, 307 (1983).
47. A. Yamagishi and A. Szabo, *Opt. Lett.* **2**, 160 (1978).
48. R. M. Macfarlane, R. M. Shelby, and R. L. Shoemaker, *Phys. Rev. Lett.* **43**, 1726 (1979).
49. R. W. Equall, R. L. Cone, and R. M. Macfarlane, *Phys. Rev. B* **52**, 3963 (1995).
50. N. Takeuchi, S. Chandra, Y. C. Chen, and S. R. Hartmann, *Phys. Lett. A* **46**, 97 (1973).
51. V. V. Samartsev and R. G. Usmanov, *Sov. Phys. Solid State*, **18**, 897 (1976).
52. T. T. Basiev, A. Ya. Karasik, V. V. Fedorov, and K. W. Ver Steeg, *JETP* **86**, 156 (1998).
53. M. Mitsunaga, R. Kachru, E. Xu, and M. K. Kim, *Phys. Rev. Lett.* **63**, 754 (1989).

54. G. . Elinn, K. W. Jang, J. Ganem, M. L. Jones, R. S. Meltzer, and R. M. Macfarlane, *Phys. Rev. B* **49**, 5821 (1994).
55. R. M. Macfarlane, T. L. Harris, R. L. Cone, and R. W. Equall, *Opt. Lett.* **22**, 871 (1997).
56. G. K. Liu and R. L. Cone, *Phys. Rev. B* **41**, 6193 (1990).
57. G. M. Wang, R. W. Equall, R. L. Cone, M. J. M. Leask, K. W. Godfrey, and F. R. Wondre, *Opt. Lett.* **21**, 818 (1996).
58. X. Wang, M. Afzelius, N. Ohlsson, U. Gustafsson, and S. Kroll, *Opt. Lett.* **25**, 945 (2000).
59. H. L. Xu, M. Nilsson, S. Ohser, N. Rauhut, S. Kroll, M. Aguilo, and F. Diaz, *Phys. Rev. B* **70**, 214115 (2004).
60. V. N. Lisin, V. V. Samartsev, A. M. Shegeda, V. A. Zuikov, and Yu. K. Rosencwage, *Laser Phys.* **17**, 87 (2007).
61. J. Hegarty, M. M. Broer, B. Golding, J. R. Simpson, and J. B. MacChesney, *Phys. Rev. Lett.* **51**, 2033 (1983).
62. Y. Silberberg, V. L. da Silva, J. P. Heritage, E. W. Chase, and M. J. Andrejco, *IEEE J. Quantum Electron.* **28**, 2369 (1992).
63. S. Zhang, Z. Sun, X. Yang, Z. Wang, J. Lin, W. Huang, Z. Xu, and R. Li, *Opt. Commun.* **241**, 481 (2004).
64. M. Koch, D. Weber, J. Feldmann, E. O. Göbel, T. Meier, A. Schulze, P. Thomas, S. Schmitt-Rink, and K. Ploog, *Phys. Rev. B* **47**, 1532 (1993).
65. T. J. Aartsma and D. A. Wiersma, *Phys. Rev. Lett.* **36**, 1360 (1976).
66. A. Nakamura, Y. Yoshikuni, Y. Ishida, S. Shiooya, and M. Aihara, *Opt. Commun.* **44**, 431 (1983).
67. L. W. Molenkamp and D. A. Wiersma, *J. Chem. Phys.* **83**, 1 (1985).
68. M. Fujiwara, R. Kuroda, and H. Nakatsuka, *J. Opt. Soc. Am. B* **2**, 1634 (1985).
69. C. A. Walsh, M. Berg, L. R. Narasimhan, and M. D. Fayer, *Chem. Phys. Lett.* **130**, 6 (1986).
70. A. Elschner, L. R. Narasimhan, and M. D. Fayer, *Chem. Phys. Lett.* **171**, 19 (1990).
71. M. van der Voort, C. W. Rella, L. F. G. van der Meer, A. V. Akimov, and J. I. Dijkhuis, *Phys. Rev. Lett.* **84**, 1236 (2000).
72. K. W. Jobson, J.-P. R. Wells, N. Q. Vinh, P. J. Phillips, C. R. Pidgeon, and J. I. Dijkhuis, *Phys. Rev. B* **74**, 165205 (2006).
73. C. L. Tang and B. D. Silverman, in *Physics of Quantum Electronics*, edited by P. L. Kelley, B. Lax, and P. E. Tannenwald (McGraw-Hill, New York, 1966), pp. 280–293.
74. H. C. Torrey, *Phys. Rev.* **76**, 1059 (1949).
75. G. B. Hocker and C. L. Tang, *Phys. Rev. Lett.* **21**, 591 (1968).
76. R. G. Brewer, in *Frontiers in Laser Spectroscopy*, vol. 1, edited by R. Balian, S. Haroche, and S. Liberman (North-Holland, Amsterdam, 1977), pp. 341–398.
77. G. B. Hocker and C. L. Tang, *Phys. Rev.* **184**, 356 (1969).
78. R. L. Shoemaker and F. A. Hopf, *Phys. Rev. Lett.* **33**, 1527 (1974).
79. M. M. T. Loy, *Phys. Rev. Lett.* **36**, 1454 (1976).
80. J. Wessel, *Appl. Phys. Lett.* **41**, 411 (1982).
81. R. G. Brewer and A. Z. Genack, *Phys. Rev. Lett.* **36**, 959 (1976).
82. V. Yu. Baranov, V. L. Borzenko, S. M. Kozochkin, K. N. Makarov, D. D. Malyuta, Yu. V. Petrushevich, Yu. A. Satov, A. N. Starostin, and A. P. Strel'tsov, *Sov. J. Quantum Electron.* **14**, 235 (1984).
83. G. Duxbury, J. F. Kelly, T. A. Blake, and N. Langford, *J. Chem. Phys.* **136**, 174317 (2012).
84. P. F. Liao, J. E. Bjorkholm, and J. P. Gordon, *Phys. Rev. Lett.* **39**, 15 (1977).
85. P. M. Farrell, W. R. MacGillivray, and M. C. Standage, *Phys. Lett. A* **107**, 263 (1985).
86. Y. S. Bai, A. G. Yodh, and T. W. Mossberg, *Phys. Rev. A* **34**, 1222 (1986).
87. Y. Q. Li, W. H. Burkett, and M. Xiao, *Opt. Lett.* **21**, 982 (1996).
88. U. Shim, S. B. Cahn, A. Kumarakrishnan, T. Sleator, and J.-T. Kim, *Jpn. J. Appl. Phys. Part 1*, **44**, 168 (2005).
89. A. V. Smith and T. D. Raymond, *Opt. Lett.* **16**, 267 (1991).

90. A. Wännström, O. Vogel, A. Arnesen, R. Hallin, A. Kastberg, C. Nordling, and S. Linnaeus, *Phys. Rev. A* **38**, 5964 (1988).
91. D. M. Burland, F. Carmona, and E. Cuellar, *Chem. Phys. Lett.* **64**, 5 (1979).
92. C. Greiner, B. Boggs, T. Loftus, T. Wang, and T. W. Mossberg, *Phys. Rev. A* **60**, R2657 (1999).
93. Y. Sun, G. M. Wang, R. L. Cone, R. W. Equall, and M. J. M. Leask, *Phys. Rev. B* **62**, 15443 (2000).
94. R. G. Brewer and R. L. Shoemaker, *Phys. Rev. A* **6**, 2001 (1972).
95. E. L. Hahn, *Phys. Rev.* **77**, 297 (1950).
96. A. Z. Genack and R. C. Brewer, *Phys. Rev. A* **17**, 1463 (1978).
97. P. Dube, M. D. Levenson, and J. L. Hall, *Opt. Lett.* **22**, 184 (1997).
98. K. Bratengeier, H. -G. Purucker, and A. Alaubereau, *Opt. Commun.* **70**, 393 (1989).
99. E. Yablonovitch and J. Goldhar, *Appl. Phys. Lett.* **25**, 580 (1974).
100. I. Coddington, W. C. Swann, and N. R. Newbury, *Opt. Lett.* **35**, 1395 (2010).
101. E. N. Chesnokov, V. V. Kubarev, P. V. Koshlyakov, and G. N. Kulipanov, *Appl. Phys. Lett.* **101**, 131109 (2012).
102. R. G. DeVoe and R. G. Brewer, *Phys. Rev. Lett.* **40**, 862 (1978).
103. U. Shim, S. Cahn, A. Kumarakrishnan, T. Sleator, and J.-T. Kim, *Jpn. J. Appl. Phys. Part 1* **41**, 3688 (2002).
104. P. L. Liao and S. R. Hartmann, *Phys. Lett. A* **44**, 361 (1973).
105. S. Nakanish., O. Tamura, T. Muramoto, and T. Hashi, *Opt. Commun.* **31**, 344 (1979).
106. A. Szabo and T. Muramoto, *Phys. Rev. A* **39**, 3992 (1989).
107. R. G. DeVoe, A. Szabo, S. C. Rand, and R. G. Brewer, *Phys. Rev. Lett.* **42**, 1560 (1979).
108. R. G. DeVoe and R. G. Brewer, *Phys. Rev. Lett.* **50**, 1269 (1983).
109. R. M. Macfarlane and R. M. Shelby, *Opt. Commun.* **42**, 346 (1982).
110. R. M. Macfarlane and R. M. Shelby, *Phys. Lett. A* **16**, 299 (1986).
111. R. M. Macfarlane and M. Zhu, *Opt. Lett.* **22**, 248 (1997).
112. R. M. Shelby, A. G. Tropper, R. T. Harley, and R. M. Macfarlane, *Opt. Lett.* **8**, 304 (1983).
113. J. Minar, B. Lauritzen, H. de Riedmatten, M. Afzelius, C. Simon, and N. Gisin, *New J. Phys.* **11**, 113019 (2009).
114. R. M. Macfarlane, A. Z. Genack, and R. G. Brewer, *Phys. Rev. B* **17**, 2821 (1978).
115. P. C. M. Planken, P. C. van Son, J. N. Hovenier, T. O. Klaassen, W. Th. Wenckebach, B. N. Murdin, and G. M. H. Knippels, *Phys. Rev. B* **51**, 9643 (1995).
116. P. de Bree and D. A. Wiersma, *Opt. Commun.* **26**, 248 (1978).

11
Optical Bistability

When the output parameter of an optical device can be a multivalued function of the input parameter, this device could be working in an optically bistable state. The most typical optical bistable device is a Fabry–Perot (F-P) cavity (etalon) containing a nonlinear optical medium, the refractive index of which depends on the light intensity. Upon the action of an intense input optical beam, the transmitted intensity or reflected intensity of this device may exhibit a nonlinear response to the incident light intensity and, especially under appropriate working conditions, the characteristic output/input curve may manifest a single (or multiple) hysteresis loop(s). This type of optical bistable device can be used as an optical differential amplifier, optical switch, optical limiter, optical clipper, optical discriminator, or an optical memory element, depending on the chosen nonlinear medium and operation conditions.[1,2]

The key requirement for a nonlinear F-P etalon device is that the light-induced refractive-index change of the intracavity medium should be as great as possible. For this purpose, certain mechanisms of resonance-enhanced refractive-index change should be employed. However, these resonant mechanisms may also lead to a certain extent of thermal effect that in many cases may dominate the induced refractive index change. This opto-thermal effect has a slow rise/decay time and poor reproducibility, and therefore should be avoid or minimized for high-speed optoelectronic applications.

11.1 Basic consideration of a nonlinear F-P etalon

11.1.1 Background of optical bistability studies

The earliest theoretical work suggesting an optical bistable device was published by Szöke et al. in 1969, which was based on the consideration of a saturable absorptive medium inside an F-P cavity.[3] Later, the first experimental observation of optical bistable effect was reported by Gibbs et al. in 1976[4]; in this work an F-P etalon containing Na vapor inside the cavity was used, and the observed bistable behavior was explained with the light-induced refractive index change instead of the early suggested absorption change. For this historical reason, optical bistable devices can be classified into two categories: the devices of absorptive type and the devices of dispersive type. For the first type of devices, the major mechanism is the light-intensity dependent absorption change of the intracavity medium; whereas for the second type of devices, the major mechanism is the light-intensity dependent refractive-index change of the intracavity medium. Following the first demonstration of the dispersive optical bistable device, several other papers were published, which were based on the same dispersive regime but used different materials as the nonlinear media.[5-9] In all these early experimental studies, plotting the transmitted intensity as a function of the incident intensity, researchers could obtain various nonlinear characteristic curves including the "hysteresis loop," which was commonly recognized as a special feature of the optical bistable process.

Although in this chapter we shall mainly consider the dispersive-type bistable device as a passive optical element, the similar optical hysteresis-loop behavior could also be observed in a certain type of laser devices with the F-P cavity, or even may be observed in a cavityless nonlinear optical system. The discussions on the latter two types of bistable devices are somewhat beyond the scope of this chapter.

11.1.2 Theory of steady-state optical bistability

Let us consider the basic behavior of the F-P cavity containing a nonlinear medium, of which the refractive-index is dependent on the intensity of the intracavity optical field. Assume that the F-P cavity consists of two identical mirrors of reflectivity R, and a homogeneous third-order nonlinear medium is inside the cavity, as shown schematically in Fig. 11.1. From the theory of F-P interferometer, we know that for a plane incident light wave the transmitted intensity, I_t, is determined by[10]

$$I_t = \frac{1}{1 + F\sin^2(\delta/2)} I_0, \qquad (11.1\text{-}1)$$

where I_0 is the incident intensity, $F = 4R/(1-R)^2$, and δ is the phase-shift factor associated with one round trip of the intracavity beam. In the present case, δ can be expressed as

$$\delta = \frac{4\pi}{\lambda} L \cos\theta \cdot (n_0 + \Delta n) = \frac{4\pi}{\lambda} L \cos\theta \cdot (n_0 + n_2 I_i), \qquad (11.1\text{-}2)$$

where λ is the free-space wavelength of the incident light wave, L is the cavity length, θ is the incident angle, n_0 and n_2 are the linear refractive index and the nonlinear refractive-index coefficient of the medium, and I_i is the overall intensity of the intracavity field that involves the forward and backward components, as shown in Fig. 11.1. In writing Eq. (11.1-2), we have assumed that the induced refractive-index change is simply proportional to the intracavity light intensity, which is a steady-state assumption. The intensity of the forward intracavity field can be expressed as[10]

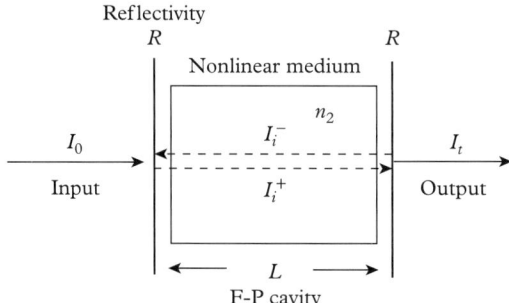

Figure 11.1 Schematic illustration of an F-P cavity containing a nonlinear medium.

$$I_i^+ = \frac{1}{1-R} \cdot \frac{1}{1+F\sin^2(\delta/2)} I_0, \quad (11.1\text{-}3)$$

and the intensity of the backward intracavity field is

$$I_i^- = \frac{R}{1-R} \cdot \frac{1}{1+F\sin^2(\delta/2)} I_0. \quad (11.1\text{-}4)$$

Thus the overall intensity of the intracavity field should be[10,11]

$$I_i = I_i^+ + I_i^- = \frac{1+R}{1-R} \cdot \frac{1}{1+F\sin^2(\delta/2)} I_0 = \frac{1+R}{1-R} I_t. \quad (11.1\text{-}5)$$

Substituting Eqs. (11.1-2) and (11.1-5) into Eq. (11.1-1) leads to

$$I_t = \frac{I_0}{1+F\sin^2\frac{1}{2}[\delta_0+\gamma I_i]}, \quad (11.1\text{-}6)$$

where

$$\left.\begin{array}{l} \delta_0 = \dfrac{4\pi}{\lambda} n_0 L \cos\theta - m2\pi \\[6pt] \gamma = \dfrac{4\pi}{\lambda} n_2 L \cos\theta \end{array}\right\}. \quad (11.1\text{-}7)$$

Here, δ_0 is the initial phase factor independent of the intensity, and m is an integer chosen for making $|\delta_0| \leq 2\pi$.

From Eq. (11.1-6) we see that I_t is an implicit function of I_0. For a given value of I_0, I_t may have multi-valued solutions, and this is the mathematical origin of the optical hysteresis loop that might be observed. Although Eq. (11.1-6) can be solved numerically, it is more straightforward and meaningful to solve this equation with a graphical method.[11,12] For this purpose, from Eqs. (11.1-1) and (11.1-5) we could give two parallel expressions for the transmissivity of a nonlinear F-P etalon:

$$\left.\begin{array}{l} T = \dfrac{I_t}{I_0} = \dfrac{1}{1+F\sin^2\dfrac{1}{2}(\delta_0+\gamma I_i)} \\[10pt] T = \dfrac{1-R}{1+R} \cdot \dfrac{1}{I_0} I_i \end{array}\right\}. \quad (11.1\text{-}8)$$

In Eq. (11.1-8), the first equation represents a periodic curve as a function of $(\delta_0 + \gamma I_i)$, as shown in Fig. 11.2, while the second equation implies a group of straight lines originating from the δ_0 position in the abscissa, the slopes of these straight lines are determined by $(1-R)/[(1+R)\gamma I_0]$. In Fig. 11.2, the straight lines labeled (1) to (5) correspond to five different increasing I_0 values. The intersections between the periodic curve and the straight lines represent the solutions of the transmissivity; for a given I_0 value, there might be more than one possible solution.

Knowing T values for different I_0 values, we can determine the corresponding I_t values, and then obtain the characteristic curve plotting I_t as a function of I_0. With shifting the straight line's

362 *Optical Bistability*

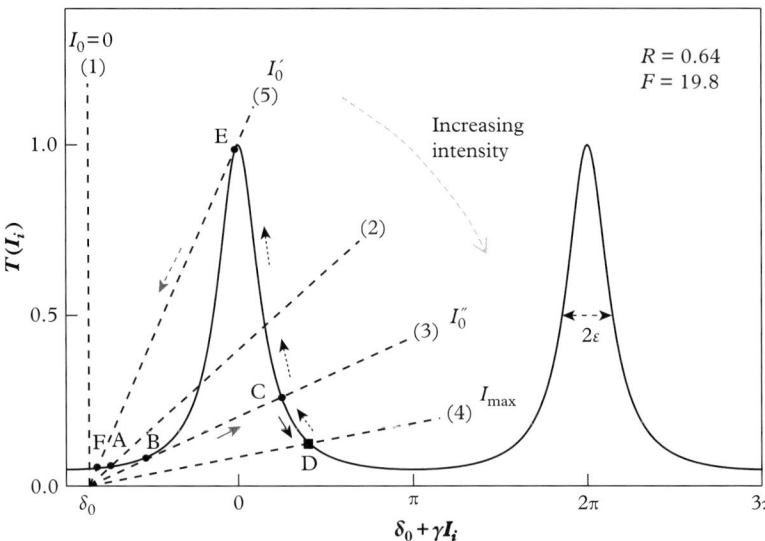

Figure 11.2 *Graphical solutions of the transmissivity T of a nonlinear F-P cavity at various incident intensity (I_0) levels.*

origin position in the abscissa of Fig. 11.2 and changing the straight line's scanning extension, various characteristic curves for a nonlinear F-P etalon could be obtained. Some typical curves are shown schematically in Fig. 11.3, which represent different functions of the bistable device, corresponding to (a) optical switching, (b) optical differential amplification, (c) optical clipping, (d) optical limiting, and (e–f) optical memory.

Now let us explain under what conditions the optical hysteresis loop shown in Fig. 11.3(e) and (f) could be obtained. For this purpose, we return to Fig. 11.2 and assume that the initial cavity detuning is arbitrarily chosen as $\delta_0 \approx -0.87\pi$. When the I_0 value increases from near zero to the level represented by the straight line (2), the transmissivity value is determined by the intersection point A. Following the subsequent increase of I_0, the T value grows continuously until the incident intensity reaches a critical value that corresponds to straight line (3); in this case, there are two intersections B and C, and the further continuous change of T is no longer possible. Thereby, the working point will jump from the point B (corresponding to a lower T value) to the point C (corresponding to a higher T value). After that transition, the transmissivity decreases following the further increase of I_0. When the I_0 value starts to decrease from its maximum level, e.g., represented by straight line (4), the working point continuously moves along the periodic curve until it reaches the point E, where the further continuous move along the same curve is impossible and the working point has to jump down from the point E (corresponding to a higher T value) to the point F (corresponding to a lower T value). After that transition, the working point will smoothly return to its initial position when I_0 approaches zero again. In describing this full cycle of the move of working point, we have assumed that the change of the T value tends to take place continuously if possible. According to this assumption, the full cycle of working point's move can be expressed by the characteristic curve shown in Fig. 11.3(e), and the two pairs of working points are indicated where the switch-up and switch-down transitions take place, respectively. From Fig. 11.3(e), one can see that in the incident intensity range between I_0' and I_0''', there is more than one

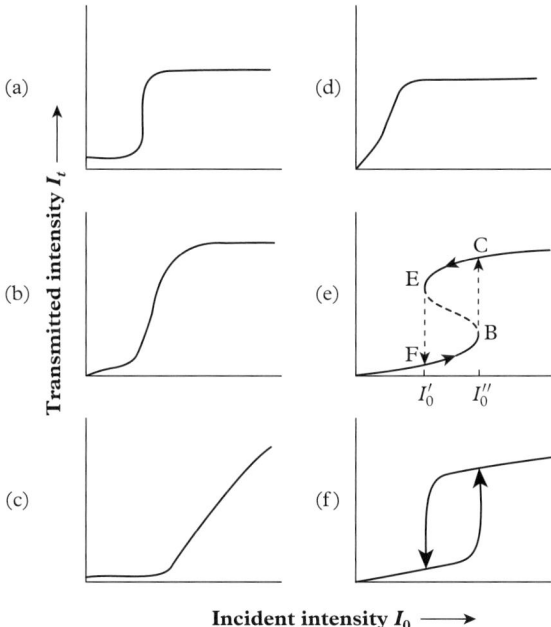

Figure 11.3 Typical characteristic curves for a nonlinear F-P bistable device, corresponding to (a) optical switching, (b) optical differential gain, (c) optical clipping, (d) optical limiting, and (e, f) optical memory.

solution of I_t, although the solutions represented by the dashed S-shape curve are not stable. As a result, an optical hysteresis loop shown in Fig. 11.3(f) could be actually observed.

In practice, the specifically observed characteristic curves of an optical bistable device depend on many working conditions, such as the values of δ_0, F, and γ, the incident intensity level and change range, the sampling manner of the transmitted optical signal, and so on.

So far, we have assumed that the intracavity medium is transparent to the incident light wave. In reality, there are always some losses due to light scattering or absorption. Assuming the linear attenuation coefficient of the nonlinear medium is α_0, we should rewrite Eq. (11.1-8) as[10,13]

$$\left. \begin{array}{c} T = \dfrac{I_t}{I_0} = \dfrac{(1-R)^2 e^{-\alpha_0 L/\cos\theta}}{(1-R')^2} \cdot \dfrac{1}{1 + F'\sin^2\dfrac{1}{2}(\delta_0 + \gamma I_i)} \\[2ex] T = \dfrac{(1-R)e^{-\alpha_0 L/\cos\theta}}{(1+R')} \cdot \dfrac{1}{I_0} I_i \end{array} \right\}, \qquad (11.1\text{-}9)$$

where

$$\left. \begin{array}{c} R' = Re^{-\alpha_0 L/\cos\theta} \\[1ex] F' = \dfrac{4R'}{(1-R')^2} \end{array} \right\}. \qquad (11.1\text{-}10)$$

From these two formulae, we see that the existence of attenuation in the nonlinear medium causes the decrease of the equivalent reflectivity of the cavity mirrors, and diminishes the maximum transmissivity to be less than one. Except for that, all the graphical description and qualitative discussions given previously remain adequate. When $\alpha_0 \Rightarrow 0$, Eq. (11.1-9) leads to Eq. (11.1-8).

11.1.3 Dynamic response of a nonlinear F-P etalon

All foregoing formulae are obtained with the steady-state assumption that the incident optical signal is a continuous-wave (cw) or intensity-slowly-changed laser beam. In this case, the optical bistable behavior of a nonlinear F-P etalon is only dependent on the instantaneous intensity of an incident laser beam. However, in practice the incident light signals are usually laser pulses with a certain pulse width. Under these circumstances, the dynamic response of the F-P etalon may exhibit more complicated features. In general, there are several parameters that affect the dynamic behavior of a nonlinear F-P device, and these are defined as follows.

(1) The characteristic variation time of the incident laser pulse is usually defined as the pulsewidth Δt_0 of the incident laser pulse.

(2) The cavity build-up time is defined as

$$\tau_c = \frac{2n_0 L}{c \cos \theta}(1 - Re^{-\alpha_0 L})^{-1} = \tau_0 \cdot (1 - Re^{-\alpha_0 L})^{-1}, \qquad (11.1\text{-}11)$$

where τ_0 is the cavity round-trip time. The physical meaning of τ_c is the average lifetime of the intracavity photons, determined by the cavity length L, mirror reflection R, and linear attenuation coefficient α_0 of the medium.

(3) Characteristic times of the induced refractive-index change—it has been mentioned in Section 5.6.4 that there are two characteristic times for describing the temporal behavior of the induced refractive-index change: the rise time τ_{rise} and the decay-time τ_{decay} of Δn. The former is used to describe how fast the refractive-index change can occur under the action of an ultrashort laser pulse, and the latter is used to describe how long the induced Δn can last after the laser pulse is terminated. It is well known that the values of τ_{rise} and τ_{decay} can vary dramatically for different mechanisms of light-induced refractive-index changes.

The requirement for the steady-state operation of a bistable F-P etalon can be expressed as

$$\Delta t_0 \gg \tau_c, \tau_{rise}, \tau_{decay}. \quad \text{(steady state)} \qquad (11.1\text{-}12)$$

In this case all formulae from Eq. (11.1-1) to Eq. (11.1-10) remain valid, and the intensities $I_0(t)$, $I_t(t)$, and $I_i(t)$ and transmissivity $T(t)$ can be treated as instantaneous functions of time t. For instance, Eq. (11.1-6) can be rewritten explicitly as

$$I_t(t) = \frac{I_0(t)}{1 + F\sin^2 \frac{1}{2}[\delta_0 + \gamma I_t(t)(1 + R)/(1 - R)]}. \qquad (11.1\text{-}13)$$

Knowing the values of F, R, δ_0, and γ, one may use this equation to calculate the transmitted pulse for a given incident pulse.

Conversely, if the input laser pulsewidth fulfills the following condition:

$$\Delta t_0 < \text{any of } \tau_c, \tau_{rise}, \text{or } \tau_{decay}, \quad \text{(transient state)} \tag{11.1-14}$$

the optical bistable device will work in a transient state. In this case, all formulae describing the steady-state behavior of a nonlinear F-P etalon are no longer adequate for the transient-state operation. The major reason is that, in the transient-state case the induced refractive-index change $\Delta n(t)$ of the nonlinear medium is not simply dependent on the instantaneous intracavity intensity $I_i(t)$; instead, $\Delta n(t)$ might be a rather complicated integral function of $I_i(t - t')$ over t'. The specific integral function may depend on the mechanisms of induced refractive-index changes as well as other working conditions.

In many experimental studies of optical bistability, the real operation could be worked in between the steady-state and the transient-state, and it may make the theoretical analysis of the observed results even more difficult.

Nevertheless, a modified formula similar to Eq. (11.1-1) can be applied to a nonlinear F-P etalon working at any operation state, which is

$$\left. \begin{aligned} I_t(t) &= \frac{1}{1 + F\sin^2\frac{1}{2}[\delta_0 + \Delta\phi(t)]} I_0(t) \\ \Delta\phi(t) &= \frac{4\pi}{\lambda} L \cos\theta \Delta n(t) \end{aligned} \right\}, \tag{11.1-15}$$

where $\Delta\phi(t)$ is the induced round-trip phase change and $\Delta n(t)$ is the associated refractive-index change of the intracavity medium. In general, it is quite difficult to predict the specific form of function $\Delta\phi(t)$ or $\Delta n(t)$; however, the real change of $\Delta\phi(t)$ or $\Delta n(t)$ as a function of t is much easier to be determined experimentally by measuring the temporal variation of $I_0(t)$ and $I_t(t)$ simultaneously.

11.2 Design of optical bistability experiments

11.2.1 Influences of spatial and spectral structures of the incident laser beam

In the preceding section, all formulae and discussions are based on the assumption that the incident field is a monochromatic and plane light wave. In a real optical bistable experiment, the incident signal is a quasi-parallel or focused laser beam with a certain spectral width. In that case, the observed optical bistable behavior of a given nonlinear F-P etalon will depend on the spectral and spatial structures of the incident laser beam.

Let us first consider the influence of spatial structure of an incident beam on the observed bistable behavior of the F-P device. According to Eq. (11.1-15), the transmissivity of a nonlinear F-P etalon is generally determined by the following phase factor:

$$\delta(t) = \frac{4\pi}{\lambda} L \cos\theta [n_0 + \Delta n(t)] = \delta_0 + \Delta\phi(t), \tag{11.2-1}$$

where δ_0 is the initial phase factor (or detuning) of the etalon when $I_0(t \to 0) = 0$, and $\Delta\phi(t)$ is the light-induced phase shift. If the incident light is an unfocused quasi-parallel laser beam and the initial transmissivity $T(t\to 0)$ is determined by the δ_0 value, then following the intensity increase of the input laser pulse the transmissivity $T(t>0)$ of the etalon will change in a certain dynamic manner. Nevertheless, throughout the full period of the input laser pulse, the transmitted light retains its quasi-parallel beam form. In that sense there is no transverse dependence of the observed bistable behavior on the sampling position of the detected signal. For this reason, we can collect the entire transmitted beam with a simple lens system, as shown schematically in Fig. 11.4(a). However, if the incident signal is a focused laser beam with a larger converging angle θ_f (aperture angle), there will be interference fringes of equal inclination associated with the transmitted beam. In such a case, the interference fringes will undergo an impulsive move due to the induced refractive index change of the nonlinear medium. This can be understood if we consider that the transmissivity $T[\Delta n(t),\theta]$ is a function of both the time t and the propagation angle $\theta \leq \theta_f$. For this reason, the observed optical bistable behavior of the F-P etalon might essentially depend on the sampling position and detecting area of the transmitted laser beam.[14–16]

To detect the impulsive fringe's shift more effectively, a small-size pinhole could be placed in an appropriate position within the transmitted beam: as shown schematically in Fig. 11.4(b), where the pinhole is in the near-field location, and in Fig. 11.4(c), where the pinhole is in the far-field location. The opening of the pinhole should match the characteristic width of a selected fringe. If the size of the pinhole is too large that several fringes can pass through it, then the light-induced fringe's shift may not be detected because of the spatial average result. In other words, the optical bistable behavior observed depends on how the signal is detected.

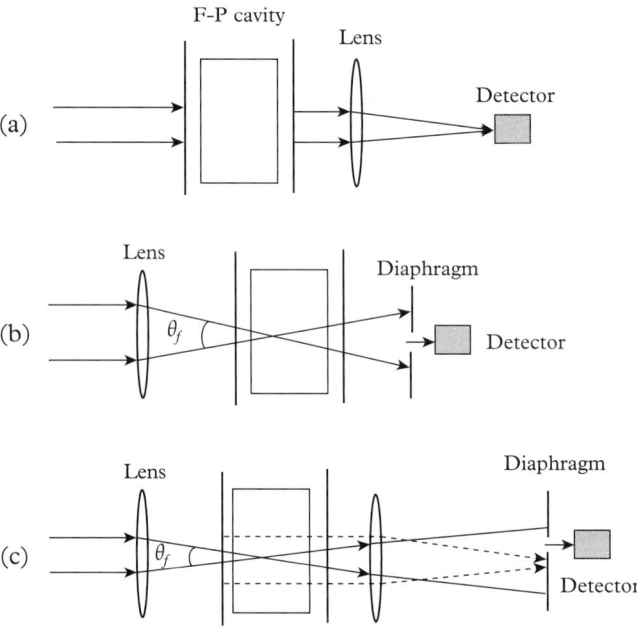

Figure 11.4 *Detection of the transmitted signal of an F-P bistable device for (a) a quasi-parallel incident beam and (b, c) a focused incident beam.*

When the aperture angle of a loosely focused incident beam is considerably smaller than the angular breadth of the highest-order interference ring, which is determined by the thickness L of the etalon, the incident beam can still be treated as a quasi-parallel beam.

Finally, let us consider the influence of spectral width of the incident laser beam on the performance of a bistable F-P device. It is well known that the spectral resolving power of an F-P etalon in near normal incidence is determined by[10]

$$\frac{\lambda_0}{\Delta\lambda_0} \approx \frac{2\zeta n_0 L}{\lambda_0}, \qquad (11.2\text{-}2)$$

where ζ is the finesse of the etalon

$$\zeta = \frac{\pi\sqrt{F}}{2} = \frac{\pi\sqrt{R}}{1-R}, \qquad (11.2\text{-}3)$$

and $\Delta\lambda_0$ is the minimum wavelength separation between two monochromatic lines around λ_0, which still can be resolved by an F-P etalon via their own interference rings. If the wavelength difference between two close spectral lines is larger than $\Delta\lambda_0$, the interference rings from these two spectral components will overlap each other and cannot be resolved. Actually, $\Delta\lambda_0$ is determined by 2ε, the full width at half maximum (FWHM) of the periodic curve shown in Fig. 11.2, i.e.,

$$\Delta\lambda_0 \approx \frac{\lambda_0^2}{4\pi n_0 L} \cdot \varepsilon = \frac{\lambda_0^2}{4\pi n_0 L} \cdot \frac{2.07\pi}{\zeta} \approx \frac{\lambda_0^2}{2\zeta n_0 L}. \qquad (11.2\text{-}4)$$

Here, 2ε determines the angular width of the interference rings from the F-P device for an input monochromatic spectral component.

From Eqs. (11.2-2) and (11.2-3) one can see that to achieve a higher spectral resolution, we need a higher reflectivity R and a larger cavity length L. The same conclusion is applicable to the consideration of the sensitivity of a nonlinear F-P etalon in the sense of the detectable refractive-index change or the interference ring shift. If the spectral linewidth of the incident laser beam is equal or less than $\Delta\lambda_0$ predicted by Eq. (11.2-2), the incident beam can be treated as an ideal monochromatic wave. When the laser spectral linewidth is considerably greater than $\Delta\lambda_0$, the angular breadth of the interference fringes becomes larger, then a small impulsive shift of these fringes will be more difficult to detect.

11.2.2 Standard setup for experimental studies

Figure 11.5 shows a standard experimental setup for observing the optical bistable effect; it consists of the following four basic units:

(1) A laser source: this can be a cw laser or a pulsed laser device. In the former case, the laser beam passes through an optical modulator and then becomes a series of pulses.
(2) A beam splitting system: it divides the laser beam into two branches, the stronger one sent to a nonlinear F-P etalon and the weaker to a photoelectric detector.
(3) A nonlinear F-P etalon: the nonlinear medium is usually placed between two dielectric coating mirrors with a higher reflectivity. If the working medium is a solid material like a semiconductor with a higher n_0 value, the two parallel and polished surfaces can form an F-P cavity with a moderate reflectivity.

368 Optical Bistability

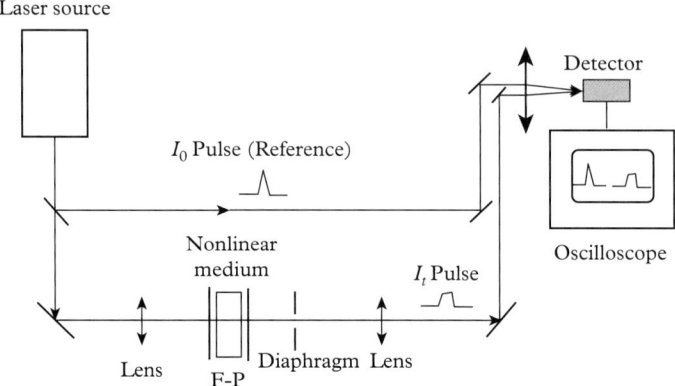

Figure 11.5 *Standard experimental setup for observing the optical bistable effect.*

(4) A system of photoelectric detector and oscilloscope: the incident and transmitted light pulses can be detected by a fast photodiode detector, and the opto-electric signals are displayed by a high-speed oscilloscope with proper relative delay between the two pulses. Comparing the digitally recorded waveforms of these two pulse signals, one may finally obtain the characteristic curve of the device under a given operation condition. The same characteristic curves can also be obtained directly, by using two identical detectors and sending the electric signals of the incident and transmitted pulses to the vertical and horizontal deflections of the oscilloscope simultaneously.

11.3 Experimental studies of optical bistability

11.3.1 Early observations of optical bistable effects

It is known that the physical origin of the dispersive bistable F-P devices is the light-induced refractive-index change (Δn) of the intracavity nonlinear medium. In general, for most optical media that are transparent at the incident light wavelengths, the induced Δn is extremely small unless the intensities of the incident beam are quite high. However, as mentioned in Chapter 5, a considerable increase of Δn could be achieved in a third-order nonlinear medium based on various resonant-enhancement mechanisms, including one-photon absorption, two-photon absorption, and Raman resonant interaction between the incident laser radiation and the nonlinear medium. On the other hand, these resonant mechanisms may also cause an undesirable attenuation of the incident signals and opto-thermal effect, which may degrade the performance of a bistable device. For these reasons, a compromise between the Δn enhancement and the signal attenuation and/or thermal effect should be made.[4,5,8,9]

The first optical bistable effect was observed in an F-P cavity filled with sodium (Na) vapor.[4] Figure 11.6 shows the measured characteristic curves under various operation conditions. These observations were made using an 11-cm F-P cavity filled with Na vapor at 10^{-4}–10^{-5} Torr in the 2.5-cm region midway between two mirrors of $\sim$90% reflectivity. The input laser beam was from a

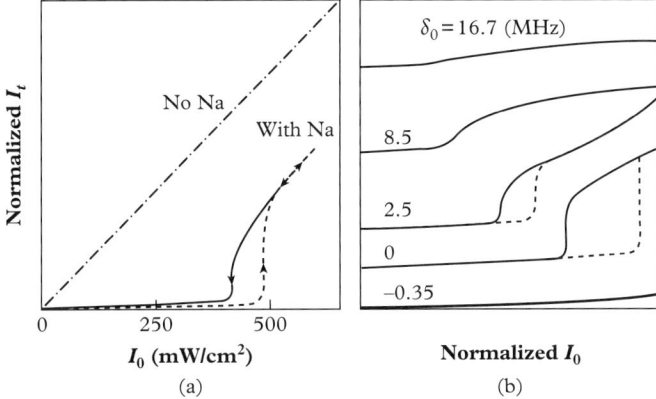

Figure 11.6 (a) Optical bistability in a Na vapor-filled F-P cavity, the oscilloscope trace is dashed for increasing input intensities and solid for decreasing. (b) Characteristic curve dependence on the initial detuning of the cavity, the free spectral range was about 1364 MHz. (Reproduced with permission from Gibbs et al.[4] Copyright 1976, American Physical Society).

tunable single-mode dye laser source. At the F-P device, the maximum input power was 12.7 mW in a 1.65-mm-diameter Gaussian beam, and the laser frequency was about 150 MHz above the Na $^2S_{1/2}$, $F=2$ to $^2P_{3/2}$ transitions (near one-photon resonance). The measured relative transmitted intensity as a function of the input intensity is shown in Fig. 11.6(a), while the observed characteristic output/input curves are shown in Fig. 11.6 (b) at different values of the initial cavity detuning δ_0 (in units of MHz). For a given F-P etalon, the free spectral range means the spacing between two adjacent peaks of the periodic curve shown in Fig. 11.2 in units of cm^{-1} or MHz, which is determined by

$$\Delta \nu_{free} = \frac{c}{2n_0 L}. \qquad (11.3\text{-}1)$$

Another example of the earliest observations of optical bistability was based on a plano-concave F-P cavity containing a 5-mm ruby crystal containing 0.03% Cr^{3+} working at 85-296 K.[5] The low-intensity room-temperature finesse of the cavity was ~18 with mirror reflectivity $R = 0.91$. The input was a 693.4-nm TEM$_{00}$ single-mode 20-mW beam from a cw ruby laser operating at 65 or 77 K. Under this condition, the wavelength of the laser emission was nearly resonant with the narrow R_1 absorption line of the ruby sample in the F-P etalon. The observed characteristic curves of optical bistable behavior are shown in Fig. 11.7. It was indicated by the authors that the thermally induced phase shift could be ruled out and the major contribution to the induced refractive-index change was associated with the laser-induced population redistribution between the ground state and the excited state of the sample. The measured switch-on and switch-off times varied from 3 to 20 ms, but did not depend on the sample temperature within a factor of 2.

The third example presented here is the early experimental observation of optical bistable effect in an F-P cavity filled with CS$_2$ liquid.[7] The measured input and output pulse shapes and corresponding characteristic curves are shown in Fig. 11.8. The F-P cavity was formed by two plane mirrors with $R = 0.98$ separated by $L = 1$ cm. The input beam was from a single-mode

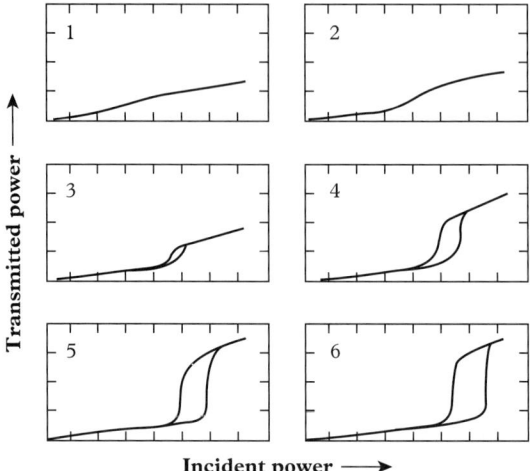

Figure 11.7 *Bistable characteristic curves for a 5-mm 0.03% Cr^{3+} ruby rod at 105 K. The horizontal scale runs from 0 to 15 mW; vertical from 0 to 1.6 mW. (Reproduced with permission from Vankatesan and MaCall.[5] Copyright 1977, AIP Publishing LLC).*

Q-switched ruby laser with a pulse width of 14 ns. Because of the attenuation due to scattering and residual absorption in the liquid sample, the effective mirror reflectivity at the ruby laser frequency was $R' \approx 0.78$, and the experimentally determined finesse of the F-P cavity was 13. It was indicated that the input laser pulsewidth was remarkably greater than the cavity build-up time as well as the characteristic times of the induced refractive-index change in CS_2 liquid.

11.3.2 Nonlinear materials for optical bistable devices

It is well known that the induced refractive-index change (Δn) of a third-order nonlinear medium is proportional to the local intensity of the applied optical field, so that a higher incident intensity or power level is necessary to create a larger Δn. On the other hand, from the viewpoint of applications, it is desirable to operate the bistable device at a lower input intensity or power level. To compromise these two requirements, researchers have to rely on various mechanisms of resonant enhancements of Δn and search for nonlinear materials with high n_2 values for optical bistable devices.

So far, a great number of materials have been utilized for experimental studies of optical bistability. The majority of those materials can be summarized as follows.

(A) semiconductors, including bulk crystals or crystal films such as GaAs,[8,17] InSb,[9,18-20] InAs,[21] PbSnSe,[22,23] PbSnTe,[24] CdHgTe,[25,26] CuCl,[27] and multiple-quantum-well (MQW) samples of GaAs-AlGaAs,[28] InGaAs-InAlAs,[29] and ZnSe–ZnS;[30]

(B) semiconductor or organic waveguides, including semiconductor waveguides of GaAs/AlGaAs,[31] GaAlAs,[32] ZnSe,[33] InGaAs/InP,[34] InP/InGaAsP,[35] and polymer film waveguides;[36,37]

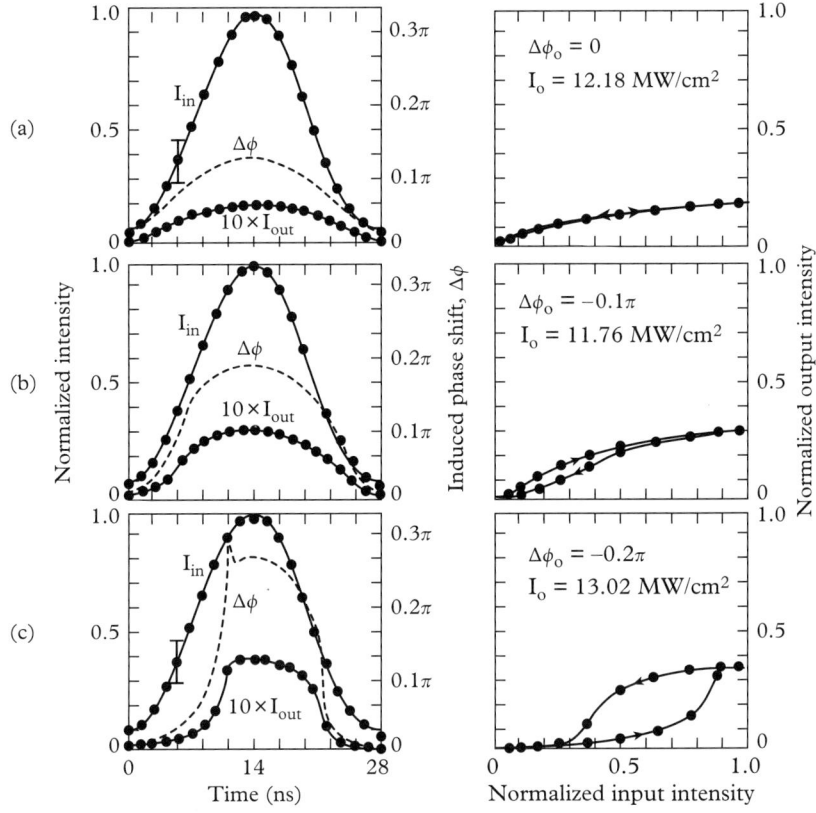

Figure 11.8 *Input and output pulse shapes (left) and characteristic curves (right) of an F-P cavity filled with CS_2 liquid at various input peak intensity levels and initial detuning values of $\Delta\phi_0$. The dashed curve denotes the calculated light-induced phase-change. (Reproduced with permission from Bischofberger and Shen.[7] Copyright 1979, American Physical Society).*

(C) optical filters, including some interference filters,[38–41] as well as semiconductor-doped glass filters;[42–44]

(D) liquid crystals, including MBBA (in isotropic phase),[6,7] PCB (in nematic phase),[45] 5CB (in nematic phase),[46,47] K15 (in nematic phase),[48,49] and MOOBB (in smectic A phase);[50]

(E) other materials, including metal vapors of Na[4,51] and Rb,[52] molecular gases of NH_3[53] and SF_6,[54] dye solutions of BDN[55,56] and fluorescein,[57] as well as optical fibers.[58]

In most cases, the above-mentioned nonlinear optical materials were placed inside an F-P cavity consisting of two separate mirrors. In some other cases, the nonlinear medium could be inside an integrated micro-cavity[59] or a microsphere,[60,61] and within a period structure[62] or photonic crystal structure.[63,64]

11.3.3 Semiconductor bistable devices

Up to now, the majority of research efforts in the field of optical bistability have been focused on developing various semiconductor materials and devices. The major advantages of utilizing semiconductors as optical bistable elements are:

(1) the ease of creating a larger refractive-index change with a lower input intensity based on the near-resonant enhancement;
(2) the superior physical and chemical stability of the materials, as well as fairly good reliability and reproducibility of their bistable performances;
(3) the capability of being compact, integrated, and multiplexed.

The earliest experimental demonstrations of semiconductor bistable devices were accomplished in 1979, in which GaAs and InSb were used as the nonlinear materials, respectively.[8,9] In the first case, the sample was grown by molecular beam epitaxy with 4.1-μm thick GaAs between 0.21-μm AlGaAs layers supported by a 150-μm GaAs substrate containing 1-2-mm-diam etched hole for the input beam.[8] A reflective coating with $R = 0.9$ was added on the two surfaces of the sample. The input beam was from a cw dye laser tunable from 770 to 870 nm, passed through an acousto-optic modulator, and focused to a 10-μm diameter on the sample at low temperature. Tuning the laser frequency to just below the exciton peak, the bistability was observed from 5 to 120 K with a switch-on time less than 1 nsec and switch-off time of 40 ns. It was indicated that the primary mechanism of Δn was light-induced changes of the exciton absorption. Figure 11.9 shows the measured waveforms of the input and output pulses and the corresponding characteristic curves.

Figure 11.10 shows the measured optical bistable characteristic curve based on a plane-parallel InSb crystal ($5 \times 5 \times 0.56$ mm^3, $N_D - N_A \sim 3 \times 10^{14}$ cm^{-3} (n-type)) at ~ 5 K.[9] The two plane-parallel faces of the crystal sample had a natural reflectivity ($R \sim 36\%$) and formed the F-P cavity. The focused input beam was provided by a cw CO laser with the output wavenumber of

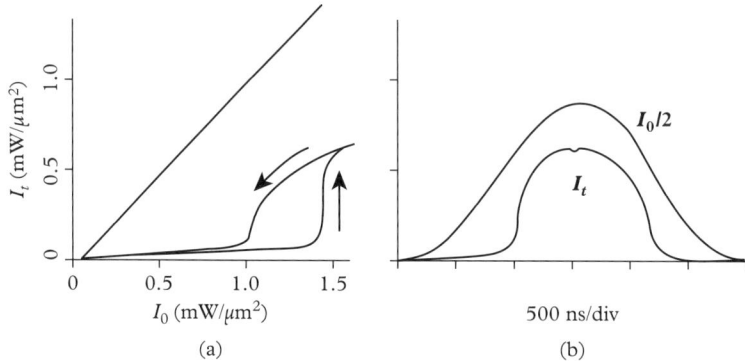

Figure 11.9 *Excitonic optical bistability in GaAs at 15 K and 819.9 nm laser wavelength. (a) Characteristic curves, in which the 45° line shows the operation at another non-resonant wavelength (∼830 nm). (b) Waveforms of the input and transmitted laser pulses. (Reproduced with permission from Gibbs et al.[8] Copyright 1979, AIP Publishing LLC).*

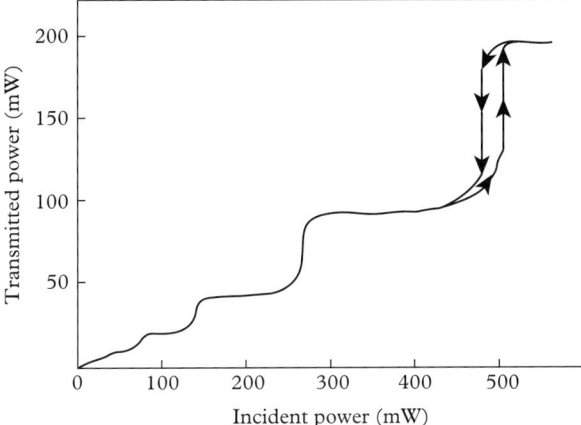

Figure 11.10 *Characteristic curve of a bistable InSb device at ~5 K for a cw CO laser beam (wavenumber 1895 cm^{-1} and incident spot size 180 µm). (Reproduced with permission from Miller et al.[9] Copyright 1979, AIP Publishing LLC).*

1895 cm^{-1}, which was near the band gap of the crystal (1899 cm^{-1} at ~5 K). The measured F value for the laser wavelength was ~0.5 due to absorption inside the crystal. From Fig. 11.10, one can see that the hysteresis loop occurred at the fifth-order interference shift; this means that the light induced round-trip phase change was about 10π. It was claimed that this phase-shift was due to electronic nonlinearity.[9]

Another example of a bistable InSb device was working at room temperature with ~10-µm CO_2 laser radiation.[20] The sample was a 174-µm thick slice of intrinsic InSb polished plane-parallel to form an F-P etalon with $R\approx 36\%$ natural reflectivity and 10 cm^{-1} loss coefficient due to free-carrier absorption. A CO_2 laser was used to generate single transverse/longitudinal mode pulses of duration 1-2 µs, and the output line was tunable from 9 to 11 µm among the P branch ($P12$-$P28$) transitions. Since adjacent CO_2 laser lines were 2 cm^{-1} apart compared with the etalon free spectral range of 8 cm^{-1}, several different laser lines could be chosen to provide different initial cavity detuning values without changing the cavity thickness or incident angle. The incident and transmitted pulse waveforms and the output/input characteristic curves for four different laser lines are shown in Fig. 11.11. The measured switch-on and switch-off times were 5 ns and 50-100 ns, respectively. It was indicated that the observed nonlinear refractive-index change was caused by two-photon absorption of the incident laser radiation; however, the thermal effect could not be totally ignored particularly with the higher peak-intensity laser pulses.[20]

11.3.4 Optical waveguide bistable devices

There is considerable interest in developing waveguide bistable devices that require much lower input light power because of (i) the transverse confinement of a guided beam and (ii) the longer interaction length between light and nonlinear guiding material.

As an example, the optical bistable behavior of a semiconductor channel waveguide used for optical bistable performance was investigated.[31] The experimental results with ~867 nm laser

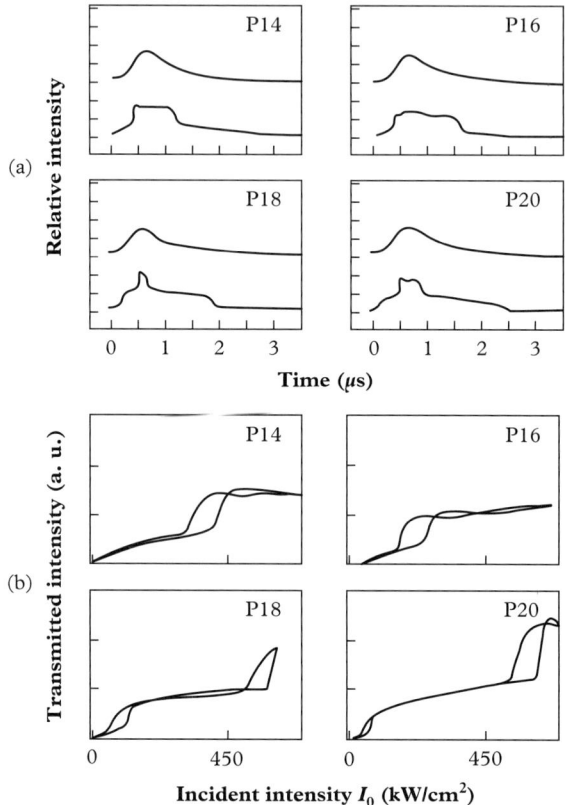

Figure 11.11 *Bistable behavior of the 174-μm thick InSb etalon at room temperature. (a) Incident (upper traces) and transmitted (lower traces) pulse shapes for four different CO_2 laser lines. (b) Corresponding characteristic output/input curves. (Reproduced with permission from Ji et al.[20] Copyright 1986, IEEE).*

wavelength and ~60 mW coupled power are shown in Fig. 11.12, which are the oscilloscope traces showing the input and the transmitted pulse shapes as well as the characteristic output/input curve. The exciton resonance of the MQW sample was measured to be at 845 nm wavelength for light incident normal to the quantum well layers. The bistable operation was observed over a range of ~10 nm around 867 nm. In Fig. 11.12, we can see two hysteresis loops indicating that the device was switching through two F-P transmission orders. It was noted that both thermal and electronic nonlinearities were present in the waveguiding materials. The Δn was negative for the electronic contribution and positive for the thermal contribution. For shorter pulse duration (e.g., 200-300 ns) the former was dominant, whereas for longer pulse duration (~8 μs) the latter was no longer negligible.[31]

Moreover, the use of a single quantum well may lead to a small filling factor and the absorption of the light wave can be low. Therefore, device operation closer to the band edge is possible,

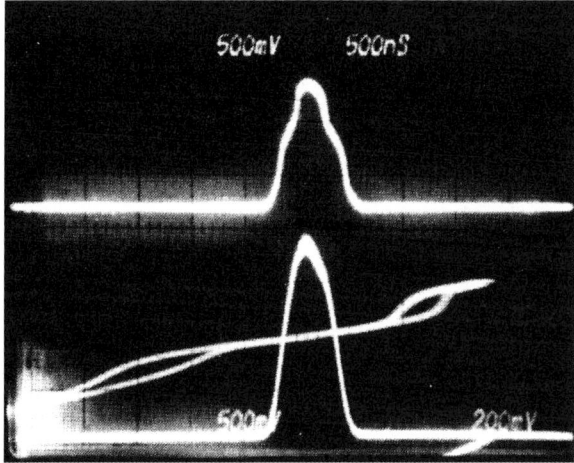

Figure 11.12 *Bistable behavior of GaAs/AlGaAs MQW strip-loaded waveguide structure: transmitted (upper trace) and incident (lower trace) pulse shapes of the 867-nm laser beam, and the characteristic output/input curve (lower left trace). Time scale is 500 ns/div. (Reproduced with permission from Warren et al.[31] Copyright 1987, AIP Publishing LLC).*

where the refractive-index change is higher. As an example, the single mode and single quantum well waveguide in a system of InP/InGaAsP was used for demonstrating optical bistability with ~1.5-μm laser wavelength at low power (~1.5 mW) and room temperature.[35] The waveguide had an ~1-μm transverse size of the guiding section and ~600-μm length. The incident beam was from a NaCl color center laser tunable in the range 1450–1700 nm with a linewidth of 5 MHz. After passing through a rotating shutter the laser beam transferred to a series of pulses with the duration of 1.5 ms and a rise time of 500 μs. The input facet was uncoated while the output facet was covered with a 40-nm-thick gold layer, forming an asymmetrical F-P cavity. The bistable behavior of the device was observed within a wavelength range 1480–1525 nm. The measured characteristic output/input curve is shown in Fig. 11.13 for the laser wavelength of 1497 nm. It was confirmed that the thermal contribution could be neglected at so low optical power level (<10 mW).

11.3.5 Transient and thermal optical bistability

As described by Eq. (11.1-14), when the characteristic variation time (Δt_0) of the input laser pulse is short than any one of the cavity build-up time (τ_c), the rise time (τ_{rise}), and the decay time (τ_{decay}) of induced refractive-index change of the intracavity medium, we will deal with transient bistable behavior because either the intracavity intensity $I_i(t)$ or the induced $\Delta n(t)$ cannot follow the change of the incident intensity $I_0(t)$ instantaneously. In comparison with the steady-state bistable operation, the features of transient bistable operations are reflected in their characteristic output/input curves and can be described as follows.

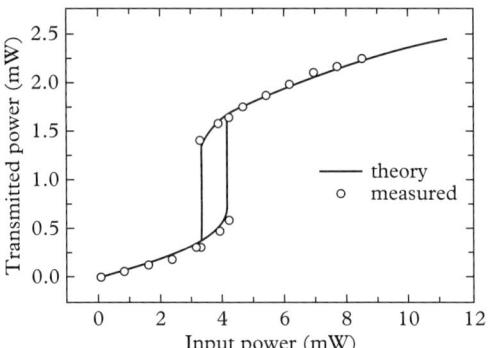

Figure 11.13 *Measured characteristic curve of a 600-μm-long single mode and single quantum well waveguide device excited with 1497-nm laser pulses. (Reproduced with permission from Hübner et al.[35] Copyright 1995, AIP Publishing LLC).*

(1) There is no vertical transition (steep switch-up or switch-down) associated with the hysteresis loop.

(2) There is no partial overlap between the intensity-increasing pathway and the intensity-decreasing pathway of the characteristic curve.

(3) The observed hysteresis loops may exhibit irregular or butterfly-like shapes.

In practice, optical bistable devices containing the following two types of materials often manifest transient behavior. The liquid crystals and photorefractive crystals belong to the first type of materials, which exhibit slow response (rise/decay) time of the induced refractive-index change.[7] The materials of the second type are those linearly absorbing at the incident lasing wavelength. In the latter case, linear absorption may easily cause the temperature increase through radiationless transitions and lead to a thermal refractive-index change that may have a relatively fast rise time but always exhibits very slow (e.g., $\geq 10^{-3}$ s) decay time.

Now let us consider an example of thermally induced bistable behavior of an F-P device, containing an linearly absorptive dye (BDN) solution, excited with 532-nm laser pulses.[56] BDN is a saturable absorption dye for Nd lasers working at 1.06 μm, whereas with this dye there is a moderate linear absorption in the broad visible spectral range. The mirror reflectivity of a 2.35-cm long F-P cavity was $R \approx 95\%$, the measured effective reflectivity $R' \approx 30\%$ was mainly due to absorption losses in the 5-mm long dye solution with a concentration of 5×10^{-6} M. The pulse duration, spectral linewidth, repetition rate, and pulse energy of the input laser were ∼6 ns, ∼0.05 cm^{-1}, 1 Hz, and a few millijoules, respectively. The oscilloscope traces of the input and output pulse shapes are shown in Fig. 11.14(a) under four different initial cavity detuning conditions.

The characteristic curve corresponding to the measurement result of case (iii) is presented in Fig. 11.14(b), where one can see a hysteresis loop of butterfly shape. Knowing the waveform of the input pulse, one can assume various trial functions of the induced phase change $\Delta\phi(t)$, and then use Eq. (11.1-15) to calculate the predicted waveform of the output pulse until the best fit with the measured waveform of the output pulse is obtained. In Fig. 11.14(c), the best-fit output pulse shape is shown by the dashed curve that is based on the following trial function of $\Delta\phi(t)$:

$$\Delta\phi(t) = bt^{3/2}, \tag{11.3-2}$$

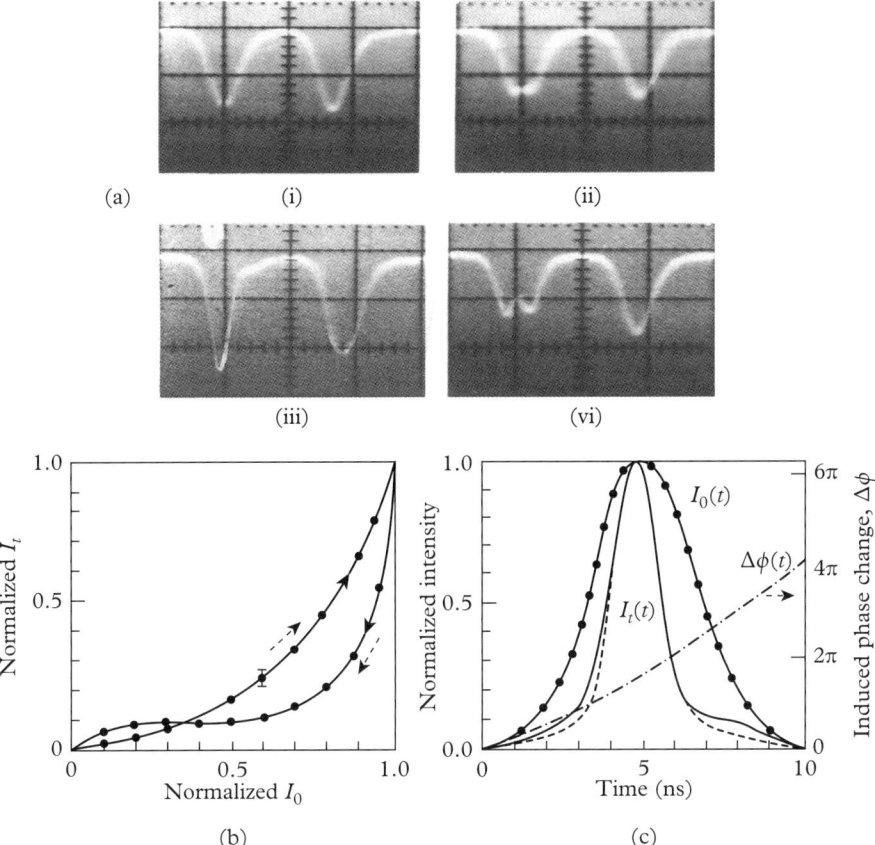

Figure 11.14 *Transient optical bistable behavior of an F-P cavity containing a linearly absorbing BDN dye solution excited with 532-nm laser. (a) Oscilloscope displays of the output pulse (left trace) and input pulse (right trace) under various initial detuning conditions, time scale is 10 ns/div. (b) Characteristic curve corresponding to the measurement (iii) in (a). (c) Assumed curve $\Delta\phi(t)$ and the fitted output pulse shape (dashed curve). (After He et al.[56]).*

where b is a fitting constant. In obtaining the fitted output pulse shape, it was assumed that the input pulse had a Gaussian shape. From Fig. 11.14(c) we can see that the fitted shape is in fairly good agreement with the measured output pulse. Eq. (11.3-2) describes the rising behavior of a thermally induced phase (and hence refractive-index) change. This is one example using optical bistable method to measure the dynamic behavior of the induced refractive index changes.

It should be noted that in Fig. 11.14, the induced phase change $\Delta\phi(t)$ or $\Delta n(t)$ still remain a high value even the input light intensity is reduced to zero, indicating the accumulating feature and slow decay of the induced refractive-index change.

Some other later experimental studies have also shown the transient behavior of thermal effect dominating some optical bistable devices.[61,63,65,66]

11.4 Recent development of bistability studies

Although studies on optical bistability still face the challenges of seeking novel materials of high n_2-value and avoiding thermal effects, research efforts, which remain active in the new century, are mainly encouraged by the potential applications of using optical bistable devices for all-optical switching and memory, optical computing, and data processing and transmission. To highlight the new trend of these efforts, some interesting results of the most recent studies are worthy of being discussed.

One development of novel bistable devices is based on the nanocavity in a photonic crystal (PhC) structure of semiconductors. An experimental example of optical bistable operation using such a nanocavity was demonstrated by Kim et al.[64] In their work, the two-dimensional PhC structure was fabricated in an InGaAsP slab containing four QWs of $In_{0.76}Ga_{0.24}As_{0.75}P_{0.25}$ in the form of a hexagonal period-hole lattice with a period of 460 nm and hole radius of 160 nm.

The nanocavity is formed in the five-cell area without holes, as shown in Fig. 11.15(a), and the cavity resonance wavelength is around 1610 nm. The calculated photon density distribution within the nanocavity is shown in Fig. 11.15(b). A tapered and strongly bent single mode fiber is utilized as an input/output coupler via evanescently coupling through the cavity surface. The input

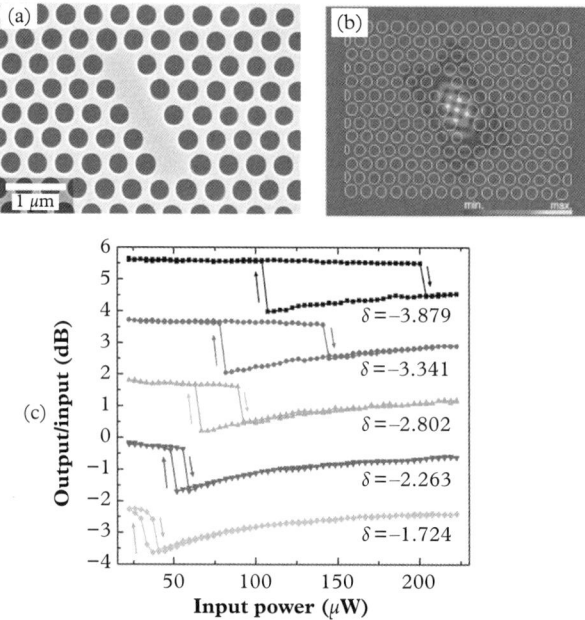

Figure 11.15 *(a) Fabricated photonic crystal structure (period-hole lattice) and five-cell nanocavity in a InGaAsP slab. (b) Calculated photon density distribution inside the cavity region. (c) Bistable output/input curves (on decibel scale) measured with different initial cavity detuning values. (Reproduced with permission from Kim et al.[64] Copyright 2007, AIP Publishing LLC).*

cw laser is tunable in the wavelength range 1608–1614 nm, and the initial normalized detuning of the cavity is defined as

$$\delta_0 \equiv (\lambda_{in} - \lambda_{res})/(\Delta\lambda_{res}/2). \tag{11.4-1}$$

Here, λ_{in} is the input laser wavelength, λ_{res} is the resonance peak wavelength of the nanocavity, and $\Delta\lambda_{res}$ is the FWHM of the cavity transmission spectrum, which is measured to be 0.732 nm. Figure 11.15(c) shows the measured bistable output/input curves under different wavelength detuning conditions, indicating the threshold bistable power is close to 35 µW and the size of the bistable loop increases with detuning. It is also indicated by the authors that the electronic band-edge nonlinearity is dominant over the thermal contribution.

The other similar example is the all-optical memory operation reported by Shinya et al., which is based on an InGaAsP core waveguide nanocavity surrounding by PhC structure of period-distributed air holes of ∼430-nm spacing.[67] The hole size and waveguide core thickness are both 200 nm, and the nanocavity is formed by shifting the three rows of side holes away from the central waveguide by distances of 9, 6, and 3 nm, respectively, as shown in Fig. 11.16(a).

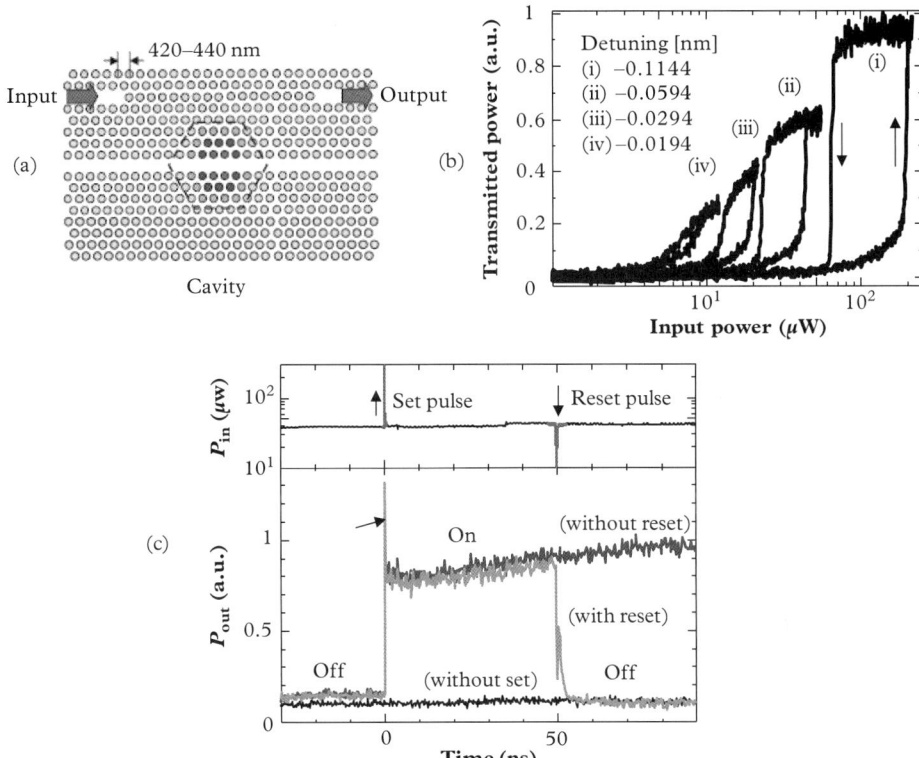

Figure 11.16 (a) Photonic crystal structure, input/output waveguide, and nanocavity on an InGaAsP substrate. (b) Hysteresis curves measured with different wavelength detuning values. (c) All-optical bistable memory operation: output status (Off/On) is controlled by the input set/reset pulses. (Reproduced with permission from Shinya et al.[67] Copyright 2008, Optical Society of America).

The input and output coupling of the nanocavity is realized through the upper waveguide sections, as also shown in Fig. 11.16(a). The resonant peak wavelength of the cavity is measured to be 1585.1151 nm, at which the linear absorption is minimized as the band gap of the waveguide core material is ~1.3 μm. The input laser is tunable around the 1585 nm wavelength, with a triangle pulse shape of 100 ns base. Figure 11.16(b) shows the measured hysteresis curves for several different input wavelengths, indicating the threshold bistable power only about several tens of μW. Moreover, the dynamic response of the on/off switching and the memory holding time are shown in Fig. 11.16(c), where an intensity increased input set pulse of 100-ps duration and an intensity decreased reset pulse of 500-ps duration are added on a below-threshold bias power level. Without applying an input set pulse the nanocavity is working in the "Off"-status with a low power output; whereas upon the action of the set pulse the cavity is working in the "On"-status with a high power output and remains in this status until the arrival of the reset pulse. The measured longest memory holding time could be as long as 150 ns. The major nonlinearity could attribute to the two-photon transition induced refractive-index change via carrier plasma effect.

Another effort on the recent optical bistable studies is to explore new mechanisms of resonant enhancement of the nonlinearity, and to utilize other types of cavity or interferometers instead of the F-P cavity. For example, EIT (electromagnetically induced transparency) induced or resonate-Raman induced nonlinear enhancements, which are based on two-beam interaction, are among these new mechanisms.[68,69]

In a recent work reported by Sheng et al., the nonlinear medium is an ^{87}Rb-vapor cell placed in a ring cavity; the cavity is injected with two laser beams of different frequencies, as shown in Fig. 11.17(a).[68] One is the probe beam working with the cavity resonance, the other is a control beam interacting only with the nonlinear medium. To achieve an EIT enhancement of the nonlinearity, the two beam frequencies are chosen to match the Λ-type three-level structure of the D_1 line of ^{87}Rb atoms (cf. Section 16.5.1). These two beams are from two laser diodes tunable around 795 nm separately. On the conditions of 105 °C vapor temperature, probe beam's frequency detuning $\Delta_p/2\pi = -137.7$ MHz, and control beam's detuning $\Delta_c/2\pi = -227.7$ MHz, the recorded bistable and multistable curves are shown in Fig. 11.17(b), measured at different control power levels and the input probe power was scanning triangularly from 0 to 15 mW. These multistable curves could be well controlled by varying the experimental parameters, which is helpful for having a better understanding of the related nonlinear optical processes.

Up to now, we have only considered and discussed the optical bistable device as a passive element in an optical logical circuit. Actually, the same functions of all-optical memory and switching can also be implemented based on some "active (gain)" optical devices, especially semiconductor microlasers.[70–72] Here we only discuss two examples of recent achievements in this research direction.

The first is an all-optical flip-flop operation of a polarization bistable vertical-cavity surface-emitting laser (VCSEL), demonstrated by Mori et al.[70] The structure of this electrically pumped laser is shown in Fig. 11.18(a). The lasing occurred around 980-nm wavelength with an output power of 520 μW and was pumped by 4.0 mA current. There were two lasing modes with orthogonal linear polarization (0° and 90°) states and a small wavelength difference (~0.014 nm), possible due to the residual optical anisotropy of the gain region. Under free running condition, the 0°-mode output was stronger than the 90°-mode. However, the lasing of these two modes can be controlled by separately injecting two light pulses each of which should have the same polarization state and nearly the same wavelength with the corresponding lasing mode. As shown schematically in Fig. 11.18(a), when a 90°-polarized set pulse is injected into the laser surface,

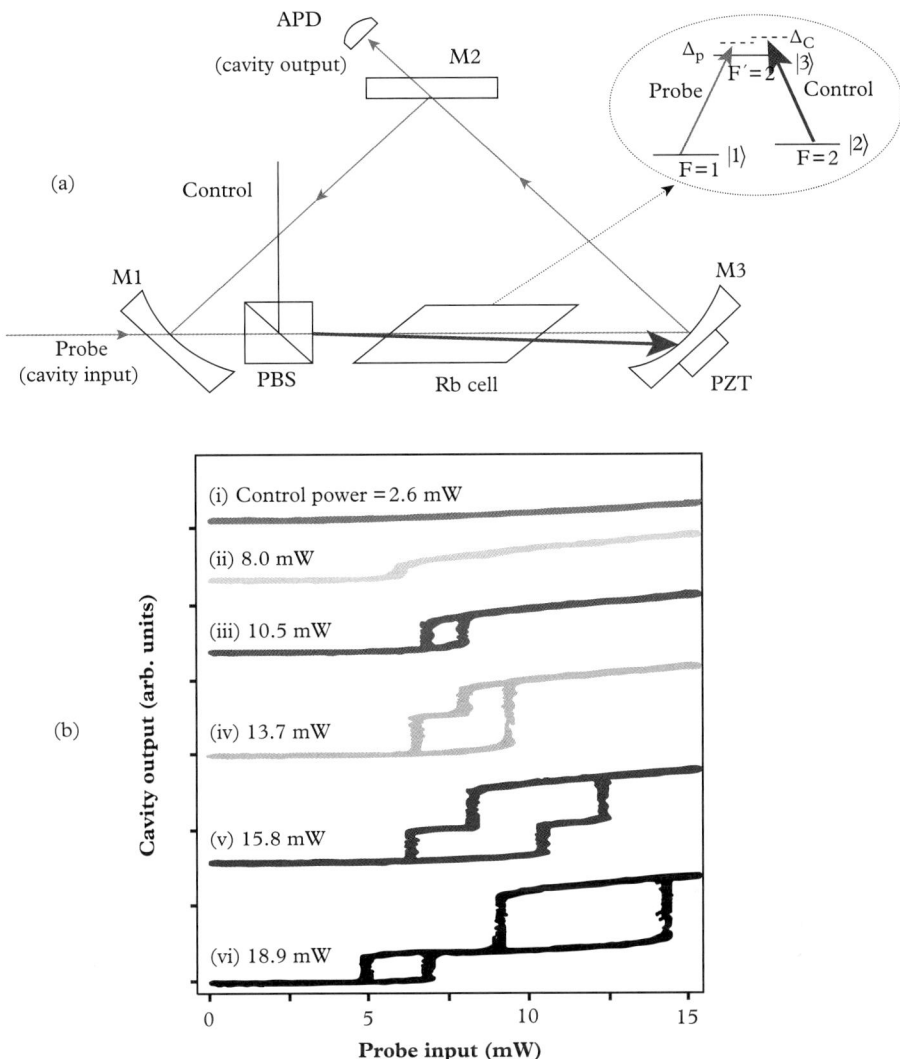

Figure 11.17 *(a) Experimental setup demonstrating optical multistability of ^{87}Rb-vapor in a ring cavity, the involved atomic state levels and the frequencies of two interacting laser fields are shown in the inset. (b) Characteristic output/input curves measured at different control beam power levels. (Reproduced with permission from Sheng et al.[68] Copyright 2012, American Physical Society).*

the 90°-lasing mode is enhanced while the 0°-mode is suppressed. The enhanced 90°-lasing could last for certain period of time (due to injection mode locking) until a 0°-polarized reset pulse is injected then the 0°-lasing mode becomes dominant. Figure 11.18(b) shows the measured waveforms of the all-optical flip-flop operation of 540 MHz switching frequency; the energies of the

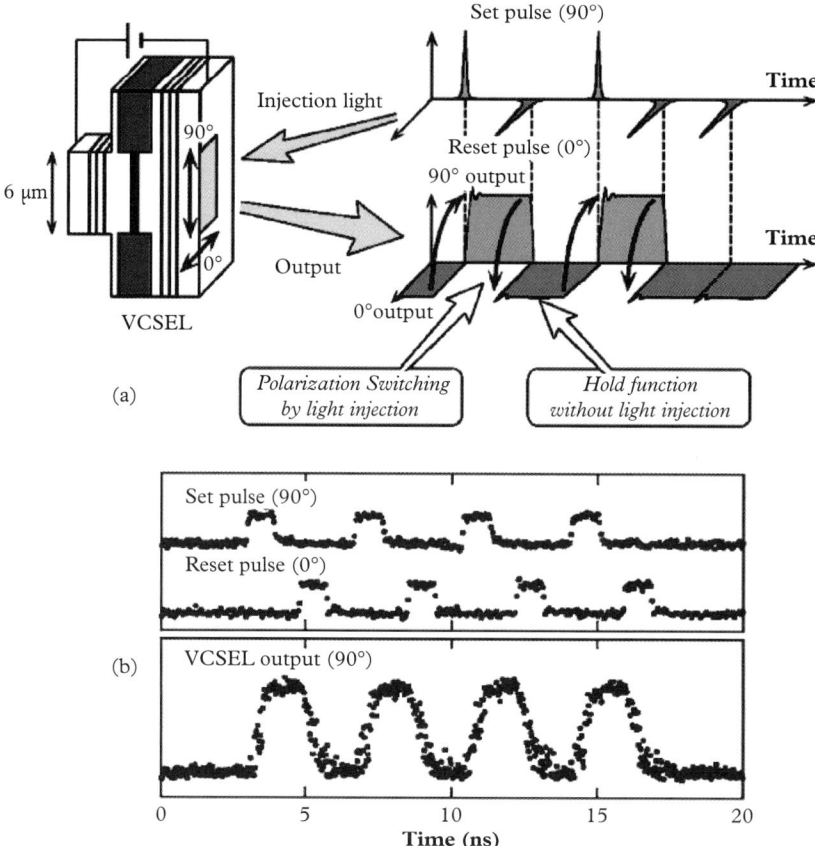

Figure 11.18 *(a) Polarization bistable vertical-cavity surface-emitting laser (VCSEL) structure and injection control. (b) All-optical flip-flop operation of 540 MHz switching frequency. (Reproduced with permission from Mori et al.[70] Copyright 2006, AIP Publishing LLC).*

set and reset injection pulses were 0.2 and 0.3 fJ, respectively. The highest switching frequency achieved in this work was 10 GHz.

The second example of all-optical memory operation is reported by Chen et al., based on a cw photonic crystal InGaAsP/InP waveguide laser working at ∼1.5-μm wavelength range and room temperature, optically pumped by 1.31-μm laser of ∼100 μW power.[71] The structure of the waveguide laser, which is shown in Fig. 11.19(a), features a $4 \times 0.3 \times 0.16$ μm³ gain volume surrounded by a periodically spaced hole structure. In the free running condition, there are two output lasing modes: a stronger Mode-B with wavelength of 1541.76 nm and a weaker Mode-A with wavelength of 1542.99 nm. In this case, shown in Fig. 11.19(b), the laser output of Mode-B can be controlled by injecting a light beam of wavelength $\lambda = \lambda_A + \delta (\geq 0.17 \text{nm})$ backward through the output waveguide of this laser. Upon the action of the injection control, the lasing of Mode-B is suppressed while the lasing of Mode-A is enhanced. Figure 11.19(c) shows the

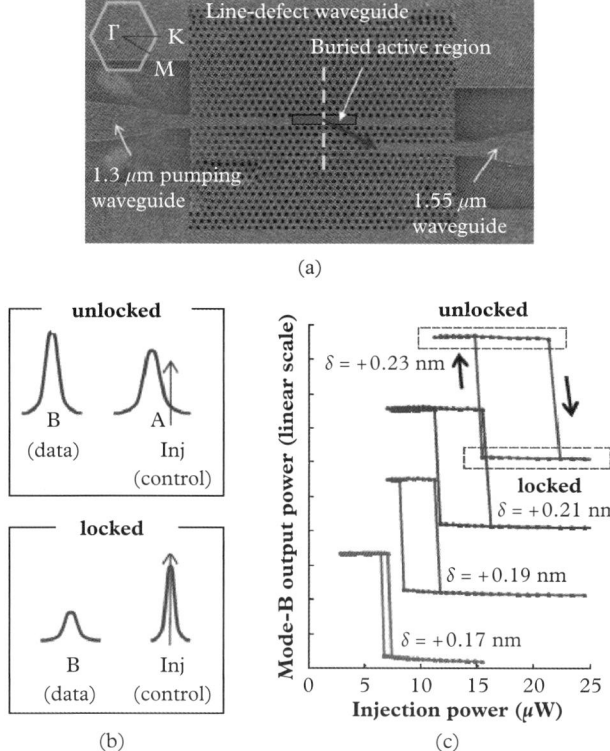

Figure 11.19 *(a) Structure of optically pumped photonic crystal waveguide laser. (b) Free-running spectra of two lasing modes and injection spectral line position; injection-controlled bistable lasing of Mode-B. (c) Measured hysteresis curves of Mode-B lasing against injection power with different injection wavelength detuning δ values. (Reproduced with permission from Chen et al.[71] Copyright 2011, Optical Society of America).*

measured hysteresis curves of the Mode-B lasing output vs. the injection power with different detuning δ values. To demonstrate the dynamic memory functions of this device, similar to that shown in Fig. 11.16(c), the set and reset pulses of 200-ns width and 0.5 MHz repetition rate could be added on a bias injection level, the fast switching time of ∼60 ps has been reached.

REFERENCES

1. H. M. Gibbs, *Optical Bistability: Controlling Light with Light* (Academic Press, New York,1985).
2. S. D. Smith, *Appl. Opt.* **25**, 1550 (1986); L. A. Lugiato, in *Progress in Optics*, edited by E. Wolf, Vol. 21 (North-Holland, Amsterdam, 1984), pp. 69–216.

3. A. Szöke, V. Daneu, J. Goldhar, and N. A. Kurnit, *Appl. Phys. Lett.* **35**, 376 (1969).
4. H. M. Gibbs, S. L. McCall, and T. N. C. Venkatesan, *Phys. Rev. Lett.* **36**, 1135 (1976).
5. T. N. C. Vankatesan and S. L. MaCall, *Appl. Phys. Lett.* **30**, 282 (1977).
6. T. Bischofberger and Y. R. Shen, *Appl. Phys. Lett.* **32**, 156 (1978).
7. T. Bischofberger and Y. R. Shen, *Phys.Rev. A* **19**, 1169 (1979).
8. H. M. Gibbs, S. L. MaCall, T. N. C. Venkatesan, A. C. Gossard, A. Passner, and W. Wiegmann, *Appl. Phys. Lett.* **35**, 451 (1979).
9. D. A. B. Miller, S. D. Smith, and A. Johnston, *Appl. Phys. Lett.* **35**, 658 (1979).
10. M. Born and E. Wolf, *Principles of Optics*, 6th ed. (Pergamon, Oxford, 1980), pp. 323–341.
11. F. S. Felber and J. H. Marburger, *Appl. Phys. Lett.* **28**, 731 (1976).
12. J. H. Marburger and F. S. Felber, *Phys. Rev. A* **17**, 335 (1978).
13. B. S. Wherrett, *IEEE J. Quantum Electron.* **20**, 646 (1984).
14. M. L. Berre, E. Ressayre, A. Tallet, K. Tai, and H. M. Gibbs, *IEEE J. Quantum Electron.* **21**, 1404 (1985).
15. J. E. Bjorkholm, P. W. Smith, W. J. Tomlinson, and A. E. Kaplan, *Opt. Lett.* **6**, 345 (1981).
16. A. V. Grigor'yants and I. N. Dynzhikhov, *J. Opt. Soc. Am. B* **7**, 1303 (1990).
17. O. Sahlén, U. Olin, E. Masseboeuf, G. Landgren, and M. Rask, *Appl. Phys. Lett.* **50**, 1559 (1987).
18. D. A. B. Miller, S. D. Smith, and C. T. Seaton, *IEEE J. Quantum Electron.* **17**, 312 (1981).
19. A. K. Kar, J. G. H. Mathew, S. D. Smith, B. Davis, and W. Prettl, *Appl. Phys. Lett.* **42**, 334 (1983).
20. W. Ji, A. K. Kar, J. G. H. Mathew, and A. C. Walker, *IEEE J. Quantum Electron.* **22**, 369 (1986).
21. C. D. Poole and E. Garmire, *Appl. Phys. Lett.* **44**, 363 (1984).
22. J. J. E. Reid, A. K. Kar, H. A. MacKenzie, P. L. Chua, R. Grisar, and H. M. Preier, *Appl. Phys. Lett.* **51**, 1215 (1987).
23. J. J. E. Reid, A. K. Kar, H. A. MacKenzie, R. Grisar, and H. M. Preier, *Opt. Commun.* **64**, 175 (1987).
24. P. P. Paskov, L. I. Pavlov, P. A. Atanasov, D. B. Kushev, and N. N. Zheleva, *Opt. Commun.* **65**, 133 (1988).
25. J. G. H. Mathew, D. Craig, and A. Miller, *Appl. Phys. Lett.* **46**, 128 (1985).
26. D. Craig, M. R. Dyball, and A. Miller, *Opt. Commun.* **54**, 383 (1985).
27. N. Peyghambarian, H. M. Gibbs, M. C. Rushford, and D. A. Weinberger, *Phys. Rev. Lett.* **51**, 1692 (1983).
28. S. S. Tarng, H. M. Gibbs, J. L. Jewell, and N. Peyghambarian, *Appl. Phys. Lett.* **44**, 360 (1984).
29. K. Nonaka, Y. Kawamura, H. Kawaguchi, and K. Kubodera, *Appl. Phys. Lett.* **56**, 2062 (1990).
30. H. Wang, L. Xu, A. Shen, Y. Chen, J. Cui, P. Qiu, Q. Li. D. Zhuang, C, Zhang, S. Yang, and W. Wang, *J. Appl. Phys.* **68**, 4338 (1990).
31. M. Warren, W. Gibbons, K. Komatsu, D. Sarid, D. Hendricks, H. M. Gibbs, and M. Sugimoto, *Appl. Phys. Lett.* **51**, 1209 (1987).
32. J. S. Aitchison, J. D. Valera, A. C. Walker, S. Ritchie, P. M. Rodgers, P. McIlroy, and G. I. Stegeman, *Appl. Phys. Lett.* **51**, 561 (1987).
33. B. G. Kim, E. Garire, N. Shibata, and S. Zembutsu, *Appl. Phys. Lett.* **51**, 475 (1987).
34. D. T. Neilson, J. E. Ehrlich, P. Meredith, A. C. Walker, G. T. Kennedy, R. S. Grant, P. D. Roberts, W. Sibbett, M. Hopkinson, and M. Pate, *Electron. Lett.* **29**, 1537 (1993).
35. B. Hübner, R. Zengerle, and A. Forchel, *Appl. Phys. Lett.* **66**, 3090 (1995).
36. B. P. Singh and P. N. Prasad, *J. Opt. Soc. Am.< **5**, 453 (1988).
37. J. Si, Y. Wang, J. Zhao, B. Zou, P. Ye, L. Qiu, Y. Shen, Z. Cai, and J. Zhou, *Opt. Lett.* **21**, 357 (1996).
38. G. R. Olbright, N. Peyghambarian, H. M. Gibbs, H. A. Macleod, and F. Van Milligen, *Appl. Phys. Lett.* **45**, 1031 (1984).
39. S. D. Smith, J. G. H. Mathew, M. R. Taghizadeh, A. C. Walker, B. S. Wherrett, and A. Hendry, *Opt. Commun.* **51**, 357 (1984).
40. Y. T. Chow, B. S. Wherrett, E. Van Stryland, B. T. McGuckin, D. Hutchins, J. G. H. Mathew, A. Miller, and K. Levis, *J. Opt. Soc. Am. B* **3**, 1535 (1986).

41. U. Eicker and J. I. B. Wilson, *J. Opt. Soc. Am. B* **8**, 614 (1991).
42. B. Daniezic, K. Nattermann, and D. von der Linde, *Appl. Phys. B* **38**, 31 (1985).
43. E. Haro-Poniatowski, M. Fernandez Guasti, E. R. Mendes, and M. Balkanski, *Opt. Commun.* **70**, 70 (1989).
44. K. Ogusu, Y. Kaneko, and K. Ishikawa, *Appl. Opt.* **34**, 3413 (1995).
45. I. C. Khoo, *Appl. Phys. Lett.* **41**, 909 (1982).
46. M. M. Cheung, S. D. Durbin, and Y. R. Shen, *Opt. Lett.* **8**, 39 (1983).
47. V. A. Gunyakov, S. A. Myslivets, V. G. Arkhipkin, V. Ya. Zyryanov, and V. F. Shabanov, *Dokl. Phys.* **58**, 219 (2013).
48. A. D. Lloyd, I. Janossy, H. A. MacKenzie, and B. S. Wherrett, *Opt. Commun.* **61**, 339 (1987).
49. A. D. Lloyd and B. S. Wherrett, *Appl. Phys. Lett.* **53**, 460 (1988).
50. C. Somerton and D. L. Tunnicliffe, *Opt. Commun.* **65**, 143 (1988).
51. F. T. Arecchi, G. Giusfredi, E. Petriella, and P. Salieri, *Appl.Phys. B* **29**, 79 (1982).
52. E. Giacobino, M. Devaud, F. Biraben, and G. Grynberg, *Phys. Rev. Lett.* **45**, 434 (1980).
53. D. M. Pepper and M. B. Klein, *IEEE J. Quantum Electron.* **15**, 1362 (1979).
54. R. G. Harrison, I. A. Al-Saidi, E. J. D. Cummins, and W. J. Firth, *Appl. Phys. Lett.* **46**, 532 (1985).
55. Z. F. Zhu and E. M. Garmire, *IEEE J. Quantum Electron.* **19**, 1495 (1983).
56. G. S. He, F. X. Zhou, and S. H. Liu, *Acta Phys. Sinica* **34**, 1241 (1985).
57. S. Speiser and F. L. Chisena, *J. Chem. Phys.* **89**, 7259 (1988); S. Speiser and F. L. Chisena, *Appl. Phys. B* **45**, 137 (1988).
58. K. Ogusu, T. Konoma, J. Yamasaki, K. Ishikawa, and M. Minakata, *Jpn. J. Appl. Phys. Part 1*, **42**, 434 (2003).
59. T. Rivera, F. R. Ladan, A. Izrael, R. Azoulay, R. Kuszelewicz, and J. L. Oudar, *Appl. Phys. Lett.* **64**, 869 (1994).
60. F. Treussart, V. S. Ilchenko, J.-F. Roch, J. Hare, V. Lefevre-Seguin, J.-M. Raimond, and S. Haroche, *Eur. Phys. J. D* **1**, 235 (1998).
61. Y. Wu, J. M. Ward, and S. Nic Chormaic, *J. Appl. Phys.* **107**, 033103 (2010).
62. C. J. Herbert and M. S. Malcuit, *Opt. Lett.* **18**, 1783 (1993).
63. M. W. Lee, C. Grillet, C. Monat, E. Mägi, S. Tomljenovic-Hanic, X. Gai, S. Madden, D.-Y. Choi, D. Bulla, B. Luther-Davies, and B. J. Eggleton, *Opt. Express* **18**, 26695 (2010).
64. M.-K. Kim, I.-K. Hwang, S.-H. Kim, H.-J. Chang, and Y.-H. Lee, *Appl. Phys. Lett.* **90**, 161118 (2007).
65. V. E. Hartwell, H. Nakajima, and N. Djeu, *Opt. Lett.* **20**, 2210 (1995).
66. M. A. Noginov, M. Vondrova, and B. D. Lucas, *Phys. Rev. B* **65**, 035112 (2002).
67. A. Shinya, S. Matsuo, Yosia, T. Tanabe, E. Kuramochi, T. Sato, T. Kakitsuka, and M. Notomi, *Opt. Express* **16**, 19382 (2008).
68. J. Sheng, U. Khadka, and M. Xiao, *Phys. Rev. Lett.* **109**, 223906 (2012).
69. I. Novikova, A. S. Zibrov, D. F. Phillips, A. Andre, and R. L. Walsworth, *Phys. Rev. A* **69**, 061802 (2004).
70. T. Mori, Y. Yamayoshi, and H. Kawaguchi, *Appl. Phys. Lett.* **88**, 101102 (2006).
71. C.-H. Chen, S. Matsuo, K. Nozaki, A. Shinya, T. Sato, Y. Kawaguchi, H. Sumikura, and M. Notomi, *Opt. Express* **19**, 3387 (2011).
72. M. Takenaka, K. Takeda, Y. Kanema, Y. Nakano, M. Raburn, and T. Miyahara, *Opt. Express* **14**, 10785 (2006).

12
Optical Temporal Solitons

12.1 Conditions for temporal soliton formation

12.1.1 Group velocity and group velocity dispersion (GVD)

It is well known that the phase velocity (v_p) of light propagating in an optical medium is defined by

$$v_p = \frac{\omega}{k} = \frac{c}{n(\omega)}, \tag{12.1-1}$$

where ω is the light frequency, c is the light velocity in free space, $n(\omega)$ is the refractive-index of the medium, $k = 2\pi n(\omega)/\lambda$ is the magnitude of the wavevector, and λ is the light wavelength in free space. The physical meaning of $v_p(\omega)$ is the propagation velocity in the medium for a continuous and monochromatic light wave with a given frequency ω.

In practice, a real optical signal launching into an optical medium always possesses a limited temporal duration and nonzero spectral bandwidth. In this case, the real optical signal propagation behavior is determined by the group velocity (v_g) that is defined by

$$v_g = \frac{d\omega}{dk} = \frac{c}{n(\omega) + \omega dn/d\omega} = \frac{c}{n(\lambda) - \lambda dn/d\lambda}. \tag{12.1-2}$$

The physical meaning of v_g (at ω_0 or λ_0) is the real velocity of a quasi-monochromatic light pulse propagating in a medium and having a central frequency ω_0 (or wavelength λ_0). The difference between the phase velocity and group velocity is that the former is only determined by the refractive-index $n(\omega$ or $\lambda)$, but the latter is determined by both $n(\lambda)$ and the refractive-index dispersion $dn(\lambda)/d\lambda$.

Assuming there are two quasi-monochromatic pulsed signals that have different central wavelengths, their propagation velocities in a medium should be generally different due to the so-called group velocity dispersion (GVD) that is described by

$$\frac{dv_g}{d\lambda} = v_g^2 \frac{\lambda}{c} \frac{d^2 n}{d\lambda^2}. \tag{12.1-3}$$

From Eqs. (12.1-2) and (12.1-3) one can see that the group velocity is determined by the first-order derivative of refractive-index on wavelength, whereas the group velocity dispersion determined by the second-order derivative of refractive-index on the wavelength. If the difference of central wavelength for these two pulses is $\Delta\lambda$ and the corresponding difference of group velocity for these two pulses is $\Delta v_g = \Delta\lambda dv_g/d\lambda$, the difference of the traveling time for these two pulses passed through a propagation distance L should be

$$\Delta t = \frac{L}{v_g} - \frac{L}{v_g + \Delta v_g} \approx \frac{L\Delta v_g}{v_g^2} = \frac{L}{v_g^2}\frac{dv_g}{d\lambda}\Delta\lambda. \quad (12.1\text{-}4)$$

Substituting $dv_g/d\lambda$ from Eq. (12.1-3) into the above equation leads to

$$\Delta t = \frac{\lambda}{c}\frac{d^2 n}{d\lambda^2}L\Delta\lambda = DL\Delta\lambda. \quad (12.1\text{-}5)$$

Here,

$$D = \frac{\lambda}{c}\frac{d^2 n}{d\lambda^2} = \frac{\lambda^2}{2\pi c}\frac{d^2 k}{d\lambda^2} \quad (12.1\text{-}6)$$

is defined as GVD coefficient or chromatic dispersion coefficient of a given propagation medium. The physical meaning of D is the pulse delay time with a unit of wavelength difference and over a unit of propagation length. For an optical fiber, D is usually characterized in units of ps/nm/km.

12.1.2 Refractive index and GVD of silica glass

Most commonly utilized optical fiber systems are made of silica (fused quartz) glass. The refractive index of this material as a function of wavelength in the visible and near-IR spectral range is shown in Fig. 12.1(a), indicating a normal refractive-index dispersion behavior as there is no linear absorption in this region. The phase velocity and the group velocity of light in this material as a function of wavelength are shown in Fig. 12.1(b), respectively. Moreover, by taking second-order derivative on the refractive-index curve shown in Fig. 12.1(a) one can readily obtain the GVD coefficient as a function of wavelength, as shown in Fig. 12.1(c); from the latter we can see that the silica glass manifests a positive (normal) GVD behavior in the spectral region <1.27 μm, a negative (anomalous) GVD behavior in the region >1.27 μm, and zero GVD at the ~1.27 μm position.

For two quasi-monochromatic light pulses that have small central-wavelength difference and propagate in a medium working on its positive GVD region, the group velocity of the pulse with a longer wavelength is greater than that of the pulse with a shorter wavelength, or otherwise if the medium is working on its negative GVD region, the pulse with a longer wavelength will propagate slower than the pulse with a shorter wavelength.

Note that for an optical fiber system made of silica glass, the GVD coefficient D could be redefined as

$$D = \frac{\lambda^2}{2\pi c}\frac{d^2 \beta}{d\lambda^2}, \quad (12.1\text{-}7)$$

where β is the fiber propagation constant. In this case, the GVD is determined not only by the materials dispersion but also by the waveguide dispersion property of the fiber system.

For optical fiber telecommunication systems, the 1.3–1.5 μm wavelength region is extremely useful as it falls into the transmission window offering minimum propagation losses. From Fig. 12.1(c) one can see that in that spectral region silica fibers exhibit a negative (anomalous) GVD; hence they are ideal candidates for fundamental studies and applications of optical temporal solitons.

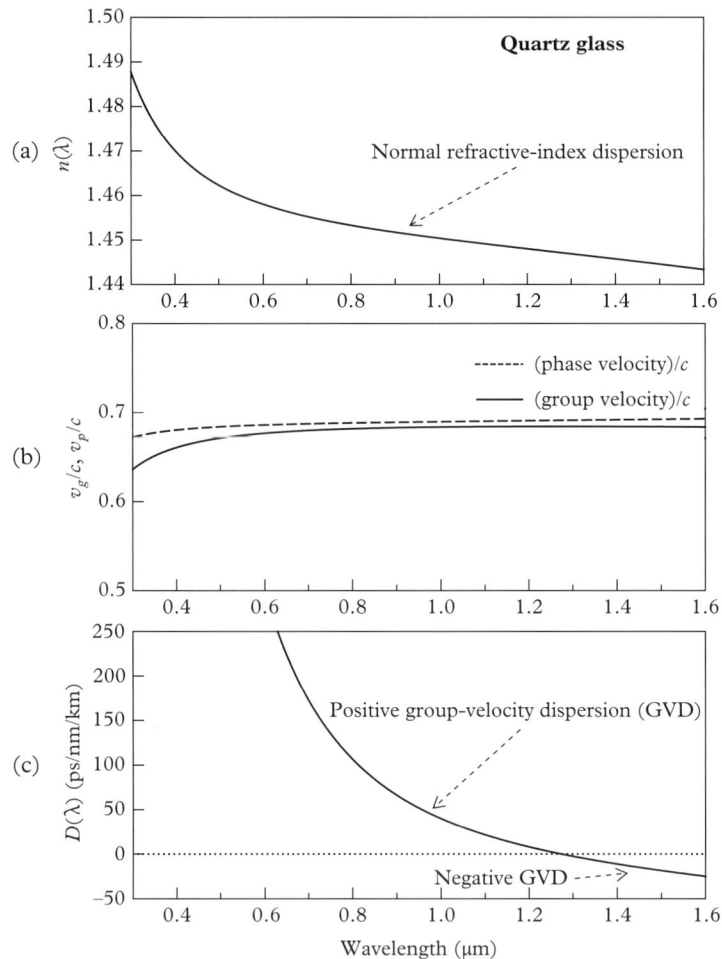

Figure 12.1 *Refractive index (a), phase velocity and group velocity (b), and GVD as functions of wavelength for silica (fused quartz) glass used for making optical fibers.*

12.1.3 Balance between GVD and self-phase modulation in a nonlinear medium

Let us consider a real light pulse having an initial pulse duration of τ_0 and spectral width of $2\delta\omega$, as shown in Fig. 12.2(a). If this pulse propagates in an optical medium in its positive GVD region, the lower-frequency spectral components of the pulse will travel faster than the higher-frequency components, as shown in Fig. 12.2(b), whereas propagating in a medium of negative GVD the light pulse undergoes a reverse situation, as shown in Fig. 12.2(c). In both cases, it is obvious that as a result of group velocity dispersion the pulse envelope will gradually become broader following the increase of the propagation distance (L) in the medium.

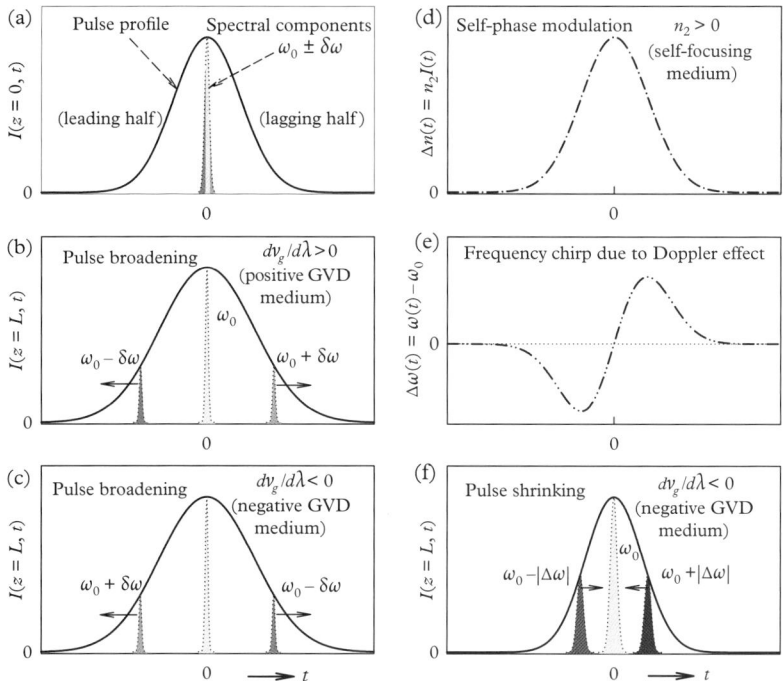

Figure 12.2 (a) Pulse intensity profile and spectral components. (b, c) Pulse broadening after passing through nonzero-GVD media. (d) Induced instantaneous refractive-index change in an $n_2 > 0$ medium. (e) Self-phase modulation induced frequency chirp. (f) Frequency chirp induced pulse shrinking in a negative GVD medium.

On the other hand, in a third-order nonlinear optical medium, there is an intensity-dependent refractive-index change expressed by

$$\Delta n(t) = n_2 |E(t)|^2, \qquad (12.1\text{-}8)$$

where n_2 is the nonlinear refractive-index coefficient of the medium and E is the electric field of the incident light field. As discussed previously in Chapter 6, in a medium with $n_2 > 0$, self-focusing may occur, whereas self-defocusing may be observed in a medium with $n_2 < 0$. If an intense laser pulse passes through such a nonlinear medium, there will be an induced phase (optical path) change experienced by pulse itself in the medium:

$$\Delta \phi(t) = \frac{\omega_0}{c} L n_2 |E(t)|^2, \qquad (12.1\text{-}9)$$

where L is the propagation distance in the nonlinear medium. Such induced time dependent phase-change is accompanied by a frequency shift described by

$$\Delta \omega(t) = -\frac{\partial}{\partial t} \Delta \phi(t). \qquad (12.1\text{-}10)$$

For a nonlinear medium with $n_2 > 0$, the induced refractive-index change is shown in Fig. 12.2(d), from which one can see that the leading half of the pulse creates an increase of the optical path length, this means that during this time period the effective thickness of the nonlinear medium is continuously increasing; according to Doppler principle the light frequency will experience a redshift as shown in Fig. 12.2(e). Based on the same consideration, the lagging half of the pulse will experience a blueshift, as shown in Fig. 12.2(e). Moreover, if the considered nonlinear medium manifests a negative GVD in the wavelength range of the input light pulse, the redshift part of the pulse will propagate slower than the blue-shifted part, so that the pulse has a tendency of becoming narrower, as shown schematically in Fig. 12.2(f). Comparing Fig. 12.2(c) with (f), one may find that in a $n_2 > 0$ nonlinear medium working in its negative GVD region, the pulse broadening effect due to GVD can be compensated by the pulse shrinking effect due to self-phase modulation, so that a stable pulse transmission without pulse width change may be achieved. This is the physical basis of formation of optical temporal solitons. The same principle is also applicable to the case of a $n_2 < 0$ nonlinear medium working in its positive GVD region. In both cases and under appropriate experimental conditions, a special nonlinear transmission of short laser pulses can be attained, which means that the pulse shape either remains unchanged over the whole propagation distance in the medium or repeats its shape periodically along the propagation distance. The pulses manifesting such nonlinear propagating properties in suitable media are called optical temporal solitons that will be further described in the following section.

12.2 Basic properties of temporal solitons

12.2.1 Wave equation governing pulse propagation in a nonlinear dispersive medium

The theoretical considerations of optical soliton formation were deduced in the early 1970s. One of the earliest suggestions of utilizing GVD effect to compress the self-phase-modulated pulse was reported by Zel'dovich and Sobel'man.[1] In the other respect, Zakharov and Shabat analyzed the wave equation describing the pulse transmission in a dispersive nonlinear medium, indicating that certain particular solitary solutions (solitons) of this equation could exist, which have a temporal envelope of hyperbolic secant shape.[2] The first clear theoretical suggestion of using dielectric optical fibers working at negative GVD region to achieve soliton formation and transmission was described by Hasegawa and Tappert in 1973.[3] In 1974, Satsuma and Yajima further theoretically indicated that there could be a series of discrete solitary solutions characterized by different positive integers of $N = 1, 2, 3\ldots$, which have predicable behavior along the propagation distance in the nonlinear medium.[4]

In a third-order nonlinear optical medium, the electric field of a quasi-monochromatic intense pulse can be expressed as

$$E(z, t) = U(z, t) \exp[i(k_0 z - \omega_0 t)]. \tag{12.2-1}$$

It is assumed that there is a uniform transverse intensity distribution and $U(z, t)$ is the slowly varying amplitude envelope function, ω_0 is the central frequency of the pulse, and k_0 is amplitude of the wavevector corresponding to ω_0. By considering the velocity dispersion and self-phase modulation, the nonlinear wave equation can be written as[2-4]

$$i\left(\frac{\partial U}{\partial z} + k'\frac{\partial U}{\partial t}\right) + \frac{k''}{2}\frac{\partial^2 U}{\partial t^2} = \frac{k_0\,n_2}{2\,n_0}|U|^2 U, \quad (12.2\text{-}2)$$

where $k' = \partial k/\partial\omega = v_g^{-1}$ and $k'' = \partial^2 k/\partial\omega^2 = -v_g^2 \partial v_g/\partial\omega$. In Eq. (12.2-2), the third term on the left-hand side describes the GVD effect, while the term on the right-hand side describes the influence of self-phase modulation. It can be mathematically proved that only when $n_2 > 0$ and $\partial v_g/\partial\omega > 0$ (anomalous GVD), or $n_2 < 0$ and $\partial v_g/\partial\omega < 0$ (normal GVD) does the above equation possess temporal solitary solutions.[2,4] After taking the following transformations

$$\begin{cases} u = \tau_0[(k_0 n_2/2n_0)/|k''|]^{1/2} U \\ \xi = |k''|\,z/\tau_0^2 \\ s = (t - k'z)/\tau_0 \end{cases}, \quad (12.2\text{-}3)$$

Eq. (12.2-2) can be reduced to the so-called nonlinear dimensionless Schrödinger equation:

$$i\frac{\partial u}{\partial \xi} = \frac{1}{2}\frac{\partial^2 u}{\partial s^2} + |u|^2 u. \quad (12.2\text{-}4)$$

Here, $u(\xi, s)$ is the dimensionless envelope function, ξ is the dimensionless form of distance z, s is the dimensionless form of time t, and τ_0 is a time scale factor whose physical meaning will become clear through the following descriptions of solitary solutions of the above equation.

12.2.2 Solitary solutions of nonlinear wave equation in optical fiber systems

Equation (12.2-4) can be perfectly applied to describe the formation and transmission of temporal solitons in a single-mode optical fiber, working in its negative GVD region.[5] Under the initial condition of an input light pulse

$$u(\xi = 0, s = s_0) = A\,\mathrm{sech}(s_0), \quad (12.2\text{-}5)$$

where A is the dimensionless amplitude maximum and $s_0 = t/\tau_0$. When $t = \tau_0$ we have sech $(s_0 = 1) = 0.648$; thus here τ_0 can be roughly recognized as the half width of the amplitude envelope of the input pulse. The solitary solutions of Eq. (12.2-4) should have the following form at the input position:

$$u(0, s_0) = N\,\mathrm{sech}(s_0), \quad (12.2\text{-}6)$$

where N can be any positive integer. There are analytical expressions for $N = 1$ and $N = 2$ solitons:[4]

$$\begin{cases} u(\xi, s) = \mathrm{sech}(s)\exp(-i\xi/2) & \text{for } N = 1 \\ u(\xi, s) = \dfrac{4[\mathrm{ch}(3s) + 3\exp(-4i\xi)\mathrm{ch}(s)]}{\mathrm{ch}(4s) + 4\mathrm{ch}(2s) + 3\cos(4\xi)}\exp(-i\xi/2) & \text{for } N = 2. \end{cases} \quad (12.2\text{-}7)$$

When $N = 1$, we deal with the fundamental temporal soliton that can propagate along the fiber length without any envelope change. For higher-order ($N = 2, 3, 4, \ldots$) solutions there is a propagation period of $\xi_0 = \pi/2$, which corresponds to a real propagation period of

$$z_0 = \frac{\pi \tau_0^2}{2|k''|} = \frac{\pi^2 c}{\lambda_0^2} \frac{\tau_0^2}{|D(\lambda_0)|}, \qquad (12.2\text{-}8)$$

where $D(\lambda)$ is the GVD coefficient of the fiber at the λ_0 position. The pulse shape and duration of the higher-order solutions remain unchanged after each propagation period of z_0. For the fundamental ($N = 1$) soliton $u_{\max} = 1$, so from Eq. (12.2-3) one can determine the required peak power for the input pulse to be (in SI units)

$$P_1 = \frac{\varepsilon_0 n_0 c \lambda_0}{4 n_2 z_0} \bar{A} = \frac{\varepsilon_0 n_0 \lambda_0^3}{4\pi^2 n_2} \frac{\bar{A}|D|}{\tau_0^2}, \qquad (12.2\text{-}9)$$

where $\bar{A}$ is the core area of the optical fiber. The physical meaning of Eq. (12.2-9) is that the required input power for fundamental soliton formation is directly proportional to the value of GVD coefficient ($|D|$) and inversely proportional to τ_0^2. It is noted that the input power ratio between the N-order soliton and the fundamental soliton is $P_N/P_1 = N^2$. For precise calculations using Eqs. (12.2-8) and (12.2-9), one should know the τ value that is the full width at half maximum (FWHM) of the input pulse power profile. If the pulse possesses a sech^2-shape of the power profile, we have $\tau = 0.669\tau'$, where τ' is the FWHM of the corresponding amplitude envelope of the sech-shape. Finally we obtain the following relation:

$$\tau_0 = \frac{\tau'/1.317}{2} = 0.568\tau. \qquad (12.2\text{-}10)$$

Figure 12.3 shows the theoretical evolution of the power envelopes for three ($N = 1$, 2, and 3) solitons at several characteristic propagation distances in a fiber.[5] For the $N = 1$ soliton, the pulse remains unchanged forever like that at the input. For $N = 2$ and $N = 3$ solitons, they exactly repeat the initial pulse shape after the same propagation period (z_0). It is noted that, for the $N = 2$ soliton at $z_0/2$ distance, the pulse narrows with a considerably increase of the peak power. For the $N = 3$ soliton, a much sharper pulse narrowing occurs at the $z_0/4$ position, whereas at $z_0/2$ distance the pulse splits into two peaks. For the $N = 4$ soliton (not shown here) there will be a threefold splitting at the $z_0/2$ position.

12.2.3 Experimental observation of temporal solitons in optical fibers

The first experimental observation of temporal solitons generated in a single-mode silica fiber was reported by Mollenauer et al. in 1980.[5] In their experiment, the input was the laser pulses of ~1.55 μm wavelength, ~7 ps duration, ~sech^2-pulse shape, and 100 MHz repetition rate, which were from a synchronously pumped mode-locked color-center laser. The fiber sample was 700 m long with a core diameter of 9.3 μm, and the GVD coefficient at 1.55 μm was $D \approx -16$(ps/nm/km). Based on Eq. (12.2-8) the calculated soliton period was $z_0 \approx 1260$ m and the fundamental soliton power was $P_1 \approx 1$ W. By varying the input power level and measuring the output pulse shape one could determine the formation of solitons at different appropriate input power levels. In this specific case, the real fiber length (700 m) was close to the half of the calculated z_0 value, such that Fig. 12.3 could be used for comparison with the experimental observation.

Shown at the bottom of Fig. 12.4 are the SHG (second-harmonic generation) autocorrelation traces of the output laser pulses measured at different input power levels. When the input power $P \ll P_1$, the output pulse width is broader than the input due to GVD influence. When $P \approx P_1$

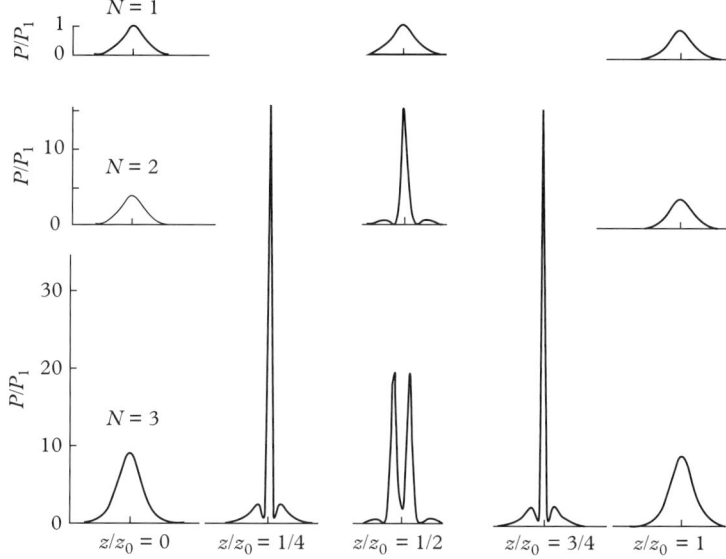

Figure 12.3 *Theoretical pulse shapes of the fundamental (N = 1) and two higher-order (N = 2, 3) temporal solitons versus the propagation distance z, where z_0 is the soliton period. (Reproduced with permission from Mollenauer et al.[5] Copyright 1980, American Physical Society).*

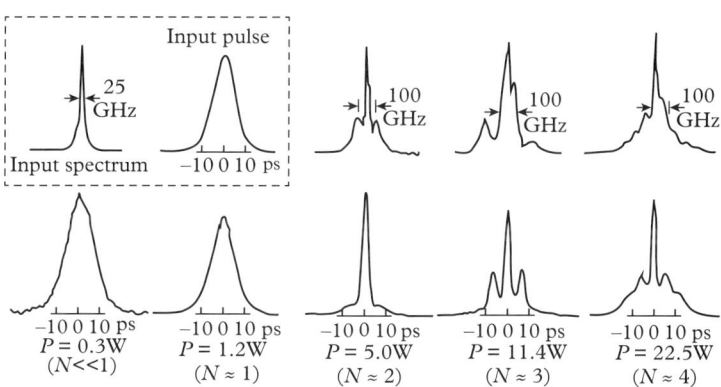

Figure 12.4 *Measured autocorrelation traces (below) and corresponding frequency spectra (above) of the fiber output as a function of input pulse peak power. Inset: spectrum and autocorrelation trace of the input laser pulse. The fiber length L = 700 m ≈ $z_0/2$, where z_0 ≈ 1260 m is the calculated soliton period. (Reproduced with permission from Mollenauer et al.[5] Copyright 1980, American Physical Society).*

or $N \approx 1$, the output pulse width is the same as the input pulse; this reveals that the GVD influence is just balanced by the self-phase modulation effect along all the propagation positions. At the input level of $P \approx 4P_1$, there is a much narrower output single pulse corresponding to $N = 2$ soliton. The $N = 3$ soliton is observed at the input level of $P \approx 9P_1$, the three peaks in the autocorrelation trace reveal the twofold splitting at $\sim z_0/2$ distance, as predicted by Fig. 12.3.

The measured frequency spectral curves of the output pulses at different input power levels are shown in the top of Fig. 12.4. One can see that comparing to the input pulse width and spectral width, the temporal narrowing of the output soliton pulses is always accompanied by the corresponding spectral broadening. The more complex behavior of the variation of temporal width and spectral width for higher-order ($N = 2, 3, 4,...$) solitons can be recognized as a result of the dynamic mutual interaction between the GVD induced pulse broadening and self-phase modulation induced pulse narrowing.

For input pulses with initial values of $A \neq N$ (cf. Eq. (12.2-5)), the pulse evolution along the propagation length has been theoretically considered,[4,6] and also studied experimentally.[7]

12.2.4 Soliton-like pulse formation in $n_2 < 0$ nonlinear media with positive GVD

As we mentioned previously, in principle temporal solitons can also be formed in a defocusing ($n_2 < 0$) nonlinear medium of positive GVD property at the working wavelength region. In this case (cf. Fig. 12.2), the leading half of the input pulse induces a decreasing refractive-index change, i.e., the effective optical path of the medium looks changed from thicker to thinner, which leads to a frequency blueshift owing to Doppler effect, whereas for the lagging half of the pulse, the induced refractive-index change makes the effective optical path of the medium changed from thinner to thicker, which causes a frequency redshift. Since the medium possesses a positive GVD property, the blue component of the pulse propagates slower than the red component; therefore an intense optical pulse may become narrowing. Furthermore, if such a pulse narrowing effect can be in balance with the pulse broadening effect owing to the pure GVD effect, the formation of temporal solitons may be achieved.

The nonlinear pulse narrowing effect has been experimentally observed in the third-order nonlinear media that manifest the properties of $n_2 < 0$ and GVD > 0 in an appropriate wavelength region[8,9]. As an example, Fig. 12.5 shows the experimental results obtained from a CdS crystal sample worked at ~ 625 μm wavelength region.[9] It is experimentally proved that at this wavelength region CdS crystal exhibits a positive GVD and a two-photon absorption enhanced negative refractive-index change. In experiment, the sample thickness was 130 μm, the input pulses were from a mode-locked dye laser with the wavelength of 625 nm, pulse duration of 105 fs, and repetition rate of 20 Hz. The input pulses of ~ 80 nJ energy were focused on the CdS sample surface with a spot diameter of ~ 33 μm. The pulse width of the transmitted laser beam was measured by SHG method through intercorrelation with a probe beam of much shorter (~ 45 fs) pulse width. As shown in Fig. 12.5, at low input intensity level (~ 0.2 GW/cm^2), the transmitted pulse width (115 fs) is broader than the input pulse width (105 fs) which is due to the pure GVD effect, whereas at high input intensity level (~ 15 GW/cm^2), the output pulse width (82 fs) is narrower than the input, clearly indicating the nonlinear pulse-narrowing effect.

In second-order nonlinear media, the magnitude and sign of the induced refractive-index change can be controlled through special three-wave mixing processes (see Section 5.7). If a second-order nonlinear medium manifests positive GVD in a given wavelength region and the induced refractive-index change is negative through a proper three-wave mixing interaction,

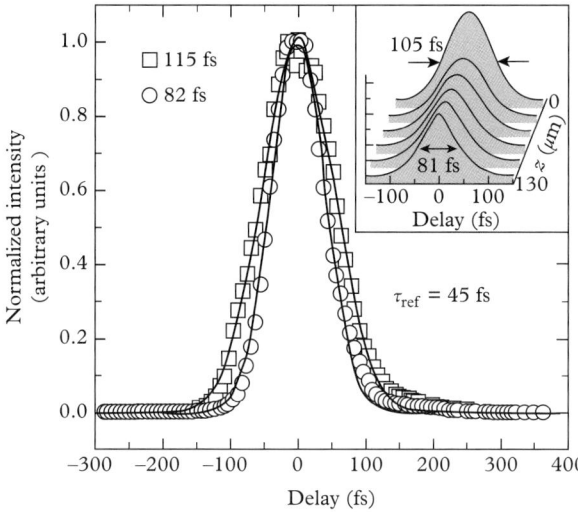

Figure 12.5 *Normalized intercorrelation traces between the reference and the transmitted probe pulses for 15-GW/cm² (open circles) and 0.2-GW/cm² (open squares) intensities. The inset shows the simulated temporal profile evolution of a 105-fs pulse propagating through a 130-µm thick CdS crystal. (Reproduced with permission from Lami et al.[9] Copyright 1999, American Physical Society).*

the temporal solitons can be formed. Some preliminary experiments performed in such type of second-order crystals have illustrated the pulse compressing effect similar to that shown in Fig. 12.5.[10,11]

12.2.5 Long-distance transmission of temporal solitons in fibers

For optical fiber-based high-bit-rate and long-distance telecommunications systems, it is important to retain the shape and width of optical pulses over a long fiber length without distortion or temporal broadening. Without considering the attenuation influence from absorption and scattering, optical solitons are the best signal carriers for this purpose. For conventional single-mode silica fibers, at ∼1.5 µm wavelength the loss rate is of ∼0.2 dB/km, which means after 10-km fiber propagation the pulse intensity will decrease to 63% of the initial level. For this reason, we must consider the loss influence on the soliton transmission over long distance, and find out solutions to overcome this problem. The issue of soliton transmission in lossy fibers was addressed by the early theoretical work.[6,12] As it is theoretically indicated that the area (the amplitude times the width) of a soliton pulse is conserved, therefore one may consider that the soliton property is retained even in the presence of loss, although the fiber loss decreases the amplitude and increases the width of the soliton. This prediction is understandable if we take the fundamental ($N = 1$) soliton as an example. The fiber loss gradually decreases the soliton power, then from Eq. (12.2-9) one can see that the soliton status can be retained but with a gradually increased τ_0 value. Kodama and Hasegawa theoretically predicted that the original soliton shape and pulse width can be retained for long-distance transmission (up to 6×10^3 km) if some spatially period amplification mechanism could be applied to the fiber system over each several tens of kilometers.[13] The suggested amplification mechanisms include stimulated Raman scattering (SRS), which should be capable to compensate for the fiber loss after each appropriate propagation distance.[14]

In experiments, two periodic amplification mechanisms have been applied to the fiber system to restore the initial soliton shape and width: one is the gain provided by SRS,[15–17] and the other is the gain offered by a rare-earth doped fiber amplifier.[18]

The first experimental demonstration of utilizing SRS gain to recover the fundamental soliton was performed in a 10-km standard single-mode silica fiber sample.[15] The input signal pulses were of 1.56 μm wavelength and ∼10 ps FWHM with a peak power level (∼375 mW) satisfying the fundamental soliton requirement. The SRS gain was created by a cw 1.46-μm laser beam counter-propagating along the same fiber. The frequency difference between the soliton beam and the pump beam was $\Delta \bar{\nu} \approx 440$ cm^{-1} that was close to the broad peak of Raman gain profile of silica glass. The loss rate of the fiber sample was ∼0.18 dB/km for 1.56 μm signals. When the 1.46-μm pump beam was turned off, the output pulse width increased to 1.5 times broader than the input due to the fiber loss, whereas when the SRS gain reached to ∼1.8 dB over the whole 10 km fiber length, the measured width of the output pulse was exactly the same as that of the input, indicating a nearly perfect recovering of the original soliton.

The further experiment demonstrated that a high-fidelity soliton transmission in fibers could be extended as far as more than 4000 km with loss periodically compensated by Raman gain.[16]

The other alternative approach to balance the fiber loss is the periodic use of conventional fiber laser amplifiers. To demonstrate this feasibility, a 400-km long fiber transmission line was tested, which was composed by eight 50-km long fiber spans; between each two spans there was a laser diode-pumped erbium-doped fiber amplifier (EDFA).[18] The input pulse was of 1.545 nm wavelength and ∼22 ps pulse duration. At the working wavelength of 1.545 nm, the fiber GVD was $D \approx -2.4$ ps/nm/km, and the fiber loss was ∼0.24 dB/km. The required pulse peak power for formatting $N = 1$ soliton was determined to be $P_1 = 6 - 10$ mW, and the gain provided by the EDFA for each fiber span should be more than 12 dB. Under these experimental conditions, the original soliton can be well retained, which was proved by the waveform measurements taken at 100, 200, 300, and 400 km distances separately.

12.3 Pulse narrowing and self-frequency shift of temporal solitons

12.3.1 Pulse narrowing of higher-order temporal solitons through a shorter fiber

From Fig. 12.3 one can see that at ∼$z_0/4$ position the $N = 3$ soliton manifests a very narrow pulse peak. The numerical simulation has indicated that for higher-order ($N > 3$) solitons this type of pulse narrowing can be further enhanced over the distances much shorter than the soliton period z_0. Based on this special property of higher-order temporal solitons, researcher may simply utilize a relatively short fiber to significantly compress the pulse duration and to increase the peak power.[19–23]

Figure 12.6 shows the experimental results of pulse compression based on higher-order solitons formation in a shorter standard single-mode fiber.[19] The input pulses had ∼1.5 μm wavelength

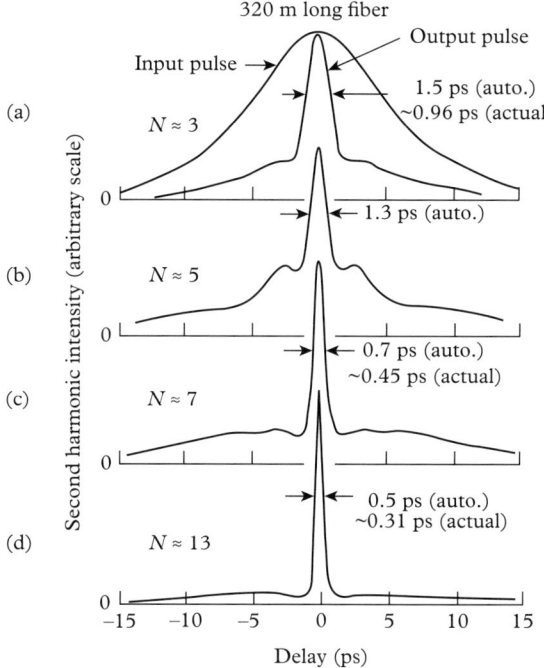

Figure 12.6 *Autocorrelation traces of pulses initially 7-ps wide emerging from a 320-m single-mode fiber at various indicated input powers and related soliton numbers. (Reproduced with permission from Mollenauer et al.[19] Copyright 1983, Optical Society of America).*

and ~7 ps duration. The fiber length was 320 m, which was nearly close to $z_0/4$. In the experiment, the input pulse peak power could be adjusted to form different-order solitons. The corresponding output pulse widths were measured by using the SHG autocorrelation method. The results clearly showed that for the $N \approx 3$ soliton the output pulse width was ~0.96 ps, whereas for the $N \approx 13$ soliton, the output width was shortened to ~0.31 ps. In the latter case, the compression ratio was ~1/23.

So far in this chapter, the most of mentioned experiments were accomplished by using ~1.5-μm input laser pulses. The similar pulse compression can also be accomplished by formatting the solitons working at ~1.3 μm wavelength. As an example of such experiments, the input laser pulses of 1.32 μm wavelength and ~1.5 ps duration were launched into a 19.5-m long single-mode fiber with $D \approx 0$ at 1.275 μm and $D \approx -5$ ps/nm/km at the 1.32 μm wavelength, respectively.[21] The calculated peak power for fundamental soliton formation was $P_1 \approx 4.16$ W, soliton period $z_0 \approx 246$ m, and $z/z_0 \approx 0.08$. Figure 12.7 shows the autocorrelation traces of the output pulses at various input peak powers and corresponding to different soliton orders. For the $N \approx 12$ soliton after 19.5 m transmission, the pulse width was compressed by more than 45 times.

If the duration of pulses from an original laser sources is quite broad (50–100 ps), the pulse compression can be accomplished by two steps. In the first step one can let these pulses pass

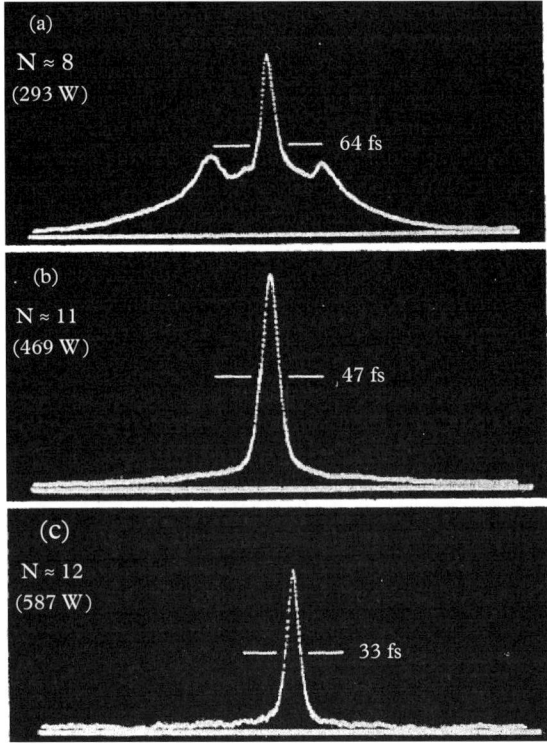

Figure 12.7 *Autocorrelation traces of 1.32 μm pulses initially ~1.5 ps wide emerging from a 19.5-m single-mode fiber at the various indicated input powers and related soliton numbers. (Reproduced with permission from Gouveia-Neto et al.[21] Copyright 1987, Optical Society of America).*

through a conventional compression system composed of a positive GVD fiber and a pair of gratings (or prisms), after which the pulse width can be reduced to 1–2 ps. In the second step, these pulses can be further compressed by passing through a short fiber with negative GVD, so that the output pulses can be finally shortened to several tens of femtoseconds.[21,22]

In practice, the input pulse peak power is restricted by either the possible optical damage of silica fiber core or by other undesirable nonlinear effects. With this respect, hollow-core photonic band-gap fibers with anomalous GVD property may be an alternative approach to generate temporal soliton pulses over a short fiber length.[24] This special type of fiber may withstand a higher peak power and possesses a smaller nonlinearity compared to the standard silica fibers.

12.3.2 Self-frequency shift of temporal solitons due to Raman gain

In Section 12.2.5, it is described how periodic Raman amplification can be adopted to balance the fiber loss for long-distance telecommunications systems. In that case, a rather small gain level

(~0.2 dB/km) is enough, which is far below the threshold requirement for generating SRS without input seed signals. Here we shall further indicate that under some experimental conditions of using relatively short fiber to compress the pulse duration of input pulses with high peak powers, the Raman gain induced frequency shift and spectral broadening must be considered. In 1985, Dianov et al. reported an interesting experiment in which a 250 m length of single mode fiber was utilized to generate higher-order solitons with ~1.54-μm and ~30-ps input pulses.[25] The input peak power was variable from 50 to 900 W, and at ~100 W input level the estimated soliton order would be $N \approx 30$. The spectral distributions of the output pulses from the fiber were measured at various input peak power levels, and the results are shown in Fig. 12.8.

From Fig. 12.8 one can see that when the input level is lower than ~100 W, there is a slightly symmetric-broadening (curves (a) and (b)), which could be due to self-modulation accompanied with soliton narrowing; when the input is higher than 100 W, there is an asymmetric spectral broadening, and a considerable pulse energy is transferred to the red-shifted wing that covers a frequency shift range up to ~200 cm^{-1} (from 1.54 to 1.59 μm). The asymmetric red-shifted broadening observed at high input power is ascribed to low-frequency Raman gain mechanism. In this case, during the higher-order soliton formation and temporal narrowing the 1.54 μm pulses significantly enhance their peak power and therefore can provide a stronger Raman gain in the fiber, while the red-shifted components due to self-modulation can be recognized as seed signals. Since the very broad Raman gain band of silica glass covers from $\Delta v \approx 0$ to ~1000 cm^{-1} position, these red-shifted seed signals can be efficiently amplified through Raman gain mechanism. The cascaded result of this elementary process may finally lead to the observed continuous Stokes shifted spectral wing. Moreover, if the fiber is not too short and the absolute value of GVD in the spectral broadening region cannot be neglected, original "pump" pulse and the newly created "Raman" pulse may be temporally separated after certain fiber propagation distance.

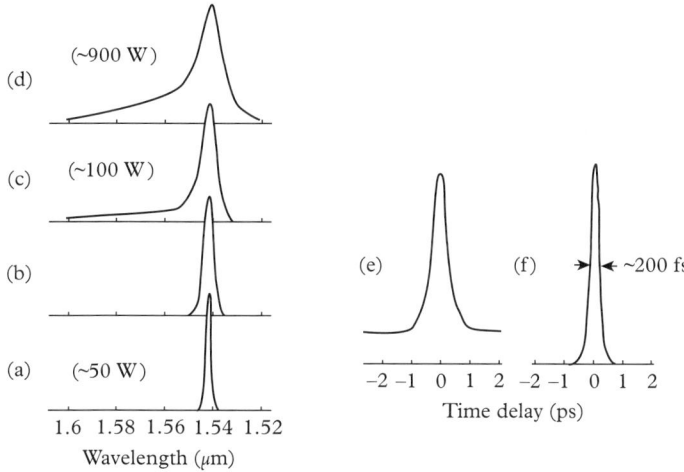

Figure 12.8 *Measured output spectra from a 250 m fiber sample at various peak power levels of the input 1.54 μm pulses of 30 ps duration (a–d). The autocorrelation traces of the output pulses measured at ~1.54 μm (e) and ~1.55 μm (f) spectral positions separately under the input power of ~900 W. (Reproduced from Dianov et al.[25]).*

Note that the pulse width is inherently related to the corresponding spectral width. In the above-mentioned situation, the measured pulse width of the Raman pulse may be different from the pump pulse. Figure 12.8(e) shows the autocorrelation trace of the output pulse measured around the 1.54 μm spectral region, which indicates a broad pedestal, whereas the trace shown in Fig. 12.8(f) reveals a sharp peak of ∼200 fs duration without background, which is measured around the 1.55 μm spectral region.

The similar self-frequency shift behavior was also observed by Mitschke and Mollenauer.[26] In their experiment, the input pulse duration was much shorter (∼0.5 ps), the peak powers were several times higher than the calculated P_1 value, and the fiber length could be varied from 50 to 392 m. It was found that the output from the fiber end contained two temporally separated pulse components, the spectral redshift was proportional to the increase of fiber length and input power; the spectral broadening could be up to ∼100 cm^{-1}, which revealed a cascaded nature of intrapulse stimulated Raman amplification processes. This type of spectral broadening behavior was also theoretically described.[27,28]

However, some other experiments also demonstrated that the spectral broadening and Raman related frequency shift could be observed in both sides of the central wavelength position of the input pulses.[29,30] As an example of those studies, Fig. 12.9 shows the measured spectral profiles of the input 1.341 μm pulses and the output pulses from the fiber samples of different length[29]. In this experiment the input pulse duration was 0.83 ps with a peak power of 531 W; the tested

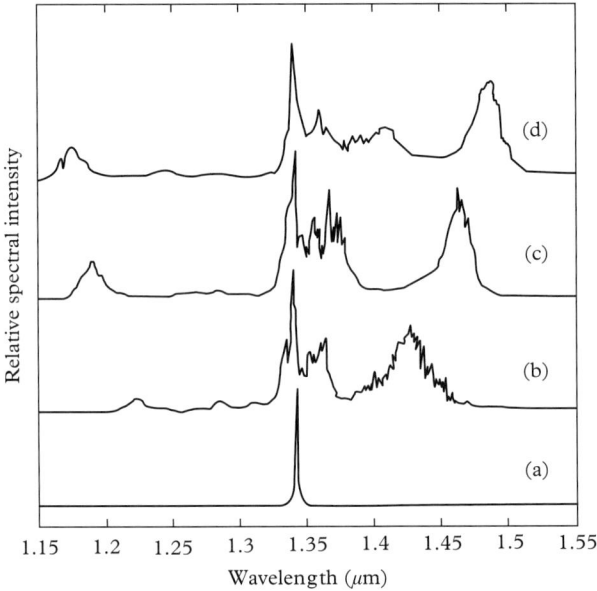

Figure 12.9 *Spectral intensity distributions of the input 1.341 μm pulse (a) and the output pulses transmitted through a fiber length of 17 m (b), 50 m (c), and 150 m (d), respectively. The input pulse duration was 0.83 ps with a peak power of 531 W. (Reproduced with permission from Zysset et al.[29] Copyright 1987, AIP Publishing LLC).*

fiber samples possessed zero GVD at $\lambda_{D=0} = 1.317\mu m$ position, their length could be varied from 17 to 150 m. From Fig. 12.9 one can see that after nonlinear transmission through fiber sample, the frequency-shifted and broadened spectral peaks appear in both Stokes and anti-Stokes sides within the Raman spectral range of the silica glass. The simultaneous generation of Stokes and anti-Stokes spectral components can be qualitatively explained by Raman resonance-enhanced four-photon parametric interaction within the fiber samples.[30] It was experimentally shown that following an increase of fiber length or input power, a higher proportion of the energy was located in the Stokes side, which could be interpreted by the attenuation of anti-Stoles components owing to the inverse-Raman effect and the simultaneous enhancement of Stokes components through the stimulated Raman gain process.

12.4 Fiber soliton lasers

12.4.1 Principles of soliton lasers

For fundamental research and especially for optical telecommunications, the stable ultra-short laser sources with controllable output repetition rate are highly desirable. The idea of temporal soliton formation in a fiber has been successfully applied to design such types of laser devices, which is based on the inherent stability of solitons and their capability offering ultra-short pulse width.

Comparing to conventional laser devices that usually contain a gain medium and two cavity mirrors, the soliton lasers manifest a unique feature that is involving a key element—a section of optical fiber necessary for soliton formation. This section of fiber is of negative GVD in working wavelengths and can be placed either in the master cavity containing the gain medium or in an external sub-cavity to control the feedback to the master cavity. The gain medium of soliton lasers can be a crystal, a semiconductor structure, and more often another shorter section of rare-earth doped fiber.

The soliton lasers have to work in mode-locked status to generate periodic and ultra-short output pulses. There are three possible ways to achieve mode-locked operation in soliton lasers: (i) active mode-locking via an active (phase, amplitude, or polarization) modulator, (ii) passive mode-locking via a saturable absorber, and (iii) self-starting mode-locking caused by nonlinearity of pulse transmission in a fiber medium. The repetition rate of laser pulses is solely determined by the effective mode-locking frequency, which can be a multiple number of the fundamental mode-locking frequency determined by $c/2L_0$, where c is the light speed in a vacuum and L_0 is the overall optical path-length of the lasing cavity. The effective-mode locking frequency can be controlled by adjusting appropriate experimental factor(s).

It should be noted that owing to the restrictions of possible optical damage and other undesirable nonlinear effects occurring the fiber systems, the soliton laser is more suitable working as an oscillator rather than an amplifier.

12.4.2 Original version of soliton lasers

The first soliton laser operation was demonstrated in 1984 by Mollenauer and Stolen.[31] In their experiment, as schematically illustrated in Fig. 12.10, the whole laser system was formed by combining two cavities: the master cavity containing a color-center crystal as lasing medium and the sub-cavity involving a section of standard single-mode fiber of length L. These two cavities were coupled through the coupling mirror M_0.

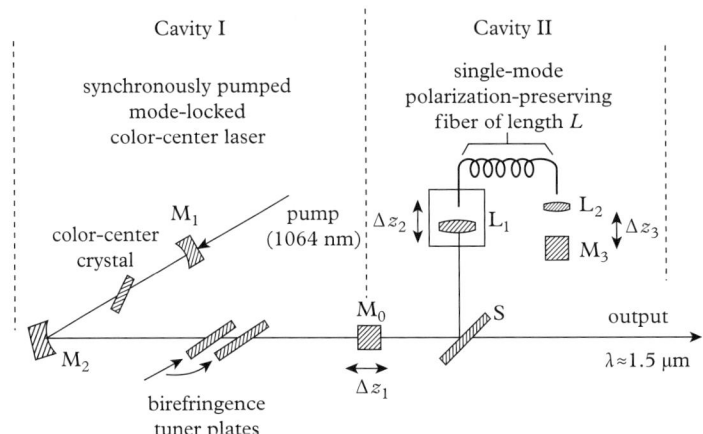

Figure 12.10 *Schematic diagram of an original soliton laser. Mirror reflectivity was 80% for M_0 and 30% for S. The main cavity length could be adjusted via Δz_1, while the sub-cavity length could be adjusted via Δz_2 or Δz_3. (Reproduced with permission from Mollenauer and Stolen.[31] Copyright 1984, Optical Society of America).*

The master laser cavity was synchronously pumped by the 1.064 μm pulses from a mode-locked Nd:YAG laser system. The lasing wavelength of the main cavity was tuned to be ~1.5 μm, and the pump synchronism is ensured by precisely adjusting the axial position of mirror M_0. Part of initial lasing pulses were launched into a single-mode fiber of length L with a negative GVD at ~1.5 μm wavelength; after double passing through this fiber section, the pulses became shorter and then re-entered the main cavity as additional feedback signals to affect the shape and width of lasing pulses. To ensure a synchronizing interaction between the feedback pulses from the sub-cavity and the pump pulses in the main cavity, the sub-cavity length could be adjusted. Under appropriate conditions, the final lasing pulses could be working in $N = 2$ soliton form satisfying the condition of $2L \approx z_0$, where z_0 was the soliton period, and the output pulse width could range from ~2.0 ps to ~0.21 ps, when the fiber length changed from ~0.5 m to ~30 m.[31] In this case, the output lasing pulses might not be stable due to the fluctuation of round-trip optical phase shift within the sub-cavity. This fluctuation could be minimized by stabilizing the sub-cavity length by controlling a piezoelectric translator of the end mirror M_3. After taking this measure and employing a 1.6-m long fiber, the very stable output pulses of ~580 fs duration were obtained.[32]

In a later but quite similar experiment, the main cavity was pumped by a cw 1064-μm laser beam of ~2 W continuous power, and the active mode-locking was created by an acousto-optic modulator working at 41 MHz.[33] The main cavity length was adjusted to match the cavity round-trip time to the second-harmonic of the modulation frequency. Without adding a sub-cavity, the main cavity generated the mode-locked ~1.5-μm lasing pulses of 6 ps duration. However, after coupling with the sub-cavity containing a 1.88-m long single-mode fiber, making the input power into the fiber to be ~30 mW, and adjusting the sub-cavity length to be an integral multiple of the main cavity length, the finally stabilized $N = \overline{2}$ soliton lasing pulses of ~380 fs width were observed.

12.4.3 Rare earth-doped fiber soliton lasers

Since the 1990s, most of the efforts for developing soliton lasers have been focused on using rare earth-doped fibers as the high gain and broad band lasing elements. Among them erbium (Er^{3+})-doped silica fibers are mostly employed for various experimental studies. The Er^{3+}-doped fiber soliton lasers usually work in mode-locked operation with lasing wavelengths tunable around the 1.55–1.57 μm region. The effective pump wavelengths and commonly adopted cw pump sources can be as follows: ~980 nm from semiconductor diode lasers[34] or from Ti:sapphire lasers[35,36], 1064 nm from Nd:YAG lasers[37], as well as 1480 nm from semiconductor diode lasers.[38] Recently, ytterbium (Yb^{3+})-doped fibers have also been utilized for soliton fiber laser devices; the lasing wavelength is tunable around the 1040–1050 nm region and the pump source is a ~980-nm cw laser beam.[39–41]

There are two basic requirements that should be considered in designing rare earth-doped fiber soliton lasers. First, the laser should always work in mode-locked status; secondly, a certain length of fiber with negative GVD is necessary to be connected with the doped gain fiber for soliton formation. The mode-locking in a fiber soliton laser under cw-pump can be achieved through three ways: (i) the active mode-locking produced by a suitable light modulator;[42–44] (ii) the passive mode-locking via an appropriate saturable absorber within the cavity;[40,41,45] (iii) the self-starting mode-locking ascribed to the initial instability and nonlinearity of the lasing loop.[46–49] In all these three cases, as we mentioned early, the effective mode-locking frequency can be an arbitrary multiple number of the fundamental mode-locking frequency that is the inverse cavity round-trip time. It has been experimentally demonstrated that the lasing pulse repetition rate could be ≥ 10 GHz. The lasing wavelength can be controlled by using intracavity pass-band tuning elements.

Figures 12.11 and 12.12 show schematic diagrams of four typical configurations for fiber soliton laser designs. In all these structures, the length for the doped gain fiber is usually around several meters, and the length for the fiber of negative GVD could be from several meters to more than a hundred meters. The pump beam is coupled into the gain fiber through a wavelength-division-multiplexing (WDM) coupler.

Figure 12.11(a) shows a linear-shape configuration with a lasing cavity comprising two mirrors (or reflectors). This type of cavity structure is equivalent to that of the early reported color-center soliton laser.[31] The recently reported Yb^{3+}-doped fiber soliton lasers are designed in this configuration, and could generate ~1040-nm output pulses of 600–800 fs duration with a lasing efficiency (the ratio between the pump and output average powers) of 2–5%.[40,41]

The configuration shown in Fig. 12.11(b) is similar to the above except that one reflecting mirror is replaced by a fiber loop that plays two roles: an equivalent reflector and an output coupler. This loop contains a modulator for active mode-locking and an optical (Faraday) isolator to ensure unidirectional optical circulation. The Er^{3+}-doped fiber soliton lasers based on this configuration usually produced laser pulses of 1–2 ps duration.[44,45] However, lasing output of ~380 fs duration was also reported with an lasing efficiency of ~0.4%[46].

Furthermore, many reported fiber soliton lasers were based on the design shown in Fig. 12.12(a), in which the whole lasing cavity manifests itself as a single close ring circuit.[42,43,47] In these cases it was assumed that the lasing worked with the $N = 1$ soliton as the output pulse width was not sensitive with the length change of the standard fiber of negative GVD.[50] In addition, when the output pulse width was $\Delta\tau = 1$–2 ps, the duration time–spectral bandwidth product ($\Delta\tau \Delta\nu$) was measured to be 0.32–0.36, i.e., very close to the value (0.315) theoretically calculated from a $sech^2$-shape soliton pulse. Sometimes the output stability of fiber soliton lasers might be influenced by mechanical vibration and temperature variations, this problem could be

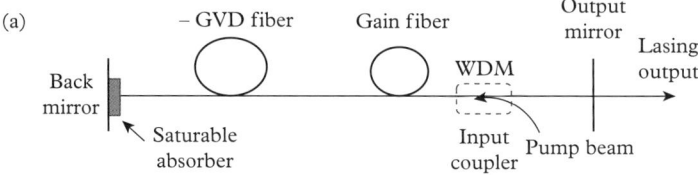

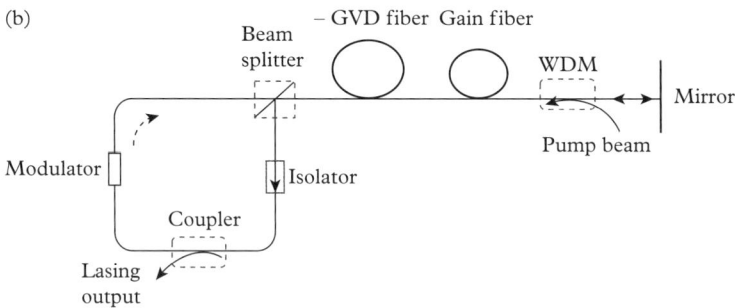

Figure 12.11 *Configurations of fiber soliton lasers with (a) a cavity composed of two mirrors and (b) a cavity composed of one mirror and one fiber feedback loop.*

solved by automatically stabilizing the cavity length or mode-locking frequency via an electronic servo-loop system.[51,52]

Figure 12.12(b) shows the soliton laser configuration comprising two ring loops with a figure-eight shape; one loop contains a doped gain fiber while the other loop provides nonlinear and unidirectional optical feedback.[53] Applying this design to erbium-fiber laser systems, it was found that the ultra-short soliton lasing output could be observed only after placing a suitable length of negative GVD fiber into the lasing circulating loop.[53,54] Base on this configuration, the measured output pulse width could be as short as 290–320 fs with a time-bandwidth product value around 0.32–0.34.[48,54] It was predicted that, in principle, the gain bandwidth of erbium-doped fiber could be ∼35 nm (or ∼150 cm^{-1} at 1.5 μm), which may allow the pulse width to be ∼80 fs.[48]

The use of optical isolators is to ensure unidirectional optical circulation, either along the whole ring cavity or within a sectional loop. The designs of isolator for fiber-optic circuits are based on the Faraday effect. They can be polarization sensitive, relying on birefringence-induced polarization rotation, or polarization insensitive, relying on birefringence-induced deflection. In addition, polarization controllers are also included in the designs shown in Figs. 12.11 and 12.12. The polarization controller utilized for fiber-optic circuits is a device to impose pressure or stress on a section of fiber to create certain birefringence and desirable polarization change. The polarization controllers are needed not only for working with polarization sensitive isolators (or modulators) but also essentially useful for optimizing mode-locking operations of fiber soliton laser systems.

Lastly, it is worth indicating that the designs shown in Fig. 12.12 can be adopted to realize all-fiber soliton lasers without involving any open bulk optical elements, which can make the whole system more compact and robust.

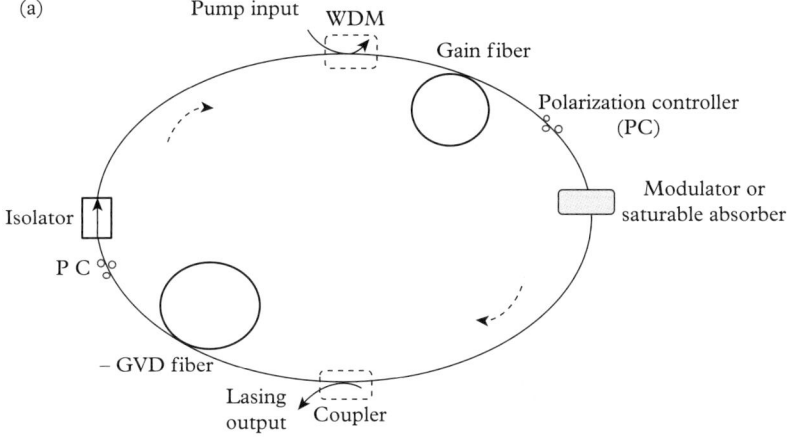

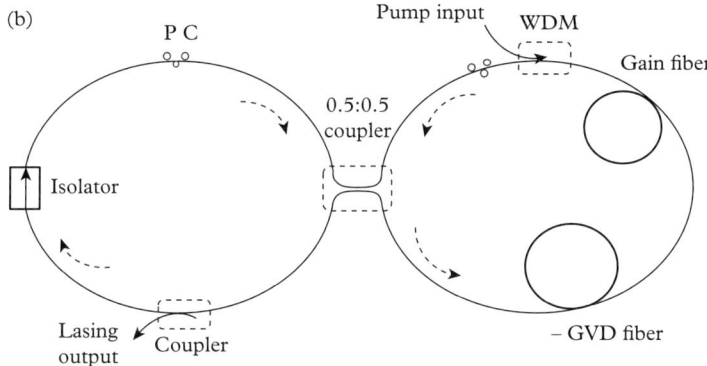

Figure 12.12 *Configurations of all-fiber soliton lasers with (a) a single ring loop and (b) a double ring loop.*

12.4.4 Fiber Raman soliton lasers

In Section 12.3 we have described the self-frequency shift and shortening of soliton pulses due to single passage through a Raman gain fiber sample. The same principle can be ready applied to achieve a fiber soliton laser based on SRS circulating within a cavity. Comparing to rare-earth doped fibers, the primary gain bandwidth of silica fiber can be greater than 150–200 cm^{-1}, so that in principle the much shorter pulse duration may be expected. On the other hand, the attainable gain level per unit of length for silica Raman fibers is considerable lower than that for rare earth-doped fibers. Therefore a longer gain fiber length is needed for Raman soliton lasers.

The early theoretical consideration of using high efficient SRS process to attain soliton lasing was given by Vysloukh and Serkin.[55] The further theoretical analysis of a synchronously pumped fiber Raman ring laser was given by Haus and Nakazawa.[56]

One of the earliest experimental demonstration of fiber Raman soliton lasing was reported by Kafka and Baer.[57] In their experiment, as shown in Fig. 12.13(a), being placed in a ring cavity

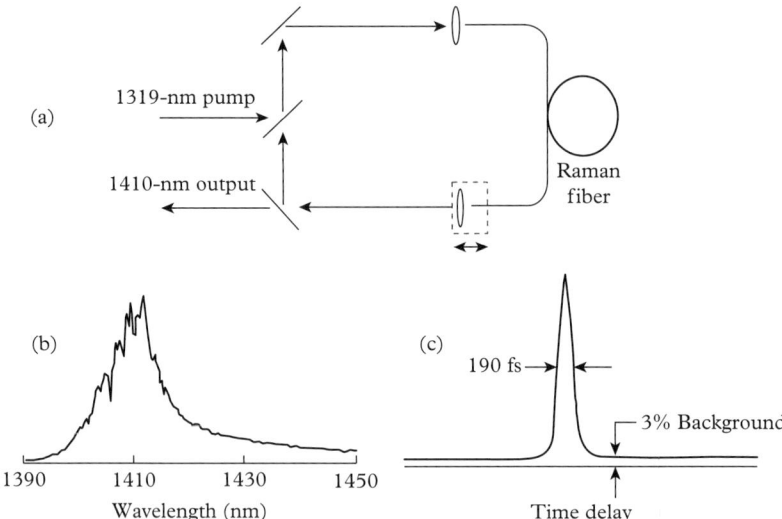

Figure 12.13 *Experimental setup (a), output spectrum (b), and autocorrelation trace (c) of a fiber Raman soliton laser. (Reproduced with permission from Kafka and Baer.[57] Copyright 1987, Optical Society of America).*

the standard single-mode fiber of 1.1 km length was synchronously pumped by a mode-locked Nd:YAG laser system with its 1319-nm pulses of 100 ps duration and 82 MHz repetition rate. In order to ensure a pump synchronism, the cavity length or mode-locking frequency could be precisely adjusted. The typical spectrum and autocorrelation trace of the output Raman soliton pulses are shown in Fig. 12.13(b) and (c), respectively. The spectral bandwidth (FWHM) was ~11 nm with a central wavelength of ~1410 nm, corresponding to a Raman frequency shift of ~489 cm^{-1}. The measured pulse width was ~190 fs, revealing a transformed-limited pulse with a time-bandwidth product of $\Delta\tau\Delta\nu \approx 0.32$. At ~300 mW average pump power level, the average Raman output power was ~50 mW, indicating a lasing efficiency of ~17%. It was indicated that under this experimental condition the effective gain length of the fiber was ~300 m limited by walk-off between the pump pulse and the Raman lasing pulse.

In a similar experimental setup, Gouveia-Neto et al. utilized a 600-m long dispersion-shifted single-mode fiber that exhibited a zero GVD at the ~1.46 μm position and ensured an efficient cascade-Raman generation and negative GVD in the required 1.5 μm regime. The pump source still was a cw mode-locked Nd:YAG laser providing 1320 nm pulses of ~100 ps duration, 100 MHz repetition rate, and 200 W peak power.[58] For single-pass pump, only one SRS component (S_1) is observed at the ~1410 nm position, as shown in Fig. 12.14(a). Upon cavity lasing, the output spectrum is as shown in Fig. 12.14(b), where three Stokes components (S_1, S_2, and S_3) are generated by cascade SRS processes. The frequency shift values for these components are ~488 cm^{-1}, ~887 cm^{-1}, and ~1326 cm^{-1}, respectively. The measured pulse width was ~200 fs for the S_2 band and ~230 fs for the S_3 band; the peak power was ~2 kW for the S_2 pulse and ~120 W for the S_3 pulse.

It was indicated by Islam et al. that the observed broadening of the Stokes SRS components could be initially facilitated by the mechanism of cross-phase modulation between the strong pump pulse and the weaker SRS pulse.[59] In another interesting experiment performed by Suzuki et al.,

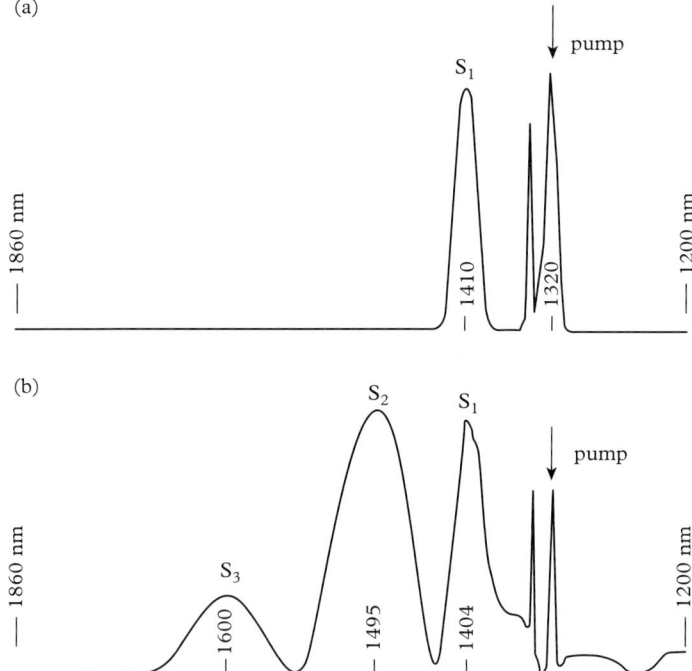

Figure 12.14 *Spectra of the SRS output from a fiber without cavity feedback (a) and the cascaded SRS output with cavity feedback. (Reproduced with permission from Gouveia-Neto et al.[58] Copyright 1987, Optical Society of America).*

both Stokes and anti-Stokes lasing components were observed due to Raman-enhanced four-photon parametric interaction[30]. In this case, however, only the former was located in the negative GVD region and could form a soliton with a shorter duration, whereas the latter was located in the positive dispersion region and could not form a soliton.[30,60]

REFERENCES

1. B. Ya. Zel'dovich and I. I. Sobel'man, *JETP Lett.* **13**, 129 (1971).
2. V. E. Zakharov and A. B. Shabat, *Sov. Phys. JETP* 1972, **34**, 62 (1972).
3. A. Hasegawa and F. Tappert, *Appl. Phys. Lett.* **23**, 142 (1973).
4. J. Satsuma and N. Yajima, *Suppl. Prog. Theor. Phys.* **55**, 284 (1974).
5. L. F. Mollenauer, R. H. Stolen, and J. P. Gordon, *Phys. Rev. Lett.* **45**, 1095 (1980).
6. N. J. Doran and K. J. Blow, *IEEE J Quantum Electron.* **19**, 1883 (1983).
7. R. H. Stolen, L. F. Mollenauer, and W. J. Tomlinson. *Opt. Lett.* **8**, 186 (1983).
8. P. Dumais, A. Villeneuve, and J. S. Aitchison, *Opt. Lett.* 1996, **21**, 260 (1996).
9. J. F. Lami, S. Petit, and C. Hirlimann, *Phys. Rev. Lett.* **82**, 1032 (1999).
10. X. Liu, L. J. Qian, and F. Wise, *Opt. Lett.* **24**, 1777 (1999).

11. S. Ashihara, J. Nishina, T. Shimura, and K. Kuroda, *J. Opt. Soc. Am. B* **19**, 2505 (2002).
12. A. Hasegawa and Y. Kodama, *Proc. IEEE* **69**, 1145 (1981).
13. Y. Kodama and A. Hasegawa, *Opt. Lett.* **7**, 339 (1982).
14. A. Hasegawa, *Opt. Lett.* **8**, 650 (1983).
15. L. F. Mollenauer, R. H. Stolen, and M. N. Islam, *Opt. Lett.* **10**, 229 (1985).
16. L. F. Mollenauer and K. Smith, *Opt. Lett.* **13**, 675 (1988).
17. A. S. Gouveianeto, P. G. J. Wigley, and J. R. Taylor, *Opt. Commun.* **72**, 119 (1989).
18. M. Nakazawa, K. Suzuki, H. Kubota, E. Yamada, and Y. Kimura, *IEEE J. Quantum Electron.* **26**, 2095 (1990).
19. L. F. Mollenauer, R. H. Stolen, J. P. Gordon, and W. J. Tomlinson, *Opt. Lett.* **8**, 289 (1983).
20. F. M. Mitschke and L. F. Mollenauer, *Opt. Lett.* **12**, 407 (1987).
21. A. S. Gouveia-Neto, A. S. L. Gomes, and J. R. Taylor, *Opt. Lett.* **12**, 395 (1987).
22. K. Tai and A. Tomita, *Appl. Phys. Lett.* **48**, 1033 (1986).
23. W. H. Xiang, S. R. Friberg, K. Watanabe, S. Machida, Y. Sakai, H. Iwamura, and Y. Yamamoto, *Appl. Phys. Lett.* **59**, 2076 (1991).
24. D. G. Ouzounov, F. R. Ahmad, D. Muller, N. Venkataraman, M. T. Gallagher, M. G. Thomas, J. Silcox, K. W. Koch, and A. L. Gaeta, *Science* **301**, 1702 (2003).
25. E. M. Dianov, A. Y. Karasik, P. V. Mamyshev, A. M. Prokhorov, V. N. Serkin, M. F. Stel'makh, and A. A. Fomichev, *JETP Lett.* **41**, 294 (1985).
26. F. M. Mitschke and L. F. Mollenauer, *Opt. Lett.* **11**, 659 (1986).
27. J. P. Gordon, *Opt. Lett.* **11**, 662 (1986).
28. G. P. Agrawal, *Opt. Lett.* **15**, 224 (1990).
29. B. Zysset, P. Beaud, and W. Hodel, *Appl. Phys. Lett.* **50**, 1027 (1987).
30. K. Suzuki, M. Nakazawa, and H. A. Haus, *Opt. Lett.* **14**, 320 (1989).
31. L. F. Mollenauer and R. H. Stolen, *Opt. Lett.* **9**, 13 (1984).
32. F. M. Mitschke and L. F. Mollenauer, *IEEE J. Quantum Electron.* **22**, 2242 (1986).
33. J. F. Pinto, C. P. Yakymyshyn, and C. R. Pollock, *Opt. Lett.* **13**, 383 (1988).
34. I. N. Duling, *Electron. Lett.* **27**, 544 (1991).
35. M. L. Dennis and I. N. Duling, *Electron. Lett.* **28**, 1894 (1992).
36. D. U. Noske, N. Pandit, and J. R. Taylor, *Electron. Lett.* **28**, 2185 (1992).
37. D. O. Culverhouse, D. J. Richardson, T. A. Birks, and P. St. J. Russell, *Opt. Lett.* **20**, 2381 (1995).
38. M. Zirngibl, L. W. Stulz, J. Stone, J. Hugi, D. DiGiovanni, and P. B. Hansen, *Electron. Lett.* **27**, 1734 (1991).
39. M. E. Fermann, A. Galvanauskas, M. L. Stock, K. K. Wong, D. Harter, and L. Goldberg, *Opt. Lett.* **24**, 1428 (1999).
40. A. Isomaki and O. G. Okhotnikov, *Opt. Express* **14**, 4368 (2006).
41. L. Orsila, R. Herda, and O. G. Okhotnikov, *IEEE Photonics Technol. Lett.* **19**, 2009 (2007).
42. J. D. Kafka, T. Baer, and D. W. Hall, *Opt. Lett.* **14**, 1269 (1989).
43. K. Smith, E. J. Greer, R. Wyatt, P. Wheatley, N. J. Doran, and M. Lawrence, *Electron. Lett.* **27**, 244 (1991).
44. T. F. Carruthers and I. N. Duling, *Opt. Lett.* **21**, 1927 (1996).
45. S. Gray and A. B. Grudinin, *Opt. Lett.* **21**, 207 (1996).
46. A. Boskovic, S. V. Chernikov, and J. R. Taylor, *J. Mod. Opt.* **42**, 1959 (1995).
47. V. J. Matsas, T. P. Newson, D. J. Richardson, and D. N. Payne, *Electron. Lett.* **28**, 1391 (1992).
48. D. J. Richardson, R. I. Laming, D. N. Payne, M. W. Phillips, and V. J. Matsas, *Electron. Lett.* **27**, 730 (1991).
49. C. J. Chen, P. K. A. Wai, and C. R. Menyuk, *Opt. Lett.* **17**, 417 (1992).
50. D. Abraham, R. Nagar, and G. Eisenstein, *Opt. Lett.* **18**, 1508 (1993).
51. X. Shan, D. Cleland, and A. Ellis, *Electron. Lett.* **28**, 182 (1992).
52. D. U. Noske and J. R. Taylor, *Electron. Lett.* **29**, 2200 (1993).
53. I. N. Duling, *Opt. Lett.* **16**, 539 (1991).

54. M. Nakazawa, E. Yoshida, and Y. Kimura, *Appl. Phys. Lett.* **59**, 2073 (1991).
55. V. A. Vysloukh and V. N. Serkin, *JETP Lett.* **38**, 199 (1983).
56. H. A. Haus and M. Nakazawa, *J. Opt. Soc. Am. B* **4**, 652 (1987).
57. J. D. Kafka and T. Baer, *Opt. Lett.* **12**, 181 (1987).
58. A. S. Gouveia-Neto, A. S. L. Gomes, J. R. Taylor, B. J. Ainslie, and S. P. Craig, *Opt. Lett.* **12**, 927 (1987).
59. M. N. Islam, L. F. Mollenauer, R. H. Stolen, J. R. Simpson, and H. T. Shang, *Opt. Lett.* **12**, 814 (1987).
60. A. S. Gouveia-Neto, M. E. Faldon, and J. R. Taylor, *Opt. Lett.* **13**, 770 (1988).

13
Optical Spatial Solitons

13.1 Definition of spatial bright and dark solitons

Spatial (bright and dark) solitons are essentially specialized applications of the studies of self-focusing and self-defocusing, which are described in Chapter 6. In this sense, one may realize that the physical origins for producing optical spatial solitons are exactly the same as that for formation of self-focusing (or self-defocusing) in nonlinear media. In other words, the coherent light beam(s) induced refractive-index change, which is dependent on the spatial intensity (or amplitude) distribution of the incident light field, is the key issue to deal with spatial optical solitons. As indicated in Chapter 5, the specific mechanisms contributing to induced refractive-index changes of nonlinear optical media can be various and dependent on the properties of the media as well as the characteristics of the incident light field.

For a third-order nonlinear medium, the induced refractive-index change can be phenomenologically expressed as

$$\Delta n = n_2 |E|^2 = n_2' I. \qquad (13.1\text{-}1)$$

Here, E and I are the electric field and intensity of the incident light, respectively, and n_2 or n_2' is the nonlinear refractive-index coefficient of the medium. If $n_2 > 0$, the medium may manifest a self-focusing or self-trapping response, whereas if $n_2 < 0$, the medium may manifest a self-defocusing response.

Under appropriate experimental conditions, the beam divergence due to natural diffraction or geometrical optical propagation may be balanced or compensated by the self-focusing response of the medium; therefore, a stationary transverse intensity distribution of the beam can remain unchanged along the whole or considerable propagation length inside the nonlinear medium. This type of process is specially termed the formation of a spatial soliton or spatial solitary wave. In this sense, the spatial soliton simply means a self-trapped beam produced in a self-focusing medium, as addressed by Chiao et al.[1] Usually, the transverse intensity distribution of a spatial soliton beam can be approximately described by a quasi-Gaussian profile or a hyperbolic-secant profile characterized by a central intensity maximum.[2] Hence, this type of spatial soliton is sometimes called a bright spatial soliton to distinguish it from the dark soliton formed in a self-defocusing medium, as described below.

When a broad laser beam, which exhibits a dark notch or narrow stripe in its center with an otherwise uniform transverse intensity distribution, is launched into a self-defocusing medium, under appropriate conditions the divergence of the dark notch or stripe due to diffraction or geometric propagation may be balanced by the self-defocusing effect of the neighboring bright field

Nonlinear Optics and Photonics. First Edition. Guang S. He. © Guang S. He 2015.
Published in 2015 by Oxford University Press.

area. Thereby, a stationary transverse distribution of the dark notch or stripe area can be retained over the whole or considerable propagation length inside the medium. This is the formation of so-called dark soliton in a self-defocusing medium. The difference between a bright soliton and a dark soliton is that the refractive index in the center of the former is higher than that at its boundary, whereas the refractive index in the center of the latter is lower than its boundary.

Theoretically, the formation, propagation, and interaction of bright (or dark) spatial solitons are generally governed by the following nonlinear wave propagation equation (or the so-called nonlinear Schrödinger equation):

$$2ik\frac{\partial E_0}{\partial z} + \nabla_\perp^2 E_0 + 2\frac{k^2}{n_0}\Delta n E_0 = 0, \qquad (13.1\text{-}2)$$

where E_0 is the slow changed amplitude function of the input beam propagating in a nonlinear medium along z-axis, k is the magnitude of the wavevector, n_0 is the linear refractive index of the medium, and Δn is the induced refractive-index change. Most results of spatial soliton experiments described in the following sections supported the simulation calculations based on the above nonlinear differential equation.

13.2 Formation of bright spatial solitons

13.2.1 Nonlinear materials for spatial soliton formation

Roughly speaking, four major types of materials have been employed for producing spatial bright solitons based on their self-focusing capabilities.[3,4] The first is isotropic third-order nonlinear media, such as gases, liquids, and solids, with an induced refractive-index change proportional to the local light intensity (generalized Kerr effect). The second is the second-order nonlinear crystals with an induced refractive-index change proportional to the amplitude of the local optical field (optical Pockels effect). The third is liquid crystals with an induced refractive-index change determined by both the applied lower-frequency electric field and the incident optical field. The fourth is photorefractive materials. The temporal response of induced refractive-index change for the first and second types of media can be either instantaneous or much faster than the duration of the input laser pulses; whereas for third and fourth types of media, the response time of the induced refractive-index change is quite slow, i.e., in the range $\geq 10^{-3}$–10^3 s.

In experiments, if the incident laser beam is pre-focused by an ordinary lens system and then launched into a bulk medium along the z-axis direction, the self-trapping may occur in both x- and y-axis directions. This is the formation of a two-dimensional (2D) bright spatial soliton. In another case, the input laser beam is pre-focused via a cylindrical lens system to generate a narrow bright stripe in the entrance of the sample medium, and the self-trapping can take place only along the direction perpendicular to the bright stripe. This is the formation of a one-dimensional (1D) bright spatial soliton.

Although self-trapping is one of beam self-acting effects related to induced refractive-index changes, the bright spatial soliton formation differs from the usually called self-focusing phenomena. In the former case, the input beam is pre-focused on the entrance of a sample medium with a very small spot size (in several tens of micrometers), and this size will be retained over a propagation length much longer (≥ 5–10 times) than the Rayleigh range or diffraction length (L_d) that is defined by

$$L_d = \pi w_0^2 n_0/\lambda_0, \qquad (13.2\text{-}1)$$

where w_0 is the focused beam's waist size (radius).

However, in usual cases of self-focusing studies, the input beam can be either a pre-focused, or a quasi-collimated, or even a loosely pre-defocused beam. Once the beam's intensity is high enough, either a single focus or multiple foci can be observed along the propagation path; or in other case, a single or multiple filaments may be formed, depending on the input intensity levels and specific transverse intensity distributions.

13.2.2 Spatial soliton formation in generalized Kerr-type nonlinear media

Two-dimensional bright spatial solitons can be produced in Kerr-like bulk nonlinear media, such as metal vapors,[5,6] liquids,[7,8] organic crystals,[9,10] and heavy metal oxide glasses.[11] In the case of metal vapors, researchers may use one-photon absorption resonance to enhance the refractive-index change and to reduce the power requirement for the input laser beam. For transparent liquid and solid media with no resonant interaction, in order to produce enough refractive-index change, pulsed input laser beams in ns- or ps-regime are needed to produce spatial soliton. As an sample, experimental and simulation results for the formation of a 2D soliton beam are shown in Fig. 13.1.[11]

The nonlinear medium was a 5.7-mm long Nb_2O_5–PbO–GeO_2 (NPG) glass rod; the input beam was 820-nm and 25-ps laser pulses with a focused spot size of ~11 μm. When the input pulse energy was low, the spot size measured at the output end was considerablly larger than the input spot size due to diffraction; whereas once the input pulse energy was higher than a level $\geq$2–3 μJ, the output spot size became the same as the input, which was the evidence of a spatial soliton formation, as shown in the bottom of Fig. 13.1. The theoretical stimulation results with taking account of the influence from three-photon absorption (3PA) at the working wavelength are shown in the top of Fig. 13.1.

One-dimensional spatial solitons can be produced in planar waveguide structures based on Kerr-type nonlinear media, such as AlGaAs slab waveguides[12,13] and glass or polymer planar waveguides.[14,15]

In addition to the passive media mentioned above, the bright spatial solitons can also formed in a gain medium such as a laser oscillator or amplifier, as well as a stimulated Raman medium.

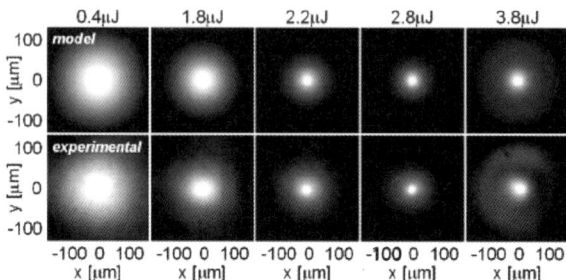

Figure 13.1 *Calculated (top) and observed (bottom) output section images of an 11-μm Gaussian beam passed through a 5.7-mm thick NPG glass rod. The fitting parameters are $n'_2 = 5.5 \times 10^{-15}$ cm^2/W and assumed 3PA coefficient $\beta_3 = 3 \times 10^{-4}$ cm^3/GW^2. (Reproduced with permission from Pasquazi et al.[11] Copyright 2007, Optical Society of America).*

In this special case, a resonance enhanced refractive-index change could be expected, which is related to a lasing transition process or a Raman interaction process.[16–18] Based on the similar consideration, spatial solitons can also be expected in some four-wave frequency mixing (such as third harmonic generation) processes with enhanced refractive-index changes.

13.2.3 Spatial soliton formation in second-order nonlinear crystals

It is well known that for the non-resonant nonlinear interaction in transparent optical media, the second-order ($\chi^{(2)}$ related) nonlinear effects are much easier to be produced than the third-order ($\chi^{(3)}$ related) nonlinear effects. In a second-order nonlinear medium, three-wave frequency mixing may cause not only the energy exchange between different frequency components, but also may lead to the induced refractice-index change and self-focusing effect.[19,20] Based on this consideration, one may expect that the spatial solitons can be efficiently produced in a second-order nonlinear crystal medium;[21,22] the spatial solitons formed are also called quadratic spatial solitons. The advatage of using second-order nonlinear media for generating spatial solitons is the instantaneous time response, because the induced refractive-index change is based on the electron-cloud distortion.

Unlike the third-order nonlinear process, in a second-order nonlinear medium the three-wave mixing-induced refractive-index change is proportional to the amplitude (not intensity) of the optical fields. As a typical example of second-order process induced refractive-index changes, we assume that a fundamental beam of frequency ω and a second-harmonic beam of 2ω are simultaneously launched into a second-order crystal and meet the phase-matcing requirement for second-harmonic generation. Based on the theoretical analyses given in Section 5.7, in type-I phase-matching conditions the induced refractive index change due to the optical Pockels effect can be written as[20]

$$\left. \begin{array}{l} \Delta n^o(\omega) = \dfrac{\chi^{(2)}_{e,(I)}}{2n_0^o(\omega)} 2A(2\omega) \cos\delta\theta \\[2mm] \Delta n^e(2\omega) = \dfrac{\chi^{(2)}_{e,(I)}}{2n_0^e(2\omega)} \dfrac{A^2(\omega)}{A(2\omega)} \cos\delta\theta \end{array} \right\}. \qquad (13.2\text{-}2)$$

Here, $\Delta n^o(\omega)$ and $\Delta n^e(2\omega)$ are the induced refractive-index changes for the fundamental beam (o-ray) and second-harmonic beam (e-ray), $A(\omega)$ and $A(2\omega)$ are the amplitude functions of the two beams, $\chi^{(2)}_{e,(I)}$ is the effective second-order nonlinear susceptibility of the medium under type-I phase-matching conditions, and $\delta\theta$ is the phase difference between these two beams. From the above equation one can see that when $\delta\theta = 0$ or $2m\pi$, where m is an integer, there will be a maximum positive refractive-index change (mutual focusing), when $\delta\theta = m'\pi$, where m' is an odd integer, there will be a maximum negative change (mutual defocusing), whereas if $\delta\theta = m'\pi/2$, there is no induced refractive-index change but there should be a most effective energy transfer between the two beams. The physical meaning of Eq. (13.2-2) is that there is a mutually induced refractive-index change between the two beams, and both beams can be self-focused or self-defocused, depending on the field amplitudes, their transverse distributions, and the value of $\delta\theta$. Similar expressions for induced refractive-index changes under type-II phase-matching conditions can also be obtained. The same considerations of induced mutual refractive-index changes are applicable to other second-order nonlinear wave-mixing processes, including the sum-frequency process and down-frequency parametric process.

The first experimental demonstration of 2D quadratic bright spatial soliton formation was reported in 1995 by Torruellas et al.[23] In this experiment, the nonlinear medium was a 1-cm long KTP crystal; a strong 1064-nm fundamental beam and a weak 532-nm second-harmonic generation (SHG) beam were colaunched into the crystal under type-II phase-matching conditions. The pulse duration of both beams was ~15 ps, and the 532-nm beam was passed through a 30-cm long gas cell filled with variable pressure nitrogen to contral the relative phase between the two beams. The pump beam was focused on the input end of the crystal with an ~20-μm beam waist and ~2-mm Rayleigh range. Without seeding the 532 nm beam and at a low intensity level, the output spot size of the 1064 nm beam at the exit end of the crystal was increased to ≥ 80 μm, as shown in Fig. 13.2 (a); when a 532 nm beam was seeded in phase with the 1064 nm beam and the input intensity level of the latter was higher than 5–10 GW/cm^2, the output spot sizes of both beams could be compressed to ~12.5 μm, as shown in Fig. 13.2(b), which is the evidence of spatial soliton formation. However, when the 532 nm beam's phase was shifted 180° with respect to the 1064 nm beam, there was no formation of self-trapping, as shown in Fig. 13.2(c). Under basically the same experimnetal conditions, by using a strong 532-nm pump beam and a weak 1064-nm seed beam, the formation of spatial solitons for both beams could also be observed.[24]

Later experimental results further showed that in quasi-phase-matching or noncritical phase-matching conditions, if even only a strong fundamental beam was focused onto the SHG crystal (without seeding a 2ω beam), a mutual self-trapping for both the fundamental beam and the generated second-harmonic beam could be observed, provided that the input intensity of the fundamental beam was higher than certain threshold levels.[25,26] In this case, the initial 2ω signals can be thought of as provided by the quantum noise of the electromagnetic fields. Thus generated second-harmonic components may have different relative phase-relations with regard to the input fundamental and, therefore, both the energy transfer and coupled refractive-index

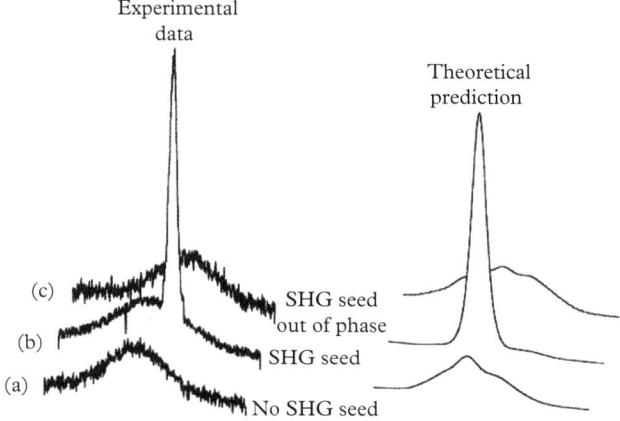

Figure 13.2 *Transverse profiles of the fundamental beam with no second harmonic seeded (a), with the second harmonic seeded in phase with the fundamental showing the formation of a solitary beam (b), and the relative phase between the two input beams is 180° (c). Left: experimental results; right: simulation results. (Reproduced with permission from Torruellas et al.[23] Copyright 1995, American Physical Society).*

changes may take place together. Similarly, based on a down-conversion single-pass parametric generation $(\omega_p \Rightarrow \omega_s + \omega_i)$ process, occurring in a 15 mm-long LBO crystal pumped by a strong 0.527 μm beam of 1.5 ps pulse duration and working in type-I temperature-tuned noncritical phase-matching conditions, the mutual self-trapping of both the pump beam of ω_p and the down-conversion signal beam of ω_s was observed.[27,28]

13.2.4 Spatial soliton formation in liquid crystals

The optical field-induced refractive-index change in a nematic liquid crystal (LC) is based on the reorientation of molecular directors of the LC medium. Owing to elastic forces between adjacent molecules and the boundary interaction with walls of the cell containing the LC, the induced refractive-index change exhibits a significant nonlocality in both temporal and spatial responses, i.e., the refractive-index change in a given location and at a given time, $\Delta n(x, y, z, t)$, is not simply determined by the local light intensity at the same time, $I(x, y, z, t)$. The response time of Δn in a LC medium for soliton formation is usually in the range of several tens of seconds.[29] However, the technical merit of utilizing a LC for a spatial soliton experiment is that, by applying a low-frequency and low voltage ac field, the nonlocal nonlinearity of a LC sample can be so large that the spatial soliton formation could be observed by using a cw laser beam of power in mW range.[30–32]

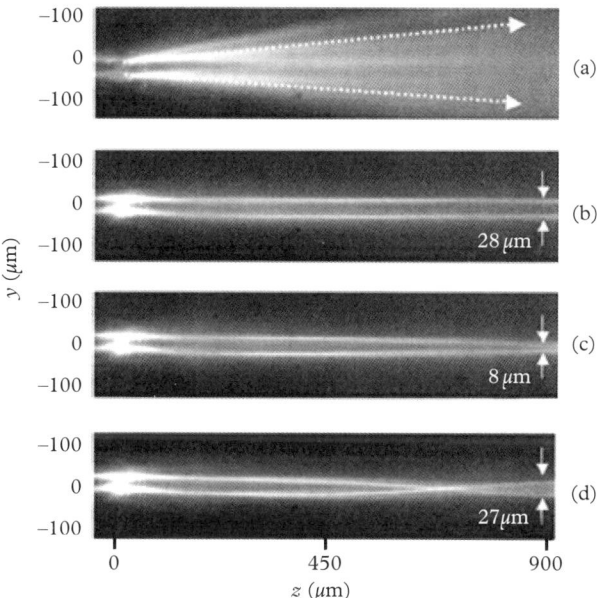

Figure 13.3 *Formation of two spatial solitons in a 0.9 mm-long planar E7 liquid crystal cell. The two beams (10 μm waist each) with a 28 μm y separation and a 5° relative angle are launched: (a) linear propagation behavior, (b) week attraction between solitons of 2.8 mW power each, (c) stronger attraction at 3.6 mW, and (d) crossing at power 4.5 mW. (Reproduced with permission from Peccianti et al.[30] Copyright 2002, Optical Society of America).*

The standard nematic E7 LC sample, which is usually employed for spatial soliton formation experiments, is filled in a planar cell with an optical entrance window. One may apply a low-frequency electric field along the perpendicular (x) direction of the thick LC layer to pre-orientate the molecules at ~45° to the z direction to maximize the effective reorientational nonlinearity.

The photographs shown in Fig. 13.3(a)–(d), which were taken above the top plate of a 75-μm thick and 0.9-mm long planar LC cell, are the scattering images of two laser beams linearly polarized along the x-axis and propagating along the z-axis through the LC layer.[30] The two equal-power input beams of 514.5 nm wavelength were from a cw Ar$^+$ laser, focused by a 20× objective onto the entrance end of the cell with a spot size of ~10 μm, y-separation of 28 μm, and a crossing angle of 5°. At a very low power level, the two beams diverge rapidly due to linear diffraction (Fig. 13.3 (a)). When the power level is increased to 2.8 mW, there is a clear formation of two parallel solitons (Fig. 13.3(b)). When the power level is increased further, the two spatial solitons attract each other and may even cross with each other (Fig. 13.3(c) and (d)).

13.2.5 Spatial soliton formation in photorefractive crystals

The origin of light-induced refractive-index change in a photorefractive (PR) medium is different from that in a Kerr-like medium or in a second-order transparent crystal. In a PR medium (usually an inorganic electro-optic crystal doped with some impurity), the mechanism of light-induced refractive-index change can be very complex, but it is basically related to the generation, separation, transport, and retrapping of space charges created by light. As a result of these processes, a light-induced space-charges field is formed inside the crystal medium, which consequently give rise to the refractive-index change due to linear electro-optic (Pockels) effect. As it involves space-charge diffusion and drift, the induced refractive-index change in a PR medium exhibits the nature of uninstantaity and nonlocality, which is similar to the case of liquid crystals.

In the first theoretical analysis of spatial soliton formation in a PR medium published in 1992 by Segev et al., a PR crystal group possessing large electro-optic nonlinearity with an externally applied dc electric-field was suggested.[33] One year later, the first experimental demonstration of beam self-trapping in a PR crystal was reported by Duree et al.[34] In this experiment, the sample medium was strontium barium niobate (SBN) crystal with an applied dc field of 400–500 V/cm, the input was a focused 457-nm cw Ar$^+$ laser beam with a spot size of 70–80 μm at the entrance end of the sample, and beam self-trapping was only observed in a short time "window" (~130 ms) along the entire temporal response process. In this sense, it was an example of transient (or so-called quasi-steady-state) spatial soliton formation in a PR medium. Furthermore, researchers found that, in addition to the externally applied dc field, an auxiliary uniform light illumination upon the sample medium could be beneficial to realize steady-state solitons and to conveniently control their behavior, because in this case the induced refractive-index change dependent on the ratio of $\eta = I_0^{max}/I_{bg}$, where I_0^{max} is the maximum intensity of the input signal beam and I_{bg} is the background illumination intensity.[35,36] Under this condition, the self-trapping of a focused input signal beam is essentially due to non-uniform screening of the external dc electric field, so that this type of steady-state soliton formation in a PR crystal is also called screening spatial soliton formation.

As an example of 2D screening soliton formation, Fig. 13.4 shows the experimental results based on a 5.5 mm-long SBN:60 crystal.[36] A focused 514-nm cw laser beam of μW power was launched into the crystal along its a-axis and linearly polarized along the c-axis, a dc field was applied along the c-axis with a value of ~3400 V/5.5 mm, the background irradiance with intensity levels at W/cm^2 was provided by a laser beam illuminating the crystal uniformly and polarized orthogonally to the soliton beam, and the intensity ratio η was chosen to be ~70. Figure 13.4 shows

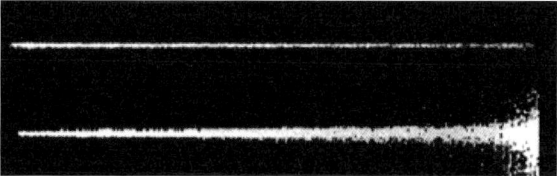

Figure 13.4 *Top-view photograph of a 10 μm soliton (upper trace with applied dc field) and a normally diffracting beam (lower trace without applied dc field) propagating along a 5.5 mm-long SBN crystal. (Reproduced with permission from Shih et al.[36] Copyright 1996, Optical Society of America).*

the top-view scattering image of the laser beam propagating through the 5.5 mm path length of the sample; the upper trace is the magnified image of the self-trapping beam with the applied dc field, while the lower trace is the image of linearly divergent beam without applying the dc field. The focused beam spot size at the input end was ~10 μm and the buildup time of steady-state soliton formation in SBN was estimated to be 0.1–1 s.[36,37] Since the response time of the PR medium is inversely proportional to the intensity of the input optical signal, a faster temporal response may be expected when using a pulsed laser beam with much higher intensities. It was experimentally shown that, by using a 532-nm and 8-ns pulsed laser of 100 MW/cm² as the focused signal beam, and a broad 532-nm beam of 10 MW/cm² as the background beam uniformly illuminating a SBN sample, the buildup time of steady-state soliton formation was estimated to be ~80 ns.[38]

It should be pointed that for a given focused input laser beam, once a stable spatial soliton configuration is formed in a PR medium, the beam's transverse phase distribution is uniform at exit face of the medium.[34,36] This fact implies that the output beam will further propagate in a linear medium as a quasi-plane wave.

13.2.6 Formation of spiraling spatial solitons

Spiraling bright spatial solitons can be generated by using the optical vortex technique. Usually, an optical vortex can be a special light beam with a doughnut- or ring-shape intensity distribution, its phase rotates around the dark center of the beam in the plane perpendicular to the propagation direction. In general, it can be describes as

$$E(r, \varphi, z) = E(r, \varphi) \exp(-ikz), \quad (13.2\text{-}3)$$

where $E(r, \varphi)$ is the amplitude function, r is the transverse radial variable, φ is the polar angle around the optical axis in the plane perpendicular to the propagation direction, and k is the modulus of the wavevector. For example, a φ-dependent amplitude function may take the form[39]

$$E(r, \varphi) = E_0 \frac{r}{w_0} \exp\left[-(r/w_0)^2\right] \exp(ip\varphi). \quad (13.2\text{-}4)$$

Here, w_0 is the waist size characterized by the Gaussian function and $p = \pm 1, 2, \ldots$ is an integer that is referred to as the topological charge of the optical vortex. A real optical vortex beam can be generated via different technical approaches. One of them is a monochromatic and quasi-plane wave passing through a special holographic mask, which can be artificially made by computer simulations of the interference fringes produced by a regular plane wave and a designed vortex wave, like that described by Eq. (13.2-4).[39] According to the holograph principle, when a plane wave passes through such a mask, the vortex wave will be restored by an appropriate diffraction order.

The early experiments using an optical vortex beam to produce bright spatial solitons were performed in a rubidium (Rb) vapor system by using a cw Ti:sapphire laser beam of ~780 nm wavelength close to the rubidium D_2 line.[40,41] In this case, the cubic nonlinearity and the induced refractive-index change can be significantly enhanced by one-photon absorption resonance. To obtain an enhanced positive refractive-index change, the input laser frequency was tuned close enough to the hyperfine transition $5S_{1/2}(F=3) \rightarrow 5P_{3/2}(F=4)$ on the high-frequency side. In addition, the nonlinearity of this system can also be controlled by changing either the temperature of the 20 cm-long Rb vapor or the input beam power. In one of these experiments, the original laser beam was linearly polarized and produced a slightly elliptical TEM_{00} mode with a ratio of the major to minor axis of ~1.15; this beam then passed through a computer-generated holographic amplitude mask to create a vortex beam with a unit topological charge ($p = 1$) and a slightly asymmetric azimuthal distribution, as shown in Fig. 13.5(a).

When the input beam power is 17 mW, frequency detuning $\Delta \nu = \nu - \nu_0 = 20$ GHz, (i.e., far from the resonance enhancement), and the cell temperature is 81 °C, the beam basically propagates linearly through the cell and the image measured at the output window is as shown in Fig. 13.5(b), which indicates the overall divergence and the more clear azimuthal asymmetry of the output beam. This image reveals an enhanced local-intensity non-uniformity due to the partial

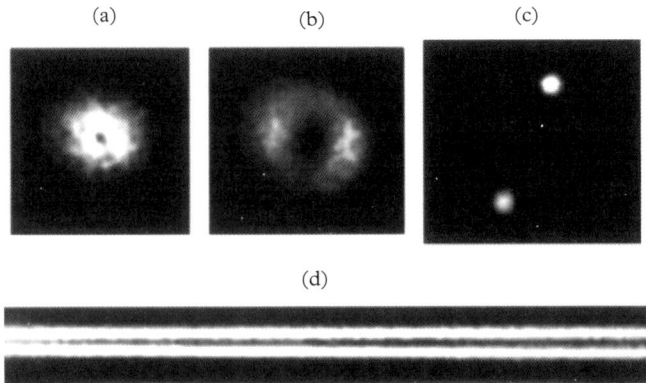

Figure 13.5 *Experimental beam profiles recorded (a) at the input to the Rb vapor cell, (b) at the output window after linear propagation (laser frequency detuning 20 GHz), and (c) at the output window after nonlinear propagation (frequency detuning 0.5 GHz). The scale, which is the same for (a)–(c), corresponds to a 1 mm × 1 mm window at the image (c), cell temperature 81 °C. (d) Image of nonlinear propagation of solitons along the cell by means of resonance fluorescence, image size 2 mm × 20 mm, offset from the input of the cell ~70 mm, cell temperature ~100 °C, other parameters as in case (c). (Reproduced with permission from Tikhonenko et al.[40] Copyright 1995, Optical Society of America).*

self-focusing response of the nonlinear medium. Under the same conditions, when the input laser frequency ν is much closer to the resonant transition frequency, i.e., $\Delta \nu = 0.5\,\text{GHz}$, the nonlinearity of the medium can be so strong that the two bright regions become two sharp bright spots, as shown in Fig. 13.5(c). The appearance of these two spots is evidence of the beam self-trapping and the formation of a pair of spatial solitons. It should be noted that the azimuthal angle of these two spots is rotated around the optical axis with respect to the position of the original two bright regions.

It has been shown experimentally that when the sign of the topological charge p changes (by choosing the opposite diffraction order from the holographic mask), the rotation direction of such a soliton pair is also reversed. To indicate the soliton feature of these two self-trapping beams, the top-view image obtained by means of resonance fluorescence of Rb vapor is shown in Fig. 13.5(d), which was taken at $\sim$70 mm distance from the input window and at a cell temperature of $\sim$100 °C.[40]

Similar experiments using an optical vortex to generate spatial solitons were also implemented in photorefractive media.[42,43]

13.3 Formation of dark spatial solitons

In the preceding section, we have described bright spatial soliton formation in different types of self-focusing media that possess a light-induced positive refractive-index change. Now, in this section, we consider dark spatial soliton formation in some self-defocusing media that possess a light-induced negative refractive-index change. In the latter case, a broad and otherwise uniform light beam, which contains a narrow dark stripe (or notch) in the beam transverse section, is launched into a self-defocusing medium. If the diffraction-induced linear divergence of the narrow and sharp dark region(s) can be balanced by the self-defocusing-induced expansion of the surrounding bright regions, the original dark region(s) can remain unchanged throughout the whole propagation length in the nonlinear medium. This is the formation of dark spatial soliton(s). The physical picture of dark spatial soliton formation is quite straightforward: the refractive-index change in the dark region is negligible, while the negative refractive-index change in the bright background regions is considerably large. As a result of self-defocusing, the bright illumination regions exhibit a tendency to expand their inner boundaries into the dark region, so that the natural diffraction of the latter can be balanced. That is, the diffraction of the original dark region is restrained by the extension of the bright region due to self-defocusing. In more general cases, the input beam(s) may contain multiple structures of dark stripes (or notches), and under appropriate conditions researchers may produce multiple dark solitons (or a dark soliton array).

Light-induced negative refractive-index change $[\Delta n = (n' - n_0) < 0]$ can be realized mainly in the following types of nonlinear optical media.

i. Third-order isotropic media with $n_2 < 0$ nonlinearity responding to the input light frequencies (see Eq. (13.1-1)). In these cases, as previously described in Chapter 5, the magnitude and sign of n_2 can be effectively controlled by using one-, two-, or multi-photon absorption resonance enhancement as well as Raman resonance enhancement, through the input frequency-tuning from the appropriate side to the resonant frequency of a given medium.

ii. Second-order anisotropic crystals with $\Delta n < 0$ through second-harmonic generation or other three-wave parametric processes (e.g., see Eq. (13.2-2)). In these cases, the sign and magnitude of the induced refractive-index change can effectively be controlled by changing the relative phase difference between the input two beams.

iii. Photorefractive (PR) media with $\Delta n < 0$ that can be obtained under appropriate arrangements of input light polarization, crystal orientation, and the polarity of applied dc voltage.

The early experiment that demonstrated spatial dark soliton formation in a third-order nonlinear medium was implemented in a sodium (Na) vapor system by using a cw dye laser beam of power ≤ 100 mW.[44] An $L = 18$ mm-long vapor cell contained $\sim 10^{12}$ atoms/cm^3 of Na vapor; the laser frequency was tuned, on the low-frequency side, close to the D_2 line of the atomic resonance to obtain an enhanced negative refractive-index change. In the experiment, the laser beam was focused on the entrance of the cell with a beam waist of $2w_0 \approx 440$ μm, and a square mesh aperture with a 160 μm wire diameter and 160 μm wire spacing was placed in the entrance plane. The output beam pattern could be measured in a distance of ≥ 1 m (far-field position), which manifested a rectangular array of spots. When the laser frequency was far from the resonance frequency of the vapor medium, $n_2 \approx 0$, and after linearly passing through the vapor cell, the measured far-field pattern was as shown in Fig. 13.6(a). When the laser frequency was tuned from the low frequency side close enough to the resonance, the measured output far-field pattern was as shown in Fig. 13.6(b). In the latter case, we can see a much clearer spot structure, separated by a series of crossing sharp dark stripes, namely, the dark spatial solitons. For comparison, Fig. 13.6(c) shows results calculated by simulating the beam propagation through a self-defocusing medium with the nonlinear parameter similar to the real experimental condition. One can see that both results are in good agreement.

Later, more studies have been implemented mainly in PR media because of the easy control of the sign and magnitude of their nonlinearity and the low requirement for the input laser power.[45-49] Once again, in these cases the dark spatial soliton formation time is still very slow, sometimes on the scale of seconds or even minutes when a cw laser beam is utilized. To effectively produce a negative Δn response to the applied optical intensity, two types of PR crystals can be employed for dark spatial soliton experiments. One is the SBN crystal that is also commonly used for producing bright spatial solitons. The other is the LiNbO$_3$ crystal, which possesses both PR and photovoltaic properties and manifests a light-intensity induced negative refractive-index change without the need to apply an external dc field.

In the case of using SBN crystal for soliton formation, the sign of the induced Δn is determined by the polarity of the external dc voltage applied along the crystalline c-axis. To provide

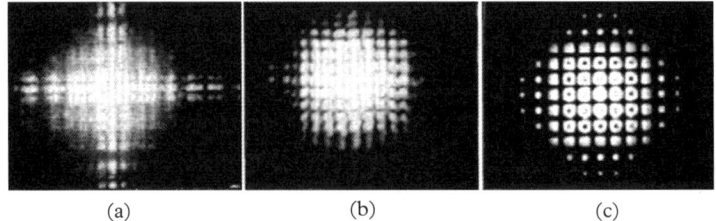

(a) (b) (c)

Figure 13.6 *(a) Measured far-field image after linearly passing through the Na-vapor cell. (b) Measured far-field image after nonlinearly passing through the cell. (c) Theoretical simulation of far-field distribution after passing through a self-defocusing medium. (Reproduced with permission from Swartzlander et al.[44] Copyright 1991, American Physical Society).*

a self-defocusing nonlinearity for dark soliton formation, a negative $(-V)$ bias voltage should be applied along the c-axis, while the input beam should be linearly polarized along the same c-axis with a dark stripe (or notch) orthogonal to this direction. In the first experiment on dark soliton formation in a SBN crystal, a cw Ar$^+$ laser beam of 457 nm wavelength was input into an ∼6 mm-long crystal sample with e-ray polarization.[46] Before launching the input beam into the sample, a thin glass slide was inserted in one half of the beam, by slightly tilting this glass plate until there was a π-phase jump between the two portions of the beam; thus a 1D dark notch could be formed through the center of the beam. The transverse intensity profiles at different propagation positions inside the PR medium could be measured by the combination of a magnifying optical system and a CCD array. In the experimental results shown in Fig. 13.7, the curves in the left column were obtained without applying a bias voltage, whereas those in the right column were recorded with a −400 V/cm voltage applied.

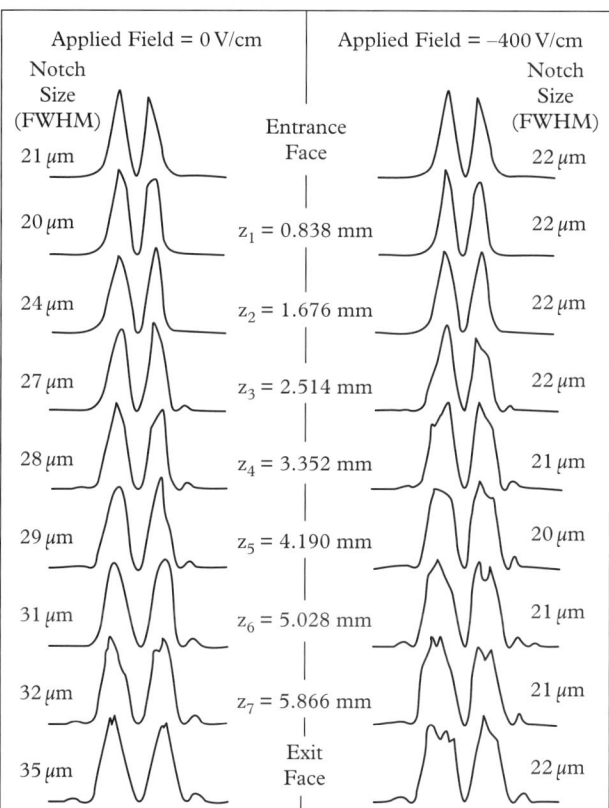

Figure 13.7 *Normalized transverse profiles of the linearly diffracting beam (left column) and the dark soliton beam (right column) versus propagation distances in a 6 mm-long SBN crystal. The size of the dark notch (FWHM) is given for each position. (Reproduced with permission from Duree et al.[46] Copyright 1995, American Physical Society).*

From Fig. 13.7 one can see that without applying an external dc field, the sizes of both the dark ridion and bright region are considerably increased after linear propagation through the crystal, whereas with applying appropriate negative bias voltage, the size of the dark notch remains unchanged throughout the whole propagation length in the medium. In addition, it should be noted that in the latter case, the size increase of the bright regions is greater than the natural diffraction; this is due to the self-defocusing of the bright region. These observations manifest a transient or quasi-steady-state feature as they can stay within a time window ≥ 2 s, and basically are independent of the input beam's intensity (from 0.3 to 30 W/cm^2).[46]

However, shortly thereafter, under similar experimental conditions but applying additional background illumination onto the same SBN crystal, researchers were able to generate steady-state dark screening PR solitons.[48,49]

Furthermore, in the case of utilizing LiNbO$_3$ crystal as a self-defocusing medium, the light field-induced refractive-index change can be written as[45,50]

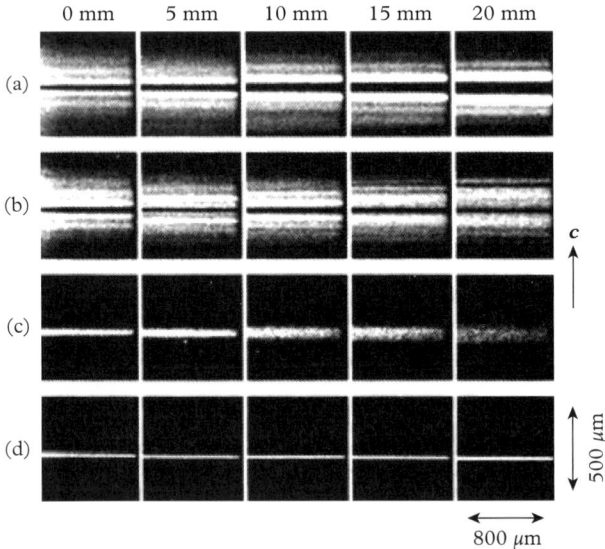

Figure 13.8 *The experimental results for dark soliton formation in a photovoltaic LiNbO$_3$ crystal. The side views of the light illumination inside the crystal at different propagation distances for the 488 nm beam of $\sim$0.6 mm diameter with a dark notch: (a) in the initial condition before soliton formation; (b) in the steady state after dark soliton formation. The corresponding side views of a separate 514 nm probe beam launched collinear to the 488 nm beam is shown in (c) before the soliton formation and (d) as guided by the waveguide left behind by the 488-nm formed dark soliton. The images were obtained at 5 mm intervals and show a width of 800 μm. (Reproduced with permission from Taya et al.[50] Copyright 1995, American Physical Society).*

$$\Delta n = -\frac{1}{2}n_0^3 r_e \gamma_p \frac{I}{I+I_d}, \tag{13.3-1}$$

where n_0 is the linear refractive index, r_e is the effective electro-optic coefficient, γ_p is the photovoltaic field constant, I is the input bright illumination intensity, and I_d is the dark irradiance intensity of the crystal. In a typical experiment conducted by using LiNbO$_3$ crystal to generate the dark spatial soliton, a cw Ar$^+$-laser beam of 488 nm wavelength, polarized along the crystalline c-axis, was input into the crystal in the direction perpendicular to the c-axis.[50] Since the induced negative refractive-index change is most sensitive to the transverse intensity gradient along the c-axis direction, the dark notch created by an inserted glass slide (π-phase jump plate) is orientated perpendicular to the c-axis.

Figure 13.8(a) shows the side-view images of the 488-nm beam after scattering, taken at different propagation distances in the initial condition before dark soliton formation. For an average bright illumination intensity of $\sim$10 W/cm^2, the buildup time is $\sim$15 min and subsequently the measured corresponding images, which are shown in Fig. 13.8(b), indicate the existence of a 1D dark soliton. It is known that after the dark soliton formation, the relative refractive-index value in the center of the dark soliton is higher than that at boundaries; therefore, an existing dark soliton can be recognized as a waveguide for another probe beam of different wavelength, whose influence on the induced refractive-index change can be neglected. In the same experiment, a 514-nm laser beam was collinearly launched into the sample to probe the existence of the dark spatial soliton formed by the above-mentioned 488 nm beam. The measured side-view images of the 514 nm probe beam, which are shown in Fig. 13.8(c), were taken in the initial condition with a much short exposure time (seconds), whereas, the corresponding images taken after formation of the dark soliton, which are shown in Fig. 13.8(d), indicate the waveguide propagation of the probe beam.

Similar experimental results were also obtained in a planar LiNbO$_3$ waveguide configuration.[51]

13.4 Spatial soliton interactions and applications

13.4.1 General features of spatial soliton interactions

So far, we have mainly considered the formation of a single spatial soliton generated by a single input beam. Now we shall further consider the interaction between two existing solitons. There is a great richness of interaction behaviors of spatial solitons, including bending, fission, and annihilation for a given soliton, as well as the pulling, repulsion, fusion, or spiraling between two (or more) solitons. For example, a single input beam with an asymmetric intensity gradient across the beam's transverse section may generate a bending spatial soliton in an appropriate nonlinear medium. Also, with a non-uniform intensity or transverse intensity modulation, the input beam may generate a bifurcating soliton. When two soliton beams in the same plane undergo a collision, they may attract or repel each other. Depending on the properties of the interaction and the nonlinear medium, the two solitons may fuse into one soliton or create an additional new soliton beam. If the interaction takes place between two soliton beams propagating not in the same plane, this may lead to the formation of a pair of spiraling spatial solitons.

There are two types of soliton interaction: coherent and incoherent. Coherent interaction means that the buildup time of the refractive-index change (Δn) is much shorter than the coherent time for each soliton beam. Therefore, the specific interaction behavior depends on the relative phase-relationship between the two involved soliton beams. In this case, one should take account of the interference effect between these two beams. On the contrary, if the response time Δn for a given medium is much longer than the phase-retaining time of the soliton beams, or the time delay between two soliton beams is longer than their coherent time, one will deal with incoherent interaction, which means that the interaction behavior between two soliton beams is independent of their relative phase-relationship. Therefore, the interference effect between them plays no role. Usually, coherent interaction occurs between spatial solitons generated in Kerr-type isotropic third-order nonlinear media and second-order nonlinear crystals, whereas incoherent interaction takes place between solitons produced in liquid crystals and photorefractive materials.

With regard to applications of spatial solitons, the first intuitive thought is to fabricate a temporary, a semi-permanent, or even a permanent waveguide structure inside a proper nonlinear medium. This method has proved to be practical and can be adopted for three-dimensional (3D) optical routing and spatial multiplexing. The other promising potential application is to use spatial soliton techniques to implement the functions of all-optical switching, steering, interconnecting, and readdressing in advanced optical circuits or multi-channel optical communication systems. In addition, spatial soliton techniques can be useful for optical data processing and optical computing.

In the remaining part of this section, we shall briefly discuss some experimental results that have shown the basic features of spatial soliton interactions as well as the feasibility for future applications.

13.4.2 Soliton interactions in Kerr-type media

The interaction (collision) between two spatial soliton beams has been experimentally investigated in planar glass waveguides. In these cases, the behavior of the two-soliton interaction depends on their relative phase relation (for coherent interaction) as well as their spatial separation, crossing angle, and intensity ratio (for incoherent interaction). As a typical example for coherent interaction, Fig. 13.9 shows the experimental results of two parallel soliton beams propagating in a 3–4 μm-thick and 5 mm-long glass planar waveguide.[52] The two input beams with the same intensity were from a colliding-pulse mode-locked dye laser system of wavelength 620 nm, pulse

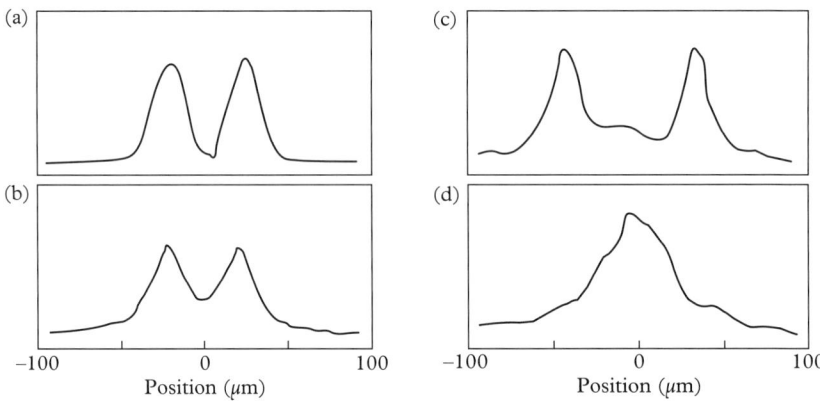

Figure 13.9 *Soliton interaction in a planar glass waveguide for two 22-μm FWHM solitons initially separated by 45 μm: (a) input light distribution, (b) output light distribution when the two beams do not overlap in time, (c) interacting solitons with π relative phase shift, (d) interacting solitons with zero phase shift. (Reproduced with permission from Aitchison et al.[52] Copyright 1991, Optical Society of America).*

duration 100 fs, and repetition rate 8.6 kHz. The input beams were focused at the entrance end of the waveguide with a spot size of 22 μm (FWHM) and peak power of 360 kW (1.5 times the soliton formation power) for each of them. The two beams were launched parallel to each other and were separated by an adjustable distance comparable to their beam size. The relative phase shift between these two beams was controlled by adjusting their relative optical-path delay.

Figure 13.9 shows the measured relative intensity distributions at the input end (a) and at the output end (b–d) under different relative phase-relation conditions. It can be seen that for a relatively small initial separation of 45 μm (left column), when the two beams are incident with zero phase shift, as a result of constructive interference, the two solitons are attracted to each other and partially fused (see (d)). When two beams have a π phase shift, as a result of destructive interference the two soliton beams repel each other and their separation at the output end is increased (see (c)). Under the condition of a larger initial separation of 55 μm, similar behavior was observed but the interaction between the two soliton beams is weaker than the former case; both the increase of separation with π phase shift and the fusion with zero phase shift are not as obvious as that observed for a small initial separation.

For incoherent interaction of two soliton beams in a planar glass waveguide structure, for example, in the case of two input beams polarized orthogonally each other, the interaction is

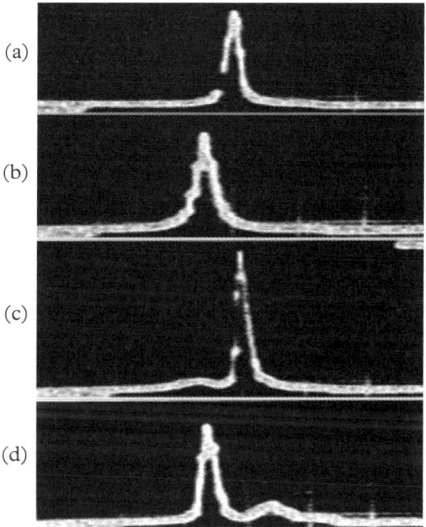

Figure 13.10 *Experimental demonstration of soliton switching in CS_2 liquid. Output beam intensity shapes for separately incident solitons are shown in (a) and (b), and for simultaneous launching with an initial phase shift of $\Delta \varphi = 2.3$ rad (c) or $\Delta \varphi = -2.3$ rad (d). (Reproduced with permission from Shalaby and Barthelemy.[56] Copyright 1991, Optical Society of America).*

phase-insensitive and always leads to an attraction (dragging) of one beam by the other.[53–55] Based on this feature, one may realize the all-optical switching and ultrafast deflection.

CS$_2$ is a widely studied Kerr liquid that possesses the largest molecular reorientational nonlinearity with a Δn buildup time $\leq 10^{-12}$ s. The coherent interaction of two soliton beams propagating in a CS$_2$ liquid sample with a small crossing angle was investigated by using two equal-intensity pulsed laser beams of 532 nm wavelength, 30 ps pulse duration, and 1 GW peak power.[56] In this experiment, two input beams were focused into two sharp elliptical spots 60 µm wide and 10 mm high. These two spots were completely superimposed at the input face of the nonlinear medium but with a small crossing angle ($\Delta\theta$). When $\Delta\theta \leq 3$ mrad and the relative phase shift $\Delta\varphi = 0$, the two soliton beams fused into one output beam. However, as shown in Fig. 13.10, for the case of $\Delta\theta = 1.8$ mrad and $\Delta\varphi = \pm 2.3$ rad, there is an efficient energy exchange between the two soliton beams, i.e., the energy is switched to one soliton or the other, depending on the sign of the relative phase shift (see Fig. 13.10(c) and (d)). Obviously, this type of phenomenon can be applied to ultrafast all-optical switching, addressing, or modulation.

Soliton interaction behaviors have also been studied experimentally in liquid crystal samples by applying an external bias electric field.[57,58] In these cases, both coherent and incoherent interaction behaviors between two soliton beams can be observed with features of slow Δn buildup time and strong nonlocal nonlinearity.

13.4.3 Soliton interactions in second-order nonlinear crystals

The mechanism of light-induced Δn in second-order nonlinear media through three-wave frequency mixing is based on the electron-cloud distortion with an extremely fast temporal response ($\leq 10^{-14}$–10^{-15} s), and the magnitude and sign of Δn depend sensitively on the phase relationship of the optical fields involved. So, here we shall deal with the typical situation of coherent interaction of two or multiple spatial solitons generated in this type of media.[59–61]

First we consider the collision between two non-coplanar soliton beams propagating with a small crossing angle in a second-order crystal.[60] The sample medium is an 11.4 mm-long KNbO$_3$ crystal, working in non-critical type-I phase-matching conditions. The two input beams of fundamental wavelength 983 nm are from a Q-switched and mode-locked Nd:YAG laser pumped optical parametric generator, and the pulse duration and repetition rate are 22 ps and 10 Hz respectively. The spot size of each focused input beam is ~18 µm, and at the input end the vertical and horizontal separations are ~10 µm and ~80 µm, respectively; the crossing angle between these two beams is ~0.9°. To create a soliton beam at the fundamental wavelength, the intensity needed is ~3 GW/cm^2, whereas for a two-beam collision experiment, the intensity for each beam is set to 4.5 GW/cm^2.

It was experimentally found that when the two beams were launched in-phase (with zero phase shift), one could see a newly created soliton beam of fundamental frequency propagating along the median direction while there was a small rotation of the two transmitted soliton beams. Figure 13.11, which shows the output patterns recorded at different phase-shift values between the two beams varying from 0 to 2π, indicates the moving behavior of the third soliton beam between the other two beams.

The other interesting example is an experimental demonstration for logical signal processing, which is based on the coherent collision between two crossing quadratic spatial solitons.[61] In such an experiment, as shown schematically in Fig. 13.12, two coplanar beams of 1548 nm fundamental wavelength, 4 ps pulse duration, 20 MHz repetition rate, and few kW peak power are launched with a small crossing angle (~0.4°) into a 63 mm-long and 10 mm-wide planar periodically

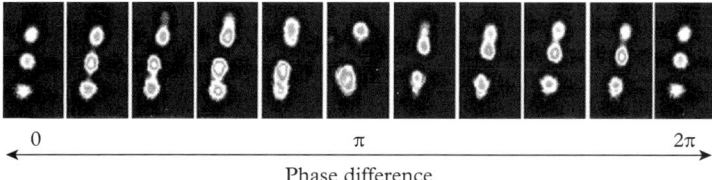

Figure 13.11 *Output soliton pattern versus the relative phase difference between two input beams. (Reproduced with permission from Jankovic et al.[60] Copyright 2003, Optical Society of America).*

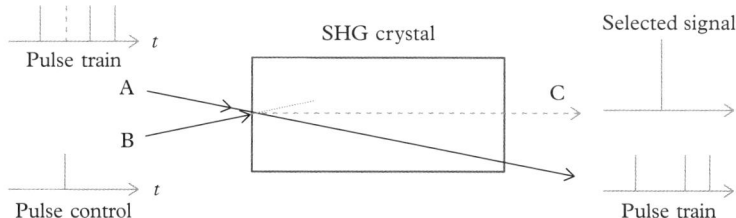

Figure 13.12 *Pulse selection by simultaneously launching beam A (containing a pulse train) and beam B (containing a single control pulse) into a second-order nonlinear crystal with a small crossing angle. A third beam C (containing the selected pulse signal) is generated along the median direction. (Adapted from Pioger et al.[61] Copyright 2004, IEEE).*

poled Ti-diffused lithium niobate (Ti:PPLN) slab waveguide, working in SHG configuration with temperature–tuning phase matching.

After being focused each beam is shaped in a highly elliptical spot, with a 60 μm size along the waveguide plane and 3.9 μm size along the perpendicular direction. For a zero phase difference between the input two beams, the constructive interference leads to a new single quadratic solitary wave propagating along the median direction between two input beams directions. In experiment, one input beam (A) contains a series of laser pulses while the other beam (B) contains only one single pulse. By adjusting the optical delay between these two beams such that the single pulse from beam B is coincident with one selected pulse from beam A, a new soliton beam with the same fundamental wavelength can be observed along the median direction provided that the phase shift between these two pulses is zero. This experiment essentially demonstrated an all-optical spatial switching with high efficiency at a repetition rate of 125 GBit/s.

13.4.4 Soliton interactions in PR media

In terms of slow response and nonlocality of the nonlinearity, the behaviors of soliton interaction in PR media are similar to that in liquid crystals, i.e., both coherent and incoherent collisions between two soliton beams can be observed. In the case of coherent interaction, two beams attract each other if they are in phase, or repel each other if they have a π phase shift. If no interference effect between the two beams should be considered, e.g., two beams are polarized orthogonally

428 *Optical Spatial Solitons*

or not spatiotemporally overlapping, we shall see an incoherent interaction between such beams. In this case, the two bright soliton beams always attract each other in a self-focusing PR medium.

As a typical experimental example, the optical setup for two-beam coherent interaction in a PR medium is shown at the top of Fig. 13.13.[62] In this experiment, two 6328 nm beams from the

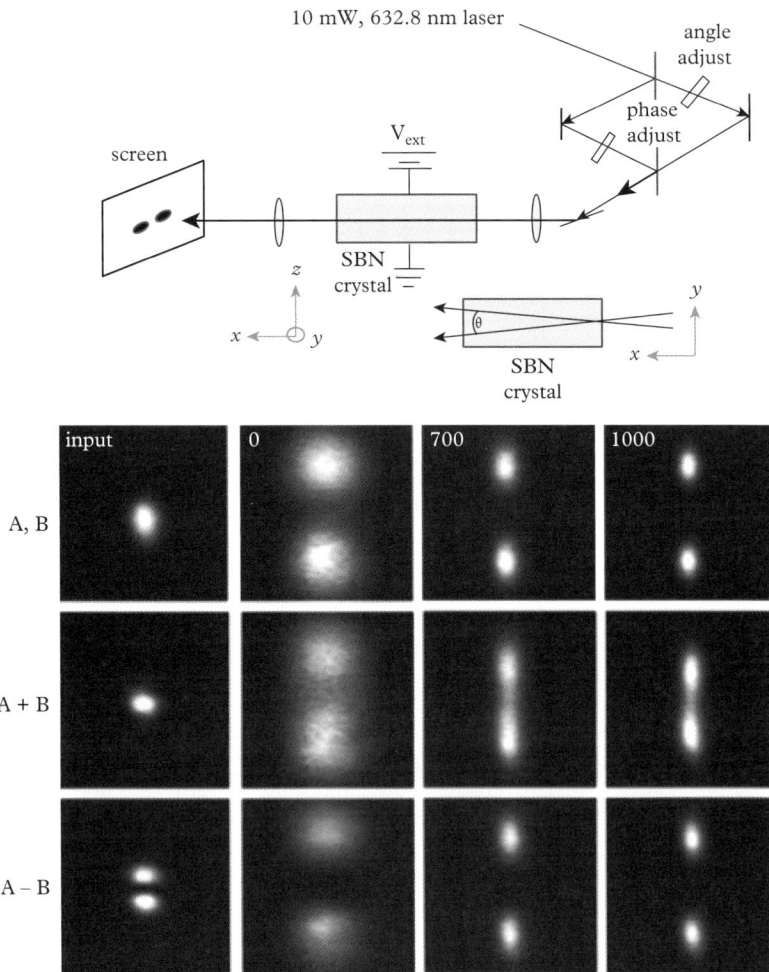

Figure 13.13 *Top: the experimental setup for two beam-induced soliton interaction in a SBN crystal. Bottom: phase-dependent soliton collisions with internal crossing angle $\theta = 6.4$ mrad. The leftmost column shows the input beams. Succeeding columns show output beams at the indicated values of external voltage. Each frame depicts a 235 μm × 235 μm region, and the spot separation is along the y-axis. (A, B): two beams incident separately; (A + B): two beams incident with zero phase shift; (A − B): two beams incident with π phase shift. (Reproduced with permission from Mamaev et al.[62] Copyright 1998, Optical Society of America).*

same He–Ne laser source propagated perpendicular to the crystalline c-axis (z-axis) of a 20 mm-long SBN:60 crystal and polarized along the same z-direction. The crystal of 5 mm thickness along the z-axis worked in the regime of steady-state screening soliton formation, with a dc-bias voltage applied along the z-axis and white-light background illumination. After being focused, the two beams intersected at the input face of the crystal with a small crossing angle of $\theta = 6.4$ mrad, a power level of 40 μW for each, and a spot size $\sim$31 μm. At the bottom of Fig. 13.13, the photos in the left column show the spot patterns of the input beams, while the succeeding columns show the output spot patterns, taken without and with applying a dc voltage, respectively. Without the applied dc field, the spot size of the output beam increased to $\sim$81 μm due to linear diffraction, whereas with applying a proper dc voltage ($\geq$ 700 V/5 mm) the output spot size reduced to nearly the same as the input. Moreover, we can see obvious attraction between two solitons when two beams are launched together with zero phase shift, or a slight repulsion when two beams are co-launched with a π phase shift. If the crossing angle is further reduced and the two input beams are in phase, the two solitons may fuse into one soliton beam.

In the regime of incoherent interaction, some other types of soliton collisions in PR media have been investigated, including 3D spiraling solitons formed by two initial non-coplanar input beams,[63] cascaded collisions of multiple solitons,[64] the collision between a 1D soliton and a 2D soliton,[65] and the interaction between two counter-propagating spatial solitons.[66]

Finally, as we mentioned previously, both the bright soliton formed in a self-focusing PR medium and the dark soliton formed in a self-defocusing PR material can provide a transient or semi-permanent waveguide structure that is suitable for beam confinement and light propagation at other wavelengths.[67,68] Such a soliton-produced waveguide structure in a PR crystal can be utilized to enhance other second-order nonlinear effects, such as SHG and optical parametric oscillation.[69,70]

REFERENCES

1. R. Y. Chiao, E. Garmire, and C. H. Townes, *Phys. Rev. Lett.* **13**, 479 (1964).
2. V. E. Zakharov and A. B. Shabat, *Sov. Phys. JETP* **34**, 62 (1972).
3. M. Segev and G. Stegeman, *Phys. Today* **51**, 42 (1998).
4. G. I. Stegeman and M. Segev, *Science* **286**, 1518 (1999).
5. D. Grischko, *Phys. Rev. Lett.* **24**, 866 (1970).
6. J. E. Bjorkholm and A. A. Ashkin, *Phys. Rev. Lett.* **32**, 129 (1974).
7. S. Maneuf, R. Desailly, and C. Froehly, *Opt. Commun.* **65**, 193 (1988).
8. R. Delafuente, A. Barthelemy, and C. Froehly, *Opt. Lett.* **16**, 793 (1991).
9. W. Torruellas, B. Lawrence, and G. I. Stegeman, *Electron. Lett.* **32**, 2092 (1996).
10. B. L. Lawrence and G. I. Stegeman, *Opt. Lett.* **23**, 591 (1998).
11. A. Pasquazi, S. Stivala, G. Assanto, J. Gonzalo, J. Solis, and C. N. Afonso, *Opt. Lett.* **32**, 2103 (2007).
12. J. S. Aitchison, K. Alhemyari, C. N. Ironside, R. S. Grant, and W. Sibbett, *Electron. Lett.* **28**, 1879 (1992).
13. V. Coda, R. D. Swain, H. Maillotte, G. J. Salamo, and M. Chauvet, *Opt. Commun.* **251**, 186 (2005).
14. J. S. Aitchison, A. M. Weiner, Y. Silberberg, M. K. Oliver, J. L. Jackel, D. E. Leaird, E. M. Vogel, and P. W. E. Smith, *Opt. Lett.* **15**, 471 (1990).
15. U. Bartuch, U. Peschel, T. Gabler, R. Waldhausl, and H. H. Horhold, *Opt. Commun.* **134**, 49 (1997).

16. G. Khitrova, H. M. Gibbs, Y. Kawamura, H. Iwamura, T. Ikegami, J. E. Sipe, and L. Ming, *Phys. Rev. Lett.* **70**, 920 (1993).
17. E. A. Ultanir, G. I. Stegeman, D. Michaelis, C. H. Lange, and F. Lederer, *Phys. Rev. Lett.* **90**, 253903 (2003).
18. G. Fanjoux, J. Michaud, M. Delque, H. Maillotte, and T. Sylvestre, *Opt. Lett.* **31**, 3480 (2006).
19. Y. N. Karamzin and A. P. Sukhorukov, *Sov. Phys. JETP* **41**, 414 (1976).
20. D. Liu, G. S. He, and S. H. Liu, *Chin. Phys. Lasers* **13**, 234 (1986).
21. K. Hayata and M. Koshiba, *Phys. Rev. Lett.* **71**, 3275 (1993).
22. L. Torner, C. R. Menyuk, W. E. Torruellas, and G. I. Stegeman, *Opt. Lett.* **20**, 13 (1995).
23. W. E. Torruellas, Z. Wang, D. J. Hagan, E. W. Vanstryland, G. I. Stegeman, L. Torner, and C. R. Menyuk, *Phys. Rev. Lett.* **74**, 5036 (1995).
24. M. T. G. Canva, R. A. Fuerst, S. Baboiu, G. I. Stegeman, and G. Assanto, *Opt. Lett.* **22**, 1683 (1997).
25. B. Bourliaguet, V. Couderc, A. Barthelemy, G. W. Ross, P. G. R. Smith, D. C. Hanna, and C. De Angelis, *Opt. Lett.* **24**, 1410 (1999).
26. R. Malendevich, L. Jankovic, S. Polyakov, R. Fuerst, G. Stegeman, C. Bosshard, and P. Gunter, *Opt. Lett.* **27**, 631 (2002).
27. S. Minardi, S. Sapone, W. Chinaglia, P. Di Trapani, and A. Berzanskis, *Opt. Lett.* **25**, 326 (2000).
28. P. Di Trapani, G. Valiulis, W. Chinaglia, and A. Andreoni, *Phys. Rev. Lett.* **80**, 265 (1998).
29. M. Peccianti, A. De Rossi, G. Assanto, A. De Luca, C. Umeton, and I. C. Khoo, *Appl. Phys. Lett.* **77**, 7 (2000).
30. M. Peccianti, K. A. Brzdakiewicz, and G. Assanto, *Opt. Lett.* **27**, 1460 (2002).
31. M. Peccianti, C. Conti, G. Assanto, A. De Luca, and C. Umeton, *Nature* **432**, 733 (2004).
32. J. F. Henninot, J. F. Blach, and M. Warenghem, *J. Opt. A: Pure Appl. Opt.* **10**, 085104 (2008).
33. M. Segev, B. Crosignani, A. Yariv, and B. Fischer, *Phys. Rev. Lett.* **68**, 923 (1992).
34. G. C. Duree, J. L. Shultz, G. J. Salamo, M. Segev, A. Yariv, B. Crosignani, P. Diporto, E. J. Sharp, and R. R. Neurgaonkar, *Phys. Rev. Lett.* **71**, 533 (1993).
35. M. D. I. Castillo, P. A. M. Aguilar, J. J. Sanchezmondragon, S. Stepanov, and V. Vysloukh, *Appl. Phys. Lett.* **64**, 408 (1994).
36. M. F. Shih, P. Leach, M. Segev, M. H. Garrett, G. Salamo, and G. C. Valley, *Opt. Lett.* **21**, 324 (1996).
37. D. Kip, M. Wesner, V. Shandarov, and P. Moretti, *Opt. Lett.* **23**, 921 (1998).
38. K. Kos, G. Salamo, and M. Segev, *Opt. Lett.* **23**, 1001 (1998).
39. N. R. Heckenberg, R. McDuff, C. P. Smith, and A. G. White, *Opt. Lett.* **17**, 221 (1992).
40. V. Tikhonenko, J. Christou, and B. Luther-Davies, *J. Opt. Soc. Am. B* **12**, 2046 (1995).
41. V. Tikhonenko, J. Christou, and B. Luther-Davies, *Phys. Rev. Lett.* **76**, 2698 (1996).
42. W. Krolikowski, E. A. Ostrovskaya, C. Weilnau, M. Geisser, G. McCarthy, Y. S. Kivshar, C. Denz, and B. Luther-Davies, *Phys. Rev. Lett.* **85**, 1424 (2000).
43. T. Carmon, R. Uzdin, C. Pigier, Z. H. Musslimani, M. Segev, and A. Nepomnyashchy, *Phys. Rev. Lett.* **87**, 143901 (2001).
44. G. A. Swartzlander, Jr., D. R. Andersen, J. J. Regan, H. Yin, and A. E. Kaplan, *Phys. Rev. Lett.* **66**, 1583 (1991).
45. G. C. Valley, M. Segev, B. Crosignani, A. Yariv, M. M. Fejer, and M. C. Bashaw, *Phys. Rev. A* **50**, R4457 (1994).
46. G. Duree, M. Morin, G. Salamo, M. Segev, B. Crosignani, P. Di-Porto, E. Sharp, and A. Yariv, *Phys. Rev. Lett.* **74**, 1978 (1995).
47. M. D. I. Castillo, J. J. Sanchezmondragon, S. I. Stepanov, M. B. Klein, and B. A. Wechsler, *Opt. Commun.* **118**, 515 (1995).
48. Z. G. Chen, M. Mitchell, M. F. Shih, M. Segev, M. H. Garrett, and G. C. Valley, *Opt. Lett.* **21**, 629 (1996).

49. Z. G. Chen, M. Segev, S. R. Singh, T. H. Coskun, and D. N. Christodoulides, *J. Opt. Soc. Am. B* **14**, 1407 (1997).
50. M. Taya, M. C. Bashaw, M. M. Fejer, M. Segev, and G. C. Valley, *Phys. Rev. A* **52**, 3095 (1995).
51. M. Chauvet, S. Chauvin, and H. Maillotte, *Opt. Lett.* **26**, 1344 (2001).
52. J. S. Aitchison, A. M. Weiner, Y. Silberberg, D. E. Leaird, M. K. Oliver, J. L. Jackel, and P. W. E. Smith, *Opt. Lett.* **16**, 15 (1991).
53. J. U. Kang, G. I. Stegeman, and J. S. Aitchison, *Opt. Lett.* **21**, 189 (1996).
54. M. Bertolotti, A. D'Andrea, E. Fazio, M. Zitelli, A. Carrera, G. Chiaretti, and N. G. Sanvito, *Opt. Commun.* **168**, 399 (1999).
55. J. Hubner, H. M. van Driel, and J. S. Aitchison, *Opt. Lett.* **30**, 3168 (2005).
56. M. Shalaby and A. Barthelemy, *Opt. Lett.* **16**, 1472 (1991).
57. M. Peccianti, C. Conti, G. Assanto, A. De Luca, and C. Umeton, *Appl. Phys. Lett.* **81**, 3335 (2002).
58. W. Hu, T. Zhang, Q. Guo, L. Xuan, and S. Lan, *Appl. Phys. Lett.* **89**, 071111 (2006).
59. V. Couderc, E. L. Lago, C. Simos, and A. Barthelemy, *Opt. Lett.* **26**, 905 (2001).
60. L. Jankovic, H. Kim, S. Polyakov, G. I. Stegeman, C. Bosshard, and P. Gunter, *Opt. Lett.* **28**, 1037 (2003).
61. P.- H. Pioger, F. Baronio, V. Couderc, A. Barthelemy, C. De Angelis, Y. Min, V. Quiring, and W. Sohler, *IEEE Photonics Technol. Lett.* **16**, 560 (2004).
62. A. V. Mamaev, M. Saffman, and A. A. Zozulya, *J. Opt. Soc. Am. B* **15**, 2079 (1998).
63. M. F. Shih, M. Segev, and G. Salamo, *Phys. Rev. Lett.* **78**, 2551 (1997).
64. C. Anastassiou, J. W. Fleischer, T. Carmon, M. Segev, and K. Steiglitz, *Opt. Lett.* **26**, 1498 (2001).
65. E. DelRe, A. D'Ercole, E. Palange, and A. J. Agranat, *Appl. Phys. Lett.* **86**, 191110 (2005).
66. M. Petrovic, D. Jovic, M. Belic, J. Schroder, P. Jander, and C. Denz, *Phys. Rev. Lett.* **95**, 053901 (2005).
67. M. Klotz, H. X. Meng, G. J. Salamo, M. Segev, and S. R. Montgomery, *Opt. Lett.* **24**, 77 (1999).
68. Z. G. Chen and K. McCarthy, *Opt. Lett.* **27**, 2019 (2002).
69. S. Lan, J. A. Giordmaine, M. Segev, and D. Rytz, *Opt. Lett.* **27**, 737 (2002).
70. C. B. Lou, J. J. Xu, H. J. Qiao, X. Z. Zhang, and Y. L. Chen, *Opt. Lett.* **29**, 953 (2004).

14
Multi-Photon Nonlinear Optical Effects

14.1 Description of multi-photon absorption (MPA)

14.1.1 Introduction to MPA studies

The concept of two-photon absorption (2PA) was first proposed in 1931 by Göppert-Mayer.[1] She theoretically predicted that simultaneous 2PA may lead to a molecular transition between two eigenstates. This prediction was based on the principles of the newly established quantum theory of radiation,[2] and the key issue to support her argument was to introduce the concept of intermediate state, which describes the quantum state of a combined system consisting of molecules and photon fields. The 2PA probability is too small to be measured with any conventional (incoherent) light for excitation. Due to this reason, no real 2PA had been observed until the advent of laser in the beginning of 1960s.

In 1961, Kaiser and Garrett reported the first observation of a 2PA-induced frequency up-conversion fluorescence in a $CaF_2:Eu^{2+}$ crystal, excited by the intense coherent radiation of 694.3 nm wavelength from a pulsed ruby-crystal laser.[3] Relying on the advantages of high directionality, high monochromaticity, high brightness, and tunability of coherent radiation from various types of laser devices, it is now possible to observe not only 2PA-related processes, but also three-photon absorption (3PA), four-photon absorption (4PA), and even higher-order multi-photon excitation (MPE) processes.

Multi-photon related studies include both fundamental research and applications of multi-photon absorption (MPA), MPE, and multi-photon active materials.[4] Here, MPA means simultaneous absorption of a number (≥ 2) of photons, accompanied by the transition of absorbing molecule from a lower energy level to a higher level. However, MPE has a much broader meaning that includes other physical, chemical, and biological effects upon simultaneous interaction of multiple photons with matter. For instance, MPE may lead not only to nonlinear attenuation of an incident coherent light beam, but also may create various other effects, such as enhanced refractive-index changes of the medium, molecular dissociation or ionization, electron emission from the materials' surface, induced conductivity change in semiconductors, and induced polymerization. All materials that can manifest such photophysical or photochemical responses upon MPE are multi-photon active. In order to realize such higher order MPE, a high local intensity (in units of MW/cm^2 or GW/cm^2) of the incident light beam is needed.

From the viewpoint of fundamental research, multi-photon related studies may greatly enrich and deepen our knowledge and understanding about interactions of intense coherent radiation with matter. The following are several specific examples that have shown how much our previous knowledge has to be renewed, and how valuable new information is obtained from these studies.

i. *Multi-photon induced surface photoelectric effect*: According to the previous theory of one-photon induced electron emission from the surface of a given photocathode material, once the wavelength of light signals is longer than the so called "red-limit," no electron emission from the surface can be observed. This is one of the key experimental facts on which Einstein based his postulate of light radiation consisting of photons. Now people know that by utilizing an intense IR coherent radiation with a wavelength longer (or much longer) than the red-limit, a two- or multi-photon excited electron emission from the same photoelectric device can be observed. A similar situation also exists for the volume photoelectric effect in a semiconductor device.

ii. *Multi-photon induced photochemical reactions*: It was previously known that some photochemical reactions, including the most familiar photographic effect, depend not only on the illumination intensity, but also on the wavelength of the light signal. For a given photographic emulsion material, no photographic response can be expected if the light wavelength is notably longer than a certain limit. Similar rules also held for some other photochemical reactions (such as photo-polymerization) that need UV or short-wavelength visible radiation. In the wake of the invention of lasers, however, it has been found that the aforementioned photochemical reactions could also be produced using intense coherent IR radiation, based on two- or multi-photon excitation processes.

iii. *New knowledge brought by two-photon-process studies*: In general, the selection rules and path-ways of molecular transitions upon MPE may be different from those of one-photon excitation. For this reason valuable new spectral information about molecular energy-level structure and transition properties of multi-photon active media can be obtained through multi-photon related studies. During the 1970s and 1980s, the main research theme in this area was focused on two-photon spectroscopy of simple atomic and molecular gaseous systems, organic solvents and compounds, as well as inorganic crystals and semiconductors. All these studies have brought a great deal of new information and knowledge on electronic, vibrational and even rotational states of the investigated atomic and molecular systems.

iv. *High-order multi-photon studies*: After the advent of laser, many researchers still thought that a two-photon excitation process was easy to observe by utilizing a pulsed and focused laser beam, but a high-order MPE would be more difficult to produce. Further development of relevant studies has shown that by excitation with nanosecond laser pulses, 3PA can be readily realized in semiconductors and organic dye solutions, and four- or five-photon absorption (4PA or 5PA) can also be easily produced by using femtosecond laser pulses in organic materials. In addition, the early investigations of multi-photon induced molecular dissociation and ionization even showed that a process involving more than ten-photon excitation was observable at quite low intensity levels of an mid-IR laser beam.[5] In other words, experimental results have shown that in many cases, multi (>2)-photon excitation processes can be produced much more easily than had been thought before. A fundamental interpretation of these experimental facts has been an interesting subject for further studies.

14.1.2 Mechanisms of MPA

In principle, there are two theoretical regimes that can be utilized to describe nonlinear optical effects, including MPA processes. The first is the semiclassical theory, and the second is the quantum electrodynamic theory.

The most essential feature of the semiclassical theory is that the media composed of atoms or molecules are described by the theory of quantum mechanics, while the light fields are described by the classical Maxwell theory. The key issue of the semiclassical theory in nonlinear optics is the explicit expressions for nonlinear electric polarization of an optical medium. It is generally known that the electric polarization vector **P** of a medium is defined by the summation of the light field-induced electric-dipole-moment vectors of all molecules within a unit volume. In the case of applying a weak light field from an incoherent light source, **P** is linearly proportional to the applied electric field **E**. If this field is a monochromatic wave with an angular frequency of ω, we have

$$\mathbf{P}(\omega) = \mathbf{P}^{(1)}(\omega) = \varepsilon_0 \chi^{(1)}(\omega) \mathbf{E}(\omega). \tag{14.1-1}$$

Here, $\chi^{(1)}(\omega)$ is the first-order (linear) susceptibility (in SI units) of a given medium, and ε_0 is the free-space permittivity. In general, $\chi^{(1)}(\omega)$ is a complex parameter in the form of a second-rank tensor: its real part determines the refractive index of the medium, while the imaginary part determines the linear (one-photon) absorption. To further consider the nonlinear absorption of a strong monochromatic coherent light of frequency ω, which penetrates through an isotropic or centrosymmetric medium, Eq. (14.1-1) has to be generalized as

$$\begin{aligned}\mathbf{P}(\omega) &= \mathbf{P}^{(1)}(\omega) + \mathbf{P}^{(3)}(\omega) + \mathbf{P}^{(5)}(\omega) + \cdots \\ &= \varepsilon_0 [\chi^{(1)}(\omega)\mathbf{E}(\omega) + \chi^{(3)}(\omega,\omega,-\omega)\mathbf{E}(\omega)\mathbf{E}(\omega)\mathbf{E}^*(\omega) + \\ &\quad \chi^{(5)}(\omega,\omega,-\omega,\omega,-\omega)\mathbf{E}(\omega)\mathbf{E}(\omega)\mathbf{E}^*(\omega)\mathbf{E}(\omega)\mathbf{E}^*(\omega) + \cdots].\end{aligned} \tag{14.1-2}$$

Here, $\mathbf{E}^*(\omega)$ is the complex conjugate of the electric-field function $\mathbf{E}(\omega)$ of the applied light wave; $\chi^{(3)}(\omega,\omega,-\omega)$ is the third-order nonlinear susceptibility of the nonlinear absorbing medium, which in general is a complex fourth-rank tensor: its real part describes induced refractive-index change depending on the light intensity $I(\omega) \propto \mathbf{E}^*(\omega)\cdot\mathbf{E}(\omega)$, while the imaginary part can phenomenologically describe the 2PA. Similarly, $\chi^{(5)}(\omega,\omega,-\omega,\omega,-\omega)$ is the fifth-order nonlinear susceptibility, which in general is a complex sixth-rank tensor: its real part describes the induced refractive-index change depending on the square of the light intensity $I^2(\omega)$, while the imaginary part can phenomenologically describe the 3PA. However, it should be noted that semiclassical theory treats light waves as classical electromagnetic fields, i.e., no quantized concept of 'photon' is introduced; therefore, the semiclassical theory cannot provide a rigorous physical description of the microscopic processes of MPA.

In contrast, the quantum theory of radiation in the regime of quantum electrodynamics treats the medium and the optical field as a combined quantized system.[6-8] In other words, both the medium and the optical field should be described quantum mechanically. As a result, the wave function of the combined system is expressed by the product of the eigenfunction of a molecular system and the eigenfunction of a quantized photon field. In this case, the key issue is to determine the probability of quantum state changes of the combined system due to interaction between the photon field and the medium. Usually, the state changes of the combined system are related to the transition of the molecular system from an initial eigenstate to a final eigenstate, accompanied by simultaneous changes of the photon numbers among different photon modes (or photon states).

The quantum theory of radiation is a more rigorous approach that, in principle, can be perfectly used to explain MPA processes in both qualitative and quantitative ways. To do so, it is necessary to introduce the concept of intermediate states co-occupied by the combined system of the photon field and a molecular system. Based on the concept of intermediate states, the principles and mechanisms of nonlinear optical effects (including MPA) can be consistently interpreted.

For simplicity, let us first consider the elementary 2PA process induced by a single laser beam of frequency v, passing through a nonlinear absorbing medium. The schematic diagram for this process in the regime of quantum theory is shown in Fig. 14.1(a) with the help of introducing the concept of intermediate state that can be schematically represented by a dashed line level between two real eigenstates of the molecule. In this case, the occurrence of 2PA inducing molecular transition between its two real states can be visualized as a "two-step" event. (i) In the first step, one photon is absorbed while the molecule leaves its initial state E_g to be excited to an intermediate state; (ii) in the second step, another photon will be absorbed while the same molecule completes its transition from the intermediate state to the final real state E_f. The key connection between these two steps is the intermediate state in which the molecular status is not certain in the sense that the molecule may stay in all its possible eigenstates (except E_g and E_f) with a certain probability distribution among these states. Because the uncertainty of energy distribution range is so large, according to the uncertainty principle, the residence time of the molecule in the intermediate state should be infinitely short. This means that the imagined "two steps" of 2PA actually occur simultaneously. In other words, the real 2PA is a single elementary process that is essentially different from the so-called *cascaded* two-step one-photon processes. In the first step of the latter case, the molecule absorbs one photon to reach a real excited state with a certain lifetime, followed by the second step when this excited molecule absorbs another photon to reach a higher real level. Obviously, in such a situation the two transition steps can take place in different time sequences.

Similarly, as shown in Fig. 14.1(b), we can describe the elementary 3PA process, in which three photons can be simultaneously absorbed by a single molecule through two intermediate states schematically represented by two dashed lines. For easy understanding, this 3PA process may be imagined by the following three "steps:" (i) in the first step, one photon is absorbed while the molecule leaves its initial state E_g to be excited to an intermediate state represented by a lower

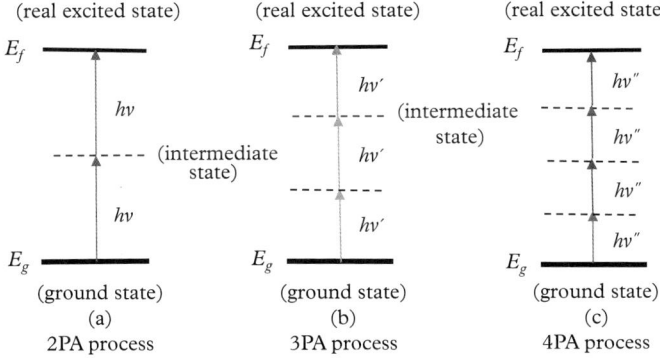

Figure 14.1 *Schematic description of an elementary process for multi-photon induced molecular transition. Solid lines represent the eigenstates that only characterize molecular system; dashed lines represent the intermediate states that characterize both molecular and photon systems.*

dashed line; (ii) in the second step, another photon is absorbed while the same molecule makes a transition to another intermediate state (a higher dashed line); (iii) in the third step, one more photon is absorbed when the excited molecule reaches its final real state E_f. Based on the same consideration of the uncertainty principle, these imagined "three steps" take place simultaneously. In this sense, the real 3PA of a molecule is also a single elementary process, resulting from the quantized interaction between radiation and matter. A similar description can be applied to multi (>3)-photon absorption by involving more intermediate states. As a further example, Fig. 14.1(c) shows the elementary process of 4PA. In all these cases, to meet the requirement of conservation of energy, the following resonance conditions should be fulfilled:

$$E_f - E_g = 2h\nu \ (2PA); \quad E_f - E_g = 3h\nu' \ (3PA); \quad E_f - E_g = 4h\nu'' \ (4PA). \tag{14.1-3}$$

So far we have only considered the absorption of multiple photons having the same frequency; which is termed *degenerate* MPA. In more general cases, the applied light field may contain two or more frequency components, and the simultaneous absorption of several photons having different frequencies is also possible. This is termed *nondegenerate* MPA. The theoretical descriptions are essentially the same for both degenerate and nondegenerate MPA processes.

14.1.3 Formulations of MPA-induced light attenuation

The attenuation of a light beam passing through an optical medium can be generally expressed by the following phenomenological expression:

$$\frac{dI(z)}{dz} = -\alpha I(z) - \beta I^2(z) - \gamma I^3(z) - \eta I^4(z) \cdots . \tag{14.1-4}$$

Here, $I(z)$ is the intensity of the incident light beam propagating along the z-axis and α, β, γ, and η are the one-, two-, three-, and four-photon absorption coefficients of the transmitting medium, respectively. For simplicity, here we have assumed that the incident light has a uniform transverse intensity distribution and the initial intensity is not dependent on time. Under the conditions that there is no linear absorption ($\alpha = 0$) at the wavelength λ of the incident light, and only 2PA satisfying Eq. (14.1-3) takes place, we have

$$\frac{dI(z)}{dz} = -\beta I^2(z). \tag{14.1-5}$$

The physical meaning of this expression is that the 2PA probability of molecules in a given position is proportional to the square of the local light intensity. The solution of Eq. (14.1-5) is

$$I(z, \lambda) = \frac{I_0(\lambda)}{1 + \beta(\lambda)I_0(\lambda)z}. \quad (2PA) \tag{14.1-6}$$

Here, $I_0(\lambda)$ is the incident light intensity with a top-hat pulse shape, z is the propagation length in the medium, and $\beta(\lambda)$ is the 2PA coefficient that is a material parameter which depends on the wavelength of the incident light. As a macroscopic parameter depending on concentration of the two-photon absorbing molecules, $\beta(\lambda)$ (in units of cm/GW) can be further expressed as

$$\beta(\lambda) = \sigma_2(\lambda)N_0 = \sigma_2(\lambda)N_A d_0 \times 10^{-3}. \tag{14.1-7}$$

Here, σ_2 is the molecular 2PA cross-section (in units of cm^4/GW), N_0 is the molecular density (in units of 1/cm^3), N_A is Avogadro's number, and d_0 is the molar concentration of the absorbing molecules (in units of M/L or simply M). Equation (14.1-7) is precisely valid only when it can be assumed that the 2PA process is not too strong and most molecules stay in their ground state. Otherwise, N_0 should be replaced by $\Delta N = N_1 - N_2$, where N_1 is the population density of the ground state and N_2 is the population density of the two-photon excited state. If N_2 is negligible we have $\Delta N \approx N_0$ for a two-state system. When two-photon excitation is strong enough and the depletion of the ground-state population is no longer negligible, the 2PA coefficient β will not be a material constant even for a given wavelength and concentration; instead it will depend on the input excitation intensity. This is the so-called 2PA saturation effect that will be discussed later.

Although the parameter $\sigma_2(\lambda)$ is a directly measurable quantity (in units of cm^4/GW) that characterizes the average two-photon absorbability per molecule, one can also use another parallel expression for the 2PA cross-section defined by

$$\sigma_2'(\lambda) = \sigma_2(\lambda) h\nu. \tag{14.1-8}$$

Here, $h\nu$ is the photon energy of the input light beam. According to this definition, $\sigma_2'(\lambda)$ is in units of cm^4/photon/s or simply cm^4 s. In practice, most of the measured values of $\sigma_2'(\lambda)$ are in the range from 10^{-51} to 10^{-46} cm^4 s. For this reason, some researchers prefer to use another informal unit (GM, the abbreviation of Göppert-Mayer) defined by

$$1\ \text{GM} = 10^{-50}\ \text{cm}^4\ \text{s}. \tag{14.1-9}$$

Based on Eq. (14.1-6), the nonlinear transmission (NLT) for a two-photon absorbing medium with optical path length l_0 can be written as

$$T(I_0, \lambda) = \frac{I(l_0, \lambda)}{I_0(\lambda)} = \frac{1}{1 + \beta(\lambda) I_0(\lambda) l_0}. \quad (2PA) \tag{14.1-10}$$

Similarly, for a pure 3PA process, Eq. (14.1-4) becomes

$$\frac{dI(z)}{dz} = -\gamma I^3(z). \tag{14.1-11}$$

The physical meaning of this expression is that the 3PA probability of molecules in a given position is proportional to the third power of the local light intensity. The solution of Eq. (14.1-11) is

$$I(z, \lambda) = \frac{I_0(\lambda)}{\sqrt{1 + 2\gamma(\lambda) I_0^2(\lambda) z}}. \quad (3PA) \tag{14.1-12}$$

Here, $\gamma(\lambda)$ is the 3PA coefficient (in units of cm^3/GW^2) of a given material, which is a macroscopic parameter depending on the concentration of the three-photon absorbing molecules, and can be further expressed as

$$\gamma(\lambda) = \sigma_3(\lambda) N_0 = \sigma_3(\lambda) N_A d_0 \times 10^{-3}. \tag{14.1-13}$$

Here, $\sigma_3(\lambda)$ is the molecular 3PA cross-section (in units of cm^6/GW^2). There is also a parallel definition of the 3PA cross-section, i.e., $\sigma_3'(\lambda) = \sigma_3(\lambda)(h\nu)^2$ (in units of cm^6 s^2). In obtaining

Eq. (14.1-13) it is assumed that the 3PA process is not too strong and most of the molecules stay in their ground state. Otherwise, N_0 should be replaced by $\Delta N = N_1 - N_2$, as described previously in connection with the 2PA process.

In general, for a sole MPA process (here M $\rightarrow$ m > 3), we have

$$\frac{dI(z)}{dz} = -\xi_m I^m(z). \qquad (14.1\text{-}14)$$

The solution of Eq. (14.1-14) is

$$I(z,\lambda) = \frac{I_0(z)}{[1 + (m-1)\xi_m(\lambda)I_0^{m-1}(\lambda)z]^{1/(m-1)}}. \quad (MPA) \qquad (14.1\text{-}15)$$

Here, $\xi_m(\lambda)$ is the MPA coefficient (in units of cm^{m+1}/GW^{m-1}), which can be further expressed as

$$\xi_m(\lambda) = \sigma_m(\lambda)N_0 = \sigma_m(\lambda)N_A d_0 \times 10^{-3}, \qquad (14.1\text{-}16)$$

where $\sigma_m(\lambda)$ is the molecular MPA cross-section (in units of cm^{2m}/GW^{m-1}). Based on Eq. (14.1-15), the NLT for a solely M-photon absorbing medium with a thickness l_0 is given by

$$T(I_0,\lambda) = \frac{1}{[1 + (m-1)\xi_m(\lambda)I_0^{m-1}(\lambda)l_0]^{1/(m-1)}}. \quad (MPA) \qquad (14.1\text{-}17)$$

In principle, the cross-section values of σ_2, σ_3, and σ_m for a given nonlinear absorbing medium can also be theoretically calculated, provided that the complete information of molecular eigenstate structure and transition parameters (including matrix elements of the dipole-moment operator for all possible transition combinations) are known. Unfortunately, for most commonly studied organic materials, the exact molecular eigenstate structure and transition properties are too complicated to calculate. However, it is much easier to directly measure MPA coefficient (a macroscopic parameter depending on concentration), and then to further determine the corresponding cross-section (a microscopic parameter of individual absorption center) for a given medium at a given wavelength by using the NLT method described later.

14.1.4 Theoretical expression of 2PA cross-section

A convenient way to derive a theoretical expression for 2PA cross-section is the one used by Göppert-Mayer in her original work. Here the intermediate state shown in Fig. 14.1 actually is related with all eigenstates of the molecular system. For two coherent monochromatic beams linearly polarized, which are of different frequencies ω_1 and ω_2 and in resonance with an absorbing isotropic medium at the two-photon transition frequency. The 2PA cross-section (in units of cm^4 s) for a molecule making a direct transition from its ground state (o) to a final excited state (t) can be expressed (in units of cm^4 s) as

$$\sigma_2'(\omega_1,\omega_2) = \left(\frac{2\pi}{c}\right)^2 \frac{\omega_1\omega_2}{6\varepsilon_0^2\hbar^2} \left|\sum_b \left[\frac{(p_1)_{ob}(p_2)_{bt}}{\omega_{bo} - \omega_1} + \frac{(p_2)_{ob}(p_1)_{bt}}{\omega_{bo} - \omega_2}\right]\right|^2 g(\omega_1 + \omega_2)$$

$$= \left(\frac{2\pi}{c}\right)^2 \frac{\omega_1\omega_2}{6\varepsilon_0^2\hbar^2} |\Lambda|^2 g(\omega_1 + \omega_2). \qquad (14.1\text{-}18)$$

Here, c is the light speed in a vacuum, h is Planck's constant, ε_0 is the permittivity of free space; p_1 and p_2 are the components of the molecular dipole-moment vector operator along the polarization directions of the two light beams; state o and state t are the initial and terminal states of molecular 2PA transition, respectively. The summation is over all one-photon transition-allowed eigenstates (b) of the molecule; $g(\omega_1 + \omega_2)$ is a normalized line shape function (with a dimension of second), and Λ represents all of the summation terms.

In more general cases, one (or two) of the incident light beams may possess more complicated polarization properties (such as being circularly or elliptically polarized). In these cases, one may introduce a generalized polarization unit vector ($\mathbf{S}$) to describe the polarization state of an input light beam. For example, if the light is linearly polarized, we have $\mathbf{S} = \{\cos\theta_x, \cos\theta_y, \cos\theta_z\}$, where θ_x, θ_y, and θ_z are the angles between the light polarization direction and the x-, y-, and z-axes of the laboratory coordinate system; for right-handed circularly polarized light we have $\mathbf{S} = (1/2)^{1/2}\{1, -i, 0\}$; and for left-handed circularly polarized light we have $\mathbf{S} = (1/2)^{1/2}\{1, i, 0\}$. Assuming two incident light beams have different polarization properties ($\mathbf{S}_1$ and $\mathbf{S}_2$), the summation terms in Eq. (14.1-18) can be rewritten as[9]

$$|\Lambda|^2 = \left| \sum_b \left[\frac{(\mathbf{S}_1 \cdot \mathbf{p}_{ob})(\mathbf{S}_2 \cdot \mathbf{p}_{bt})}{\omega_{bo} - \omega_1} + \frac{(\mathbf{S}_2 \cdot \mathbf{p}_{ob})(\mathbf{S}_1 \cdot \mathbf{p}_{bt})}{\omega_{bo} - \omega_2} \right] \right|^2. \quad (14.1\text{-}19)$$

Here, $\mathbf{p}_{ob}$ and $\mathbf{p}_{bt}$ are the corresponding matrix elements of molecular dipole moment. From this expression one can clearly see that the magnitude of 2PA cross-section and the related selection rules are dependent on the combination of the two input light fields. On the other hand, the direction of molecular dipole-moment vector is generally dependent on molecular orientation; therefore, for a system consisting of a great number of randomly orientated molecules, it is necessary to take an orientation average over molecular assembly. Based on the above expression, researchers may investigate the details of molecular energy structures, selection rules, and other 2PA related properties by changing polarization states of the input light fields.

In the simplest case for a degenerate 2PA process, the input is a single linearly polarized light beam, and Eq. (14.1-18) of molecular 2PA cross-section can be simplified as

$$\sigma'_2(\omega) = \left(\frac{2\pi}{c}\right)^2 \frac{\omega^2}{6\varepsilon_0^2 h^2} \left| \sum_b \left[2\frac{(p_s)_{ob}(p_s)_{bt}}{\omega_{bo} - \omega} \right] \right|^2 g(2\omega). \quad (14.1\text{-}20)$$

Here, $(p_s)_{ob}$ is the transition matrix element of the electric dipole-moment component along the polarization direction of the input light field. Once again, an orientational average is needed if the light induced dipole-moment vector is dependent on the molecular orientation.

The selection rules for one- and two-photon transitions are mutually exclusive, in the case of a centrosymmetric molecule. Since the dipole operator is ungerade, only an ungerade state can be reached from the usually gerade ground state. For the products of matrix elements that appear in Eqs. (14.1-18)–(14.1-20) the state b must be ungerade; otherwise $(p)_{ob}$ vanishes. To make the second matrix element $(p)_{bt}$ nonzero, the final state t must be gerade. Transitions from the ground state to such a state are one-photon forbidden. The mutual exclusion between one- and two-photon transitions in centrosymmetric molecules is similar to the one known for IR and Raman transitions.

Finally, it should be noted that the same expression (such as Eq. (14.1-20)) of 2PA cross-section can also be obtained through semi-classical theory by considering $\chi^{(3)}(\omega, \omega, -\omega)$ related complex refractive-index change ($\Delta n(\omega)$) that is induced by $I(\omega)$ in a given medium, the real part of $\Delta n(\omega)$ represents the intensity-dependent real refractive-index change, while the imaginary part of $\Delta n(\omega)$ determines the 2PA coefficient (see Section 9.3.2). For the case of degenerate 2PA, the relationship between β and $\chi^{(3)}(\omega,\omega,-\omega)$ can be obtained as follows:

$$\beta \approx \frac{2\pi}{n_0 \lambda} \text{Im} \chi_e^{(3)} \frac{|E_0|^2}{I(0)} = \frac{4\pi}{\varepsilon_0 c n_0^2 \lambda} \text{Im} \chi_e^{(3)}(\omega,\omega,-\omega), \tag{14.1-21}$$

where n_0 is the linear refractive index of the medium, λ is the incident light wavelength, and $\text{Im}\chi_e^{(3)}$ is the imaginary part of the effective third-order nonlinear susceptibility characterizing the 2PA process.

14.2 Highly multi-photon absorbing materials

14.2.1 Need for highly multi-photon active materials

Since the mid-1990s, research interests in the multi-photon area have shifted from fundamental spectroscopic studies to the development of new materials and exploration of their technical applications.[4] In principle, all sorts of known optical media are linearly (one-photon) absorbing in certain spectral ranges; therefore they should be also nonlinear (multi-photon) absorbing in corresponding other spectral ranges. For most commonly utilized and visibly transparent optical materials (such as glasses, crystals, polymers, solvents, and optical fibers), they have very strong linear absorption in the UV range and, therefore, can be multi-photon active in the visible or near-IR spectral ranges. However, the experimental measurements indicated that the 2PA two-photon (or beyond two-photon) absorption cross-section values for these types of conventional optical materials are quite small.

On the other hand, even in their early studies, researchers found that a much stronger 2PA could be easily found in various dye solutions and some bulk semiconductor crystals. For this reason, the efforts of synthesizing new dyes (or chromophores) have been greatly stimulated since the late 1990s, promoted by the emerging multi-photon related applications, including optical limiting and stabilization, bio-imaging, photodynamic therapy, frequency up-converted lasing, three-dimensional (3D) data storage, and microfabrication. These specifically designed and synthesized organic chromophores are much more efficient in MPA than those commercially available dyes.

As we mentioned previously, 2PA and MPA are high-order nonlinear optical effects that require high intensities of input laser pulses. From the viewpoint of applications, the first requirement for the candidates is that they should exhibit very small linear absorption at working wavelengths, while have a large nonlinear (MPA) coefficient at the same wavelengths. Secondly, the nonlinear materials should exhibit very good physical/chemical stabilities and can withstand high irradiation power density without photobleaching, degradation, and optical damage.

As shown in Fig. 14.2, there are two large categories of multi-photon active materials for MPA studies and applications. Generally speaking, organic materials manifest larger nonlinear absorption coefficients than inorganic materials, whereas the inorganic materials may exhibit better physical/chemical stabilities and anti-damage capabilities.

14.2.2 Basic structures of multi-photon active chromophores

With linear absorption generally in the visible region, most organic π-conjugated molecules are naturally promising materials for 2PA applications in the near-IR, where they are usually linearly

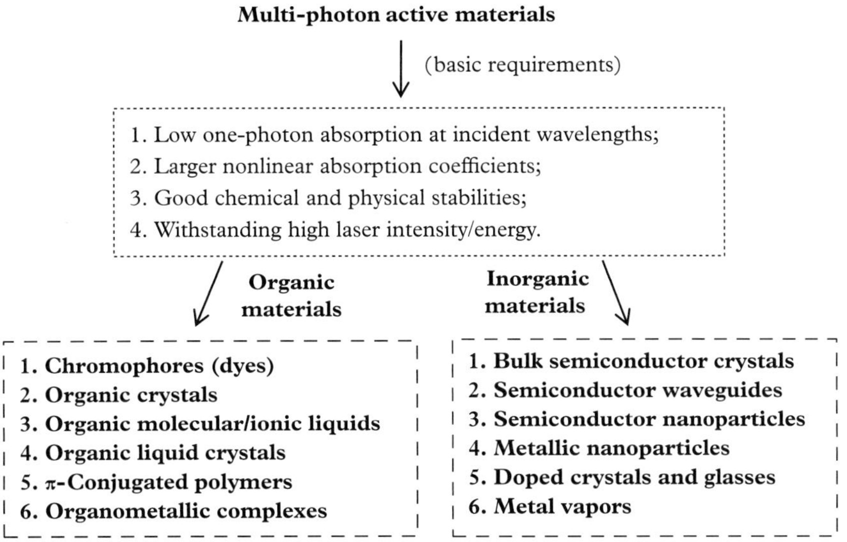

Multi-photon active materials

↓ (basic requirements)

1. Low one-photon absorption at incident wavelengths;
2. Larger nonlinear absorption coefficients;
3. Good chemical and physical stabilities;
4. Withstanding high laser intensity/energy.

↙ Organic materials   Inorganic materials ↘

Organic:
1. Chromophores (dyes)
2. Organic crystals
3. Organic molecular/ionic liquids
4. Organic liquid crystals
5. π-Conjugated polymers
6. Organometallic complexes

Inorganic:
1. Bulk semiconductor crystals
2. Semiconductor waveguides
3. Semiconductor nanoparticles
4. Metallic nanoparticles
5. Doped crystals and glasses
6. Metal vapors

Figure 14.2 *Schematic description of various types of multi-photon active materials.*

transparent. Theoretical calculations manifest that there is a strong correlation between intramolecular charge-transfer processes and two-photon absorptivity of a molecule. The chromophores with large 2PA usually have a number of key molecular features. With an intramolecular charge-transfer process as the driving force, the presence of a strong π-electron donor (D), a strong π-electron acceptor (A), and a polarizable π-conjugated bridge are essentially important for molecules with large 2PA. In addition, the extent (i.e., length) of π-conjugation is found to be particularly important to enhance the 2PA cross-section of a molecule, as it leads to states with extended charge separation. Hence, the co-planarity of the π-conjugation is also critical in boosting the efficiency of an intramolecular charge transfer.

Figure 14.3 show the basic molecular structures of one-dimensional multi-photon active chromophores, which can be asymmetrical with an arrangement such as A-π-D, or symmetrical with arrangements such as D-π-D, A-π-A, A-D-π-D-A, and D-A-π-A-D (where π refers to a π-conjugated bridge).

It has been theoretically and experimentally proved that realizing two- or three-dimensional multi-branched structures can greatly enlarge the MPA cross-section of a chromophore. This enhancement in molecular multi-photon absorbability is based on the increase of the effective multi-photon absorbing centers (active units) within the organic molecule as well as the cooperative interaction between its multi-branches. Figure 14.4 shows some schematic structures of multi-branched two- and three-dimensional two-photon absorbing molecules.

14.2.3 Features of novel multi-photon active materials

So far, several thousand research articles have been published in the area of multi-photon active materials, and most of them are related to the molecular design, synthesis, and characterization of organic or organic/inorganic hybrid systems.[4] A detailed description of various multi-photon active materials is far beyond the scope of this subsection. Here we can only outline general features

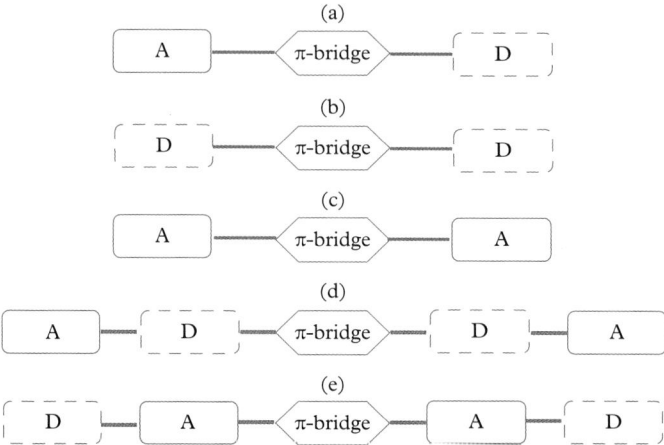

Figure 14.3 Molecular-structure motifs for one-dimensional multi-photon absorbing chromophores: asymmetrical (a) and symmetrical (b–e). D: π-electron donor; A: π-electron acceptor; π-bridge: π-conjugated bridge.

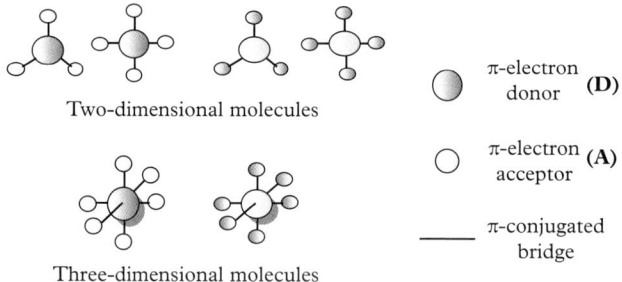

Figure 14.4 Basic structural motifs for two and three-dimensional two-photon absorbing molecules.

of several major types of novel materials. Some specific materials or compounds will be mentioned in the following sections regarding their characterizations and applications.

14.2.3.1 Novel chromophores (dyes)

Since the mid-1990s, considerable research efforts have been focused on designing and synthesizing novel organic chromophore (dye) systems whose MPA cross-section values are much larger than ordinary and commercial available dyes. In general, excellent multi-photon active chromophores will exhibit high MPA cross-section values. They also may provide quite high fluorescence quantum yields upon a MPE, which is essentially advantageous for frequency-upconversion imaging and sensing applications. In addition, these chromophores can be easily dissolved in some organic solvents or doped into transparent polymer matrixes while retain reasonable opto-chemical stabilities. When chromophore-doped polymer rods are utilized for power limiting or frequency-upconversion lasing, opto-bleaching or opto-damaging may occur upon the irradiation of high-power (energy) and/or high-repetition laser pulses.

Regarding the structural designs of chromophores, there were two parallel early efforts to achieve a high 2PA activity. One was pursued by the research group led by Reinhardt and Tan, whose strategy was more focused on developing some chromophores with asymmetrical structures and multi-branched structures.[10,11] Meanwhile the other effort was pursued by Brédas, Marder, Perry, and co-workers with emphasis on some symmetrical structures.[12,13]

Experimental results have shown that the novel chromophores with three-branched structure (such as AF350, AF380, and PRL701) are among the best chromophores in the respect of high 2PA cross-section values, reasonable solubility and good opto-stability.[14–16]

Design of dendritic structures is another direction to increase the effective number of multi-photon absorbing centers within a macromolecule.[17–19] For the non-conjugated dendritic chromophores, only a linear enhancement in two photon absorption was found in going from a generation to the next.[18] However, with use of a π-conjugated dendritic architecture, a cooperative increase in 2PA was achieved from a generation to the next.[19]

14.2.3.2 *Polymers, liquid crystals, and ionic liquids*

For some applications (such as optical power limiting), a larger MPA coefficient is desired, which is proportional to the density of chromophore molecules. However, the solubility of those MPA active chromophores in solutions or polymeric matrices is usually limited to be less than 0.03–0.1 M.

To overcome the solubility limitation of multi-photon absorbing molecules, one of the best solutions is to employ a modified MPA-active chromophore as the basic repeating unit to form an intrinsic MPA polymer.[20,21] The other alternative solutions are to synthesize intrinsic organic liquid crystals[22,23] or ionic liquids[24], which consist of pure MPA active chromophore molecules. In these cases, the molar concentration of MPA active chromophores can easily exceed 1 M.

14.2.3.3 *Coordination and organometallic compounds*

The coordination compounds and organometallics are hybrid systems comprised of a metal center covalently surrounded by organic ligands, which are the above-mentioned π-conjugated chromophores with certain functional groups for metal ion binding. Here, the metal ions can either serve as a non-interacting center (or a multidimensional template) for increasing the molecular density of two-photon active components or serve as an important part of the structure to tune the charge-transfer process. When the metal ions are used as templates, they can assemble simple ligands to more sophisticated complexes, which will have tunable optical and electronic properties, depending on the nature of metal center and the energetics of the metal-ligand interactions. In principle, metal ions including main group metals, transition metals and lanthanides can be used for the construction of multi-photon absorbing organometallic complexes. However, the multi-photon active ligands should have metal ion chelatable group such as phenanthrolines, terpyridines, crown ethers, etc.[25–27] The introduction of metal ions into an organic π-conjugated system can enhance its MPA if there is a favorable intramolecular charge transfer. In some instances, materials with quenched emissions could be obtained by the introduction of metal ions, which is a favorable feature for the applications where strong fluorescence is hazardous.[25]

Metallophophyrins are another important class of coordination and organometallic compounds with large two-photon absorption.[28] In combination with their unique excited state absorption capability, metallophophyrins have been frequently used for a nonlinear absorbing medium in optical limiting, as well as a photosensitizer in photodynamic therapy.

14.2.3.4 Nanoparticles

Similar to the broad interest in nanostructured materials for the material community, there is an increasing attention focused on the nonlinear optical properties of metal and semiconducting nanoparticles, especially on the aspect of surface-plasmon enhancement that is important for fluorescence imaging-based applications. The other nonlinear optical applications related with multi-photon active nanoparticles are drug delivery, cancer targeting, and photodynamic therapy.

The intrinsic MPA of semiconductor nanoparticles and their upconversion emission properties depend on their compositions, shapes, sizes, coatings, and environments.[29–31]

It has been experimentally shown that there is an obvious enhancement of the multi-photon induced upconversion emission when the organic chromophores are incorporated into some metal nanaoparticles.[32,33] Moreover, it was reported that a strong upconversion photoluminescence could be observed from gold nanoparticles with diameters as small as a few nanometers upon an excitation with fs-laser pulses of 790-nm wavelength.[34]

14.2.3.5 Optical fibers and waveguides

For optical limiting, optical switching, multi-photon pumped lasing, and stimulated scattering in a multi-photon absorbing medium, the long interaction length (effective thickness) and high local light intensity are beneficial for device performances. To meet these two requirements, multi-photon active materials can be a core-medium adopted into an optical fiber or waveguide configuration. In some special cases, after coupling a focused laser beam into the waveguide structures, a high local light intensity can be retained for a much longer interaction length.

14.3 Characterizations of MPA materials

14.3.1 Selection of excitation wavelengths

For a given nonlinear absorbing material, the choice of excitation wavelength depends on the purpose of specific studies or applications. For novel materials, it is important to know their nonlinear absorption spectral structures and then to choose suitable excitation wavelengths. Generally speaking, it is not an easy task to measure the complete MPA spectra for a given material.

For simplicity, Fig. 14.5 shows a schematic diagram of the linear (one-photon) absorption spectral band centered on the wavelength position of λ_0, as well as the corresponding 2PA and 3PA bands for a given medium. Here, it should be noted that the selection rules, pathways of molecular transitions, resonance enhancements, and magnitudes of relevant matrix elements for one- and multi-photon processes may be generally different; therefore even for a given material, the relative shapes of MPA spectra may differ from the corresponding one-photon absorption (1PA) spectrum. For instance, for centrosymmetric molecules the 2PA spectrum is inherently different from the 1PA spectrum because of the different selection rules (see Section 14.1.4). For 3PA processes, the selection rules are the same as for 1PA irrespective of the molecular symmetry, and a similar band-shape may appear in the 3PA spectrum. However, the intensity and relative distribution of different spectral components may still differ from the linear spectrum. For these reasons, the peak (or central) wavelengths for the corresponding 2PA and 3PA bands may not be exactly located at $2\lambda_0$ and $3\lambda_0$ positions. Based on quite limited results of spectral comparison between 1PA and 2PA spectra for a given sample medium, it could be roughly assumed that 2PA peak wavelength is close to or slightly shorter than the $2\lambda_0$ position. On the other hand, there is still lack of reliable experimental comparisons between the linear spectrum and MPA spectra for given

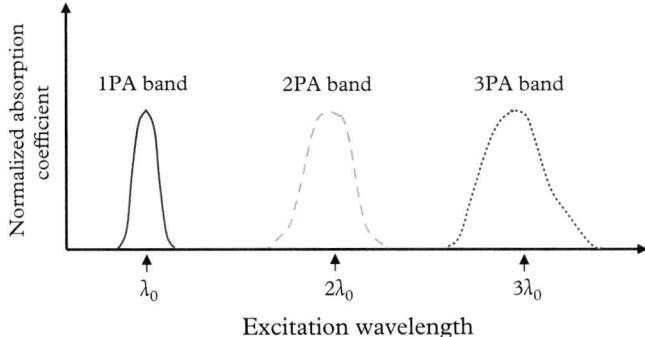

Figure 14.5 *Schematic description of relative spectral positions and bandwidths for 1PA, 2PA, and 3PA bands. The relative shapes of these three absorptive bands do not have to be the same.*

materials. Nevertheless, for a given sample medium without knowing its MPA spectral structures, one may choose those wavelengths located in the following spectral ranges for MPA excitation studies:

$$\begin{aligned} \lambda_0 \ll \lambda_{exc} &\leq 2\lambda_0 \quad &\text{(for 2PA)} \\ 2\lambda_0 \ll \lambda_{exc} &\leq 3\lambda_0 \quad &\text{(for 3PA)} \\ 3\lambda_0 \ll \lambda_{exc} &\leq 4\lambda_0 \quad &\text{(for 4PA)} \end{aligned} \qquad (14.3\text{-}1)$$

Several different types of coherent light sources can provide intense coherent radiation with suitable wavelengths for multi-photon related studies:

1. pulsed or mode-locked lasers working at single (or several discrete) wavelength(s), such as Nd:YAG lasers (1064 nm, 532 nm), Ti:sapphire lasers (~800 nm), excimer lasers, and Raman lasers;
2. pulsed lasers and optical parametric generators (OPGs) working in certain tunable spectral ranges, such as tunable dye lasers, Ti:sapphire lasers, and laser-pumped OPGs that are based on second-order nonlinear crystals;
3. coherent white-light continuum (WLC) generators, in which an intense coherent white-light beam can be generated in various transparent nonlinear optical media (such as heavy water, quartz crystal, and optical glasses and fibers), pumped by an input laser beam of high-peak power. This type of coherent light source with super-broad continuous spectral distribution is especially useful for MPA spectral measurements.

14.3.2 Measurement of MPA cross-section at discrete wavelengths

14.3.2.1 *Nonlinear transmission (NLT) method*

The most simple and straightforward method to experimentally determine values of MPA coefficient and/or cross-section is to measure the NLT of the sample as a function of the input intensity of an applied laser beam at a given wavelength. For example, if the linear (one-photon) absorption of a sample medium at the incident wavelength is negligible, the 2PA-induced NLT for an input laser beam passing through the sample medium can be expressed as (cf. Eq. (14.1-10)):

$$T(I_0) = \frac{1}{1 + \beta(\lambda)I_0 l_0}. \quad (2PA) \quad (14.3\text{-}2)$$

Here, I_0 is the initial intensity of the incident light beam with wavelength λ, β is the 2PA coefficient, and l_0 is the optical path length (thickness) of the sample. In above equation it was assumed that (i) the incident light beam possesses a uniform transverse intensity distribution and (ii) the beam's section is unchanged within the sample's path length. However, these two assumptions are not exactly held in most experimental conditions under which the incident beam is usually focused on the center of a sample with thickness of about 0.2–2 cm. In this case, the focused laser beam has a nearly Gaussian transverse intensity distribution near the sample center. With this correction, Eq. (14.3-2) should be modified as[35]

$$T(I_0) = \frac{\ln(1 + \beta(\lambda)I_0 l_0)}{\beta(\lambda)I_0 l_0}. \quad (2PA) \quad (14.3\text{-}3)$$

For the same measured NLT values, the estimated β value from Eq. (14.3-2) will be ~0.43 times smaller than that estimated from Eq. (14.3-3), although this difference will be much less than the experimental uncertainty that is mainly related with the estimation of a spatiotemporally averaged value of light intensity (I_0) at the center position of the sample medium.

Based on the known values of I_0, l_0, and the measured NLT, $T(I_0)$, the value of the 2PA coefficient $\beta(\lambda)$ (in units of cm/GW) can be readily determined through Eq. (14.3-2) or Eq. (14.3-3). Moreover, if the concentration of the two-photon absorbing molecules is also known, the 2PA cross-section σ_2 (in units of cm^4/GW) or σ_2' (in units of cm^4 s or GM) can be finally determined through Eq. (14.1-7) or Eq. (14.1-8).

Mentioned above is a simplest one-point measurement. To reduce the experimental uncertainty, one may conduct a multi-point measurement, i.e., to repeat the same measurement at several different I_0 levels, and then average over the measured values of $\beta(\lambda)$. If there are enough measured points, one may also make experimental curve of NLT as a function of incident intensity, and then fit the experimental data by using Eq. (14.3-2) or Eq. (14.3-3) with a best-fitting β value. In a multi-point measurement, the variation of the input intensity level can be done by two different methods. One is to fix the sample position and vary the input intensity (or pulse energy) by using a variable optical attenuator; the other is to fix the input pulse energy level of a given focused laser beam and change the distance from the sample to the focal point position. The second approach is also called (open-aperture) Z-scan method.[36]

The same experimental setup and procedures mentioned above can also be applied for measuring 3PA and MPA coefficients by using Eqs. (14.1-12) and (14.1-15), respectively. Then the 3PA and MPA cross-section values can be finally determined based on the relationships expressed by Eqs. (14.1-13) and (14.1-16).

In general, MPA coefficient and cross-section values for a given sample medium are the function of wavelength of the input light. If there is a tunable coherent light source, one may repeat the NLT measurement with deferent wavelengths and finally obtain multi-points (wavelengths) information of the MPA spectral distribution.

14.3.2.2 Two-photon excited fluorescence (2PEF) method

In some case, when an expensive ultra-short and high peak power laser source is not available, or the concentration of the tested samples cannot be high enough, the 2PA-induced NLT change may be too small to measure. In such a case, an alternative method can be used to determine the 2PA cross-section at a given excitation wavelength if the sample is highly fluorescent. This method is called two-photon excited fluorescence (2PEF) technique,[37] which is based on a 2PA-induced fluorescence intensity measurement that is compared to a standard fluorescent sample, of which the 2PA cross-section value $\sigma_2^s(\lambda)$ and the quantum yield of 2PA-induced fluorescence $\eta_{2PEF}^s(\lambda)$ at a given excitation wavelength (λ) are known. In this case, the measured 2PEF signal intensity of the standard sample solution can be simply expressed as

$$S_{2PEF}^s(\lambda) = A_\Sigma \eta_{2PEF}^s(\lambda) \sigma_2^s(\lambda). \qquad (14.3\text{-}4)$$

Here, A_Σ is a phenomenological proportional coefficient that depends on the excitation light intensity and many other experimental factors, including the optical setup, spectral sensitivity and aperture of the detector, sample concentration and thickness, etc. However, under the same experimental conditions except replacing the standard sample by a tested sample, the measured 2PEF signal intensity emitted from the latter can be expressed as:

$$S_{2PEF}^{test}(\lambda) = A_\Sigma \eta_{2PEF}^{test}(\lambda) \sigma_2^{test}(\lambda). \qquad (14.3\text{-}5)$$

Comparing Eq. (14.3-5) to Eq. (14.3-4) we have

$$\eta_{2PEF}^{test}(\lambda) \sigma_2^{test}(\lambda) = (S_{2PEF}^{test}/S_{2PEF}^s) \eta_{2PEF}^s(\lambda) \sigma_2^s(\lambda). \qquad (14.3\text{-}6)$$

The physical meaning of Eq. (14.3-6) is that if the four quantities on the right-hand side of the equation are known, we can determine the product of the two quantities on the left-hand side. If it can be assumed that $\eta_{2PEF}^{test} \approx \eta_{2PEF}^s \approx 1$, then we will have

$$\sigma_2^{test}(\lambda) = (S_{2PEF}^{test}/S_{2PEF}^s) \sigma_2^s(\lambda). \qquad (14.3\text{-}7)$$

The major advantage of 2PEF method is that in conjunction with a highly sensitive photo-electric detector (photomultiplier or CCD array), even very weak fluorescence signals can be measured. Therefore, no high-power or high-energy pulsed laser source is needed, and a sample of low concentration ($\leq 10^{-3}$–10^{-4} M) or with a small path length can be employed. The disadvantage of this method is that in general it can only determine the product of $\eta_{2PEF}^{test}(\lambda)\sigma_2^{test}(\lambda)$, not $\sigma_2^{test}(\lambda)$ alone. Logically, $\eta_{2PEF}^{test}(\lambda)$ could be determined only after $\sigma_2^{test}(\lambda)$ is known. From an experimental viewpoint, the absolute measurements of quantum yield value for either $\eta_{2PEF}^s(\lambda)$ or $\eta_{2PEF}^{test}(\lambda)$ are quite complicated; besides, some MPA materials are weakly fluorescent or even non-fluorescing.

14.3.2.3 Time-regime dependence of measured cross-section values

During the late 1990s, researchers found that even for the same nonlinearly absorbing medium excited at the same wavelength, the 2PA cross-section values measured by using the NLT method were dependent on the pulse duration of the applied laser beam. In particular, the apparent 2PA cross-section values measured in the sub-picosecond or femtosecond regime for a given dye–solution sample could be two orders of magnitude smaller than that measured in the 5–10 ns regime.[15,38,39] This huge difference of 2PA cross-section values for the same molecular system

448 *Multi-Photon Nonlinear Optical Effects*

might puzzle some researchers, as the cross-section should be a molecular constant depending only on the wavelength, not the time duration of the applied light pulses. Although time regime-dependent results of MPA cross-section measurements can strongly depend on specific experimental conditions and sample materials, some proposed explanations described below can be helpful for a better understanding in this specific issue.

MPA-induced excited state absorption is proposed to be one of the major mechanisms leading to a much greater effective (or apparent) MPA cross-section value measured in the nanosecond regime, comparing to the true cross-section values measured in the sub-picosecond regime. For simplicity, Fig. 14.6 shows the schematic diagram of energy-state structures typical for most multi-photon active organic molecular systems.[40]

With two-photon (or three-photon) excitation, molecules in the lowest vibrational sublevel of ground electronic singlet state (S_0) can be excited to a higher electronic singlet state (S_1 or S_n). These excited molecules have a tendency to rapidly relax to the lowest vibrational sublevel of S_1 through the internal conversion and/or vibrational relaxation. For commonly investigated fluorescent organic chromophores, this fast relaxation process takes place within a time period of $\leq$1–10 ps. The lowest vibrational sublevel of S_1 is a metastable state exhibiting a longer lifetime around 0.2–2 ns values. Once molecules relaxed to this state, they have four possible pathways

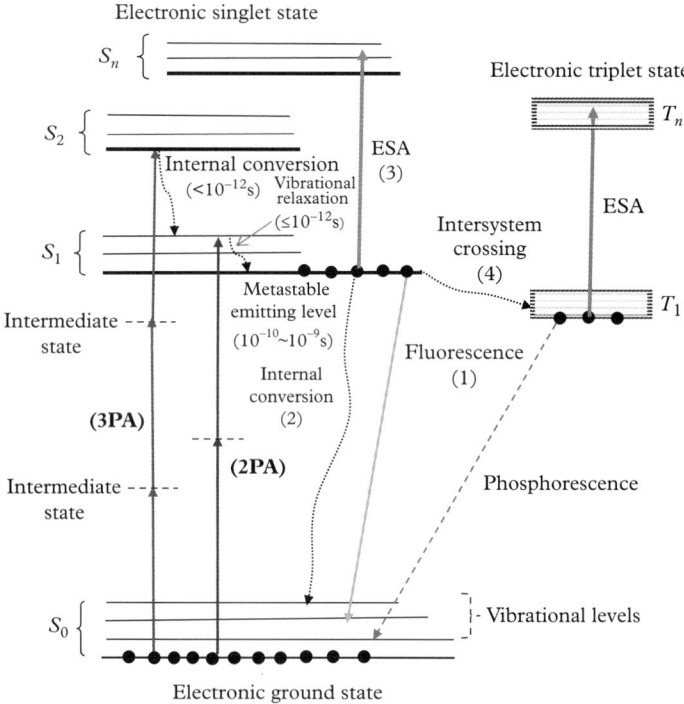

Figure 14.6 *Schematic state-level structures and MPA-induced transition pathways of an organic chromophore system. It is noted that there are four possible channels for the excited molecules to leave from the metastable emitting level. The dotted-line arrows represent radiationless relaxations.*

to leave it: (1) to emit fluorescence and return to the ground state, (2) to directly return to the ground state through a radiationless transition via internal conversion, (3) to be further excited to a higher excited state through 1PA, and (4) to relax to the triplet state T_1 that usually exhibits a much longer lifetime (e.g., 10^2 ns) than S_1. If the medium is highly fluorescent, i.e., the quantum yield of the fluorescence is close to 1, the second and fourth processes can be neglected. The third process is called MPA induced excited-state absorption (ESA).

It should be noted that the ESA probability of a molecule making the transition from a lower excited state to a higher excited state is proportional to the lifetime in the former. If the molecular population in the metastable excited state is non-negligible and the 1PA cross-section for the excited molecules in that state is much larger than the MPA cross-section of the ground-state molecules, ESA caused attenuation of the input light beam may become comparable to or even greater than the MPA-induced attenuation. In that case, the apparent MPA coefficient (or cross-section) values determined by measuring the NLT of the sample medium may be considerably or even significantly larger than the true values contributed from a sole MPA process.

It is obvious that the extent of ESA influence on the NLT strongly depends on the state structures and transition properties of the sample medium, and also depends on the parameters (wavelength, intensity level, and pulse duration) of the input laser beam. For given sample medium and excitation wavelength, the ESA contribution to the measured nonlinear attenuation is merely determined by the pulse duration and the intensity level of the input laser beam. To elucidate this statement, we take two-photon excitation as an example and only consider two extreme situations. In the first case, we assume that the pulse duration is around 100–160 fs, such as the pulses from an ultra-short Ti:sapphire laser device. In this situation, the pulse duration is so short that most of the two-photon excited molecules have not relaxed yet to the lowest sublevel of S_1. Only after a time period of approximately 1–2 ps (internal conversion and vibrational relaxation time), those excited molecules can be finally accumulated in this metastable sublevel, while the input laser pulse is already passed away. For this reason the major ESA influence from the metastable state can be neglected, and the 2PA cross-section values measured with sub-picosecond pulses are more close to the true values. In the second case, we assume that the input pulse duration (e.g., 5–10 ns) is considerably longer than the lifetime of the lowest sublevel of S_1, thus a large number of excited molecules cab be accumulated in that level, and therefore the ESA process from this metastable state can take place with a great effectiveness. In addition, when the input laser pulse duration is less than a picosecond, once the excited molecules have left the ground state, they are unable to return to it within the pulse duration, thus the depletion of ground state is an unmitigated process. In contrast, for the input pulses of nanoseconds, the excited molecules can return to the ground state for many circles and repeatedly contribute to the MPA and ESA processes. At least, these two considerations mentioned above can partially explain why the effective or apparent 2PA cross-section values measured in the nanosecond regime are significantly greater than that measured in the femtosecond regime.

There are some experimental examples of 2PA- and 3PA-induced ESA studies.[41–44] Sutherland et al. have reported their quantitative results of ESA influence on the NLT behavior of two-photon absorbing solutions measured in the nanosecond regime.[45]

14.3.3 Saturation effect of MPA in the sub-picosecond regime

In the sub-picosecond regime, a saturation effect of pure MPA process may be observed, which means that when the input laser intensity is higher than a certain level, the measured MPA coefficient value will decrease with further increase of the input intensity levels.[46–48] To explain the origin of saturation effect, we should return to the expression for MPA coefficient given in

Eq. (14.1.16), in which it is assumed that the MPA coefficient is simply proportional to the molecular density N_0. This assumption is acceptable only when the input light intensity is not too high and the depletion of the molecular population in the ground state can be neglected during the laser pulse duration. In a more rigorous manner for a two-state system, N_0 should be replaced by $\Delta N = N_1 - N_2$, where N_1 is the population density in the ground state, while N_2 is the population density in an upper state that becomes populated via MPA. For a weak excitation, we may approximately assume that $N_2 \approx 0$ and $N_1 \approx N_0$. However, in the case of very strong excitation, the depletion of N_1 is no longer negligible, which means that $(N_1 - N_2)$ will actually depend on the light intensity I_0. Thus for 2PA processes, Eq. (14.1-7) should be rewritten as[49]

$$\beta(I_0) = \frac{\sigma_2 N_0}{1 + (I_0/I_{s,2PA})^2} = \frac{\beta_0}{1 + (I_0/I_{s,2PA})^2}. \qquad (14.3\text{-}8)$$

Here, β_0 is the non-saturation 2PA coefficient that is independent of I_0, and $I_{s,2PA}$ is a material parameter called saturation intensity of a given two-photon absorbing medium at which the effective β value decreases to the half of β_0. By combining this expression with the original Eq. (14.3-2), the NLT becomes

$$T(I_0) = \frac{1}{1 + \beta_0 I_0 l_0 / \left[1 + (I_0/I_{s,2PA})^2\right]}. \qquad (2PA) \qquad (14.3\text{-}9)$$

As an example, Fig. 14.7 shows the measured NLT for three 2PA chromophore solutions as a function of the input intensity of 775-nm and 160-fs laser pulses. From this figure one can see that at low input intensity levels, the experimental data can be well fitted without saturation (Eq. (14.3-2)), whereas at high input levels the data can be fitted only by a theory (Eq. (14.3-9)) that includes saturation. Based on this type of measurements and theoretical fitting, both non-saturation value of β_0 and saturation intensity $I_{s,2PA}$ can be determined.

In the 3PA case, if the input light intensity is high enough, the same saturation effect may also be observed in NLT measurements. With inclusion of saturation effect, the original expression for 3PA coefficient, Eq. (14.1-13) turns into[49]

$$\gamma(I_0) = \frac{\sigma_3(\lambda) N_0}{1 + (I_0/I_{s,3PA})^3} = \frac{\gamma_0}{1 + (I_0/I_{s,3PA})^3}. \qquad (14.3\text{-}10)$$

Here, γ_0 is the non-saturation 3PA coefficient, and $I_{s,3PA}$ is the saturation intensity for a given three-photon absorbing medium. In this case, based on Eq. (14.1-12) the general expression of 3PA induced NLT will be

$$T(I_0) = \frac{1}{\sqrt{1 + 2\gamma_0 I_0^2 l_0 / \left[1 + (I_0/I_{s,3PA})^3\right]}}. \qquad (3PA) \qquad (14.3\text{-}11)$$

14.3.4 Measurements of MPA spectra

Hitherto, we have only considered the measurement of MPA coefficient or cross-section at a single or several discrete wavelengths by using different methods. In order to obtain a complete MPA spectrum for a given sample medium, researchers have to repeat this type of measurement at

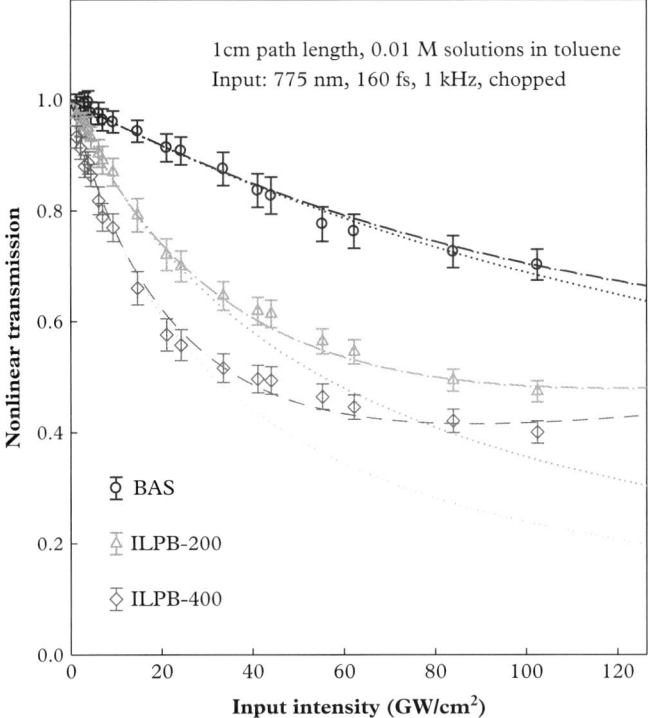

Figure 14.7 *2PA-induced NLT versus the input light intensity for three chromophores (BAS, ILPB-200, and ILPB-400) in toluene. The dotted curves are the best fits given by the non-saturation theory, and the dash-dotted curves are the best fits given by theory that includes saturation. (After He et al.[49] Copyright 2007, AIP Publishing LLC).*

a sufficient number of different wavelengths covering a broad spectral range. To do so, a tunable coherent light source can be utilized in conjunction with different approaches, such as direct NLT measurement,[50] open-aperture Z-scan measurement,[51] multi-photon excited fluorescence measurement,[52] and the pump–probe two-beam approach.[53] Obviously, this kind of measurement is very sophisticated and highly time-consuming.

There is another more efficient approach by using an intense coherent WLC (white-light continuum) beam to replace a tunable monochromatic laser source. This new approach may provide two major advantages. First, it is a continuous-spectrum measurement instead of a discrete multi-wavelength measurement. Second, it is a time-saving approach based on which a complete 2PA spectrum for a given sample medium can be recorded within a very short time. The disadvantage of using a WLC is that the spectral intensity at a given wavelength is still lower than that can be provided by a tunable coherent monochromatic light source (such as an OPG).

WLC generation is a nonlinear optical effect that can be efficiently produced in many transparent liquids and solids by using high peak-power ultrashort laser pulses of picosecond or femtosecond duration. The laser generated continuum is a white-light coherent emission that features high directionality, super-broad spectral band (usually covering entire visible range and near

IR range), coherent phase relationship among different spectral components, and quite high spectral intensity. Even in the early stage of continuum-generation with picosecond laser pulses, this new type of coherent white light source was employed to measure the transient linear absorption spectra of sample materials. The major advantage of this application is the elimination of the need for spectral tunability and wavelength-scanning.

14.3.4.1 Nondegenerate 2PA spectral measurement

In 1999, Belfield et al. reported a novel approach to measure nondegenerate 2PA spectra by utilizing a pump-probe two-beam configuration, in which the strong pump was a monochromic IR laser beam and the weak probe was a WLC beam.[54] As shown schematically in Fig. 14.8, these two pulsed beams were overlapping in a two-photon absorbing sample with a small crossing angle. Assuming that the very weak probe beam cannot produce degenerate 2PA by itself, but it can be nonlinearly attenuated through the simultaneous absorption of one photon from the monochromatic pump beam and one photon from the white-light probe beam. By recording the relative spectrum change of the probe beam with and without overlapping with the pump beam, one could finally obtain the relative spectrum of nondegenerate 2PA for the sample medium.

Assuming that the duration values of the pump pulses and the probe pulses are both in sub-picosecond regime, a temporal overlap is as important as the spatial overlap between these two beams to ensure a nondegenerate 2PA process. For this reason, a variable optical delay line was needed. In experiments, the pump beam was a 1220-nm output from an OPG, while the probe beam (WLC) was generated in a 1.5-mm-thick fused silica plate on which a 1400-nm laser beam from another OPG was focused. These two OPGs were pumped separately by 775-nm, 120-fs laser pulses from the same Ti:sapphire device.

In this case, there is a difficulty related to the chirping effect taking place in continuum generation, and also related to the group velocity dispersion (GVD) effect on the probe pulse propagation of different spectral components through optical elements and sample medium. Owing to these two effects, the arriving time in the center of sample medium will be different for different spectral components of the continuum pulse. This means that only part of the probe pulse can be temporally overlapping with the pump pulse. To overcome this difficulty, one may repeat multi-point measurements with different time-delay values between these two beams, to ensure a sequential temporal overlapping with all spectral components of the probe pulse.[55]

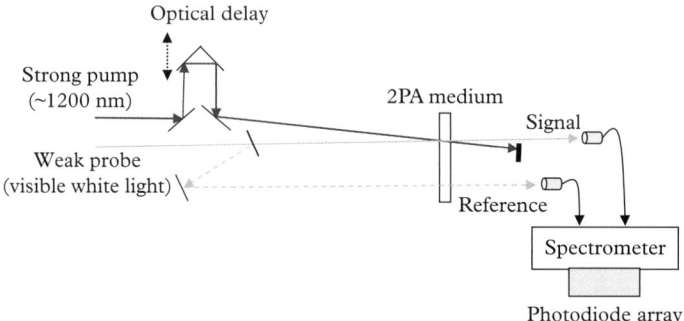

Figure 14.8 *Optical layout for measuring nondegenerate 2PA spectra by using a two-beam configuration: a strong IR pump beam and a weak visible continuum beam.*

14.3.4.2 Degenerate 2PA spectral measurement

In 2002, He and co-workers reported another new approach to directly measure the degenerate 2PA spectra by using only one intense continuum beam.[56] The unique feature of this method is that a collimated and powerful WLC beam passes through a spectral-dispersion element (prism or grating), and then is focused by a lens on the sample medium. Under this special arrangement, different spectral components of the WLC pulse will be spatially separated inside the sample, nondegenerate 2PA between those different spectral components can be excluded, and only degenerate 2PA induced nonlinear attenuation of these spectral components will be recorded. The major advantages of this method are its simplicity and reliability, as well as the chirping effect and GVD effect related to continuum generation and propagation have no influence on the 2PA spectral measurement. In addition, there is no critical requirement for spatial and temporal overlapping as needed in the two-beam (probe-pump) approach.

Figure 14.9 shows the optical set-up for degenerate 2PA spectral measurements using an intense and spectrally dispersed WLC beam. Here, heavy water (D_2O) was chosen as the non-linear transparent medium for providing continuum generation because of its high efficiency and stability. The pump source for this continuum generation was a focused ultra-short pulsed laser beam provided by a Ti:sapphire laser oscillator/amplifier system with the wavelength ~790 nm and pulse duration ~140 fs. The intensity distribution of different spectral components of the continuum at the sample position can be further imaged through a camera lens-set on the surface of a CCD array controlled by a computer system. By comparing the recorded continuum spectral distribution passed through a chromophore solution to that passed through a pure solvent sample, the dynamic attenuation of different spectral components due to the investigated chromophores can be readily determined. Furthermore, if the linear absorption of the chromophores in the measured spectral range is known or negligible, the relative nonlinear absorption spectrum due to degenerate 2PA can be obtained. After getting the relative degenerate 2PA spectral distribution curve for a given sample medium, one may measure the absolute value of the 2PA coefficient $\beta(\lambda_0)$ at a given wavelength λ_0 via the NLT method using a monochromatic laser beam, so that after calibration the absolute 2PA spectral curve can be finally obtained.

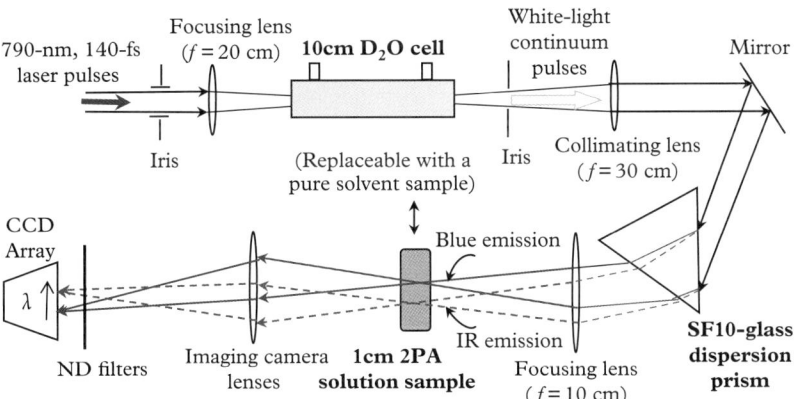

Figure 14.9 *Experimental setup for degenerate 2PA spectral measurement, using a single and spectrally dispersed WLC beam. (After He et al.[56] Copyright 2002, Optical Society of America).*

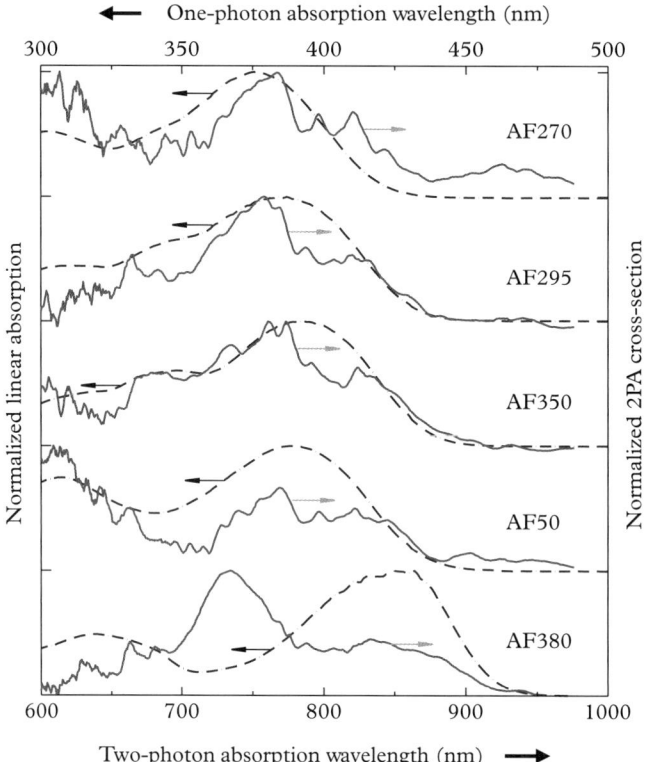

Figure 14.10 *Normalized spectra of 2PA (solid-line curves) and 1PA (dash-dotted line curves) of five AFX chromophores in THF. (After He et al.[57] Copyright 2004, AIP Publishing LLC).*

Figure 14.10 shows the normalized degenerate 2PA spectra for five AFX chromophores (AF270, AF295, AF350, AF50, and AF380) in solution phase in tetrahydrofuran (THF),[57] measured with the single WLC beam configuration shown in Fig. 14.9. The concentration and path length for all samples were 0.02 M and 1 cm, respectively. In the same figure, the solid-line curves are the normalized linear absorption spectra measured with samples of a much lower concentration (0.0001 M). From Fig. 14.10 one can see two noticeable features: (a) for AF-270, −295, and −350 samples, there is a similarity between 2PA and linear absorption spectra; (b) for AF380, the 2PA spectrum is obviously different from the corresponding linear absorption spectrum, which may exemplify in this specific case the different selection rules for one- and 2PA processes.

14.3.4.3 *3PA spectral measurement results*

In principle, all methods used for 2PA spectral measurements are also applicable to 3PA or even 4PA spectral measurements. In the latter two cases, of course, a much higher intensity (or peak power) of the coherent light source is required. We present in what follows two examples of 3PA spectral measurement that were obtained with two different technical approaches.

Suo et al. reported their 3PA spectral measurement of three chromophores (TFA-01, TFA-02, and TFA-03), which belong to a series of fluorophores based on the tri(9,9-diethyl-9H-fluorenyl)amine unit.[58] These results were obtained by using the three-photon excited fluorescence (3PEF) method. The samples were chromophores in dichloromethane of concentration $\sim 10^{-4}$ M; the excitation source was output beam from an OPG with a wavelength tunable from 1100 to 1600 nm, pulse duration $\sim$100 fs and repetition rate of 1 kHz. Through comparison with a standard sample medium whose absolute value of 3PA cross-section at a given wavelength was known, the absolute 3PA spectral data could be finally determined. As an example, the complete 3PA spectral data for a TFA-03 sample solution are shown in Fig. 14.11 by full squares. The solid-line curve is the linear absorption spectrum. It can be seen that in this case the measured 3PA spectrum is generally following the 1PA spectrum.

Zheng et al. reported a cooperative enhancement of the 3PA cross-section in a series of chromophores, going from a one-branched to a three-branched and then to a dendritic structure.[59] The results were obtained by using the direct NLT method. In 3PA measurement, the concentrations for the one-branched, three-branched, and the dendritic chromophores were 0.02 M, 0.0067 M, and 0.0033 M in chloroform ($CHCl_3$), respectively. The pump source for the 3PA measurement was a focused ultra-short pulsed laser beam from an OPG with the wavelength tunable from 1100 to 1600 nm. The pulse duration and the repetition rate of the pump beam were 160 fs and 1 kHz, respectively. The output laser energy from the OPG was kept at $\sim$1.0 µJ by using neutral density filters. As an example, the 3PA cross-section values as a function of the input light wavelength for the dendritic chromophore are shown in Fig. 14.12. The dashed line is a

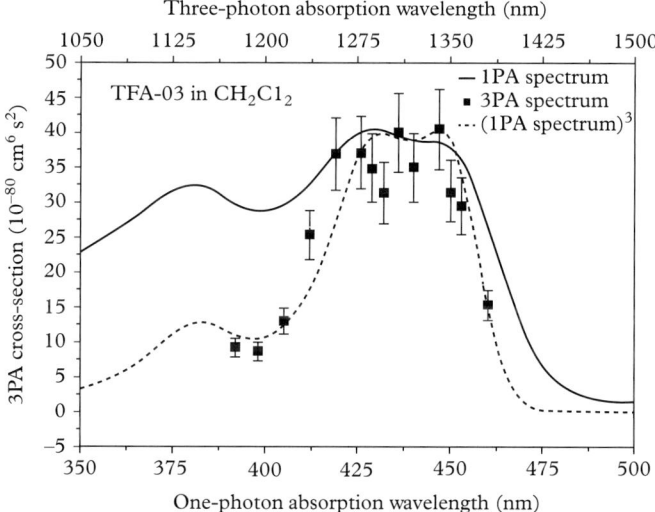

Figure 14.11 Measured 3PA spectral data of TFA-03 solution in CH_2Cl_2 by using the three-photon excited fluorescence (3PEF) method. The solid-line curve is the relative linear absorption spectrum, and the dotted-line curve is the cube of the relative linear absorption spectrum. (Reproduced with permission from Suo et al.[58] Copyright 2005, American Chemical Society).

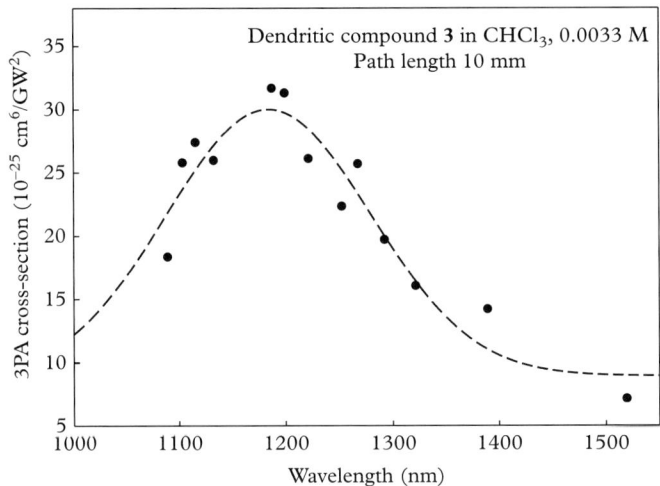

Figure 14.12 *Measured 3PA cross-section values of a π-conjugated dendritic chromophore solution in $CHCl_3$ versus the excitation wavelength by using the NLT method. The dashed line is a Gaussian fitting curve. (After Zheng et al.[59] Copyright 2006, American Chemical Society).*

Gaussian fitting curve for the 3PA spectrum. As shown in the figure, the dendritic chromophore exhibits a peak 3PA cross-section value of 30.0×10^{-25} cm^6/GW^2 at ~1200 nm wavelength that is approximately three-times longer than the corresponding linear absorption peak wavelength.

14.3.5 Characterization of MPA-induced fluorescence emission

The majority of multi-photon active organic compounds investigated so far are fluorescent materials that can emit visible fluorescence excited by IR coherent radiation. Studies on properties of this type of fluorescence emission may help researchers to have a better understanding about the transition pathways, selection rules, relaxation rates, excited-state lifetimes, and other useful information of MPE processes. Characterizations of MPA induced fluorescence from a given material usually include the following measurements: emission spectrum, quantum yield, temporal (rising and decaying) behavior, excitation intensity dependence, concentration dependence and polarization dependence.

14.3.5.1 Excitation intensity dependence of fluorescence emission

The measurement of fluorescence emission intensity as a function of the input excitation intensity is particularly essential to determine how many simultaneously absorbed photons are involved in the elementary excitation process. In many published papers, the authors presented a square dependence of the fluorescence on the input intensity to prove the 2PA nature, or a cubic dependence to prove the 3PA nature. Strictly speaking, even for a pure 2PA or 3PA process, the simple square or cubic law between the observed fluorescence signals and the input excitation intensity does not hold automatically. Instead, only under some specially arranged conditions, one may observe a simple square law or cubic law. To explain this, we have to return to some original equations given in Section 14.1. For a pure 2PA process and according to Eq. (14.1-6), the total change of the intensity of the input beam will be

$$\Delta I = I_0 - I(l_0) = \frac{I_0^2 \beta l_0}{1 + I_0 \beta l_0}. \tag{14.3-12}$$

It can be assumed that the molecular population number (ΔN) excited to the emitting level is proportional to ΔI; therefore the signal intensity (I_{fluor}) of the detected fluorescence from the sample cell will be proportional to ΔN, and thus we have

$$I_{fluor} \propto \Delta N \propto \Delta I \propto \frac{I_0^2 \beta l_0}{1 + I_0 \beta l_0}. \quad (2PA) \tag{14.3-13}$$

It is obvious that in general there is no simple quadratic relationship between the fluorescence intensity and the input excitation light intensity. Only when $I_0 \beta l_0 \ll 1$, the following quadratic relationship is approximately valid,

$$I_{fluor} \propto I_0^2 \beta l_0. \quad (2PA) \tag{14.3-14}$$

In practice, to meet this requirement, the concentration and thickness (l_0) of the sample medium should be quite small, and the input intensity should not be too high. Similarly, for a pure 3PA-induced fluorescence process, based on Eq. (14.1-12), we can write the total input intensity change of the input beam as

$$\Delta I = I_0 - I(l_0) = I_0 \left(1 - \frac{1}{\sqrt{1 + 2I_0^2 \gamma l_0}}\right). \quad (3PA) \tag{14.3-15}$$

Only when the condition of $2I_0^2 \gamma l_0 \ll 1$ is met, we have the following approximate expression:

$$I_{fluor} \propto \Delta I \approx I_0 [1 - (1 - I_0^2 \gamma l_0)] \approx I_0^3 \gamma l_0. \quad (3PA) \tag{14.3-16}$$

In practice, to observe a cubic dependence, the values of sample concentration, thickness, and input intensity should not be too high.

As a typical example, Fig. 14.13 shows the fluorescence spectra and intensity dependence of the silicon 3-nm quantum dots solution in $CHCl_3$, excited by 339-nm, 778-nm, and 1335-nm laser pulses of ~160 fs duration.[31] From the results shown in Fig. 14.13 one can see that the emission spectra remain basically the same under three different excitation conditions; the 2PA-induced fluorescence intensity basically follows the square low, whereas the 3PA-induced fluorescence intensity follows the cubic law.

14.3.5.2 Spectral profile and lifetime of MPA-induced fluorescence

For a given multi-photon absorbing and fluorescent organic chromophore sample, if the molecules excited via MPA are finally relaxed to the same metastable emitting state, i.e., the lowest vibrational sublevel of S_1 as shown in Fig. 14.6, the fluorescence spectral and decaying behaviors should be the same no matter how many photons are simultaneously absorbed for a single molecular transition. Many experimental results showed that the fluorescence emission spectra for a given medium excited by 1PA, 2PA, and 3PA are nearly the same.[15,49,60]

For some organic chromophores in solution phase, it was experimentally shown that the temporal decay and lifetime of one-, two-, and three-photon induced fluorescence emission were also nearly the same for a given sample medium.[15,49,60]

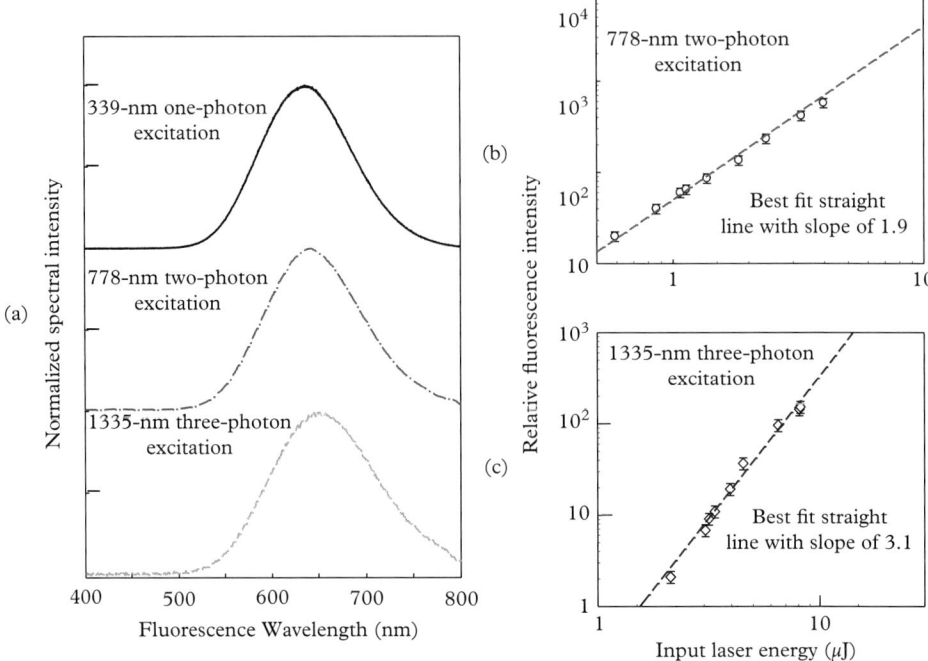

Figure 14.13 *(a) One-, two-, and three-photon excited fluorescence spectra of silicon 3-nm quantum dots solution in CHCl$_3$. (b) Quadratic dependence of two-photon induced emission on the excitation intensity. (c) Cubic dependence of three-photon induced emission on the excitation intensity. Sample concentration is 1.43 mg/mL, path length ~1 mm. (After He et al.[31] Copyright 2008, American Chemical Society).*

14.4 Multi-photon pumped (MPP) frequency upconversion lasing

The significance of multi-photon pumped (MPP) lasing studies is that they are a new approach to accomplish frequency upconversion of coherent light. The other commonly used approaches of frequency upconversion are based on either sum-frequency or second-harmonic generation (SHG) in second-order nonlinear crystals, or third-harmonic generation (THG) in third-order nonlinear media. In all those cases, phase-matching requirements have to be fulfilled, which cannot be easily achieved except if very special techniques (such as birefringence in crystals or anomalous dispersion in isotropic materials) are employed. In contrast, MPP lasing can be realized in liquids, films, solid rods, optical fibers, and waveguides without a phase-matching requirement. In addition, MPP lasing in dye-based systems exhibits the advantage of tunability within quite a broad spectral range.

From a long-term viewpoint, there is a definite trend to extend the lasing wavelengths into a much shorter spectral range (even to the vacuum UV and soft X-ray range), because extremely short lasing wavelengths will bring new advantages and promotions to various laser-based applications. In this direction, MPP lasing can also play a vital role in the future, provided that the number of photons simultaneously absorbed from the pump beam (in near IR or visible range) can be considerably increased.

14.4.1 General features of MPP lasing materials and devices

It is well known that dye solutions and dye-doped solid materials are among the best lasing materials from the viewpoint of low pump threshold requirement, high lasing efficiency, broad wavelength tunability, low cost, and ease of modification. The low pump threshold and high lasing efficiency are based on the so-called four-level lasing scheme and high quantum efficiency for the majority of commercial lasing dyes; while the lasing tunability is ensured by the relatively broad spectral band of fluorescence emission for dye materials. Shortly after the advent of lasers, it was found that two- and even three-photon absorption-induced frequency-up-converted fluorescence in organic dye materials could be observed by using pulsed laser excitation. Naturally, upon the pump action of strong input laser pulses, the population inversion and stimulated emission in an appropriate dye system can be achieved through MPE.

In practice, the early observations of two-photon pumped (2PP) lasing were reported during the 1960s–1970s in several semiconductor crystals operating at low temperature.[61–63] Also, there were a few papers reporting two-photon excited stimulated emission in gas and metal vapor systems during the 1970s–1990s.[64–66] On the other hand, between the 1970s and 1993, there were several experimental reports on two-photon excited stimulated emission for commercial organic dye systems.[67–71] In those cases with no cavity enhancement, the observed output was mostly the so-called amplified spontaneous emission (ASE) that usually was only characterized by spectral narrowing, not by high directionality and brightness. In 1995, He et al. reported 2PP cavity lasing with narrow spectral linewidth and high directionality in solutions of a novel dye and in the dye-doped solid matrix.[72,73] Since then, an increasing number of novel 2PP lasing dyes and cavity lasing configurations have been reported. The first three-photon pumped (3PP) lasing in a dye solution was reported in 2002.[74] In 2005, four-photon pumped (4PP) lasing is achieved in a series of new dye solutions.[75] More recently, five-photon pumped (5PP) lasing was reported in 2013.[76]

To date, the number of reported novel MPP lasing dyes is much smaller than that of the reported novel two-photon absorbing chromophores. However, it should be noted that the major requirements for MPP lasing dyes are different from that for those chromophores designed for other applications where a larger MPA cross-section is more important. Here, for the MPP lasing purpose, the most essential requirements for lasing dyes are the ease of establishing population inversion and a higher lasing efficiency. Thus, it is noted that the best two-photon absorbing materials may not necessarily produce multi-photon pumped lasing. This may be the reason why only a small number of the reported novel multi-photon active chromophores can be employed for MPP lasing.

So far, researchers do not know exactly what specific parameters, molecular structures and dynamic processes are the criteria to judge the MPP lasing capability of an organic material. However, based on the reported experimental results, some common features of MPP lasing dyes can be briefly discussed. Hitherto, most efficient MPP lasing dyes are salt-type asymmetric molecules consisting of a π-conjugated framework containing an electron donor on one end, and an electron acceptor on the other end. Figure 14.14 shows a general structure for this type of pull–push organic dye. In this example, a dialkylamino group is used as an electron donor and a pyridinium cation is used as an electron acceptor. Actually, other electron donors (alkyloxy groups, alkylthio groups, etc.)[75] or π-excessive heterocycles (substituted pyrroles, etc.)[77] can also be used for constructing a MPP lasing dye. As to the electron-acceptor end, one may replace the

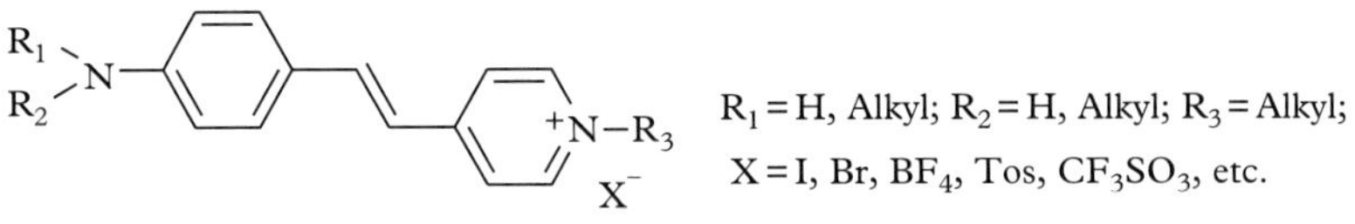

460 Multi-Photon Nonlinear Optical Effects

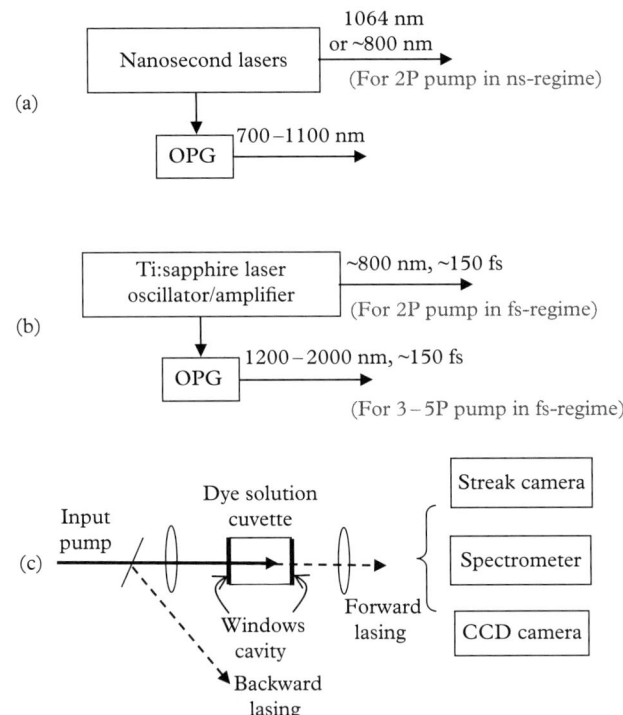

Figure 14.14 Generalized molecular structures of MPP lasing dyes.

Figure 14.15 (a) Pump sources for 2PP lasing in the nanosecond regime. (b) Pump sources for MPP lasing in the femtosecond regime. (c) Experimental setup for MPP lasing and output measurements.

pyridinium cation by other π-deficient heterocycles (substituted thiazoliums, benzothiazoliums, etc.)[78] or strong electron-withdrawing groups (alkylsulfonylphenyl groups, etc.).[74]

Figure 14.15 shows schematic diagrams of pump sources and experimental setups for MPP lasing studies. Figure 14.15(a) illustrates the pump sources for 2PP lasing in the nanosecond regime, which can be either a Q-switched Nd:YAG laser with 1064-nm output, a frequency-doubled Q-switched Nd:YAG laser pumped dye laser system with ∼800-nm output, or the same Nd:YAG laser pumped OPG with coherent light output, tunable in a range from 700 to 1100 nm. In these cases, both the pump pulses and 2PP lasing pulses are in duration of nanoseconds. Moreover, Fig. 14.15(b) show the pump sources working in the femtosecond regime, which can be either a Ti:sapphire laser (oscillator/amplifier) system with ~800-nm and ~150-fs output pulses, or an OPG system pumped by this laser, producing 1200–2000 nm tunable coherent light pulses. Among these two choices, the former can be utilized to achieve 2PP lasing, and the latter can be employed for 3PP, 4PP, and 5PP lasing. In the case of femtosecond pulse excitation, the upconverted lasing pulses will have a pulse duration of picoseconds, because the dye population inversion can last for a much longer time period (>10 ps) than the input pump pulse duration. Finally, Fig. 14.15(c) shows a typical setup for MPP lasing and characterization.

The input pump beam is typically focused into the center of a dye-solution filled cuvette (or a dye-doped rod) with a path length from several millimeters to one centimeter. Once the local pump intensity exceeds a certain threshold, an intense one-pass stimulated emission will be observed in both forward and backward directions, featuring drastically spectral narrowing as well as high directionality that is different from fluorescence emission or ordinary ASE. In this case, the latter feature is automatically ensured by the extremely large geometrical ratio between the gain length and the transverse size of the focused pump beam. In the nanosecond pump cases, the lasing can be further enhanced by using two parallel optical windows of the cuvette or two end surfaces of a solid gain rod as a cavity to provide optical feedback to form a cavity lasing. However, in the case of femtosecond excitation, the photon round time within this cavity is much longer than the pump pulse duration as well as the remaining time of the population inversion; therefore no multi-pass cavity lasing would be expected, and only cavityless lasing can be effectively generated. In the latter case, the high directionality of the output lasing beam is ensured by the focused pump beam geometry inside the gain medium.

14.4.2 Two-photon pumped (2PP) cavity lasing

Since the mid-1990s, a sizable number of novel organic chromophores have been developed which can be used as highly efficient 2PP lasing dyes.[4] The commonly used near-IR wavelengths for 2PP lasing in dye-based materials are 1064 nm and ~800 nm. Both 2PP cavity lasing and cavity-less lasing can be easily achieved using various optical configurations of the gain medium, which can be a dye-solution cell,[79] a dye-doped solid rod,[80] a dye-solution filled liquid-core fiber,[72] a dye-doped polymer-core fiber,[81] a thin film waveguide structure,[82] a photonic crystal structure,[83,84] etc. Recently, the 2PP stimulated emission was also generated in undoped polymers,[85,86] a semiconductor quantum well structure,[87] and in nanoparticle systems.[88–90]

ASPI is one of the best 2PP lasing dyes reported so far, exhibiting the advantages of high solubility in polar solvents, low pump threshold, high lasing efficiency, and high photo-chemical stability. Figure 14.16 shows the measured 2PP cavity lasing output energy values as a function of the input pump energy, from three gain media of the same length and concentration: ASPI in PHEMA (solid), ASPI in benzyl alcohol (solution), and ASPI in HEMA (solution).[79] From the data shown in this figure, one can see that at the same pump level, a dye-doped polymer rod could provide a considerably high lasing efficiency. In this specific experiment, at a pump energy level of 3.6 mJ, the laser output energy from the ASPI-doped PHEMA rod was ~0.15 mJ, and the overall energy transfer efficiency from the input pump to the output lasing was $\eta \approx 4.2\%$. On the other hand, at this pump level, the measured one-pass nonlinear absorption ratio was $\Delta \approx 0.25$; therefore the net conversion efficiency from the absorbed pump energy to the output lasing energy was $\eta' = \eta/\Delta \approx 17\%$.

In some specific cases, it is desirable to make the lasing output wavelength tunable with a spectral width as narrow as possible. In principle, various conventional methods used for one-photon pumped tunable dye lasers can also be applied to 2PP laser devices. There is an example in which a holographic polymer-dispersed liquid-crystal (H-PDLC) grating was employed as a

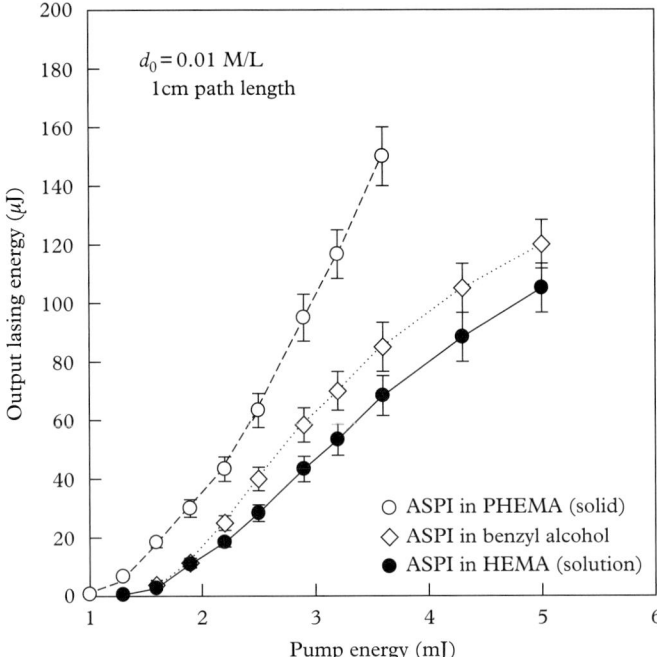

Figure 14.16 *2PP cavity lasing output energy from three ASPI-based gain media of same concentration versus the 1064-nm pump laser input energy. (After He et al.[79] Copyright 1997, AIP Publishing LLC).*

feedback mirror and a lasing tuning element as well.[91] The gain medium was a 2-cm long APSS dye solution in dimethyl sulfoxide (DMSO), pumped with ~815-nm and ~8-ns laser pulses. Shown in Fig. 14.17 are the measured spectral curves of the two-photon induced fluorescence, the cavity-less lasing, the cavity lasing with window reflection, and the tunable cavity lasing with a H-PDLC grating reflection, respectively. In this case, the spectral width for fluorescence is ~82 nm, for cavity-less lasing ~20 nm, for window cavity lasing ~13 nm, for grating reflection lasing at $\theta = 0°$ is ~5 nm, and finally for grating reflection lasing at $\theta = 20°$ is 3 nm only. By changing the reflection angle of the grating, the output lasing wavelength can be smoothly tuned over a 25 nm range.

14.4.3 Three- to five-photon pumped lasing

In 2002 and 2005, He and co-workers reported the unambiguous three-photon and 4PP lasing with drastically narrowed spectral width and highly directionality from novel organic dye solutions.[74,75] As mentioned previously, in these cases a powerful pulsed coherent light source working in 1.3–2.0 μm wavelength range with femtosecond duration was used to create population inversion and to provide sufficient single-pass gain of stimulated emission. If the path length and the refractive index of a dye-based gain medium are 1 cm and 1.5, respectively, the photon round-time within the cavity formed by two end-windows will be ~100 ps, which is much longer than the pump pulse duration (e.g., 100–500 fs) as well as the remaining period of the

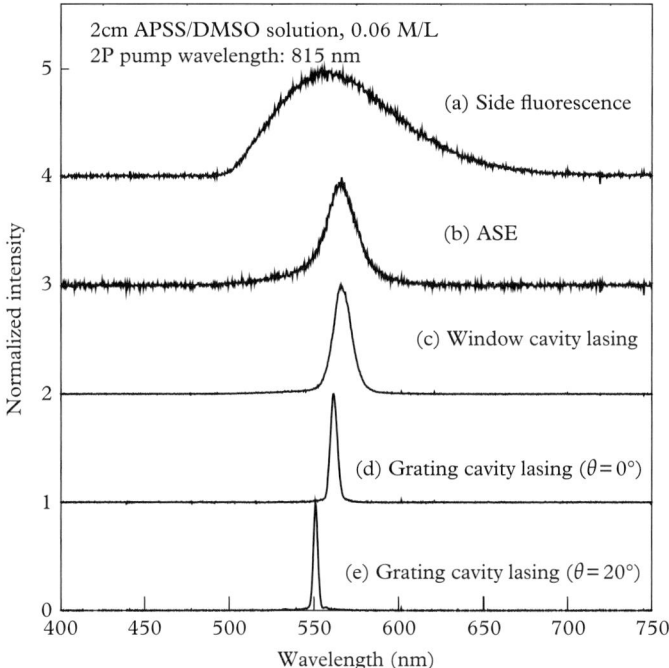

Figure 14.17 *Spectral distribution for (a) two-photon induced fluorescence, (b) 2PP cavity-less stimulated emission, (c) cavity lasing formed by two windows of the cell, (d) cavity lasing with a feedback from a PDLC grating of $\theta = 0°$, and (e) cavity lasing with a feedback from a PDLC grating of $\theta = 20°$. The gain medium was APSS solution in DMSO pumped by ~815-nm and ~8-ns laser pulses. (After He et al.[91] Copyright 2003, AIP Publishing LLC).*

peak population inversion. For this reason, a multi-pass based cavity enhancement is not available. Nevertheless, the high directionality of the one-pass cavityless lasing can still be obtained owing to the focused pump beam configuration, only the forward and backward stimulated emission can have the maximum gain length compared to stimulated emission propagating along all other directions.

In general, good organic chromophores and materials for 2PP lasing are also good candidates for 3PP and 4PP lasing purposes. For instance, APSS and ASPI are among the best 2PP lasing dyes, at the same time they are also good chromophores for 3PP lasing. Figure 14.18 shows a photograph of the 553-nm 3PP lasing output beam from a 1-cm long APSS solution cell pumped by ~1.3-μm and ~150-fs pump pulses.

Furthermore, Fig. 14.19 shows (a) the measured temporal profiles of the pump pulse, (b) the forward 3PP lasing pulse, and (c) the three-photon induced fluorescence decay behavior, respectively, by using a streak camera of 2-ps resolution. From Fig. 14.19 one can see that there is a certain delay between the input pump pulse and the burst of forward stimulated emission. The forward stimulated emission pulse can last for a period of 50–70 ps, which is much longer

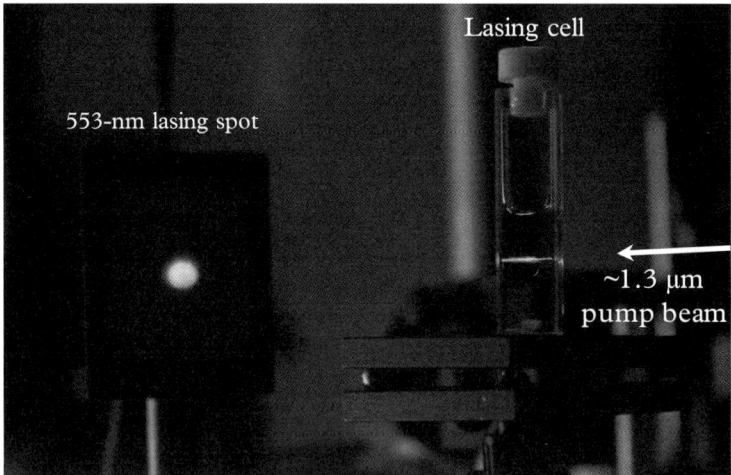

Figure 14.18 *Photographs of 3PP forward visible lasing of 553-nm wavelength from a 1-cm APSS solution in DMSO pumped by ~1.3-μm and ~150-fs IR laser pulses, showing the high directionality of the forward cavityless lasing beam. (After He et al.[74] Copyright 2002, Nature Publishing Group) (See also Color fig. 13.).*

than the pump pulse while much shorter than the spontaneous fluorescence lifetime of ~720 ps. At an input pump energy level of ~1.5 μJ, the output lasing energy was ~17 nJ from a 1-cm dye solution of 0.06 M concentration, indicating an overall lasing efficiency of $\eta \approx 1.1\%$. At this pump level, the measured NLT of the gain medium for the pump beam was ~0.47; therefore the net 3PP lasing efficiency in this case should be $\eta' = \eta/0.47 \approx 2.1\%$. In another 3PP lasing experiment based on ASPI solution in DMSO of 0.08 M concentration, pumped by ~1.5 mm laser pulses, the measured overall lasing efficiency was $\eta \approx 6.7\%$, and the corresponding net efficiency was $\eta' \approx 12\%$.[92]

A comprehensive experimental study of a series of newly synthesized chromophores (PRL-L1 to -L10) for two-, three- and four-photon pumped lasing properties in solutions have been reported.[75] These chromophores are stilbazolium dyes having basically the same molecular backbone, but differ either in their electron-donors or their electron-acceptors. Dialkylamino, alkyloxyl, and hydroxyl groups were chosen as electron donors. Various lasing wavelengths can be obtained from these dyes by changing the electron donating ability of the terminal groups on the stilbazolium backbone. Multi-photon pumped lasing can be achieved based on these dyes in solutions. At the same time, different electron donors or acceptors may influence dye interactions with a surrounding medium (solvent), which may affect their multi-photon pumped lasing properties.

During the lasing measurements of these dye solutions, an unexpected phenomenon was observed: there was a wavelength difference between the forward and the backward lasing output under 3PP and 4PP conditions. In these two cases, the forward lasing wavelength is found to be shorter than the backward lasing wavelength by an amount of 17–21 nm for the PRL-L5 solution, and 26–31 nm for the PRL-L10 solution. The other related fact is that at the same pump level, the forward lasing output energy is always significantly greater than the backward lasing output

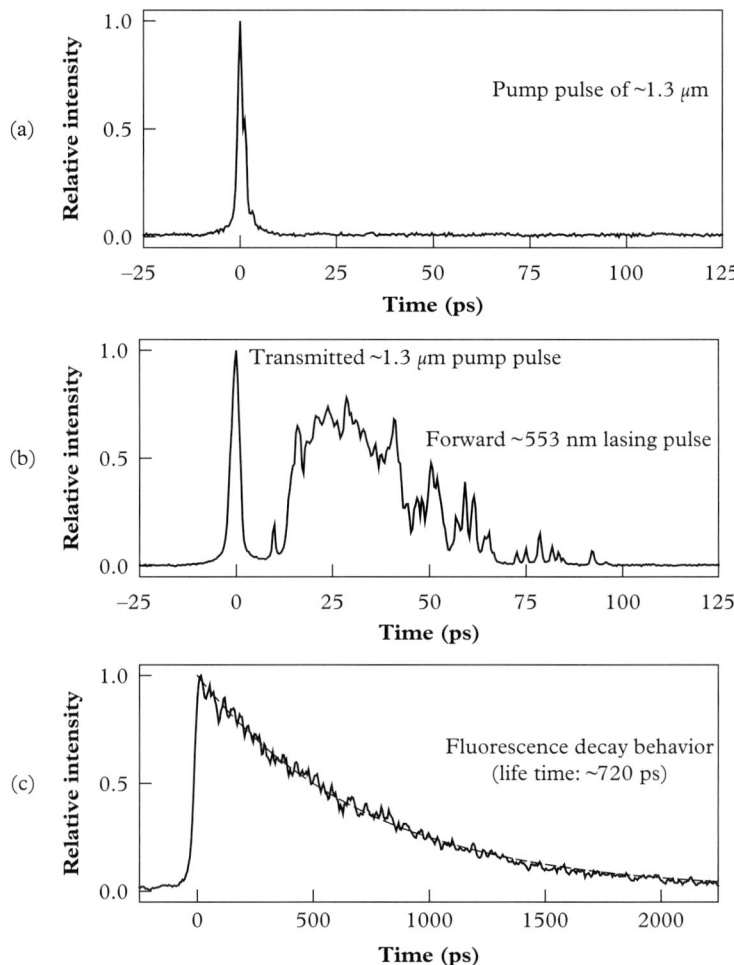

Figure 14.19 *Temporal profiles of the 1.3-μm pump pulse (a), the transmitted pump pulse together with the 553-nm forward lasing pulse (b), and the three-photon induced fluorescence decay (c). (After He et al.[74] Copyright 2002, Nature Publishing Group).*

energy under 3PP and 4PP conditions. These two effects which reflect the asymmetry between the two lasing output beams, as well as other dynamic lasing behavior can be well explained by the subsequent studies.[44,93]

Most recently, 5PP lasing was reported by Zheng et al,[76] based on a novel dye (IPPS) solution in DMSO-d_6 pumped by 2100-nm fs-laser pulses; the upconversion lasing is observed at 501 nm wavelength with the features of threshold behavior, obvious spectral narrowing, and high directionality. This work is the first real application for 5PA, which is a ninth-order nonlinear process.

14.5 MPA-based optical limiting, stabilization, and reshaping

14.5.1 Principles of optical limiting

An optical limiter is a special device by which the optical transmissivity decreases when the input signal intensity increases. This kind of device can be used for protection of human eyes or optical sensors against high-intensity laser radiation-induced damage.[94] Several nonlinear optical effects can be employed for the design and performance of optical limiting devices, which can be generally classified into two categories: one is an energy-spreading type of device, and the other is an energy-absorbing type of device. The limiting function of the first type of device is based on intensity-dependent change in spatial structure of a laser beam passing through a nonlinear medium. At low input intensity levels, this change can be neglected and the whole laser beam can be detected through a properly placed aperture in front of a detector. At high intensity levels, this change becomes so severe that only a small fraction of the transmitted beam can pass through the same aperture and be finally detected. In contrast, the operation of the second type of device is based on intensity-dependent nonlinear attenuation of the laser energy in a given nonlinear medium, whereas the beam-structure change is not so important. In the remaining part of this section, we only consider the second type of optical power limiting device, for which the intensity-dependent transmission change in a nonlinearly absorbing medium is most important.

The ideal requirements for choosing a proper nonlinearly absorbing medium for optical limiting can be described as follows: (i) there should be no linear absorption at the working wavelength range, so that the medium is highly transparent for weak input light signals; (ii) there is a strong dependence of the NLT on the input light intensity, so that the medium is highly absorptive for intense input laser signals; (iii) there should be a very fast temporal response of NLT change following the intensity change of the input signals; and (iv) the optical limiting behavior should be reversible and reproducible for the input laser pulses.

Three major nonlinear absorption mechanisms can be employed for optical limiting: the first is linear (one-photon) absorption-initiated ESA or so-called reverse saturable absorption (RSA); the second is MPA and MPA-initiated ESA; and the third is backward stimulated scattering. By working with the first mechanism, the initial linear transmission has to be considerably lower than 100%; while with the third mechanism, certain requirements for time-duration, spectral width, and peak intensity level have to be fulfilled. In contrast, the second mechanism may fulfil all requirements for power limiting performance, such as high transmissivity (∼100%) for weak input signal, almost instantaneous response of transmission change to the intense laser signal, and superior reproducibility.

14.5.2 MPA-based optical limiting

Semiconductor crystals were the solid materials initially used for 2PA-based optical limiting studies in the mid-1980s by Van Stryland et al.[95,96] For these studies, the wavelengths of input laser beams were usually in the IR range, and the measured optical limiting effects might also be due to other mechanisms, such as induced free carriers absorption or nonlinear refraction.

Although it was known for a long time that many organic compounds exhibited the property of 2PA or 2PA-induced fluorescence emission, the intense effort of using 2PA organic compounds for optical limiting purpose started only in the mid-1990s.[97,98] Since then, various newly synthesized organic materials for optical limiting performance have been reported by many research groups around the world.[4] In most cases, the investigated nonlinear absorbing materials are organic chromophores solutions and dye-doped solid rods. In some other cases, the two-photon absorbing materials for optical limiting measurements could also be organic crystals,[99] inorganic crystals,[100] a hybrid organic-inorganic crystal,[101] organic-liquid-core fiber array systems,[102] or a polycrystalline solid film.[103] Recently, there have been some encouraging developments in new directions: one is the use of a neat liquid-crystal dye in its isotropic phase,[22] a neat liquid-dye salt,[24] as well as an intrinsic polymer[104] because of their high molar concentrations; the other is the use of various nanoparticle systems in solutions or in solid matrixes.[105,106]

In the general case, for a 2PA medium with small linear absorption at the working wavelength the overall transmission T can be expressed as[107]

$$T = T_0 T' = T_0 \frac{1}{1 + \beta I_0 l_0}. \quad (2PA) \qquad (14.5\text{-}1)$$

Here, T_0 is the linear transmission for very weak light, T' is the NLT due to 2PA, I_0 is the input light intensity, and l_0 and β are the optical path length and 2PA coefficient of the medium, respectively. From Eq. (14.5-1) one can see that the NLT will decrease following the increase of the input laser intensity, which is the basis of 2PA-based optical limiting. It is obvious that a larger β value means a better limiting behavior. However, sometimes in optical limiting experiments, the measured intensity-dependent decrease of apparent transmission is not only due to real nonlinear absorption (2PA or 2PA plus ESA), but also may be due to other 2PA initiated processes, such as induced nonlinear scattering, nonlinear refraction, self-focusing or self-defocusing, and thermal-lens effect occurring in nonlinearly absorbing media. In all these situations, an apparent huge and much overestimated β (or σ_2) value may be artificially obtained.

For organic chromophore solutions or doped solid matrices used for optical limiting studies, the nonlinear absorptivity is directly proportional to the concentration of absorbing molecules. In practice, the dye concentration values are usually limited to be less than 10^{-2}–10^{-1} M in solutions as well as in doped solid matrices due to dye aggregation at high concentration. Therefore, new types of organic nonlinear materials are desirable, which exhibit a much higher effective concentration of multi-photon absorbing molecules. This type of materials can be organic crystals consisting of molecules with nonlinearly absorbing chromophores, neat liquid crystals consisting of dye molecules, and neat dye liquid salts.

For example, shown in Fig. 14.20 are the measured NLT data and the output/input characteristic curve of an optical limiter using a neat dye liquid crystal as 2PA medium, interacted with $\sim$815-nm, $\sim$5-ns, and 10-Hz laser pulses.[22] The 1 cm path length cuvette was filled with the dye liquid crystal and heated to $\sim$100 °C to keep the nonlinear medium in its isotropic phase, and an $F/100$ focusing configuration was adopted. In Fig. 14.20 the solid lines are the best fitting curves given by 2PA theory with a 2PA coefficient of $\beta = 6.25$ cm/GW. The NLT changes from $\sim$100% at low input levels to $\sim$30% at high input levels.

Another example of recent progress in 2PA-based limiting studies is the use of semiconductor CdTe quantum dots suspended in an organic solvent ($CHCl_3$) as the 2PA medium, working with $\sim$1.3-μm and $\sim$160-fs laser pulses.[108] The experimental results, which are shown in Fig. 14.21, indicate that the NLT changes from $\sim$100% at low input intensity to $\sim$20% at high input intensity ($\sim$600 GW/cm^2). It should be noted that the $\sim$1.3 μm wavelength is one of the most useful wavelengths for optical telecommunications and optical network systems.

The 3PA mechanism can also be employed for optical limiting purposes. Moreover, owing to the cubic dependence of the nonlinear absorption on the local light intensity, the material may produce a very sharp change of NLT with respect to the input intensity change. The other advantage

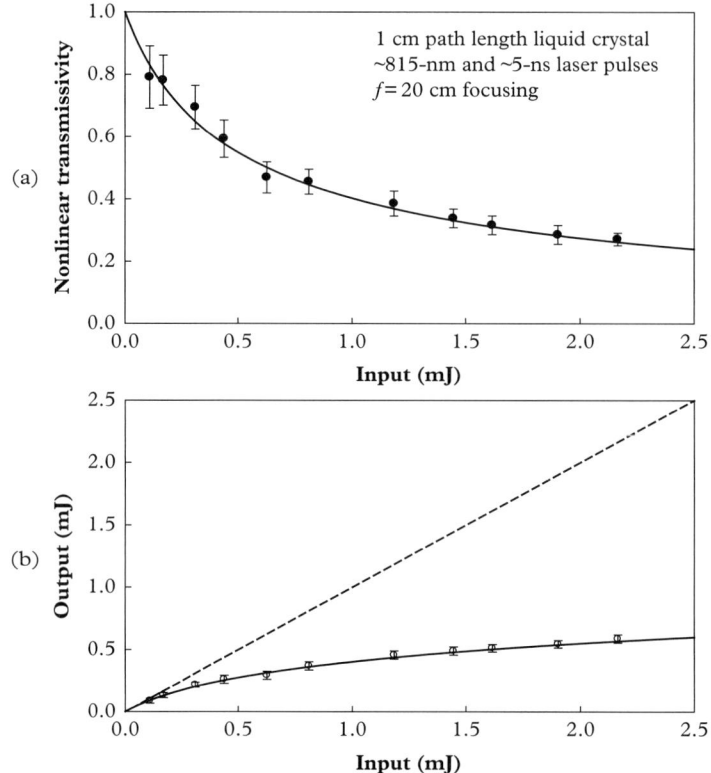

Figure 14.20 *Measured NLT data (a) and output pulse energy (b) versus the input pulse energy through a liquid crystal sample. The solid lines are the best fitting curves with a 2PA coefficient $\beta = 6.25$ cm/GW. (After He et al.[22] Copyright 2003, AIP Publishing LLC).*

of using 3PA mechanism to execute optical limiting is that the working wavelengths can be much longer and far away from the linear absorption band; therefore the harmful linear absorption influence can be avoided. For a pure 3PA process, the NLT is given by (cf. Eq. (14.1-12))

$$T' = \frac{1}{\sqrt{1 + 2\gamma I_0^2 l_0}}. \quad (3PA) \qquad (14.5\text{-}2)$$

Here, I_0 is the input intensity and γ and l_0 are the 3PA coefficient and thickness of the medium, respectively.

The earliest experimental result of 3PA-based optical limiting in a chromophore solution was reported in 1995 and the samples were dye solutions operating with 1064-nm, 10-ns laser pulses.[109] In the recent decade, there have been more studies on 3PA-based optical limiting.[110–113] Input laser wavelengths often used for 3PA optical limiting experiments are within the spectral range from ~1.0 to ~1.3 μm, the pulse duration is either in the 30–40 ps range or in the 100–160 fs range, and the peak intensity of the focused laser beam varies from 10 to 400 GW/cm^2.

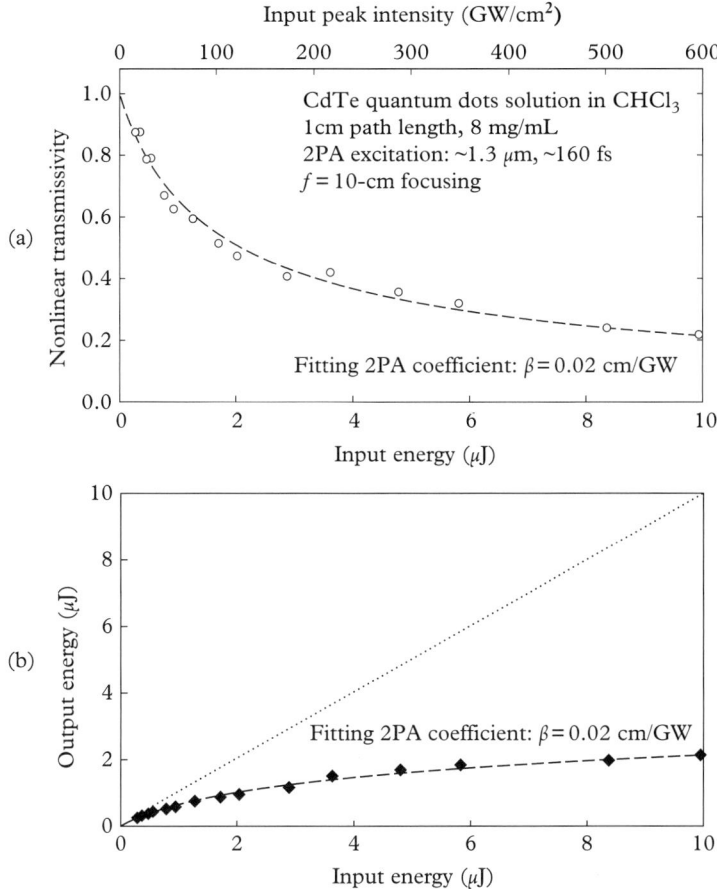

Figure 14.21 *(a) Measured NLT of CdTe quantum dots in CHCl$_3$ as a function of the input peak intensity of ~1.3-μm and ~160-fs laser pulses. (b) Measured output pulse energy versus input energy of the laser pulse. The dashed-line curve is the theoretical curve with a best fitting parameter of β = 0.02 cm/GW. (After He et al.[108] Copyright 2007, AIP Publishing LLC).*

As an example of 3PA-based optical limiting performed in the femtosecond regime,[112] the nonlinear absorbing medium was a neat liquid dye salt slightly diluted with benzyl alcohol in volume ratio of ~1/1, with an estimated molecular concentration of $d_0 \approx 1$ M. This sample medium exhibits a strong linear absorption band centered at ~500 nm, and a linearly transparent window in the ~1.3 μm range; therefore, one may expect that an efficient 3PA excitation could be realized by using a strong IR radiation in that window region. In experiment, the input ~1.3-μm and ~160-fs laser pulses were from an OPG device pumped by a Ti:sapphire laser system. Figure 14.22 shows the measured values of output pulse intensity of the laser beam passed through a 1-cm thick sample as a function of the input intensity, the dash-dotted line is the theoretical curve given by Eq. (14.5-2) with a best fitting parameter of $\gamma = 1.6 \times 10^{-5}$ cm^3/GW^2.

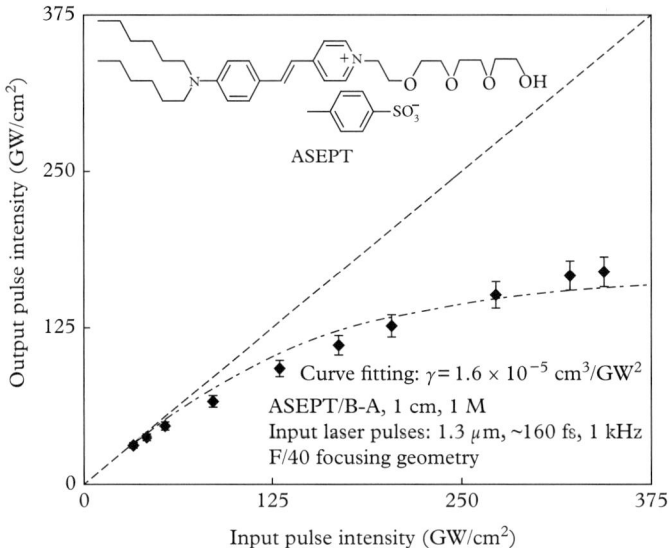

Figure 14.22 *Measured 3PA-based optical limiting behavior of a 1-cm thick liquid dye salt sample (ASEPT diluted by benzyl alcohol) using ∼1.3-µm, ∼160-fs laser pulses; the dash-dotted line is the theoretical fitting curve with a 3PA coefficient value of $\gamma = 1.6 \times 10^{-5}$ cm³/GW². An F/40 focusing geometry was used and the chemical structure of ASEPT is shown in the left top corner. (After He et al.[112] Copyright 2005, IEEE).*

There is another example of 3PA-based optical limiting performance that is based on CdSe quantum dots (QDs) suspended in hexane, working with ∼1.3-µm and ∼160-fs laser pulses focused by an $f = 10$ cm lens.[106] The average QD size is 3.9 nm and the QD weight concentration is 70 mg/mL. The measured data of NLT and optical limiting behavior are shown in Fig. 14.23(a) and (b), respectively. In Fig. 14.23(a), the dotted line is the fitting curve based on an unsaturated 3PA process, whereas the dashed line is the theoretical fitting with considering saturation effect (cf. Eq. (14.3-11)). When the input laser intensity is higher than ∼200 GW/cm², the decrease of NLT becomes slower due to the 3PA saturation effect occurring in this specific system.

14.5.3 MPA-based optical stabilization

From Figs. 14.20(b), 14.21(b), 14.22, and 14.23(b), which show optical limiting behavior based on the 2PA or 3PA mechanism, we can see that in the input intensity range where the output/input characteristic curve becomes flattened, a considerably large input intensity change only leads to a much smaller output intensity change. This feature of 2PA- and 3PA-based optical limiting can be cleverly utilized to stabilize spatiotemporal fluctuations of intense coherent optical fields. For example, there is always a peak intensity fluctuation of a series of laser pulses; this kind of fluctuation is harmful for optical telecommunications and optical data processing. A simple multi-photon absorbing medium can be adopted to effectively reduce such pulse fluctuation, because

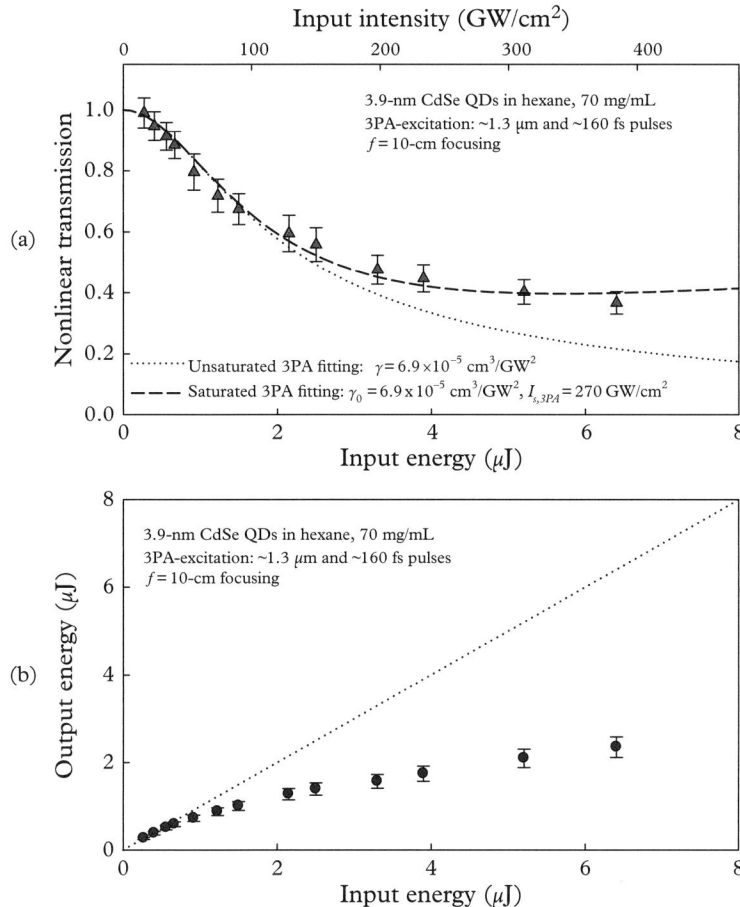

Figure 14.23 *(a) Measured 3PA-induced NLT change of CdSe QDs solution versus the input 1.3-μm laser pulse energy (intensity). (b) Measured output pulse energy versus the input pulse energy. (After He et al.[106] Copyright 2007, Optical Society of America).*

the nonlinear medium manifests a lower transmission for the pulses of a higher intensity, while providing a higher transmission to the pulses of a lower intensity. As result, the peak intensity fluctuation of the transmitted pulses will be much smaller than that of the input pulses.

In the first experimental demonstration of optical stabilization based on a 2PA process, which was reported in 1995 by He et al., a 2.4-cm long dye (BBTDOT)-doped epoxy rod of $d_0 = 0.09$ M concentration was employed as the 2PA medium, and the input 602-nm, 0.5-ps pulsed laser beam was focused via an F/100 optical system onto the center of the epoxy rod.[114] The relative peak intensity fluctuation of the input laser pulses with a repetition rate of 30 Hz was $\Delta \approx \pm 0.11$, whereas after passing through the two-photon absorbing rod, the relative peak intensity fluctuation was reduced to $\Delta' \approx \pm 0.038$.

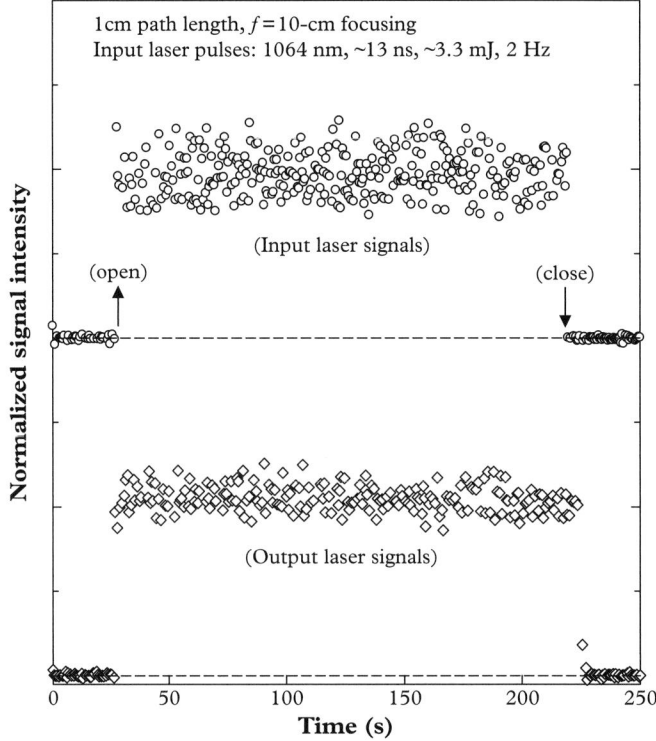

Figure 14.24 *Measured relative laser pulse fluctuation for the input signals (upper trace), and for the output signals (lower trace) passed through a 1-cm long dye liquid salt (ASDPT) sample. Each point represents the measured pulse energy value. (After He et al.[24] Copyright 2005, Optical Society of America).*

In later studies of optical stabilization experiments, the nonlinear absorbing media employed could be a chromophore solution,[107] a neat liquid-crystal dye system,[22] liquid dye salt systems,[24,112] as well as a nanoparticles system.[108,113]

Figure 14.24 shows an example of 2PA-based optical stabilization using a 1-cm long dye liquid salt (ASDPT) sample as the nonlinear absorbing medium, operated with 1064-nm, ~13-ns laser pulses of 2-Hz repetition rate.[24] To obtain these results, the input laser beam from a Nd:YAG laser device was focused via an F/50 optical system onto the sample cell. As the shape of the input laser pulses roughly remained the same, the peak intensity fluctuation was mainly due to random variation of the pulse energy around its average value. In the figure, each point represents the measured relative energy value of a single pulse, which was recorded by a gated integrator (Boxcar), and the average input energy level was ~3.3 mJ. From this figure one can see that the input energy (intensity) fluctuation is $\Delta \approx \pm 0.25$, whereas the output fluctuation is $\Delta' \approx \pm 0.13$. There is a twofold reduction in the relative fluctuation for the output pulse signals.

The 3PA mechanism can also be applied to optical limiting and stabilization with an additional advantage of a cubic dependence of the nonlinear response of the material on the local intensity change of the beam. The first experimental demonstration of 3PA-based optical stabilization was reported in 2005,[112] in which a 2-cm long liquid dye salt (ASEPT) sample was employed as a 3PA medium, the input 1.3-μm and ~160-fs laser pulses of 1-kHz repetition rate and 2.5-mm beam size were from a OPG device pumped by a Ti:sapphire laser system. Figure 14.25 shows the measured intensity fluctuation profiles of the input and the output pulses recorded by a Boxcar, where each measured point represents the average value over 8 pulses, and the total measured points were ~7500 over an exposure period of ~60 s. The average input pulse energy (or intensity) level was ~4 μJ (or ~500 GW/cm^2). From Fig. 14.25 one can see that after averaging eight pulses, the recorded relative fluctuation of the input signals is $\Delta \approx \pm 0.33$, whereas for the output signals

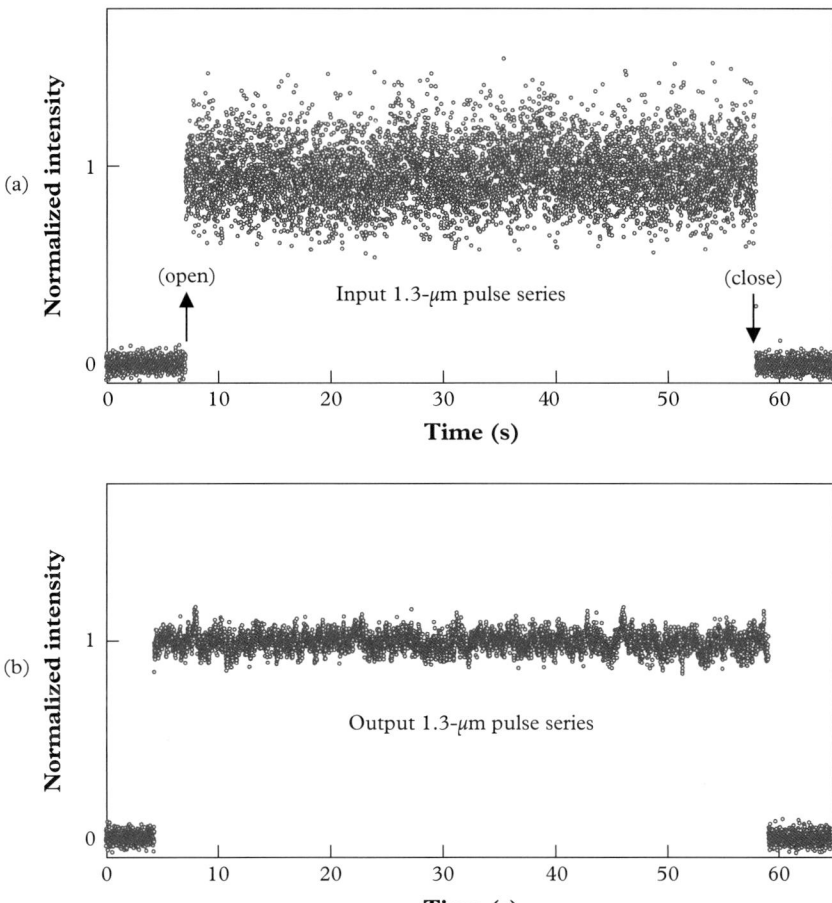

Figure 14.25 *(a) Measured intensity-fluctuation profile for the input 1.3-μm, ~160-fs laser pulses of 1-kHz repetition rate. (b) Measured intensity-fluctuation profile for the output laser pulses passed through a 2-cm liquid dye salt (ASEPT) as the 3PA medium. Each point represents measured average energy value over eight pulses. (After He et al.[112] Copyright 2005, IEEE).*

the fluctuation is reduced to $\Delta' \approx \pm 0.08$. This is strong experimental evidence of a tremendous improvement of intensity stability of laser pulses after passing through a 3PA medium.

14.5.4 MPA-based optical reshaping

The same MPA mechanisms can also be utilized for reshaping the temporal and/or spatial profiles of a pulsed laser beam. For all these types of applications, the only requirement is to let a high intensity laser beam pass through a multi-photon absorbing medium.

In the earliest experiment showing a 2PA-based optical reshaping effect, the nonlinear absorbing medium was a solution of ASPI dye in DMSO with a concentration of $d_0 = 0.05$ M, filled into a 20-cm long hollow fiber system to increase the effective interaction length.[115] The input laser beam consisted of a series of 1064-nm pulse trains, generated by a Q-switched and mode-locked Nd:YAG laser operating at a mode-locked frequency of 100 MHz and a Q-switched frequency of 500 Hz; each pulse train contained ~30 sub-pulses of ~130-ps duration each. Figure 14.26 shows

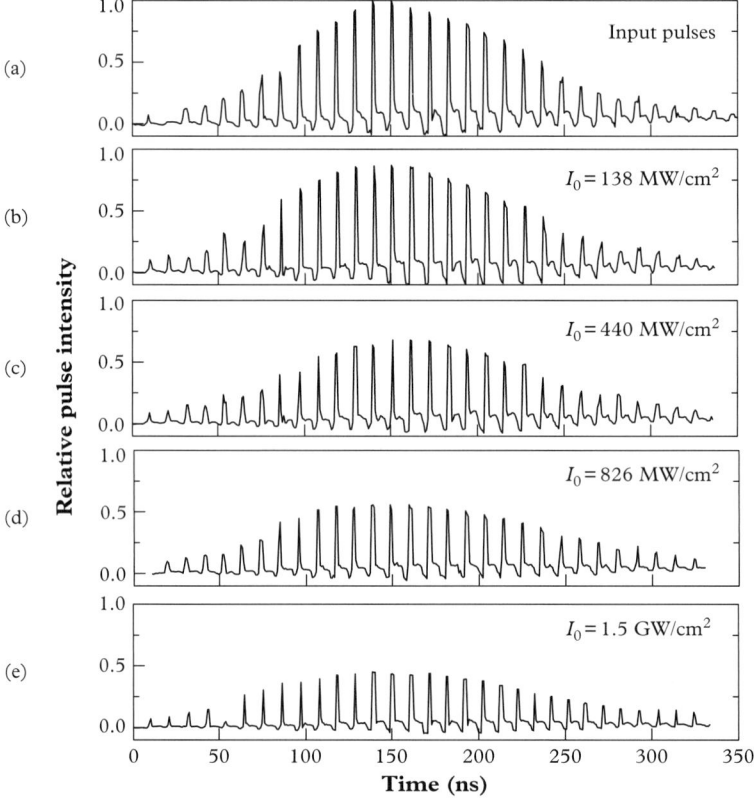

Figure 14.26 Measured temporal profiles of (a) the input pulse train and (b–e) the output pulse trains passed through a 20-cm long ASPI/DMSO filled hollow fiber at different average input peak intensity levels. (After He et al.[115] Copyright 1997, Optical Society of America).

the measured profiles for the input pulse train and for the output pulse train after passing through the 2PA liquid at different average input peak intensity (I_0) levels.

From Fig. 14.26 one can easily see that when the input peak intensity is increased from 138 MW/cm^2 to 1.5 GW/cm^2, the envelope of the transmitted pulse train becomes flatter and broader in the middle region of the train. This is a straightforward demonstration of optical reshaping effect on the temporal profile of a laser pulse train, based on the 2PA mechanism.

It is easy to imagine that the same MPA mechanism can also be employed to reshape or smooth the spatial intensity distribution of an intense coherent light field. The first experimental demonstration of 2PA-based spatial reshaping was reported in 2000.[15] In that experiment, an 810-nm and 7-ns laser beam from a pulsed dye laser system was first split into two beams and then focused with a small crossing angle onto the center of a 1-cm path-length liquid cell to produce a transversely modulated intensity distribution via two-beam interference. The two-photon absorbing medium used for this experiment was an AF350 dye solution in THF of concentration $d_0 = 0.033$ M. By measuring the transverse intensity distributions of interference pattern at the focal plane position after passing through a 1-cm dye solution and passing through a 1-cm pure solvent separately, it was found that the modulation depth of the interference pattern changed from ~67% for passing through the solvent to ~39% for passing though the 2PA dye solution. Under this condition, the ratio of $(I_{max} - I_{min})/I_{max}$ was reduced by a factor of two.

A more recent experiment that shows superior performance of optical spatiotemporal reshaping is based on a highly two-photon absorbing intrinsic polymer film.[116] In that experiment, a chemically modified poly(fluorene-*alt*-benzothiadiazole) (PFBT) polymer film of ~70-μm thickness was prepared as the highly two-photon absorbing medium. To demonstrate spatial reshaping

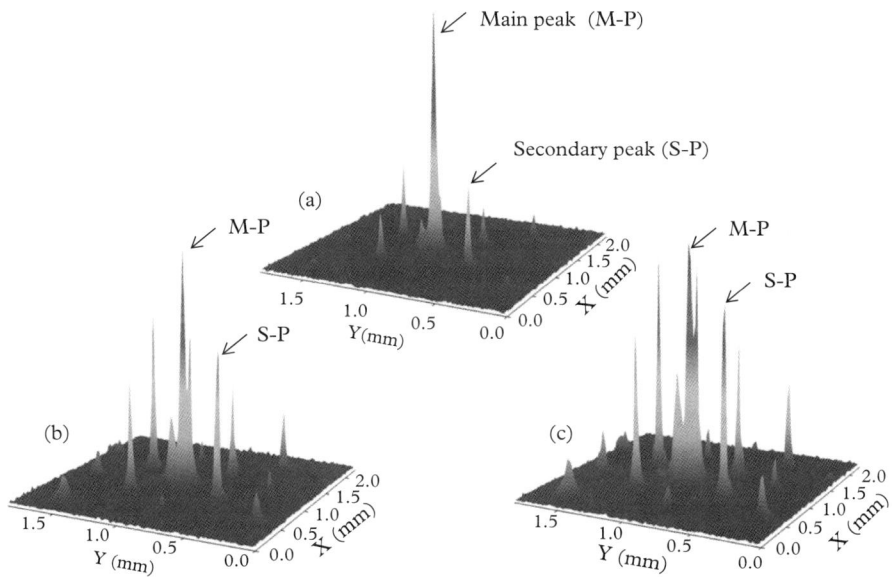

Figure 14.27 *Far-field intensity distributions of a transmitted grids-diffracted laser field passed through an ~70 μm-thick PFBT polymer film, measured at three different input pulse energy levels: (a) ~0.07 μJ, (b) ~2.6 μJ, (c) ~4.5 μJ. Input laser pulses: ~780 nm and ~160 fs. (After He et al.*[116] *Copyright 2011, Optical Society of America).*

feasibility, the ~780-nm and ~160-fs pulse laser beam was first passed through a 300-mesh copper grids, the diffracted laser beam was then focused by an $f_1 = 10$ cm lens on the surface of the polymer film to form a characteristic far-field pattern determined by grids diffraction. The relative intensity distribution of such a far-field pattern of the laser field on the polymer film was recorded by a CCD-array camera. Figure 14.27 shows the relative spatial intensity distributions of the far-field on the polymer film recorded by the transmitted laser beam, under three different input-level conditions. When the overall input pulse energy measured before the film is ~0.07 µJ, the recorded far-field pattern is shown in Fig. 14.27(a) where the nonlinear attenuation for all spatial peaks are nearly negligible, and the intensity ratio between the main peak and the secondary peak is near to 1:0.3. As shown in Fig. 14.27(b), when the overall input pulse energy is increased to ~2.6 µJ, the nonlinear absorption becomes considerably strong for the central main spatial peak; as a result the measured ratio is raised to 1:0.6. Furthermore, as shown in Fig. 14.27(c), when the input energy is increased to ~4.5 µJ, the intensity ratio between the main and the secondary peak reaches to 1:0.82. Comparing Fig. 14.27(a) to (c) indicates that after simply passing through a highly nonlinearly absorbing polymer film, the relative intensity distributions of a laser field can be drastically reshaped or smoothed.

In conclusion, it is obvious that MPA-based optical stabilization and reshaping will be especially useful for those applications such as optical telecommunications, optical data recording and processing.

14.6 3D data storage and microfabrication based on multi-photon excitation (MPE)

14.6.1 Common features of MPE for data storage and microfabrication

For several decades, a focused laser beam has been utilized for data recording and reading, as well as for material processing and fabrication. In most of these applications, the basic mechanism concerning the interaction between laser radiation and materials is 1PA or other linear optical processes. In all these cases, the available data storage density or the precision of optical machining is finally limited by the resolution of a laser focusing system (usually an objective of microscopy). It is known that the resolution of a focusing system is estimated by the minimum size of the focal spot, which is simply the product of the beam divergence angle and the focal length. In the best situation, the beam divergence angle reaches its diffraction limit, and the available minimum focal spot size is determined by

$$a_0 = 1.22\lambda/(n\sin\theta_0) = 1.22\lambda/NA. \qquad (14.6\text{-}1)$$

Here, λ is the illumination wavelength, n is the refractive index in the object space, θ_0 is the acceptance angle of the lens, and $n\sin\theta_0 = NA$ is the numerical aperture of the objective. One may achieve a larger NA value by using an oil-immersion objective, although the available NA still is around the values of 0.8–1.4. For this reason, the resolution of a laser focusing system is limited roughly by the applied light wavelength.

It is easy to make materials that are highly one-photon absorbing; hence a low intensity laser beam is enough for optical data recording or storage. In this case, the intensity attenuation of the laser beam along the propagation (z) direction inside a recording medium follows an exponential law. According to this law, if the recording medium is highly one-photon absorptive, the incident focused laser beam can only penetrate into a shallow layer near the material surface, in which two-dimensional (2D) information (in the x–y plane) can be recorded. Owing to the nature of the one-photon interaction, the material response to the incident local laser intensity manifests a linear relationship and, therefore, the lateral reactive range of the material is basically determined by the spot size of the focused laser beam. Based on this fact, the maximum 2D data-storage density in a linear absorbing medium surface is limited to $\sim 10^8$ bits/cm^2 for laser wavelengths in the visible or near-IR range. Furthermore, if the recording medium is a poorly one-photon absorbing thick film or plate, a focused laser beam will easily penetrate the whole thickness of the medium, which means that a multi-layer recording along the beam propagation direction inside the medium is not available. For the same reason, a focused laser beam can be employed to fabricate a 2D structure in the linear medium but not a 3D structure.

The principle of MPA-based optical data storage and fabrication is remarkably different from that mentioned above for one-photon interaction. In this new case, a much higher focused laser intensity is needed while the materials used for recording or fabrication should be multi-photon active. The response of a nonlinear material is proportional to the square of local light intensity for a 2PA mechanism, or proportional to the cubic of local intensity for a 3PA mechanism. This nonlinear response implies a very unique feature of the interaction between a tightly focused laser beam and a multi-photon active medium. Only in the focal region where the laser beam exhibits the maximum local intensity, will the medium undergo MPA-induced physical and/or chemical changes. To illustrate this special feature, we can consider normalized lateral intensity profiles in the focal plane for 1PA, 2PA, and 3PA reaction, respectively.

In Fig. 14.28, the solid-line curve is the lateral intensity distribution function $I(r)$ in the focal plane of a Gaussian laser beam with an assumed characteristic spot radius of $r_0 = 0.5$ μm, while the square and cube of $I(r)$ are expressed by a dashed-line curve and a dot-dashed-line curve,

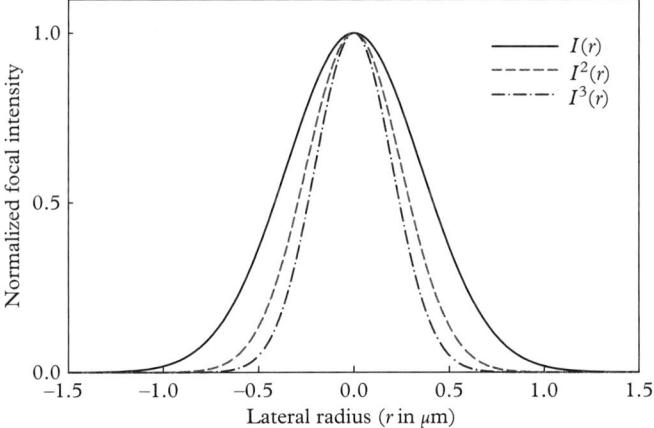

Figure 14.28 Profiles of the normalized transverse focal intensity function $I(r)$ (solid line), function $I^2(r)$ (dashed line), and function $I^3(r)$ (dot-dashed line). The solid-line curve represents a Gaussian lateral intensity distribution at the focal plane with a characteristic focal spot size of 1 μm.

respectively. The linear response of the material in the focal plane is determined by the curve $I(r)$, whereas the nonlinear 2PA or 3PA response will be determined by the function $I^2(r)$ or $I^3(r)$, respectively. It is obvious from Fig. 14.28 that the full-width of the solid-line curve measured at 1/e points is $2r_0 = 1$ µm, whereas the full-width by the same definition is ~0.7 µm for the dashed-line curve and ~0.6 µm for the dot-dashed line curve, respectively. Based on this comparison one may realize that in a multi-photon medium the reactive region can be smaller than the incident laser focal spot size in the recoding (x–y) plane.

Furthermore, the same conclusion is also applicable to the material response along the z-axial direction. It is commonly known that for a focusing optical system, a shorter focal length can provide a smaller focal depth of the beam. In the cases of optical data storage and microfabrication, a microscope objective with very short focal length is utilized and therefore, a much smaller focal depth of the tightly focused laser beam is expected along the z-direction. Just for illustration purposes, in Fig. 14.29 we depict the spatial profiles of two focused beams as a function of the z-distance from the focal plane, where the solid-line curves show a focal spot size of 1.5 µm while the dashed-line curves show a focal spot size of 1.0 µm, respectively. In obtaining these curves, a $NA = 0.8$ value of the focusing objective and a hyperbolic shape of the beam configuration near the focal region are assumed.

From Fig. 14.29 one can see that the beam with smaller focal spot size possesses a shorter focal depth. More specifically, if we define the focal depth as the distance between two positions along the z-axis at which the transverse beam size is two times larger than the focal spot size, the focal depth value is estimated to be $\Delta z_f \approx 2.0$ µm for the beam having a focal spot size of $2r_0 = 1.5$ µm, and $\Delta z_f \approx 1.4$ µm for the beam having a focal spot size of $2r_0 = 1.0$ µm, respectively. These estimates mean that under an optimum focusing condition, the focal depth can be controlled at a value slightly larger than the focal spot size. Moreover, in the case of multi-photon interaction, the effective reactive range in 3D (x–y–z) space can be further confined due to the nonlinear dependence of material response on the local light intensity.

Finally, another important feature of using multi-photon techniques to record optical data is that the incident laser beam can be focused on any designed depth of the recording medium. This

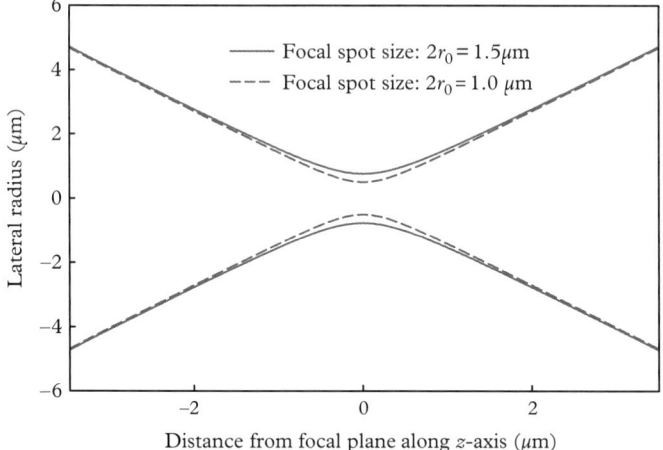

Figure 14.29 *Variation of the beam size along the propagation direction for two beams with different focal spot sizes.*

advantage is ensured by two facts: (i) there is no linear absorption of the medium for the laser wavelength chosen for multi-photon interaction, and (ii) only the local intensity near the focal point region will be high enough to cause nonlinear absorption or other nonlinear responses inside the medium. Based on this feature, multi-layer optical data recording or 3D microfabrication can be easily achieved in a thick film or a bulk that is made of multi-photon active material. Assuming that the bit spacing in a single recorded layer is ~1 µm and the distance between two neighboring layers is ~1.5 µm, one may reach a volume density of ~ 6.7×10^{11} bits/cm^3 for 3D data storage that is not possible by one-photon technique.

14.6.2 3D data storage in two-photon active materials

Recently, there has been significant progress in multi-photon based 3D data storage.[117,118] However, the earliest experimental demonstration of 3D data storage was presented by Parthenopoulos and Rentzepis in 1989.[119,120] In their pioneering work, a photochromic organic molecule-doped polymer bulk was used as the recording medium, and two perpendicularly crossed laser beams (532 nm + 532 nm, or 532 nm + 1064 nm, 25-ps pulse duration) were used as writing beams to create a two-photon induced chemical change of the organic molecules at the destined location within the polymer sample. After two-photon excitation in the recording spot, the chromatic molecules changed their energy structure and absorption–emission properties. By using other two crossed laser beams with longer wavelength (1064 nm + 1064 nm) as reading beams, the recorded spot could be read out by detecting the two-reading-beams induced fluorescence emission. Shortly after this work, in 1991 Strickler and Webb reported another technical approach in which a single tightly focused laser beam (~620 nm wavelength and ~100 fs pulse duration) was used to enhance the polymerization process via 2PA mechanism inside a UV-pre-irradiated 100-µm thick photopolymer film.[121] In the focal position (recorded bit location), the density and refractive index were different from the un-irradiated area. This writing beam induced refractive-index change could be read out by using another reading beam (488 nm with low intensity) through the so-called differential interference contrast microscopy technique. The significance of this work was to provide an example for realizing volume (25 layers) data storage by using only one intense laser beam. Since the mid-1990s, advancement in the research area of 3D data storage has been driven by Kawata's group,[122] Gu's group,[123] Prasad's group,[124] Rentzepis' group,[125] Belfield's group,[126] as well as some other groups.

Usually, a complete 3D data storage system consists of the following: (i) a multi-photon active recording medium, (ii) a data-writing laser beam in conjunction with focusing optics, (iii) x–y–z scanning mechanisms, and (iv) a data-reading laser beam in conjunction with related optics. The recording medium can be a thick film or a thin slab made of solid matrix (polymer, sol–gel glass, or crystal) containing multi-photon active centers (dye molecules, impurities, or nanoparticles). The writing beam is usually a pulsed laser beam of ~800 nm wavelength and 100–150 fs duration from a mode-locked Ti:sapphire laser system for two-photon excitation. The focusing optics comprises a microscopic objective with a larger *NA* value (0.8–1.4). The 3D recording is ensured by a computer controlled scanning system to change the focal position of the writing beam inside the recording medium. In a simple case, the sample medium can be mounted upon an x–y–z translation piezoelectric stage. The choices of reading-beam wavelength, power level, and reading manner depend on the specific mechanism of the writing beam induced physical/chemical changes of the nonlinear medium in the recorded bits positions.

Figure 14.30 shows a typical experimental setup for a 2PA-based 3D data writing and reading system, based on which a 2D picture can be recorded in a designed thin layer within the nonlinear medium by scanning the tightly focused laser beam along x–y plane. After slightly changing the

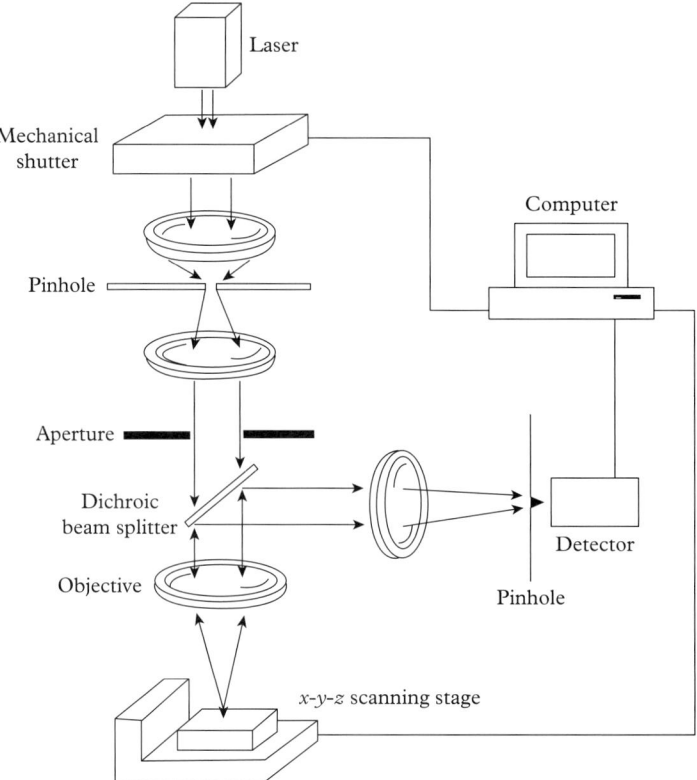

Figure 14.30 *Schematic diagram of a two-photon confocal microscope used for recording and reading 3D data bits in a photobleaching polymer. (Reproduced with permission from Gu and Day.[123] Copyright 1999, Optical Society of America).*

depth of the scanning plane, another 2D picture can be recorded in the next thin layer, and so on. In their early experiment, Gu and Day recorded a picture in four different layers within a photobleaching polymer doped with a 2PA dye.[123] The information was recorded as bleached areas by a stronger writing laser beam, and those areas will not produce fluorescence by another weaker reading beam. Each picture consists of 24 × 24 bits recorded in a single layer, the bit separation is 4.3 μm, and the exposure time is 200 ms.

In most cases reported to date, the recording media have been two-photon active organic chromophore-doped polymers, in which the intense writing laser beam could produce some type of property change of the material in the focal spot region via different mechanisms. The following are the major mechanisms used for 2PA-based 3D data storage: (a) photochromic reaction that may lead to induced changes of absorption, fluorescence, or refractive index;[127,128] (b) photobleaching reaction that may lead to a change of absorption or fluorescence;[129,130] (c) photorefractive reaction in a photorefractive polymer or crystal,[122,131] (d) photo-polymerization

Figure 14.31 *Nine frames of an animated movie stored at different depths in a two-photon dye doped polymer block. The spacing between adjacent layers is 5 µm, and the transverse scale bar is 50 µm. (After Dagani.[136]) (See also Color fig. 15.).*

reaction[121,132] and some other photo-chemical reactions leading to optical property changes. In some cases, researchers may also utilize the anisotropy of certain optical property induced by two-photon excitation to record and readout the optical data.[130,133,134] More recently, some novel materials have been reported for 2PA-based data storage, such as densely packed semiconductor nanocrystalline solid films and Au nanoparticle-doped sol–gel glass films.[133,135]

Figure 14.31 shows the early experimental achievement of multi-layer 3D-data storage in a two-photon active dye-doped polymer block, working via the photobleaching mechanism. The two-photon absorbing dye was APSS, doped into a transparent polymer (poly(hydroxyethyl methacrylate)) block of $3 \times 3 \times 3$ mm size. The writing laser beam of $\sim$800 nm wavelength was from a Ti:sapphire laser working with a confocal laser scanning microscope.[136] In the writing beam focused bit position, the 2PA processes could bleach the fluorescence capability of the dye molecules, and then produced a "dark" spot compared to un-irradiated areas when scanned by a reading beam of the same wavelength.

The key issue to increase 3D data storage density is to reduce both the lateral bit spacing within one layer and the axial layer spacing in a given recording medium. The reported density value was 33 Gbits/cm^3 in a photorefractive crystal medium,[122] 205 Gbits/cm^3 in a polymer dispersed liquid crystals medium,[137] and $\sim$300 Gbits/cm^3 in a photo-polymerization medium.[121] Moreover, based on the lateral and axial sizes of the recorded bits, researchers estimate that a 10^{12} bits/cm^3 or 1000 Gbits/cm^3 density value could be achieved.

14.6.3 Two-photon polymerization-based 3D microfabrication

The features of high spatial confinement of multi-photon interactions and penetrating capability of a focused laser beam can be used to create 3D microstructures or to fabricate micro-machines with a sub-diffraction-limit spatial resolution. The emergence of so-called 3D microfabrication is one of the latest achievements of multi-photon based applications. This new technique is essentially a multi-photon-interaction-based 3D photolithography. In recent years, a number of excellent studies in this new area have been done by Kawata's group,[138,139] Marder and Perry's group,[140,141] Sun's group,[142] Belfield and Van Stryland's group,[118] Baldeck's group,[143] Lee's group,[144] and many others.

The first experimental demonstration of using a two-photon-induced polymerization process to produce a 3D polymer part with microstructure was reported by Kawata et al. in 1997.[138] In this work, the sample medium was a commercial resin (SCR-500) containing a UV-sensitive photoinitiator, urethane acrylate monomer, and urethane acrylate oligomers; the incident was a 790-nm, 200-fs pulsed laser beam focused by a $NA = 0.85$ objective. After laser exposure and 2PA-induced polymerization, the resultant solidified polymer structures were extracted by removing the unsolidified resin with ethanol. Another important year for two-photon microfabrication was 1999, in which several groups published their research works with different features. Among them, Cumpston et al. presented their high-quality 3D microstructure from two-photon polymer resins films; the photos of their products are shown in Fig. 14.32.[140] Sun et al. reported their 3D photonic crystal structures through two-photon induced polymerization in a resin medium; the microstructures consisted of 20 layers of solidified micro-rods with 0.6–2.0 μm diameter and 1.2–1.4 μm spacing.[142] Prasad et al. reported fabrication of 3D optical waveguide circuitry in a two-photon polymerizable optical resin.[145]

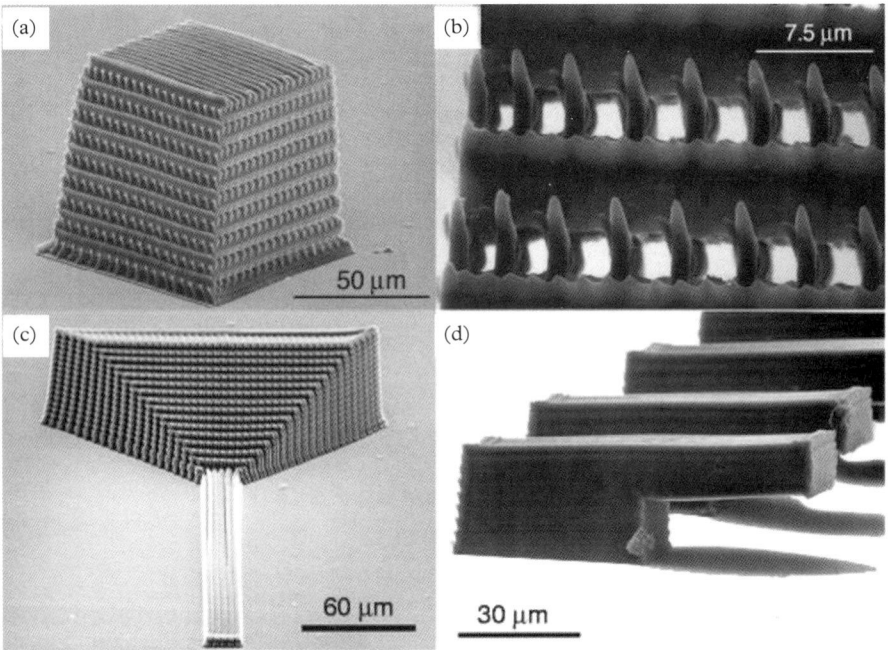

Figure 14.32 *3D microstructures produced by two-photon-initiated polymerization. (a) Photonic bandgap structure. (b) Magnified top-view of structure in (a). (c) Tapered waveguide structure. (d) Array of cantilevers. (Reproduced with permission from Cumpston et al.[140] Copyright 1999, Nature Publishing Group).*

To avoid the drift and distortion of components in microfabrication processes, it was proposed to utilize a pre-exposure technique (by a xenon lamp) to increase the viscosity of resins and let the solidified elements be tightly confined at the laser exposed sites.[146]

In 2001, Kawata et al. reported the 3D fabrication of several microstructures with a much higher spatial resolution (~0.12 μm) in a commercially available resin, consisting of urethane acrylate monomers and oligomers as well as 2PA-active photoinitiators.[139] By pinpoint-scanning the laser focus according to pre-programmed patterns, designs can be faithfully replicated to form real structures. One of the microstructures they made was a computer-designed 10-μm long and 7-μm high micro-bull. Figure 14.33 shows the scanning electron microscope (SEM) micrographs of a "micro-bull" sculpture through 2PA-induced photo-polymerization processes.

From a chemical point of view, one of the key issues for 3D microfabrication is the design and synthesis of highly 2PA-active photoinitiators in appropriate photo-polymerizable systems, which may lead to higher polymerization efficiency while requiring a lower laser beam intensity or power level.

Recently, Teh et al. have demonstrated that even by using low numerical aperture optics (10×, 0.3 NA objective), the rapid microfabrication of photoplastic pillars, planes, and cage structures with a ultrahigh aspect ratio could be achieved, based on SU-8 resin/resist films of thickness from

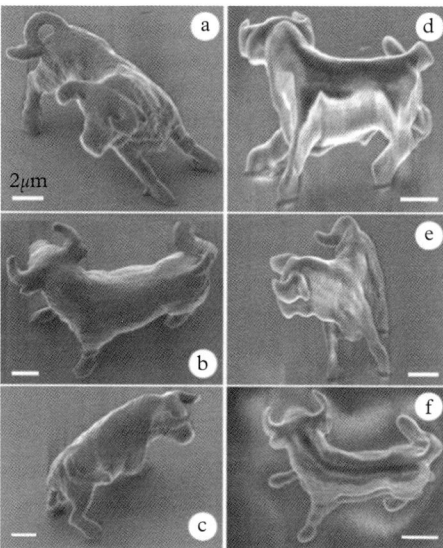

Figure 14.33 *(a–c) Bull sculpture produced by raster scanning; the process took 180 min; (d–f) Surface of the bull was defined by 2PA (that is a surface-profile scanning) and was then solidified internally by illumination under a mercury lamp, reducing the 2PA-scanning time to 13 min. Scale bars: 2 μm. (Reproduced with permission from Kawata et al.[139] Copyright 2001, Nature Publishing Group).*

500 to 975 nm.[147] The designable 3D nanofabrication technique by femtosecond laser direct writing has been reported by Sun's group.[148]

REFERENCES

1. M. Göppert-Mayer, *Ann. Phys. (Berlin)* **401**, 273 (1931).
2. P. A. M. Dirac, *Proc. Roy. Soc. London A* **114**, 243 (1927).
3. W. Kaiser and C. G. B. Garrett, *Phys. Rev. Lett.* **7**, 229 (1961).
4. G. S. He, L.-S. Tan, Q. Zheng, and P. N. Prasad, *Chem. Rev.* **108**, 1245 (2008).
5. Y. R. Shen, *The Principle of Nonlinear Optics* (Wiley, New York, 1984).
6. W. Heitler, *The Quantum Theory of Radiation* (Oxford University, London, 1954).
7. D. Marcuse, *Principles of Quantum Electronics* (Academic, New York, 1980).
8. D. P. Craig and T. Thirunamachandran, *Molecular Quantum Electrodynamics* (Academic, London, 1984).
9. P. R. Monson and W. M. Mcclain, *J. Chem. Phys.* **53**, 29 (1970).
10. B. A. Reinhardt, *Photonics Sci. News* **4**, 21(1999); B. A. Reinhardt, L. L. Brott, S. J. Clarson, A. G. Dillard, J. C. Bhatt, R. Kannan, L. Yuan, G. S. He, and P. N. Prasad, *Chem. Mater.* **10**, 1863(1998).
11. R. Kannan, B. A. Reinhardt, and L.-S. Tan, U.S.Patent 6,300,502, October 9, 2001.
12. M. Albota, D. Beljonne, J. L. Bredas, J. E. Ehrlich, J. Y. Fu, A. A. Heikal, S. E. Hess, T. Kogej, M. D. Levin, S. R. Marder, D. McCord-Maughon, J. W. Perry, H. Rockel, M. Rumi, G. Subramaniam, W. W. Webb, X. L. Wu, and C. Xu, *Science* **281**, 1653 (1998).
13. S. R. Marder, W. E. Torruellas, M. Blanchard-Desce, V. Ricci, G. I. Stegeman, S. Gilmour, J. L. Bredas, J. Li, G. U. Bublitz, and S. G. Boxer, *Science* **276**, 1233 (1997).
14. M. P. Joshi, J. Swiatkiewicz, F. Xu, P. N. Prasad, B. A. Reinhardt, and R. Kannan, *Opt. Lett.* **23**, 1742 (1998).
15. G. S. He, J. Swiatkiewicz, Y. Jiang, P. N. Prasad, B. A. Reinhardt, L.-S. Tan, and R. Kannan, *J. Phys. Chem. A* **104**, 4805 (2000).
16. S.-J. Chung, K.-S. Kim, T.-C. Lin, G. S. He, J. Swiatkiewicz, and P. N. Prasad, *J. Phys. Chem. B* **103**, 10741 (1999).
17. M. Drobizhev, A. Karotki, A. Rebane, and C. W. Spangler, *Opt. Lett.* **26**, 1081 (2001); P. Wei, X. Bi, Z. Wu, and Z. Xu, *Org. Lett.* **7**, 3199 (2005).
18. A. Adronov, J. M. J. Frechet, G. S. He, K. S. Kim, S. J. Chung, J. Swiatkiewicz, and P. N. Prasad, *Chem. Mater.* **12**, 2838 (2000); G. S. He, T.-C. Lin, Y. Cui, P. N. Prasad, D. W. Brousmiche, J. M. Serin, and J. M. J. Frechet, *Opt. Lett.* **28**, 768 (2003).
19. Q. Zheng, G. S. He, and P. N. Prasad, *Chem. Mater.* **17**, 6004 (2005).
20. M. J. Dalton, R. Kannan, J. E. Haley, G. S. He, D. G. McLean, T. M. Cooper, P. N. Prasad, and L.-S. Tan, *Macromolecules* **44**, 7195 (2011); A. Samoc, M. Samoc, M. Woodruff, and B. Lutherdavies, *Opt. Lett.* **20**, 1241 (1995).
21. G. Banfi, D. Fortusini, P. Dainesi, D. Grando, and S. Sottini, *J. Chem. Phys.* **108**, 4319 (1998); L. De Boni, A. A. Andrade, D. S. Correa, D. T. Balogh, S. C. Zilio, L. Misoguti, and C. R. Mendonca, *J. Phys. Chem. B* **108**, 5221 (2004).
22. G. S. He, T.-C. Lin, P. N. Prasad, C.-C. Cho, and L.-J. Yu, *Appl. Phys. Lett.* **82**, 4717 (2003).
23. I. C. Khoo, A. Diaz, M. V. Wood, and P. H. Chen, *IEEE J. Sel. Top. Quantum Electron.* **7**, 760 (2001); F. Lincker, P. Bourgun, P. Masson, P. Didier, L. Guidoni, J. Y. Bigot, J. F. Nicoud, B. Donnio, and D. Guillon, *Org. Lett.* **7**, 1505 (2005).
24. G. S. He, Q. Zheng, P. N. Prasad, R. Helgeson, and F. Wudl, *Appl. Opt.* **44**, 3560 (2005).

25. Q. Zheng, G. S. He, and P. N. Prasad, *J. Mater. Chem.* **15**, 579 (2005).
26. S. Righetto, S. Rondena, D. Locatelli, D. Roberto, F. Tessore, R. Ugo, S. Quici, S. Roma, D. Korystov, and V. I. Srdanov, *J. Mater. Chem.* **16**, 1439 (2006).
27. S. J. K. Pond, O. Tsutsumi, M. Rumi, O. Kwon, E. Zojer, J. L. Brédas, S. R. Marder, and J. W. Perry, *J. Am. Chem. Soc.* **126**, 9291 (2004).
28. T. C. Wen, L. C. Hwang, W. Y. Lin, C. H. Chen, and C. H. Wu, *Chem. Phys.* **286**, 293 (2003).
29. P. P. Paskov, P. O. Holtz, B. Monemar, J. M. Garcia, W. V. Schoenfeld, and P. M. Petroff, *Appl. Phys. Lett.* **77**, 812 (2000).
30. D. R. Larson, W. R. Zipfel, R. M. Williams, S. W. Clark, M. P. Bruchez, F. W. Wise, and W. W. Webb, *Science* **300**, 1434 (2003).
31. G. S. He, Q. Zheng, K.-T. Yong, F. Erogbogbo, M. T. Swihart, and P. N. Prasad, *Nano Lett.* **8**, 2688 (2008).
32. W. Wenseleers, F. Stellacci, T. Meyer-Friedrichsen, T. Mangel, C. A. Bauer, S. J. K. Pond, S. R. Marder, and J. W. Perry, *J. Phys. Chem B* **106**, 6853 (2002).
33. I. Cohanoschi and F. E. Hernandez, *J. Phys. Chem. B* **109**, 14506 (2005).
34. R. A. Farrer, F. L. Butterfield, V. W. Chen, and J. T. Fourkas, *Nano Lett.* **5**, 1139 (2005).
35. T. F. Boggess, K. M. Bohnert, K. Mansour, S. C. Moss, I. W. Boyd, and A. L. Smirl, *IEEE J. Quantum Electron.* **22**, 360 (1986).
36. M. Sheik-Bahae, A. A. Said, T.-H. Wei, D. J. Hagan, and E. W. Vanstryland, *IEEE J. Quantum Electron.* **26**, 760 (1990).
37. C. Xu and W. W. Webb, *J. Opt. Soc. Am. B* **13**, 481(1996).
38. J. E. Ehrlich, X. L. Wu, I. Y. S. Lee, Z. Y. Hu, H. Rockel, S. R. Marder, and J. W. Perry, *Opt. Lett.* **22**, 1843(1997).
39. J. Swiatkiewicz, P. N. Prasad, and B. A. Reinhardt, *Opt. Commun.* **157**, 135 (1998).
40. J. R. Lakowicz, *Principles of Fluoresent Spectroscopy* (Kluwer Academic/Plenum, New York, 1999).
41. S. Polyakov, F. Yoshino, M. Liu, and G. Stegeman, *Phys. Rev. B* **69**, 115421 (2004).
42. D. A. Oulianov, I. V. Tomov, A. S. Dvornikov, and P. M. Rentzepis, *Opt. Commun.* **191**, 235 (2001).
43. J. Thomas, M. Anija, J. Cyriac, T. Pradeep, and R. Philip, *Chem. Phys. Lett.* **403**, 308 (2005).
44. G. S. He, C. G. Lu, Q. D. Zheng, A. Baev, M. Samoc, and P. N. Prasad, *Phys. Rev. A* **73**, 033815 (2006).
45. R. L. Sutherland, M. C. Brant, J. Heinrichs, J. E. Rogers, J. E. Slagle, D. G. McLean, and P. A. Fleitz, *J. Opt. Soc. Am. B* **22**, 1939 (2005).
46. J. F. Lami, P. Gilliot, and C. Hirlimann, *Phys. Rev. Lett.* **77**, 1632 (1996).
47. S. M. Kirkpatrick, R. R. Naik, and M. O. Stone, *J. Phys. Chem. B* **105**, 2867 (2001).
48. R. Schroeder and B. Ullrich, *Opt. Lett.* **27**, 1285 (2002).
49. G. S. He, Q. D. Zheng, A. Baev, and P. N. Prasad, *J. Appl. Phys.* **101**, 083108 (2007).
50. H. Lei, Z. L. Huang, H. Z. Wang, X. J. Tang, L. Z. Wu, G. Y. Zhou, D. Wang, and Y. B. Tian, *Chem. Phys. Lett.* **352**, 240 (2002).
51. M. Samoc, A. Samoc, M. G. Humphrey, M. P. Cifuentes, B. Luther-Davies, and P. A. Fleitz, *Mol. Cryst. Liq. Cryst.* **485**, 894 (2008).
52. M. Drobizhev, A. Karotki, M. Kruk, Y. Dzenis, A. Rebane, Z. Suo, and C. W. Spangler, *J. Phys. Chem. B* **108**, 4221(2004).
53. P. Feneyrou, O. Doclot, D. Block, P. L. Baldeck, S. Delysse, and J. M. Nunzi, *Opt. Lett.*, **22**, 1132 (1997).
54. K. D. Belfield, D. J. Hagan, E. W. Van Stryland, K. J. Schafer, and R. A. Negres, *Org. Lett.* **1**, 1575 (1999).
55. R. A. Negres, J. M. Hales, D. J. Hagan, and E. W. Van Stryland, *IEEE J. Quantum Electron.* **38**, 1205 (2002).

56. G. S. He, T. C. Lin, and P. N. Prasad, *Opt. Express* **10**, 566 (2002).
57. G. S. He, T. C. Lin, J. M. Dai, P. N. Prasad, R. Kannan, A. G. Dombroskie, R. A. Vaia, and L. S. Tan, *J. Chem. Phys.* **120**, 5275 (2004).
58. Z. Y. Suo, M. Drobizhev, C. W. Spangler, N. Christensson, and A. Rebane, *Org. Lett.* **7**, 4807 (2005).
59. Q. D. Zheng, G. S. He, A. Baev, and P. N. Prasad, *J. Phys. Chem. B* **110**, 14604 (2006).
60. T. C. Lin, G. S. He, Q. D. Zheng, and P. N. Prasad, *J. Mater. Chem.* **16**, 2490 (2006).
61. N. G. Basov, A. Z. Grasyuk, I. G. Zubarev, and V. A. Katulin, *JETP Lett.* **1**, 118 (1965); N. G. Basov, A. Z. Grasyuk, I. G. Zubarev, V. A. Katulin, and O. N. Krokhin, *Sov. Phys. JETP* **23**, 366 (1996).
62. C. K. N. Patel, P. A. Fleury, R. E. Slusher, and H. L. Frisch, *Phys. Rev. Lett.* **16**, 971 (1966).
63. S. K. Manlief and E. D. Palik, *Appl. Phys. Lett.* **22**, 443 (1973).
64. D. M. Bloom, J. T. Yardley, J. F. Young, and S. E. Harris, *Appl. Phys. Lett.* **24**, 427 (1974).
65. G. D. Willenberg, C. O. Weiss, and H. Jones, *Appl. Phys. Lett.* **37**, 133 (1980).
66. A. D. Tserepi, E. Wurzberg, and T. A. Miller, *Chem. Phys. Lett.* **265**, 297 (1997).
67. W. Rapp and B. Gronau, *Chem. Phys. Lett.* **8**, 529 (1971).
68. V. I. Prokhorenko, E. Λ. Tikhonov, and M. T. Shpak, *Kvantovaya Elektron.* **8**, 229 (1981).
69. P. Qiu and A. Penzkofer, *Appl. Phys. B* **48**, 115 (1989).
70. A. S. Kwok, A. Serpenguzel, W. F. Hsieh, R. K. Chang, and J. B. Gillespie, *Opt. Lett.* **17**, 1435 (1992).
71. A. Mukherjee, *Appl. Phys. Lett.* **62**, 3423 (1993).
72. G. S. He, J. D. Bhawalkar, C. F. Zhao, C. K. Park, and P. N. Prasad, *Opt. Lett.* **20**, 2393 (1995).
73. G. S. He, C. F. Zhao, J. D. Bhawalkar, and P. N. Prasad, *Appl. Phys. Lett.* **67**, 3703 (1995).
74. G. S. He, P. P. Markowicz, T. C. Lin, and P. N. Prasad, *Nature* **415**, 767 (2002).
75. G. S. He, T. C. Lin, S. J. Chung, Q. D. Zheng, C. G. Lu, Y. P. Cui, and P. N. Prasad, *J. Opt. Soc. Am. B* **22**, 2219 (2005).
76. Q. Zheng, H. Zhu, S.-C. Chen, C. Tang, E. Ma, and X. Chen, *Nat. Photonics* **7**, 234 (2013).
77. A. Abbotto, L. Beverina, R. Bozio, S. Bradamante, C. Ferrante, G. A. Pagani, and R. Signorini, *Adv. Mater.* **12**, 1963 (2000).
78. X. J. Tang, L. Z. Wu, L. P. Zhang, and C. H. Tung, *Chem. Phys. Lett.* **356**, 573 (2002).
79. G. S. He, L. X. Yuan, Y. P. Cui, M. Li, and P. N. Prasad, *J. Appl. Phys.* **81**, 2529 (1997).
80. G. S. He, J. D. Bhawalkar, C. F. Zhao, and P. N. Prasad, *IEEE J. Quantum Electron.* **32**, 749 (1996).
81. G. S. He, J. D. Bhawalkar, C. F. Zhao, C. K. Park, and P. N. Prasad, *Appl. Phys. Lett.* **68**, 3549 (1996).
82. G. S. He, R. Signorini, and P. N. Prasad, *IEEE J. Quantum Electron.* **34**, 7 (1998).
83. K. Shirota, H. B. Sun, and S. Kawata, *Appl. Phys. Lett.* **84**, 1632 (2004).
84. C. Bauer, B. Schnabel, E. B. Kley, U. Scherf, H. Giessen, and R. F. Mahrt, *Adv. Mater.* **14**, 673 (2002).
85. G. Tsiminis, A. Ruseckas, I. D. W. Samuel, and G. A. Turnbull, *Appl. Phys. Lett.* **94**, 253304 (2009).
86. F. Scotognella, D. P. Puzzo, M. Zavelani-Rossi, J. Clark, M. Sebastian, G. A. Ozin, and G. Lanzani, *Chem. Mater.* **23**, 805 (2011).
87. C. Moro, M. Lepore, R. Cingolani, R. Tommasi, M. Ferrara, I. M. Catalano, K. Ploog, and A. Fischer, *Phys. Rev. B* **44**, 8384 (1991).
88. M. H. Nayfeh, N. Barry, J. Therrien, O. Akcakir, E. Gratton, and G. Belomoin, *Appl. Phys. Lett.* **78**, 1131 (2001).
89. C. Zhang, F. Zhang, A. Cheng, B. Kimball, A. Y. Wang, and J. Xu, *Appl. Phys. Lett.* **95**, 183109 (2009).
90. C. Zhang, F. Zhang, T. Zhu, A. Cheng, J. Xu, Q. Zhang, S. E. Mohney, R. H. Henderson, and Y. A. Wang, *Opt. Lett.* **33**, 2437 (2008).

91. G. S. He, T. C. Lin, V. K. S. Hsiao, A. N. Cartwright, P. N. Prasad, L. V. Natarajan, V. P. Tondiglia, R. Jakubiak, R. A. Vaia, and T. J. Bunning, *Appl. Phys. Lett.* **83**, 2733 (2003).
92. G. S. He, J. M. Dai, T. C. Lin, P. P. Markowicz, and P. N. Prasad, *Opt. Lett.* **28**, 719 (2003).
93. G. S. He, H.-Y. Qin, Q. Zheng, P. N. Prasad, S. Jockusch, N. J. Turro, M. Halim, D. Sames, H. Agren, and S. He, *Phys. Rev. A* **77**, 013824 (2008).
94. L. W. Tutt and T. F. Boggess, *Prog. Quantum Electron.* **17**, 299 (1993).
95. V. V. Arsenev, V. S. Dneprovskii, D. N. Klyshko, and A. N. Penin, *Sov. Phys. JETP* **29**, 413 (1969).
96. E. W. Vanstryland, H. Vanherzeele, M. A. Woodall, M. J. Soileau, A. L. Smirl, S. Guha, and T. F. Boggess, *Opt. Eng.* **24**, 613 (1985); E. W. Vanstryland, Y. Y. Wu, D. J. Hagan, M. J. Soileau, and K. Mansour, *J. Opt. Soc. Am. B* **5**, 1980 (1988).
97. G. S. He, G. C. Xu, P. N. Prasad, B. A. Reinhardt, J. C. Bhatt, R. McKellar, and A. G. Dillard, *Opt. Lett.* **20**, 435 (1995).
98. G. S. He, J. D. Bhawalkar, C. F. Zhao, and P. N. Prasad, *Appl. Phys. Lett.* **67**, 2433 (1995).
99. Y. Morel, A. Ibanez, C. Nguefack, C. Andraud, A. Collet, J. F. Nicoud, and P. L. Baldeck, *Synth. Met.* **115**, 265 (2000).
100. R. A. Ganeev, A. I. Ryasnyansky, R. I. Tugushev, M. K. Kodirov, F. R. Akhmedjanov, and T. Usmanov, *Opt. Quantum Electron.* **36**, 807 (2004).
101. Y. Morel, J. Zaccaro, A. Ibanez, and P. L. Baldeck, *Opt. Commun.* **201**, 457 (2002).
102. I. C. Khoo, A. Diaz, and J. W. Ding, *J. Opt. Soc. Am. B* **21**, 1234 (2004).
103. M. Nakazawa, Y. Watanabe, T. Tsuchiya, and S. Fujitsu, *J. Ceram. Soc. Jpn.* **104**, 918 (1996).
104. Y.-H. Jiang, Y.-C. Wang, J.-B. Yang, J.-L. Hua, B. Wang, S.-Q. Qian, and H. Tian, *J. Polym. Sci. A* **49**, 1830 (2011).
105. N. Venkatram, D. N. Rao, and M. A. Akundi, *Opt. Express* **13**, 867 (2005).
106. G. S. He, K.-T. Yong, Q. Zheng, Y. Sahoo, A. Baev, A. I. Ryasnyanskiy, and P. N. Prasad, *Opt. Express* **15**, 12818 (2007).
107. G. S. He, L. X. Yuan, N. Cheng, J. D. Bhawalkar, P. N. Prasad, L. L. Brott, S. J. Clarson, and B. A. Reinhardt, *J. Opt. Soc. Am. B* **14**, 1079 (1997).
108. G. S.He, Q. Zheng, K.-T. Yong, A. I. Ryasnyanskiy, P. N. Prasad, and A. Urbas, *Appl. Phys. Lett.* **90**, 181108 (2007).
109. G. S. He, J. D. Bhawalkar, P. N. Prasad, and B. A. Reinhardt, *Opt. Lett.* **20**, 1524 (1995).
110. G. S. Maciel, N. Rakov, C. B. de Araujo, A. A. Lipovskii, and D. K. Tagantsev, *Appl. Phys. Lett.* **79**, 584 (2001).
111. W. B. Ma, Y. Q. Wu, J. H. Han, D. H. Gu, and F. X. Gan, *Chem. Phys. Lett.* **410**, 282 (2005).
112. G. S. He, Q. D. Zheng, C. G. Lu, and P. N. Prasad, *IEEE J. Quantum Electron.* **41**, 1037 (2005).
113. G. S. He, K.-T. Yong, J. Zhu, H.-Y. Qin, and P. N. Prasad, *IEEE Quantum Electron.* **46**, 931 (2010).
114. G. S. He, R. Gvishi, P. N. Prasad, and B. A. Reinhardt, *Opt. Commun.* **117**, 133 (1995).
115. G. S. He, L. X. Yuan, J. D. Bhawalkar, and P. N. Prasad, *Appl. Opt.* **36**, 3387 (1997).
116. G. S. He, H. S. Oh, and P. N. Prasad, *Opt. Lett.* **36**, 4431 (2011).
117. S. Kawata and Y. Kawata, *Chem. Rev.* **100**, 1777 (2000).
118. K. D. Belfield, K. J. Schafer, Y. U. Liu, J. Liu, X. B. Ren, and E. W. Van Stryland, *J. Phys. Org. Chem.* **13**, 837 (2000).
119. D. A. Parthenopoulos and P. M. Rentzepis, *Science* **245**, 843 (1989).
120. D. A. Parthenopoulos and P. M. Rentzepis, *J. Appl. Phys.* **68**, 5814 (1990).
121. J. H. Strickler and W. W. Webb, *Opt. Lett.* **16**, 1780 (1991).
122. Y. Kawata, H. Ishitobi, and S. Kawata, *Opt. Lett.* **23**, 756 (1998).
123. M. Gu and D. Day, *Opt. Lett.* **24**, 288 (1999).
124. H. E. Pudavar, M. P. Joshi, P. N. Prasad, and B. A. Reinhardt, *Appl. Phys. Lett.* **74**, 1338 (1999).
125. A. S. Dvornikov and P. M. Rentzepis, *Opt. Commun.* **136**, 1 (1997).
126. K. D. Belfield and K. J. Schafer, *Chem. Mater.* **14**, 3656 (2002).

127. A. Toriumi, S. Kawata, and M. Gu, *Opt. Lett.* **23**, 1924 (1998).
128. Y. J. Zhou, H. H. Tang, W. H. Huang, A. D. Xia, F. Sun, and F. S. Zhang, *Opt. Eng.* **44**, 035202 (2005).
129. U. Kubitscheck, M. TschodrichRotter, P. Wedekind, and R. Peters, *J. Microsc. (Oxford)* **182**, 225 (1996).
130. I. Polyzos, G. Tsigaridas, M. Fakis, V. Giannetas, P. Persephonis, and J. Mikroyannidis, *Chem. Phys. Lett.* **369**, 264 (2003).
131. D. Day and M. Gu, *Opt. Lett.* **24**, 948 (1999).
132. S. M. Kirkpatrick, J. W. Baur, C. M. Clark, L. R. Denny, D. W. Tomlin, B. R. Reinhardt, R. Kannan, and M. O. Stone, *Appl.Phys. A* **69**, 461 (1999).
133. J. W. M. Chon, P. Zijlstra, M. Gu, J. van Embden, and P. Mulvaney, *Appl. Phys. Lett.* **85**, 5514 (2004).
134. S. O. Konorov, D. A. Sidorov-Biryukov, I. Bugar, D. Chorvat, D. Chorvat, and A. M. Zheltikov, *Chem. Phys. Lett.* **381**, 572 (2003).
135. M. Fukushima, H. Yanagi, S. Hayashi, H. B. Sun, and S. Kawata, *Physica E* **21**, 456 (2004).
136. R. Dagani, *Chem. Eng. News* **74**(39), 68 (1996).
137. D. McPhail and M. Gu, *Appl. Phys. Lett.* **81**, 1160 (2002).
138. S. Maruo, O. Nakamura, and S. Kawata, *Opt. Lett.* **22**, 132 (1997).
139. S. Kawata, H. B. Sun, T. Tanaka, and K. Takada, *Nature* **412**, 697 (2001).
140. B. H. Cumpston, S. P. Ananthavel, S. Barlow, D. L. Dyer, J. E. Ehrlich, L. L. Erskine, A. A. Heikal, S. M. Kuebler, I. Y. S. Lee, D. McCord-Maughon, J. Q. Qin, H. Rockel, M. Rumi, X. L. Wu, S. R. Marder, and J. W. Perry, *Nature* **398**, 51 (1999).
141. W. H. Zhou, S. M. Kuebler, K. L. Braun, T. Y. Yu, J. K. Cammack, C. K. Ober, J. W. Perry, and S. R. Marder, *Science* **296**, 1106 (2002).
142. H. B. Sun, S. Matsuo, and H. Misawa, *Appl. Phys. Lett.* **74**, 786 (1999).
143. I. Wang, M. Bouriau, P. L. Baldeck, C. Martineau, and C. Andraud, *Opt. Lett.* **27**, 1348 (2002).
144. H. K. Yang, M. S. Kim, S. W. Kang, K. S. Kim, K. S. Lee, S. H. Park, D. Y. Yang, H. J. Kong, H. B. Sun, S. Kawata, and P. Fleitz, *J. Photopolym. Sci. Technol.* **17**, 385 (2004).
145. M. P. Joshi, H. E. Pudavar, J. Swiatkiewicz, P. N. Prasad, and B. A. Reianhardt, *Appl. Phys. Lett.* **74**, 170 (1999).
146. H. B. Sun, T. Kawakami, Y. Xu, J. Y. Ye, S. Matuso, H. Misawa, M. Miwa, and R. Kaneko, *Opt. Lett.* **25**, 1110 (2000).
147. W. H. Teh, U. Durig, G. Salis, R. Harbers, U. Drechsler, R. F. Mahrt, C. G. Smith, and H. J. Guntherodt, *Appl. Phys. Lett.* **84**, 4095 (2004).
148. Y. L. Zhang, Q. D. Chen, H. Xia, and H. B. Sun, *Nano Today.* **5**, 435 (2010).

15
Nonlinear Photoelectric Effects

Conventional photoelectric effects are related to two types of optical and electrical phenomena: one is free electron emission from a solid (mostly a metal) surface upon light irradiation, which is termed the photoemission effect; the other is the light-induced electrical conductivity change of a solid material (mostly a semiconductor), which is termed the photoconductive effect. Before the advent of lasers, both effects could be well described with a one-photon excitation mechanism. The common features of these two effects are: (i) the input light wavelength should be shorter than a definite cut-off wavelength; and (ii) the photoemitted electron number or the change of photoconductivity is linearly proportional to the input photon number or the light intensity. In the sense of the second feature, conventional photoelectric effects can also be called linear photoelectric effects.

After the advent of laser technology, it was soon found that the same photoelectric effects can be observed by using laser radiation with wavelengths much longer than the cut-off wavelength for a given material, and that the measured photoemitted electron numbers or the induced photoconductive response manifest a nonlinear relation with the input laser intensity. These new observations can be well interpreted based on multi-photon excitation mechanisms, and thereby are termed multi-photon (or nonlinear) photoelectric effects.

15.1 Introduction to photoelectric effects

15.1.1 One-photon photoemission effect

In 1887, Hertz reported the first demonstration of free electron emission from a solid (metal) surface illuminated by UV light.[1] This phenomenon is termed the photoemission effect or external photoelectric effect. In 1905, Einstein published his famous scientific article about the quantum (photon) nature of light, based on which the essence of the photoemission effect could be well interpreted.[2] According to his hypothesis, a delocalized electron near to the solid surface can absorb a single photon energy of $h\nu$ and be excited to escape from the solid surface. The escaped electron could have a certain kinetic energy E_k that is determined by

$$E_k = h\nu - \phi, \qquad (15.1\text{-}1)$$

where ϕ is the work function of the solid and $h\nu \geq \phi$ is the necessary photon energy. The physical meaning of ϕ (a material parameter) is the minimum energy needed by an electron to escape from the surface into vacuum; if the incident light frequency is lower than the threshold value of $\nu_0 = \phi/h$, no photoemission effect can be observed. Equation (15.1-1) indicates that when $\nu > \nu_0$ the kinetic energy of the emitted photoelectrons is only determined by the incident photon energy

---

Nonlinear Optics and Photonics. First Edition. Guang S. He. © Guang S. He 2015.
Published in 2015 by Oxford University Press.

(frequency) but not the number of photons. Since the photon model was able to well explain all early results of photoelectric experiments, Einstein was awarded the Nobel Prize in Physics in 1921.

For a very long time, the photoemission effect has been applied for designing various phototubes and photomultipliers as photodetectors. Under proper working conditions, the photocurrent (I_{cur}) created by a given photodetector will be linearly proportional to the incident light intensity (I_0), i.e.,

$$I_{\text{cur}} \propto I_0. \tag{15.1-2}$$

15.1.2 Electronic band structures of solids

According to the quantum-mechanical theory of solids, the electric and optical properties of solid materials are primarily determined by the energy band structure of the valence (outer-shell) electrons. As shown schematically in Fig. 15.1, there are three types of solids distinguished by their band structures, where the upper energy band represents the unfilled conduction band while the lower represents the valence band.

For monovalent metals, the valence band is only half-filled, as shown in Fig. 15.1(a); therefore all valence electrons can be conduction electrons. For divalent metals, the valence and conduction bands partially overlap, as shown in Fig. 15.1(b), so that they also have excellent electrical conductivity.

For semiconductors and insulators, shown in Fig. 15.1(c) and (d), there is an energy bandgap (E_g) between the empty conduction band and the full valence band. Usually, insulators have a large bandgap (e.g., $E_g > 5$–6 eV), and it is difficult to excite a valence electron into the conduction band by applying an electric field or using optical excitation without breaking down the materials. In contrast, semiconductors exhibit a moderate or even small bandgap (e.g., $E_g < 4$–5 eV); the valence electrons are much more easily excited into the conduction band by applying an electric field or optical excitation.

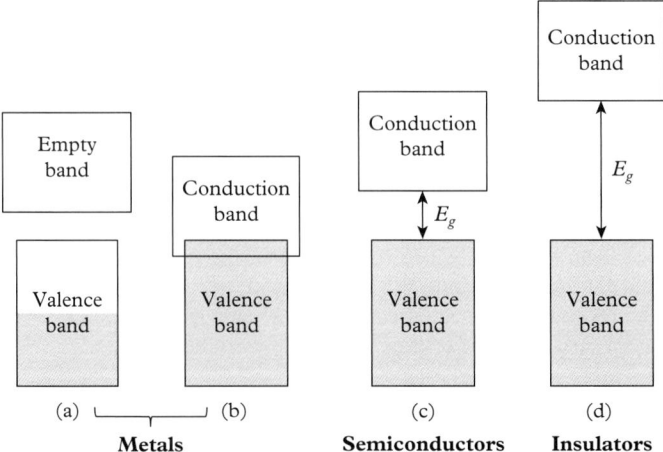

Figure 15.1 *Schematic diagram of electron energy band structures of solids: metals (a, b), semiconductors (c), and insulators (d). E_g is the bandgap between the valence band and conduction band.*

15.1.3 One-photon induced photoconductivity in semiconductors

When a semiconductor is illuminated by light radiation of suitable wavelength, the electric conductivity of a semiconductor sample can be increased. This is the photoconductivity effect or the so-called internal photoelectric effect. If the input illumination is from a conventional light source or a low-intensity laser beam, the light-induced conductivity change is a result of the one-photon-excited creation of electric carriers (electron and hole), and usually there is a linear relation between the conductivity change and the input light intensity. This is the linear photoconductivity effect via one-photon excitation. If the one-photon absorption (1PA) is accompanied by an interband transition of one electron from the valence band to the conduction band (while leaving behind a hole in the valence band), the incident photon energy must be greater than the band gap, i.e., $h\nu \geq E_g$, and this is the so-called intrinsic photoconductivity effect. In another case, if the sample is a doped semiconductor, the 1PA may also occur through the electronic transition from the valence band to a trap (acceptor) level or from a trap (donor) level to the conduction band; this corresponds to the so-called extrinsic photoconductivity effect. Obviously, in the latter case the incident photon energy can be smaller than the band gap. In the rest of this chapter, we only consider the intrinsic photoconductivity effect.

15.1.4 Image-potential states (IPSs) of an electron at a metal surface

Let us consider an electron outside but very close to the surface of a metal; this electron will be attracted by the surface electric field originating from an induced equal charge of opposite sign inside the metal. Under equilibrium conditions of the interaction between the electron and the induced surface field, the metal surface should be an equipotential surface, which means that the electric lines of force characterizing this interaction should everywhere be perpendicular to the metal surface. Based on this consideration, one may assume that this type of interaction is equivalent to the interaction between the outside electron and an equal image charge of opposite sign inside the metal, as shown schematically in Fig. 15.2(a). Upon this assumption, the outside electron can be trapped in an image potential described by

$$V(z) = -\frac{e^2}{16\pi\varepsilon_0 z}, \qquad (15.1\text{-}3)$$

where e is the charge of electron and z is the distance from the electron to the metal surface. This classical potential function is quite similar to that experienced by an electron inside a hydrogen atom.

Starting from Eq. (15.1-3) and after quantum mechanical treatment, one can obtain the solutions of Schrödinger's equation describing the above-mentioned interaction. Eigenenergies of these solutions are described by a Rydberg-like series:[3]

$$E_n = -\frac{0.85\text{eV}}{(n+a)^2}, \qquad (n = 1, 2, 3, \ldots) \qquad (15.1\text{-}4)$$

where n is the quantum number and a is quantum defect ($0 \leq a \leq 0.5$) depending on the material. These hydrogen-like electronic states are called the image-potential states (IPSs) that characterize the bound states of an outside electron very close to a conductor surface. As shown in Fig. 15.2(b),

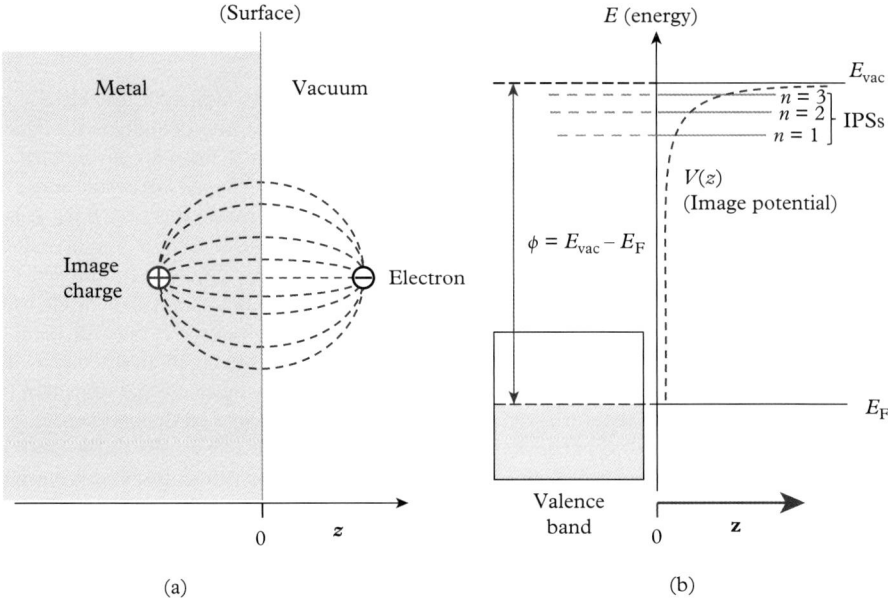

Figure 15.2 *Schematic diagrams of (a) the interaction between an outside electron and induced image-charge inside a metal, and (b) the classical image-potential curve (right-hand half) and quantized image-potential-state energy levels. E_{vac}: vacuum level energy, E_F: Fermi level energy, and Φ: work function. Above E_{vac}, the electron becomes free.*

the eigenvalues of high-order IPSs converge to the vacuum level of E_{vac}, above which the electron becomes entirely free from the surface of a metal.

In Fig. 15.2(b), there is an energy level (labeled by E_F) called Fermi level that is a concept introduced in quantum statistics. It is known that the conduction electrons in metals satisfy the Fermi–Dirac statistical law, i.e., the probability of a quantum statistical state (j) occupied by one electron is determined by

$$f_j = \frac{1}{e^{(E_j - \mu)/k_b T} + 1}, \tag{15.1-5}$$

where k_b is the Boltzmann constant, T the absolute temperature, and E_j and μ are the energy and chemical potential of the electron, respectively. It can be proved that at temperature $T = 0$, $f_j = 1$ for $E_j^0 \leq \mu^0$, and $f_j = 0$ for $E_j^0 \geq \mu^0$. Therefore the energy level of $E_F = E_j^0 = \mu^0$ is defined as the Fermi level. Here, μ^0 is the chemical potential at $T = 0$, and E_F means the highest energy level occupied by a valence electron at $T = 0$. At arbitrary temperature of $T \neq 0$, Fermi level is defined by $E_F = \mu(T)$ at which and from Eq. (15.1-5) we have

$$f(E_j = E_F) = 1/2. \tag{15.1-6}$$

Although chemical potential μ is a function of T, we approximately have $\mu(T) \approx \mu^0$ if T is not too high in comparison with room temperature. For this reason, the two ways of defining E_F are identical, and roughly speaking, the meaning of Fermi level for a metal is the highest energy level that could be occupied by a conduction electron.

Referring to Fig. 15.2(b), one can see that the work function of a given metal is defined by

$$\phi = E_{\text{vac}} - E_{\text{F}}. \tag{15.1-7}$$

If there is a conventional light radiation of $h\nu \geq \phi$ incident on the metal surface, the valence electrons, located near to or below the E_{F} level can be emitted into vacuum, which is the ordinary photoemission effect. On the other hand, a valence electron below the E_{F} level may also be excited to one of the image-potential states by absorbing a photon of appropriate energy.

15.2 Multi-photon photoemission (MPPE) effects

15.2.1 Early observations of MPPE phenomena

The advent of the laser opened a new research area of multi-photon excited photoemission and photoconductivity. In these cases, the linear relationship between the photocurrent and the input light intensity (see Eq. (15.1-2)) no longer holds. Instead, the following relation between the peak photocurrent and the peak laser intensity can be applied:

$$I_{\text{cur}} \propto I_0^m. \quad (m = 2, 3, 4, \ldots) \tag{15.2-1}$$

Here, m is the number of photons that are involved in an elementary photoemission process, and it is assume that the primary mechanism leading to the observed photoelectric effect is due to simultaneous m-photon absorption through one or more intermediated states (see Section 14.1), as shown schematically in Fig. 15.3. For m-photon excited photoemission, the energy conservation requires that the incident photon energy should be in the range

$$(m-1)h\nu < \phi \text{ and } mh\nu > \phi, \tag{15.2-2}$$

where ϕ is the work function of a metal; meanwhile the extra energy,

$$E_{\text{kin}} = mh\nu - \phi - \Delta, \tag{15.2-3}$$

will become the kinetic energy of the emitted electron in vacuum. Here, Δ is the energy spacing from an initial occupied state to the Fermi level. Since the initial states may have a continuous distribution below Fermi level, even for a given excitation photon energy $(h\nu)$ one will still observe a continuous kinetic energy distribution of the emitted electrons in vacuum.

By using a pulsed laser beam of suitable wavelength, the multi-photon photoemission effect could be observed even in commercially available (or modified) phototubes and photomultipliers. In 1964, the two-photon photoemission (2PPE) was observed in a RCA1P28 photomultiplier (with Cs_3Sb photocathode) using 1.06 μm laser pulses ($h\nu = 1.17$ eV),[4] and in a photomultiplier with a vapor-deposited sodium (Na, $\Phi = 2.3$ eV) as the cathode irradiated by 840-nm laser pulses ($h\nu = 1.48$ eV).[5] In both cases, a quadratic dependence of photocurrent on the laser intensity was observed.

Later on, the effect of three-photon photoemission (3PPE) was observed in a photomultiplier with the gold (Au, $\Phi = 4.8$ eV) cathode irradiated by the Q-switched 694.3-nm laser pulses ($h\nu = 1.79$ eV).[6] Furthermore, in the 1970s, researchers reported the four-photon photoemission

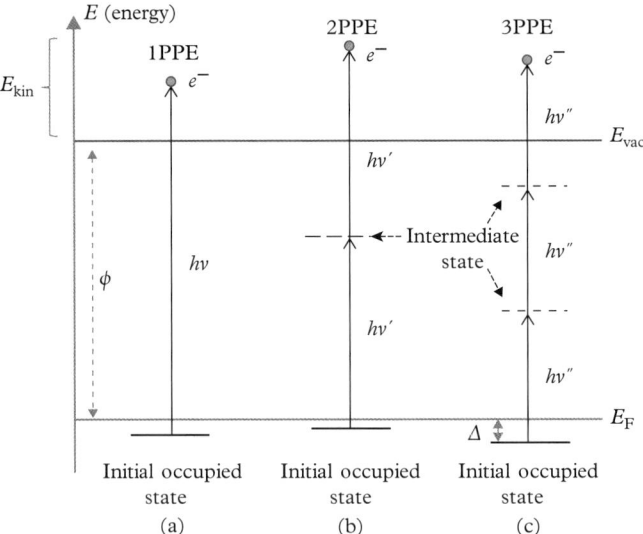

Figure 15.3 *Electron-transition diagrams of photoemission excited by (a) one photon absorption, (b) simultaneous two-photon absorption, and (c) simultaneous three-photon absorption. mPPE (m = 1, 2, 3, . . .) means m-photon photoemission.*

(4PPE) from a tungsten (W, Φ = 4.49 eV) surface by using mode-locked 1064-nm laser pulses ($h\nu$ = 1.17 eV),[7] and the five-photon photoemission (5PPE) from a 1-mm thick Au cathode irradiated by 30-ps and 1064-nm laser pulses.[8] In all these early studies, the nonlinear relation between the photocurrent and the laser intensity predicted by Eq. (15.2-1) was proved experimentally.[4–9]

15.2.2 Resonance-enhanced MPPE effects

In Chapter 14, we discussed the multi-photon absorption (MPA) properties of materials consisting of molecules or atoms. In those cases, the MPA behavior for a given medium is actually determined by outer-shell valence electron(s); thereby the basic concept and theoretical description of MPA are also applicable to multi-photon-induced photoelectric processes. Here the concept of *intermediate state* is particularly important through which *simultaneous* absorption of multiple photons by a valence electron can really take place. It can be theoretically proved that when one of the intermediate states involved in a specific MPA process approaches a real eigenenergy level of the excited electron, the probability of this process can be significantly increased.

Now taking 2PPE as an example, three possible situations may occur, which are shown schematically in Fig. 15.4. In the case shown in Fig. 15.4(a), there is only a single laser beam incident on a metal surface with a photon energy $h\nu$ much smaller than the work function (Φ), and the intermediate state for any allowed two-photon absorption (2PA) transitions is much far from any real excited-state levels of the electron; therefore we will observe a smooth kinetic energy distribution of the photoemitted electrons based on non-resonant 2PPE processes. When the incident photon energy is still lower than Φ but quite close to one of the IPSs described in Section 15.1.4, the 2PA cross-section will be enhanced and we shall see a near-resonant 2PPE process, as shown

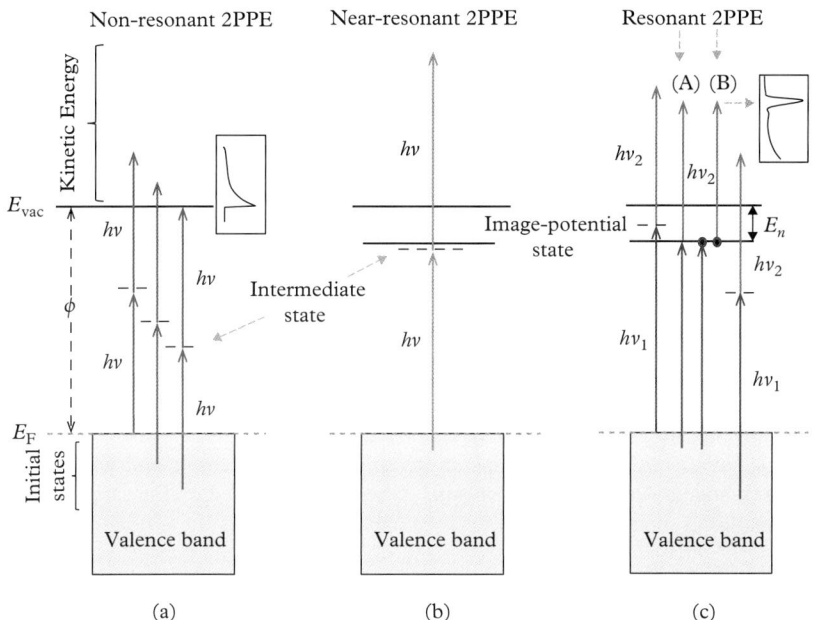

Figure 15.4 *Schematic diagrams showing (a) non-resonant two-photon photoemission (2PPE), (b) near-resonant 2PPE, and (c) two types of resonant 2PPE processes: (A) simultaneous two-photon excitation and (B) two sequential one-photon excitations.*

in Fig. 15.4(b). Furthermore, as shown in Fig. 15.4(c), if there are two incident laser beams of different frequencies and the intermediate state related to a nondegenerate 2PA process (absorbing one $h\nu_1$ photon from beam 1 and another $h\nu_2$ photon from beam 2) is precisely overlapping with one of the electron's IPSs, the resonance-enhancement will reach to its maximum. This so-called resonant 2PPE process leads to a resonance peak in the observed electron kinetic energy spectrum. It can be seen in Fig. 15.4(c) that the resonance peak in an energy spectrum will be located at the position of

$$E_{kin}^{peak} = h\nu_2 - E_n, \tag{15.2-4}$$

where E_n is the involved IPS energy position below the vacuum level. In such a way, the energy locations of various IPSs can be readily determined by measuring the specific positions of different resonant peaks in a recorded electron kinetic-energy spectral curve.

In the case of resonant 2PPE, however, two different mechanisms may contribute to an observed resonant peak in the kinetic energy spectrum. One mechanism still is the simultaneous 2PA, while the other is a stepwise 2PA mechanism. The latter essentially involves two sequential 1PA processes: (i) a valence electron to be excited to one IPS level via absorbing one $h\nu_1$ photon, and (ii) the excited electron populated in the IPS level can be further emitted to the vacuum by absorbing another $h\nu_2$ photon. These two events do not have to take place at the same time, which means that even the second (probe) laser pulse of $h\nu_2$ comes later than the first (pump) pulse of $h\nu_1$, one may still see the resonant 2PPE peak signal because the excited electron in the IPA level can stay for finite lifetime. In contrast, once the relative delay of these two laser pulses is larger than the pulse width, the contribution from the first mechanism vanishes. Since the excited electron can stay in an IPS for limited but nonzero lifetime, it can be expected that the resonant 2PPE peak signal can remain observable when we gradually change the relative delay of the hv_2 beam from zero to a certain limit. Actually, based on this principle, researchers can determine the lifetime of a given IPS by measuring its corresponding 2PPE resonance peak value as a function of the relative time delay between the pump and probe laser pulses.

Figure 15.5 shows a schematic diagram of a typical experimental setup for energy- and time-resolved resonant 2PPE studies, which consists of following three basic parts:

1. *Ultra-short pulsed laser sources*: They usually can provide two pulsed laser beams with a pulse duration about or less than 100 fs; the wavelength of one (pump) beam is in the UV range and the wavelength of the other (probe) beam is in the near-IR or visible range.
2. *Optical delay control system*: With a suitable optical-delay design the relative time delay between the pump pulse and probe pulse can be well controlled within a desired time range for the lifetime measurements of IPSs.
3. *Vacuum chamber system*: It contains an investigated metal sample and an electron energy analyzer, and allows the two laser beams to be loosely focused on the same surface area of the sample.

In most of the later experimental studies, the primary laser source is a mode-locked and Q-switched Ti:sapphire laser oscillator/amplifier system with a fundamental pulsed output of wavelength ~800 nm and pulse duration ≤ 100 fs. The fundamental IR beam or its second-harmonic generation (SHG) beam can be employed as the probe beam, while its third-harmonic generation (THG) beam can be used as the pump beam.

The metal sample is placed inside a specially designed ultra-high-vacuum (UHV) chamber with a background pressure lower than 10^{-10} Torr. The surface of the metal sample should be

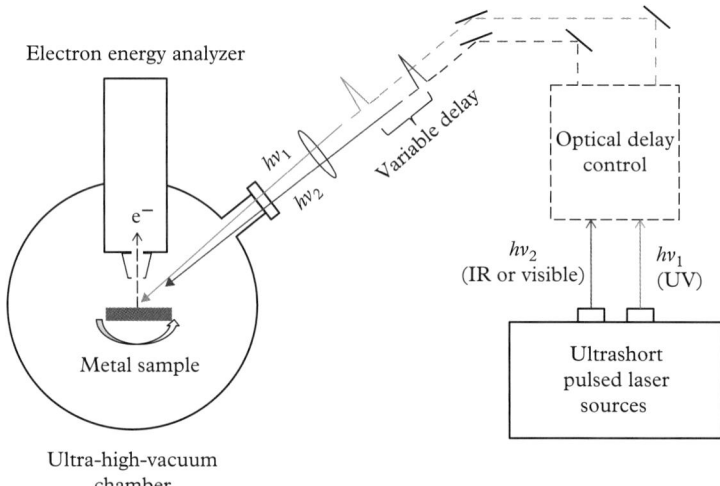

Figure 15.5 *Schematic diagram of the experimental setup for energy- and time-resolved 2PPE studies, utilizing the ultrashort pulsed laser sources with two different output frequencies as well as an adjustable optical delay device.*

mechanically and chemically polished and cleaned by cycles of sputtering and annealing, and the sample temperature can be controlled according to the specific requirement. The kinetic energy distribution and dynamics of photoemitted electrons from the sample surface can be measured by using an electron energy analyzer (such as a hemispherical-type or a time-of-flight type energy analyzer) attached to the vacuum chamber. Usually, the laser beams of p-polarization (electric vector parallel to the plane of incidence) irradiate the sample surface with an incident angle of ∼45°, while the photoelectrons are detected along the normal direction of the surface. If someone needs to investigate the angular dispersion behavior of the photoemitted electrons, the angle between the detection direction and the normal of the sample surface can be changed accordingly by rotating the sample. At the sample surface, the laser intensity should be not too high to avoid the opto-thermal effect and space-charge effect; the latter may lead to a distortion of the energy distribution of photoelectrons.

15.2.3 MPPE studies on clean and/or adsorbing metal surfaces

Energy- and time-resolved resonant MPPE studies have been conducted in a series of single-crystal metal surfaces, such as Ag(111) and Cu(111) (2PPE),[10] Ag(100) and Cu(100) (2PPE),[11–14] Ni(110, 111) (2PPE),[15] Co(0001) and Fe(110) (2PPE),[16] Cu(001) (3PPE),[17] and Pt(111) (3PPE).[18] Based on these studies, researchers can determine the binding energy, lifetime, and linewidth of various IPSs for a given metal sample.[19]

As a typical example, Fig. 15.6 shows kinetic energy spectra of the electrons photoemitted from Cu(100) and Ag(100) samples, excited by 2PA from a single dye laser beam of which the photon energy $h\nu$ can be changed.[11] These spectral curves were measured at the surface normal direction along which $k_{//} = 0$, where $k_{//}$ is the component parallel to the sample surface of the wavevector of electron wave function. At higher photon energies, both $n = 1$ and $n = 2$ IPSs can be clearly resolved, as shown in Fig. 15.6. The specific energy-level position (below E_{vac}) for the nth IPS can be directly determined by (see Eq. (15.2-4))

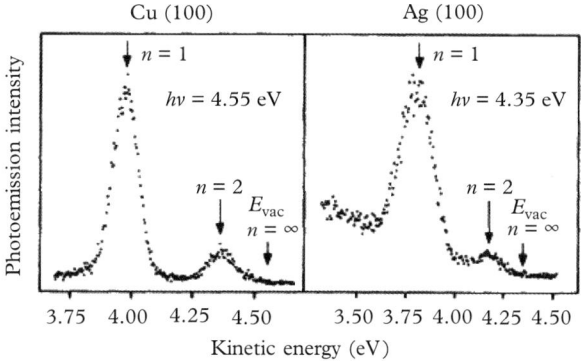

Figure 15.6 *Two-photon photoemission spectra from Cu(100) and Ag(100) in the $k_{//} = 0$ direction. Both $n = 1$ and $n = 2$ IPSs are clearly resolved. (Reproduced with permission from Giesen et al.[11] Copyright 1987, American Physical Society).*

$$E_n = h\nu - E_{kin}^n, \tag{15.2-5}$$

where E_{kin}^n is the measured position of the resonant peak corresponding to the nth IPS.

Moreover, the population lifetime in IPSs can be measured by using a two-beam (pump-probe) configuration with a tunable time delay between the pump and probe pulses. Figure 15.7 shows the 2PEE spectra of Ag(100) measured at different time delays.[13] To obtain these results, a 285–nm laser beam of $h\nu_1 = 4.35$ eV and $\sim$90 fs pulse width was used as the pump, while another 620-nm laser beam of $h\nu_2 = 2.0$ eV and $\sim$55 fs pulse width was the probe. It is clear that following the increase of delay, the $n = 1$ and $n = 2$ resonant peak values are gradually decreased. Based on this type of time-resolved measurement, the lifetime of various IPSs can be determined accordingly.[3,12,14]

The multi-photon photoemission technique can be effectively employed to investigate the influence of an atomic/molecular monolayer (or overlayers) on the photoemission behavior of a pure metal surface. The selected atoms or molecules can be adsorbed physically or chemically on a clean metal surface. Some of the investigated systems are: O/Cu(111),[20] Ag/Pd(111),[21] Xe/Ag(111),[22] Na/Cu(111),[23] CO/Cu(111),[24] and Cs/Cu(111).[25] The adsorbed layer(s) may modify the intrinsic photoemission properties or produce new interfacial states. As an example, Fig. 15.8 shows the decay behavior of $n = 3$ and $n = 4$ IPSs of a clean Cu(100) surface compared to the surface covered with different amounts of CO monolayer (ML).[26] In that experiment, the fundamental $\sim$800-nm and $\sim$45-fs output from a Ti:sapphire laser was used as the probe beam while its THG beam of $\sim$75 fs duration was the pump beam. As can be seen in Fig. 15.8, the exponential decay of the $n = 3$ 2PPE peak from the clean Cu(100) surface experiences a periodic modulation that is due to the quantum beat effect,[3]

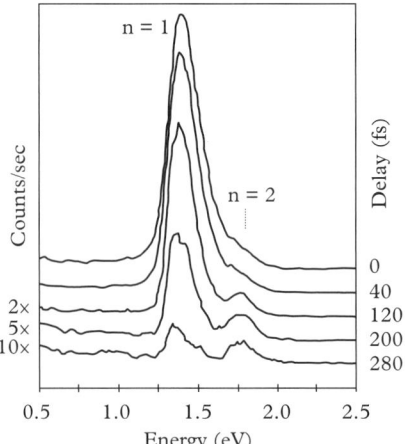

Figure 15.7 *Two-photon excited photoelectron kinetic energy spectra from Ag(100) measured with different time delays between the UV and visible laser pulses. (Reproduced with permission from Schoenlein et al.[13] Copyright 1990, American Physical Society).*

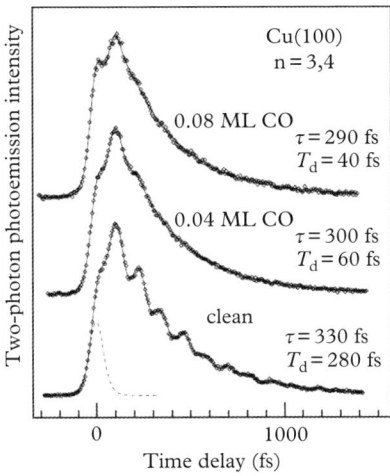

Figure 15.8 *Decay behavior of the $n = 3$ 2PPE peak from clean and CO-adsorbed Cu(100) surfaces, where the modulation superimposed on an exponential decay is due to the quantum beat between the $n = 3$ and $n = 4$ IPSs. The beat decay time (T_d) of the induced modulation decreases rapidly upon CO adsorption, whereas the exponential lifetime remains almost unchanged. The dashed line shows the cross correlation between the pump and probe pulses. (Reproduced with permission from Reuß et al.[26] Copyright 1999, American Physical Society).*

resulting from the wave-function interference between the coherently excited $n = 3$ and $n = 4$ states. The beat decay time (T_d) of that modulation decreases from $T_d \approx 280$ fs for the clean surface to $T_d \approx 40$ fs for the 0.08-ML CO adsorbed surface, which implies that the adsorption of CO on Cu(100) leads to a strong decrease of the phase coherence between two wave functions; meanwhile the exponential decay time of the $n = 3$ state remains almost the same ($\tau \approx 300$ fs).

The time-resolved two-color pump–probe technique has been employed to investigate the dynamic photoemission behavior of other surface systems, such as 2PPE processes in a thin metal film on a semiconductor substrate,[27] as well as the 3PPE processes in metal–insulator–metal junctions.[28]

To acquire the specific information about the electron wavevector dependence of relevant energy-band structures, some angle-dependent 2PPE spectroscopic studies have been done.[22–25,29] If θ is the angle between the normal of sample surface and the detecting direction of electron kinetic energy, the detected energy will be a function of angle θ, i.e.,[30]

$$E_{kin} = \hbar^2 k_\parallel^2(\theta)/2m^* + E_{kin}^0. \tag{15.2-6}$$

Here, E_{kin}^0 is the kinetic energy along the normal direction, m^* is the effective mass of the interfacial electron, $k_\parallel$ is the electron wavevector component parallel to the surface and given by

$$k_\parallel = (2m_e E_{\text{kin}}/\hbar^2)^{1/2} \sin\theta, \qquad (15.2\text{-}7)$$

where m_e is the free electron mass. Based on the experimental measurement of $E_{\text{kin}}(\theta)m$, one can determine the effective electron mass associated with various IPSs.

Furthermore, there are some other subjects of MPPE studies based on the same physical approaches mentioned above but with extended research coverage. The following are some of those issues.

i. *Spin-resolved MPPE studies*: Combining an electron spin detector with the electron energy analyzer in a MPPE experimental system, researchers can directly measure the spin polarization of photoelectrons and be able to explore the spin dependence of the relevant electronic states involved in MPPE processes.[31–34]

ii. *MPPE studies in nanoparticles systems*: Some studies have been done on Pt and Pd clusters (4PPE),[35] Ag nanoparticles (2PPE),[36,37] and silver-coated spherical nanoparticles.[38] In these cases, surface plasmon enhancement may take place in the related MPPE processes.

iii. *Imaging applications of MPPE*: Imaging of nanoparticles and/or nanostructures of patterned films has been demonstrated, based on MPPE mechanisms and the use of a photoemission electron microscope with the lateral resolution of ~20 nm.[38–42]

15.3 Multi-photon photoconductivity (MPPC) effects

15.3.1 Mechanisms of multi-photon induced photoconductivity

Upon the action of high-intensity pulsed laser radiation, an electron in the valence band of a semiconductor can be excited into the conduction band via MPA accompanied by the simultaneous creation of a hole in the valence band. The excited electron is mobile in the conduction while the hole is mobile in the valence band, and both will contribute to an increase of conductivity when a dc electric voltage is applied to the sample. Figure 15.9(a) shows three possible ways to excite an electron from the valence band into the conduction band: (i) by simultaneously absorbing two identical incident photons (degenerate 2PA); (ii) by simultaneously absorbing two incident photons with different frequencies (nondegenerate 2PA); and (iii) by simultaneously absorbing three identical incident photons (degenerate 3PA). In these cases, it is assumed that any single photon energy is smaller than the band gap (E_g) while the sum of the simultaneously absorbed multi-photon energy is greater than E_g. For a degenerate MPA process, the photon energy requirement will be

$$mh\nu \geq E_g \text{ and } (m-1)h\nu < E_g. \ (m = 2, 3, 4, \ldots) \qquad (15.3\text{-}1)$$

A typical experimental setup for the study of multi-photon photoconductivity (MPPC), which is shown schematically in Fig. 15.9(b), involves the following three parts: (i) a short or ultra-short pulsed laser source, (ii) a test semiconductor sample connected with an electric circuit, and (iii) a suitable electric detecting system to measure the multi-photon induced conductivity change of a sample medium.

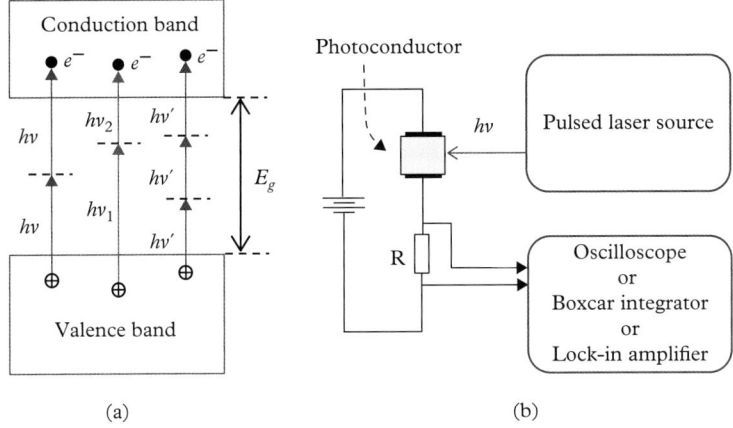

Figure 15.9 *(a) Electronic transitions in a semiconductor excited by degenerate 2PA, nondegenerate 2PA, and degenerate 3PA, respectively. (b) Schematic diagram of a typical experimental setup for the multi-photon-induced photoconductivity study.*

Let us first consider two-photon photoconductivity (2PPC) in a semiconductor sample. If I_0 denotes the input laser intensity, the sample thickness along the light propagation direction is l_0, the intensity decrease inside the sample due to 2PA should be (cf. Eq. (14.3-12))

$$\Delta I = I_0 \frac{I_0 \beta l_0}{1 + I_0 \beta l_0}, \qquad (15.3\text{-}2)$$

where β is the 2PA coefficient of the sample medium. In practice, it is usually true that $I_0 \beta l_0 \ll 1$ and Eq. (15.3-2) can be simplified to

$$\Delta I \approx I_0^2 \beta l_0. \qquad (15.3\text{-}3)$$

Then the 2PA-induced photocurrent change can be expressed by

$$\Delta \mathcal{J} \propto \frac{eS}{2h\nu} I_0^2 \beta l_0. \quad (2PPC) \qquad (15.3\text{-}4)$$

Here, e is the electron charge and S is the illumination area on the sample.

For three-photon excitation, we have (cf. Eq. (14.3-15))

$$\Delta I = I_0 - I(l_0) = I_0 \left(1 - \frac{1}{\sqrt{1 + 2I_0^2 \gamma l_0}}\right), \qquad (15.3\text{-}5)$$

where γ is the three-photon absorption coefficient of the sample. In experiments on three-photon photoconductivity (3PPC), usually $2I_0^2 \gamma l_0 \ll 1$ and Eq. (15.3-5) can be simplified to

$$\Delta I \approx I_0 [1 - (1 - I_0^2 \gamma l_0)] \approx I_0^3 \gamma l_0. \qquad (15.3\text{-}6)$$

Therefore, the induced photocurrent change will be

$$\Delta \mathcal{J} \propto \frac{eS}{3h\nu} I_0^3 \gamma l_0. \qquad (3PPC) \qquad (15.3\text{-}7)$$

Similarly, for multi-photon excited photoconductive processes, the following relation between the induced photocurrent change and the input pulsed laser intensity is generally valid:

$$\Delta \mathcal{J} \propto I_0^m. \qquad (MPPC) \qquad (15.3\text{-}8)$$

In practice, measuring the photoconductive response of a given sample medium as a function of the input laser intensity, one may determine the number m and what is the major mechanism contributing to the experimental results.

15.3.2 Observations of MPPC effects in semiconductors and dielectric media

The early observations of multi-photon excited photoconductivity were performed by using Q-switched pulsed laser beams. The examined nonlinear photoconductive samples were mainly single-crystal semiconductors as well as some dielectric optical crystals, including (but not limited to) anthracene (organic semiconductor, 2PPC),[43] PbTe and InSb (2PPC),[44] AlN (3PPC),[45] ZnS and CdS (2PPC),[46,47] GaAs and InP (2PPC),[48] GaP (3PPC),[49] etc. The multi-photon-induced nonlinear photoconductive behavior of some dielectric crystals, such as NaCl (5PPC) and Al$_2$O$_3$ (3PPC),[50] KCl, KBr, and NaBr (4–6PPC),[51] crystal quartz (4PPC),[52] as well as organic crystal CHI$_3$ (2PPC),[53] has also been reported.

The key issue of those early studies was to measure the pulsed laser radiation-induced photocurrent (or charge) change as the function of the input laser intensity at a given wavelength, to compare the experimental results with the relation predicted by Eq. (15.3-8), and finally to determine the dominant mechanism (the number m) responsible to the observed results.

As one of the typical examples of such early studies, the specific mechanisms of MPPC in the crystals of KI, NaCl, and KCl were investigated, using 694.3-nm and 20-ns ruby pulsed laser and its frequency-doubled radiation of 347 nm wavelength.[54] The bandgaps of these three crystals are 5.8, 8.1, and 8 eV, respectively, whereas the photon energies of the fundamental ruby laser and its second harmonic are 1.78 and 3.56 eV, respectively. The measured results of the induced photoconductive charge as a function of the input laser intensity plotted on a double-logarithmic scale confirmed the 2PPC, 3PPC, 4PPC, and 5PPC processes that occur in these crystals.

The later studies on MPPC have been pursued in more semiconductor materials, such as Hg$_{1-x}$Cd$_x$Te (2PPC),[55] CdI$_2$ (2PPC, 3PPC),[56,57] ZnSe and CdSe (2PPC, 3PPC),[58,59] PbI$_2$ (2PPC, 3PPC),[59] GaN (2PPC),[60] and BN (3PPC, 4PPC).[61]

15.3.3 2PPC-based spectroscopic studies on semiconductors

For a degenerate 2PA excitation process in a photoconductive semiconductor sample, the absorbed two-photon energy must equal the energy difference between the initial level in a valence band and the final level in the conduction band, which can be written as

$$2h\nu = E_{\text{fin}}^{\text{con}} - E_{\text{ini}}^{\text{val}}. \qquad (15.3\text{-}9)$$

At low temperature (e.g., ≤20 K), near the bandgap top within the conduction band there can be a series of discrete energy levels. These levels may belong to free excitons or to Landau energy level splitting; the latter exists only when a strong magnetic field is applied to the sample. Assuming the input laser beam is frequency tunable, when the incident two-photon energy is equal to the electronic transition energy from an initial level in the valence band to one of the discrete levels in the conduction band, the free electrons in the conduction band and free holes in the valence band can be significantly increased; subsequently a resonant enhancement of the photoconductivity signal will be observed. Based on this principle, researchers have developed a nonlinear photoconductive spectroscopic technique.

Some 2PPC spectroscopic experiments are related to the study of free exciton energy structures and dynamics. When the two-photon energy is close to but not too high above the bandgap energy, an electron (in the conduction band) and a hole (in the valence band) can be generated. Under appropriate conditions, this electron–hole pair of charges with opposite sign can form an electrically neutral quasiparticle bound by electrostatic Coulomb interaction. This corresponds to the formation of a so-called free exciton that exhibits the energy level structure of a hydrogenic atom. Without applying an electric field, the dissociation or breakup of a free exciton can occur either by the interaction with an impurity or defect center, or by the interaction with optical phonons. Owing to exciton dissociation, the free electrons and/or free holes will increase. By applying a dc-bias electric field to the semiconductor, the above-mentioned exciton dissociation process can be further expedited as the electron and hole have a tendency to move in opposite directions. This is the mechanism whereby an observed 2PA-induced photoconductive signal peak can correspond to a resonant transition frequency of the free exciton.

In practice, one may utilize a frequency-tunable laser beam to illuminate a semiconductor sample at low temperature by applying a suitable bias voltage; by recording the two-photon-induced photoconductive signal as a function of the input two-photon energy, the final spectral curve could indicate the discrete peak structures that are associated with the corresponding free exciton transitions.

Multi-photon photoconductive spectroscopy can be employed to investigate the Zeeman splitting of both the exciton transitions and the interband Landau transitions for a sample in a magnetic field. Compared to 1PA spectroscopy, two-photon spectroscopy may reveal much valuable information about the selection rules and transition details.

Studies using 2PPC spectroscopy have been focused on single crystal semiconductor samples at low temperature and in a strong magnetic field, such as PbTe and InSb,[44,62,63] CdS,[64,65] GaAs,[66,67] and $Hg_{1-x}Cd_xTe$.[68] Taking the CdS crystal as an example, this crystal has a hexagonal wurtzite structure and exhibits three (A, B, and C) valence bands.[64] The transitions from each valence band to the conduction band may produce free excitons, occupying different discrete excited states. The excitons initiated from the A valence band are named A excitons, while B excitons are initiated from the B valence band. The lowest bandgap is due to the A valence band at the Brillouin-zone center with E_g = 2.582 eV. In an experiment, a CdS single crystal at 1.8 K was subjected to a low bias voltage and a high magnetic field; the sample was irradiated by two nanosecond-pulsed laser beams: one was a tunable visible dye laser ($h\nu_{vis}$ = 1.7–1.834 eV) and another was an IR laser ($h\nu_{IR}$ = 0.784 eV).[65] Figure 15.10 shows the spectral curve recorded at the 10-T magnetic field level, from which the spectral peak positions of different A and B excitonic transitions could be identified with a high resolution, which was only limited by the laser spectral width (≤0.3 meV).

Another example is the 1- and 2PPC spectroscopic study of a GaAs/AlGaAs multiple quantum well sample consisting of 10 unit cells of double quantum wells: a wide (155 Å) well and a narrow (75 Å) well, separated by an undoped $Al_{0.3}Ga_{0.7}As$ barrier layer with a thickness of 70 Å.[69]

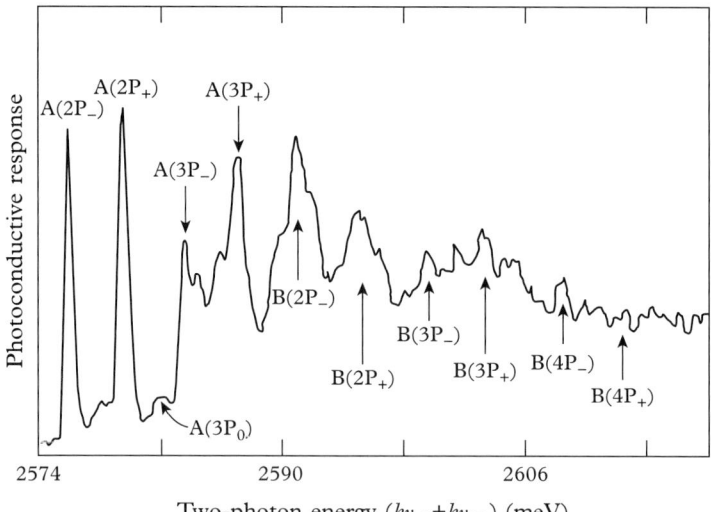

Figure 15.10 *High-resolution 2PPC spectra of a CdS crystal obtained at a magnetic field of 10 Tesla, showing the complexity of the A and B exciton structures. The arrows indicate the resonant peaks corresponding to specific excitonic transitions. (Reproduced with permission from Seiler et al.[65] Copyright 1983, American Physical Society).*

The sample was at 20 K without applying a magnetic field, illuminated by a single pulsed laser beam with tunable wavelength and at intensity level of ~100 MW/cm². Figure 15.11(a) and (b) depict the recorded two-photon-induced photoconductivity spectra of the same sample but corresponding to two different well widths; from these two spectral curves one could identify two-photon excited 2PHH (2P exciton due to HH1 subband) and 2PLH (2P exciton due to LH1 subband) excitonic transitions, respectively. For GaAs, the 1S transitions are not allowed for 2PA due to selection rules; therefore it is concluded that the 1S peaks shown in Fig. 15.11(a) and (b) are actually produced by one-photon excitation from the SHG (with frequency $2h\nu$) of the input laser radiation in the crystal. Later on, the magneto-absorption spectroscopic study of a GaAs/AlGaAs multiple quantum wells sample in a magnetic field has also been implemented.[70]

15.3.4 MPPC-based autocorrelation measurements of ultrashort laser pulses

SHG-based auto- or cross-correlation is a commonly used method to measure the time-width of ultrashort laser pulses. In this case, the phase-matching requirement of the utilized second-order nonlinear crystals has to be met, and the generated SHG signal as a function of the time delay between the two input fundamental pulses is usually detected by a photomultiplier. The intensity of such a detected SHG signal can be expressed as

$$I_{2\nu}(\Delta t) \propto \int_{-\infty}^{+\infty} I_\nu^{(1)}(t) I_\nu^{(2)}(t + \Delta t) dt, \qquad (15.3\text{-}10)$$

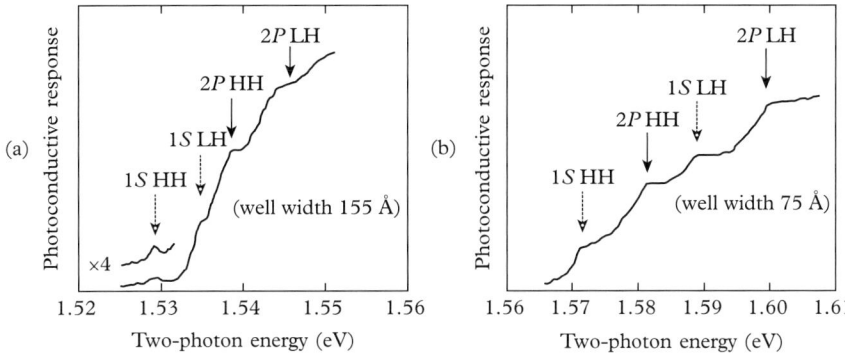

Figure 15.11 *(a, b) 2PPC spectra of the same sample but corresponding to the different well widths. The arrows indicate the spectral peak positions of the identified excitonic transitions, whereas the 1S transitions are actually produced by the second-harmonic radiation of the input laser beam. (Reproduced with permission from Nithisoontorn et al.[69] Copyright 1989, American Physical Society).*

where $I_\nu^{(1)}$ and $I_\nu^{(2)}$ are the intensity functions of the two fundamental pulses, and Δt is the time delay between these two pulses. Assuming that the temporal shape of the fundamental pulse is known and symmetric on the time scale, its real pulse width can be readily determined by the corresponding width of the measured $I_{2\nu}(\Delta t)$ curve. However, if the input fundamental pulse has an asymmetric temporal shape, such a shape asymmetry cannot be revealed by using this second-order intensity correlation method.

On the other hand, there is an alternative technique to measure the time-duration of ultrashort laser pulses without any phase-matching requirement, which is based on 2PPC or 3PPC effect in a semiconductor photodetector. In this case, a laser beam of ultrashort pulses is split into two relatively delayed beams; those two beams then focused onto a semiconductor detector with spatially overlapping each other. For the 2PPC mechanism, the recorded photocurrent (or voltage) change $\mathcal{J}_{PC}$ as a function of the relative delay between the two beams can be written as

$$\mathcal{J}_{pc}(\Delta t) \propto \int_{-\infty}^{+\infty} I_\nu^{(1)}(t) I_\nu^{(2)}(t + \Delta t) dt. \tag{15.3-11}$$

The same second-order intensity correlation is also applicable to the nondegenerate 2PPC mechanism, where two input laser beams have different frequencies and we shall deal with a cross-correlation measurement.

In practice, various commercial semiconductor photodetectors have been adopted as auto-correlators to measure the laser pulse width, such as photodiodes of Si (2PPC),[71] GaAsP (2PPC),[71,72] SiC (2PPC),[73] GaN (3PPC),[74] SiC (2PPC),[75] AlGaAs,[76] and diamond.[77] In addition, some other types of semiconductor photodetectors with a plate or waveguide shape, and made from CdS,[71] ZnSe,[78] GaAs,[79,80] or GaN,[81] have also employed to implement the autocorrelation measurements. Moreover, by using ultrashort ultraviolet laser pulses of 267 nm wavelength, the autocorrelation measurement based on 2PPC in a biased fused silica sampler of $E_g \approx 9$ eV has been demonstrated.[82]

Let us first examine an example of the nondegenerate 2PPC-based cross-correlation measurement reported by Schade et al., utilizing a GaAsP diffusion photodiode of $E_g \approx 1.8$ eV.[83] One of the two input pulsed laser beams had a wavelength of $\lambda_1 = 775$ nm (from a Ti:sapphire laser oscillator) and the other had $\lambda_2 = 1300$ nm (from an optical parametric generator). The measured 2PPC-induced photocurrent data versus the time delay of the two beams are shown in Fig. 15.12(a); these data can be well fitted by a Gaussian function with a time width of 299 fs. Moreover, at a fixed input power level of 28 mW for the $\lambda_1 = 775$ nm beam, the measured maximum photocurrent as a function of the input power of the $\lambda_2 = 1300$ nm beam, which is plotted in Fig. 15.12(b), shows a linear relationship; this is understandable because the maximum correlation photocurrent should be proportional to the product of power values of these two beams.

In the next example, we shall consider a 3PPC-based autocorrelation experiment accomplished by Streltsov et al. utilizing a GaN photodiode of $E_g \approx 3.4$ eV irradiated by 60-fs and 820-nm (1.51 eV) laser pulses.[74] Figure 15.13(a) shows the measured peak photocurrent data as a function of the input pulse energy, indicating a cubic dependence characterized for the 3PPC mechanism, while Fig. 15.13(b) shows the recorded third-order autocorrelation trace produced by the two relatively delayed 820-nm laser beams with slightly different intensity values. In the latter case, unequal intensities in two laser beams may reveal even a slight asymmetry of the real pulse shape.

To explain why using unequal intensities in two beam arms can break the time-direction ambiguity that is found in the second-order intensity correlation, for a 3PA-based photoconductive detector employed for autocorrelation measurement, the photocurrent change as a function of the time delay (Δt) between two pulsed laser beams can be written as:

$$Q_{pc}(\Delta t) \propto \langle I_1^3(t) + 3I_1^2(t)I_2(t+\Delta t) + 3I_1(t)I_2^2(t+\Delta t) + I_2^3(t+\Delta t) \rangle, \quad (15.3\text{-}12)$$

where I_1 and I_2 are the intensity functions of the two pulsed laser beams, and $\langle \rangle$ denotes a time-average. Assuming $a = I_1/I_2 \ll 1$, the above expression turns to

$$Q_{pc} \propto \langle a^3 I_2^3(t) + 3a^2 I_2^2(t) I_2(t+\Delta t) + 3I_1(t) I_2^2(t+\Delta t) + I_2^3(t+\Delta t) \rangle.$$

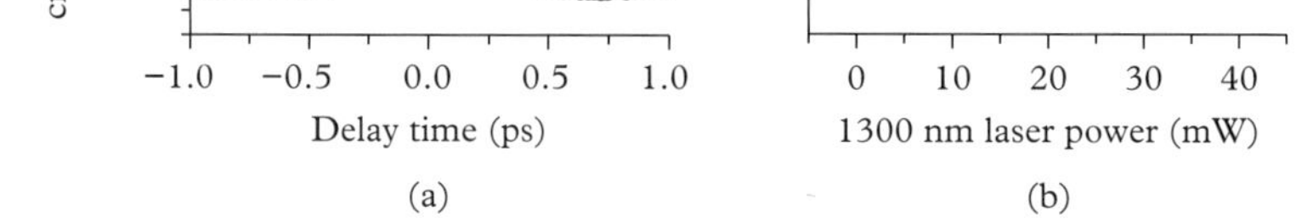

Figure 15.12 (a) GaAsP photodiode measured nondegenerate 2PPC cross-correlation data with $\lambda_1 = 775$ nm and $\lambda_2 = 1300$ nm laser pulses, fitted by a Gaussian function of 259 fs width. (b) Power dependence of the two-photon induced cross-correlation photocurrent on the changed λ_2 average power; the λ_1 average power is fixed at 28 mW. (Reproduced with permission from Schade et al.[83] Copyright 1998, Elsevier).

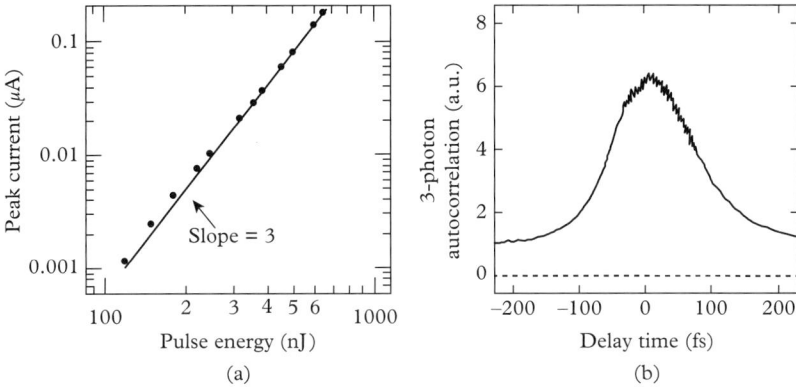

Figure 15.13 *(a) 3PPC-induced peak photocurrent produced by the GaN photodiode as a function of the input energy for an 820-nm laser pulse of 60 fs duration, indicating a cubic relationship. (b) Measured autocorrelation trace of the 820-nm laser pulse, the intensities of the two beams are slightly different. (Reproduced with permission from Streltsov et al.[74] Copyright 1999, AIP Publishing LLC).*

If $a \ll 1$, the first and second terms can be neglected and we finally have[84]

$$Q_{pc} \propto \langle 3I_1(t)I_2^2(t+\Delta t) + I_2^3(t+\Delta t)\rangle. \tag{15.3-13}$$

In Eq. (15.3-13), the second term contributes to a dc-background contribution, while the first term determines the third-order intensity correlation that measures the pulse asymmetry, since it is no longer compensated for by the addition of its time-reversed copy.

The above-mentioned capability of using the third-order intensity correlation to measure the laser pulse's asymmetry has been further demonstrated by Langlois and Ippen, utilizing a GaAsP photodiode as the third-order autocorrelator and a Si photodiode as the second-order autocorrelator, both working with ~1.5-μm laser pulses.[84] In this experiment, a double-pulse, composed of two sub-pulses of ~138-fs duration and 320-fs separation with a peak intensity ratio of 14, was employed as a test asymmetrical pulsed laser signal. This double-pulse laser beam was split into two arms with an intensity ratio of $a = 0.04$. Such measured autocorrelation traces obtained by using two photodiodes are depicted in Fig. 15.14, indicating a symmetric second-order trace (based on 2PPC) and a third-order trace (based on 3PPC), respectively. It is evident that only the latter can reveal the asymmetry (two-peak structure) of the input pulsed signal.

15.3.5 Other related studies

In addition to those applications mentioned in the preceding subsection, some other studies related to nonlinear photoconductive processes and materials have been pursued since the 2000s. For example, MPPC effects in organic materials, such as poly(*p*-phenylene vinylene),[85] poly(9,9-dioctylfluorene),[86] as well as Ga–Ge–S–S–Se:Er^{3+} chalcogenide glasses,[87] have been studied. Researchers have also reported the 2PPC effect in the $Hg_{1-x}Cd_xTe$ and GaAs/AlGaAs quantum well photodetectors working with the far-IR coherent radiation from a free-electron laser

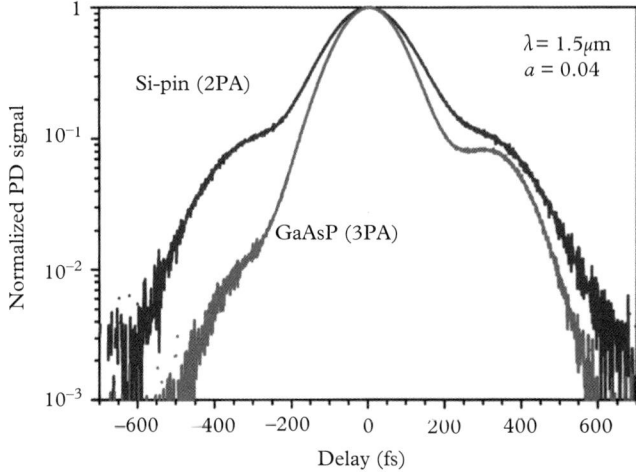

Figure 15.14 *Autocorrelation traces of the 1.5-μm double-pulse laser signal with asymmetrical temporal profile, measured by the two photodiodes with an intensity ratio of $a = I_1/I_2 = 0.04$ between the two laser arms, showing a strong asymmetry of the third-order intensity correlation. (Reproduced with permission from Langlois and Ippen.[84] Copyright 1999, Optical Society of America).*

source.[88,89] Nonlinear photoconductive performances based on an all-silicon photonic crystal photoconductor[90] and on a GaAs photomultiplier tube[91] have also been demonstrated.

In recent decades, a new research area has been the generation and applications of coherent terahertz (THz) radiation produced by ultrashort laser pulses. Briefly, researchers can utilize two primary techniques to generate ultrashort coherent THz radiation via optical excitation. One is based on the optical rectification (or difference-frequency mixing) effect in a second-order nonlinear optical crystal, which will be described in detail in Chapter 17. The other is based on the use of a photoconductive semiconductor antenna system.[92,93] In the latter case, if the input pulsed pump laser wavelength is shorter than the cutoff wavelength of the photoconductor used, the one-photon excitation induced transient photocurrent change will be the physical source of coherent THz radiation. However, if the input pump intensity is high enough, the two-photon excitation can contribute to the induced photoconductivity change and the corresponding coherent THz wave generation.[94,95]

Holzman and Elezzabi have demonstrated that with two-photon excitation the THz radiation can be effectively generated from a suitable photoconductor.[96] They employed polycrystalline ZnSe as the nonlinear photoconductive medium having a bandgap of $E_g \approx 2.67$ eV, working with the ultrashort ~800-nm ($h\nu \approx 1.55$ eV) laser pulses from a Ti:sapphire laser oscillator. The pump laser beam was focused as a 3 μm spot onto the region next to the anode with an ~9 μm photoconductive gap. The THz radiation field (amplitude) variation is shown in Fig. 15.15, measured in the conditions of ~160 mW pump power and 230 V bias voltage. The experimental results also indicate that the peak to peak amplitude of the generated THz field is nearly proportional to the pump laser power, revealing the two-photon excitation mechanism.

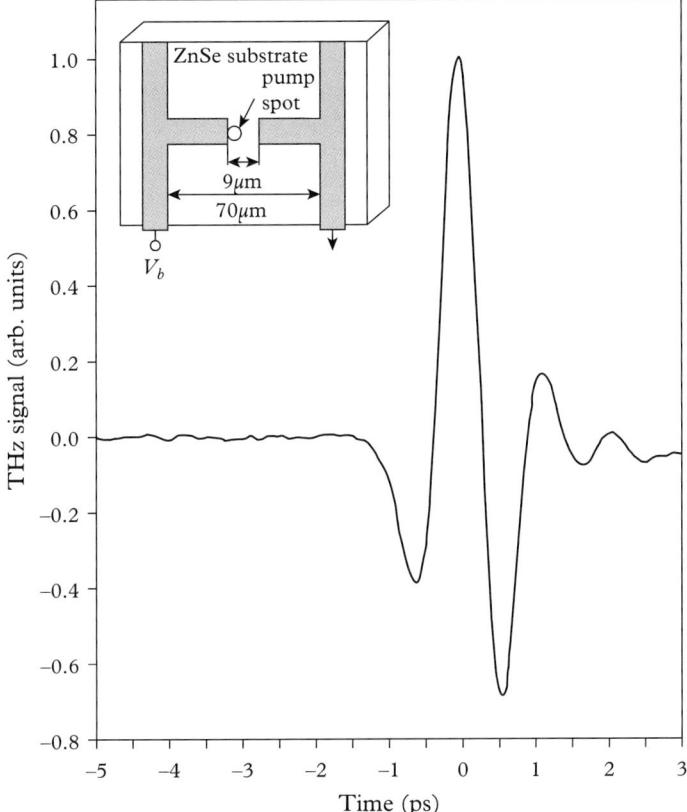

Figure 15.15 *Temporal variation of the two-photon excited THz radiation field generated from a polycrystalline ZnSe emitter with a 160-mW pump power and a $V_b = 230$ V bias voltage. The edge-illuminated PC gap and structure of the dipole emitter are shown in the inset. (Reproduced with permission from Holzman and Elezzabi.[96] Copyright 2003, AIP Publishing LLC).*

REFERENCES

1. H. Hertz, *Ann. Phys. (Berlin)* **267**, 983 (1887).
2. A. Einstein, *Ann. Phys. (Berlin)* **322**, 132 (1905).
3. U. Höfer, I. L. Shumay, Ch. Reuß, U. Thomann, W. Wallauer, and T. Fauster, *Science* **277**, 1480 (1997).
4. H. Sonnenberg, H. Heffner, and W. Spicer, *Appl. Phys. Lett.* **5**, 95 (1964).
5. M. C. Teich, J. M. Schroeer, and G. J. Woga, *Phys. Rev. Lett.* **13**, 611 (1964).
6. E. M. Logothetis and P. L. Hartman, *Phys. Rev. Lett.* **18**, 581 (1967).
7. J. H. Bechtel, W. L. Smith, and N. Bloembergen, *Opt. Commun.* **13**, 56 (1975).

8. L. A. Lompre, J. Thebault, and Gy. Farkas, *Appl. Phys. Lett.* **27**, 110 (1975).
9. J. H. Bechtel, W. L. Smith, and N. Bloembergen, *Phys. Rev. B* **15**, 4557 (1977).
10. K. Giesen, F. Hage, F. J. Himpsel, H. J. Riess, and W. Steinmann, *Phys. Rev. Lett.* **55**, 300 (1985).
11. K. Giesen, F. Hage, F. J. Himpsel, H. J. Riess, and W. Steinmann, *Phys. Rev. B* **35**, 971 (1987).
12. R. W. Schoenlein, J. G. Fujimoto, G. L. Eesley, and T. W. Capehart, *Phys. Rev. Lett.* **61**, 2596 (1988).
13. R. W. Schoenlein, J. G. Fujimoto, G. L. Eesley, and T. W. Capehart, *Phys. Rev. B* **41**, 5436 (1990).
14. I. L. Shumay, U. Höfer, Ch. Reuß, U. Thomann, W. Wallauer, and Th. Fauster, *Phys. Rev. B* **58**, 13974 (1998).
15. H. W. Rudolf, D. Rieger, and W. Steinmann, *Solid State Commun.* **34**, 427 (1980).
16. R. Fischer, N. Fischer, S. Schuppler, Th. Fauster, and F. J. Himpsel, *Phys. Rev. B* **46**, 9691 (1992).
17. A. Winkelmann, W.-C. Lin, C.-T. Chiang, F. Bisio, H. Petek, and J. Kirschner, *Phys. Rev. B* **80**, 155128 (2009).
18. I. Kinoshita, T. Anazawa, and Y. Matsumoto, *Chem. Phys. Lett.* **259**, 445 (1996).
19. T. Fauster and W. Steinmann, in *Electromagnetic Waves: Recent Developments in Research, Vol. 2: Photonic Probe of Surfaces*, edited by P. Halevi (Elsevier, Amsterdam, 1995), Chapter 8, pp. 347–411.
20. D. Rieger, T. Wegehaupt, and W. Steinmann, *Phys. Rev. Lett.* **58**, 1135 (1987).
21. R. Fischer, S. Schuppler, N. Fischer, Th. Fauster, and W. Steinmann, *Phys. Rev. Lett.* **70**, 654 (1993).
22. D. F. Padowitz, W. R. Merry, R. E. Jordan, and C. B. Harris, *Phys. Rev. Lett.* **69**, 3583 (1992).
23. X. Y. Wang, R. Paiella, and R. M. Osgood, Jr., *Phys. Rev. B* **51**, 17035 (1995).
24. D. Velic, E. Knoesel, and M. Wolf, *Surf. Sci.* **424**, 1 (1999).
25. M. Bauer, S. Pawlik, and M. Aeschlimann, *Phys. Rev. B* **60**, 5016 (1999).
26. C. Reuß, I. L. Shumay, U. Thomann, M. Kutschera, M. Weinelt, Th. Fauster, and U. Höfer, *Phys. Rev. Lett.* **82**, 153 (1999).
27. S. Schramm, S. Dantscher, C. Schramm, O. Autzen, C. Wesenberg, E. Hasselbrink, and W. Pfeiffer, *Appl. Phys. B* **88**, 459 (2007).
28. D. Diesing, M. Merschdorf, A. Thon, and W. Pfeiffer, *Appl. Phys. B* **78**, 443 (2004).
29. G. Ferrini, C. Giannetti, S. Pagliara, F. Banfi, G. Galimberti, and F. Parmigiani, *J. Electron Spectrosc. Relat. Phenom.* **144–147**, 565 (2005).
30. N.-H. Ge, C. M. Wong, R. L. Lingle, Jr., J. P. McNeill, K. J. Gaffney, and C. B. Harris, *Science* **279**, 202 (1998).
31. M. Aeschlimann, M. Bauer, S. Pawlik, W. Weber, R. Burgermeister, D. Oberli, and H. C. Siegmann, *Phys. Rev. Lett.* **79**, 5158 (1997).
32. A. Scholl, L. Baumgarten, R. Jacquemin, and W. Eberhardt, *Phys. Rev. Lett.* **79**, 5146 (1997).
33. A. B. Schmidt, M. Pickel, M. Wiemhöfer, M. Donath, and M. Weinelt, *Phys. Rev. Lett.* **95**, 107402 (2005).
34. W.-C. Lin, A. Winkelmann, C.-T. Chiang, F. Bisio, and J. Kirschner, *New J. Phys.* **12**, 083022 (2010).
35. N. Pontius, P. S. Bechthold, M. Neeb, and W. Eberhardt, *J. Electron Spectrosc. Relat. Phenom.* **106**, 107 (2000).
36. W. Pfeiffer, C. Kennerknecht, and M. Merschdorf, *Appl. Phys. A* **78**, 1011 (2004).
37. M. Sipilä, A. A. Lushnikov, L. Khriachtchev, M. Kulmala, H. Tervahattu, and M. Räsänen, *New. J. Phys.* **9**, 368 (2007).
38. S. J. Peppernick, A. G. Joly, K. M. Beck, and W. P. Hess, *J. Chem. Phys.* **134**, 034507 (2011).
39. O. Schmidt, G. H. Fecher, Y. Hwu, and G. Schönhense, *Surf. Sci.* **482–485**, 687 (2001).
40. G. H. Fecher, O. Schmidt, Y. Hwu, and G. Schönhense, *J. Electron Spectrosc. Relat. Phenom.* **126**, 77 (2002).
41. G. Lilienkamp, F. Lindla, C. Senft, and W. Daum, *Surf. Sci.* **602**, 2658 (2008).
42. R. C. Word, J. Fitzgerald, and R. Koenenkamp, *Appl. Phys. Lett.* **99**, 041106 (2011).
43. K. Hasegawa and S. Yoshimura, *Phys. Rev. Lett.* **14**, 689 (1965).

44. K. J. Button, B. Lax, M. H. Weiler, and M. Reine, *Phys. Rev. Lett.* **17**, 1005 (1966).
45. K. Kawabe, K. Yoshino, and Y. Inuishi, *J. Phys. Soc. Japan*, **21**, 1604 (1966).
46. A. Cingolani, F. Ferrero, A. Minafra, and D. Trigiante, *Nuovo Cimento B* **4**, 217 (1971).
47. G. Koren and Y. Yacoby, *Phys. Rev. Lett.* **30**, 920 (1973).
48. C. C. Lee and H. Y. Fan, *Appl. Phys. Lett.* **20**, 18 (1972).
49. I. M. Catalano, A. Cingolani, and A. Minafra, *Solid State Commun.* **16**, 417 (1975).
50. V. S. Dneprovskii, D. N. Klyshko, and A. N. Penin, *JETP Lett.* **3**, 251 (1966).
51. G. I. Aseev, M. L. Kats, and V. K. Nikol'skii, *JETP Lett.* **8**, 103 (1968).
52. D. D. Venable and R. B. Kay, *Appl. Phys. Lett.* **27**, 48 (1975).
53. M. Samoc, A. Samoc, and D. F. Williams, *Mol. Cryst. Liq. Cryst.* **78**, 15 (1981).
54. I. M. Catalano, A. Cingolani, and A. Minafra, *Phys. Rev. B* **5**, 1629 (1972).
55. D. G. Seiler, S. W. McClure, R. J. Justice, M. R. Loloee, and D. A. Nelson, *J. Vac. Sci. Technol. A* **4**, 2034 (1986).
56. N. V. Unnikrishnan, R. D. Singh, and M. Matera, *Solid State Commun.* **71**, 1001 (1989).
57. M. I. Miah, *Opt. Mater.* **25**, 353 (2004).
58. A. Gaur, D. K. Sharma, D. S Ahlawat, and N. Singh, *J. Opt. A* **9**, 260 (2007).
59. A. Gaur, D. K. Sharma, K. S. Singh, and N. Singh, *Mod. Phys. Lett. B* **23**, 2783 (2009).
60. J. Miragliotta and D. K. Wickenden, *Appl. Phys. Lett.* **69**, 2095 (1996).
61. V. Ageev, S. Klimentov, M. Ugarov, E. Loubnin, A. Bensaoula, N. Badi, A.Tempez, and D. Starikov, *Appl. Surf. Sci.* **138–139**, 364 (1999).
62. S. K. Manlief and E. D. Palik, *Solid State Commun.* **12**, 1071 (1973).
63. M. H. Weiler, R. W. Bierig, and B. Lax, *Phys. Rev.* **184**, 709 (1969).
64. D. G. Seiler, D. Heiman, R. Feigenblatt, R. L. Aggarwal, and B. Lax, *Phys. Rev. B* **25**, 7666 (1982).
65. D. G. Seiler, D. Heiman, and B. S, Wherrett, *Phys. Rev. B* **27**, 2355 (1983).
66. D. G. Seiler, C. L. Littler, and D. Heiman, *J. Appl. Phys.* **57**, 2191 (1985).
67. J. S. Michaelis, K. Unterrainer, E. Gornik, and E. Bauser, *Phys. Rev. B* **54**, 7917 (1996).
68. D. G. Seiler, J. R. Lowney, C. L. Littler, and M. R. Loloee, *J. Vac. Sci. Technol. A* **8**, 1237 (1990).
69. M. Nithisoontorn, K. Unterrainer, S. Michaelis, N. Sawaki, E. Gornik, and H. Kano, *Phys. Rev. Lett.* **62**, 3078 (1989).
70. M. Nithisoontorn, R. Dum, K. Unterrainer, J. S. Michaelis, E. Gornik, N. Sawaki, and H. Kano, *Surf. Sci.* **267**, 505 (1992).
71. Y. Takagi, T. Kobayashi, K. Yoshihara, and S. Imamura, *Opt. Lett.* **17**, 658 (1992).
72. J. K. Ranka, A. L. Gaeta, A. Baltuska, M. S. Pshenichnikov, and D. A. Wiesma, *Opt. Lett.* **22**, 1344 (1997).
73. T. Feurer, A. Glass, and R. Sauerbrey, *Appl. Phys. B* **65**, 295 (1997).
74. A. M. Streltsov, K. D. Moll, A. L. Gaeta, P. Kung, D. Walker, and M. Razeghi, *Appl. Phys. Lett.* **75**, 3778 (1999).
75. S. Lochbrunner, P. Huppmann, and E. Ricdle, *Opt. Commun.* **184**, 321 (2000).
76. A. K. Sharma, M. Raghuramaiah, P. A. Naik, and P. D. Gupta, *Opt. Commun.* **246**, 195 (2005).
77. N. F. Kleimeier, T. Haarlammert, H. Witte, U. Schuehle, J.-F. Hochedez, A. BenMoussa, and H. Zacharias, *Opt. Express* **18**, 6945 (2010).
78. W. Rudolph, M. Sheik-Bahae, A. Bernstein, and L. F. Lester, *Opt. Lett.* **22**, 313 (1997).
79. J. Montoya and Q. Hu, *J. Appl. Phys.* **95**, 2230 (2004).
80. F. Lacassie, D. Kaplan, Th. De Saxce, and P. Pignolet, *Eur. Phys. J. Appl. Phys.* **11**, 189 (2000).
81. V. Pacebutas, A. Krotkus, T. Suski, P. Perlin, and M. Leszczynski, *J. Appl. Phys.* **92**, 6930 (2002).
82. A. M. Streltsov, J. K. Ranka, and A. L. Gaeta, *Opt. Lett.* **23**, 798 (1998).
83. W. Schade, D. L. Osborn, J. Preusser, and S. R. Leone, *Opt. Commun.* **150**, 27 (1998).
84. P. Langlois and E. P. Ippen, *Opt. Lett.* **24**, 1868 (1999).
85. C. Soci, D. Moses, and A. J. Heeger, *Synth. Met.* **153**, 145 (2005).
86. X. Zhang, Y. Xia, and R. H. Friend, *Phys. Rev. B* **75**, 245128 (2007).

87. T. Y. Ivanova, A. A. Manshina, A. V. Povolotskiy, Y. S. Tver'yanovich, S.-K. Liaw, and Y.-S. Hsieh, *J. Phys. D: Appl. Phys.* **41**, 175110 (2008).
88. X. Yuan, W. Lu, J. Wei, G. Xu, X. Shen, M. Wang, X. Yang, G. Wu, and Y. Li, *Sci. China A* **44**, 1579 (2001).
89. J. Jiang, Y. Fu, N. Li, X. S. Chen, H. L. Zhen, W. Lu, M. K. Wang, X. P. Yang, G. Wu, Y. H. Fan, and Y. G. Li, *Appl. Phys. Lett.* **85**, 3614 (2004).
90. L.-D. Haret, X. Checoury, Z. Han, P. Boucaud, S. Combrie, and A. De Rossi, *Opt. Express* **18**, 23965 (2010).
91. F. Boitier, J.-B. Dherbecourt, A. Godard, and E. Rosencher, *Appl. Phys. Lett.* **94**, 081112 (2009).
92. D. H. Auston, K. P. Cheung, and P. R. Smith, *Appl. Phys. Lett.* **45**, 284 (1984).
93. J. T. Darrow, X.-C. Zhang, and D. H. Auston, *Appl. Phys. Lett.* **58**, 25 (1991).
94. J. B. Baxter and C. A. Schmuttenmaer, *Phys. Rev. B* **80**, 235206 (2009).
95. C.-K. Lee, C.-S. Yang, S.-H. Lin, S.-H. Huang, O. Wada, and C.-L. Pan, *Opt. Express* **19**, 23689 (2011).
96. J. F. Holzman and A. Y. Elezzabi, *Appl. Phys. Lett.* **83**, 2967 (2003).

16
Fast and Slow Light

Fast or slow light propagation means that the group velocity of a quasi-monochromatic light pulse propagating in a resonant medium can be significantly faster or slower than the light speed c in a vacuum. Since the end of the last century and start of this century, studies on this subject have attracted a great deal of interest, and remarkable progress has been achieved based on various nonlinear optical interactions in resonant media. The fundamental research on this subject can enrich our understanding of special relativity and causality in physics.[1-4] On the other hand, the mechanisms and techniques developed for fast and slow light propagation have provided the potentials for many future applications, such as optical buffers and variable delay lines for optical networks, optical computing, optical telecommunications, and optical data storage and signal processing.[5,6]

16.1 Definitions of light speeds

There are several different scientific terms that have been introduced in the literature to describe the speed of light propagation, including the phase velocity, group velocity, signal velocity, and energy transport velocity.[4,7,8] Among them, the concepts of phase velocity and group velocity have clear mathematical definitions and also unambiguous physical interpretations. The definitions of other concepts of light velocity are conditional and sometimes may cause dispute due to their physical ambiguity.[9-11]

16.1.1 Phase velocity of a monochromatic light

Phase velocity, $v_p(\omega)$, is defined as the propagation speed of the co-phasal surface of a monochromatic light wave with an angular frequency of $\omega = 2\pi\nu$. A given monochromatic plane wave propagating along the z-axis direction is expressed by

$$E(z,t) = E_0(z)e^{-i(\omega t - kz)}, \qquad (16.1\text{-}1)$$

where $E_0(z)$ is the amplitude function independent of time t, $k = 2\pi n(\omega)/\lambda$ is the wavenumber (magnitude of the wavevector), $n(\omega)$ is the refractive index of the medium in which the wave propagates, and λ is the light wavelength in a vacuum. In above expression, the spatiotemporal variation of the wave phase is determined by the phase factor $\phi(t,z) = (\omega t - kz)$, and the motion of the co-phasal plane is determined by the condition

$$\phi(t,z) = (\omega t - kz) = \text{constant}. \qquad (16.1\text{-}2)$$

Taking differential operation on both sides of the above equation leads to $k\Delta z = \omega \Delta t$ (as k and ω are fixed), and the phase velocity is obtained by

$$v_p(\omega) = \frac{\Delta z}{\Delta t} = \frac{\omega}{k} = \frac{\omega \lambda}{2\pi n(\omega)} = \frac{c}{n(\omega)}, \qquad (16.1\text{-}3)$$

where c is the light velocity in a vacuum. In practice, it is impossible to create an ideal monochromatic light wave with infinitesimal spectral linewidth; besides, one cannot use such a monochromatic light wave with infinite time extension to do light speed measurement. Even though the phase velocity is not a directly measureable quantity of speed, the refractive index $n(\omega)$ for a given medium can be directly measured by using various optical methods (such as prismatic refraction or interference refractometer).

It is generally acknowledged that the frequency of a light wave remains unchanged during the propagation in vacuum ($n = 1$) or in a medium of $n \neq 1$. According to the definition of phase velocity by Eq. (16.1-3), the wavelength of a monochromatic light wave in a propagation medium will change to $\lambda' = \lambda/n(\lambda)$, and this may be the only physical meaning of the phase velocity. Most dielectric optical media in the optical frequency region exhibit values of $n(\omega) \geq 1$, and correspondingly there should be $v_p \leq c$ or $\lambda' \geq \lambda$. For some metallic substances like gold and silver, the refractive index in the optical frequency region can be smaller than 1. Moreover, the refractive index of glass for X-rays can also be smaller than 1. In such cases, the phase velocity of light can be greater than c, which means that the light wavelength in those media become shorter than its value in vacuum, i.e., $\lambda' < \lambda$.

16.1.2 Group velocity of a quasi-monochromatic light pulse

In practice, one can measure the real speed of a light pulse or wave packet passing through a given optical medium. In this case, a pulsed and quasi-monochromatic light wave can be expressed as

$$E(z, t) = E_0(z, t) e^{-i(\omega_0 t - kz)}, \qquad (16.1\text{-}4)$$

where $E_0(z, t)$ is a time-dependent amplitude function that may have a quasi-Gaussian or hyperbolic-secant shape with a certain pulse width, and ω_0 is the central frequency of the pulse's frequency spectrum. According to the uncertainty principle or Fourier theorem, a light pulse with a finite time duration should have a corresponding finite frequency spread, as shown in Fig. 16.1(a) and (b), respectively. In this case, the light pulse can be recognized as a wave packet that is composed of a great number of monochromatic components with different frequencies and amplitudes, as shown schematically in Fig. 16.1(c). Only at the peak position (t_0) of the wave packet, all monochromatic components are perfectly in phase to give a maximum of temporal interference; whereas at other time positions apart from t_0, owing to partial diphasing between those different monochromatic components the overall amplitude is declining following the increase of $|t - t_0|$. The constitution behavior of a quasi-monochromatic wave packet can be more clearly illustrated in Fig. 16.1(d) by means of vector addition of contributions from all spectral components. Over there the contribution from each individual spectral component is represented by a vector, the length of which is proportional to the amplitude of each component and the direction is determined by the phase status relative to the neighboring spectral components. The sequential addition of the vectors from all components determines the amplitude of the wave packet at different time positions. The magnitude of the sum vector is determined by the connection length from the starting point of the first vector to the end of the last vector. For instance, as shown in Fig. 16.1(d), at an arbitrarily chosen point (i) of the pulse front, sum vector of all contributing

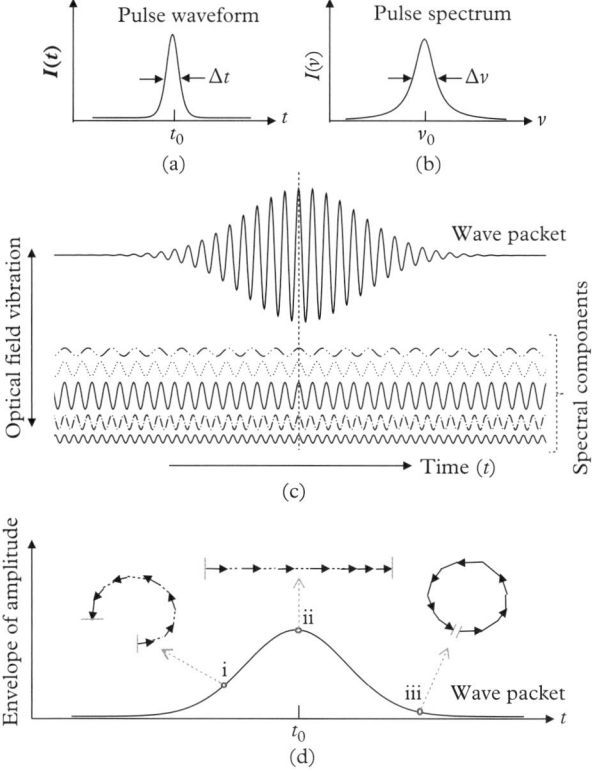

Figure 16.1 *(a) Pulse waveform as a function of time. (b) Pulse spectrum as a function of frequency. (c) Pulse as a wave packet that is composed of a great number of monochromatic components, only in the peak position all components are in phase and result in a maximum of interference. (d) Wave-packet amplitude as a function of time determined by summation of contributing vectors from all components.*

vectors from different components leads to moderate amplitude of the wave packet. At the peak point (ii) of the wave packet, all contributing vectors are added up in the same direction and give a maximum overall amplitude; finally, at a point (iii) located in the far tail part of the wave packet, the magnitude of the sum vector approaches to zero due to destructive interference.

Here we define the moving speed of the peak of an optical pulse (wave packet) as the group velocity of this pulse propagating in a medium. As explained above and shown schematically in Fig. 16.1, the peak position of the pulse in time scale is characterized by the co-phase behavior of superposition of all spectral components with different frequencies, which means that for this particular position the phase factor $\phi(t, z) = (\omega t - kz)$ in Eq. (16.1-1) should be independent of ω. Let

$$\frac{d\phi(t, z)}{d\omega} = 0, \qquad (16.1\text{-}5)$$

we have

$$t - z\frac{dk}{d\omega} = 0,$$

and the group velocity is defined by

$$v_g = \frac{z}{t} = \frac{d\omega}{dk}. \tag{16.1-6}$$

As $k(\omega) = \frac{2\pi}{\lambda_0}n(\omega) = \frac{\omega}{c}n(\omega)$, we have

$$\frac{dk}{d\omega} = \frac{1}{c}\left[n(\omega) + \omega\frac{dn(\omega)}{d\omega}\right].$$

Substituting the above expression into Eq. (16.1-6) leads to

$$v_g(\omega) = \frac{d\omega}{dk} = \frac{c}{n(\omega) + \omega\frac{dn(\omega)}{d\omega}} = \frac{c}{n_g(\omega)}, \tag{16.1-7}$$

and $n_g(\omega)$ is the group refractive index of the medium, defined by

$$n_g(\omega) = n(\omega) + \omega\frac{dn(\omega)}{d\omega}. \tag{16.1-8}$$

From Eqs. (16.1-3) and (16.1-7) we see that the phase velocity $v_p(\omega)$ is simply determined by the refractive index $n(\omega)$, whereas the group velocity $v_g(\omega)$ is determined by both $n(\omega)$ and $\frac{dn(\omega)}{d\omega}$.

In vacuum for all frequencies, $n(\omega) = 1$, $\frac{dn(\omega)}{d\omega} = 0$, we always have

$$v_g(\omega) = c. \tag{16.1-9}$$

In a dispersion-less spectral range medium, $n(\omega) \neq 1$, $\frac{dn(\omega)}{d\omega} \Rightarrow 0$, and we will see

$$v_g(\omega) = \frac{c}{n(\omega)} = v_p(\omega). \tag{16.1-10}$$

In a linearly absorbing medium and very near to resonant frequencies, $\frac{dn(\omega)}{d\omega} < 0$ (anomalous dispersion), $n_g(\omega)$ could be less than 1 or even be negative; thus we have

$$v_g(\omega) > c \text{ or } v_g(\omega) < 0. \quad \text{(Fast light)} \tag{16.1-11}$$

This corresponds to so-called fast light or superluminal light propagation.

On the contrary, in a gain medium or in an absorbing medium with an induced narrow-band transparency, it may take place that $\omega\frac{dn(\omega)}{d\omega} \gg 1$ (steep normal dispersion) and $n_g(\omega) \gg 1$, then we shall have

$$v_g(\omega) \ll c. \quad \text{(Slow light)} \tag{16.1-12}$$

This corresponds to so-called slow light propagation. We shall further discuss the physical meanings of slow and fast light propagation in the following sections.

16.2 Group velocity in a resonant medium

In this section, we briefly describe the group velocity properties in an absorbing medium and in a gain medium, respectively. Here it is assumed that the central frequency of the input light pulse is very close to the absorption frequency or gain frequency of the resonant medium. In those cases, an abrupt variation of the refractive index over a very narrow frequency region will lead to a huge change of the values of group velocity.

16.2.1 Complex refractive index of an absorbing medium

As described in Section 5.1, for an isotropic linearly absorptive medium the linear susceptibility $\chi^{(1)}(\omega)$ is a complex quantity, and the generalized linear refractive index can also be expressed as a complex quantity, i.e.,

$$\tilde{n} = (n + in') = \sqrt{1 + \chi^{(1)}}, \tag{16.2-1}$$

where n and n' are the real and imaginary parts of $\tilde{n}$, respectively. In this case, the linear susceptibility $\chi^{(1)}$ can be expressed as

$$\chi^{(1)} = \chi^{(1)}_{NR} + \chi^{(1)}_R = \chi^{(1)}_{NR} + \chi^{(1)}_{R,re} + i\chi^{(1)}_{R,im}, \tag{16.2-2}$$

where $\chi^{(1)}_{NR}$ is the non-resonant contribution (real) and $\chi^{(1)}_R$ is the resonant contribution that contains a real part $\chi^{(1)}_{R,re}$ and an imaginary part $\chi^{(1)}_{R,im}$, respectively. According to Eq. (5.1-10), we have

$$\left. \begin{array}{l} \chi^{(1)}_{R,re} = \dfrac{\Delta N(p_0)^2}{\varepsilon_0 \hbar} \dfrac{\omega_0 - \omega}{(\omega_0 - \omega)^2 + \Gamma_0^2} \\ \chi^{(1)}_{R,im} = \dfrac{\Delta N(p_0)^2}{\varepsilon_0 \hbar} \dfrac{\Gamma_0}{(\omega_0 - \omega)^2 + \Gamma_0^2} \end{array} \right\}, \tag{16.2-3}$$

where ΔN is the population density difference between the low- and high-level of a considered absorptive transition, p_0 the matrix element of that dipole-transition, ω_0 the central frequency of the transition, and 2Γ is the full width of the absorption spectral line. Substituting Eq. (16.2-2) into Eq. (16.2-1) leads to

$$(n + in') = \sqrt{1 + \chi^{(1)}_{re}} \sqrt{1 + i\frac{\chi^{(1)}_{R,im}}{1 + \chi^{(1)}_{re}}}, \tag{16.2-4}$$

where $\chi^{(1)}_{re} = \chi^{(1)}_{NR} + \chi^{(1)}_{R,re}$ is the real part of $\chi^{(1)}$. Since $\chi^{(1)}_{R,re}$ and $\chi^{(1)}_{R,im}$ are of the same order of magnitude and both are considerably smaller than $(1 + \chi^{(1)}_{re})$, we approximately have

$$(n + in') \approx \sqrt{1 + \chi^{(1)}_{re}} \left[1 + i\frac{1}{2}\frac{\chi^{(1)}_{R,im}}{1 + \chi^{(1)}_{re}} \right]. \tag{16.2-5}$$

Equalizing the real part and imaginary part on both sides of Eq. (16.2-5) leads to the following expression for the real refractive index:

$$n(\omega) = \sqrt{1 + \chi_{re}^{(1)}(\omega)} = \sqrt{1 + \chi_{NR}^{(1)}(\omega) + \chi_{R,re}^{(1)}(\omega)}$$

$$\approx n_0(\omega) + \frac{\Delta N(p_0)^2}{2n_0(\omega)\varepsilon_0\hbar} \frac{\omega_0 - \omega}{(\omega_0 - \omega)^2 + \Gamma_0^2} \quad (16.2\text{-}6)$$

$$= n_0(\omega) + \delta_0^{\max} \frac{2(\omega_0 - \omega)\Gamma_0}{(\omega_0 - \omega)^2 + \Gamma_0^2}.$$

Here, $n_0(\omega) = \sqrt{1 + \chi_{NR}^{(1)}(\omega)}$ is the refractive index originating from the non-resonant susceptibility, and $\delta_0^{\max}$ is the maximum deviation of the resonant contribution from non-resonant contribution to the refractive index, which is determined by

$$\delta_0^{\max} = \frac{\Delta N(p_0)^2}{4n_0(\omega_0)\varepsilon_0\hbar\Gamma_0}. \quad (16.2\text{-}7)$$

From Eq. (16.2-5) one can also obtain the following expression for the imaginary part of the complex refractive index:

$$n'(\omega) = \frac{1}{2} \frac{\chi_{R,im}^{(1)}(\omega)}{\sqrt{1 + \chi_{re}^{(1)}(\omega)}} = \frac{1}{2n(\omega)} \chi_{R,im}^{(1)} \approx \frac{1}{2n_0(\omega)} \chi_{R,im}^{(1)}. \quad (16.2\text{-}8)$$

Here it is assumed that $\delta_0^{\max}$ is considerably smaller than the value of n_0. Substituting the second expression of Eq. (16.2-3) into Eq. (16.2-8) leads to

$$n'(\omega) = \frac{\Delta N(p_0)^2}{2n_0(\omega)\varepsilon_0\hbar} \frac{\Gamma_0}{(\omega_0 - \omega)^2 + \Gamma_0^2} = \delta_0^{\max} \frac{2\Gamma_0^2}{(\omega_0 - \omega)^2 + \Gamma_0^2}. \quad (16.2\text{-}9)$$

On the other hand, a monochromatic plane wave propagating in an absorbing medium along the z direction can be expressed as

$$E(\omega, z) = E(\omega, 0) e^{-i(\omega t - \tilde{k}z)}, \quad (16.2\text{-}10)$$

where $\tilde{k}$ is the magnitude of a generalized wavevector, which is defined by the complex refractive index:

$$\tilde{k}(\omega) = \frac{2\pi}{\lambda_0} \tilde{n}(\omega) = \frac{2\pi}{\lambda_0} [n(\omega) + in'(\omega)] = k(\omega) + ik'(\omega). \quad (16.2\text{-}11)$$

Here, λ_0 is the wavelength in vacuum corresponding to the frequency of ω. Substituting Eq. (16.2-11) into Eq. (16.2-10) leads to

$$E(\omega, z) = E(\omega, 0) e^{-i(\omega t - kz)} e^{-k'z}. \quad (16.2\text{-}12)$$

The intensity attenuation after propagation distance z will be

$$I(\omega, z) = \frac{1}{2}\varepsilon_0 c n_0 |E(\omega, z)|^2 = I(\omega, 0) e^{-2k'z} = I(\omega, 0) e^{-\alpha(\omega)z}, \quad (16.2\text{-}13)$$

where $\alpha(\omega)$ is the linear absorption coefficient of the medium, defined by

$$\alpha(\omega) = 2k' = \frac{4\pi}{\lambda_0} n'(\omega). \qquad (16.2\text{-}14)$$

Substituting Eq. (16.2-9) into Eq. (16.2-14) leads to the final expression of $\alpha(\omega)$

$$\alpha(\omega) = \frac{4\omega\delta_0^{max}}{c} \frac{\Gamma_0^2}{(\omega_0-\omega)^2+\Gamma_0^2}. \quad \text{(Absorption)} \qquad (16.2\text{-}15)$$

Comparing Eq. (16.2-15) with Eq. (16.2-6), we should note that in a resonant medium the linear absorption coefficient $\alpha(\omega)$ and the refractive index $n(\omega)$ are both determined by the key parameter δ_0^{max} of the medium. As a quantitative example, Fig. 16.2(a) and (b) show the curves of $\alpha(\omega)$ and $n(\omega)$ as functions of $(\omega-\omega_0)/\Gamma_0$, respectively, with the assumed values of $\delta_0^{max} = 0.15$, $n_0 = 1.2$, $\omega_0 = 2\pi \times 5 \times 10^{14}\text{s}^{-1}$, and $\omega_0/\Gamma_0 = 200$.

Figure 16.2(a) shows that the absorption coefficient $\alpha(\omega)$ reaches its maximum value $(4\omega_0\delta_0^{max}/c)$ at the $\omega = \omega_0$ position, while Fig. 16.2(b) indicates that the refractive index $n(\omega)$ attains its maximum deviation from n_0 at $|\omega-\omega_0| = \Gamma_0$ positions.

16.2.2 Group refractive index of an absorbing medium

According to Eqs. (16.1-8) and (16.2-6), the group refractive index of an absorbing medium is given by

$$n_g(\omega) = n(\omega) + 2\omega\Gamma_0\delta_0^{max} \frac{d}{d\omega}\left[\frac{(\omega_0-\omega)}{(\omega_0-\omega)^2+\Gamma_0^2}\right]$$

$$= n(\omega) + 2\omega\Gamma_0\delta_0^{max} \frac{(\omega_0-\omega)^2-\Gamma_0^2}{[(\omega_0-\omega)^2+\Gamma_0^2]^2}. \qquad (16.2\text{-}16)$$

The group refractive index curve in the vicinity of central absorption frequency ω_0 is depicted in Fig. 16.2(c), using the same assumed values of other parameters.

From Eq. (16.2-16) and Fig. 16.2(c) one can find that for the curve of $n_g(\omega)$ there are three extremum positions located at $(\omega-\omega_0) = 0$ and $(\omega-\omega_0) = \pm\sqrt{3}\Gamma_0$, respectively, with the following extremum values:

$$n_g = n_0 + \begin{cases} -\dfrac{2\omega_0}{\Gamma_0}\delta_0^{max} & (\text{at } \omega = \omega_0) \\ +\dfrac{\omega_0}{4\Gamma_0}\delta_0^{max} & (\text{at } |\omega-\omega_0| = \sqrt{3}\Gamma_0) \end{cases}. \qquad (16.2\text{-}17)$$

Here the most important conclusion is that around the central absorption frequency ω_0, the group refractive index can indeed take negative values, which means that the corresponding group velocity can also be negative in that frequency region.

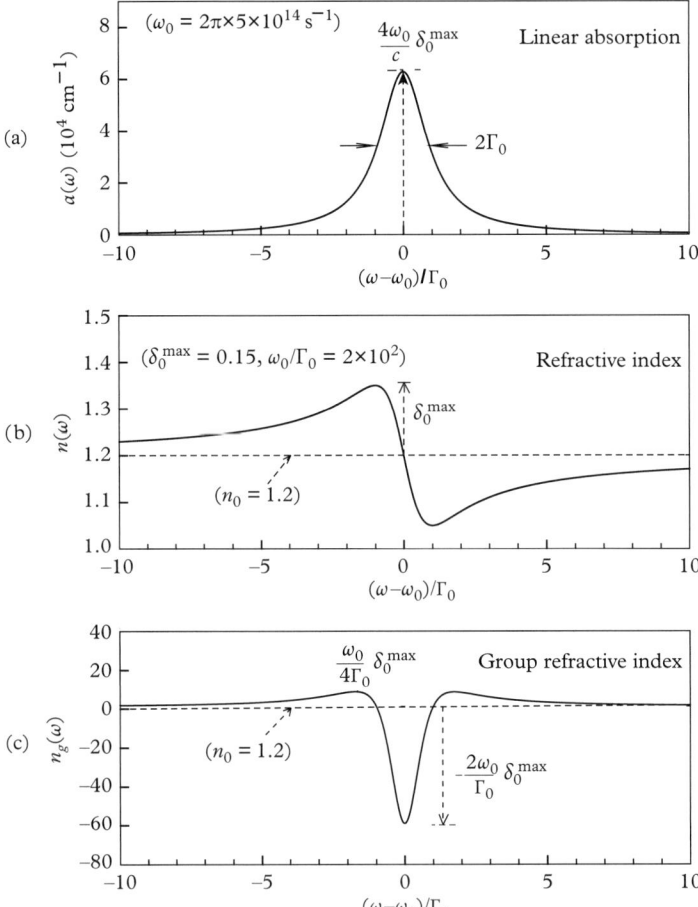

Figure 16.2 *Linear absorption curve (a), real refractive index dispersion curve (b), and group refractive index curve (c) of a resonant medium. It is assumed that non-resonant refractive index $n_0 = 1.2$, maximum resonant contribution $\delta_0^{max} = 0.15$, central frequency of absorption $\omega_0 = 2\pi \times 5 \times 10^{14} s^{-1}$, and absorption line width $2\Gamma = \omega_0/100$.*

16.2.3 Group velocity of a light pulse in an absorbing medium

Knowing the group refractive index $n_g(\omega)$ given by Eq. (16.2-16), we can write the expression for group velocity as

$$v_g(\omega) = \frac{c}{n_g(\omega)} = c\left[n(\omega) + 2\omega\Gamma_0\delta_0^{max}\frac{(\omega_0-\omega)^2 - \Gamma_0^2}{[(\omega_0-\omega)^2 + \Gamma_0^2]^2}\right]^{-1}. \quad (16.2\text{-}18)$$

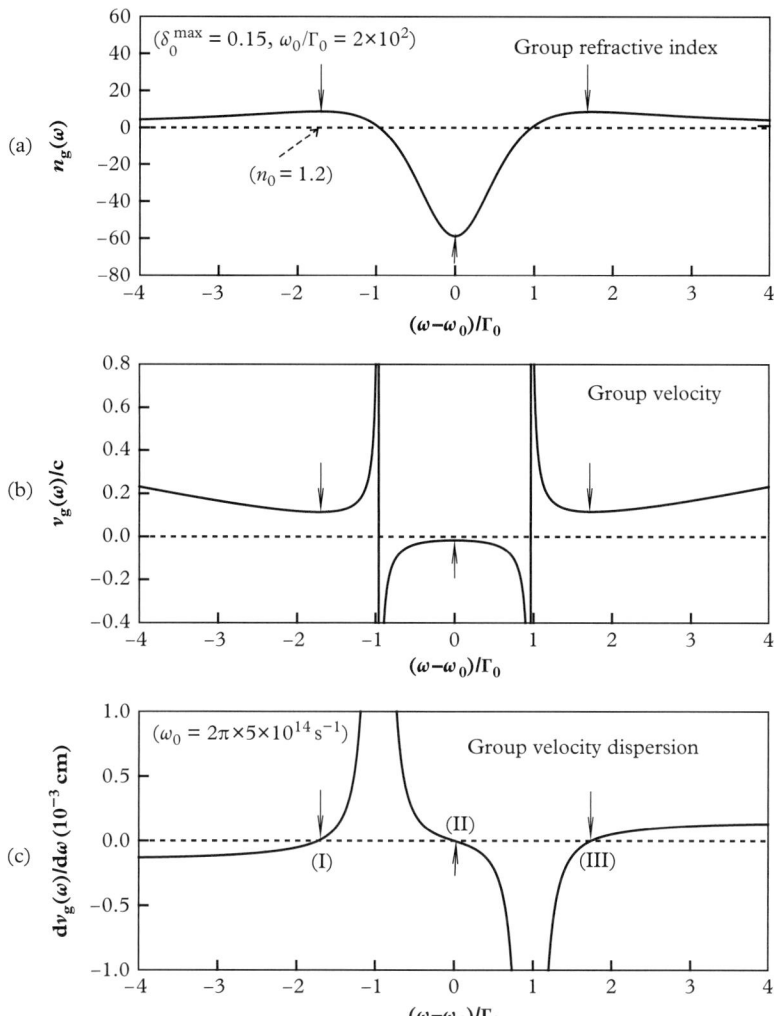

Figure 16.3 *Group refractive index curve (a), group velocity curve (b), and group velocity dispersion curve (c) of a linearly absorbing medium. The assumed medium parameters are the same as shown in Fig. 16.2.*

In Fig. 16.3(a) we redraw the group refractive index curve, while the corresponding group velocity curve is shown in Fig. 16.3(b). At three extremum positions (indicated by vertical solid-line arrows) we have

$$v_g = \begin{cases} c/[n_0 - \dfrac{2\omega_0}{\Gamma_0}\delta_0^{\max}] & (\text{at } \omega = \omega_0) \\ c/[n_0 + \dfrac{\omega_0}{4\Gamma_0}\delta_0^{\max}]. & (\text{at } |\omega - \omega_0| = \sqrt{3}\Gamma_0) \end{cases} \qquad (16.2\text{-}19)$$

For a quasi-monochromatic input light pulse with finite spectral linewidth, in order to avoid severe distortion of the pulse shape, the group velocity dispersion (GVD) of the medium should remain as small as possible. The GVD for a resonant medium is determined by

$$\frac{dv_g}{d\omega} = c\frac{d}{d\omega}(n_g)^{-1} = -\frac{c}{n_g^2}\left[2\frac{dn}{d\omega} + \omega\frac{d^2n}{d\omega^2}\right]. \tag{16.2-20}$$

Substituting the first- and second-order derivatives of $n(\omega)$ expressed by Eq. (16.2-6) into Eq. (16.2-20) leads to

$$\frac{dv_g}{d\omega} = -\frac{c}{n_g^2\omega}\left\{2n_g - 2n + \frac{4\omega^2\Gamma_0\delta_0^{max}(\omega_0-\omega)[(\omega_0-\omega)^2 - 3\Gamma_0^2]}{[(\omega_0-\omega)^2 + \Gamma_0^2]^3}\right\}. \tag{16.2-21}$$

As a specific example, in the case of $n_0 = 1.2$, $\delta_0^{max} = 0.15$, and $\omega_0/\Gamma = 2 \times 10^2$, the GVD, $\frac{dv_g}{d\omega}$, in the vicinity of an absorption central frequency ω_0 is depicted in Fig. 16.3(c).

In Fig. 16.3(c), we can see that the GVD approaches to zero only in three particular frequency positions that correspond to three extrema of the group refractive index curve shown in Fig. 16.3(b). From an experimental viewpoint, it is suitable to choose the input light pulse working in such three frequency positions to minimize the GVD influence on the pulse's shape and width.

16.2.4 Group velocity in a gain medium

For the sake of simplicity, here we consider a two-level gain system with population inversion $(-\Delta N)$ between the low- and high-energy level of a resonant transition. In such a case, the expression of Eq. (16.2-3) for the real and imaginary parts of the linear resonant susceptibility should change to

$$\begin{aligned}\chi_{R,re}^{(1)} &= \frac{-\Delta N(p_0)^2}{\varepsilon_0\hbar}\frac{\omega_0-\omega}{(\omega_0-\omega)^2 + \Gamma_0^2}\\ \chi_{R,im}^{(1)} &= \frac{-\Delta N(p_0)^2}{\varepsilon_0\hbar}\frac{\Gamma_0}{(\omega_0-\omega)^2 + \Gamma_0^2}\end{aligned}. \tag{16.2-22}$$

The corresponding refractive index is (see Eq. (16.2-6))

$$\begin{aligned}n(\omega) &= n_0(\omega) - \frac{\Delta N(p_0)^2}{2n_0(\omega)\varepsilon_0\hbar}\frac{\omega_0-\omega}{(\omega_0-\omega)^2 + \Gamma_0^2}\\ &= n_0(\omega) - \delta_0^{max}\frac{2(\omega_0-\omega)\Gamma_0}{(\omega_0-\omega)^2 + \Gamma_0^2},\end{aligned} \tag{16.2-23}$$

while the exponential gain coefficient (in units of cm^{-1}) should be (see Eq. (16.2-15))

$$\alpha'(\omega) = g(\omega) = \frac{4\omega\delta_0^{max}}{c}\frac{\Gamma_0^2}{(\omega_0-\omega)^2 + \Gamma_0^2}, \quad \text{(Gain)} \tag{16.2-24}$$

which determines the following gain behavior of the input light intensity:

$$I(\omega,z) = I(\omega,0)e^{g(\omega)z}. \tag{16.2-25}$$

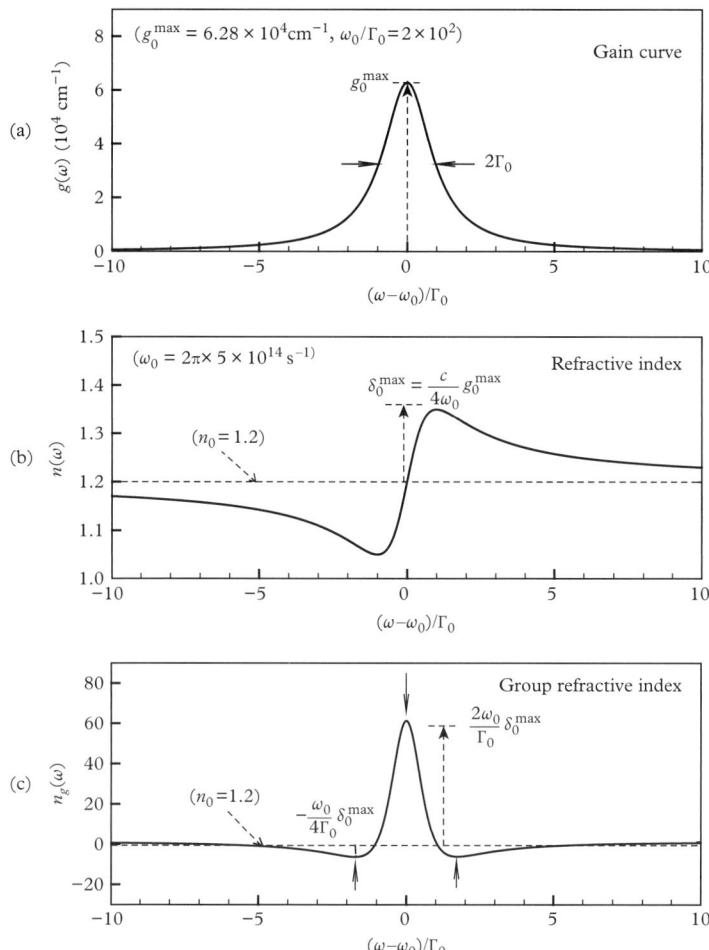

Figure 16.4 Exponential gain coefficient curve (a), refractive index curve (b), and group refractive index curve (c) of a resonant gain medium. The assumed values of n_0, ω_0, and 2Γ are the same as shown in Fig. 16.2, and the maximum gain coefficient is assumed to be $g_0^{max} = 6.28 \times 10^4$ cm^{-1}.

Similarly, the group refractive index of a gain medium can be written as (see Eq. (16.2-16))

$$n_g(\omega) = n(\omega) - 2\omega\Gamma_0 \delta_0^{max} \frac{d}{d\omega}\left[\frac{(\omega_0 - \omega)}{(\omega_0 - \omega)^2 + \Gamma_0^2}\right]$$

(16.2-26)

$$= n(\omega) - 2\omega\Gamma_0 \delta_0^{max} \frac{(\omega_0 - \omega)^2 - \Gamma_0^2}{[(\omega_0 - \omega)^2 + \Gamma_0^2]^2}.$$

Figure 16.4 shows the calculated curves for (a) gain coefficient $g(\omega)$, (b) refractive index $n(\omega)$, and (c) group refractive index $n_g(\omega)$, respectively, with the indicated parameters of the gain medium.

Comparing Fig. 16.4 with Fig. 16.3 one can see that the essential difference between a gain medium and absorbing medium is that at the central frequency ω_0 of a resonant transition; the former may have a maximum positive $n_g^{\max}$ value, whereas the latter may have a maximum negative $n_g^{\max}$ value. In other words, one may achieve fast light (or superluminal) propagation in an absorbing medium or slow light propagation in a gain medium, provided that the saturation effect on the absorption or gain does not take place. In both cases, the maximum deviation of group velocity v_g from c is determined by the ratio of $\delta_0^{\max}/\Gamma_0$, where $\delta_0^{\max}$ is the maximum deviation of the resonant refractive index contribution from the non-resonant contribution, and $2\Gamma_0$ is the full linewidth of the resonant transition.

16.3 Fast/slow light propagation in a resonant medium

16.3.1 Features of light pulse propagation in a resonant medium

In general, the group velocity of a quasi-monochromatic light pulse propagating in an absorbing or gain medium is determined by the group refractive index of the medium. In practice, the experimental setup schematically shown in Fig.16.5 is commonly adopted to measure the group velocity for laser pulses passing through a given resonant medium. In this type of experiment, an input pulsed laser beam is split into two beams; one is used to pass through the test medium while another is directly sent to the detector as a reference beam. It is ensured that before entering the detectors, the two beams have passed through the same optical pathlength. By using a two-channel digital oscilloscope, the relative time delay (Δt) between the transmitted pulse passed through the test medium and the reference pulse can be precisely measured.

The measured time delay is actually determined by

$$\Delta t = \frac{L}{v_g} - \frac{L}{c} = \frac{L}{c}\left(\frac{c}{v_g} - 1\right) = \frac{L}{c}(n_g - 1). \tag{16.3-1}$$

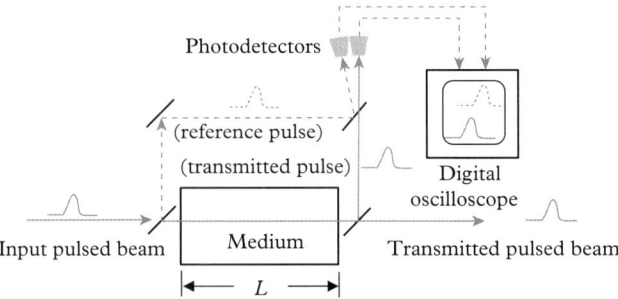

Figure 16.5 *Schematic diagraph of the experimental setup for measuring the relative delay between the transmitted light pulse passed through a resonant medium and the reference pulse.*

Table 16.1 *Value ranges of n_g, v_g, and Δt.*

Group refractive index	Group velocity	Relative time delay	Features (medium)
$n_g \gg 1$	$v_g \ll c$	$\Delta t \gg L/c$	Slow light (gain)
$n_g = 1$	$v_g = c$	$\Delta t = 0$	
$0 < n_g < 1$	$v_g > c$	$-L/c < \Delta t < 0$	Fast light (absorbing)
$n_g = 0$	$v_g = \infty$	$\Delta t = -L/c$	Fast light (absorbing)
$n_g < 0$	$v_g < 0$	$\Delta t < -L/c$	Fast light (absorbing)

Here, v_g, n_g, and L are the group velocity, group refractive index, and path length, respectively, of the tested medium. Based on Eq. (16.3-1), one can find that the value range of Δt is directly determined by the value range of n_g for a given resonant medium. Table 16.1 summarizes the possible ranges of the values for n_g, v_g, and Δt. It is noted that once the experimental values of delay time Δt are known, one may readily estimate the values of n_g and v_g.

Figure 16.6 schematically shows several possible situations of light propagation through a resonant (absorbing or gain) medium. Here we have neglected the change of pulse amplitude due to attenuation or amplification from the medium. The situations shown in Fig. 16.6(a–c) are easy to understand, whereas the situation shown in Fig. 16.6(d) seems not so straightforward to perceive by intuition. The pulse propagation with $v_g < 0$ is accompanied by two unusual features: (i) the transmitted pulse peak could be observed even before the input pulse peak enters the resonant medium; (ii) inside the resonant medium the pulse peak (indicated by the arrow) will propagate

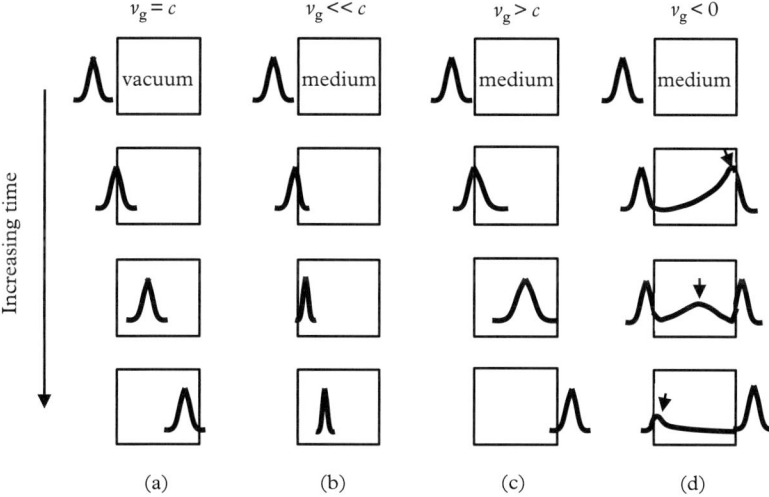

Figure 16.6 *Schematic diagrams showing the pulse propagation in vacuum (a), in a slow light medium (b), and in a fast light medium (c, d). For $v_g < 0$, as shown in (d), inside the medium the pulse peak is moving backward. (Adapted from Boyd and Gauthier.[2]).*

backward following the time increase. These two predictions can be proved by either experimental measurement or numerical simulation for a given medium of negative n_g values.

16.3.2 Light propagation versus causality and special relativity

In vacuum or free space, any types of light pulse or radiation signal always have the same $v_p = v_g \equiv c$, and the light velocity is not dependent on the moving speed of either the light source or the receiver. This is one of Einstein's postulates to establish his special theory of relativity. Our preceding analyses of fast or slow light propagation (i.e., v_g behavior) in a resonant medium do not violate the above postulate. On the other hand, the special theory of relativity concludes that in any case, the moving speed of a substance or matter should always be less than or equal to the light speed c in a vacuum owing to the request of causality. In some situations we shall deal with in this chapter, the group velocity v_g of light pulses could be faster than c. This fact is not in contradiction with the causality, because here v_g is not the moving speed of a substance; instead it is the moving speed of the peak position of a light pulse or wave packet.

As we mentioned before, a wave packet is the superposition of a great number of monochromatic plane wave components of differing frequencies with certain phase relationship to each other. According to Einstein's definition, a photon corresponds to a monochromatic plane wave of the quantized electromagnetic fields; the energy and momentum of a photon are determined by $\hbar\omega$ and $\hbar k$, where ω and k are the angular frequency and the wavevector magnitude, respectively, of a corresponding monochromatic plane wave. Thus, a wave packet can be regarded as a collection (group) of photons. The moving speed of intensity peak for a given photon group can be much slower or faster than c depending on the refractive index dispersion property of the medium. In this case it should be noted that each individual photon of frequency ω always propagates with a speed of $c/n(\omega)$.

To further clarify the physical meaning of group velocity of a wave packet, we may rely on Fig. 16.7, which depicts the possible waveforms of a wave packet after passing through a resonant medium in comparison with the waveform after passing through the same path length in vacuum. Here we assume that the intensity peak is originally located in the center of the original wave packet, as shown in Fig. 16.7(a). After passing through a resonant medium, one possible situation is that the peak is forward shifted along the wave packet, as shown in Fig. 16.7(b), and the other possibility is that the peak is backward shifted, as shown in Fig. 16.7(c). Obviously, these two possibilities correspond to fast and slow light propagation, respectively.

As we mentioned before, the peak position of a light wave packet is determined by the phase relationship among different spectral components of which the wave packet is composed, so that in a resonant medium, the relative peak position may be drastically shifted (forward or backward) within the wave packet duration, owing to the huge dispersion effect of medium's refractive index $n(\omega)$ around a central resonant frequency ω_0. Based on this mechanism, researchers could achieve the control of group velocity of the laser pulses propagating in various resonant media.

16.3.3 Methods of creating fast and slow light propagation

So far we already know that the key to control the group velocity is to achieve a huge value of $\omega \left| \dfrac{dn(\omega)}{d\omega} \right|$. For the case of $\omega \dfrac{dn(\omega)}{d\omega} \gg 1$ (steep normal dispersion), we could realize slow light propagation, whereas for $\omega \dfrac{dn(\omega)}{d\omega} \ll 0$ (steep anomalous dispersion) we could attain fast light

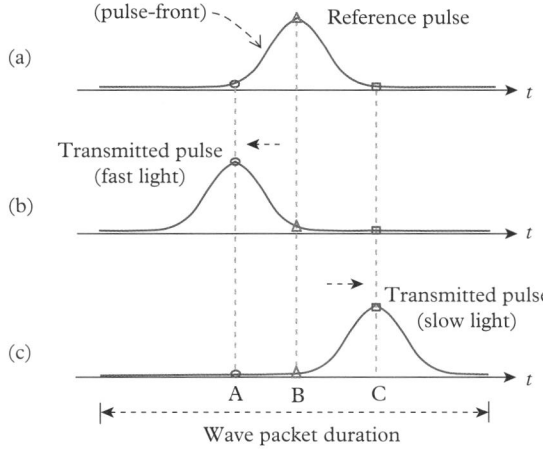

Figure 16.7 *(a) Pulse waveform of an original (reference) pulse. (b) Peak-advanced pulse waveform after passing through a fast light medium. (c) Peak-delayed pulse waveform after passing through a slow light medium.*

propagation. In both cases, it is imperative to create an abrupt refractive index variation over a very narrow frequency interval.

From Fig. 16.2 one can see that a steep anomalous dispersion can be achieved in a absorbing medium with large linear absorption coefficient $\alpha(\omega_0)$ and narrow spectral linewidth $2\Gamma_0$. However, a large $\alpha(\omega_0)$ value means strong attenuation and short penetration distance for the input light signal, which makes the fast light measurement rather difficult, if not impossible.

Until now, we have only considered the light propagation in an absorbing or gain medium with a simple two energy-level structure; besides, only the influence of refractive index dispersion on the group velocity has been taken into account. In practical experimental studies of light propagation, the situations may be much more complicated because of the following two reasons. First, the resonant medium may not be a linearly absorbing or gain medium, once the input light intensity is high enough a saturation effect for absorption or gain may occur. Second, under some specific experimental conditions, the energy transfer between a resonant medium and light pulse may also lead to a shape change or apparent shift of the pulse peak along the wavepacket. For instance, according to our aforementioned discussions, a slow light propagation should be expected by sending a signal pulse passing through a gain medium with population inversion. On the other hand, however, if the input signal pulse meets the condition of 2π-pulse, the measured moving speed of the apparent pulse peak may greater than c (see Section 10.2). This is because the first half of the pulse is fully amplified while the second half is absorbed. The other situation of apparent peak shift may take place when there is strong gain saturation of a light amplifier.

The breakthrough in fast and slow light studies should be based on extending the previous theoretical analyses to some special cases, in which the medium is not a simple absorbing or gain system involving only two levels and interacting with only one incident light beam. In those special cases, for example, the medium can be a Raman or Brillouin gain medium and two incident laser beams could also be involved, where the frequency difference of the two beams should be tuned to match the Raman or Brillouin frequency of the medium. In this sense, the medium is transparent for each individual beam incidence, but it could provide a resonant response upon the coupling interaction of the two beams. Under these circumstances, the medium's resonant response may appear as a rapid variation of refractive index over a narrow spectral range for at least one of those two beams. Moreover, even for a conventional absorbing medium, by using the method of two-beam (pump–probe) coupling interaction, one may create a narrow and steep refractive index variation around the frequency of one beam; meanwhile the strong attenuation effect could be avoided.

For a simple two-level system, it is already known that the refractive index variation near the central resonant frequency of the medium is directly related to the absorption or gain curve (see Figs. 16.2 and 16.4). In other words, knowing the former one can determine the latter, and vice versa. Moreover, this conclusion is applicable to more generalized cases concerning resonant interaction between matter and radiation. Under resonant interaction conditions, the medium response to the incident light field(s) can be generally described by a complex susceptibility

$$\chi(\omega) = \chi_{re}(\omega) + i\chi_{im}(\omega). \tag{16.3-2}$$

Kramers–Kronig relations uniquely specify the following relationships between the real part and imaginary part of the complex susceptibility:

$$\chi_{re}(\omega) = \frac{2P}{\pi} \int_0^\infty \frac{\omega' \chi_{im}(\omega')}{\omega'^2 - \omega^2} d\omega', \quad \chi_{im}(\omega) = -\frac{2\omega P}{\pi} \int_0^\infty \frac{\chi_{re}(\omega')}{\omega'^2 - \omega^2} d\omega', \tag{16.3-3}$$

where P denotes the Cauchy principal value at $\omega' = \omega$. It is well known that the real part and imaginary part of the susceptibility determine the refractive index and absorption (or gain) of the medium, respectively. Equation (16.2-3) is a special example of the Kramers–Kronig relation applied to an absorbing medium with a Lorentzian line shape.

In practice, it is much easier to measure the attenuation (or gain) of a light signal as a function of frequency than to measure the refractive index variation versus frequency. Thus, once the spectral curve of attenuation (or gain) as a function of frequency is obtained, one can readily extract the refractive index curve by invoking the Kramers–Kronig relations. Knowing the refractive index dispersion curve near a given resonant frequency, we may further predict or analyze the group velocity behavior of the investigated medium.

The typical mechanisms to create fast or slow light propagation in a resonant medium are demonstrated schematically in Fig. 16.8. The common feature of these schemes is to create a steep variation of attenuation or gain within a very narrow spectral region, thereby to ensure an abrupt variation of refractive index over the same narrow spectral region. As depicted in Fig. 16.8(a), in order to realize fast light propagation, a steep attenuation increase or gain decrease is imperative to ensure a narrow anomalous dispersion with $\frac{dn(\omega)}{d\omega} < 0$ and $\omega \left| \frac{dn(\omega)}{d\omega} \right| \gg 1$. Conversely, the key requirement for schemes shown in Fig. 16.8(b) is to create a steep gain increase or absorption decrease over a narrow spectral region, where a huge $\omega \frac{dn(\omega)}{d\omega} \gg 1$ value may maintain slow light propagation.

Figure 16.8(a) shows three approaches to realize fast light propagation based on (i) a medium exhibiting very narrow linear absorption line or induced absorption line, (ii) a medium exhibiting very narrow gain saturation hole in a broad gain band, and (iii) a medium with two gain lines that are close to each other and forming a very narrow gain drop between them. These three approaches are actually equivalent, and the frequency of probe light signal should be tuned to the central position of the spectral curves shown in Fig. 16.8(a).

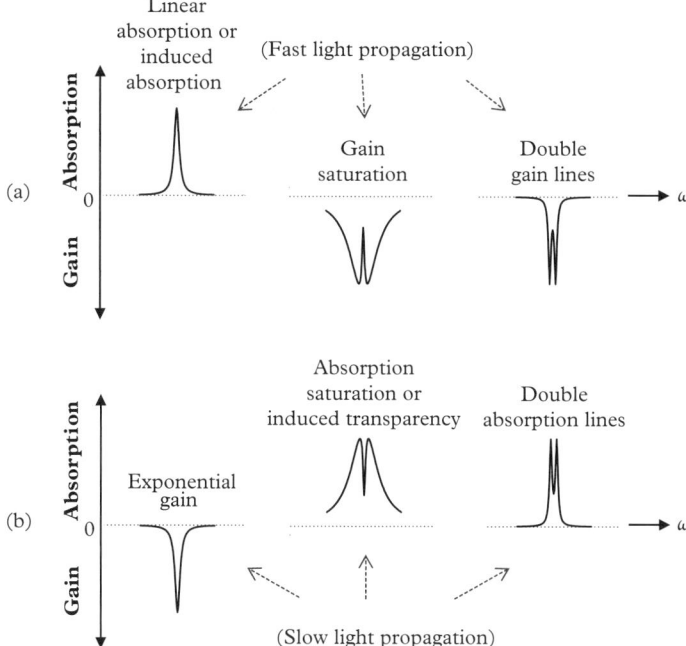

Figure 16.8 *Typical schemes for achieving fast light propagation (a) and slow light propagation (b) in a resonant medium.*

Moreover, Fig. 16.8(b) shows three typical schemes to implement slow light propagation, which are based on (i) a medium with very narrow spectral gain line, (ii) a medium with very narrow absorption drop hole due to absorption saturation or induced transparency, and (iii) a medium with two absorption lines that are close to each other and form a very narrow absorption drop between them. It is obvious that the spectral curves shown in Fig. 16.8(b) are the reverse versions of those shown in Fig. 16.8(a).

16.4 Studies on fast light effects

16.4.1 Fast light in linear absorbing media

The earliest theoretical work on fast light propagation in a linear absorbing medium was reported by Garrett and McCumber in 1970.[12] They considered the propagation of light pulses of Gaussian shape with frequency spread less than the resonance linewidth of the medium. The theoretical conclusions were that the propagation speed of pulse peak could be greater than c or negative, while the pulse shape may remain unchanged.

The early indirect observation of fast (or slow) light propagation in an absorbing (or gain) gas cell, placed inside the cavity of a mode-locked gaseous laser system, was based on measuring the repetition-rate change of the output laser pulses from the cavity.[13]

The first straightforward demonstration of fast light propagation in an absorbing solid medium was accomplished by Chu and Wong in 1982.[14] The samples were epitaxially grown GaP:N films with different thicknesses and nitrogen concentrations, working at 1.7 K to ensure a narrow absorption line width. This material exhibited a well-isolated bound A-exciton absorption line around the 534 nm position. A pulsed laser beam of 48 ps duration, generated from a dye laser tunable around the 534 nm region, was split into two beams by a beam splitter; as shown in Fig. 16.9(a), one passed through the sample and the other propagated along an adjustable optical delay line. Then the two beams were focused on a KDP crystal to generate a non-collinear second-harmonic generation (SHG) signal. The thickness of the epilayer was adjusted to ensure that the peak absorption never exceeded ~6 absorption lengths.

Figure 16.9(b) shows the cross-correlation SHG waveforms measured at different values of laser detuning, which was $\delta = h(\nu - \nu_0)$ in units of meV, where ν was the laser frequency while

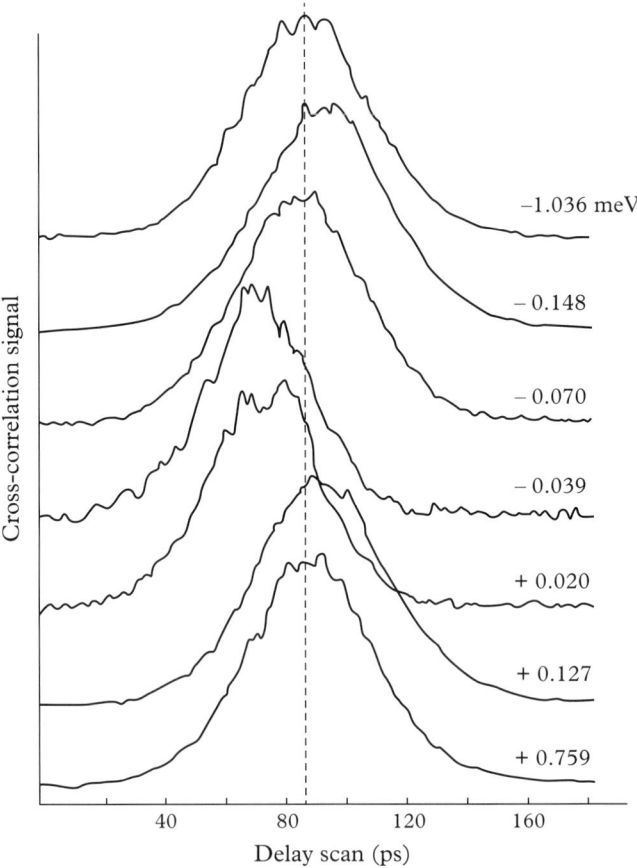

Figure 16.9 *Experimental result for fast light measurement in a cooled GaP:N epilayer sample: SHG signal waveforms measured at different frequency-detuning values (in units of meV) and sample nitrogen concentration [N] = 1.5 × 10^{17} cm^{-3}. (Reproduced with permission from Chu and Wong.[14] Copyright 1982, American Physical Society).*

v_0 was the central absorption frequency of the sample medium. From this figure, one can see a maximum negative time delay under the precisely resonant ($\delta \approx 0$) condition. The pulse delays measured were independent of peak intensities between 100 and 3 W/cm².

Moreover, the measured data (hollow circles) of pulse delay (Δt) as a function of laser detuning are shown in Fig. 16.10. It can be clearly seen that on tuning the laser frequency from off-resonance to the central frequency, the time delay smoothly changes from a near-zero value to a positive value range, then into a negative value range, and finally reaches its negative maximum value at the exact resonance position. Correspondingly, the group velocity value changes from the off-resonance value (8.6×10^9 cm/s) to a reduced value range, then passes through $\pm\infty$ and further becomes negative, as shown in the same figure by the solid-line curve. The experimental results are in good agreement with the theoretical calculation. However, it was indicated that the cross-correlation curves shown in Fig. 16.9(b), could not reveal any possible distortions of the transmitted laser pulses.[15]

A similar fast light experiment in a laser-cooled ^{85}Rb atomic system has been implemented by using the laser pulses of 35 ns duration and sub-megahertz linewidth.[16] A ^{85}Rb atom cloud was confined by a magneto-optical trap that allowed some control over the atom density and absorption length to ensure a transmission between 20% and 50%. The laser frequency was scanned across the $5S_{1/2}\,(F=3) - 5P_{2/3}\,(F'=2,3,4)$ transitions of rubidium atoms. The measured maximum (anomalous) dispersion was $dn/d\nu = -1.3 \times 10^{-12}\,\text{Hz}^{-1}$ with an atomic density of $0.6 \times 10^{10}\,\text{cm}^{-3}$ in the trap. The observed negative group velocity could reach to $-c/360$.

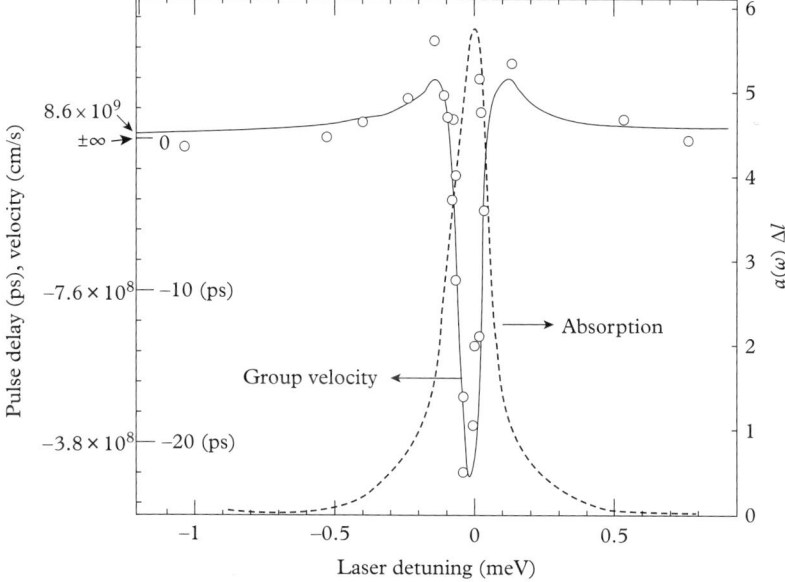

Figure 16.10 *Measured data of pulse delay (Δt) as a function of laser frequency detuning in units of meV. Sample thickness $\Delta l = 76\,\mu m$, nitrogen concentration $[N] = 1.5 \times 10^{17}\,cm^{-3}$; solid line is the calculated group velocity curve, while the dashed line is the absorption curve. (Reproduced with permission from Chu and Wong.[14] Copyright 1982, American Physical Society).*

16.4.2 Fast light in double-line gain media

In 1994, Steinberg and Chiao theoretically proposed that if there were two nearly positioned gain lines in a population inversed medium, a narrow gain dip formed between the two gain maxima could provide a huge anomalous dispersion, which might support fast light propagation (cf. Fig. 16.8(a)).[17] Meanwhile, it was also proposed that fast light propagation should be observable in the two-wing regions of a single gain line (cf. Fig. 16.4(c)).[18] Obviously, the advantage of realizing fast light in a gain medium is that the attenuation effect on the light pulse signal can be essentially avoided.

The experimental demonstration of fast light propagation in a stimulated Raman gain doublet system was accomplished by Wang et al. in 2000.[19] In their work, the gain medium was atomic caesium (Cs) vapor at 30 °C filled in a 6-cm long glass cell that was placed in a small (1.0 G) uniform magnetic field parallel to the light propagation direction. The working mechanism is shown in Fig. 16.11(a), where the Cs atoms are optically prepared to stay in the hyperfine magnetic sublevel $|F = 4, m = -4\rangle$ of the $6S_{1/2}$ state. This populated sublevel (marked as $|1\rangle$) serves as the initial state of Raman transition, while another empty sublevel $|F = 4, m = -2\rangle$ (marked as $|2\rangle$) acts as the final state of the Raman transition.

Two strong cw pump laser beams were applied to the Cs vapor system with a frequency difference of $2\Delta = 2.7$ MHz, and their average frequency was far enough from the high sublevel $|F = 4, m = -3\rangle$ of the $6P_{3/2}$ by an amount Δ_0. A third weak laser beam working either in cw or in pulsed mode was tunable around the Raman gain region provided by the two pump beams. Figure 16.11(b) shows the frequency-dependent curves of Raman gain and refractive index of such a system, from which one can see a steeper anomalous dispersion of refractive index in the spectral region between these two gain peaks.

Figure 16.12(a) shows the measured data of Raman gain and refractive index as a function of the probe detuning by using a cw probe beam. There is a $\Delta n = \pm |1.8 \times 10^{-6}|$ refractive-index change over a narrow frequency range of $\Delta \nu = 1.9$ MHz between the two gain lines, which means a group refractive index value of $n_g = -330$ in that frequency region.

In the same experiment, by using a modulated probe beam of ~ 2 μs pulse duration, the flight time delay of the probe pulse could be directly measured. Figure 16.12(b) shows the normalized

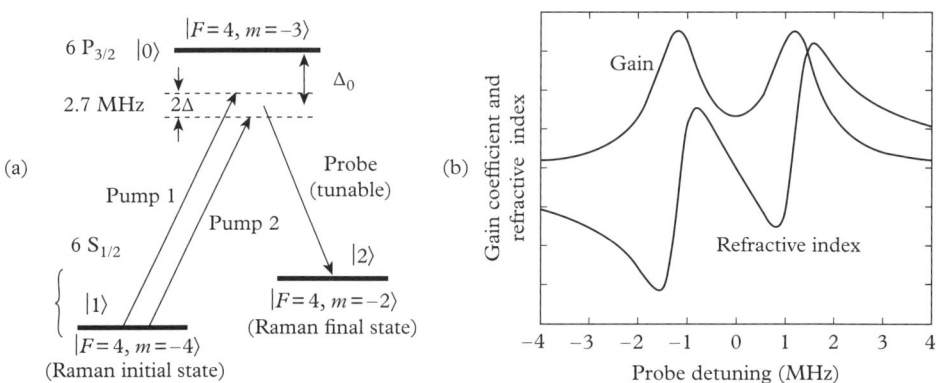

Figure 16.11 *(a) Double-line Raman gain scheme in Cs atomic gas system; (b) theoretical curves of gain and refractive index versus the frequency detuning of probe beam. (Reproduced with permission from Wang et al.[19] Copyright 2000, Nature Publishing Group).*

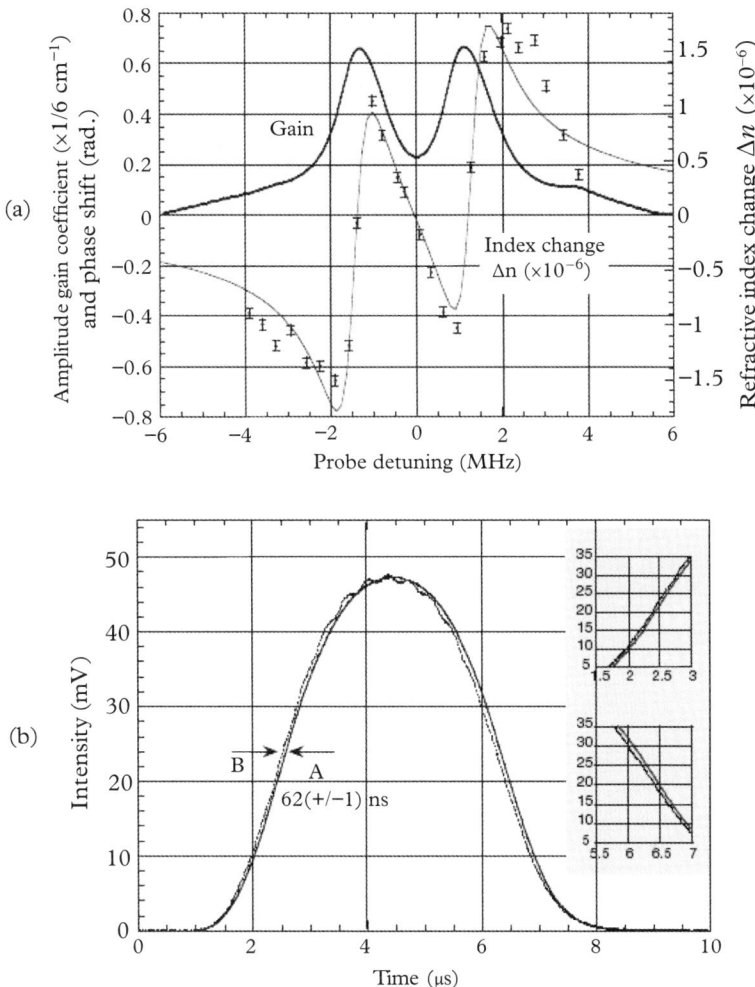

Figure 16.12 *(a) Measured gain and refractive index as a function of the probe detuning. (b) Normalized waveforms of the probe pulse A (passed through the medium in off-resonance manner) and the probe pulse B (in zero-detuning manner), indicating a 62 ns advancement of the latter or a group velocity of $-c/310$. Insets show the front and trailing portions of the pulses (Reproduced with permission from Wang et al.[19] Copyright 2000, Nature Publishing Group).*

waveforms measured for the transmitted pulse A, which was passed through the medium under off-resonance condition, and for the pulse B that was passed through the medium under near-zero detuning condition. There is a 62 ns time advancement for the latter, which means a negative group velocity $v_g = -c/310$.

The aforementioned principle with gain doublet is also applicable to a stimulated Brillouin scattering (SBS) medium. Song et al. demonstrated such an experiment by using a 5-km long conventional optical fiber with a Brillouin frequency shift of $\nu_B = 10.834$ GHz.[20] Their experimental setup is shown in Fig. 16.13. The 1552-nm laser beam from a DFB (distributed feed-back) laser diode was launched into an electro-optic modulator (EOM-1) to create two sidebands while the carrier wave was suppressed. The low-frequency sideband beam was reflected by a narrow-band fiber Bragg grating and then optically gated to form a forward probe pulse of 37 ns duration. To create a single gain line, the high-frequency sideband beam was amplified through an Er-doped fiber amplifier (EDFA) and then launched into the optical fiber as a backward pump beam. The frequency difference between these two beams was set to $\nu_B + \Delta\nu$, where $\Delta\nu$ could be tuned from 2 MHz to 50 MHz by changing the modulation frequency of EOM-1. Furthermore, to provide a gain doublet, the modulation frequency of EOM-1 was fixed to ν_B and another modulator (EOM-2) was inserted in the setup to modulate the high-frequency sideband beam from the EOM-1. After EOM-2 and EDFA the two newly created sidebands with a tunable frequency difference of $\Delta\nu$ played the roles of two backward pump beams.

The schematic curves of gain, refractive index change Δn, and group refractive index change Δn_g as a function of detuning of the probe beam for the cases of single peak and double peaks are shown in Fig. 16.14(a) and (b), respectively, from which one can see that the fast light propagation could be realized either in the wing region of the Δn_g curve with a single peak, or in the middle region of the two peaks.

The measured gain and time delay of the probe pulse as a function of detuning frequency for the cases of single frequency pump and double frequency pump are shown in Fig. 16.14(c)

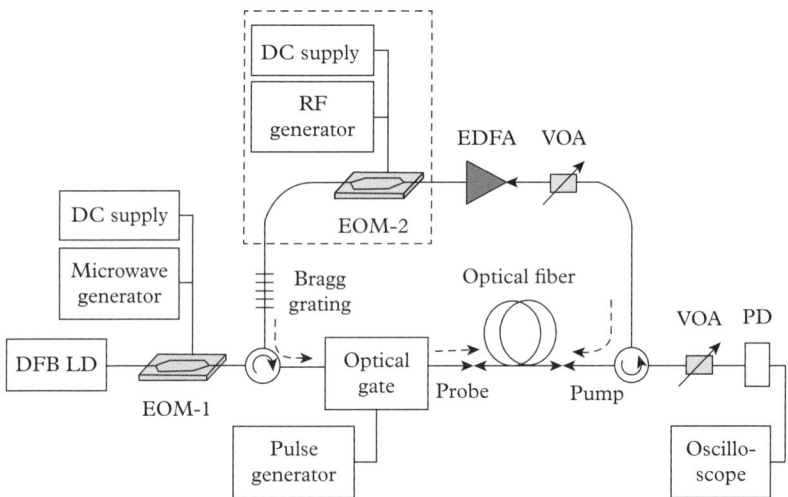

Figure 16.13 *Experimental setup for fast and slow light measurement by using an optical fiber as Brillouin gain medium. Device marked by the dashed box are only inserted for the double gain–peak configuration: LD, laser diode; VOA, variable optical attenuator; EOM, electro-optic modulator; PD, photodiode; EDFA, erbium-doped fiber amplifier. (Reproduced with permission from Song et al.[20] Copyright 2005, Optical Society of America).*

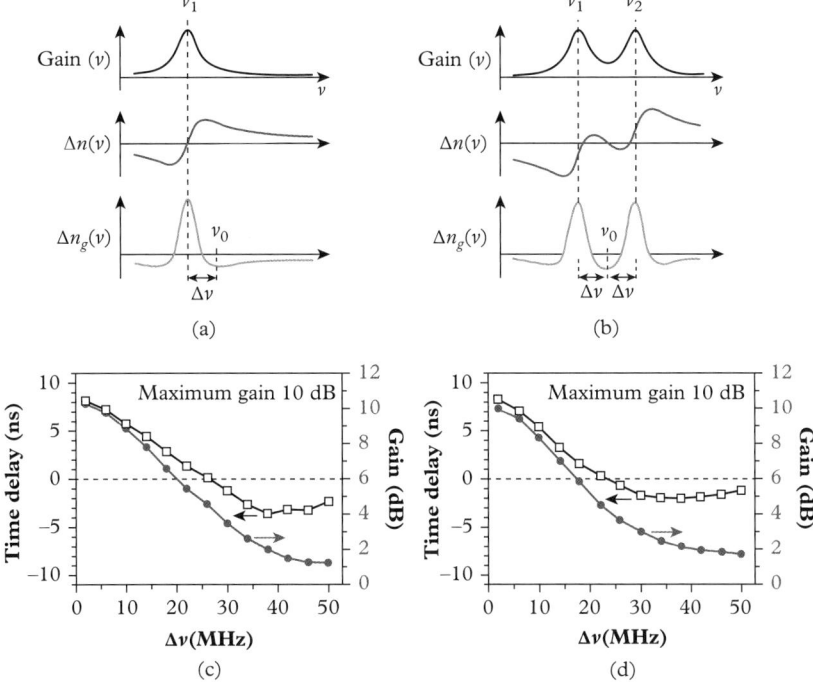

Figure 16.14 Gain, refractive index change Δn, and group refractive index change Δn_g in the vicinity of Brillouin resonance of an optical fiber in cases of (a) single gain peak and (b) double gain peaks. The measured gain and time delay of the probe pulse as a function of detuning are shown in (c) and (d) in corresponding to the cases (a) and (b), respectively. (Reproduced with permission from Song et al.[20] Copyright 2005, Optical Society of America).

and (d), respectively. In both cases, a negative time delay was observed within the detuning range from ∼ 25 MHz to 50 MHz. However, the waveform measurement indicated that in the latter case, the pulse shape of the probe beam experienced a less distortion than that in the former case.

16.4.3 Fast light in induced absorption systems

In slow/fast light studies, the mechanism of so-called electromagnetically induced transparency (EIT) was first employed to achieve slow light propagation in an absorbing medium without severe attenuation (see Fig. 16.8(b) and the details in the next section). After that, researchers naturally thought that a reverse effect, i.e., electromagnetically induced absorption (EIA) should support fast light propagation (see Fig. 16.8(a)). Actually, Akulshin et al. did investigate the EIA in a coherently prepared degenerate two-level atomic system, and observed a very narrow induced absorption line with an associated steep anomalous dispersion.[21,22] In their experiment, the medium was a 5-cm long ^{87}Rb vapor, excited at its D_2 line by a strong drive laser beam and probed by another tunable weak laser beam. The highest anomalous dispersion value was measured to be $dn/d\nu \approx -6 \times 10^{-11}\,\text{Hz}^{-1}$, corresponding to a group velocity of $v_g = -c/2300$.[22]

On the other hand, an early study of intensity-dependent absorption behavior of alexandrite crystal showed that, depending on specific wavelengths of a strong exciting laser beam, a narrow hole (dip) due to saturated absorption or a narrow anti-hole due to excited-state absorption (ESA) could be observed on the broad absorption profile by using a weak tunable probe laser beam.[23] Here one is dealing with a strong drive beam-induced absorption reduction (or increase) over a narrow frequency region in the medium; therefore, these two cases can be employed to achieve slow (or fast) light propagation, respectively. In 2003, Bigelow et al. accomplished such an experiment to demonstrate the real propagation of slow light and fast light at two different excitation wavelengths.[24] They have indicated that in the alexandrite crystal ($BeAl_2O_4:Cr^{3+}$) working at room temperature, there are two groups of Cr^{3+} ions which are separately located in the mirror sites and the inversion sites of the $BeAl_2O_4$ lattice. These two groups of Cr^{3+} ions have different intensity-dependent absorption responses to the incident strong drive beam. Specifically, as shown in Fig. 16.15, under strong excitation at 476 nm wavelength the Cr^{3+} ions in mirror sites exhibit dominant ESA. In the experiment, the 476 nm drive beam was from an Ar-ion laser, after an electro-optic modulator, the frequency-shifted sideband acted as a probe beam, and the frequency difference between the drive and probe beams was tunable by changing the modulation frequency.

The measured attenuation and time delay of probe pulses for a 4-cm long crystal sample are shown in Fig. 16.16, where the relative modulation attenuation is defined as the attenuation of the modulated signal (probe) relative to the attenuation of the drive beam. One can see that the attenuation becomes smaller following the increase of the detuning of the probe beam, which indicates the narrow resonant feature of induced absorption. At a transmission level of $\sim 3.5\%$ for the probe pulses, the measured maximum delay time is $\Delta t = -50\,\mu s$, which corresponds to a group velocity of $v_g = -800$ m/s.

Moreover, in a later article Mikhailov et al. reported that in a 2.5-cm long ^{87}Rb vapor cell added with a buffer gas (Ne) and excited by a strong drive laser beam tunable across the $5S_{1/2} \rightarrow 5P_{1/2}$ (D1) transition, a two-photon resonance-induced absorption and large negative time delay of $\Delta t \approx -300\,\mu s$ for the probe pulses could be obtained, which corresponded to a group velocity of $v_g = -84$ m/s.[25]

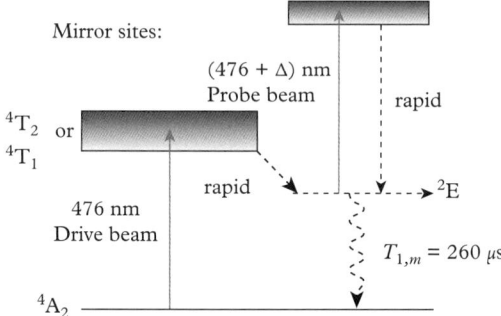

Figure 16.15 *Energy levels and excited state absorption of Cr^{3+} ions in mirror sites of alexandrite crystal excited with strong drive beam of 476 nm wavelength. (Reproduced with permission from Bigelow et al.[24] Copyright 2003, AAAS).*

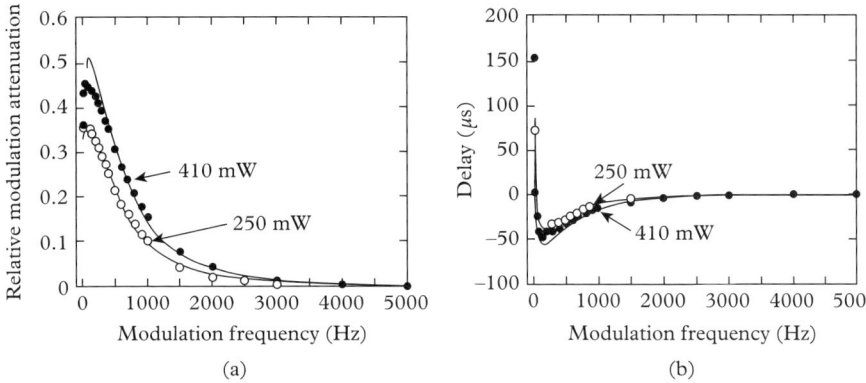

Figure 16.16 *Relative modulation attenuation (a) and time delay measured for a 4-cm-long alexandrite crystal at a wavelength of 476 nm with drive powers of 250 and 410 mW. The negative time delay corresponds to fast light propagation, and the solid lines indicate the theoretical curves. (Reproduced with permission from Bigelow et al.[24] Copyright 2003, AAAS).*

16.4.4 Backward motion of a pulse peak inside a fast light medium

Negative group velocity of light pulse propagation implies two physical meanings: (i) the peak of the transmitted pulse can leave a resonant medium before the peak of the incident pulse enters the medium; (ii) the pulse peak appears to move in the backward direction within the medium. Although these two features of the negative group velocity are somewhat unintelligible on our first thought, they are both experimentally observable. Many experimental results have proved the first feature. Next, we shall give two examples that can prove the second feature.

A straightforward demonstration of laser pulse peak's backward propagation within an erbium-doped optical amplifier (EDOA) was accomplished in 2006 by Gehring et al.[26] In this case, the EDOA sample was used as a fast-light sustaining medium, based on a narrow dip in the broad gain profile of the EDOA through the mechanism of the so-called coherent population oscillation (CPO) or induced gain saturation. The experimental setup is shown schematically in Fig. 16.17. A 9-m long erbium-doped optical fiber (EDOF) was pumped by a 980-nm laser beam of 128 mW to provide a broad gain band. Another modulated or pulsed 1550-nm probe laser beam of 0.5 mW was coupled together with the pump beam into the EDOF sample to create a narrow dip of the gain profile as well as to undergo the fast-light propagation with a negative group velocity. The time evolution of the probe pulse within the fiber amplifier was monitored by successively cutting back the length of the fiber and recording the corresponding output waveform; each cutting length was about 25 cm and the procedure was continued until only several centimeters of fiber remained. In this way, the time evolution of the pulse could be determined at many points along the length of the fiber sample.

As an example, Fig. 16.18(a) shows the measured normalized waveforms and the relative delay of the input probe pulse and the transmitted pulse passed through a 6-m long EDOF sample, which indicates an evident negative time delay. The measured group velocity was −75 km/s and the group index was −4000.

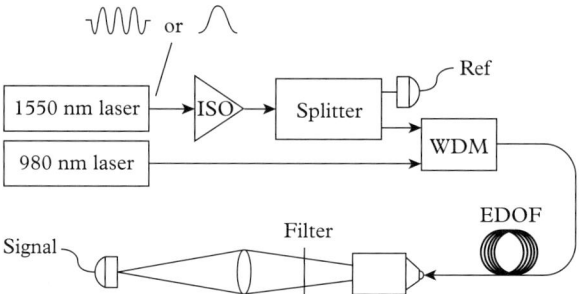

Figure 16.17 *Experimental setup for measuring the time evolution of the probe pulse within the fiber amplifier by successively changing the fiber length. ISO: fiber optical isolator; WDM: wavelength-division-multiplexing coupler. (Reproduced with permission from Gehring et al.[26] Copyright 2006, AAAS).*

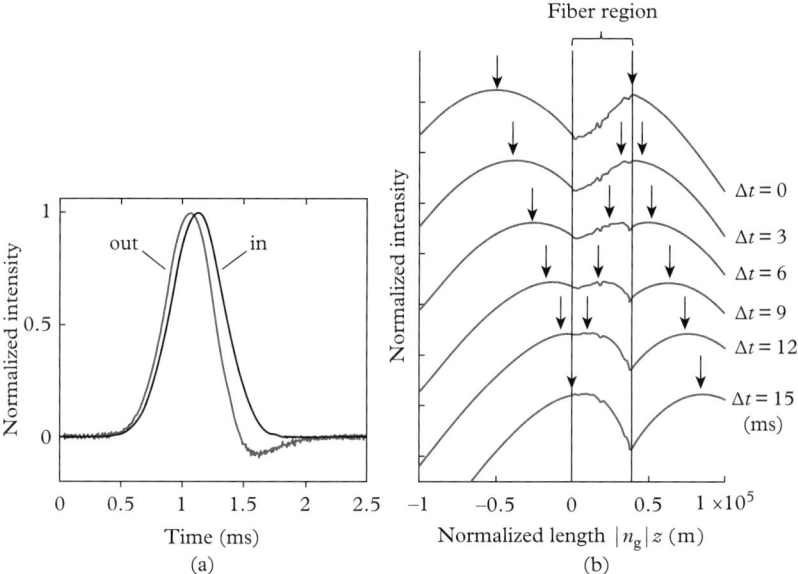

Figure 16.18 *(a) Measured waveforms for input and output probe pulses, indicating a negative time delay. (b) Time evolution of the probe pulse's spatial waveforms measured at different time intervals. The arrows indicate the pulse peak positions inside and outside the fiber medium. (Reproduced with permission from Gehring et al.[26] Copyright 2006, AAAS).*

Figure 16.18(b) shows the more interesting result of the time evolution of the probe pulse inside and outside the fiber region. To obtain these results, the measured data have been normalized at each point within the fiber to remove the effects of gain. From this figure, one can indeed see that the peak of the transmitted pulse exits from the fiber before the peak of the incident pulse enters the fiber, and the pulse peak moves backward inside the fiber as time increases.

It is worth indicating that though there is apparent backward motion of the pulse peak inside the resonant medium, the complete light pulse with its carrying energy always keeps propagating along the forward direction. In the same experiment, a specially designed test has proved that there is no backward pulse signal or energy flow exiting from the EDOA sample.[26]

A theoretical (analytical and numerical) study of pulse propagation in a resonant medium with negative group velocity was performed by Clader et al.[27] They also demonstrated the backward motion feature of the pulse peak inside the medium, by examining a short 2π-pulse incident to an initially population inversed two-levels amplifier system. Choosing a Rb vapor system working at its D_2 line as the example, for the input pulse duration of $\tau = 0.1$ ns and gain coefficient of 8.15 cm^{-1}, the calculated group velocity was $v_g = -3.27c$.

Figure 16.19 shows the normalized spatial waveforms of the 2π-pulse propagating through the gain medium, which are calculated at different instants. Once again, from this figure one can see that the outgoing peak leaves the gain medium before the input peak enters it, and the pulse peak appears to be moving backward inside the medium.

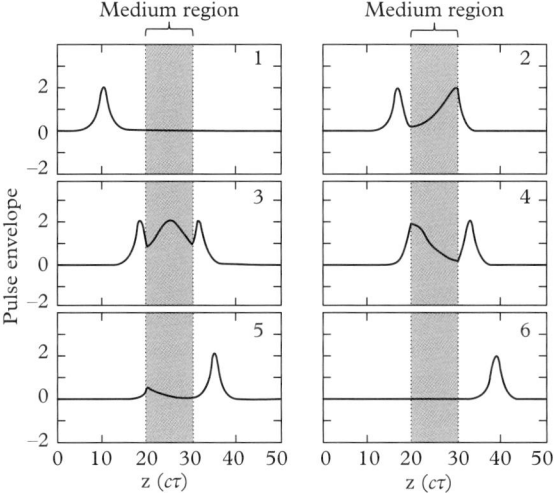

Figure 16.19 *Time evolution of the spatial waveforms for a 2π-pulse propagating through a two-level gain medium, indicating the apparent backward movement of the pulse peak inside the medium. (Reproduced with permission from Clader et al.[27] Copyright 2006, Optical Society of America).*

16.5 Studies on slow light effects

16.5.1 Slow light based on electromagnetically induced transparency (EIT)

The principles of creating slow light propagation in resonant media are opposite to that of creating fast light. In other words, to provide a slow light propagation, there should be a narrow gain line, or two neighboring absorption lines, or a narrow saturation dip on a broad absorption profile for a given medium (see Fig. 16.8(b)). In slow light studies, the first breakthrough was using the mechanism of EIT to realize slow light propagation in a highly absorbing medium.

The basic idea and physical interpretations of EIT have been well described by Harris et al.[28–30] The EIT mechanism in an absorbing medium can be explained by referring to Fig. 16.20(a), where three energy levels of the atoms are involved. The level $|1\rangle$ is mostly populated while level $|2\rangle$ is partially populated or nearly vacant. The transitions from $|1\rangle$ to $|3\rangle$ with frequency ω_{31} and from $|2\rangle$ to $|3\rangle$ with frequency ω_{32} are allowed. When a light of frequency $\omega \approx \omega_{31}$ is incident alone upon the medium, there will be a strong linear absorption. However, if there are two monochromatic laser beams acting on the medium at the same time, one is a strong coupling beam with frequency of $\omega_c \approx \omega_{32}$, while another is a weak probe beam with frequency $\omega_p \approx \omega_{31}$. Under these particular conditions, the medium may react to the probe beam as a "transparent" medium. This is the EIT effect that is due to coherent population trapping of the atoms in state $|1\rangle$, caused by the presence of the strong coupling beam, so that the absorption transition at $\omega_p \approx \omega_{31}$ for the probe beam is "prohibited."[30]

In this specific case, the essential consequence of the co-action of these two beams is that the atoms are driven into a coherent superposition of states $|1\rangle$ and $|2\rangle$, while the state $|3\rangle$ remains empty. In other words, EIT can be interpreted as the quantum interference effect of a two-beam resonant interaction with the medium. The probability amplitude of the atom's transition from

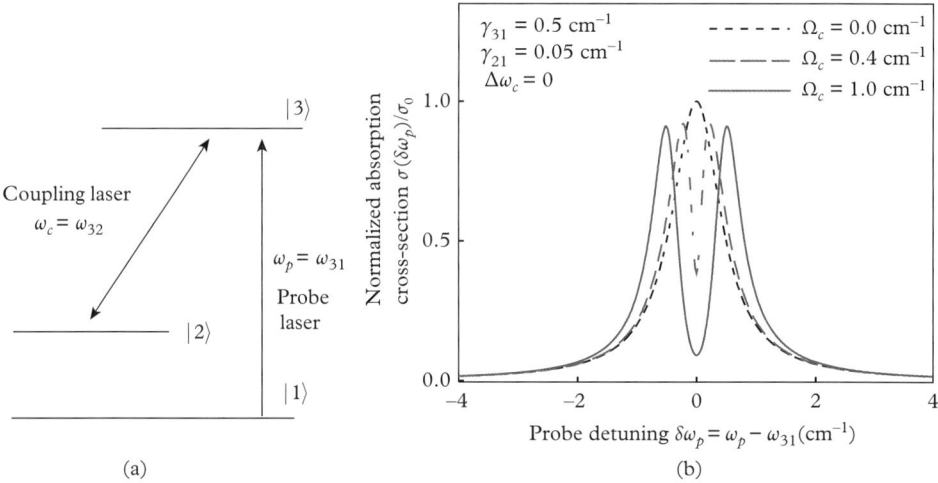

Figure 16.20 (a) Three-state structure for an EIT atomic system. (b) Normalized absorption curves of the probe beam as a function of detuning $\delta\omega_p = \omega_p - \omega_{31}$, at three different Rabi frequency (Ω_c) values of the coupling beam with $\Delta\omega_c = \omega_c - \omega_{32} = 0$.

|1> to |3> can be comparable to that of the transition from |2> to |3>, but with opposite phase factor to each other. Because of destructive quantum interference, the net probability of the atom being excited from |1> to |3> can be zero, provided that the intensity of the coupling beam is strong enough and precise resonance conditions of $\omega_c \approx \omega_{32}$ and $\omega_p \approx \omega_{31}$ are satisfied.

Based on the EIT theory and assuming $\Delta\omega_c = \omega_c - \omega_{32} = 0$, the absorption cross-section of probe beam as a function of detuning $\delta\omega_p = \omega_p - \omega_{31}$ can be written as[29]

$$\sigma(\delta\omega_p) = \sigma_0 \frac{(\delta\omega_p \gamma_{31})^2 + \gamma_{21}\gamma_{31}(\Omega_c^2/4 + \gamma_{21}\gamma_{31})}{[\Omega_c^2/4 + \gamma_{21}\gamma_{31} - (\delta\omega_p)^2]^2 + [\delta\omega_p(\gamma_{21} + \gamma_{31})]^2}. \qquad (16.5\text{-}1)$$

Here, σ_0 is the peak absorption cross-section of the $|1\rangle \to |3\rangle$ transition in the absence of the coupling beam, γ_{21} and γ_{31} are the linewidth factors associated with $|1\rangle \to |2\rangle$ and $|1\rangle \to |3\rangle$ transitions, respectively, and Ω_c is the coupling beam's Rabi frequency defined by

$$\Omega_c = p_0 E_c/\hbar, \qquad (16.5\text{-}2)$$

where E_c is the amplitude of the coupling laser field and p_0 is the dipole matrix element of the $|2\rangle \to |3\rangle$ transition. Figure 16.20(b) shows the normalized curves of probe absorption as a function of the detuning $\delta\omega_p$, as predicted by Eq. (16.5-1) with three different Ω_c values. From this figure, one can see that at appropriate values of coupling beam intensity, there is a deep and narrow drop near the center frequency of the probe absorption curve. In this particular case, the induced transparency within a narrow frequency region could provide a rapid refractive index variation with a huge positive dispersion that is suitable for slow light propagation.

In an early theoretical estimation, Harris et al. specifically considered a 10-cm long ^{208}Pb vapor sample with an atom density of 7×10^{15} atoms/cm^3.[31] The coupling transition would be $6s^26p^2\,^3P_2 \to 6s^26p7s\,^3P_1$ with a wavelength of 405.9 nm, the probe transition would be $6s^26p^2\,^3P_0 \to 6s^26p7s\,^3P_1$ with a wavelength of 283 nm, and the required Rabi frequency of the coupling beam would be $\Omega_c \approx 0.7$ cm^{-1}. Under these conditions, the calculated group velocity of the probe pulse would be $v_g \approx c/250$. Subsequently, in 1995 Kasapi et al. accomplished an experimental study on the previously proposed ^{208}Pb vapor system, using 406-nm laser pulses of ~100 ns duration as the coupling beam and 283-nm laser pulses of ~12 ns duration as the probe beam.[32] For a 10-cm long vapor sample of atom density $N_0 \approx 2 \times 10^{14}$ atoms/cm^3 and linear absorption coefficient $\alpha_0 = 600$ cm^{-1} at probe wavelength, the measured transmission curves are shown in Fig. 16.21, under the conditions with and without the coupling laser beam. Based on the measured pulse-delay results, the group velocity for the probe pulse was $v_g = c/165$ with a transmission value of 55%.

In the meantime, Xiao et al. reported their dispersion measurement results in a 76-mm long Rb vapor sample working at 20 °C; the EIT occurred along a cascade three energy-level configuration. At the central frequency of the probe transmission, the measured maximum positive dispersion was 19.4×10^{-9} s, which would yield a group velocity of $v_g = c/13.2$.[33]

In 1999, Hau et al. reported an extremely low group velocity of $v_g = 17$ m/s, which was achieved in an ultracold gas (cloud) of Na atoms cooled to nanokelvin temperatures by laser/evaporative cooling techniques.[34] The density and thickness of the sample medium were $N_0 \approx 5 \times 10^{12}$ atoms/cm^3 and $L = 229$ μm, respectively, and the medium was in the state of Bose–Einstein condensation. The weak probe laser was in resonance with the D_2 line of wavelength 589 nm, corresponding to the transition from the hyperfine level $|1\rangle = |F = 1, M_F = -1\rangle$ of the ground state $3^2S_{1/2}$ to the hyperfine level $|3\rangle = |F = 2, M_F = -2\rangle$ of the upper state $3^2P_{3/2}$.

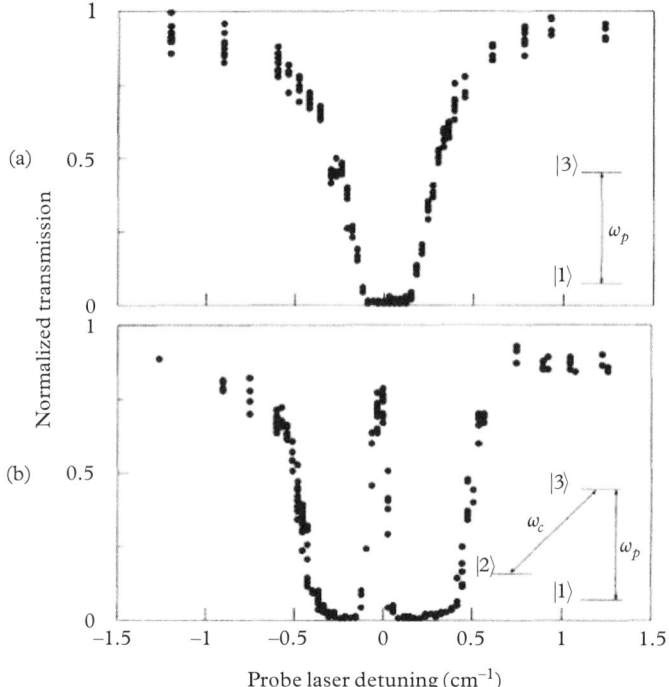

Figure 16.21 Measured transmission versus probe detuning in ^{208}Pb vapor: (a) probe alone and (b) with coupling laser of $\Omega_c = 0.4$ cm^{-1}. Sample length $L = 10$ cm; density $N_0 \approx 2 \times 10^{14}$ atoms/cm^3; linear absorption coefficient $\alpha_0 = 600$ cm^{-1}. (Reproduced with permission from Kasapi et al.[32] Copyright 1995, American Physical Society).

The strong coupling laser was tuned in resonance with the transition from another hyperfine level $|2\rangle = |F = 2, M_F = -2\rangle$ of the ground state to the same upper level of $|3\rangle = |F = 2, M_F = -2\rangle$. Figure 16.22(a) shows the partial energy level diagram of the Na atoms, and indicates the coupling and probe transitions, respectively.

Figure 16.22(b) and (c) show the calculated transmission and refractive index, respectively, as a function of the probe detuning, under the conditions of $T = 450$ nK, $N_0 = 3.3 \times 10^{12}$ atoms/cm^3, $L = 229\,\mu$m, and coupling laser intensity $I_c = 52$ mW/cm^2. Due to the very small Doppler broadening of the $|1\rangle \rightarrow |2\rangle$ transition in this ultracold medium, even a low coupling intensity can produce a probe transparency peak with a width much smaller than the natural linewidth of the $|1\rangle \rightarrow |3\rangle$ transition, and correspondingly leads to a very steep positive dispersion.

Figure 16.23 shows the experimental results of time delay between the reference pulse and the transmitted probe pulse, which indicates a time delay of $\Delta t = 7.05\,\mu$s and the group velocity of $v_g = 32.5$ m/s. Under optimum conditions, the measured lowest group velocity was 17 m/s.

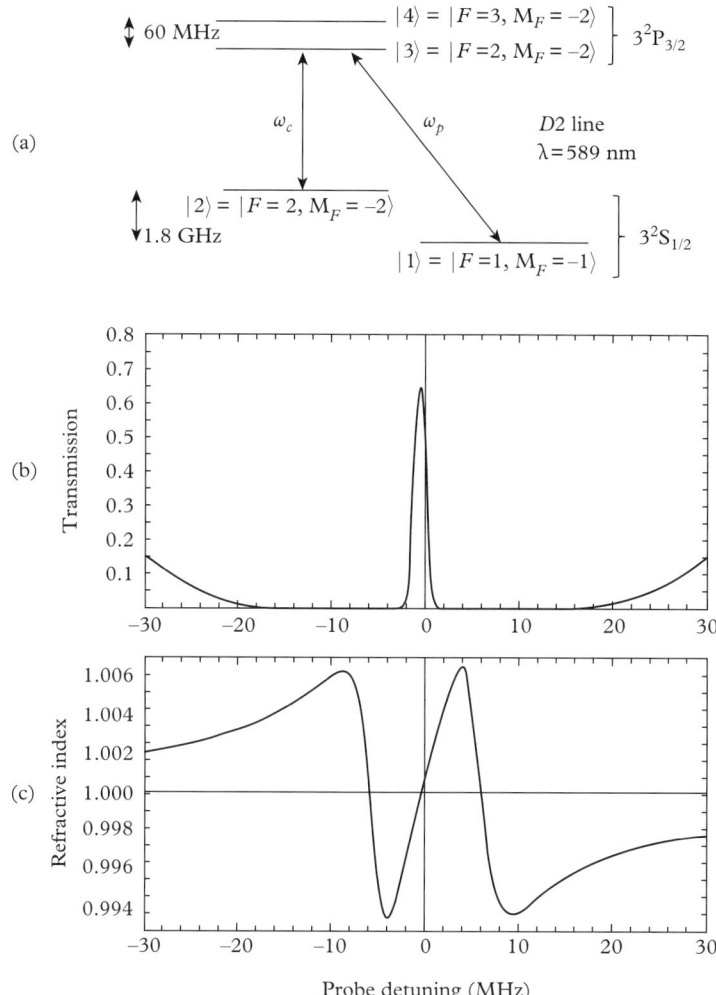

Figure 16.22 (a) Partial energy level diagram of Na atoms and the probe and coupling transitions for EIT. (b) Calculated probe transmission versus its detuning. (c) Calculated probe refractive index versus its detuning. The shift of transmission peak from the zero-detuning position is due to an ac Stark shift of level $|2\rangle$ caused by the coupling laser field. (Reproduced with permission from Hau et al.[34] Copyright 1999, Nature Publishing Group).

Based on the same EIT mechanism, Kash et al. reported a measured pulse delay of $\Delta t = 0.26$ ms and a group velocity of $v_g = 90$ m/s, in a 2.5-cm long hot (360 K) vapor sample of ^{87}Rb with added Ne buffer gas.[35]

Moreover, Budker et al. reported an even slower group velocity of $v_g = 8$ m/s in a ^{85}Rb vapor system working via a mechanism similar to EIT.[36]

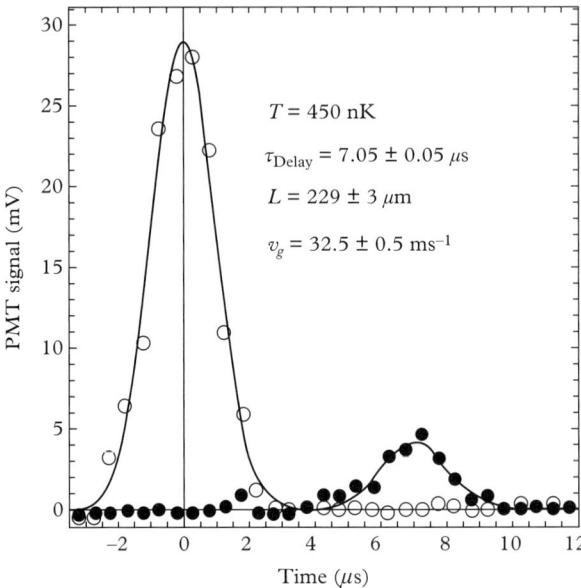

Figure 16.23 *Measured time delay for the probe pulse, which corresponds to a group velocity of $v_g = 32.5 \pm 0.5$ m/s. The open circles represent the reference probe pulse (recorded in the absence of Na atoms), the filled circles represent the transmitted probe pulse, and the solid lines indicate the Gaussian-fit curves. (Reproduced with permission from Hau et al.[34] Copyright 1999, Nature Publishing Group).*

16.5.2 Slow light based on absorption saturation or coherent population oscillations

Absorption saturation is a well-known nonlinear optical effect. It is commonly utilized for passive Q-switching and mode locking of lasing, as well as for Doppler-free saturation spectroscopy (cf. Section 9.2). Consider a monochromatic laser beam passing through an absorbing medium with a broader absorbing band; if the light frequency falls into the central region of medium's absorption band, the absorption of the incident beam becomes lower following the increase of the incident intensity. In other words, at appropriate incident intensity levels, a narrow dip or hole burning can be observed on the broad absorption profile of the medium; this narrow absorption reduction can be detected either by a weak probe beam of the same frequency or by the incident strong beam itself.[37,38] According to Kramers–Kronig relations, one may use this effect to realize slow light propagation.

If a strong laser beam induced absorption saturation is detected by a weak probe beam with a slightly different frequency, a theoretical model of coherent population oscillations was proposed for a simplified two-level system.[38] According to that model, the population in the ground state and in the excited state will undergo a periodic modulation at the beat frequency between the pump beam and the probe beam, and the absorption coefficient change detected by the probe beam can be expressed as[39]

$$\alpha(\delta) = \frac{\alpha_0}{1 + I_0}\left(1 - \frac{I_0(1 + I_0)}{(T_1\delta)^2 + (1 + I_0)^2}\right), \tag{16.5-3}$$

where α_0 is the unsaturated absorption coefficient, $I_0 = I_p/I_s$ (I_p the pump intensity and I_s the saturation intensity parameter of the medium), δ is the frequency difference between the pump and the probe beam, and T_1 is the ground state recovery time.

The first experimental demonstration of using the absorption saturation effect to implement slow light propagation was reported by Bigelow et al. in 2003.[39] The saturable absorbing medium was a ruby crystal working at room temperature, the driving beam was an argon-ion laser beam of 514.5 nm wavelength that was falling into the strong absorption band ($4A_2 \rightarrow 4F_2$ transition) of the ruby crystal. The input laser beam was first passed through an electro-optic modulator and then focused via an $f = 40$ cm lens onto a 7.25-cm long crystal sample; the modulator applied a 6% sinusoidal amplitude modulation on the input laser beam, so that the dc component of the modulated beam played the role of a strong pump beam to create the saturation of absorption, whereas the side-band frequency components played the role of a weak probe beam. The detuning between the pump beam and the modulated probe beam could be achieved by simply changing the modulation frequency.

Figure 16.24 shows the experimental data of time delay between a reference modulated signal and the transmitted signal passed through the ruby sample as a function of the modulation frequency, measured at two different input pump power levels. The solid lines represent the theoretical fit curves. From these data and curves, one could determine the spectral width of

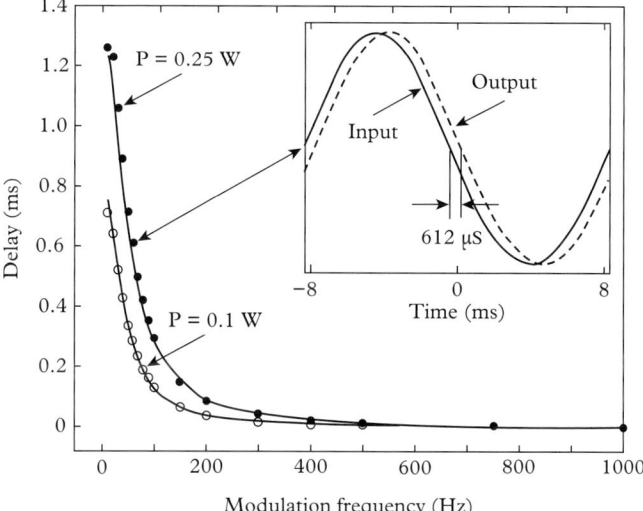

Figure 16.24 *Experimental data of time delay for the modulated probe signal as a function of modulation frequency, measured at input pump powers of 0.1 and 0.25 W, respectively. The inset shows the normalized 60-Hz input and output signal at input level of 0.25 W, indicating a signal delay of 612 μs and a group velocity of 118 m/s. (Reproduced with permission from Bigelow et al.[39] Copyright 2003, American Physical Society).*

saturation-induced absorption dip at a given pump level. As shown in the inset of this figure, with 60 Hz modulation frequency and at 0.25-W pump power level, the measured time delay is 612 μs that corresponds to an average group velocity of v_g = 118 m/s. The largest observed delay was 1.26 ms and the inferred group velocity was 57.5 m/s.[39] The authors also indicated that it is not necessary to have separate pump and probe beams; the intense pump pulse alone is able to create the saturation and to experience the slow light propagation.

In 2005, Baldit et al. reported the similar experiment based on the same mechanism but using Er^{3+}:Y_2SiO_5 crystal at $T = 1.5$ K as the saturable absorber.[40] The pump radiation was a cw fiber laser beam of 1536.1 nm wavelength corresponding to the strong absorption transition of Er^{3+} from $^4I_{15/2}$ (ground state) to $^4I_{13/2}$ (exited state). After passing through an acousto-optic modulator, the frequency unshifted pump beam and the frequency shifted probe components were focused into a 3-mm thick crystal sample. Under the conditions of pump intensity $I_p = 2.1$ mW/cm^2 and $I_p/I_s \approx 1/2$, where I_s was the saturation intensity parameter of the crystal sample, the measured delay time was $\Delta t = 1.1$ ms at the modulation frequency of 10 Hz, which corresponded to a group velocity of $v_g = 2.7$ m/s.

16.5.3 Light pulses halted (stored) in an EIT medium

In Section 16.5.1, we have discussed the slow light effect in an EIT medium under the steady-state condition, i.e., the strong coupling field is constant or has time duration much longer than the propagation time of the probe signal within the resonant medium. Now we further consider the dynamic behavior of interaction among the coupling field, probe field, and the resonant medium, when the coupling field can be controllably varied in time scale.

In 2001, Liu et al. accomplished an experiment to demonstrate how to halt a light pulse temporarily and then release it in an EIT medium[41]; the experimental conditions and the employed medium were basically the same as those described in Hau et al.[34] The medium is a 339-μm thick cloud of Na atoms cooled to 0.9 μK, working at 589.6 nm wavelength of the D_1 line. The energy-level structures and involved EIT transitions are depicted in Fig. 16.25(a). The cooled atoms are initially magnetically trapped in the state of $|3S, F = 1, M_F = -1\rangle$, the coupling beam is in resonance with $|3S, F = 2, M_F = +1\rangle \rightarrow |3P, F = 2, M_F = 0\rangle$ transition, while the probe laser is tuned to the transition from $|3S, F = 1, M_F = -1\rangle$ to $|3P, F = 2, M_F = 0\rangle$.

As shown in Fig. 16.25(b), under ordinary conditions of EIT with a stationary coupling field, the probe pulse is delayed by 11.8 μs corresponding to a group velocity of $v_g = 28$ m/s, and the time period for probe pulse propagating within the resonant medium is 6.3 μs. However, as shown in Fig. 16.25(c) and (d), when the coupling field is abruptly turned off at $t = 6.3$ μs and then turned back on at $t = 44.3$ μs and $t = 839.3$ μs, a retrieved probe pulse can be observed at the moment that the coupling field is turned back on. Within the time interval of turning off the coupling field there is no output probe signal, so that the probe pulse seems to be halted or stored inside the slow light medium. Then the probe pulse can be retrieved in a controllable way by again turning on the coupling field. It is found that the energy of the retrieved probe pulse is decreased with an increase of the time interval between the turning off and turning back on of the coupling field.

There is a theoretical model that can be employed to explain the experimental results mentioned above.[41,42] According to such a theory, under perfect EIT conditions, the atomic system is driven into a dark-state superposition of state $|1\rangle$ and state $|2\rangle$, and the probability of atoms being excited to state $|3\rangle$ approaches zero. When the coupling field is rapidly turned off while the probe pulse is contained within the medium, the probe energy is completely transferred into the

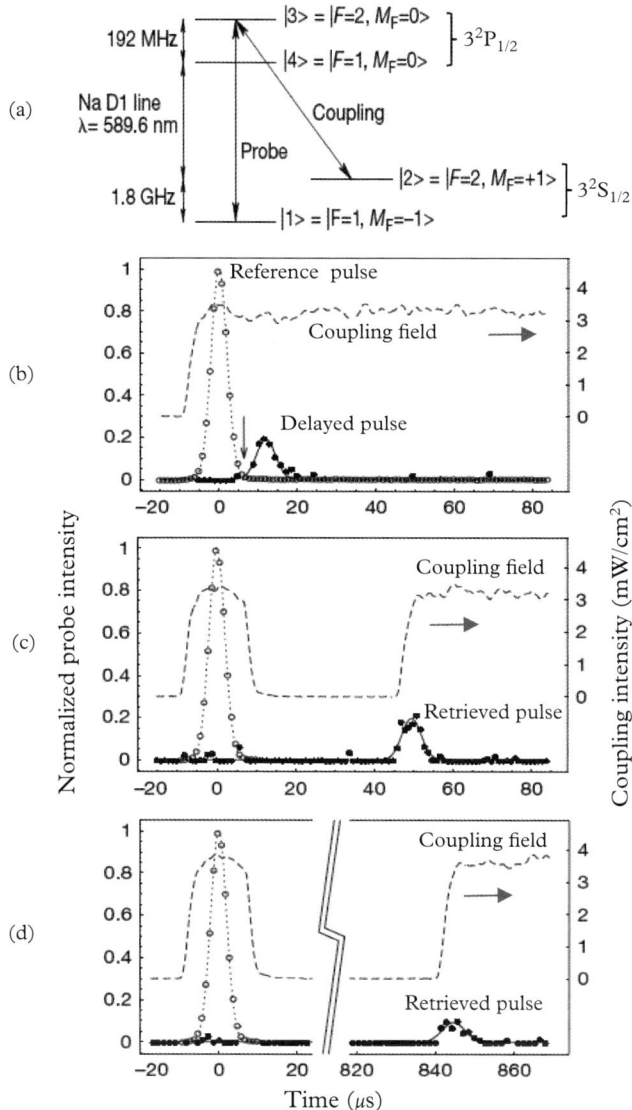

Figure 16.25 *(a) Energy-level structures and EIT involved transitions of Na atoms. (b) Probe pulse is delayed by 11.8 μs with a stationary coupling field, the arrow at 6.3 μs indicates the time when the probe pulse is contained completely within the atomic cloud. (c, d) Revival of the probe pulse after the coupling field is turned off at t = 6.3 μs and turned back on at t = 44.3 μs and t = 839.3 μs, respectively. (Reproduced with permission from Liu et al.[41] Copyright 2001, Nature Publishing Group).*

coupling field and atomic system via stimulated and resonance-enhanced (Stokes) Raman transition processes; a certain amount of atoms in state $|1\rangle$ has been coherently excited into state $|2\rangle$. The unique feature of this process is that the information (amplitude, duration, and phase) of the probe field is stored in the atomic system and can remain for a certain period of time; within this period once coupling field is turned on again, the process is reversed and a revived probe pulse can be regenerated through a stimulated (anti-Stokes) Raman process driven by the strong coupling field. In the latter case, the regenerated probe energy is from both the coupling field and the pre-excited atomic system. The intensity decrease of the retrieved probe pulse with the increasing storage time is due to various decoherence processes of the coherently excited atomic system. The effective pulse storage time measured in the experiment described above could be up to 0.9 ms.

Almost at the same time, Phillips et al. reported their independent experimental result showing laser pulse storage in the ^{87}Rb vapor system, working at 70–90 °C with D_1-line resonance of ~795 nm wavelength.[43] The probe pulse duration was 10–30 μs, coupling beam power ~1 mW, probe beam power ~0.1 mW; the stationary EIT-induced ultraslow group velocity was $v_g = 1$ km/s, and the effective storage intervals could be up to 0.5 ms.

In 2012, Turukhin et al. reported their experimental demonstration of pulse storage in a Pr:Y$_2$SO$_5$ crystal, working at 5 K with ~606-nm transitions.[44] Under stationary EIT conditions, for a 3-mm thick sample the measured delay time could be 65 μs corresponding to $v_g = 45$ m/s, and this delay time can be varied by changing the intensity of the coupling beam. It was indicated that the pulse storage time could be up to ~0.5 ms that would be determined by the spin homogenous lifetime of the involved ground-state sublevels.

16.5.4   Slow light effect in a Raman gain medium

As shown in Figs. 16.4 and 16.8(b), optical pulses with a frequency within the gain line of a resonant medium will experience the influence of slow light propagation.

The first experimental demonstration of using stimulated Raman scattering gain for observing a slow light effect was reported by Lee and Lawandy in 2001.[45] The Raman gain medium was a 5-cm long Ba(NO$_3$) crystal working at 300 K, with a Raman shift frequency of 1048.6 cm^{-1} and gain linewidth of 0.4 cm^{-1}.

As shown in Fig. 17.26 (a), the pump beam was 1064-nm and 7-ns laser pulses from a Q-switched Nd:YAG laser; the signal beam was 1197-nm and 90-ps pulses generated in another Ba(NO$_3$) Raman crystal pumped by a Q-switched and mode-locked Nd:YAG laser.

The pump and signal beams were co-propagating along the forward direction inside the crystal sample, providing a simulated Raman scattering gain coefficient of 23 cm/GW. A KTP crystal was employed to measure the signal pulse delay by the using non-collinear SHG method. As can be seen in Fig. 16.26(b), the pumped gain signal pulse shows a time delay of 57 ps with respect to the correlation curve without pump action.

In 2005, Sharping et al. reported their measurement of slow light propagation by using a 1-km long optical fiber as the Raman gain medium.[46] The pump beam was 1535-nm and 500-ps pulses from a diode laser, while the signal beam was 430-fs laser pulses with wavelength tunable around the range from 1590 nm to 1643 nm. The Raman gain peak was located near to the 1640 nm wavelength position, and the maximum signal gain was ~35 dB for a peak pump power of 2.6 W. The achieved maximum signal pulse delay was 370 fs.

In above-mentioned two experiments, the time delay of the signal pulses is mainly limited by the spectral linewidth of Raman transition.

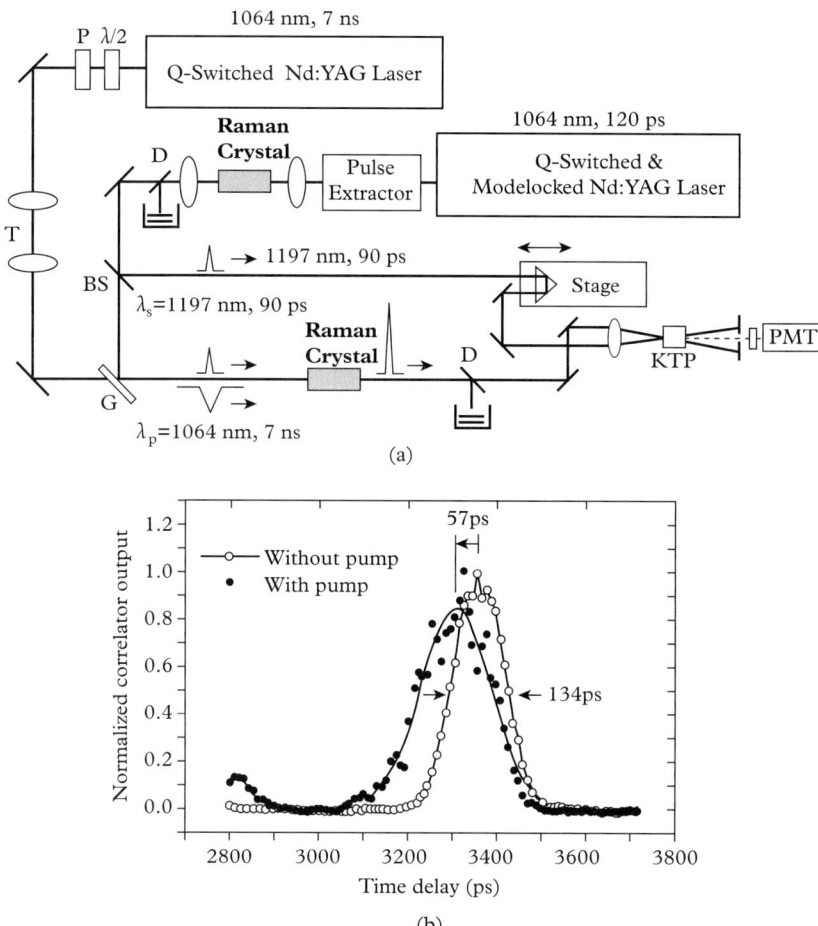

Figure 16.26 (a) Optical layout for observing the slow light effect in a Raman gain crystal; a KTP crystal is utilized to generating non-collinear SHG and to measure the pulse width and relative delay. (b) Measured correlation curves as a function of the relative scan time delay for SHG under the condition with and without pump action on the Raman crystal. (Reproduced with permission from Lee and Lawandy.[45] Copyright 2001, AIP Publishing LLC).

Furthermore, Pati et al. accomplished an experiment in which a 10-cm long rubidium vapor at 70 °C was employed as a Raman gain medium pumped with a relatively strong (20 mW) laser beam; another weak signal beam was co-propagating with the pump beam to obtain a Raman gain.[47] The experiment showed that the signal pulses experienced a slow light propagation due to the gain effect, while the pump pulses experienced a fast light propagation due to an attenuation effect.

16.5.5 Slow light effect in a Brillouin gain medium

The first demonstration of the slow light effect in a Brillouin gain optical fiber system was reported by Song et al. in 2005.[48] Their experimental setup, which is shown in Fig. 16.27, is quite similar to that shown in Fig. 16.13.

A distributed feed-back (DFB) laser diode provided a 1552-nm cw laser beam, which was launched to an electro-optic modulator (EOM) to generate two sidebands with a frequency difference equal to the Brillouin frequency of a tested fiber sample. To observe slow light effect, the low-frequency sideband was used as a probe beam while the high-frequency sideband was employed as a pump beam. The unshifted component of the carrier wave was suppressed by controlling the dc bias voltage delivered into the EOM with a feedback circuit. After passing through EOM the laser beam containing two sideband-components was sent to a fiber Bragg grating, which reflected the low-frequency component to pass through an optical gate to form a pulsed probe beam forward entering a long fiber sample, while the high-frequency component transmitted through the Bragg grating was amplified via an EDFA and then sent to the same fiber sample as a backward pump beam. Two samples were tested: one was a 11.8-km long standard fiber with $v_B \approx 10.844$ GHz, and the other was a 6.7-km long dispersion-shifted fiber (DSF) of $v_B \approx 10.42$ GHz. The Brillouin gain linewidth was estimated to be $\Delta v_B \approx 30 - 40$ GHz and the duration of the signal pulse was ~ 100 ns.

Figure 16.28(a) shows the measured time delay (slow light effect) of the signal pulse passed through the standard fiber sample at different Brillouin gain levels. By exchanging the roles of the two spectral components, i.e., using the high-frequency component as a weak signal beam and the low-frequency component as a strong pump beam, the time advance (fast light effect) of the signal pulse could also be observed due to Brillouin attenuation mechanism.[49] Figure 16.28(b) shows the measured delay time of the signal pulse as a function the gain (or attenuation) levels of the two tested fiber samples. In both cases the measured time delay (or advance) of the weak signal pulse is approximately proportional to the gain (or attenuation) in dB units.

In the mean time, a similar slow-light experiment using optical fiber as Brillouin gain medium was implemented independently by Okawachi et al.[50] The feature of their setup was that two

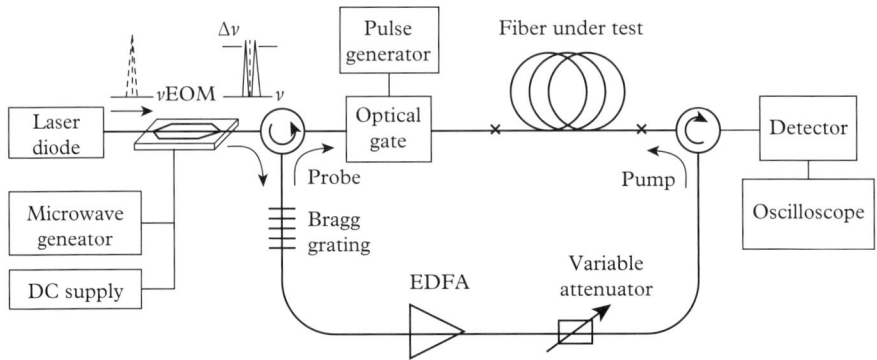

EOM: electro-optic modulator; EDFA: erbium-doped fiber amplifier.

Figure 16.27 *Experimental setup for observing the slow light effect in a Brillouin gain fiber system. (Reproduced with permission from Song et al.[48] Copyright 2005, Optical Society of America).*

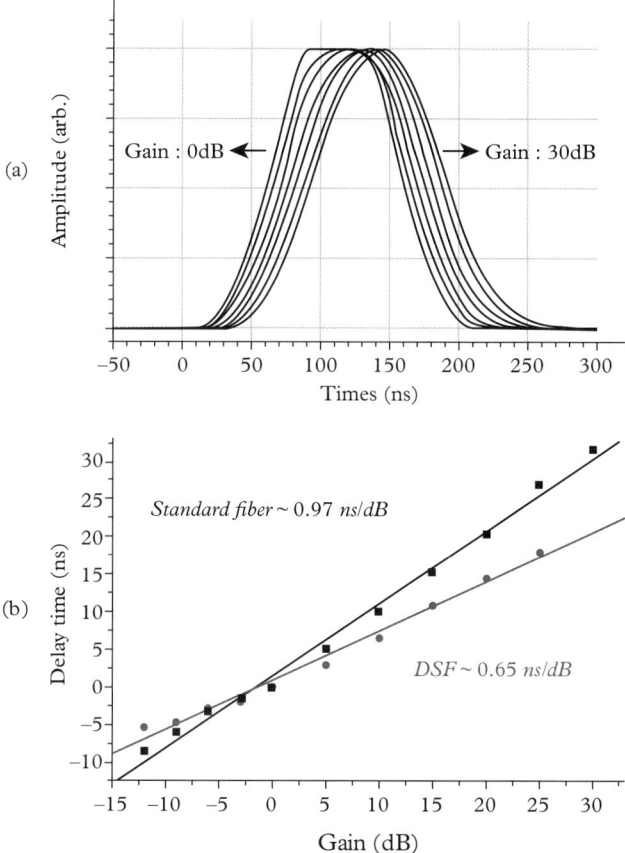

Figure 16.28 *(a) Traces of the probe pulses for different Brillouin gains (standard fiber), showing a clear delay due to the modified group velocity. (b) Delay time of the probe pulse as a function of the Brillouin gain; in a gain situation the pulse is delayed while it is accelerated in a loss configuration. (Reproduced with permission from Song et al.[48] Copyright 2005, Optical Society of America).*

separate sections of fiber (SMF-28e) with different lengths were employed: a 1-km long fiber pumped by a 1550-nm laser was used for Stokes-shifted SBS generation; another 500-m long fiber was used as a slow-light medium. Both the Stokes-shifted SBS beam (signal) and the 1550-nm laser beam (pump) were sent into the second fiber section. After modulation, the signal pulse duration could be 15 ns or 63 ns. The measured time-delay data versus Brillouin gain parameter G of the second fiber are shown in Fig. 16.29, where $G = g_b I_p L$ (g_b = gain factor, I_p = pump intensity, L = fiber length). Once again, this figure shows a linear relationship between the signal time delay and the Brillouin gain.

From Fig. 16.29 one can find another fact that under the same pump intensity the time delay is different for the signal pulses with different pulse duration. Since a pulse-duration change may

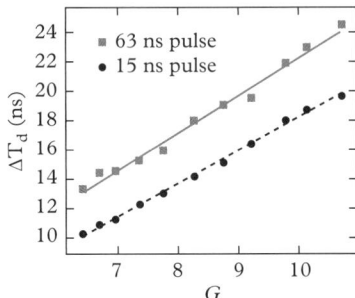

Figure 16.29 *Signal pulse delays as a function of the Brillouin gain parameter G for 63 ns long (square) and 15 ns long (circle) input signal (Stokes) pulses. (Reproduced with permission from Okawachi et al.[50] Copyright 2005, American Physical Society).*

lead to a change of the corresponding spectral linewidth of the signal beam, the above-mentioned fact reveals the influence of spectral linewidth of the signal pulse on the slow light propagation in a Brillouin gain medium. When the signal spectral linewidth is comparable or larger than the Brillouin gain width, the effective gain and slow light effect may be reduced. For this reason some techniques are reported for broadening the Brillouin gain width of fiber medium, such as to modulate the pump laser beam or to employ a pump laser source with a reasonable bandwidth.[51–53] The another benefit provided by a broader gain width is to reduce the distortion of the output signal pulse shape after passing through a Brillouin fiber amplifier.

16.5.6 Slow/fast light effects in a semiconductor absorber/amplifier or a fiber amplifier

As previously indicated in Fig. 16.8, a strong laser beam may induce a narrow dip (hole burning) in the broad absorption curve of a saturable absorbing medium, that narrow absorption hole is equivalent to a narrow "gain" spectral peak, which can be detected by a weak probe beam working at that narrow spectral region and experiencing a slow light effect. In an opposite case, if there is a saturable gain medium with a broader gain band, a strong laser beam may induce a narrow gain dip on the spectral gain curve, such a gain dip can be seen as a narrow "attenuation" spectral peak by a weak probe beam, and renders the latter a fast light propagation.

Semiconductor waveguide devices can be used as either a saturable absorber for slow light propagation or a saturable amplifier for fast light propagation, depending on their working conditions. Comparing with other resonant media aforementioned in this chapter, semiconductor devices exhibit obvious advantages of compactness, direct current control, large bandwidth, and easy integration with electrical/optical circuits. The disadvantage of semiconductor waveguide devices for slow/fast light application is that the propagation length is relatively short.

The earliest experiment for demonstrating the slow light effect in a semiconductor saturable absorber was reported by Ku et al. in 2004.[54] They used a GsAs/AlGaAs quantum well sample working at 10 K; the strong saturation beam was from a Ti:sapphire laser with a variable intensity of 0.01–1.5 kW/cm^2, while the weak probe beam was from a tunable diode laser with an intensity of 1.5 W/cm^2. The saturation (control) beam-induced absorption dip detected by the probe beam as a function of the frequency detuning between these two beams, which is shown in Fig. 16.30(a), was measured at different saturation beam intensity levels. Figure 16.30(b) shows the measured phase and absorption changes of the probe beam as a function of the detuning, with a 1 kW/cm^2 control intensity and 0.09 kW/cm^2 probe intensity. Note that the slope of the phase change curve determines the group velocity of the probe beam. The measured group velocity was 9600 m/s, and the induced absorption dip width could be as large as 2 GHz.

Later on, other similar slow light experiments were implemented in semiconductor saturable absorbers working at room temperature.[55,56] As an example, Mørk et al. utilized a commercial InGaAsP bulk p–i–n electro-absorption modulator of 100 μm thickness as the saturable absorber, the absorption behavior of which was controllable by a reverse voltage.[55] A single input 1550-nm laser beam, modulated at a frequency of 16.7 GHz, was used to saturate the absorption and also to achieve slow light propagation. Figure 16.31 shows the optical phase shift (time delay) curves as a function of the applied reverse voltage, measured at different input optical power levels (in units of dBm). The minimum group velocity observed was 3.1 times lower than that for an unsaturated sample.

On the other hand, a saturable semiconductor optical amplifier (SOA) can be employed to realize fast light propagation; its gain behavior can be optically and electrically controlled.[57,58] In an experiment performed by Su and Chuang, the fast light device was an InAs/GaAs quantum dot

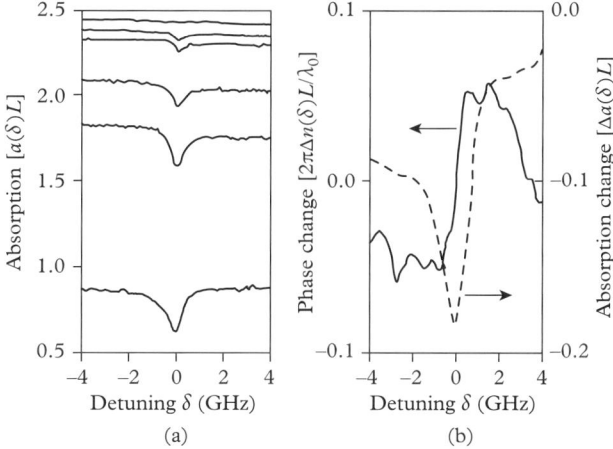

Figure 16.30 *(a) Absorption spectra as a function of detuning δ, obtained at different control (saturation) beam's intensities (from top to bottom): I_c = 0.01, 0.05, 0.1, 0.5, 1.5 kW/cm^2. The probe beam's intensity is 1.5 W/cm^2, δ is the frequency difference between these two beams. (b) Phase and absorption changes experienced by the probe beam as a function of detuning, the control and probe intensities are 1 and 0.09 kW/cm^2, respectively, and the slope of the phase change curve gives the group velocity. (Reproduced with permission from Ku et al.[54] Copyright 2004, Optical Society of America).*

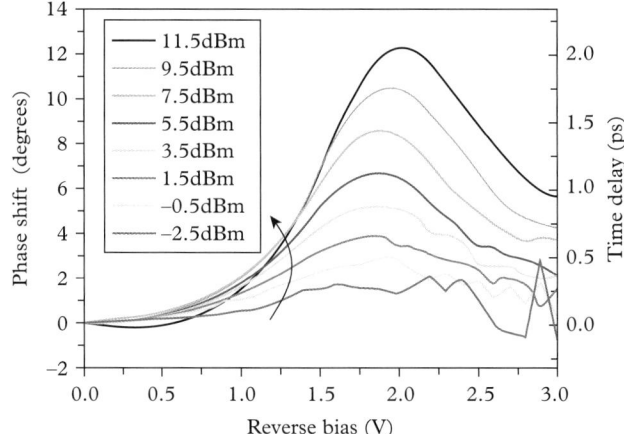

Figure 16.31 *Phase shift and time delay curves versus reverse voltage for the 16.7 GHz-modulated optical signal, measured at different input power levels: from −2.5 dBm (bottom) up to 11.5 dBm (top). (Reproduced with permission from Mørk et al.[55] Copyright 2005, Optical Society of America).*

SOA working at room temperature with an injection current of 300 mA and waveguide length of ~2 mm.[57] A strong saturation beam of 1290 nm wavelength and a weak tunable probe beam modulated at 130 MHz were launched into the OSA from two opposite directions; the spectral transmission (gain) and phase delay curves of the probe beam as a function of the frequency detuning between these two beams were measured at different input power levels of the saturation beam. The results clearly indicated the gain saturation of the SOA and associated fast light propagation (phase advancement).

In a later experiment, the fast light effect was tested in an InGaAsP/InP quantum well SOA of high gain capability; with a single input 1.3-µm laser beam modulated at 1 GHz, the measured time advancement could vary from ~0 to ~70 ps by changing the injection current from 100 to 400 mA.[58]

Another advantage of using SOA devices for fast/slow light purpose is that through controlling injection current levels these devices can work as an absorber or an amplifier, and therefore ensure an electrically switching from slow to fast light propagation.

To demonstrate such a feasibility, Pesala et al. implemented an experiment based on two concatenated quantum-well SOAs with small signal gain of 28 dB.[59] In their experimental setup, which is shown schematically in Fig. 16.32(a), a 600-fs pulsed laser beam is split into two branches: one is used as a reference for performing cross-correlation measurements; the other is passed through two SOAs in series. The current of both SOAs is controlled independently. At low injection current levels the SOAs behave as saturable absorbers to allow slow light propagation, whereas at high injection current levels they behave as gain saturable amplifiers to allow fast light propagation. Figure 16.32(b) shows the cross-correlation traces indicating the pulse advance/delay measured at different current levels (in mA). With a current of ~100 mA for both SOAs, the maximum tuning range of the time shift is ~2 ps, which corresponds to a time shift of 3.3 pulses.

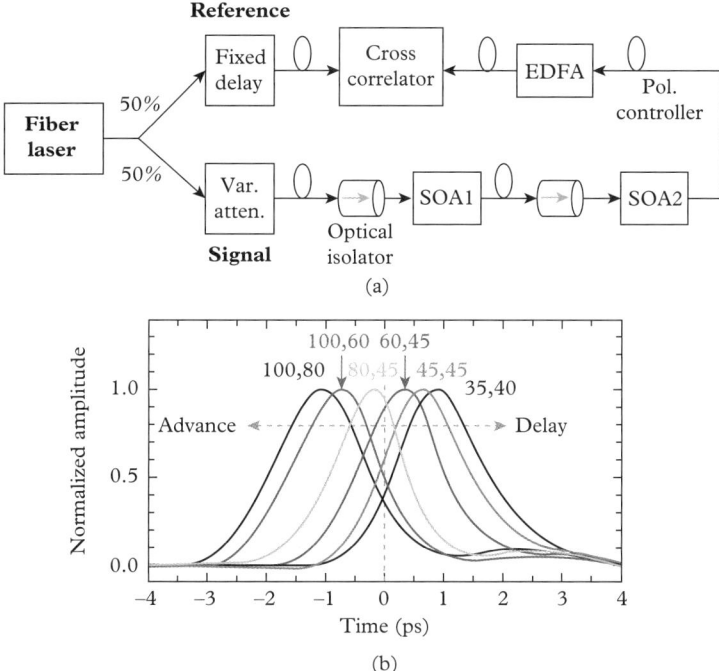

Figure 16.32 *(a) Experimental layout to realize fast/slow light propagation of 600-fs pulses passing through two concatenated SOAs. The cross correlator is adopted to measure the time shift with respect to the reference pulse. (b) Normalized cross-correlation traces showing the time advance/delay of the signal pulses. The numbers indicated are the currents (in mA) applied to each of the amplifiers. (Reproduced with permission from Pesala et al.[59] Copyright 2007, Optical Society of America).*

In another experiment using a device consisting of two concatenated semiconductor electro-absorbers working with a 1555-nm laser beam modulated at 10 GHz, the measured time delay could reach 14 ps.[60] Additionally, by using a 2-mm long InAs/InP quantum dash SOA device, working with a 1.55-μm single modulated laser beam of −17 dBm power, Martinez et al. reported a maximum optical time shift of ∼136 ps at 250 MHz (modulation frequency) and of ∼55 ps at 2 GHz by means of control of the injection current.[61]

The same physical mechanisms described above are also applicable to other types of saturable absorber/amplifier systems for slow/fast light propagation. Schweinsberg et al. utilized a 13-m long Er-doped fiber (EDF) as an optical saturable absorber (without pump) or an amplifier (with pump) to demonstrate slow and fast light effects.[62] Figure 16.33(a) shows the measured optical delay for a signal pulse of 1550 nm wavelength and 3.2 ms duration (FWHM) after passing through the un-pumped EDF sample; the input peak power of the signal beam is 0.8 mW. Figure 16.33(b) shows the optical advance of the signal pulse with 1.8-ms duration after passing through the same EDF section pumped with a 980-nm cw laser beam of 12 mW.

Detailed studies of slow/fast light in EDF absorber/amplifier systems under various beam-modulation conditions have also been reported.[63–65]

556 *Fast and Slow Light*

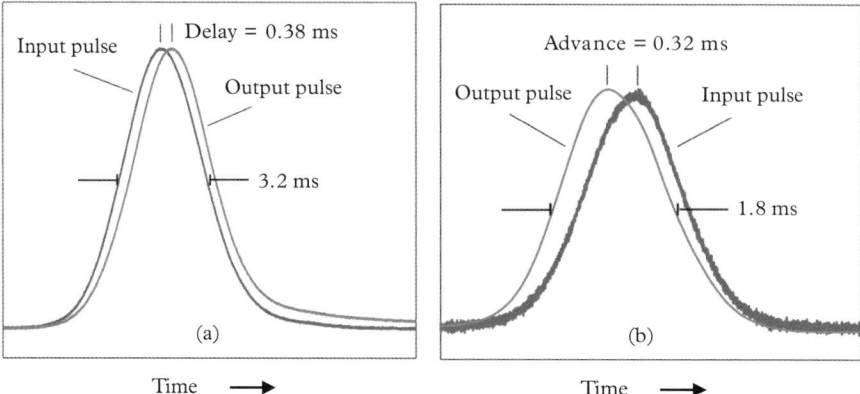

Figure 16.33 *(a) Slow light propagation of the 1550-nm and 3.2-ms pulse in an un-pumped Er-doped fiber of 13 m length; input peak power was 0.8 mW. (b) Fast light propagation of the 1.8-ms signal pulse in an Er-doped fiber amplifier pumped with a 980-nm and 12-mW cw laser. (Reproduced with permission from Schweinsberg et al.[62] Copyright 2006, IOP Publishing).*

REFERENCES

1. L. Brillouin, *Wave Propagation and Group Velocity* (Academic Press, New York, 1960).
2. R. W. Boyd and D. J. Gauthier, in *Progress in Optics*, Vol. 43, edited by E. Wolf (North-Holland, Amsterdam, 2002), pp.497–530.
3. R. Y. Chiao and P. W. Milonni, *Opt. Photonics News* **13**, 26 (2002).
4. P. W. Milonni, *J. Phys. B* **35**, R31 (2002); P. W. Milonni, *Fast Light, Slow Light, Left-Handed Light* (Institute of Physics, Bristol, UK, 2005).
5. L. Thevenaz, *Nat. Photonics* **2**, 474 (2008).
6. U. Bortolozzo, S. Residori, and J.-P. Huignard, *Laser Photonics Rev.* **4**, 483 (2010).
7. R. L. Smith, *Am. J. Phys.* **38**, 978 (1970).
8. M. Ware, S. Glasgow, and J. Peatross, *Opt. Express* **9**, 506 (2001); M. Ware, S. Glasgow, and J. Peatross, *Opt. Express* **9**, 519 (2001).
9. A. Kuzmich, A. Dogariu, L. J. Wang, P. W. Milonni, and R. Y. Chiao, *Phys. Rev. Lett.* **86**, 3925 (2001).
10. M. D. Stenner, D. J. Gauthier, and M. A. Neifeld, *Nature* **425**, 695 (2003).
11. G. Nimtz, *Nature* **429**, 40 (2004).
12. C. G. B. Garrett and D. E. McCumber, *Phys. Rev. A* **1**, 305 (1970).
13. F. R. Faxvog, C. N. Y. Chow, T. Bieber, and J. A. Carruthers, *Appl. Phys. Lett.* **17**, 192 (1970); L. Casperson and A. Yariv, *Phys. Rev. Lett.* **26**, 293 (1971).
14. S. Chu and S. Wong, *Phys. Rev. Lett.* **48**, 738 (1982).
15. A. Katz and R. R. Alfano, *Phys. Rev. Lett.* **49**, 1292 (1982); S. Chu and S. Wong, *Phys. Rev. Lett.* **49**, 1293 (1982).
16. W. G. A. Brown, R. McLean, A. Sidorov, P. Hannaford, and A. Akulshin, *J. Opt. Soc. Am. B* **25**, c82 (2008).
17. A. M. Steinberg and R. Y. Chiao, *Phys. Rev. A* **49**, 2071 (1994).

18. R. Y. Chiao, *Phys. Rev. A* **48**, R34 (1993); E. L. Bolda, J. C. Garrison, and R. Y. Chiao, *Phys. Rev. A* **49**, 2938 (1994).
19. L. J. Wang, A. Kuzmich, and A. Dogariu, *Nature* **406**, 277 (2000).
20. K. Y. Song, M. G. Herraes, and L. Thevenaz, *Opt. Express* **13**, 9758 (2005).
21. A. M. Akulshin, S. Barreiro, and A. Lezama, *Phys. Rev. A* **57**, 2996 (1998).
22. A. M. Akulshin, S. Barreiro, and A. Lezama, *Phys. Rev. Lett.* **83**, 4277 (1999).
23. M. S. Malcuit, R. W. Boyd, L. W. Hillman, J. Krashinski, and C. R. Stroud, Jr., *J. Opt. Soc. Am. B* **1**, 73 (1984).
24. M. S. Bigelow, N. N. Lepeshkin, and R. W. Boyd, *Science* **301**, 200 (2003).
25. E. E. Mikhailov, V. A. Sautenkov, Y. V. Rostovtsev, and G. R. Welch, *J. Opt. Soc. Am. B* **21**, 425 (2004).
26. G. M. Gehring, A. Schweinsberg, C. Barsi, N. Kostinski, and R. W. Boyd, *Science* **312**, 895 (2006).
27. B. D. Clader, Q.-H. Park, and J. H. Eberly, *Opt. Lett.* **31**, 2921 (2006).
28. S. E. Harris, J. E. Field, and A. Imanoglu, *Phys. Rev. Lett.* **64**, 1107 (1990).
29. K.-J. Boller, A. Imamoglu, and S. E. Harris, *Phys. Rev. Lett.* **66**, 2593 (1991).
30. S. E. Harris, *Phys. Today* **50**(7), 36 (1997).
31. S. E. Harris, J. E. Field, and A. Kasapi, *Phys. Rev. A* **46**, R29 (1992).
32. A. Kasapi, M. Jain, and S. E. Harris, *Phys. Rev. Lett.* **74**, 2447 (1995).
33. M. Xiao, Y.-Q. Li, S.-Z, Jin, and J. Gea-Banacloche, *Phys. Rev. Lett.* **74**, 666 (1995).
34. L. V. Hau, S. E. Harris, Z. Dutton, and C. H. Behroozi, *Nature* **397**, 594 (1999).
35. M. M. Kash, V. A. Sautenkov, A. S. Zibrov, L. Hollberg, G. R. Weich, M. D. Lukin, Y Rostovtsev, E. S. Fry, and M. O. Scully, *Phys. Rev. Lett.* **82**, 5229 (1999).
36. D. Budker, D. F. Kimball, S. M. Rochester, and V. V. Yashchuk, *Phys. Rev. Lett.* **83**, 1767 (1999).
37. S. E. Schwarz and T. Y. Tan, *Appl. Phys. Lett.* **10**, 4 (1967).
38. R. W. Boyd, M. G. Raymer, P. Narum, and D. J. Harter, *Phys. Rev. A* **24**, 411 (1981); L. W. Hillman, R. W. Boyd, J. Krasinski, and C. R. Stroud, Jr., *Opt. Commun.* **45**, 416 (1983).
39. M. S. Bigelow, N. N. Lepeshkin, and R. W. Boyd, *Phys. Rev. Lett.* **90**, 113903 (2003).
40. E. Baldit, K. Bencheikh, P. Monnier, J. A. Levenson, and V. Rouget, *Phys. Rev. Lett.* **95**, 143601 (2005).
41. C. Liu, Z. Dutton, C. H. Behroozi, and L. V. Hau, *Nature* **409**, 490 (2001).
42. M. Fleischhauer and M. D. Lukin, *Phys. Rev. Lett.* **84**, 5094 (2000).
43. D. F. Phillips, A. Fleischhauer, A. Mair, R. L. Walsworth, and M. D. Lukin, *Phys. Rev. Lett.* **86**, 783 (2001).
44. A. V. Turukhin, V. S. Sudarshanam, M. S. Shahriar, J. A. Musser, B. S. Ham, and P. R. Hemmer, *Phys. Rev. Lett.* **88**, 023602 (2002).
45. K. Lee and N. M. Lawandy, *Appl. Phys. Lett.* **78**, 703 (2001).
46. J. E. Sharping, Y. Okawachi, and A. L. Gaeta, *Opt. Express* **13**, 6092 (2005).
47. G. S. Pati, M. Salit, K. Salit, and M. S. Shahriar, *Opt. Express* **17**, 8775 (2009).
48. K. Y. Song, M. G. Herraez, and L. Thevenaz, *Opt. Express* **13**, 82 (2005).
49. M. Gonzales-Herraez, K.-Y. Song, and L. Thevenaz, *Appl. Phys. Lett.* **87**, 081113 (2005).
50. Y. Okawachi, M. S. Bigelow, J. E. Sharping, Z. Zhu, A. Schweinsberg, D. J. Gauthier, R. W. Boyd, and A. L. Gaeta, *Phys. Rev. Lett.* **94**, 153902 (2005).
51. M. G. Herraez, K. Y. Song, and L. Thevenaz, *Opt. Express* **14**, 1395 (2006).
52. A. Minardo, R. Bernini, and L. Zeni, *Opt. Express* **14**, 5866 (2006).
53. Z. Zhu, A. M. C. Dawes, D. J. Gauthier, L. Zhang, and A. E. Willner, *J. Lightwave Technol.* **25**, 201 (2007).
54. P.-C. Ku, F. Sedgwick, C. J. Chang-Hasnain, P. Palinginis, T. Li, H. Wang, S.-W. Chang, and S.-L. Chuang, *Opt. Lett.* **29**, 2291 (2004).
55. J. Mørk, R. Kjær, M. van der Poel, and K. Yvind, *Opt. Express* **13**, 8136 (2005).
56. H. Su and S. L. Chuang, *Opt. Lett.* **31**, 271 (2006).
57. H. Su and S. L. Chuang, *Appl. Phys. Lett.* **88**, 061102 (2006).

58. H. Su, P. Kondratko, and S. L. Chuang, *Opt. Express* **14**, 4800 (2006).
59. B. Pesala, F. Sedgwick, A. V. Uskov, and C. Chang-Hasnain, *Opt. Express* **15**, 15863 (2007).
60. F. Ohman, K. Yvind, and J. Moerk, *Opt. Express* **14**, 9955 (2006).
61. A. Martinez, G. Aubin, F. Lelarge, R. Brenot, J. Landreau, and A. Ramdane, *Appl. Phys. Lett.* **93**, 091116 (2008).
62. A. Schweinsberg, N. N. Lepeshkin, M. S. Bigelow, R. W. Boyd, and S. Jarabo, *Europhys. Lett.* **73**, 218 (2006).
63. S. Stepanov and M. Plata Sanchez, *Phys. Rev. A* **80**, 053830 (2009).
64. S. Melle, O. G. Calderon, and M. Moreno, *J. Phys. B* **43**, 215401 (2010).
65. K. Qian, L. Zhan, L. Zhang, Z. Q. Zhu, J. S. Peng, Z. C. Gu, X. Hu, S. Y. Luo, and Y. X. Xia, *Opt. Lett.* **36**, 2185 (2011).

17
Terahertz Nonlinear Photonics

Terahertz (THz) radiation, which lies between the microwave and far-infrared regions in the electromagnetic spectrum, is loosely defined as the spectral range from 0.1 to 10 THz, which corresponds to the wavelength range from 3 mm to 30 μm. It is inaccessible by either conventional electronics or photonics, leading to the term "THz gap." In the recent decades, a variety of techniques has been developed to generate and detect THz radiation, and this THz gap has begun to be filled. Recently, THz technology has undergone tremendous growth with strong interests for spectroscopy, remote sensing, imaging, and optoelectronics.[1-7]

Nonlinear optical techniques have played important roles in the development of THz technology. For example, optical rectification (or difference-frequency generation) has become one of the main mechanisms for the generation of broadband single-cycle or near single-cycle THz pulses; electro-optic (EO) sampling based on the Pockels effect has been widely used to detect THz waveforms. In recent years, with advances in ultrafast lasers and nonlinear optical techniques, it is now possible to produce intense single-cycle THz pulses with peak electric fields exceeding 10^7 V/m. The strong THz electric field associated with these ultrashort electromagnetic pulses provides unique opportunities, not only to investigate new nonlinear optical effects in the THz spectral range but also to probe novel ultrafast nonlinear phenomena in the single-cycle regime. In this chapter, we shall primarily describe how to generate and measure THz pulses using nonlinear optical methods, and then we will discuss how to utilize the intense THz pulses to study nonlinear optical effects.

17.1 THz generation via optical rectification

17.1.1 Principle of generating THz radiation in a second-order nonlinear crystal

THz radiation is typically generated via optical rectification and/or difference-frequency generation of intense laser pulses in a second-order nonlinear optical medium. Before starting the major discussion of this section, it is worth reviewing the principle of optical rectification and difference-frequency generation, which are based on the second-order nonlinear frequency mixing described in Chapter 3. Suppose there is a monochromatic coherent light of frequency ω passing through a second-order nonlinear crystal that lacks centrosymmetry; a dc-field may be induced in the crystal through the following second-order nonlinear polarization process:

$$\mathbf{P}^{(2)}(\omega' = \omega - \omega = 0) = \varepsilon_0 \chi^{(2)}(\omega, -\omega)\mathbf{E}(\omega)\mathbf{E}^*(\omega). \qquad (17.1\text{-}1)$$

Here, $\mathbf{E}(\omega)$ is the electric field (in complex form) of the input light, $\chi^{(2)}(\omega,-\omega)$ is the second-order nonlinear susceptibility describing the optical rectification effect, and $\mathbf{P}^{(2)}(\omega' = 0)$ is the medium's nonlinear polarization responsible for the generation of the corresponding dc-field $\mathbf{E}(\omega' = 0)$. Furthermore, if there are two monochromatic laser waves with frequencies ω_1 and ω_2 passing through the same second-order nonlinear medium, the nonlinear polarization component

$$\mathbf{P}^{(2)}(\omega' = \omega_1 - \omega_2) = \varepsilon_0 \chi^{(2)}(\omega_1,-\omega_2)\mathbf{E}(\omega_1)\mathbf{E}^*(\omega_2) \qquad (17.1\text{-}2)$$

may lead to the generation of a difference-frequency field $\mathbf{E}(\omega_1 - \omega_2)$. Here $\omega_1 > \omega_2$ and $\chi^{(2)}(\omega_1,-\omega_2)$ is the medium's second-order susceptibility tensor describing the optical difference-frequency generation effect.

In practice, a femtosecond laser pulse is mostly utilized to generate THz emission from the second-order nonlinear medium. In this case, the spectrum of the laser pulse exhibits a broader bandwidth owing to the uncertainty principle. For the pulse-duration range 150–20 fs, the corresponding spectral bandwidth is about 7–50 THz. The temporal waveform and the corresponding spectrum of an ultrashort laser pulse are shown schematically in Fig. 17.1(a).

One may realize that upon the action of such a broadband intense laser field, a dc or low-frequency electromagnetic radiation could be generated through the interaction between any two frequency components within the spectral bandwidth and with very small frequency difference, while a high frequency (in THz range) electromagnetic radiation could be simultaneously created through the interaction between any two frequency components within the spectral bandwidth of the input laser pulse but with a greater frequency difference. The overall result is the generation of pulsed THz radiation with a spectrum ranging from low frequencies to THz frequencies. It is

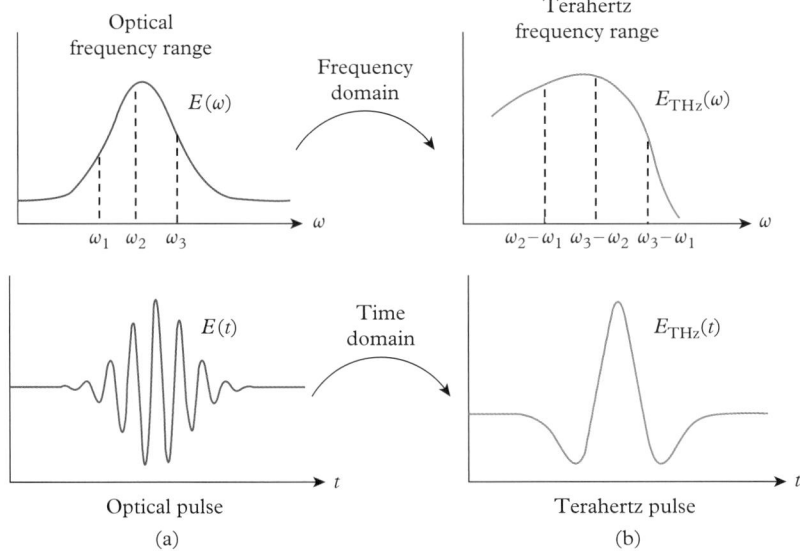

Figure 17.1 *An illustration of THz generation via optical rectification. Any frequency pairs, such as (ω_1, ω_2), (ω_1, ω_3), and (ω_2, ω_3) within the spectrum of a laser pulse can generate induced electric polarization components at corresponding difference frequencies, i.e., ($\omega_2 - \omega_1$), ($\omega_3 - \omega_1$), and ($\omega_3 - \omega_2$), respectively.*

obvious that the highest obtainable THz frequency is limited by the spectral linewidth of the input laser pulse. The temporal waveform and corresponding spectrum of the generated THz radiation pulse are shown schematically in Fig. 17.1(b), where ω_1, ω_2, and ω_3 are three arbitrarily chosen frequency components within the spectrum of the input optical pulse.

From the above description one would conclude that THz generation in a second-order nonlinear medium is based on both the optical rectification and difference-frequency generation effect due to interactions between different frequency components within the spectral linewidth of the input ultrashort laser pulse. For the sake of conciseness, it is customarily said that THz generation is simply based on the optical rectification effect of an intense ultrashort laser pulse interacting with a second-order nonlinear medium.

With regard to THz generation in a second-order nonlinear medium, the nonlinear electric polarization as a function of time, $P^{(2)}_{\text{THz}}(t)$ (ignoring its vector nature), is related to its spectral components, $P^{(2)}_{\text{THz}}(\omega')$, by

$$P^{(2)}_{\text{THz}}(t) = \int_{-\infty}^{\infty} P^{(2)}_{\text{THz}}(\omega') e^{-\omega' t} d\omega'; \qquad (17.1\text{-}3)$$

the THz radiation field as a function of time, $E_{\text{THz}}(t)$, is governed by the following wave equation:

$$\nabla^2 E_{\text{THz}} - \frac{1}{c^2} \frac{\partial^2 E_{\text{THz}}(t)}{\partial t^2} = \frac{1}{\varepsilon_0 c^2} \frac{\partial^2 P_{\text{THz}}(t)}{\partial t^2}. \qquad (17.1\text{-}4)$$

Theoretically, knowing the temporal profile and associated spectral structure of an input ultrashort pump pulse, one may determine the spectral distribution of $P^{(2)}_{\text{THz}}(\omega')$ and further predict the temporal behavior of $P^{(2)}_{\text{THz}}(t)$ and $E^{(2)}_{\text{THz}}(t)$ via Eqs. (17.1-3) and (17.1-4). Fortunately, the temporal profile $E^{(2)}_{\text{THz}}(t)$ of a generated THz radiation pulse from a second-order nonlinear medium can be directly measured based on the Pockels effect by using another second-order nonlinear sample, and the spectral distribution of the associated THz radiation $E^{(2)}_{\text{THz}}(\omega')$ can be determined through a Fourier transform.

As shown in Fig. 17.1, a broadband THz pulse can be generated via optical rectification in a second-order nonlinear optical medium. For example, with the laser pulse durations of $\sim$100 fs, THz radiation of a bandwidth $\sim$10 THz could be produced. However, there are several factors that limit the bandwidth of THz pulses. The most important of these is phase matching between the optical pulse and the generated THz pulse. Phase matching requires conservation of energy and momentum in the nonlinear process, which is described by $\omega_2 - \omega_1 = \omega_{\text{THz}}$ and $\mathbf{k}_2 - \mathbf{k}_1 = \mathbf{k}_{\text{THz}}$. This phase matching is satisfied when the group velocity of the optical pulse equals the phase velocity of the THz pulse. As a result, all three waves participating in the optical rectification process can keep in phase, leading to maximum energy conversion. Since the optical pulse has a much higher frequency than the THz pulse, the THz pulse only sees the profile of the optical pulse rather than the oscillations. If the refractive-index dispersion properties of second-order nonlinear material for both the optical and THz beams are taken into account, the coherence length (effective interaction length) can be written as[8]:

$$l_c = \frac{\pi}{\Delta k} = \frac{\pi c}{\Delta \omega_{\text{THz}} \left| n_g - n_{\text{THz}} \right|} = \frac{\pi c}{\Delta \omega_{\text{THz}} \left| n_{\text{opt}} - \lambda_{\text{opt}} (dn_{\text{opt}}/d\lambda) - n_{\text{THz}} \right|}. \qquad (17.1\text{-}5)$$

Here, $\Delta\omega_{THz}$ is the spectrum width of THz pulse, c is the speed of light in a vacuum, λ_{opt} is the optical wavelength in vacuum, n_{opt} and n_{THz} are the optical and THz refractive indices, respectively, and n_g is the group refractive index of the optical pulse. The coherence length, which limits the usable thickness of second-order nonlinear crystals, is thus inversely proportional to the THz spectral width. To obtain higher THz frequencies from shorter optical pulses, very thin EO crystals must be used, although this reduces the magnitude of the THz signal.

17.1.2 Experiments on THz generation in second-order nonlinear materials

When selecting nonlinear materials for THz wave generation via optical rectification, three major factors need to be considered: the second-order nonlinearity of the materials, absorption of both optical and THz waves in the materials, and the phase matching that allows effective conversion of the optical wave to THz wave. THz generation has been reported from a variety of second-order nonlinear materials, including semiconductor crystals (e.g., ZnTe[8,9], GaAs,[10] GaSe[11]), inorganic crystals (e.g., LiNbO$_3$, LiTaO$_3$[12,13]), organic crystals (e.g., DAST[14,15], DSTMS[16]), and EO polymers.[17,18]

Monochromatic THz emission at ~3 THz was first demonstrated by optical rectification of near-infrared laser radiation in quartz in 1965.[19] Broadband THz radiation (~0.06–0.36 THz) through optical rectification was obtained using picosecond Nd:glass laser pulses in LiNbO$_3$ in 1971,[20] and THz radiation with ~1 THz bandwidth was later demonstrated using femtosecond CPM dye laser pulses in LiNbO$_3$ in 1990.[21] Nowadays, the most often used material for THz generation via optical rectification is ZnTe, because it fulfills the phase matching condition for the pump wavelength around 800 nm, i.e., within the tuning range of the widely used Ti:sapphire femtosecond lasers, and the THz frequencies around ~1 THz. However, ZnTe exhibits strong two-photon absorption at ~800 nm, which limits the pump intensity and restricts the effective THz generation. The highest THz energy ever obtained from ZnTe to date is ~1.5 µJ with a spectral range extending to ~3 THz by using 30-fs, 48-mJ, and 800-nm excitation pulses with ~40 µm large focused spot size,[22] which corresponds to energy conversion efficiency of ~3.1 × 10^{-5}. GaSe is another promising semiconductor crystal that has been exploited recently for THz generation with femtosecond-fiber lasers at telecommunication wavelengths. The organic crystalline salts, such as DAST and DSTM, and poled EO polymers possess much higher nonlinear coefficients, but their relatively low damage threshold limits the possible pump intensity.

Compared to ZnTe, the nonlinear coefficient of LiNbO$_3$ is about three times larger. In addition, it has a bandgap much larger than ZnTe, which allows no two-photon absorption at ~800 nm excitation, and makes it possible for high pump intensities and high-energy THz pulse generation. The challenge for LiNbO$_3$ as a THz emitter is that it does not have sufficient birefringence to match the index between the optical and the THz waves. The typical values of the group index of LiNbO$_3$ in the near-infrared is n_g = 2.3, and the refractive index at a frequency of 1 THz is n_{THz} = 5.17. Therefore, the THz and optical waves "walk away" from each other in a short distance, and the energy conversion coefficient is limited. The large angle between the THz and optical beams also brings difficulty in coupling the THz wave out of the nonlinear crystal.

However, THz wave generation from LiNbO$_3$ can satisfy the phase matching condition by tilting the pulse front of the excitation beam.[23] The basic idea is shown in Fig. 17.2. Assume that the pulse intensity front of the pump light can be tilted with respect to the phase front, which is perpendicular to the propagation direction of the excitation light. According to Huygens' principle, the THz radiation is excited impulsively along this tilted pulse front and will propagate

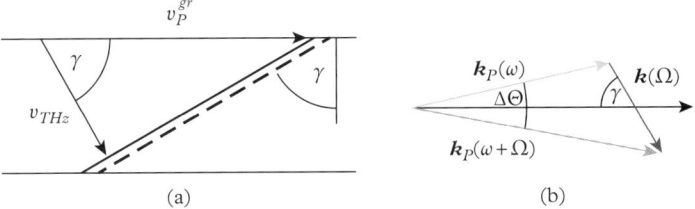

Figure 17.2 *Velocity matching using tilted-pulse-front excitation. (a) The THz wave generated in a LiNbO$_3$ crystal by the tilted intensity front of the pump pulse (dashed bold line) propagates perpendicularly to the THz phase front (bold line). (b) Wavevector diagram for difference-frequency generation between the spectral components of an ultrashort pump pulse, where Ω is the generated THz frequency. (Reproduced with permission from Hebling et al.[23] Copyright 2008, Optical Society of America).*

perpendicularly to this front with a phase velocity, v_{THz}. Although the pump pulse moves with a group velocity, v_p^{gr}, the projection of this velocity in the propagation direction of the generated THz radiation satisfies the relationship $v_p^{gr} \cos \gamma = v_{\text{THz}}$, which determines the phase matching between the optical and the THz waves. For LiNbO$_3$, $\gamma \approx 63\text{--}65°$. Figure 17.3 shows the schematic setup for generating THz waves from LiNbO$_3$ by optical pulses with a tilted pulse front.[23] The pump source can be a femtosecond Ti:sapphire oscillator/amplifier laser system; the pump pulses of horizontal polarization are incident on a grating of 2000 lines/mm, which is used to tilt the intensity front of the optical pump pulses. After passing through a $\lambda/2$-waveplate the pump pulses become vertically polarized and aligned with the optical axis of a stoichiometric LiNbO$_3$ crystal. The average power of the generated THz pulses can be measured by a Si bolometer. At an optical pump pulse energy level of $\sim$100 μJ, the measured THz pulse energy could be up to $\sim$50 nJ, indicating a conversion efficiency of $\sim 0.5 \times 10^{-3}$.

In another recent experiment using a tilted-pulse-front arrangement, a large-scale LiNbO$_3$ crystal was pumped with 25-mJ and 50-fs laser pulses of 800 nm wavelength; the generated THz pulse energy was $\sim$30 μJ, corresponding to a conversion efficiency of 1.2×10^{-3}.[24]

17.1.3 THz generation via four-wave mixing in plasmas

Another nonlinear optical method for THz generation is based on the interaction of a fundamental (ω) laser beam and the second harmonic (SH, frequency 2ω) beam in a plasma system generated in air or other gases by the same laser pulses. An accurate theoretical description of this process requires simultaneous treatments of plasma formation, generation of THz radiation, and propagation of the THz radiation through the plasma. However, a complete theory is presently not available. If we consider the laser-induced plasma as a third-order nonlinear optical medium, a four-wave mixing (FWM)-based optical rectification model can give an easier understanding of this particular process of THz generation. According to this model, in a third-order nonlinear optical medium (plasma) excited by an intense ultrashort pulsed laser beam and its SH beam simultaneously, the THz radiation can be generated by a FWM (or equivalent four-photon parametric) process described by the following third-order nonlinear polarization:

$$\mathbf{P}^{(3)}(\Omega = \omega_1 + \omega_2 - 2\omega_3) = \varepsilon_0 \chi^{(3)}(\omega_1, \omega_2, -2\omega_3)\mathbf{E}(\omega_1)\mathbf{E}(\omega_2)\mathbf{E}^*(2\omega_3). \qquad (17.1\text{-}6)$$

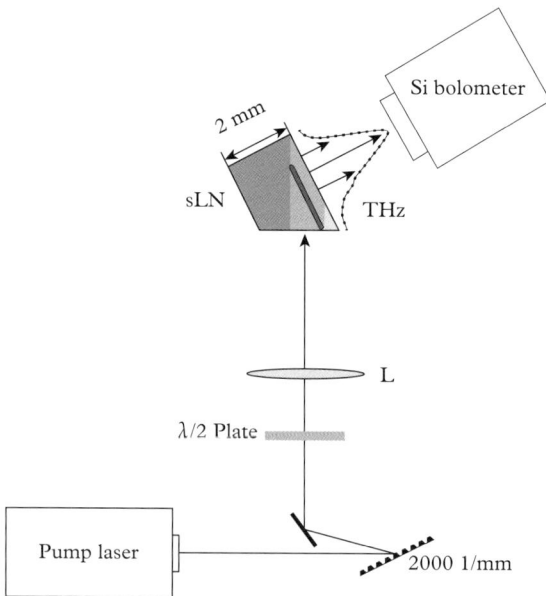

Figure 17.3 *Schematic diagram to generate THz waves from a stoichiometric $LiNbO_3$ crystal by using a tilted-pulse-front excitation beam via a 2000 lines/mm grating. The optical pump source is a Ti:sapphire laser system, and the average power of generated THz radiation can be measured by a Si bolometer. (Reproduced with permission from Hebling et al.[23] Copyright 2008, Optical Society of America).*

Here, $\mathbf{E}(\omega_1)$ and $\mathbf{E}(\omega_2)$ are two arbitrary Fourier components of higher frequency values within the fundamental (~800 nm) laser pulse's spectral bandwidth (cf. Fig. 17.1), $\mathbf{E}(2\omega_3)$ is a Fourier component of the SH beam (~400 nm) generated in a second-order nonlinear crystal by $\mathbf{E}(\omega_3)$ component of a lower frequency value within the spectral band of the same fundamental beam, $\mathbf{P}^{(3)}(\Omega)$ gives the coherent radiation at frequency Ω in THz region, and $\chi^{(3)}(\omega_1, \omega_2, -2\omega_3)$ is the third-order nonlinear susceptibility tensor of the medium corresponding to this process. This physical process can be more clearly described by the equivalent four-photon parametric process, schematically depicted in Fig. 17.4(a), which involves the annihilation of two photons of frequencies ω_1 and ω_2, and the simultaneous creation of two photons of frequency $2\omega_3$ and Ω, respectively (see Section 4.1). The efficient generation of this process requires the phase-matching requirement to be met. In practice, the relatively weak input SH beam plays the role of initial seed signal of $\mathbf{E}(2\omega_3)$, which may significantly enhance the efficiency of the above-mentioned FWM rectification process.

Figure 17.4(b) shows the schematic diagram of this approach to generate THz radiation. The intense pulsed pump beam of femtosecond duration is focused by a lens into air (or other gas) to achieve intensities sufficient for plasma generation. Part of the fundamental radiation is converted into its SH by a nonlinear crystal, such as a BBO (β-barium borate). Because of the different

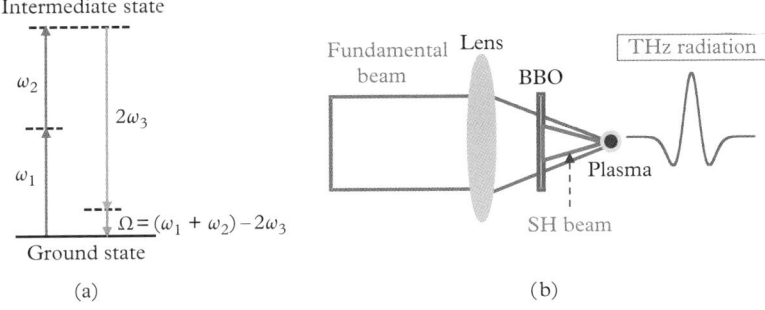

Figure 17.4 *(a) Transition diagram for four-wave parametric interaction. (b) Experimental setup for THz generation via FWM in third-order nonlinear plasma medium.*

phase velocities of the ω and 2ω fields in air, the phase matching between these two components at focus depends on the distance d between the BBO crystal and focus, which is given by $\phi = d(2n_{2\omega} - n_\omega)\omega/c$, where n_ω and $n_{2\omega}$ are the refractive indices of air at the fundamental and the SH frequencies, respectively. The temporal retardation between the SH and the fundamental waves can be compensated by inserting an appropriate birefringent plate. If type-I phase matching is used for second-harmonic generation (SHG), the polarization direction of the THz radiation will be parallel to the polarization direction of the second harmonic wave.

Compared to THz generation using second-order solid state or organic materials, third-order nonlinear gaseous media do not have a damage threshold or experience phonon absorption, which make them ideal THz emitters and sensors. Their broadband spectral response is only limited by the laser pulse width. Such a method was first demonstrated by Cook and Hochstrasser, using $\sim$800-nm and $\sim$65-fs laser pulses of $\sim 5 \times 10^{14}$ W/cm^2 intensity focused onto ambient air; the generated THz radiation field was estimated to be 2 kV/cm with the spectrum peak located at 2 THz.[25] Based on the same method, later studies have further improved the earlier result and the generated THz field strength could be higher than 10^2 kV/cm in ionized air.[26] Moreover, it has been shown experimentally that the polarity and strength of the THz field can be controlled by changing the relative phase between the ω and 2ω waves, and the plasma-enhanced $\chi^{(3)}$ ensures the measured high efficiency of THz generation.[27] A broadband THz spectrum ranging from 0.2 to over 30 THz has been demonstrated, which indicates potential applications in THz remote sensing and spectroscopy.

17.2 THz detection using nonlinear optical methods

17.2.1 THz detection via electro-optic (EO) sampling

Free-space EO sampling is based on the Pockels effect in second-order nonlinear crystals.[28,29] In EO sampling, a THz pulse and an ultrashort probe laser pulse co-propagate through an EO medium, such as a ZnTe crystal. In the presence of the THz field, it induces birefringence and refractive-index change through the Pockels effect. As a result, the polarization of the laser pulse is rotated, and two orthogonal polarized components of the laser pulse, which are oriented along the principal axes of the EO crystal, have a phase retardation

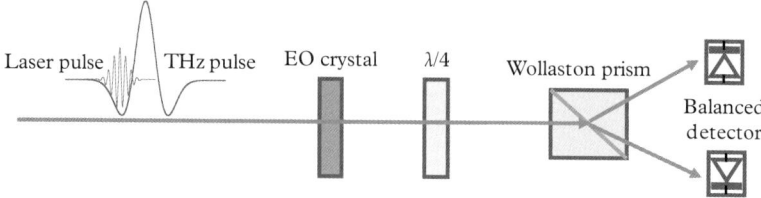

Figure 17.5 *Schematic diagram of THz waveform measurement using EO sampling.*

$$\Gamma = \frac{\omega}{c}(n_1 - n_2)L, \qquad (17.2\text{-}1)$$

where n_1 and n_2 are the refractive indices along the principal axes and L is crystal thickness. The phase shift is proportional to the THz electric field, i.e., $(n_1 - n_2) \propto E_{\text{THz}}$. This birefringence and polarization change can be measured by a system consisting of a quarter waveplate, a polarization beam splitter, and a pair of balanced detectors, as shown in Fig. 17.5. The intensity measured at the two detectors are given by

$$I_1 = I_0 \sin^2(\Gamma + \Phi) \quad \text{and} \quad I_2 = I_0 \cos^2(\Gamma + \Phi), \qquad (17.2\text{-}2)$$

where I_0 is the intensity incident on the analyzer and is a constant; Φ is the initial phase difference between two polarization components. As the two detectors measure orthogonal polarization components, here we have $\Phi = \pi/2$, then the differential signal gives the THz electric field through the following relation:

$$\frac{I_1 - I_2}{I_1 + I_2} = \sin \Gamma \approx \Gamma \propto E_{\text{THz}}. \qquad (17.2\text{-}3)$$

As an example of the early experimental study of using this method, Fig. 17.6(a) depicts the optical setup for generating THz radiation in a ZnTe crystal pumped by ~800-nm and ~130-fs laser pulses, and the detection of the THz pulses via another ZnTe crystal as the EO sampling element. The measured waveform of the THz signal amplitude and the corresponding spectrum obtained via Fourier transform are shown in Fig. 17.6(b) and (c), respectively.[8]

In experiments of laser pulse-excited THz wave generation, the laser pulse (intensity) width is shorter than the THz pulse (amplitude) width, so only a small part of the latter can overlap with the former. By varying the time delay between these two pulses, the electric field of the THz pulse is sampled.

As with the generation of THz pulses, the bandwidth of the THz detection depends on the sampling laser pulse width, EO crystal properties, and its thickness. The materials used for generation of THz radiation by optical rectification, such as ZnTe, GaSe, DAST, and poled polymers,[9,30–32] can also be used for THz waveform measurements.

17.2.2 THz detection via FWM

In a centrosymmetric crystal or isotopic medium, the direct optical SHG is prohibited in the dipole-transition approximation. However, if a low-frequency (Ω) electric field is applied on it,

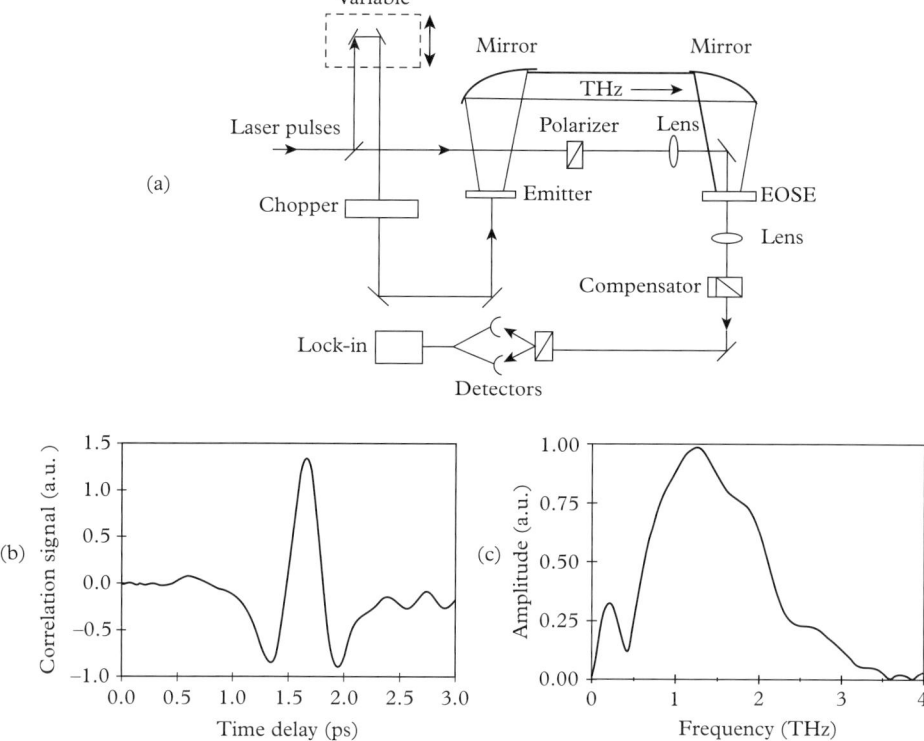

Figure 17.6 *(a) Experimental setup for THz generation via a ZnTe emitter and detection via an EO sampling element. (b) THz waveform detected by EO sampling, and (c) corresponding THz spectrum. (Reproduced with permission from Nahata et al.[8] Copyright 1996, AIP Publishing LLC).*

the SH wave of an incident optical frequency wave can be generated through a FWM process described by the following third-order nonlinear polarization vector:

$$\mathbf{P}(\omega' = 2\omega + \Omega) = \varepsilon_0 \chi^{(3)}(\omega, \omega, \Omega) \mathbf{E}(\omega) \mathbf{E}(\omega) \mathbf{E}(\Omega). \tag{17.2-4}$$

Here, $\chi^{(3)}(\omega, \omega, \Omega)$ is the corresponding third-order nonlinear susceptibility tensor of the medium, ω is the input light frequency, Ω is the frequency of the applied low-frequency (bias) electric field, and ω' is the frequency of newly generated optical wave. Since $\Omega \ll \omega$ and $\omega' \approx 2\omega$, this process is named electric field-induced second harmonic (EFISH) generation. As a third-order nonlinear process, the energy-conversion efficiency from the fundamental beam to the SH beam is usually quite small; in this condition the SH field can be approximately written as

$$E(\omega' \approx 2\omega) \propto \chi_e^{(3)}(\omega, \omega, \Omega) E(\omega) E(\omega) E(\Omega), \tag{17.2-5}$$

where $\chi_e^{(3)}(\omega, \omega, \Omega)$ is the effective third-order susceptibility value of the medium characterizing the EFISH process.

The aforementioned principle can be employed to measure a given THz field generated by ultrashort femtosecond laser pulses. Using a centrosymmetric silicon sample as the third-order nonlinear medium that was irradiated with a laser pulsed beam of ω and a THz pulsed beam of Ω, Nahata and Heinz accomplished their early experimental demonstration of THz signal sampling, based on measuring the THz-field induced SHG.[33]

Moreover, Dai et al. have demonstrated that an ionized gas (plasma) can be employed as the third-order nonlinear sensing medium to detect a given incident THz pulse field, based on the EFISH method.[34] In this case, according to Eq. (17.2-5) the THz field-induced SH signal intensity ($I_{2\omega}^{\text{THz}}$) should be

$$I_{2\omega}^{\text{THz}} \propto |E_{2\omega}^{\text{THz}}|^2 \propto |\chi_e^{(3)} E_\omega E_\omega E_\Omega|^2 \propto (\chi_e^{(3)} I_\omega)^2 I_\Omega, \qquad (17.2\text{-}6)$$

where E_ω and E_Ω are the electric field amplitudes of the input probe laser pulse and the THz pulse, respectively. The physical meaning of this equation is that at a given intensity level of the input probe laser beam, the detected SH signal intensity is proportional to the intensity of the input THz wave; thus the phase information of the latter is lost.

On the other hand, in the laser-induced plasma there could be local coherent light components at frequency $\omega' = 2\omega + \Omega \approx 2\omega$, which are from the white-light supercontinuum emission through the self-phase modulation mechanism. In this situation, one should consider the interference effect between the local SH field, $E_{2\omega}^{\text{loc}}$, and the THz-induced SH field, $E_{2\omega}^{\text{THz}}$, and then the overall intensity of the detected SH signal will be

$$I_{2\omega} \propto \langle E_{2\omega}^2 \rangle = \langle (E_{2\omega}^{\text{THz}} + E_{2\omega}^{\text{loc}})^2 \rangle = \langle (E_{2\omega}^{\text{THz}})^2 \rangle + 2 \langle E_{2\omega}^{\text{THz}} E_{2\omega}^{\text{loc}} \rangle + \langle (E_{2\omega}^{\text{loc}})^2 \rangle. \qquad (17.2\text{-}7)$$

With a zero or a weak local SH signal, the first term is dominant, leading to $I_{2\omega}^{\text{THz}} \propto I_\Omega$. The second term is the interference term that is proportional to E_Ω, and provides the basis of coherent detection of the amplitude and the phase of the THz field. The last term is the contribution from the local SH signal; it as a background signal can be easily deducted from the $I_{2\omega}$, or filled out by modulating the THz beam. Thus, Eq. (17.2-7) can be simplified to

$$I_{2\omega} \propto [(\chi_e^{(3)} I_\omega)^2 I_\Omega + 2\chi_e^{(3)} I_\omega E_{2\omega}^{\text{loc}} E_\Omega \cos\phi], \qquad (17.2\text{-}8)$$

where φ is the phase difference between $E_{2\omega}^{\text{THz}}$ and $E_{2\omega}^{\text{loc}}$. With probe laser intensity less than the air ionization threshold, the $E_{2\omega}^{\text{loc}}$ is negligible, and the total SH is dominated by the first term in Eq. (17.2-8); as a result, the measured $I_{2\omega}$ is unipolar, and the detection is incoherent. With the probe intensity much higher than the plasma generation threshold, the second term may be dominant and $I_{2\omega}$ is proportional to the THz electric field with a bipolar waveform. In the latter case, the detection will be coherent.

A schematic diagram of the experimental setup is shown in Fig. 17.7.[34] The THz wave is generated by mixing the ω pump beam and its SH beam (from a type-I BBO crystal) at the first air plasma point (see Section 17.1.3). The first parabolic mirror collimates the THz beam. A high resistivity silicon wafer blocks the residual 800- and 400-nm beams. The second parabolic mirror focuses the collimated THz beam. A $\lambda/2$ wave plate controls the polarization of the probe beam. The THz wave is detected by measuring the time-resolved SH signal produced by mixing the ω beam and the THz field at the second plasma point.

Figure 17.8 shows the measured results of THz waveform obtained using this method. As the probe pulse energy increases, the THz waveform changes from unipolar ($\propto I_\Omega$) to bipolar ($\propto E_\Omega$),

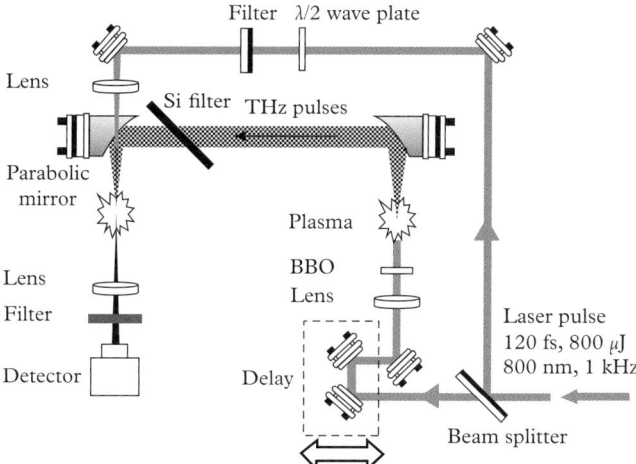

Figure 17.7 *Schematic diagram of the experimental setup for coherent THz generation and detection in plasmas. (Reproduced with permission from Dai et al.[34] Copyright 2006, American Physical Society).*

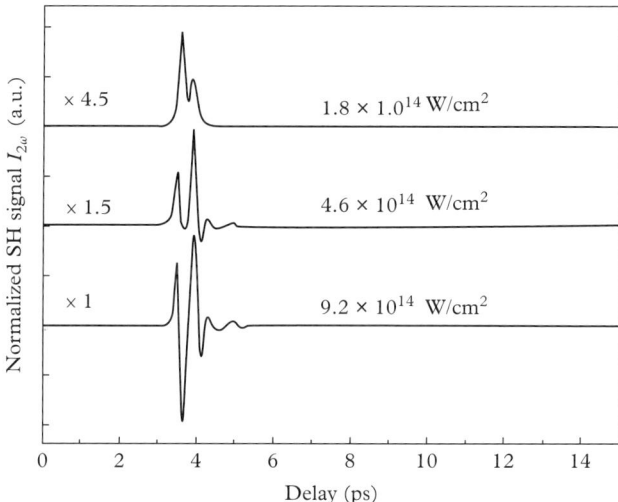

Figure 17.8 *THz waveforms measured with a gas sensor at three different levels of probe laser intensity I_ω: 1.8×10^{14} W/cm² (upper), 4.6×10^{14} W/cm² (middle), and 9.2×10^{14} W/cm² (lower), respectively. (Reproduced with permission from Dai et al.[34] Copyright 2006, American Physical Society).*

indicating that coherent detection of THz waves can be realized when sufficient optical probe energy (or intensity) is used.

In another similar experiment, the local SH wave was efficiently created by applying an external dc (bias) field across the focal volume of the probe laser beam based on the same EFISH mechanism.[35] In this case, the intensity requirement for probe intensity can be considerably reduced, and the measured signal-to noise ratio can be improved by using lock-in detection at the modulation frequency of the applied external bias field.

17.3 Nonlinear optical applications of strong THz fields

17.3.1 Nonlinear phase modulation with intense single-cycle THz pulses

When an intense optical pulse propagates through a nonlinear medium it causes a change of the refractive index, which in turn induces a phase shift to the optical pulse. The associated optical phenomenon is referred to as the Kerr effect (third-order effect) or Pockels effect (second-order effect). The Kerr effect in the temporal domain results in self-phase modulation, but it can also occur in the spatial domain, where it produces self-focusing or self-defocusing, which is known as Kerr lensing effect. When a weak probe pulse co-propagates with the intense pulse through a nonlinear medium, the probe pulse undergoes a phase modulation due to the temporal and spatial variation of the induced refractive-index change originating from the intense primary pulse, leading to a substantial modification of the temporal, spectral and spatial profile of the probe pulse. Such a process, which is defined as cross-phase modulation, has been extensively studied via the third-order Kerr effect. The same concepts apply to the second-order Pockels effect, except that now the induced index change is driven by the electric field instead of the intensity of the driving optical field. However, one may think what will happen if an intense THz field and a probe laser pulse interact through a second-order medium. Such type of study is essentially important and interesting for ultrafast and nonlinear optics. Here we shall describe that an intense single-cycle THz pulse may induce nonlinear phase modulation to a probe laser pulse in solid and liquid materials via Pockels effect and/or Kerr effect.

Consider an intense THz pulse and a weak laser probe pulse, of complex amplitude E_{THz} and E_{probe}, respectively, co-propagating through a nonlinear medium, as shown schematically in Fig. 17.9. The overall nonlinear polarization (ignoring its vector nature) at light pulse frequency ω is given by

$$P(\omega) = \varepsilon_0 (\chi_e^{(1)} E(\omega) + \chi_e^{(2)} E_{THz} E(\omega) + \chi_e^{(3)} |E_{THz}|^2 E(\omega) + \ldots), \quad (17.3\text{-}1)$$

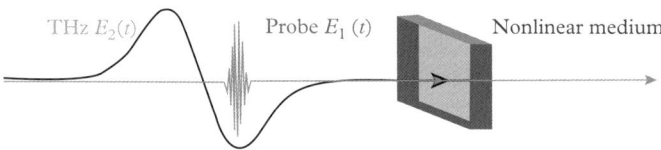

Figure 17.9 *Schematic diagram of single-cycle THz-induced nonlinear phase modulation on a co-propagating ultrashort laser pulse through Pockels and Kerr effects. Here it is assumed that the pulse width of the probe laser pulse is much narrower than that of the THz pulse.*

where $\chi_e^{(i)}$ (here $i = 2$ or 3) is the effective ith order susceptibility value of the sample medium for induced refractive-index change. The change in refractive index seen by the weak probe can be expressed as

$$\Delta n(\omega) = n' - n_0 = \Delta n_1 + \Delta n_2, \qquad (17.3\text{-}2)$$

where $n_0 = \sqrt{1 + \chi^{(1)}}$ is the linear refractive index of the nonlinear medium, and

$$\Delta n_1 = \chi_e^{(2)} E_{\text{THz}}/n_0 \quad \text{and} \quad \Delta n_2 = \chi_e^{(3)} |E_{\text{THz}}|^2/(2n_0) \qquad (17.3\text{-}3)$$

are the refractive-index changes induced by Pockels and Kerr effects, respectively. For a ZnTe crystal, $\chi_e^{(2)} = 9.0 \times 10^{-11}$ m/V and $\chi_e^{(3)} = 3 \times 10^{-19}$ m^2/V^2.[36] With $E_{\text{THz}} = 5 \times 10^7$ V/m, $\Delta n_2/\Delta n_1 \approx 0.2$, indicating that the refractive-index change in ZnTe is governed by the Pockels effect when $E_{\text{THz}} < 5 \times 10^7$ V/m. This refractive-index change affects the phase of the probe pulse, resulting in a phase shift of

$$\Delta\varphi(t) = \frac{2\pi \chi_e^{(2)}}{n_0 \lambda_0} \int_0^L E_{\text{THz}}(t - \xi z)\, dz, \qquad (17.3\text{-}4)$$

where λ_0 is the central wavelength of probe pulse, L is the thickness of the ZnTe crystal, $\chi_e^{(2)}$ is the effect second-order susceptibility coefficient charactering the Pockels effect of the crystal medium, $\xi = 1/v_{\text{probe}} - 1/v_{\text{THz}}$ is the walk-off parameter and is calculated to be 150 fs/mm for ZnTe, where v_{probe} and v_{THz} are the group velocity and phase velocity of the probe and THz pulses, respectively. The induced phase change means a temporal variation of the effective optical thickness of the nonlinear medium and, therefore, leads to a time-dependent frequency shift of the probe light wave due to the Doppler effect. Such a frequency shift (chirping) is determined by

$$\delta\omega(t) = \omega(t) - \omega_0 = -\frac{\partial}{\partial t}\Delta\varphi(t). \qquad (17.3\text{-}5)$$

As an example, Fig. 17.10(a) shows the calculated phase shift from Eq. (17.3-4), which is basically proportional to a given THz electric(E)-field, although the walk-off leads to an overall

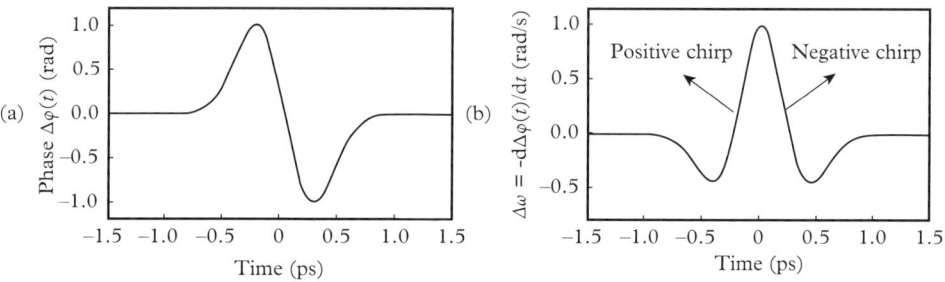

Figure 17.10 Probe pulse's phase change and frequency shift versus time induced by a single-cycle THz pulse: (a) normalized phase change and (b) normalized frequency shift. (Adapted from Shen et al.[38] Copyright 2008, American Physical Society).

reduction in the accumulated phase shift. Figure 17.10(b) shows the corresponding temporal variation of the induced frequency shift $\delta\omega(t)$. Comparing these two curves of induced phase change and frequency shift change, one may find several interesting features. First, when the probe pulse interacts with different segments of the THz pulse, $\delta\omega$ is negative (red-shifted) near the leading and trailing edges, and $\delta\omega$ is positive (blue-shifted) in the region between the two amplitude peak positions. Second, the probe acquires a positive chirp when scanned across the leading edge to the zero crossing of the THz pulse. The maximum spectral shift occurs when the probe overlaps with the middle (zero-crossing) of the THz E-pulse curve, and around this region the frequency chirp is positive and linear. Third, the frequency chirp remains linear but reverses sign from positive to negative when the probe pulse is scanned from the zero-crossing to the trailing edge. These features are basically consistent with the experimental results described below.

This type of study has been demonstrated by Shen et al. using an accelerator-based coherent THz source, which can produce up to ~100 µJ single-cycle THz pulses with the peak electric-field of ~7 × 10^7 V/m and pulse duration of ~0.7 ps.[37,38] A wire mesh attenuator is used to limit the THz energy on the ZnTe crystal to ~35 µJ. The corresponding E-field strength at the ZnTe crystal is estimated to be ~4 × 10^7 V/m for which the refractive-index change in ZnTe is dominated by the Pockels effect while the Kerr effect can be neglected. A ~120-fs and 795-nm pulse from a Ti:sapphire laser amplifier is used as a probe. Experiments are conducted to measure the intensity, phase and spectrum of the probe as a function of the time delay between the THz and the probe pulses with a SHG frequency-resolved optical gating (FROG) device.

Figure 17.11 shows the measurement results, where the column (a) illustrates that the probe pulse interacts with different parts of the single-cycle THz pulse by changing the time delay. The retrieved pulses of FROG measurements in the time and frequency domains, including intensity and phase, are shown in columns (b) and (c), respectively. The first row of Fig. 17.11 shows that the probe phase in the absence of THz is constant over the pulse duration, and the spectrum represents a nearly transform-limited pulse. As the probe is scanned across the single-cycle THz E-field, the spectrum is blue-shifted when the probe overlaps the central zero-crossing of the THz E-field, and red-shifted at the leading and trailing edges, as shown in the second, fourth, and sixth rows of Fig. 17.11. The wavelength shift is accompanied by a small asymmetric spectral broadening and the probe pulse is slightly chirped. The central wavelength of the probe is shifted by as much as ~10 nm toward the blue side and ~6 nm toward the red.

Since the distribution of the THz E-field is both time and space dependent, this leads to a phase modulation not only in time, but also in space, which affects the spatial profile of the co-propagating probe laser beam. Consider a THz pulse, with a beam radius a propagating through ZnTe. If the THz E-field's transverse profile is approximated by a parabola, the resulting phase shift is given by

$$\Delta\varphi(r,t) = \frac{2\pi \chi_e^{(2)} E_{\text{THz}}(t) L}{n_0 \lambda}(1 - (r/a)^2). \tag{17.3-6}$$

This quadratic phase variation is equivalent to a lens of focal length $f = n_0 a^2 \big/ 2\chi_e^{(2)} E_{\text{THz}}(t) L$. The focal length of the induced lens is controllable by changing the THz E-field strength and can be as short as ~35 cm with $E_{\text{THz}} = 4 \times 10^7$ V/m. As the probe co-propagates with the single-cycle THz pulse through the ZnTe and coincides with the crest of the THz pulse, $E_{\text{THz}} > 0$, and, therefore, $\Delta\varphi(r, t) > 0$. The THz pulse, together with the EO crystal, acts as a positive lens, imposing a converging wavefront on the probe and thereby focusing it. If the time delay is adjusted to overlap the probe with the THz trough, then $E_{\text{THz}} < 0$ and $\Delta\varphi(r, t) < 0$, and the

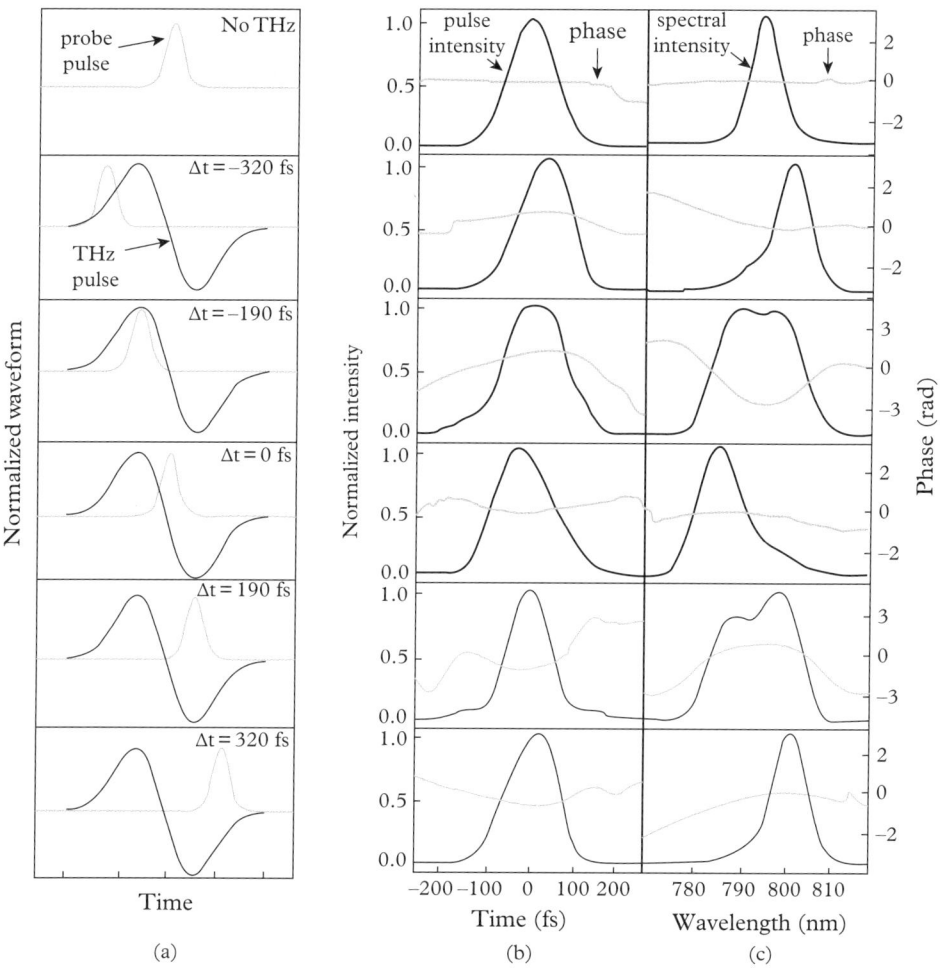

Figure 17.11 *Temporal and spectral effects of cross-phase modulation with an intense single-cycle THz pulse. (a) By changing the time delay, an ultrashort probe pulse (thin light line) interacts with different parts of a single-cycle THz pulse (thick line). (b, c) Retrieved intensity (thick line) and phase (light line) of the probe pulse in time and frequency domains. (Reproduced with permission from Shen et al.[38] Copyright 2008, American Physical Society).*

induced positive lens switches to a negative lens. In analogy to Kerr lensing, such a lensing effect can be called "Pockels lensing." It is different from the conventional Kerr lensing effect in that it is able to focus or defocus a light beam in the same nonlinear medium by simply switching the relative phase between THz and laser pulses.

In the experiment, the intensity profiles of the probe laser are measured for different THz E-field strengths and for different time delays between the THz and probe pulses. Under the experimental conditions, the effects of self-focusing of the probe itself and induced focusing by third-order Kerr effects can be neglected. Figure 17.12(a) shows the spatial profile of the probe

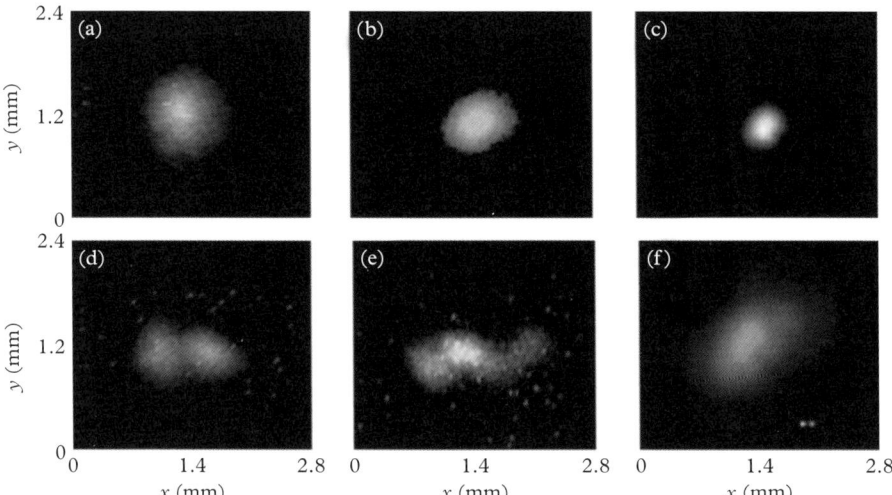

Figure 17.12 *Measured spatial profile of the probe laser beam: (a) without THz field; (b) $E_{THz} = 2.5 \times 10^7$ V/m; (c) $E_{THz} = 4.0 \times 10^7$ V/m; (d) $E_{THz} = 4.5 \times 10^7$ V/m; (e) $E_{THz} = 5.0 \times 10^7$ V/m; and (f) $E_{THz} = -4.0 \times 10^7$ V/m. (Reproduced with permission from Shen et al.[38] Copyright 2008, American Physical Society).*

laser when the THz field is not present. By varying the time delay such that the probe coincides with the crest of the THz field, there is a clear contraction of the beam indicating focusing. The beam size decreases with increasing THz E-field, as shown in Fig. 17.12(b) and (c). With a further increase of the THz E-field strength, the probe beam breaks up into two spots and then collapses into a complex pattern with several hot spots, as illustrated in Figs. 17.12(d) and (e). However, when the probe overlaps with the trough of the THz pulse, the probe beam size is broadened and defocused, as shown in Fig. 17.12(f). The beam break-up cannot be explained by the thin lens approximation, which is valid only if the THz induced phase shift is small. For a large phase shift, this approximation breaks down, and the evolution of the probe and THz beams needs to be described by the nonlinear coupled equations.

The Kerr effect is another well-known third-order nonlinear phenomenon, which is distinct from the Pockels effect in that the induced change of refractive index and the phase modulation are proportional to the square of the applied electric field. Recently, the intense single-cycle THz pulses have been used to study the Kerr effect in various liquids (CS_2, CH_2I_2, CCl_4, $CHCl_3$, and benzene),[39] in which a strong single-cycle THz pulse with peak electric field up to 150 kV/cm is used to induce the birefringence in the liquid samples, and the variation of the polarization of a co-propagating 100 fs, 800 nm laser pulse is controlled and measured simply by varying the magnitude of the applied THz electric field. Figure 17.13(a) shows the quadratic dependence of the incident THz electric field measured with CS_2, which indicates the nonlinear Kerr effect. The study found that, for most of the liquids, the THz Kerr constants were close to the values reported for the dc Kerr measurements. For example, a polarization rotation of 0.27% was obtained for CS_2 with THz electric field of 150 kV/cm, which corresponds to a refractive-index change, $\Delta n = 4.3 \times 10^{-6}$. This leads to a Kerr constant of $K = 2.4 \times 10^{-14}$ m/V^2, which is close to the value of $K = 2.8 \times 10^{-14}$ m/V^2 for the dc measurement.

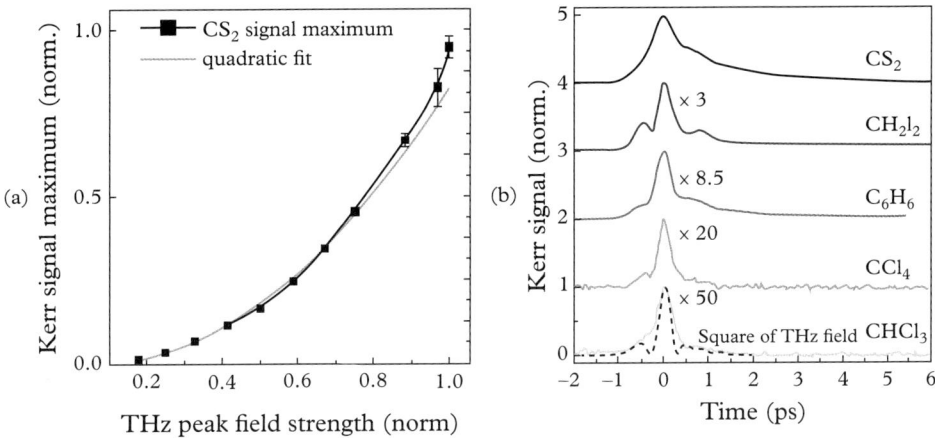

Figure 17.13 *(a) Square dependence of Kerr signal with the applied THz electric field, and (b) THz-induced time-dependent birefringence in several liquid samples. (Reproduced with permission from Hoffmann et al.[39] Copyright 2009, AIP Publishing LLC).*

The nonlinear temporal response is recorded by varying the time delay between the strong THz pump pulse and the weak near-infrared laser pulse. The measurements results are shown in Fig. 17.13(b). The study reveals both instantaneous and non-instantaneous response of the liquid to the applied single-cycle THz pulse. The fast response follows the THz electric field and is due to the re-arrangement of the molecular electronic cloud; and the slow exponential response is associated with the molecular reorientation. This study indicates that the THz Kerr effect may reveal polarizability dynamics associated with electronic, vibrational and structural responses in different kinds of materials.

17.3.2 Strong-field THz-induced nonlinear absorption in semiconductors

Strong-field THz radiation is ideal for probing the nonlinear optical response of low-energy excitation in materials. For example, in recent studies, free-carrier saturable absorption or absorption bleaching in n-doped InGaAs thin film has been investigated using intense near-single-cycle THz pulses with peak electric field of ~200 kV/cm.[40,41] An open aperture Z-scan technique is employed in the study, in which the transmittance of the sample is measured as the sample is moved along the propagation path z of a focused THz beam. Therefore, THz intensity is at maximum at the focus and is progressively reduced as the sample is moved away from the focus in both directions. Significant enhancement in the transmission near the focus is observed, in contrast to the transmission away from the focus. Figure 17.14(a) shows that the absorption bleaching effect is not observed when the same scan is performed on a bare InP substrate. The temporal profile of the transmitted THz electric field at different z-positions is measured, and Fig. 17.14(b) shows no significant temporal shift between the transmitted pulses, which indicates that the imaginary part of the conductivity is not appreciably changing in this measurement. Figure 17.14(c) shows the modulus squared of the traces in Fig 17.14(b), which is proportional to the transmitted energy and is consistent with the direct energy measurement shown in Fig. 17.14(a). The dynamics of the

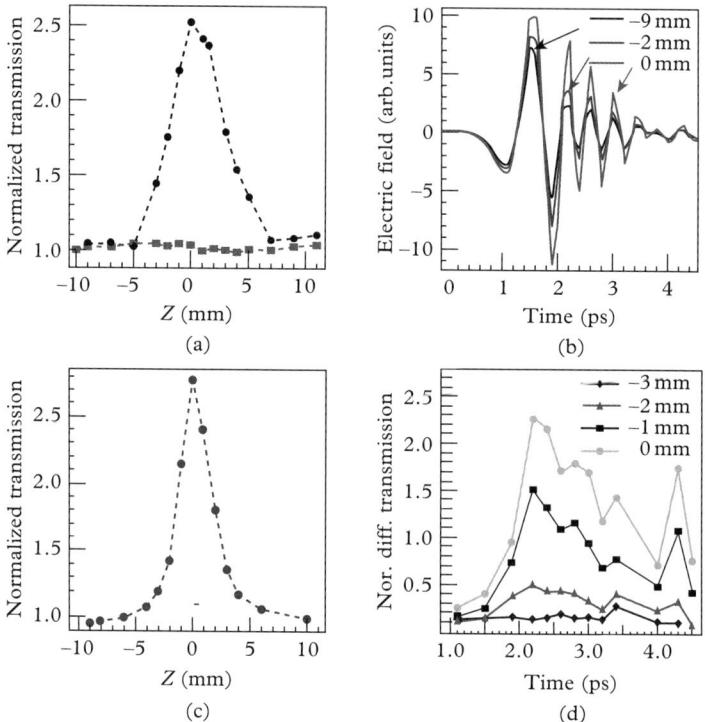

Figure 17.14 *(a) Open-aperture Z-scan normalized transmission of the total THz pulse energy measured with a pyroelectric detector after the sample (the solid circle refers to the InGaAs epilayer on an InP substrate and the solid square to the InP substrate only). (b) Transmitted THz pulse electric field measured at different positions of the Z-scan. (c) Normalized time integral associated with the modulus squared of the transmitted electric field as a function of the z position along the scan. (d) Normalized electric field differential transmission as a function of time measured at different z positions. (Reproduced with permission from Razzari et al.[40] Copyright 2009, American Physical Society).*

bleaching process is more evident in Fig. 17.14(d), where the normalized electric field differential transmission (defined as the transmitted electric field difference between the actual position in the Z-scan and a position far away from the focus, for each peak of the THz pulse) is plotted as a function of time for different z-positions along the scan. The curves show an initial increase in transmission over a period of 1 ps followed by a slower decay.

The absorption bleaching observed during the interaction of intense THz pulses with the electrons in the conduction band of the InGaAs sample can be attributed to THz-induced inter-valley scattering between non-equivalent valleys of the conduction band, in particular, the Γ and the L valleys, which are the closest upper valleys in InGaAs. The onset of conductivity in InGaAs occurs when electrons are in the high-mobility Γ valley. The applied THz electric field can promote free electrons in the Γ valley to overcome the nearest intervalley separation and scatter into

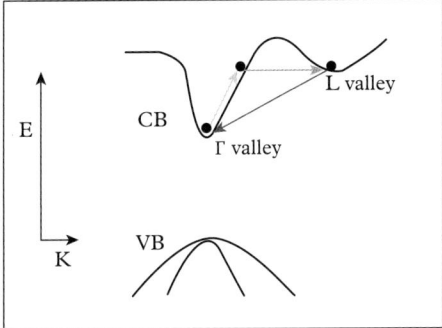

Figure 17.15 *Mechanism of intense THz pulse-induced inter-valley scattering. The electrons in the conduction band are accelerated by the THz electric field (upward arrow); after acquiring enough kinetic energy, they may scatter into a satellite valley (L valley in this case, horizontal arrow) from which they then have a finite probability to scatter back to the bottom of the conduction band (downward arrow). (Reproduced with permission from Sharma et al.[41] Copyright 2010, IEEE).*

an upper valley (i.e., the L valley), as shown in Fig. 17.15.[41] Since the effective mass of electrons is higher in the L valley, and the carrier mobility is lower, thus reducing the overall conductivity of the semiconductor film. Since the transmission of the THz pulse increases with a decrease in the conductivity of the sample, the transmission is enhanced at high THz electric fields. The electrons in this upper valley will then have a finite probability of scattering back to the Γ valley, where the effective mass is smaller. Therefore, the conductivity of the film will eventually increase back to its original value, which will eventually lead to a decrease in the THz transmission. Saturable absorption of the free carriers in n-doped semiconductors, such as Ge, Si, and GaAs, in the THz frequency region has also been observed recently.[42,43]

REFERENCES

1. Y. R. Shen, *Prog. Quantum Electron.* **4**, 207 (1976).
2. D. Dragoman and M. Dragoman, *Prog. Quantum Electron.* **28**, 1 (2004).
3. S. L. Dexheimer (Ed.), *Terahertz Spectroscopy: Principles and Applications* (CRC Press, Boca Raton, FL, 2008).
4. C. A. Schmuttenmaer, *Chem. Rev.* **104**, 1759 (2004).
5. K. Reimann, *Rep. Prog. Phys.* **70**, 1597 (2007).
6. M. Tonouchi, *Nat. Photonics* **1**, 97 (2007).
7. A. Redo-Sanchez and X. C.Zhang, *IEEE J. Sel. Top. Quantum Electron.* **14**, 260 (2008).
8. A. Nahata, A. S. Weling, and T. F. Heinz, *Appl. Phys. Lett.* **69**, 2321 (1996).

9. Q. Wu and X.-C. Zhang, *Appl. Phys. Lett.* **67**, 3523 (1995).
10. A. Rice, Y. Jin, X. F. Ma, X.-C. Zhang, D. Bliss, J. Larkin, and M. Alexander, *Appl. Phys. Lett.* **64**, 1324 (1994).
11. R. Huber, A. Brodschelm, F. Tauser, and A. Leitenstorfer, *Appl. Phys. Lett.* **76**, 3191 (2000).
12. L. Xu, X.-C. Zhang, and D. H. Auston, *Appl. Phys. Lett.* **61**, 15 (1992).
13. T. J. Carrig, G. Rodriguez, T. S. Clement, and A. J. Taylor, and Kevin R. Stewart, *Appl. Phys. Lett.* **66**, 121 (1995).
14. P. Y. Han, M. Tani, F. Pan, and X.-C. Zhang, *Opt. Lett.* **25**, 675 (2000).
15. A. Schneider, M. Neis, M. Stillhart, B. Ruiz, R. U. A. Khan, and P. Günter, *J. Opt. Soc. Am. B* **23**, 1822 (2006).
16. M. Stillhart, A. Schneider, and P. Günter, *J. Opt. Soc. Am. B* **25**, 1914 (2008).
17. X. Zheng, A. Sinyukov, and L. M. Hayden, *Appl. Phys. Lett.* **87**, 081115 (2005).
18. C. V. McLaughlin, L. M. Hayden, B. Polishak, S. Huang, J. Luo, T.-D. Kim, and A. K.-Y. Jen, *Appl. Phys. Lett.* **92**, 151107 (2008).
19. F. Zernike Jr. and P. R. Berman, *Phys. Rev. Lett.* **15**, 999 (1965).
20. K. H. Yang, P. L. Richards, and Y. R. Shen, *Appl. Phys. Lett.* **19**, 320 (1971).
21. B. B. Hu, X.-C. Zhang, D. H. Auston, and P. R. Smith, *Appl. Phys. Lett.* **56**, 506 (1990).
22. F. Blanchard, L. Razzari, H. C. Bandulet, G. Sharma, R. Morandotti, J.-C. Kieffer, T. Ozaki, M. Reid, H. F. Tiedje, H. K. Haugen, and F. A. Hegmann, *Opt. Express* **15**, 13212 (2007).
23. J. Hebling, K.-L. Yeh, M. C. Hoffmann, B. Bartal, and K. A. Nelson, *J. Opt. Soc. Am. B* **25**(7), B6 (2008).
24. A. G. Stepanov, L. Bonacina, S. V. Chekalin, and J.-P. Wolf, *Opt. Lett.* **33**, 2497 (2008).
25. D. J. Cook and R. M. Hochstrasser, *Opt. Lett.* **25**, 1210 (2000).
26. T. Bartel, P. Gaal, K. Reimann, M. Woerner, and T. Elsaesser, *Opt. Lett.* **30**, 2805 (2005).
27. X. Xie, J. Dai, and X.-C. Zhang, *Phys. Rev. Lett.* **96**, 075005 (2006).
28. G. Gallot and D. Grischkowsky *J. Opt. Soc. Am. B* **16**, 1204 (1999).
29. J. A. Valdmanis, G. Mourou, and C. W. Gabel, *Appl. Phys. Lett.* **41**, 211 (1982).
30. C. Kubler, R. Huber, S. Tubel and A. Leitenstorfer, *Appl. Phys. Lett.* **85**, 3360 (2004).
31. A. Schneider, I. Biaggio and P. Gunter, *Appl. Phys. Lett.* **84**, 2229 (2004).
32. H. Cao, T. F. Heinz, and A. Nahata, *Opt. Lett.* **27**, 775 (2002).
33. A. Nahata and T. F. Heiz, *Opt. Lett.* **23**, 67 (1998).
34. J. Dai, X. Xie, and X.-C. Zhang, *Phys. Rev. Lett.* **97**, 103903 (2006).
35. J. Dai, J. Liu, and X.-C. Zhang, *IEEE J. Sel. Top. Quantum Electron.* **17**, 182 (2011).
36. J-P. Caumes, L. Videau, C. Rouyer, and E. Freysz, *Phys. Rev. Lett.* **89**, 047401 (2002).
37. Y. Shen, T. Watanabe, D. A. Arena, C.-C. Kao, J. B. Murphy, T.Y. Tsang, X. J. Wang, and G. L. Carr, *Phys. Rev. Lett.* **99**, 043901 (2007).
38. Y. Shen, G. L. Carr, J. B. Murphy, T. Y. Tsang, X. J. Wang, and X. Yang, *Phys. Rev. A* **78**, 043813 (2008).
39. M. C. Hoffmann, N. C. Brandt, H. Y. Hwang, K.-L. Yeh, and K. A. Nelson, *Appl. Phys. Lett.* **95**, 231105 (2009).
40. L. Razzari, F. H. Su, G. Sharma, F. Blanchard, A. Ayesheshim, H.-C. Bandulet, R. Morandotti, J.-C. Kieffer, T. Ozaki, M. Reid, and F. A. Hegmann, *Phys. Rev. B* **79**, 193204 (2009).
41. G. Sharma, L. Razzari, F. H. Su, F. Blanchard, A. Ayesheshim, T. L. Cocker, L. V. Titova, H. C. Bandulet, T. Ozaki, J.-C. Kieffer, R. Morandotti, M. Reid, and F. A. Hegmann, *IEEE Photonics J.* **2**, 578 (2010).
42. J. Hebling, M. C. Hoffmann, H. Y. Hwang, K.-L. Yeh, and K. A. Nelson1, *Phys. Rev. B* **81**, 035201 (2010).
43. M. C. Hoffmann and D. Turchinovich, *Appl. Phys. Lett.* **96**, 151110 (2010).

18

Detailed Theory of Nonlinear Susceptibilities

Nonlinear susceptibilities of optical media are the key parameters to describe various nonlinear optical effects and phenomena within the framework of semiclassical theory. In this chapter, explicit expressions for various orders of nonlinear susceptibilities are derived in the electric-dipole moment approximation by using the density matrix method. The basic properties of nonlinear susceptibilities are discussed in detail, including their symmetry properties, local-field correction, and resonance enhancement.[1-5] Strictly speaking, the formulations of various orders of susceptibilities described here are applicable only to those cases where (i) the electron-cloud distortion and the intramolecular motion are the major sources of the nonlinear polarization of the medium, and (ii) the nonlinear polarization response of the medium is instantaneous and localized with respect to an applied optical field. Therefore, these quantitative formulations are not valid for other cases where one of those mechanisms, such as molecular reorientation, electrostriction, slow population change, and the opto-thermal effect, plays a major role, or a transient process is involved.

18.1 Density matrix and interaction energy

18.1.1 Basic equations of the density matrix

In principle, the quantitative properties of the nonlinear polarization of any type of medium can be described by using the method of the density matrix, which is a combination of the perturbation theory of quantum mechanics and the method of statistical mechanics. Based on the density matrix approach, we can investigate the macroscopic optical properties of the medium, starting from the microscopic response of an individual molecule to an applied external optical field.[1-3]

Considering the medium and the applied classical optical field as a coupled system, and assuming that the Hamiltonian (energy) operator of the whole system is H and the density-matrix operator is ρ, which describes the physical state of the system, the time variation of the density matrix is governed by the following equation without considering damping:

$$\frac{d\rho}{dt} = -\frac{i}{\hbar}[H\rho - \rho H]. \tag{18.1-1}$$

Here, $\hbar$ is the Planck constant divided by 2π. For resonant interaction, the damping effect should be considered and the above equation should be modified to

Nonlinear Optics and Photonics. First Edition. Guang S. He. © Guang S. He 2015.
Published in 2015 by Oxford University Press.

$$i\hbar \frac{d\rho(t)}{dt} = [H, \rho(t)] + i\hbar \left[\frac{d\rho}{dt}\right]_{relax}, \qquad (18.1\text{-}2)$$

where the first term on the right-hand side of the equation is expressed by Poisson brackets and the second term is phenomenologically introduced to address the relaxation influence of the damping effect on the density matrix.

On the other hand, the Hamiltonian of the whole system is composed of two parts, i.e.,

$$H = H_0 + H'(t), \qquad (18.1\text{-}3)$$

where H_0 is the unperturbed Hamiltonian of the medium without an applied field and $H'(t)$ is the interaction Hamiltonian between the medium and the applied optical field. The latter can be viewed as a time-dependent perturbation. In the case of perturbation, the density-matrix operator can be expressed in the form of a series:

$$\rho(t) = \rho^{(0)} + \rho^{(1)}(t) + \rho^{(2)}(t) + \cdots + \rho^{(r)}(t). \qquad (18.1\text{-}4)$$

Here, $\rho^{(0)}$ is the initial value of density matrix when there is no external field, and $\rho^{(r)}(t)$ is the rth term that is assumed to be proportional to the rth power of $H'(t)$.

If the medium is in thermal equilibrium, we have

$$\rho^{(0)} = Ze^{-H_0/k_B T}, \qquad (18.1\text{-}5)$$

where Z is a normalization factor and k_B is the Boltzmann constant. Substituting Eqs. (18.1-4) and (18.1-3) into Eq. (18.1-2) leads to the following equations:

$$\begin{aligned} i\hbar \frac{d\rho^{(1)}}{dt} &= [H_0, \rho^{(1)}] + [H', \rho^{(0)}] + i\hbar\Gamma\rho^{(1)} \\ &\quad \cdots\cdots \\ i\hbar \frac{d\rho^{(r)}}{dt} &= [H_0, \rho^{(r)}] + [H', \rho^{(r-1)}] + i\hbar\Gamma\rho^{(r)} \end{aligned} \qquad (18.1\text{-}6)$$

Here, $i\hbar\Gamma\rho^{(r)}$ is the relaxation term representing the damping effect and Γ is a phenomenologically introduced constant, which has a definite meaning when we later examine the relaxation behavior of the off-diagonal elements of the density matrix.

From Eq. (18.1-6) one can see that if $\rho^{(0)}$ and $H'(t)$ are known, the solutions from $\rho^{(1)}$ to $\rho^{(r)}$ can be obtained by using a step-by-step method; finally the overall density-matrix operator ρ can be determined.

On the other hand, once the density-matrix operator ρ is known, the macroscopic average of an operator ξ, which represents an observable physical quantity, can be determined by the following general formulation:

$$\bar{\xi} = \sum_n (\xi\rho)_{nn} = \text{Tr}(\xi\rho). \qquad (18.1\text{-}7)$$

Here, symbol $\text{Tr}(A)$ denotes the trace of a matrix A.

According to Eq. (18.1-7), if we know the operator of the sum of the induced electric-dipole moment in a unit volume of the medium and the density matrix ρ, the macroscopic polarization and the relevant susceptibilities of the medium can be finally determined. Therefore, the first issue for us is to obtain the solutions of various orders of the density matrix components.

18.1.2 Expression for interaction energy

From Eq. (18.1-6) we can see that to obtain solutions of various-order components of the density matrix, we must know the specific expression for $H'(t)$, the operator of interaction energy between the medium and an applied optical field. For simplicity, here we only consider the interaction between an outer-shell electron and the applied optical electromagnetic field. In this simplest case, the interaction energy operator can be written as[2,5,6]

$$H'(t) = -\frac{e}{2m_e}(\mathbf{p}_e \cdot \mathbf{A} + \mathbf{A} \cdot \mathbf{p}_e) + \frac{e^2}{2m_e}\mathbf{A} \cdot \mathbf{A} + e\tilde{V}, \qquad (18.1\text{-}8)$$

where e and m_e are the charge and mass of the electron, $\mathbf{p}_e$ is the momentum operator of the electron, and $\mathbf{A}$ and $\tilde{V}$, respectively, are the vector and the scalar potential of the applied optical field. Choosing Coulomb gauge so that $\nabla \cdot \mathbf{A} = 0$ and $\tilde{V} = 0$, and neglecting the smaller term of $\mathbf{A} \cdot \mathbf{A}$, we have[6]

$$H'(t) = -\frac{e}{m_e}\mathbf{A} \cdot \mathbf{p}_e. \qquad (18.1\text{-}9)$$

On the other hand, the vector potential $\mathbf{A}$ can be expressed as a Fourier integral, i.e.,

$$\mathbf{A}(\mathbf{r}, t) = \int_{-\infty}^{\infty} \mathbf{A}_0(\omega) e^{-i(\omega t - \mathbf{k} \cdot \mathbf{r})} d\omega, \qquad (18.1\text{-}10)$$

where $\mathbf{k}$ is the wavevector of a monochromatic plane wave of frequency ω, and $\mathbf{r}$ is the position vector of the electron in a molecular coordinate system. Similarly, the electric field $\mathbf{E}$ can also be expressed by the Fourier integral. Based on the following relation between $\mathbf{E}$ and $\mathbf{A}$:

$$\mathbf{E}(\mathbf{r}, t) = -\frac{\partial}{\partial t}\mathbf{A}(\mathbf{r}, t), \qquad (18.1\text{-}11)$$

we obtain the amplitude of the Fourier component of the electric field

$$\mathbf{E}_0(\omega) = i\omega \mathbf{A}_0(\omega). \qquad (18.1\text{-}12)$$

Substituting Eq. (18.1-10) into Eq. (18.1-9) we obtain the expression for the Fourier component of $H'(t)$

$$H'(\omega_{mn}) = -\frac{e}{im_e\omega_{mn}}\mathbf{E}_0(\omega_{mn}) \cdot \mathbf{p}_e e^{i(\mathbf{k} \cdot \mathbf{r})}. \qquad (18.1\text{-}13)$$

In order to derive the quantitative expression for the susceptibility of a medium, we must know the matrix elements of $H'(\omega_{mn})$. The latter is determined by

$$H'_{mn}(\omega_{mn}) = -\frac{e}{im_e\omega_{mn}}\mathbf{E}_0(\omega_{mn}) \cdot \int \varphi_m^* \mathbf{p}_e e^{i(\mathbf{k} \cdot \mathbf{r})} \varphi_n d\tau, \qquad (18.1\text{-}14)$$

where $\varphi_m(\tau)$ and $\varphi_n(\tau)$ are the unperturbed eigenfunctions of the molecular system with no external optical field and τ denotes the complete molecular coordinates. In Eq. (18.1-14) we can replace the exponential term approximately by

$$e^{i(\mathbf{k} \cdot \mathbf{r})} \approx 1 + i(\mathbf{k} \cdot \mathbf{r}) + \cdots. \qquad (18.1\text{-}15)$$

Taking only the first two terms in the above expression and substituting them into Eq. (18.1-14), we obtain

$$H'_{mn}(\omega_{mn}) = -\mathbf{E}(\omega_{mn}) \cdot (\mathbf{D}^{(1)}_{mn} + \mathbf{D}^{(2)}_{mn}). \tag{18.1-16}$$

Here, $\mathbf{D}^{(1)}_{mn}$ comes from the first term of Eq. (18.1-15), which refers to the matrix element of the molecular electric-dipole moment operator and is determined by

$$\mathbf{D}^{(1)}_{mn} = \frac{e}{im_e\omega_{mn}} \int \phi^*_m \mathbf{p}_e \phi_n d\tau = \frac{e}{im_e\omega_{mn}} (\mathbf{p}_e)_{mn} = (e\mathbf{r})_{mn}, \tag{18.1-17}$$

whereas $\mathbf{D}^{(2)}_{mn}$ comes from the second term of Eq. (18.1-15), which refers to the matrix element of the molecular electric-quadrupole and magnetic-dipole operator and is determined by

$$\mathbf{D}^{(2)}_{mn} = \frac{e}{m_e\omega_{mn}} \int \phi^*_m (\mathbf{k} \cdot \mathbf{r}) \mathbf{p}_e \phi_n d\tau = \frac{e}{\omega_{mn}} \left[(\mathbf{k} \cdot \mathbf{r}) \frac{d\mathbf{r}}{dt} \right]_{mn}. \tag{18.1-18}$$

According to the time-derivative rule of an operator, it follows that

$$(\mathbf{k} \cdot \mathbf{r}) \frac{d\mathbf{r}}{dt} = \frac{1}{2} \frac{d}{dt}[(\mathbf{k} \cdot \mathbf{r})\mathbf{r}] - \frac{1}{2}\left[\mathbf{k} \times \left(\mathbf{r} \times \frac{d\mathbf{r}}{dt}\right)\right].$$

Since

$$\left(\mathbf{r} \times \frac{d\mathbf{r}}{dt}\right) = \frac{1}{m_e}(\mathbf{r} \times \mathbf{p}_e) = \frac{1}{m_e}\mathbf{M}, \tag{[18.1-19}$$

where $\mathbf{M}$ is the angular momentum operator of the electron, Eq. (18.1-18) can be rewritten as

$$\mathbf{D}^{(2)}_{mn} = \frac{e}{\omega_{mn}} \left\{ \frac{1}{2} \frac{d}{dt}[(\mathbf{k} \cdot \mathbf{r})\mathbf{r}] \right\}_{mn} - \frac{e}{2m_e\omega_{mn}} (\mathbf{k} \times \mathbf{M})_{mn}$$

$$= \frac{ie}{2}[(\mathbf{k} \cdot \mathbf{r})\mathbf{r}]_{mn} - \frac{e}{2m_e\omega_{mn}} (\mathbf{k} \times \mathbf{M})_{mn}.$$

If we further define the electric quadrupole (tensor) operator by

$$\hat{q} = \frac{e}{2}\mathbf{rr} = \frac{e}{2}\begin{bmatrix} x^2 & xy & xz \\ yx & y^2 & yz \\ zx & zy & z^2 \end{bmatrix}, \tag{18.1-20}$$

and define the magnetic dipole operator by

$$\mathbf{m} = \frac{e}{2m_e}\mathbf{M}, \tag{18.1-21}$$

then the matrix element of $\mathbf{D}^{(2)}_{mn}$ can be finally expressed as

$$\mathbf{D}^{(2)}_{mn} = i(\mathbf{k}\hat{q})_{mn} - \frac{1}{c}(\mathbf{n}_0 \times \mathbf{m})_{mn}, \tag{18.1-22}$$

where $\mathbf{n}_0$ is the unit vector along the direction of the wavevector and c is the speed of light in free space.

Substituting Eqs. (18.1-17) and (18.1-22) into Eq. (18.1-16) leads to

$$H'_{mn}(\omega) = -\mathbf{E}_0(\omega) \cdot [\mathbf{p} + i(\mathbf{k}\hat{q}) - \frac{1}{c}(\mathbf{n}_0 \times \mathbf{m})]_{mn}. \qquad (18.1\text{-}23)$$

Here the previous notation $\mathbf{D}_{mn}^{(1)}$ of the electric-dipole moment operator is replaced by a customary notation $\mathbf{p} = e\mathbf{r}$. It should be noted that when the position vector $\mathbf{r}$ of the outer-shell electron reverses its direction, the operator $\mathbf{p}$ changes its sign, whereas the signs of the electric-quadrupole operator $\hat{q}$ and magnetic-dipole operator $\mathbf{m}$ remain unchanged.

Based on Eq. (18.1-23) we can also present the interaction-Hamiltonian operator in the following form:

$$H'(\omega) = -\mathbf{p} \cdot \mathbf{E}_0(\omega) - i(\mathbf{k}\hat{q}) \cdot \mathbf{E}_0(\omega) + \frac{1}{c}(\mathbf{n}_0 \times \mathbf{m}) \cdot \mathbf{E}_0(\omega). \qquad (18.1\text{-}24)$$

Using the following relation between the electric field $\mathbf{E}$ and the magnetic induction $\mathbf{B}$:

$$\mathbf{B}_0 = \frac{1}{c}(\mathbf{n}_0 \times \mathbf{E}_0),$$

Eq. (18.1-24) can be rewritten as

$$H'(\omega) = -\mathbf{p} \cdot \mathbf{E}_0(\omega) - i\mathbf{k} \cdot \hat{q}\mathbf{E}_0(\omega) - \mathbf{m} \cdot \mathbf{B}_0(\omega). \qquad (18.1\text{-}25)$$

In the above expression, the first term represents the contribution from the electric-dipole interaction, the second term represents the contribution from the electric-quadrupole interaction, and the third term represents the magnetic-dipole interaction.

Now we can see that in Eq. (18.1-15) the first term corresponds to the electric-dipole interaction, and the second term represents both the electric-quadrupole and the magnetic-dipole interactions. $H'(\omega)$ and its matrix elements $H'_{mn}(\omega)$ are most essential for calculating transition probabilities of a molecular system, and for determining susceptibilities of the medium. In these cases, the magnitude of the position vector of the outer-shell electron is roughly equal to the radius a_0 of an individual molecule (or atom); therefore, the ratio of the first term to the second term in Eq. (18.1-15) is determined approximately by

$$\frac{1}{\mathbf{k} \cdot \mathbf{r}} \approx \frac{1}{ka_0} \approx \frac{\lambda}{2\pi a_0}. \qquad (18.1\text{-}26)$$

For an optical field, the wavelength λ is much larger than the molecular radius a_0; hence, the second term (electric-quadrupole and magnetic-dipole contributions) in Eq. (18.1-15) can be neglected. Nevertheless, in some special cases, the contribution from the electric-dipole interaction is zero; therefore, the contribution from the electric-quadrupole and magnetic-dipole interaction may have to be considered. Throughout this chapter, all discussions are conducted in the electric-dipole approximation unless noted otherwise.

18.2 Expressions for susceptibilities based on density matrix solutions

18.2.1 Solutions of density matrix equations

When we consider the interaction of light with matter, it is often convenient to treat the applied optical field as a combination of a limited number of monochromatic plane electromagnetic waves, and the general incident field can be expressed in the form of a Fourier series:

$$\mathbf{E}(\mathbf{r}, t) = \mathbf{E}(t) = \sum_n \mathbf{E}(\omega_n) e^{i\omega_n t}. \tag{18.2-1}$$

Here we neglect the spatial variation of the field within a molecular scale because the wavelength of the field is much greater than the molecular radius. In the electric-dipole approximation, the interaction Hamiltonian operator for a single molecule can also be written as a Fourier series,

$$H'(t) = -\mathbf{E}(t) \cdot \mathbf{p} = -\sum_n \mathbf{p} \cdot \mathbf{E}(\omega_n) e^{i\omega_n t} = \sum_n H'(\omega_n) e^{i\omega_n t}. \tag{18.2-2}$$

Here, $\mathbf{p}$ is the electric-dipole operator of an individual molecule.

Accordingly, various-order components of the density-matrix operator can also be expressed by a Fourier series

$$\rho^{(r)}(t) = \sum_n \rho^{(r)}(\omega_n) e^{i\omega_n t}. \tag{18.2-3}$$

Substituting Eq. (18.2-2) and

$$\rho^{(1)}(t) = \sum_n \rho^{(1)}(\omega_n) e^{i\omega_n t},$$

into the first equation of Eq. (18.1-6), we obtain

$$-\hbar \omega_n \rho^{(1)}(\omega_n) = [H_0, \rho^{(1)}(\omega_n)] + [H'(\omega_n), \rho^{(0)}] + i\hbar \Gamma \rho^{(1)}(\omega_n). \tag{18.2-4}$$

Next, we use the unperturbed eigenfunctions of the molecular system to perform a matrix-element manipulation on both sides of the above equation, and notice that

$$[H_0, \rho^{(1)}(\omega_n)]_{ab} = \int \phi_a^* [H_0, \rho^{(1)}(\omega_n)] \phi_b d\tau$$
$$= (E_a - E_b)[\rho^{(1)}(\omega_n)]_{ab}$$
$$= \hbar \omega_{ab} [\rho^{(1)}(\omega_n)]_{ab},$$
$$[H'(\omega_n), \rho^{(0)}]_{ab} = (\rho_{bb}^{(0)} - \rho_{aa}^{(0)})[H'(\omega_n)]_{ab},$$
$$[\Gamma \rho^{(1)}(\omega_n)]_{ab} = \Gamma_{ab} [\rho^{(1)}(\omega_n)]_{ab},$$

where E_a and E_b are the eigenenergy values of the unperturbed molecular system, $\omega_{ab} = (E_a - E_b)/\hbar$ is the transition frequency from the state a to the state b, $\rho_{aa}^{(0)}$ and $\rho_{bb}^{(0)}$ are the diagonal elements of the zero-order density matrix, and Γ_{ab} is the damping factor related to the off-diagonal element of the density matrix. Then we obtain the following equation:

$$-\hbar \omega_n [\rho^{(1)}(\omega_n)]_{ab} = \hbar \omega_{ab} [\rho^{(1)}(\omega_n)]_{ab} + (\rho_{bb}^{(0)} - \rho_{aa}^{(0)})[H'(\omega_n)]_{ab} + i\hbar \Gamma_{ab} [\rho^{(1)}(\omega_n)]_{ab},$$

and the solution of the Fourier component of the first-order density matrix is

$$\rho_{ab}^{(1)}(\omega_n) = \frac{H'_{ab}(\omega_n)(\rho_{bb}^{(0)} - \rho_{aa}^{(0)})}{\hbar(\omega_{ba} - \omega_n - i\Gamma_{ab})} = \frac{H'_{ab}(\omega_n)(\rho_{bb}^{(0)} - \rho_{aa}^{(0)})}{D_{ab}(\omega_n)}, \qquad (18.2\text{-}5)$$

where

$$\left. \begin{array}{l} \rho_{ab}^{(1)}(\omega_n) = [\rho^{(1)}(\omega_n)]_{ab} \\ H'_{ab}(\omega_n) = [H'(\omega_n)]_{ab} \\ D_{ab}(\omega_n) = \hbar(\omega_{ba} - \omega_n - i\Gamma_{ab}) \\ \omega_{ba} = -\omega_{ab} \end{array} \right\}.$$

To find the solution of the second-order density matrix, we can substitute the expression

$$\rho^{(2)}(t) = \sum_n \rho^{(2)}(\omega_n) e^{i\omega_n t}$$

into the second equation of Eq. (18.1-6) and perform a matrix-element manipulation on both sides of this equation; then we obtain

$$\sum_n (-\hbar\omega_n - \hbar\omega_{ab} - i\hbar\Gamma_{ab}) \rho_{ab}^{(2)}(\omega_n) e^{i\omega_n t} = [H'(t), \rho^{(1)}(t)]_{ab}. \qquad (18.2\text{-}6)$$

Assuming that the applied field involves only two monochromatic waves of different frequencies, the interaction Hamiltonian can be written as

$$H'(t) = [\mathbf{E}(\omega_1) e^{i\omega_1 t} + \mathbf{E}(\omega_2) e^{i\omega_2 t}] \cdot \mathbf{p} = H'(\omega_1) e^{i\omega_1 t} + H'(\omega_2) e^{i\omega_2 t}.$$

Substituting the above equation and the solution of $\rho^{(1)}(t)$,

$$\rho^{(1)}(t) = \sum_m \rho^{(1)}(\omega_m) e^{i\omega_m t},$$

into the right-hand side of Eq. (18.2-6) leads to

$$\sum_n D_{ab}(\omega_n) \rho_{ab}^{(2)}(\omega_n) e^{i\omega_n t} = \sum_m [H'(\omega_1), \rho_1(\omega_m)]_{ab} e^{i(\omega_1 + \omega_m)t}$$
$$+ \sum_m [H'(\omega_2), \rho_1(\omega_m)]_{ab} e^{i(\omega_2 + \omega_m)t}.$$

It will become evident that the second-order density matrix can be used to describe the second-order nonlinear polarization processes. Considering only the sum-frequency generation and equating the terms with the same sum-frequency argument on both sides of the above equation, we obtain

$$D_{ab}(\omega_1 + \omega_2) \rho_{ab}^{(2)}(\omega_1 + \omega_2) = [H'(\omega_1), \rho^{(1)}(\omega_2)]_{ab} + [H'(\omega_2), \rho^{(1)}(\omega_1)]_{ab}. \qquad (18.2\text{-}7)$$

Remembering the manipulation rule of the product of two operators,

$$[AB]_{ab} = \sum_c (A)_{ac} (B)_{cb},$$

and substituting the elements of $\rho^{(1)}(\omega_1)$ and $\rho^{(1)}(\omega_2)$ given by Eq. (18.2-5) into Eq. (18.2-7) leads to a solution of the element of the second-order density matrix:

$$\rho^{(2)}_{ab}(\omega_1+\omega_2) = \frac{1}{D_{ab}(\omega_1+\omega_2)} \sum_c \left\{ \frac{H'_{ac}(\omega_1)H'_{cb}(\omega_2)[\rho^{(0)}_{bb}-\rho^{(0)}_{cc}]}{D_{cb}(\omega_2)} \right.$$
$$+ \frac{H'_{ac}(\omega_2)H'_{cb}(\omega_1)[\rho^{(0)}_{bb}-\rho^{(0)}_{cc}]}{D_{cb}(\omega_1)} - \frac{H'_{ac}(\omega_2)H'_{cb}(\omega_1)[\rho^{(0)}_{cc}-\rho^{(0)}_{aa}]}{D_{ac}(\omega_2)} \quad (18.2\text{-}8)$$
$$\left. - \frac{H'_{ac}(\omega_1)H'_{cb}(\omega_2)[\rho^{(0)}_{cc}-\rho^{(0)}_{aa}]}{D_{ac}(\omega_1)} \right\}.$$

Here, $D_{ab}(\omega_1+\omega_2) = \hbar(\omega_{ba}-\omega_1-\omega_2-i\Gamma_{ab})$, $D_{cb}(\omega_2) = \hbar(\omega_{bc}-\omega_2-i\Gamma_{cb})$, etc., and the subscripts a, b, and c denote different unperturbed eigenstates of the molecular system. Thus, Eq. (18.2-8) can be rewritten in a more compact form

$$\rho^{(2)}_{ab}(\omega_1+\omega_2) = \frac{1}{D_{ab}(\omega_1+\omega_2)}$$
$$\cdot \sum_{(\omega_1\omega_2)} \sum_c \left[H'_{ac}(\omega_1)H'_{cb}(\omega_2) \left(\frac{\rho^{(0)}_{bb}-\rho^{(0)}_{cc}}{D_{cb}(\omega_2)} + \frac{\rho^{(0)}_{aa}-\rho^{(0)}_{cc}}{D_{ac}(\omega_1)} \right) \right]. \quad (18.2\text{-}9)$$

Here, the first summation symbol $\sum_{(\omega_1\omega_2)}$ denotes that the expression that follows it is to be summed over the permutation of ω_1 and ω_2.

After going through a similar procedure, we could also obtain the following solution for the Fourier component of the third-order density matrix:

$$\rho^{(3)}_{ab}(\omega_1+\omega_2+\omega_3) = \frac{1}{D_{ab}(\omega_1+\omega_2+\omega_3)}$$
$$\cdot \sum_{(\omega_1\omega_2\omega_3)} \sum_{c,d} \left\{ \frac{H'_{ac}(\omega_1)H'_{cd}(\omega_2)H'_{db}(\omega_3)}{D_{cb}(\omega_2+\omega_3)} \left[\frac{\rho^{(0)}_{bb}-\rho^{(0)}_{dd}}{D_{db}(\omega_3)} + \frac{\rho^{(0)}_{cc}-\rho^{(0)}_{dd}}{D_{cd}(\omega_2)} \right] \right.$$
$$\left. + \frac{H'_{ad}(\omega_2)H'_{dc}(\omega_3)H'_{cb}(\omega_1)}{D_{ac}(\omega_2+\omega_3)} \left[\frac{\rho^{(0)}_{dd}-\rho^{(0)}_{cc}}{D_{cd}(\omega_3)} + \frac{\rho^{(0)}_{dd}-\rho^{(0)}_{aa}}{D_{da}(\omega_2)} \right] \right\}$$
$$\quad (18.2\text{-}10)$$

Here the first summation symbol denotes that the expression that follows it is to be summed over all possible permutations of ω_1, ω_2, and ω_3.

In principle, we could find the further solutions for the higher-order density-matrix elements. However, in most cases it is enough to find the solutions up to the third order.

18.2.2 Explicit formulations of various-order susceptibilities

As mentioned in the first section of this chapter, the density matrix is an operator depending on the Hamiltonian operator for a given system, and can be used to determine the statistical average of any observable physical quantity based on Eq. (18.1-7).

Now let us consider a macroscopic sub-ensemble of molecules (or atoms). The dimension of this sub-ensemble should be much larger than that of each individual particle but much shorter than the wavelength of the applied optical field. For such a specially chosen system, the average of a given physical quantity over the whole sub-ensemble represents a specific macroscopic property of the medium. In the meantime, we can neglect the influence of the spatial variation of the optical field within the volume of this sub-ensemble. The latter feature may greatly simplify the theoretical treatment.

Assuming that the sub-ensemble of volume V consists of ζ identical and independent molecules, the sum of the electric-dipole operators of molecules within a unit volume is

$$\mathbf{R} = \frac{\zeta}{V}\mathbf{p} = N\mathbf{p}, \qquad (18.2\text{-}11)$$

where $\mathbf{p}$ is the electric-dipole-moment operator of an individual molecule and N is the number of molecules in a unit volume.

It is well known that the electric polarization of a medium is a macroscopic physical quantity defined as the average value of the sum of the induced electric-dipole moment per unit volume. According to Eq. (18.1-7), the electric polarization of this sub-ensemble should be

$$\mathbf{P}^{(r)}(t) = \text{Tr}[\mathbf{R}\rho^{(r)}(t)] = N\,\text{Tr}[\mathbf{p}\rho^{(r)}(t)]. \qquad (18.2\text{-}12)$$

Here, $\rho^{(r)}(t)$ is the rth-order density operator.

18.2.2.1 First-order susceptibility

Based on Eq. (18.2-12) it is easy to write the expression for the first-order polarization of the considered system,

$$\mathbf{P}^{(1)}(t) = N\,\text{Tr}[\mathbf{p}\rho^{(1)}(t)]. \qquad (18.2\text{-}13)$$

According to Eq. (18.2-3), $\rho^{(1)}(t)$ can be expanded as a Fourier series, so the above expression becomes

$$\mathbf{P}^{(1)}(t) = N\sum_n \text{Tr}[\mathbf{p}\rho^{(1)}(\omega_n)]e^{i\omega_n t},$$

and the Fourier component of the first-order polarization is

$$\mathbf{P}^{(1)}(\omega) = N\,\text{Tr}[\mathbf{p}\rho^{(1)}(\omega)] = N\sum_a [\mathbf{p}\rho^{(1)}(\omega)]_{aa}$$

$$= N\sum_{a,b} (\mathbf{p})_{ab}\rho^{(1)}_{ba}(\omega) \qquad (18.2\text{-}14)$$

$$= N\sum_{a,b} (\mathbf{p})_{ab} H'_{ba}(\omega)\frac{\rho^{(0)}_{aa} - \rho^{(0)}_{bb}}{D_{ba}(\omega)}.$$

In obtaining the above expression, we utilized Eq. (18.2-5). On the other hand, from Eq. (18.2-2) we know that in the electric-dipole approximation, the element of the Fourier component of the perturbed Hamiltonian operator can be written as

$$H'_{ba}(\omega) = -[\mathbf{p}\cdot\mathbf{E}(\omega)]_{ba} = -\sum_j [p_j E_j(\omega)]_{ba} = -(p_j)_{ba} E_j(\omega), \qquad (18.2\text{-}15)$$

where $j = x, y, z$, and, for the sake of conciseness, we omit the summation symbol $\sum_j$ over the repeated subscript j.

Substituting Eq. (18.2-15) into Eq. (18.2-14), we obtain the linear electric polarization in terms of its Cartesian components:

588 *Detailed Theory of Nonlinear Susceptibilities*

$$P_i^{(1)}(\omega) = -N \sum_{a,b} \frac{(p_i)_{ab}(p_j)_{ba}}{D_{ba}(\omega)} [\rho_{aa}^{(0)} - \rho_{bb}^{(0)}] E_j(\omega). \tag{18.2-16}$$

On the other hand, we already know that (see Eq. (2.3-3))

$$P_i^{(1)}(\omega) = \varepsilon_0 \chi_{ij}^{(1)} E_j(\omega), \tag{18.2-17}$$

Here we also omit the summation symbol over the repeated subscript j; therefore we can obtain the explicit expression for the first-order susceptibility,

$$\begin{aligned}\chi_{ij}^{(1)}(\omega) &= -\frac{N}{\varepsilon_0} \sum_{a,b} \frac{(p_i)_{ab}(p_j)_{ba}}{D_{ba}(\omega)} [\rho_{aa}^{(0)} - \rho_{bb}^{(0)}] \\ &= \frac{N}{\varepsilon_0 \hbar} \sum_{a,b} \rho_{aa}^{(0)} \left[\frac{(p_i)_{ab}(p_j)_{ba}}{\omega_{ba} + \omega + i\Gamma_{ba}} + \frac{(p_j)_{ab}(p_i)_{ba}}{\omega_{ba} - \omega - i\Gamma_{ba}} \right].\end{aligned} \tag{18.2-18}$$

Here it is assumed that $\omega_{ab} = -\omega_{ba}$ and $\Gamma_{ab} = \Gamma_{ba}$.

For non-resonant interaction, the frequency of the applied field is far away from the Bohr frequency of the molecule, i.e., $(\omega_{ba} - \omega) \gg \Gamma_{ba}$. Thus the terms of $i\Gamma_{ab}$ in Eq. (18.2-18) can be neglected and we have

$$\chi_{ij}^{(1)}(\omega) = \frac{N}{\varepsilon_0 \hbar} \sum_{a,b} \rho_{aa}^{(0)} \left[\frac{(p_i)_{ab}(p_j)_{ba}}{\omega_{ba} + \omega} + \frac{(p_j)_{ab}(p_i)_{ba}}{\omega_{ba} - \omega} \right]. \tag{18.2-19}$$

Since the electric-dipole moment **p** is a Hermitian operator, it follows that $(p_i)_{ab} = [(p_i)_{ba}]^*$. On the other hand, **p** is a real operator, and the unperturbed eigenfunctions ϕ_a and ϕ_b of the molecule can also be chosen as real functions; therefore the elements of **p** should be real, i.e.,

$$(p_i)_{ab} = (p_i)_{ba}. \tag{18.2-20}$$

Based on this property, Eq. (18.2-19) can be further simplified as

$$\chi_{ij}^{(1)}(\omega) = \frac{N}{\varepsilon_0 \hbar} \sum_{a,b} \rho_{aa}^{(0)} (p_i)_{ab}(p_j)_{ba} \frac{2\omega_{ba}}{\omega_{ba}^2 - \omega^2}. \tag{18.2-21}$$

Here the subscripts a and b denote all possible eigenstates of the molecule.

18.2.2.2 Second-order susceptibility

After going through a procedure similar to that described above, we can obtain the expression for the second-order electric polarization of the medium as follows:

$$\begin{aligned}P_i^{(2)}(\omega_1 + \omega_2) &= N \sum_{a,b} (p_i)_{ab} \rho_{ba}^{(2)}(\omega_1 + \omega_2) \\ &= N \sum_{a,b} (p_i)_{ab} \sum_{(\omega_1 \omega_2)} \sum_c \frac{H'_{bc}(\omega_1) H'_{ca}(\omega_2)}{D_{ba}(\omega_1 + \omega_2)} \left[\frac{\rho_{aa}^{(0)} - \rho_{cc}^{(0)}}{D_{ca}(\omega_2)} + \frac{\rho_{bb}^{(0)} - \rho_{cc}^{(0)}}{D_{bc}(\omega_1)} \right] \\ &= N \sum_{(\omega_1 \omega_2)} \sum_{a,b,c} \frac{(p_i)_{ab}(p_j)_{bc}(p_k)_{ca}}{D_{ba}(\omega_1 + \omega_2)} \left[\frac{\rho_{aa}^{(0)} - \rho_{cc}^{(0)}}{D_{ca}(\omega_2)} + \frac{\rho_{bb}^{(0)} - \rho_{cc}^{(0)}}{D_{bc}(\omega_1)} \right] E_j(\omega_1) E_k(\omega_2).\end{aligned} \tag{18.2-22}$$

In obtaining this result, Eq. (18.2-9) and the electric-dipole approximation have been used.

On the other hand, from Eq. (2.3-6) we know that the second-order polarization can be expressed in a more conventional form, i.e.,

$$P_i^{(2)}(\omega_1 + \omega_2) = \varepsilon_0 \chi_{ijk}^{(2)}(\omega_1 + \omega_2) E_j(\omega_1) E_k(\omega_2). \tag{18.2-23}$$

Here we omit the summation symbol over the repeated subscripts j and k. Comparing the above expression with Eq. (18.2-22) one can immediately obtain the explicit expression for $\chi_{ijk}^{(2)}$. Considering that in Eq. (18.2-22) the permutation between $E(\omega_1)$ and $E(\omega_2)$ does not cause any physical difference, we have

$$\chi_{ijk}^{(2)}(\omega_1, \omega_2) = \frac{1}{2}[\chi_{ijk}^{(2)}(\omega_1, \omega_2) + \chi_{ikj}^{(2)}(\omega_2, \omega_1)]$$

$$= \frac{N}{\varepsilon_0 2} \sum_{(jk)}^{(\omega_1 \omega_2)} \sum_{a,b,c} \frac{(p_i)_{ab}(p_j)_{bc}(p_k)_{ca}}{D_{ba}(\omega_1 + \omega_2)} \left[\frac{\rho_{aa}^{(0)} - \rho_{cc}^{(0)}}{D_{ca}(\omega_2)} + \frac{\rho_{bb}^{(0)} - \rho_{cc}^{(0)}}{D_{bc}(\omega_1)} \right]$$

$$= \frac{N}{\varepsilon_0 2} \sum_{(jk)}^{(\omega_1 \omega_2)} \sum_{a,b,c} \rho_{aa}^{(0)} \left[\frac{(p_i)_{ab}(p_j)_{bc}(p_k)_{ca}}{D_{ba}(\omega_1 + \omega_2) D_{ca}(\omega_1)} - \frac{(p_i)_{cb}(p_j)_{ba}(p_k)_{ac}}{D_{bc}(\omega_1 + \omega_2) D_{ac}(\omega_2)} \right.$$

$$\left. + \frac{(p_i)_{ba}(p_j)_{ac}(p_k)_{cb}}{D_{ab}(\omega_1 + \omega_2) D_{ac}(\omega_1)} - \frac{(p_i)_{cb}(p_j)_{ba}(p_k)_{ac}}{D_{bc}(\omega_1 + \omega_2) D_{ba}(\omega_1)} \right]. \tag{18.2-24}$$

Here, the first summation symbol $\sum_{(jk)}^{(\omega_1 \omega_2)}$ denotes that the following expression should be summed over all possible permutation of the pairs of $j\omega_1$ and $k\omega_2$, whereas the subscripts a, b, and c denote all possible unperturbed eigenstates of the molecule. Substituting the explicit expressions for all terms in denominators into the above equation and noticing that the terms involved in the square brackets remain unchanged for the simultaneous permutation of the pairs of $j\omega_1$ and $k\omega_2$, we obtain the final explicit expression for the second-order susceptibility:

$$\chi_{ijk}^{(2)}(\omega_1, \omega_2) = \frac{N}{\varepsilon_0 2\hbar^2} \sum_{(jk)}^{(\omega_1 \omega_2)} \sum_{a,b,c} \rho_{aa}^{(0)} \left[\frac{(p_i)_{ab}(p_j)_{bc}(p_k)_{ca}}{(\omega_{ba} - \omega_1 - \omega_2 - i\Gamma_{ba})(\omega_{ca} - \omega_2 - i\Gamma_{ca})} \right.$$

$$+ \frac{(p_j)_{ab}(p_i)_{bc}(p_k)_{ca}}{(\omega_{cb} - \omega_1 - \omega_2 - i\Gamma_{bc})} \left(\frac{1}{\omega_{ac} + \omega_2 + i\Gamma_{ac}} + \frac{1}{\omega_{ba} + \omega_1 + i\Gamma_{ba}} \right) \tag{18.2-25}$$

$$\left. + \frac{(p_k)_{ab}(p_j)_{bc}(p_i)_{ca}}{(\omega_{ca} + \omega_1 + \omega_2 + i\Gamma_{ca})(\omega_{ba} + \omega_2 + i\Gamma_{ba})} \right].$$

For non-resonant interaction, the frequencies ω_1 and ω_2 and their combination $(\omega_1 \pm \omega_2)$ are far away from the molecular transition frequencies ω_{ba}, ω_{ca}, ω_{cb}, etc., so that all the damping factors can be neglected and Eq. (18.2-25) is simplified as

$$\chi_{ijk}^{(2)}(\omega_1, \omega_2) = \frac{N}{\varepsilon_0 2\hbar^2} \sum_{(jk)}^{(\omega_1 \omega_2)} \sum_{a,b,c} \rho_{aa}^{(0)} \left[\frac{(p_i)_{ab}(p_j)_{bc}(p_k)_{ca}}{(\omega_{ba} - \omega_1 - \omega_2)(\omega_{ca} - \omega_2)} \right.$$

$$\left. + \frac{(p_j)_{ab}(p_i)_{bc}(p_k)_{ca}}{(\omega_{ba} + \omega_1)(\omega_{ca} - \omega_2)} + \frac{(p_k)_{ab}(p_j)_{bc}(p_i)_{ca}}{(\omega_{ba} + \omega_2)(\omega_{ca} + \omega_1 + \omega_2)} \right]. \tag{18.2-26}$$

18.2.2.3 Third-order susceptibility

Similarly, the third-order nonlinear susceptibility can be finally obtained as

$$\chi^{(3)}_{ijkl}(\omega_1,\omega_2,\omega_3) = \frac{N}{\varepsilon_0 6\hbar^3} \sum_{(jkl)}^{(\omega_1\omega_2\omega_3)} \sum_{a,b,c,d} \rho^{(0)}_{aa}$$

$$\cdot \left[\frac{(p_i)_{ab}(p_j)_{bc}(p_k)_{cd}(p_l)_{da}}{(\omega_{ba}-\omega_1-\omega_2-\omega_3-i\Gamma_{ba})(\omega_{ca}-\omega_2-\omega_3-i\Gamma_{ca})(\omega_{da}-\omega_3-i\Gamma_{da})} \right.$$

$$+ \frac{(p_j)_{ab}(p_i)_{bc}(p_k)_{cd}(p_l)_{da}}{(\omega_{ba}+\omega_1+i\Gamma_{ba})(\omega_{ca}-\omega_2-\omega_3-i\Gamma_{ca})(\omega_{da}-\omega_3-i\Gamma_{da})}$$

$$+ \frac{(p_j)_{ab}(p_k)_{bc}(p_i)_{cd}(p_l)_{da}}{(\omega_{ba}+\omega_1+i\Gamma_{ba})(\omega_{ca}+\omega_1+\omega_2+i\Gamma_{ca})(\omega_{da}-\omega_3-i\Gamma_{da})}$$

$$\left. + \frac{(p_j)_{ab}(p_k)_{bc}(p_l)_{cd}(p_i)_{da}}{(\omega_{ba}+\omega_1+i\Gamma_{ba})(\omega_{ca}+\omega_1+\omega_2+i\Gamma_{ca})(\omega_{da}+\omega_1+\omega_2+\omega_3+i\Gamma_{da})} \right].$$

(18.2-27)

Here the first summation symbol denotes the permutation among the pairs of $j\omega_1$, $k\omega_2$, and $l\omega_3$ and the sequential summation of all terms.

If ω_1, ω_2, and ω_3 (as well as their arbitrary linear combinations) are far from the molecular resonant frequencies and all damping terms can be neglected, then Eq. (18.2-27) becomes

$$\chi^{(3)}_{ijkl}(\omega_1,\omega_2,\omega_3) = \frac{N}{\varepsilon_0 6\hbar^3} \sum_{(jkl)}^{(\omega_1\omega_2\omega_3)} \sum_{a,b,c,d} \rho^{(0)}_{aa}$$

$$\cdot \left[\frac{(p_i)_{ab}(p_j)_{bc}(p_k)_{cd}(p_l)_{da}}{(\omega_{ba}-\omega_1-\omega_2-\omega_3)(\omega_{ca}-\omega_2-\omega_3)(\omega_{da}-\omega_3)} \right.$$

$$+ \frac{(p_j)_{ab}(p_i)_{bc}(p_k)_{cd}(p_l)_{da}}{(\omega_{ba}+\omega_1)(\omega_{ca}-\omega_2-\omega_3)(\omega_{da}-\omega_3)} + \frac{(p_j)_{ab}(p_k)_{bc}(p_i)_{cd}(p_l)_{da}}{(\omega_{ba}+\omega_1)(\omega_{ca}+\omega_1+\omega_2)(\omega_{da}-\omega_3)}$$

$$\left. + \frac{(p_j)_{ab}(p_k)_{bc}(p_l)_{cd}(p_i)_{da}}{(\omega_{ba}+\omega_1)(\omega_{ca}+\omega_1+\omega_2)(\omega_{da}+\omega_1+\omega_2+\omega_3)} \right].$$

(18.2-28)

Here, the subscripts a, b, c, and d denote all the possible eigenstates of the molecule.

18.3 Properties of nonlinear susceptibilities

18.3.1 Local-field corrections

In the preceding subsection, all the explicit expressions for the first- to the third-order susceptibilities have been obtained on the assumption that the medium is composed of a great number of identical and independent molecules. Under this condition, the effective local electric field **E'** imposed on an individual molecule can be recognized as equal to the applied external field **E**. For gaseous media, the average distance between the neighboring molecules is relatively large and the interaction between molecules is negligible; hence the difference between **E'** and **E** can also be neglected.

However, for a liquid or solid medium the interaction between the neighboring molecules can no longer be ignored, and the local electric field imposed on the individual molecule might differ considerably from the macroscopic average field in the medium. In this case, we must consider the so-called local-field corrections of various-order susceptibilities. In general, it is a rather complicated issue, and only for some liquids and ionic crystals with cubic lattice, a simple analytic correction relation could be obtained based on the so-called Lorentz model. According to this model, the local effective field $\mathbf{E}'$ imposed on an individual particle involves two parts: one is the average macroscopic field $\mathbf{E}$; the other is the field generated by the electric dipoles of the neighboring particles. Thus we have the following simple relation:[3-5]

$$\mathbf{E}' = \mathbf{E} + \frac{1}{3\varepsilon_0}\mathbf{P}' = \mathbf{E} + \frac{1}{3\varepsilon_0}(\mathbf{P}'^{(1)} + \mathbf{P}'^{(2)} + \cdots), \qquad (18.3\text{-}1)$$

where $\mathbf{P}'$ is the local polarization field created by the neighboring particles, which can be expanded as a series. Assuming the linear polarizability of the particle is $\alpha^{(1)}$, the linear local polarization field is determined by

$$\mathbf{P}'^{(1)} = \varepsilon_0 N \alpha^{(1)} \mathbf{E}',$$

where N is the number of particles in a unit volume. Substituting the above expression into Eq. (18.3-1) leads to

$$\mathbf{P}'^{(1)} = \frac{\varepsilon_0 N \alpha^{(1)}}{1 - N\alpha^{(1)}/3}\mathbf{E} + \frac{1}{3}\frac{N\alpha^{(1)}}{1 - N\alpha^{(1)}/3}\mathbf{P}'^{(2)}. \qquad (18.3\text{-}2)$$

On the other hand, in the first-order approximation the following relation holds:

$$\mathbf{D} = \varepsilon_0 \mathbf{E} + \mathbf{P}'^{(1)} = \varepsilon_0 \mathbf{E} + \frac{\varepsilon_0 N \alpha^{(1)}}{1 - N\alpha^{(1)}/3}\mathbf{E} = \varepsilon_0 \varepsilon_r \mathbf{E},$$

where $\mathbf{D}$ is the electric displacement and

$$\varepsilon_r = \frac{1 + 2N\alpha^{(1)}/3}{1 - N\alpha^{(1)}/3} \qquad (18.3\text{-}3)$$

is the linear dielectric constant of the medium. Based on the above expression for $\mathbf{D}$, Eq. (18.3-2) can be rewritten as

$$\mathbf{P}'^{(1)} = \varepsilon_0(\varepsilon_r - 1)\mathbf{E} + \frac{(\varepsilon_r - 1)}{3}\mathbf{P}'^{(2)}. \qquad (18.3\text{-}4)$$

Furthermore, in the second-order approximation we have

$$\mathbf{D} = \varepsilon_0 \mathbf{E} + (\mathbf{P}'^{(1)} + \mathbf{P}'^{(2)}).$$

Substituting Eq. (18.3-4) into the above expression results in

$$\mathbf{D} = \varepsilon_0 \varepsilon_r \mathbf{E} + \frac{\varepsilon_r + 2}{3}\mathbf{P}'^{(2)} = \varepsilon_0 \varepsilon_r \mathbf{E} + \mathbf{P}^{(2)},$$

which implies that

$$\mathbf{P}^{(2)} = \frac{\varepsilon_r + 2}{3}\mathbf{P}'^{(2)}. \qquad (18.3\text{-}5)$$

In order to find the explicit relation between $\mathbf{E}'$ and $\mathbf{E}$, we can neglect the second term in Eq. (18.3.4) and then have

$$\mathbf{P}'^{(1)} \approx \varepsilon_0(\varepsilon_r - 1)\mathbf{E}.$$

Substituting this expression into Eq. (18.3-1) and neglecting the third terms leads to

$$\mathbf{E}' = \mathbf{E} + \frac{1}{3\varepsilon_0}\mathbf{P}'^{(1)} = \frac{\varepsilon_r + 2}{3}\mathbf{E}. \tag{18.3-6}$$

Then $\mathbf{P}'^{(2)}$ can be expressed as

$$\mathbf{P}'^{(2)}(\omega_3) = \varepsilon_0 N\alpha^{(2)}\mathbf{E}'(\omega_1)\mathbf{E}'(\omega_2) = \varepsilon_0 N\alpha^{(2)}\left(\frac{\varepsilon_r(\omega_1)+2}{3}\right)\left(\frac{\varepsilon_r(\omega_2)+2}{3}\right)\mathbf{E}(\omega_1)\mathbf{E}(\omega_2), \tag{18.3-7}$$

where $\alpha^{(2)}$ is the second-order nonlinear polarizability tensor of the individual particle. On the other hand, from Eq. (18.3-5) we know that

$$\mathbf{P}^{(2)}(\omega_3 - \omega_1 + \omega_2) = \frac{\varepsilon_r(\omega_3)+2}{3}\mathbf{P}'^{(2)}(\omega_3) = \varepsilon_0 \chi^{(2)}(\omega_1,\omega_2)\mathbf{E}(\omega_1)\mathbf{E}(\omega_2). \tag{18.3-8}$$

Substituting Eq. (18.4-7) into Eq. (18.4-8), we finally obtain the expression for the second-order nonlinear susceptibility with the local-field correction, i.e.,

$$\chi^{(2)}(-\omega_3;\omega_1,\omega_2) = N\alpha^{(2)}\left(\frac{\varepsilon_r(\omega_3)+2}{3}\right)\left(\frac{\varepsilon_r(\omega_1)+2}{3}\right)\left(\frac{\varepsilon_r(\omega_2)+2}{3}\right). \tag{18.3-9}$$

Here, $N\alpha^{(2)}$ is the second-order nonlinear susceptibility tensor calculated in the preceding section; its elements are given by Eq. (18.2-25) for the general case and by Eq. (18.2-26) for the non-resonant case.

The same procedure can be applied to the high-order nonlinear susceptibilities. For instance, the third-order nonlinear susceptibility tensor with local-field correction should be

$$\chi^{(3)}(-\omega_4;\omega_1,\omega_2,\omega_3) = N\alpha^{(3)}\left(\frac{\varepsilon_r(\omega_4)+2}{3}\right)\left(\frac{\varepsilon_r(\omega_1)+2}{3}\right)\left(\frac{\varepsilon_r(\omega_2)+2}{3}\right)\left(\frac{\varepsilon_r(\omega_3)+2}{3}\right). \tag{18.3-10}$$

Here, $N\alpha^{(3)}$ is the third-order nonlinear susceptibility tensor calculated in the previous section; its elements are given by Eq. (18.2-27) for the general case and by Eq. (18.2-28) for the non-resonant case.

18.3.2 Spatial symmetry

Based on the explicit expressions for the elements of nonlinear susceptibilities given in Section 18.2, we could further discuss the influences of the microscopic and macroscopic spatial symmetries of the particle system on the polarization properties of the medium.[1-4]

First, let us consider the second-order nonlinear susceptibility. In the general case, the element of the second-order susceptibility tensor described by Eq. (18.2-25) is determined by the summation of terms proportional to the product of three elements of the electric-dipole operator, such as $(p_i)_{ab}(p_j)_{bc}(p_k)_{ca}$, etc. Here p_i is the Cartesian component of the electric-dipole operator, the matrix element of which is determined by

$$(p_i)_{ab} = \int \phi_a^*(x,y,z)(p_i)\phi_b(x,y,z)dxdydz,$$

where ϕ_a and ϕ_b are the unperturbed eigenfunctions of the individual molecule. If the molecular eigenfunctions possess a centrosymmetry, we have

$$\phi_a(x,y,z) = \phi_a(-x,-y,-z). \tag{18.3-11}$$

This implies that when the Cartesian coordinates are reversed, the molecular eigenfunctions remain unchanged, but the electric-dipole moment vector changes its sign, i.e., $(p'_i) = -(p_i)$, so that we have

$$(p'_i)_{ab} = \int \phi_a^*(-x,-y,-z)(-p_i)\phi_b(-x,-y,-z)\,dxdydz$$
$$= -\int \phi_a^*(x,y,z)(p_i)\phi_b(x,y,z)\,dxdydz = -(p_i)_{ab}$$

and

$$(p'_i)_{ab}(p'_j)_{bc}(p'_k)_{ca} = -(p_i)_{ab}(p_j)_{bc}(p_k)_{ca}. \tag{18.3-12}$$

This indicates that before the inversion of the Cartesian coordinates we have

$$P_i^{(2)}(\omega_1 + \omega_2) = \varepsilon_0 \chi_{ijk}^{(2)}(\omega_1,\omega_2) E_j(\omega_1) E_k(\omega_2),$$

and after the inversion of Cartesian coordinates we have

$$P_i^{(2)}(\omega_1 + \omega_2) = -\varepsilon_0 \chi_{ijk}^{(2)}(\omega_1,\omega_2) E_j(\omega_1) E_k(\omega_2).$$

However, the physical property of the medium should not depend on how we choose the Cartesian coordinates. For this reason, we can conclude that

$$\chi_{ijk}^{\prime(2)} = -\chi_{ijk}^{(2)} \equiv 0. \tag{18.3-13}$$

This relation means that, if the eigenfunctions of the particles (molecule or atoms) possess a centrosymmetry, the second-order nonlinear susceptibility vanishes. In the more general cases, this conclusion is valid for all even-order nonlinear susceptibilities.

It is important to point out that the above conclusion is obtained in the electric-dipole approximation. This conclusion does not hold if we consider the higher-order (electric-quadrupole and magnetic-dipole) approximation. For example, assuming that the molecular eigenfunctions are centrosymmetric, and considering the electric-quadrupole contribution for the second-order nonlinear susceptibility, we can simply replace the electric-dipole operator $\mathbf{p}$ by the operator $i(\mathbf{k}\hat{q})$ (see Eq. (18.1-24)). Here $\mathbf{k}$ is the wavevector of the optical field and $\hat{q}$ is the electric-quadrupole operator defined by Eq. (18.1-20). From that definition, one can see that after the inversion of the Cartesian coordinates, $\hat{q}$ remains unchanged. Therefore, we could get the following relation:

$$(\mathbf{k}\hat{q}')_{ab} = \int \phi_a^*(-x,-y,-z)(\mathbf{k}\hat{q})\phi_b(-x,-y,-z)\,dxdydz = (\mathbf{k}\hat{q})_{ab}. \tag{18.3-14}$$

After replacing $\mathbf{p}$ by $i(\mathbf{k}\hat{q})$, all derivations in Section 18.2 can be applied to the electric-quadrupole approximation. Especially, from Eq. (18.2-25) one can see that the element of the second-order nonlinear susceptibility is determined by the summation of terms proportional to the product of three elements of operator $(\mathbf{k}\hat{q})$, such as $[(\mathbf{k}\hat{q})_i]_{ab}[(\mathbf{k}\hat{q})_j]_{bc}[(\mathbf{k}\hat{q})_k]_{ca}$, etc. Based on this consideration and Eq. (18.3-14), we obtain

$$\chi'^{(2)}_{ijk} = \chi^{(2)}_{ijk} \neq 0. \tag{18.3-15}$$

This means that, in the electric-quadrupole approximation, the second-order susceptibility does not vanish even for the molecular system of centrosymmetry. This conclusion is valid for all even-order susceptibilities.

Regarding the third-order nonlinear susceptibility, from Eq. (18.2-27) one can see that the element of the third-order susceptibility is determined by the summation of terms proportional to the product of four elements of the electric-dipole operator $\mathbf{p}$, such as $(p_i)_{ab}(p_j)_{bc}(p_k)_{cd}(p_l)_{da}$, etc. This means that the property of the third-order nonlinear polarization of the medium remains unchanged after the inversion of the Cartesian coordinates. Therefore, we can conclude that even in the electric-dipole approximation, the third-order susceptibility does not vanish for the assembly of molecules possessing a centrosymmetry. This conclusion is applicable to all odd-order (including the first-order) susceptibilities.

So far, in this subsection we have only discussed the influence of the microscopic symmetry of the molecular system on the susceptibilities of the medium. Now we further examine the influence of the macroscopic symmetry of the medium on various-order susceptibilities.

First, let us consider gases, liquids, and amorphous solids consisting of anisotropic molecules. For these types of media, although the eigenfunctions for each individual molecule may not be centrosymmetric, the whole system containing a great number of molecules still exhibits a macroscopic isotropy because of the random distribution of molecular orientations. In Section 18.2.2, all formulations are obtained on the assumption that the medium consists of a great number of identical and independent particles. Therefore, the status of a given sub-ensemble can be described in terms of the eigenfunctions of the individual particle. However, for many real optical media the microscopic status of different particles (such as their orientations) may not be identical. For this reason the formulae given in Section 18.2.2 cannot be simply applied. Nevertheless, for these types of real media we still can chose such a sub-ensemble, the dimension of which should be much larger than that of the particle but much shorter than the wavelength of the light. In this case, the operator of the sum of the molecular electric-dipole moments over the whole sub-ensemble can be expressed by

$$\mathbf{R} = \sum_{i=1}^{N} \mathbf{p}_i, \tag{18.3-16}$$

where $\mathbf{p}_i$ is the electric-dipole operator of the ith molecule and N is the density of the particles. According to Eq. (18.1-7), the rth-order electric polarization of the real medium should be

$$\mathbf{P}^{(r)} = \mathrm{Tr}[\mathbf{R}\rho^{(r)}] = \sum_n [\mathbf{R}\rho^{(r)}]_{nn}. \tag{18.3-17}$$

Here, $\rho^{(r)}$ is the rth-order component of the density operator of the sub-ensemble. Comparing the above two expressions with Eqs. (18.2-11) and (18.2-12), one can find that after going through the same mathematical procedures, we will obtain similar formulae for various-order susceptibilities, except that the factor ($N\mathbf{p}$) in the previous formulae is replaced by $\mathbf{R}$, and the elements of operators are determined by the eigenfunctions of the whole sub-ensemble. As we mentioned previously, owing to random orientation distributions of a great number of particles, the eigenfunctions of the sub-ensemble should still be centrosymmetric, so that all the even-order nonlinear susceptibilities should vanish in the electric-dipole approximation.

Finally, let us turn to a crystalline medium that consists of a great number of elementary cells with an identical and regular spatial orientation. The previous conclusions about the influence of spatial symmetry of the medium on the susceptibilities are also valid for crystalline media. Specifically, in the electric-dipole approximation, there is no second-order nonlinear polarization effect for the crystals possessing a centrosymmetry. Among all 32 crystal classes, only 21 crystal classes lack the inversion symmetry, and therefore are suitable for the observation of second-order nonlinear optical effects. In addition, for a given crystal the various-order susceptibilities should remain unchanged upon all allowed symmetry operations. Suppose that one of the allowed symmetry operations is denoted by a matrix (A) with 3×3 elements; after such an operation the following relation for the second-order susceptibility should hold:

$$\chi^{(2)}_{\alpha\beta\gamma} = A_{\alpha i}A_{\beta j}A_{\gamma k}\chi^{(2)}_{ijk} = \chi^{(2)}_{ijk}. \tag{18.3-18}$$

Generalizing this relation to the rth-order susceptibility we have

$$\chi^{(r)}_{\alpha\beta\cdots\eta} = A_{\alpha i}A_{\beta j}\cdots A_{\eta s}\chi^{(r)}_{ij\cdots s} = \chi^{(r)}_{ij\cdots s}. \tag{18.3-19}$$

Based on the restriction imposed by this requirement, the number of independent and nonvanishing elements of various-order susceptibility tensors can be greatly reduced.

In Appendices 3–6 of this book, the distribution and number of nonvanishing elements of various-order susceptibility tensors are given for different symmetry groups of crystals as well as for isotropic media.

18.3.3 Permutation symmetry and time-reversal symmetry of susceptibilities

18.3.3.1 Intrinsic permutation symmetry

The permutation symmetry of nonlinear susceptibilities implies the invariance of the element of a susceptibility tensor under the operation of simultaneous interchange among its dummy indices and the corresponding frequency arguments. For instance, from Eq. (18.2-25), the general expression for the second-order nonlinear susceptibility, we can see that a specially defined summation symbol $\sum\limits_{(jk)}^{(\omega_1\omega_2)}$ is involved. Therefore, the following relation holds:

$$\chi^{(2)}_{ijk}(\omega_1,\omega_2) = \chi^{(2)}_{ikj}(\omega_2,\omega_1). \tag{18.3-20}$$

Similarly, for the element of the third-order nonlinear susceptibility, considering the invariance under the permutation indicated in the first summation symbol in Eq. (18.2-27), we find that

$$\chi^{(3)}_{ijkl}(\omega_1,\omega_2,\omega_3) = \chi^{(3)}_{ikjl}(\omega_2,\omega_1,\omega_3) = \chi^{(3)}_{iljk}(\omega_3,\omega_1,\omega_2) = \cdots. \tag{18.3-21}$$

The above two equations indicate that the elements of various-order susceptibility tensors remain unchanged when we permute the positions of the coordinate indices (except i) with the corresponding frequency arguments together. This property is known as the intrinsic permutation symmetry of the nonlinear susceptibilities; it applies to both resonant and non-resonant interactions.

18.3.3.2 Overall permutation symmetry

For a non-resonant interaction, the permutation symmetry can be further extended. In this case, the second-order susceptibility element is given by Eq. (18.2-26)

$$\chi^{(2)}_{ijk}(\omega_1,\omega_2) = \frac{N}{\varepsilon_0 2\hbar^2} \sum_{(jk)}^{(\omega_1\omega_2)} \sum_{a,b,c} \rho^{(0)}_{aa} \left[\frac{(p_i)_{ab}(p_j)_{bc}(p_k)_{ca}}{(\omega_{ba}-\omega_1-\omega_2)(\omega_{ca}-\omega_2)} \right.$$
$$\left. + \frac{(p_j)_{ab}(p_i)_{bc}(p_k)_{ca}}{(\omega_{ba}+\omega_1)(\omega_{ca}-\omega_2)} + \frac{(p_k)_{ab}(p_j)_{bc}(p_i)_{ca}}{(\omega_{ba}+\omega_1)(\omega_{ca}+\omega_1+\omega_2)} \right].$$

In this expression, we see that the positions of Cartesian indices j and k correspond to the frequencies ω_1 and ω_2. If the first Cartesian index i corresponds to a formally introduced frequency argument $\omega' = -(\omega_1 + \omega_2)$, then the second term in the square brackets can be obtained from the first term by taking a simultaneous interchange between (i, ω') and (j, ω_1), and the third term can be obtained from the first by the permutation between (i, ω') and (k, ω_2). Thus the above expression can be rewritten in a more compact form:

$$\chi^{(2)}_{ijk}(\omega';\omega_1,\omega_2) = \frac{N}{\varepsilon_0 2\hbar^2} \sum_{(ijk)}^{(\omega'\omega_1\omega_2)} \sum_{a,b,c} \rho^{(0)}_{aa} \frac{(p_i)_{ab}(p_j)_{bc}(p_k)_{ca}}{(\omega_{ba}+\omega')(\omega_{ca}-\omega_2)}, \tag{18.3-22}$$

and it follows that

$$\chi^{(2)}_{ijk}(\omega';\omega_1,\omega_2) = \chi^{(2)}_{jik}(\omega_1;\omega',\omega_2) = \chi^{(2)}_{jki}(\omega_1;\omega_2,\omega') = \cdots. \tag{18.3-23}$$

The physical meaning of this identity is that the second-order susceptibility element is invariant under any of the permutations among the pairs (i, ω'), (j, ω_1), and (k, ω_2).

The same procedure can be followed to simplify the expression for the third-order susceptibility under the non-resonant condition, and Eq. (18.2-28) can be rewritten in a more compact form:

$$\chi^{(3)}_{ijkl}(\omega';\omega_1,\omega_2,\omega_3) = \frac{N}{\varepsilon_0 6\hbar^3} \sum_{(ijkl)}^{(\omega'\omega_1\omega_2\omega_3)} \sum_{a,b,c,d} \rho^{(0)}_{aa} \frac{(p_i)_{ab}(p_j)_{bc}(p_k)_{cd}(p_l)_{da}}{(\omega_{ba}+\omega')(\omega_{ca}-\omega_2-\omega_3)(\omega_{da}-\omega_3)}. \tag{18.3-24}$$

Here, $\omega' = -(\omega_1 + \omega_2 + \omega_3)$. Then the following relations hold

$$\chi^{(3)}_{ijkl}(\omega';\omega_1,\omega_2,\omega_3) = \chi^{(3)}_{jikl}(\omega_1;\omega',\omega_2,\omega_3) = \chi^{(3)}_{lkij}(\omega_3;\omega_2,\omega',\omega_1) = \cdots. \tag{18.3-25}$$

The above-mentioned overall permutation symmetry can be generalized to the rth-order susceptibility under the condition of non-resonant interaction.

18.3.3.3 Time-reversal symmetry

For non-resonant interaction, there is an additional symmetry property for the nonlinear susceptibilities. To describe this property, we return to Eq. (18.3-22) and take a complex-conjugation operation on both sides of this equation. On the right-hand side of this equation, there are those terms that involve the dipole matrix elements of $(p_i)_{ab}$, $(p_j)_{bc}$, and $(p_k)_{ca}$ in the numerator. From the derivation of Eq. (18.2-20), we already know that all these dipole matrix elements are real quantities. On the other hand, the zero-order density matrix element $\rho_{aa}^{(0)}$ is also a real quantity determined by Eq. (18.1-5):

$$\rho_{aa}^{(0)} = Ze^{-E_a/k_B T}, \tag{18.3-26}$$

where E_a is the eigenenergy of the molecule in the eigenstate a and Z is a normalization factor. Based on the above considerations, we realize that the second-order susceptibility is real for the non-resonant case; therefore, we have

$$[\chi_{ijk}^{(2)}(\omega_1, \omega_2)]^* = \chi_{ijk}^{(2)}(\omega_1, \omega_2).$$

Based on the same consideration, for the ith-order susceptibility we could obtain

$$[\chi_{ij\cdots s}^{(r)}(\omega_1, \omega_2, \cdots \omega_r)]^* = \chi_{ij\cdots s}^{(r)}(\omega_1, \omega_2, \cdots \omega_r). \tag{18.3-27}$$

This is valid only for the non-resonant interaction.

Remember that in Chapter 2 for the general case, we have given the following relation to describe the complex conjugation for the rth-order susceptibility (see Eq. (2.4-7)):

$$[\chi_{ij\cdots s}^{(r)}(\omega_1, \omega_2, \cdots \omega_r)]^* = \chi_{ij\cdots s}^{(r)}(-\omega_1, -\omega_2, \cdots -\omega_r). \tag{18.3-28}$$

Comparing Eq. (18.3-27) with Eq. (14.3-28) leads to

$$\chi_{ij\cdots s}^{(r)}(\omega_1, \omega_2, \cdots \omega_r) = \chi_{ij\cdots s}^{(r)}(-\omega_1, -\omega_2, \cdots -\omega_r). \tag{18.3-29}$$

This relation indicates that for non-resonant interaction, the nonlinear susceptibility element is invariant when all the frequency arguments change their signs. This is the so-called time-reversal symmetry.

18.3.3.4 Kleinman symmetry

For non-resonant interaction, the second-order susceptibility is described by Eq. (18.3-22) and the overall permutation symmetry holds. If we can assume that the frequencies ω_1, ω_2, and $-\omega' = (\omega_1 + \omega_2)$ are much smaller than the molecular resonance frequencies ω_{ba} and ω_{ca}, any interchange among ω_1, ω_2, and ω' brings no significant change of the susceptibility element. Therefore, according to the property of overall permutation symmetry we have

$$\chi_{ijk}^{(2)}(\omega'; \omega_1, \omega_2) = \chi_{jik}^{(2)}(\omega'; \omega_1, \omega_2) = \chi_{kji}^{(2)}(\omega'; \omega_1, \omega_2) = \cdots. \tag{18.3-30}$$

This relation indicates that if we could neglect the dependence of the second-order susceptibility on the specific involved frequency components, the susceptibility element is invariant under all permutation of the coordinate subscripts i, j, and k. Similarly, under the same condition for the third-order susceptibility, we obtain

$$\chi_{ijkl}^{(3)}(\omega'; \omega_1, \omega_2, \omega_3) = \chi_{jikl}^{(3)}(\omega'; \omega_1, \omega_2, \omega_3) = \chi_{klji}^{(3)}(\omega'; \omega_1, \omega_2, \omega_3) = \cdots. \tag{18.3-31}$$

This additional property expressed by the above two equations is the so-called Kleinman symmetry.[7]

In summary, knowing all these symmetry properties of various-order nonlinear susceptibilities, we could considerably reduce the number of the nonzero and independent tensor elements involved in a given nonlinear optical process and, therefore, greatly simplify the theoretical analyses.

18.4 Resonance enhancement of nonlinear susceptibilities

18.4.1 Introduction

In the preceding section, we have discussed the overall permutation symmetry, time-reversal symmetry, and Kleinman symmetry of the nonlinear susceptibilities. All these descriptions are valid only for the non-resonant interaction.

Now in this section we shall consider the resonant interaction when the frequencies (including their linear combinations) of the applied optical field are quite close to the molecular (absorptive or Raman) resonant frequencies of the medium. In this case, the various-order susceptibilities are complex quantities and the effective susceptibility values for a given nonlinear optical process could be significantly enhanced. This is often a favorable property of the nonlinear susceptibilities.

On the other hand, however, in some cases an exact resonance may cause considerable attenuation of the incident optical wave or a newly generated wave. As a result of such attenuation due to resonant absorption, the efficiency of the desired nonlinear optical effect might be reduced. For this reason, in many practical studies researchers have to make a choice of compromise. For instance, the most commonly used second-order nonlinear crystals are highly transparent for the involved optical frequencies, and a high efficiency could be achieved even without utilizing any resonance enhancement. However, for some third-order nonlinear processes (such as four-wave mixing (FWM) and light-induced refractive-index change), the efficiency is much lower than that of the second-order nonlinear effects. Therefore, the resonance enhancement of the third-order nonlinear susceptibility becomes essentially important.

18.4.2 Resonance enhancement of the first- and second-order susceptibilities

The general expression for the first-order susceptibility tensor is given by Eq. (18.2-18). In the case of resonant interaction, the frequency ω of the applied monochromatic field is very close to the frequency ω_{to} of one allowed absorptive transition of the molecules. If the subscripts o and t denote the ground state and the excited state of this transition, and in the summation of Eq. (18.2-18) only those terms related with $(a, b) = (o, t)$ should be considered, then we obtain

$$\chi_{ij}^{(1)}(\omega) = \frac{N}{\varepsilon_0 \hbar} \left[\rho_{oo}^{(0)} \left(\frac{(p_i)_{ot}(p_j)_{to}}{\omega_{to} + \omega + i\Gamma_{to}} + \frac{(p_j)_{ot}(p_i)_{to}}{\omega_{to} - \omega - i\Gamma_{to}} \right) \right.$$
$$\left. + \rho_{tt}^{(0)} \left(\frac{(p_i)_{to}(p_j)_{ot}}{\omega_{ot} + \omega + i\Gamma_{ot}} + \frac{(p_j)_{to}(p_i)_{ot}}{\omega_{ot} - \omega - i\Gamma_{ot}} \right) \right]. \tag{18.4-1}$$

Since $\omega_{to} = -\omega_{ot}$ and $\Gamma_{to} = \Gamma_{ot}$, the first and fourth terms in the above equation can be neglected, and we have the final expression for the resonant contribution for $\chi^{(1)}$

$$\chi_{ij}^{(1)}(\omega) = \frac{N}{\varepsilon_0 \hbar} [\rho_{oo}^{(0)} - \rho_{tt}^{(0)}] \frac{(p_i)_{to}(p_j)_{ot}}{\omega_{to} - \omega - i\Gamma_{to}} = \frac{N_o - N_t}{\varepsilon_0 \hbar} \frac{(p_i)_{to}(p_j)_{ot}}{\omega_{to} - \omega - i\Gamma_{to}}. \tag{18.4-2}$$

Here, N_o and N_t are, respectively, the numbers of molecules in the ground state a and the excited state b within the unit volume of the medium.

It is well known in conventional optics that the real part of $\chi_{ij}^{(1)}(\omega)$ in Eq. (18.4-2) determines the anomalous dispersion behavior of the refractive index, while the imaginary part determines the linear (one-photon) absorption property and Γ_{to} determines the spectral linewidth of the absorptive transition.

In the same way, we can also consider the resonant effect of the second-order nonlinear susceptibility generally expressed by Eq. (18.2-25). For simplicity, we consider here the two-photon absorption resonance effect, i.e., $(\omega_1 + \omega_2) \approx \omega_{to}$. In the summation operation of Eq. (18.2-25), letting $(a, b, c) = (o, t)$ and only retaining the resonant terms related to ω_{to}, we obtain

$$\chi^{(2)}_{ijk}(\omega_1,\omega_2) = \frac{N/(\varepsilon_0 2\hbar^2)}{\omega_{to} - (\omega_1 + \omega_2) - i\Gamma_{to}} \sum_b \left\{ \rho^{(0)}_{oo}(p_i)_{ot} \left[\frac{(p_j)_{tb}(p_k)_{bo}}{\omega_{bo} - \omega_2 - i\Gamma_{bo}} \right. \right.$$
$$\left. + \frac{(p_k)_{tb}(p_j)_{bo}}{\omega_{bo} - \omega_1 - i\Gamma_{bo}} \right] - \rho^{(0)}_{tt}(p_i)_{ot} \left[\frac{(p_k)_{tb}(p_j)_{bo}}{\omega_{bt} + \omega_2 + i\Gamma_{bt}} + \frac{(p_j)_{tb}(p_k)_{bo}}{\omega_{bt} + \omega_1 + i\Gamma_{bt}} \right] \bigg\}. \tag{18.4-3}$$

Furthermore, if we assume that the one-photon resonance is negligible, i.e., $i\Gamma_{bo} \approx i\Gamma_{bt} \approx 0$ and $\omega_{bt} = \omega_{bo} + \omega_{ot} = \omega_{bo} - \omega_{to} \approx \omega_{bo} - (\omega_1 + \omega_2)$, then Eq. (18.4-3) can be further simplified as

$$\chi^{(2)}_{ijk}(\omega_1,\omega_2) = \frac{(N_o - N_t)/(\varepsilon_0 2\hbar^2)}{\omega_{to} - (\omega_1 + \omega_2) - i\Gamma_{to}}(p_i)_{ot} \sum_b \left[\frac{(p_j)_{tb}(p_k)_{bo}}{\omega_{bo} - \omega_2} + \frac{(p_k)_{tb}(p_j)_{bo}}{\omega_{bo} - \omega_1} \right]. \tag{18.4-4}$$

18.4.3 Resonance enhancement of the third-order susceptibility

Let us consider the general FWM process in which three frequency components ω_1, ω_2, and ω_3 are mixed to generate a new frequency component $\omega_4 = \omega_1 + \omega_2 + \omega_3$ through the third-order susceptibility expressed by Eq. (18.2-27). In this expression we can see that the resonance enhancement of $\chi^{(3)}_{ijkl}(\omega_1, \omega_2, \omega_3)$ can be achieved in three possible ways: (i) one-photon resonance when the frequency of one wave approaches a molecular resonant frequency ω_{to}, (ii) two-photon resonance when the sum- or difference-frequency of two waves approaches ω_{to}, and (iii) three-photon resonance when the frequency combination of three waves is close to ω_{to}. For simplicity, in this subsection we only consider the first and second ways.

18.4.3.1 One-photon resonance enhancement of $\chi^{(3)}$

Assume that the frequency of one incident wave is close to a molecular resonant transition frequency, e.g., $\omega_3 \approx \omega_{to}$, and in the summation operation of Eq. (18.2-27) let $(a, c) = (o, t)$ and keep only those terms related to the one-photon resonance; we then obtain

$$\chi^{(3)}_{ijkl}(\omega_1,\omega_2,\omega_3) = \frac{(N_o - N_t)/(\varepsilon_0 6\hbar^3)}{\omega_{to} - \omega_3 - i\Gamma_{to}}(p_l)_{to}$$
$$\sum_{b,c} \left[\frac{(p_i)_{ob}(p_j)_{bc}(p_k)_{ct}}{(\omega_{bo} - \omega_1 - \omega_2 - \omega_3)(\omega_{co} - \omega_2 - \omega_3)} + \frac{(p_j)_{ob}(p_i)_{bc}(p_k)_{ct}}{(\omega_{bo} + \omega_1)(\omega_{co} - \omega_2 - \omega_3)} \right.$$
$$+ \frac{(p_j)_{ob}(p_k)_{bc}(p_i)_{ct}}{(\omega_{bo} + \omega_1)(\omega_{co} + \omega_1 + \omega_2)} + \frac{(p_i)_{ob}(p_k)_{bc}(p_j)_{ct}}{(\omega_{bo} - \omega_1 - \omega_2 - \omega_3)(\omega_{co} - \omega_1 - \omega_3)}$$
$$\left. + \frac{(p_k)_{ob}(p_i)_{bc}(p_j)_{ct}}{(\omega_{bo} + \omega_2)(\omega_{co} - \omega_1 - \omega_3)} + \frac{(p_k)_{ob}(p_j)_{bc}(p_i)_{ct}}{(\omega_{bo} + \omega_2)(\omega_{co} + \omega_1 + \omega_2)} \right]. \tag{18.4-5}$$

In this expression it is obvious that when $\omega_3 \Rightarrow \omega_{to}$, there is a resonance enhancement for the third-order susceptibility.

18.4.3.2 Two-photon sum-frequency resonance enhancement of $\chi^{(3)}$

Assume that the sum of frequencies ω_2 and ω_3 of the two incident waves is close to a molecular resonant frequency, i.e., $(\omega_2 + \omega_3) \approx \omega_{to}$, and in the summation operation of Eq. (18.2-27) let $(a, c) = (o, t)$ and keep only those terms related to the two-photon sum-frequency resonance; we then obtain

$$\chi_{ijkl}^{(3)}(\omega_1, \omega_2, \omega_3) = \frac{(N_o - N_t)/(\varepsilon_0 6 \hbar^3)}{\omega_{to} - (\omega_2 + \omega_3) - i\Gamma_{to}} \sum_b \left[\frac{(p_i)_{ob}(p_j)_{bt}}{\omega_{ba} - \omega_1 - \omega_2 - \omega_3} \right.$$
$$\left. + \frac{(p_j)_{ob}(p_i)_{bt}}{\omega_{ba} + \omega_1} \right] \cdot \sum_b \left[\frac{(p_k)_{tb}(p_l)_{bo}}{\omega_{ba} - \omega_3} + \frac{(p_l)_{tb}(p_k)_{bo}}{\omega_{ba} - \omega_2} \right]. \qquad (18.4-6)$$

In obtaining the above expression we neglected the contribution from all non-resonant terms and utilized the relation: $\omega_{bt} = \omega_{bo} + \omega_{ot} = \omega_{bo} - \omega_{to} \approx \omega_{bo} - (\omega_2 + \omega_3)$. From Eq. (18.4-6) we see that when $(\omega_2 + \omega_3) \Rightarrow \omega_{to}$, $\chi_{ijkl}^{(3)}$ can be significantly enhanced. This enhancement effect can be applied to third-order harmonic generation as well as other FWM processes.

Moreover, if the three waves are mixed in such a way that two waves have the same frequency, the frequency combination is $(-\omega_2 + \omega_1 + \omega_2)$, and the two-photon sum-frequency resonance $(\omega_1 + \omega_2) \approx \omega_{to}$ is fulfilled, then from Eq. (18.4-6) we obtain

$$\chi_{ijkl}^{(3)}(-\omega_2, \omega_1, \omega_2) = \frac{(N_o - N_t)/(\varepsilon_0 6 \hbar^3)}{\omega_{to} - (\omega_1 + \omega_2) - i\Gamma_{to}} \sum_b \left[\frac{(p_i)_{ob}(p_j)_{bt}}{\omega_{bo} - \omega_1} \right.$$
$$\left. + \frac{(p_j)_{ob}(p_i)_{bt}}{\omega_{bo} - \omega_2} \right] \cdot \sum_b \left[\frac{(p_k)_{tb}(p_l)_{bo}}{\omega_{b0} - \omega_2} + \frac{(p_l)_{tb}(p_k)_{bo}}{\omega_{bo} - \omega_1} \right]. \qquad (18.4-7)$$

This resonant third-order susceptibility is a complex quantity that can be used to describe the two-photon (sum-frequency) resonance enhancement for a general FWM process. In addition, the real part of this resonant susceptibility can be used to describe the resonance-enhanced refractive-index change, whereas the imaginary part can be used to describe the two-photon absorption process. To explain the latter, we should realize that in the present case the third-order susceptibility, $\chi_{ijkl}^{(3)}(-\omega_2, \omega_1, \omega_2) = \chi_{ijkl}^{(3)}(-\omega_1; -\omega_2, \omega_1, \omega_2)$, implies that the resulting frequency (ω_1) of the nonlinear frequency mixing $(-\omega_2 + \omega_1 + \omega_2)$ is identical with one of the incident waves, and the imaginary part of this susceptibility determines the attenuation of the ω_1 wave owing to the two-photon ($\omega_1 + \omega_2$) absorption. In other words, there is energy transfer from ω_1 and ω_2 waves to the nonlinear medium. It is obvious that the imaginary part of another resonant susceptibility,

$$\chi_{ijkl}^{(3)}(-\omega_1, \omega_1, \omega_2) = \chi_{ijkl}^{(3)}(-\omega_2; -\omega_1, \omega_1, \omega_2),$$

can be used to describe the attenuation of the ω_2 wave owing to the same two-photon absorption process.

18.4.3.3 Two-photon difference-frequency resonance enhancement of $\chi^{(3)}$

Now let us consider another special resonant effect in the FWM process, wherein the frequency difference of two incident waves is close to a molecular resonant frequency, e.g., $(\omega_2 - \omega_3) \approx \omega_{to}$. In this case, from Eq. (18.2-27) we can finally obtain

$$\chi^{(3)}_{ijkl}(\omega_1,-\omega_2,\omega_3) = \frac{(N_o - N_t)/(\varepsilon_0 6\hbar^3)}{\omega_{to} - (\omega_2 - \omega_3) + i\Gamma_{to}} \sum_b \left[\frac{(p_i)_{ob}(p_j)_{bt}}{\omega_{bo} + \omega_1 - \omega_2 + \omega_3} \right.$$
$$\left. + \frac{(p_j)_{ob}(p_i)_{bt}}{\omega_{bo} - \omega_1} \right] \cdot \sum_b \left[\frac{(p_k)_{tb}(p_l)_{bo}}{\omega_{bo} + \omega_3} + \frac{(p_l)_{tb}(p_k)_{bo}}{\omega_{bo} - \omega_2} \right].$$
(18.4-8)

In most practical cases, the frequency difference between two waves is tuned in resonance with a Raman transition frequency $\Delta\omega_r$ of the medium, i.e., $(\omega_2 - \omega_3) \approx \Delta\omega_r$. If one incident beam is a stronger pump wave with frequency ω_p and the other is a weaker signal beam with frequency ω_s, then the following four different effects can be observed under the Raman resonant condition.

18.4.3.3.1 Raman gain effect This is described by

$$\chi^{(3)}_{ijkl}(\omega_p,-\omega_p,\omega_s) = \frac{(N_o - N_t)/(\varepsilon_0 6\hbar^3)}{\Delta\omega_r - (\omega_p - \omega_s) + i\Gamma_r} \sum_b \left[\frac{(p_i)_{ob}(p_j)_{bt}}{\omega_{bo} + \omega_s} \right.$$
$$\left. + \frac{(p_j)_{ob}(p_i)_{bt}}{\omega_{bo} - \omega_p} \right] \cdot \sum_b \left[\frac{(p_k)_{tb}(p_l)_{bo}}{\omega_{bo} + \omega_s} + \frac{(p_l)_{tb}(p_k)_{bo}}{\omega_{bo} - \omega_p} \right],$$
(18.4-9)

where $\omega_p > \omega_s$ and Γ_r is the spectral width of the Raman transition. The imaginary part of this expression determines the gain of the signal beam at frequency ω_s. This process is the basis of the Raman-gain spectroscopy.

18.4.3.3.2 Raman attenuation effect This is described by

$$\chi^{(3)}_{ijkl}(\omega_p,\omega_s,-\omega_p) = [\chi^{(3)}_{ijkl}(-\omega_p,-\omega_s,\omega_p)]^* = \frac{(N_o - N_t)/(\varepsilon_0 6\hbar^3)}{\Delta\omega_r - (\omega_s - \omega_p) - i\Gamma_r}$$
$$\sum_b \left[\frac{(p_i)_{ob}(p_j)_{bt}}{\omega_{bo} - \omega_s} + \frac{(p_j)_{ob}(p_i)_{bt}}{\omega_{bo} + \omega_p} \right] \cdot \sum_b \left[\frac{(p_k)_{tb}(p_l)_{bo}}{\omega_{bo} + \omega_p} + \frac{(p_l)_{tb}(p_k)_{bo}}{\omega_{bo} - \omega_s} \right],$$
(18.4-10)

where $\omega_s > \omega_p$. The imaginary part of this expression determines the attenuation of the signal beam at the higher frequency ω_s. That process is the basis of the inverse Raman spectroscopy.

18.4.3.3.3 Coherent Stokes Raman emission This is described by

$$\chi^{(3)}_{ijkl}(\omega_s,-\omega_p,\omega_s) = \frac{(N_o - N_t)/(\varepsilon_0 6\hbar^3)}{\Delta\omega_r - (\omega_p - \omega_s) + i\Gamma_r} \sum_b \left[\frac{(p_i)_{ob}(p_j)_{bt}}{\omega_{bo} + 2\omega_s - \omega_p} \right.$$
$$\left. + \frac{(p_j)_{ob}(p_i)_{bt}}{\omega_{bo} - \omega_s} \right] \cdot \sum_b \left[\frac{(p_k)_{tb}(p_l)_{bo}}{\omega_{bo} + \omega_s} + \frac{(p_l)_{tb}(p_k)_{bo}}{\omega_{bo} - \omega_p} \right],$$
(18.4-11)

where $\omega_p > \omega_s$. This resonant susceptibility is used to describe the generation of coherent Raman emission at a new frequency of $\omega = \omega_s - (\omega_p - \omega_s) = \omega_s - \Delta\omega_r$ through phase-matched FWM; the intensity of this newly generated signal is proportional to $\left|\chi^{(3)}_{ijkl}(\omega_s,-\omega_p,\omega_s)\right|^2$. This process is the basis of the coherent Stokes Raman spectroscopy (CSRS).

18.4.3.3.4 Coherent anti-Stokes Raman emission This is described by

$$\chi_{ijkl}^{(3)}(\omega_p, \omega_p, -\omega_s) = [\chi_{ijkl}^{(3)}(-\omega_p, -\omega_p, \omega_s)]^* = \frac{(N_o - N_t)/(\varepsilon_0 6\hbar^3)}{\Delta\omega_r - (\omega_p - \omega_s) - i\Gamma_r}$$

$$\cdot \sum_b \left[\frac{(p_i)_{ob}(p_j)_{bt}}{\omega_{bo} - 2\omega_p + \omega_s} + \frac{(p_j)_{ob}(p_i)_{bt}}{\omega_{bo} + \omega_p} \right] \cdot \sum_b \left[\frac{(p_k)_{tb}(p_l)_{bo}}{\omega_{bo} + \omega_s} + \frac{(p_l)_{tb}(p_k)_{bo}}{\omega_{bo} - \omega_p} \right],$$

(18.4-12)

where $\omega_p > \omega_s$. This resonant susceptibility is used to describe the generation of the coherent Raman emission at a new frequency of $\omega = \omega_p + (\omega_p - \omega_s) = \omega_p + \Delta\omega_r$ through phase-matched FWM; the intensity of this newly generated signal is proportional to $\left|\chi_{ijkl}^{(3)}(\omega_p, \omega_p, -\omega_s)\right|^2$. This process is the basis of the coherent anti-Stokes Raman spectroscopy (CARS).

In addition, the real parts of Eqs. (18.4-9) and (18.4-10) can also be used to describe the Raman resonance-enhanced refractive-index change at the frequencies of the two incident light beams.

18.5 Quantum-mechanical expressions for nonlinear susceptibilities

18.5.1 Validity of quantum-mechanical expressions for nonlinear susceptibilities

Up to now, we have completed the rigorous derivations of the quantum-mechanical expressions for various-order susceptibilities within the framework of semiclassical theory. However, it should be pointed out that to obtain these explicit expressions for nonlinear susceptibilities, the following preconditions have been assumed: (i) there is an instantaneous relationship between the electric polarization $\mathbf{P}(t)$ of a medium and the applied optical field $\mathbf{E}(t)$; (ii) the electric polarization $\mathbf{P}(t, r)$ in a given position of the medium is only a function of the optical field $\mathbf{E}(t, r)$ in the same position; (iii) the microscopic polarization response of an individual particle (molecule or atom) can be described in the quantum-mechanical way. For polarization processes arising from electron-cloud distortion, all these three preconditions are fulfilled. For polarization processes arising from intramolecular motion, these requirements can also be fulfilled if the variation of the amplitude function of the optical field is much slower than the frequency of the intramolecular motion. Thus, we can conclude that the quantum-mechanical expressions for susceptibilities presented in this chapter are valid only for those polarization responses of the medium, which come from the electron-cloud distortion or intramolecular motion (Raman vibration).

On the other hand, these quantum-mechanical expressions for susceptibilities cannot be simply applied to those polarization processes caused by other mechanisms. For instance, in Kerr liquids the optical field-induced molecular reorientation leads to a refractive-index change; however, this molecular reorientation does not follow a pure quantum-mechanical rule, owing to the existence of the macroscopic damping (viscosity) field. For the polarization response associated with the electrostriction process, the applied optical field may induce a macroscopic variation of the density distribution in the medium; this density variation is irrelevant to the quantum states of the individual particles, but can be described by classical elastic-mechanics. As for other mechanisms, such as population change or opto-thermal effect, the related polarization change of the medium is no longer an instantaneous function of the amplitude variation of the optical field. Moreover, the nonlinear polarization theory specifically presented in this chapter is not suitable to describe various coherent transient effects discussed in Chapter 10, these effects have to be elucidated by employing the Bloch equation theory, though the latter is still within the semiclassical regime.

18.5.2 Born–Oppenheimer approximation for nonlinear susceptibilities of a molecular medium

In principle, for a given nonlinear medium if we know the molecular eigenfunctions and the macroscopic constitutional property, various susceptibilities can be theoretically calculated. In practice, however, even for the simplest case, i.e., a medium consisting of identical and independent molecules, the computation task is still extremely difficult because of the lack of specific knowledge of the complete eigenfunctions of the molecule. For these reasons, when researchers intend to conduct such theoretical computation or simulation, they have to use various approximation methods based on some greatly simplified physical models. Among those, the Born–Oppenheimer (B-O) approximation is one of the most useful approaches for the theoretical study of nonlinear susceptibilities of molecular systems.

The B-O approximation, which is a well-known approximation method in quantum mechanics, can be used to simplify the descriptions of quantum states and energy structures of a molecular system.[8] This approximation is based on the following two steps of theoretical treatments.

Step I: Assuming that the nuclei of a molecule are fixed in certain positions, the eigenfunctions $\phi_n(x, X)$ and eigenenergy $U_n(X)$ of the electron of the molecule can be obtained by solving the electronic wave equation alone. Here, n labels different quantum states of the electron, x denotes the coordinates of the electron with respect to the nuclei, and X denotes the relative positions of nuclei with respect to a laboratory coordinate system. The physical meaning of this assumption is that the electronic eigenfunctions and eigenenergy are functions of the coordinates of the nuclei.

Step II: Substituting the electronic eigenenergy $U_n(X)$ as a potential function into the wave equation of the nuclear motion, then the solutions of eigenfunctions $\Phi_{n,v}(X)$ and eigenenergy $E_{n,v}^{nuc}$ of the nuclei can be obtained.

For any given value of electronic quantum number $n = 0, 1, 2, \ldots$, there is a series of quantum numbers v corresponding to different vibrational (or vibration–rotational) states of the nuclei. Therefore, the complete eigenfunctions of the whole molecule can be written as

$$\Psi_{n,v}(x, X) = \phi_n(x, X)\Phi_{n,v}(X). \tag{18.5-1}$$

From this expression one can see that in the B-O approximation, we could consider the electronic motion and nuclear motion separately; the connection between these two parts is the use of electronic eigenenergy function as the potential function of the wave equation of the nuclei.

The B-O approximation can also be used to simplify the theoretical description of the nonlinear susceptibilities for molecular medium.[2,9] In this case, use of the B-O approximation can provide two advantages: (i) the contributions to nonlinear susceptibilities from the electronic motion and from the nuclear motion can be separated, and (ii) the number of nonzero and independent elements of various-order susceptibilities can be further reduced based on the additional permutation symmetry properties in this approximation. The nonlinear susceptibility derivations of a molecular system in the B-O approximation are given in detail by Hellwarth.[9]

Based on the above-mentioned additional permutation symmetry, the number of nonzero and independent elements of the third-order nuclear susceptibility can be further reduced.

In Appendix 7, the distribution and the number of nonzero and independent elements of the third-order nuclear susceptibility in the B-O approximation are given for various types of media.[9]

REFERENCES

1. P. N. Butcher, *Nonlinear Optical Phenomena* (Ohio State University Press, Columbus, 1965).
2. C. Flytzanis, In *Quantum Electronics: A Treatise*, Vol. 1, Part A, edited by H. Rabin and C. L. Tang (Academic Press, New York, 1975), pp.9–278.
3. J. Ducuing, In *Nonlinear Optics*, edited by P. G. Harper and B. S. Wherrett (Academic Press, New York, 1977), pp.11–46.
4. P. N. Butcher and D. Cotter, *The Elements of Nonlinear Optics* (Cambridge University Press, 1990), Chapters 2, 4, and 5.
5. M. Schubert and B. Wilhelmi, *Nonlinear Optics and Quantum Electronics* (Wiley, New York, 1986), Chapters 1, 3, and 4.
6. W. Heitler, *The Quantum Theory of Radiation*, 3rd ed. (Oxford University Press, London, 1954), Chapter 3.
7. D. A. Kleinman, *Phys. Rev.* **126**, 1977 (1962).
8. L. I. Schiff, *Quantum Mechanics*, 3rd ed. (McGraw-Hill, New York, 1968), pp.446–449.
9. R. W. Hellwarth, *Prog. Quantum Electron.* **5**, Part I, 1 (1977).

Appendices

Appendix 1
Physical Constants Commonly Used in Nonlinear Optics

Name	Symbol	Value	SI System♣	Gaussian System♦
Speed of light in vacuum	c	2.99793	10^8 m/s	10^{10} cm/s
Permittivity of free space	ε_0	8.85419	10^{-12} F/m	1
Permeability of free space	μ_0	1.25664	10^{-6} H/m	1
Planck constant	h	6.62608	10^{-34} J s	10^{-27} erg s
	$\hbar$	1.05457	10^{-34} J s	10^{-27} erg s
Boltzmann constant	k_B	1.38066	10^{-23} J/K	10^{-16} erg/K
Avogadro constant	N_A	6.02214	10^{23} mol^{-1}	10^{23} mol^{-1}
Electron charge	e	1.60218	10^{-19} C	
		4.803		10^{-10} esu
Electron rest mass	m_e	9.10939	10^{-31} kg	10^{-28} g
Electron radius	r_e	2.81794	10^{-15} m	10^{-13} cm
Bohr radius	a_0	5.29177	10^{-11} m	10^{-9} cm
Proton rest mass	m_p	1.67262	10^{-27} kg	10^{-24} g
Bohr magneton	μ_B	9.27402	10^{-24} J/T	10^{-21} erg/G
Nuclear magneton	μ_n	5.05079	10^{-27} J/T	10^{-24} erg/G
Rydberg constant	R_∞	1.09737	10^7 m^{-1}	10^5 cm^{-1}
Fine structure constant	α	7.29735	10^{-3}	10^{-3}
Electronvolt	1 eV	1.60218	10^{-19} J	10^{-12} erg

Abbreviations: C = coulomb, F = farad = coulomb/volt, G = gauss, H = henry, J = joule, K = kelvin, T = tesla, V = volt, s = second, m = meter, cm = centimeter, kg = kilogram, g = gram, mol = mole.
♣D. R. Lide, *CRC Handbook of Chemistry and Physics*, 71st ed. (CRC Press, Boston, 1990–1991), p. 1-1.
♦R. W. Boyd, *Nonlinear Optics* (Academic Press, New York, 1992), pp. 433–434.

Appendix 2
Numerical Estimates and Conversion of Units

A2.1 Estimate of the atomic electric field strength

As an example, we consider the atomic electric field strength exerted on the electron in a hydrogen atom. In the SI system of units, it can be calculated by

$$E_{atom} = \frac{e}{4\pi\varepsilon_0 a_0^2} \quad \text{(V/m)}. \tag{A2-1}$$

Here, $e = 1.60218 \times 10^{-19}$ C is the electron charge, $\varepsilon_0 = 8.85419 \times 10^{-12}$ F/m (=8.85419 $\times$ 10^{-12} C/V m) is the permittivity of free space, and $a_0 = 5.29177 \times 10^{-11}$ m is the Bohr radius. By substituting these values into Eq. (A2-1), we obtain

$$E_{atom} \approx 5.14 \times 10^{11} \text{ V/m}. \tag{A2-2}$$

A2.2 Estimate of the electric field strength of a laser beam

For a quasi-monochromatic and directional light beam, the relation between the light intensity I and the electric field strength E_0 is determined by Eq. (1.3-8), i.e.,

$$I = \frac{1}{2}\varepsilon_0 n_0 c E_0^2 \quad \text{(W/m}^2\text{)}, \tag{A2-3}$$

where n_0 is the relative refractive index of the medium and $c = 2.99793 \times 10^8$ m/s is the speed of light in a vacuum. If the light intensity is known, the electric field strength of the light wave can be calculated by

$$E_0 = \left(\frac{2I}{\varepsilon_0 n_0 c}\right)^{1/2} \quad \text{(V/m)}. \tag{A2-4}$$

The commonly employed intensity levels of a focused laser beam for nonlinear optical studies are around $I \approx 10^8$–10^{10} W/cm^2 = 10^{12}–10^{14} W/m^2. Assuming $n_0 \approx 1$, according to Eq. (A2-4) we obtain

$$E_0 \approx 0.28 - 2.7 \times 10^8 \text{ V/m}. \tag{A2-5}$$

Comparing these values with the value indicated by Eq. (A2-2), we can see that the electric field strength of a focused laser beam is quite close to the atomic electric field strength in order of magnitude. For this reason, many nonlinear optical effects can be observed.

A2.3 Units of various-order susceptibilities

According to Eq. (5.1-4) the relative dielectric constant of the medium is defined by

$$\varepsilon_r(\omega) = 1 + \chi^{(1)}(\omega). \tag{A2-6}$$

This definition indicates that the linear susceptibility $\chi^{(1)}$ is dimensionless in the system of SI units. Consequently, the linear electric polarization of the medium and its unit can be expressed as

$$\mathbf{P}^{(1)} = \varepsilon_0 \chi^{(1)} \mathbf{E} \quad (C/m^2). \tag{A2-7}$$

Since the overall electric polarization is expressed by

$$\mathbf{P} = \mathbf{P}^{(1)} + \mathbf{P}^{(2)} + \mathbf{P}^{(3)} + \cdots,$$

the second- and third-order polarization should have the same units as the linear polarization has, i.e.,

$$\left.\begin{array}{l} \mathbf{P}^{(2)} = \varepsilon_0 \chi^{(2)} \mathbf{E}^2 \quad (C/m^2) \\ \mathbf{P}^{(3)} = \varepsilon_0 \chi^{(3)} \mathbf{E}^3 \quad (C/m^2) \end{array}\right\}. \tag{A2-8}$$

Based on this expression we can further determine the units for various-order susceptibilities:

$$\left.\begin{array}{l} \chi^{(1)} \to \text{(dimensionless)} \\ \chi^{(2)} \to (m/V) \\ \chi^{(3)} \to (m/V)^2 \\ \cdots \\ \chi^{(r)} \to (m/V)^{r-1} \end{array}\right\}. \tag{A2-9}$$

A2.4 Conversion of susceptibilities among different systems of units

In some early papers of nonlinear optical studies, the Gaussian system of units (esu) was used. In this system, the electric field strength has the following unit:

$$\mathbf{E} \to \text{(statvolt/cm)}, \tag{A2-10}$$

and $\mathbf{E}$ is related to the electric displacement $\mathbf{D}$ through the following equation:

$$\mathbf{D}(\omega) = \mathbf{E}(\omega) + 4\pi \mathbf{P}(\omega). \tag{A2-11}$$

Therefore, the polarization $\mathbf{P}$ should possess the same unit as $\mathbf{E}$, i.e.,

$$\mathbf{P} = \mathbf{P}^{(1)} + \mathbf{P}^{(2)} + \mathbf{P}^{(3)} + \cdots \quad \text{(statvolt/cm)}. \tag{A2-12}$$

On the other hand, in the Gaussian system various-order susceptibilities are defined by

$$\left.\begin{array}{l} \mathbf{P}^{(1)} = \chi^{(1)} \mathbf{E} \\ \mathbf{P}^{(2)} = \chi^{(2)} \mathbf{E}^2 \\ \mathbf{P}^{(3)} = \chi^{(3)} \mathbf{E}^3 \\ \cdots \end{array}\right\}. \tag{A2-13}$$

Based on these expressions we can conclude that

$$\left.\begin{array}{l}\chi^{(1)} \rightarrow \text{(dimensionless)} \\ \chi^{(2)} \rightarrow \text{(cm/statvolt)} \\ \chi^{(3)} \rightarrow \text{(cm/statvolt)}^2 \\ \cdots \\ \chi^{(r)} \rightarrow \text{(cm/statvolt)}^{r-1}\end{array}\right\}. \qquad \text{(A2-14)}$$

We know that 1 statvolt/cm for electric field strength in the esu system is equivalent to 3×10^4 V/m in the SI system and the conversion of various-order susceptibilities in these two systems of units is governed by the following relations:

$$\left.\begin{array}{l}\chi^{(1)}(\text{SI})/\chi^{(1)}(\text{esu}) = 4\pi \\ \chi^{(2)}(\text{SI})/\chi^{(2)}(\text{esu}) = 4\pi/(3 \times 10^4) \\ \chi^{(3)}(\text{SI})/\chi^{(3)}(\text{esu}) = 4\pi/(3 \times 10^4)^2 \\ \cdots \\ \chi^{(r)}(\text{SI})/\chi^{(r)}(\text{esu}) = 4\pi/(3 \times 10^4)^{r-1}\end{array}\right\}. \qquad \text{(A2-15)}$$

Appendix 3
Tensor Elements of the Linear Susceptibility for Crystals and other Media

The matrix element $\chi_{ij}^{(1)}$ is denoted by its Cartesian indices. The number of the independent nonzero elements is given in parentheses.

Crystal System	$\chi^{(1)}$	
Triclinic	$\begin{bmatrix} xx & xy & zx \\ xy & yy & yz \\ zx & yz & zz \end{bmatrix}$	(6)
Monoclinic	$\begin{bmatrix} xx & 0 & zx \\ 0 & yy & 0 \\ zx & 0 & zz \end{bmatrix}$	(4)
Orthorhombic	$\begin{bmatrix} xx & 0 & 0 \\ 0 & yy & 0 \\ 0 & 0 & zz \end{bmatrix}$	(3)
Trigonal, Tetragonal, Hexagonal	$\begin{bmatrix} xx & 0 & 0 \\ 0 & xx & 0 \\ 0 & 0 & zz \end{bmatrix}$	(2)
Cubic, Isotropic	$\begin{bmatrix} xx & 0 & 0 \\ 0 & xx & 0 \\ 0 & 0 & xx \end{bmatrix}$	(1)

After P. N. Butcher and D. Cotter, *The Elements of Nonlinear Optics* (Cambridge University Press, Cambridge, 1990).

Appendix 4
Tensor Elements of the Second-Order Susceptibility for Various Crystal Classes

The matrix element $\chi_{ijk}^{(2)}$ is denoted by its Cartesian indices. The number of the independent nonzero elements is given in parentheses. A bar denotes the negative. For other unlisted crystal classes, all matrix elements are identically zero. Note that the arrangement of the elements shown here in each row is different from that shown in Eq. (2.3-5).

Crystal System	Crystal Class										
Triclinic	1	$\begin{bmatrix} xxx & xyy & xzz & xyz & xzy & xzx & xxz & xxy & xyx \\ yxx & yyy & yzz & yyz & yzy & yzx & yxz & yxy & yyx \\ zxx & zyy & zzz & zyz & zzy & zzx & zxz & zxy & zyx \end{bmatrix}$									(27)
Monoclinic	2	$\begin{bmatrix} 0 & 0 & 0 & xyz & xzy & 0 & 0 & xxy & xyx \\ yxx & yyy & yzz & 0 & 0 & yzx & yxz & 0 & 0 \\ 0 & 0 & 0 & zyz & zzy & 0 & 0 & zxy & zyx \end{bmatrix}$									(13)
	m	$\begin{bmatrix} xxx & xyy & xzz & 0 & 0 & xzx & xxz & 0 & 0 \\ 0 & 0 & 0 & yyz & yzy & 0 & 0 & yxy & yyx \\ zxx & zyy & zzz & 0 & 0 & zzx & zxz & 0 & 0 \end{bmatrix}$									(14)
Orthorhombic	222	$\begin{bmatrix} 0 & 0 & 0 & xyz & xzy & 0 & 0 & 0 & 0 \\ 0 & 0 & 0 & 0 & 0 & yzx & yxz & 0 & 0 \\ 0 & 0 & 0 & 0 & 0 & 0 & 0 & zxy & zyx \end{bmatrix}$									(6)
	mm2	$\begin{bmatrix} 0 & 0 & 0 & 0 & 0 & xzx & xxz & 0 & 0 \\ 0 & 0 & 0 & yyz & yzy & 0 & 0 & 0 & 0 \\ zxx & zyy & zzz & 0 & 0 & 0 & 0 & 0 & 0 \end{bmatrix}$									(7)
Trigonal	3	$\begin{bmatrix} xxx & \overline{xxx} & 0 & xyz & xzy & xzx & xxz & \overline{yyy} & \overline{yyy} \\ \overline{yyy} & yyy & 0 & xxz & xzx & \overline{xzy} & \overline{xyz} & \overline{xxx} & \overline{xxx} \\ zxx & zxx & zzz & 0 & 0 & 0 & 0 & zxy & \overline{zxy} \end{bmatrix}$									(9)
	32	$\begin{bmatrix} xxx & \overline{xxx} & 0 & xyz & xzy & 0 & 0 & 0 & 0 \\ 0 & 0 & 0 & 0 & 0 & \overline{xzy} & \overline{xyz} & \overline{xxx} & \overline{xxx} \\ 0 & 0 & 0 & 0 & 0 & 0 & 0 & zxy & \overline{zxy} \end{bmatrix}$									(4)

continued

Crystal System	Crystal Class	Matrix	
	3m	$\begin{bmatrix} 0 & 0 & 0 & 0 & 0 & xzx & xxz & \overline{yyy} & \overline{yyy} \\ \overline{yyy} & yyy & 0 & xxz & xzx & 0 & 0 & 0 & 0 \\ zxx & zxx & zzz & 0 & 0 & 0 & 0 & 0 & 0 \end{bmatrix}$	(5)
Tetragonal	4	$\begin{bmatrix} 0 & 0 & 0 & xyz & xzy & xzx & xxz & 0 & 0 \\ 0 & 0 & 0 & xxz & xzx & \overline{xzy} & \overline{xyz} & 0 & 0 \\ zxx & zxx & zzz & 0 & 0 & 0 & 0 & zxy & \overline{zxy} \end{bmatrix}$	(7)
	$\bar{4}$	$\begin{bmatrix} 0 & 0 & 0 & xyz & xzy & xzx & xxz & 0 & 0 \\ 0 & 0 & 0 & \overline{xxz} & \overline{xzx} & xzy & xyz & 0 & 0 \\ zxx & \overline{zxx} & 0 & 0 & 0 & 0 & 0 & zxy & zxy \end{bmatrix}$	(6)
	422	$\begin{bmatrix} 0 & 0 & 0 & xyz & xzy & 0 & 0 & 0 & 0 \\ 0 & 0 & 0 & 0 & 0 & \overline{xzy} & \overline{xyz} & 0 & 0 \\ 0 & 0 & 0 & 0 & 0 & 0 & 0 & zxy & \overline{zxy} \end{bmatrix}$	(3)
	4mm	$\begin{bmatrix} 0 & 0 & 0 & 0 & 0 & xzx & xxz & 0 & 0 \\ 0 & 0 & 0 & xxz & xzx & 0 & 0 & 0 & 0 \\ zxx & zxx & zzz & 0 & 0 & 0 & 0 & 0 & 0 \end{bmatrix}$	(4)
	$\bar{4}2m$	$\begin{bmatrix} 0 & 0 & 0 & xyz & xzy & 0 & 0 & 0 & 0 \\ 0 & 0 & 0 & 0 & 0 & xzy & xyz & 0 & 0 \\ 0 & 0 & 0 & 0 & 0 & 0 & 0 & zxy & zxy \end{bmatrix}$	(3)
Hexagonal	6	$\begin{bmatrix} 0 & 0 & 0 & xyz & xzy & xzx & xxz & 0 & 0 \\ 0 & 0 & 0 & xxz & xzx & \overline{xzy} & \overline{xyz} & 0 & 0 \\ zxx & zxx & zzz & 0 & 0 & 0 & 0 & zxy & \overline{zxy} \end{bmatrix}$	(7)
	$\bar{6}$	$\begin{bmatrix} xxx & \overline{xxx} & 0 & 0 & 0 & 0 & 0 & \overline{yyy} & \overline{yyy} \\ \overline{yyy} & yyy & 0 & 0 & 0 & 0 & 0 & \overline{xxx} & \overline{xxx} \\ 0 & 0 & 0 & 0 & 0 & 0 & 0 & 0 & 0 \end{bmatrix}$	(2)
	622	$\begin{bmatrix} 0 & 0 & 0 & xyz & xzy & 0 & 0 & 0 & 0 \\ 0 & 0 & 0 & 0 & 0 & \overline{xzy} & \overline{xyz} & 0 & 0 \\ 0 & 0 & 0 & 0 & 0 & 0 & 0 & zxy & \overline{zxy} \end{bmatrix}$	(3)
	6mm	$\begin{bmatrix} 0 & 0 & 0 & 0 & 0 & xzx & xxz & 0 & 0 \\ 0 & 0 & 0 & xxz & xzx & 0 & 0 & 0 & 0 \\ zxx & zxx & zzz & 0 & 0 & 0 & 0 & 0 & 0 \end{bmatrix}$	(4)
	$\bar{6}m2$	$\begin{bmatrix} 0 & 0 & 0 & 0 & 0 & 0 & 0 & \overline{yyy} & \overline{yyy} \\ \overline{yyy} & yyy & 0 & 0 & 0 & 0 & 0 & 0 & 0 \\ 0 & 0 & 0 & 0 & 0 & 0 & 0 & 0 & 0 \end{bmatrix}$	(1)

continued

Crystal System	Crystal Class		
Cubic	432	$\begin{bmatrix} 0 & 0 & 0 & xyz & \overline{xyz} & 0 & 0 & 0 & 0 \\ 0 & 0 & 0 & 0 & 0 & xyz & \overline{xyz} & 0 & 0 \\ 0 & 0 & 0 & 0 & 0 & 0 & 0 & xyz & \overline{xyz} \end{bmatrix}$	(1)
	$\bar{4}3m$	$\begin{bmatrix} 0 & 0 & 0 & xyz & xyz & 0 & 0 & 0 & 0 \\ 0 & 0 & 0 & 0 & 0 & xyz & xyz & 0 & 0 \\ 0 & 0 & 0 & 0 & 0 & 0 & 0 & xyz & xyz \end{bmatrix}$	(1)
	23	$\begin{bmatrix} 0 & 0 & 0 & xyz & xzy & 0 & 0 & 0 & 0 \\ 0 & 0 & 0 & 0 & 0 & xyz & xzy & 0 & 0 \\ 0 & 0 & 0 & 0 & 0 & 0 & 0 & xyz & xzy \end{bmatrix}$	(2)

After P. N. Butcher and D. Cotter, *The Elements of Nonlinear Optics* (Cambridge University Press, Cambridge, 1990).

Appendix 5
Tensor Elements of the Susceptibility of Second-Harmonic Generation for Various Crystal Classes

Here $d_{il} = \chi^{(2)}_{ijk}(\omega,\omega)$ when $j = k$, or $d_{il} = \frac{1}{2}[\chi^{(2)}_{ijk}(\omega,\omega) + \chi^{(2)}_{ikj}(\omega,\omega)]$ when $j \neq k$. The relations between the new subscript l and the original subscripts jk are:

$$\begin{cases} l = 1 & 2 & 3 & 4 & 5 & 6 \\ jk = xx & yy & zz & yz, zy & xz, zx & xy, yx \end{cases}$$

The polarization of the second-harmonic generation can be expressed as

$$\begin{bmatrix} P^{(2)}_x(2\omega) \\ P^{(2)}_y(2\omega) \\ P^{(2)}_z(2\omega) \end{bmatrix} = \varepsilon_0 \begin{bmatrix} d_{11} & d_{12} & d_{13} & d_{14} & d_{15} & d_{16} \\ d_{21} & d_{22} & d_{23} & d_{24} & d_{25} & d_{26} \\ d_{31} & d_{32} & d_{33} & d_{34} & d_{35} & d_{36} \end{bmatrix} \begin{bmatrix} E^2_x(\omega) \\ E^2_y(\omega) \\ E^2_z(\omega) \\ 2E_yE_z \\ 2E_xE_z \\ 2E_xE_y \end{bmatrix}.$$

The distributions of nonzero elements of d_{il} are the same as those of the piezoelectric modulus tensors of crystals. The number of the independent nonzero elements of d_{il} is given in parentheses. A bar denotes the negative.

Crystal System	Crystal Class							
Triclinic	1	$\begin{bmatrix} d_{11} & d_{12} & d_{13} & d_{14} & d_{15} & d_{16} \\ d_{21} & d_{22} & d_{23} & d_{24} & d_{25} & d_{26} \\ d_{31} & d_{32} & d_{33} & d_{34} & d_{35} & d_{36} \end{bmatrix}$							(18)
Monoclinic	2	$\begin{bmatrix} 0 & 0 & 0 & d_{14} & 0 & d_{16} \\ d_{21} & d_{22} & d_{23} & 0 & d_{25} & 0 \\ 0 & 0 & 0 & d_{34} & 0 & d_{36} \end{bmatrix}$							(8)
	m	$\begin{bmatrix} d_{11} & d_{12} & d_{13} & 0 & d_{15} & 0 \\ 0 & 0 & 0 & d_{24} & 0 & d_{26} \\ d_{31} & d_{32} & d_{33} & 0 & d_{35} & 0 \end{bmatrix}$							(10)

continued

Crystal System	Crystal Class		
Orthorhombic	222	$\begin{bmatrix} 0 & 0 & 0 & d_{14} & 0 & 0 \\ 0 & 0 & 0 & 0 & d_{25} & 0 \\ 0 & 0 & 0 & 0 & 0 & d_{36} \end{bmatrix}$	(3)
	mm2	$\begin{bmatrix} 0 & 0 & 0 & 0 & d_{15} & 0 \\ 0 & 0 & 0 & d_{24} & 0 & 0 \\ d_{31} & d_{32} & d_{33} & 0 & 0 & 0 \end{bmatrix}$	(5)
Trigonal	3	$\begin{bmatrix} d_{11} & \overline{d_{11}} & 0 & d_{14} & d_{15} & 2\overline{d_{22}} \\ \overline{d_{22}} & d_{22} & 0 & d_{15} & \overline{d_{14}} & 2\overline{d_{11}} \\ d_{31} & d_{31} & d_{33} & 0 & 0 & 0 \end{bmatrix}$	(6)
	32	$\begin{bmatrix} d_{11} & \overline{d_{11}} & 0 & d_{14} & 0 & 0 \\ 0 & 0 & 0 & 0 & \overline{d_{14}} & 2\overline{d_{11}} \\ 0 & 0 & 0 & 0 & 0 & 0 \end{bmatrix}$	(2)
	3m	$\begin{bmatrix} 0 & 0 & 0 & 0 & d_{15} & 2\overline{d_{22}} \\ \overline{d_{22}} & d_{22} & 0 & d_{15} & 0 & 0 \\ d_{31} & d_{31} & d_{33} & 0 & 0 & 0 \end{bmatrix}$	(4)
Tetragonal	4	$\begin{bmatrix} 0 & 0 & 0 & d_{14} & d_{15} & 0 \\ 0 & 0 & 0 & d_{15} & \overline{d_{14}} & 0 \\ d_{31} & d_{31} & d_{33} & 0 & 0 & 0 \end{bmatrix}$	(4)
	$\overline{4}$	$\begin{bmatrix} 0 & 0 & 0 & d_{14} & d_{15} & 0 \\ 0 & 0 & 0 & \overline{d_{15}} & d_{14} & 0 \\ d_{31} & \overline{d_{31}} & 0 & 0 & 0 & d_{36} \end{bmatrix}$	(4)
	422	$\begin{bmatrix} 0 & 0 & 0 & d_{14} & 0 & 0 \\ 0 & 0 & 0 & 0 & \overline{d_{14}} & 0 \\ 0 & 0 & 0 & 0 & 0 & 0 \end{bmatrix}$	(1)
	4mm	$\begin{bmatrix} 0 & 0 & 0 & 0 & d_{15} & 0 \\ 0 & 0 & 0 & d_{15} & 0 & 0 \\ d_{31} & d_{31} & d_{33} & 0 & 0 & 0 \end{bmatrix}$	(3)
	$\overline{4}2m$	$\begin{bmatrix} 0 & 0 & 0 & d_{14} & 0 & 0 \\ 0 & 0 & 0 & 0 & d_{14} & 0 \\ 0 & 0 & 0 & 0 & 0 & d_{36} \end{bmatrix}$	(2)
Hexagonal	6	$\begin{bmatrix} 0 & 0 & 0 & d_{14} & d_{15} & 0 \\ 0 & 0 & 0 & d_{15} & \overline{d_{14}} & 0 \\ d_{31} & d_{31} & d_{33} & 0 & 0 & 0 \end{bmatrix}$	(4)
	$\overline{6}$	$\begin{bmatrix} d_{11} & \overline{d_{11}} & 0 & 0 & 0 & 2\overline{d_{22}} \\ \overline{d_{22}} & d_{22} & 0 & 0 & 0 & 2\overline{d_{11}} \\ 0 & 0 & 0 & 0 & 0 & 0 \end{bmatrix}$	(2)

continued

Crystal System	Crystal Class		
	622	$\begin{bmatrix} 0 & 0 & 0 & d_{14} & 0 & 0 \\ 0 & 0 & 0 & 0 & \overline{d_{14}} & 0 \\ 0 & 0 & 0 & 0 & 0 & 0 \end{bmatrix}$	(1)
	6mm	$\begin{bmatrix} 0 & 0 & 0 & 0 & d_{15} & 0 \\ 0 & 0 & 0 & d_{15} & 0 & 0 \\ d_{31} & d_{31} & d_{33} & 0 & 0 & 0 \end{bmatrix}$	(3)
	$\bar{6}m2$	$\begin{bmatrix} 0 & 0 & 0 & 0 & 0 & 2\overline{d_{22}} \\ \overline{d_{22}} & d_{22} & 0 & 0 & 0 & 0 \\ 0 & 0 & 0 & 0 & 0 & 0 \end{bmatrix}$	(1)
Cubic	$\bar{4}3m$	$\begin{bmatrix} 0 & 0 & 0 & d_{14} & 0 & 0 \\ 0 & 0 & 0 & 0 & d_{14} & 0 \\ 0 & 0 & 0 & 0 & 0 & d_{14} \end{bmatrix}$	(1)
	23	$\begin{bmatrix} 0 & 0 & 0 & d_{14} & 0 & 0 \\ 0 & 0 & 0 & 0 & d_{14} & 0 \\ 0 & 0 & 0 & 0 & 0 & d_{14} \end{bmatrix}$	(1)

See P. N. Butcher and D. Cotter, *The Elements of Nonlinear Optics* (Cambridge University Press, Cambridge, 1990); F. F. Nye, *Physical Properties of Crystals* (Oxford University Press, London, 1957).

Appendix 6
Tensor Elements of the Third-Order Susceptibility for Crystals and other Media

The matrix element $\chi^{(3)}_{ijkl}$ is denoted by its Cartesian indices.

Triclinic

For both classes, 1 and $\bar{1}$, there are **81** independent nonzero elements.

Monoclinic

For all three classes, 2, *m*, and 2/*m*, there are **41** independent nonzero elements:

- 3 elements with indices all equal;
- 18 elements with indices equal in pairs;
- 12 elements with indices having two *y*'s, one *x*, and one *z*;
- 4 elements with indices having three *x*'s and one *z*;
- 4 elements with indices having three *z*'s and one *x*.

Orthorhombic

For all three classes, 222, *mm*2, and *mmm*, there are **21** independent nonzero elements:

- 3 elements with indices all equal;
- 18 elements with indices equal in pairs.

Trigonal

For the two classes 3 and $\bar{3}$, there are 73 nonzero elements, of which only **27** are independent. They are:

$$zzzz$$

$$xxxx = yyyy = xxyy + xyyx + xyxy \begin{cases} xxyy = yyxx \\ xyyx = yxxy \\ xyxy = yxyx \end{cases}$$

$yyzz = xxzz$ $\quad\quad\quad xyzz = -yxzz$
$zzyy = zzxx$ $\quad\quad\quad zzxy = -zzyx$
$zyyz = zxxz$ $\quad\quad\quad zxyz = -zyxz$
$yzzy = xzzx$ $\quad\quad\quad xzzy = -yzzx$
$yzyz = xzxz$ $\quad\quad\quad xzyz = -yzxz$
$zyzy = zxzx$ $\quad\quad\quad zxzy = -zyzx$

$$xxxy = -yyyx = yyxy + yxyy + xyyy \begin{cases} yyxy = -xxyx \\ yxyy = -xyxx \\ xyyy = -yxxx \end{cases}$$

$yyyz = -yxxz = -xyxz = -xxyz$
$yyzy = -yxzx = -xyzx = -xxzy$
$yzyy = -yzxx = -xzyx = -xzxy$
$zyyy = -zyxx = -zxyx = -zxxy$
$xxxz = -xyyz = -yxyz = -yyxz$
$xxzx = -xyzy = -yxzy = -yyzx$
$xzxx = -yzxy = -yzyx = -xzyy$
$zxxx = -zxyy = -zyxy = -zyyx$

For the three classes $3m$, $\bar{3}m$ and 32, there are 37 nonzero elements, of which only **14** are independent. They are:

$zzzz$

$$xxxx = yyyy = xxyy + xyyx + xyxy \begin{cases} xxyy = yyxx \\ xyyx = yxxy \\ xyxy = yxyx \end{cases}$$

$yyzz = xxzz$ $\quad\quad\quad xxxz = -xyyz = -yxyz = -yyxz$
$zzyy = zzxx$ $\quad\quad\quad xxzx = -xyzy = -yxzy = -yyzx$
$zyyz = zxxz$ $\quad\quad\quad xzxx = -xzyy = -yzxy = -yzyx$
$yzzy = xzzx$ $\quad\quad\quad zxxx = -zxyy = -zyxy = -zyyx$
$yzyz = xzxz$
$zyzy = zxzx$

Tetragonal

For the three classes 4, $\bar{4}$ and $4/m$, there are 41 nonzero elements, of which only **21** are independent. They are:

$xxxx = yyyy$ $\quad\quad zzzz$
$zzxx = zzyy$ $\quad\quad xyzz = -yxzz$ $\quad\quad xxyy = yyxx$ $\quad\quad xxxy = -yyyx$
$xxzz = yyzz$ $\quad\quad zzxy = -zzyx$ $\quad\quad xyxy = yxyx$ $\quad\quad xxyx = -yyxy$
$zxzx = zyzy$ $\quad\quad xzyz = -yzxz$ $\quad\quad xyyx = yxxy$ $\quad\quad xyxx = -yxyy$
$xzxz = yzyz$ $\quad\quad zxzy = -zyzx$ $\quad\quad\quad\quad\quad\quad\quad\quad\quad yxxx = -xyyy$
$zxxz = zyyz$ $\quad\quad zxyz = -zyxz$
$xzzx = yzzy$ $\quad\quad xzzy = -yzzx$

For the four classes 422, 4*mm*, 4/*mmm*, and $\bar{4}2m$, there are 21 nonzero elements, of which only **11** are independent. They are:

$xxxx = yyyy$ $zzzz$
$yyzz = xxzz$ $yzzy = xzzx$ $xxyy = yyxx$
$zzyy = zzxx$ $yzyz = xzxz$ $xyxy = yxyx$
$zyyz = zxxz$ $zyzy = zxzx$ $xyyx = yxxy$

Hexagonal

For the three classes 6, $\bar{6}$ and 6/*m*, there are 41 nonzero elements, of which only **19** are independent. They are:

$zzzz$

$xxxx = yyyy = xxyy + xyyx + xyxy \begin{cases} xxyy = yyxx \\ xyyx = yxxy \\ xyxy = yxyx \end{cases}$

$yyzz = xxzz$ $xyzz = -yxzz$
$zzyy = zzxx$ $zzxy = -zzyx$
$zyyz = zxxz$ $zxyz = -zyxz$
$yzzy = xzzx$ $xzzy = -yzzx$
$yzyz = xzxz$ $xzyz = -yzxz$
$zyzy = zxzx$ $zxzy = -zyzx$

$xxxy = -yyyx = yyxy + yxyy + xyyy \begin{cases} yyxy = -xxyx \\ yxyy = -xyxx \\ xyyy = -yxxx \end{cases}$

For the four classes 622, 6*mm*, 6/*mmm* and $\bar{6}m2$, there are 21 nonzero elements, of which only **10** are independent. They are:

$zzzz$

$xxxx = yyyy = xxyy + xyyx + xyxy \begin{cases} xxyy = yyxx \\ xyyx = yxxy \\ xyxy = yxyx \end{cases}$

$yyzz = xxzz$ $yzzy = xzzx$
$zzyy = zzxx$ $yzyz = xzxz$
$zyyz = zxxz$ $zyzy = zxzx$

Cubic

For the two classes 23 and $m3$, there are 21 nonzero elements, of which only 7 are independent. They are:

$$xxxx = yyyy = zzzz$$
$$yyzz = zzxx = xxyy$$
$$zzyy = xxzz = yyxx$$
$$yzyz = zxzx = xyxy$$
$$zyzy = xzxz = yxyx$$
$$yzzy = zxxz = xyyx$$
$$zyyz = xzzx = yxxy$$

For the three classes 432, $\bar{4}3m$ and $m3m$, there are 21 nonzero elements, of which only 4 are independent. They are:

$$xxxx = yyyy = zzzz$$
$$yyzz = zzyy = zzxx = xxzz = xxyy = yyxx$$
$$yzyz = zyzy = zxzx = xzxz = xyxy = yxyx$$
$$yzzy = zyyz = zxxz = xzzx = xyyx = yxxy$$

Isotropic

There are 21 nonzero elements, of which only 4 are independent. They are:

$$xxxx = yyyy = zzzz = xxyy + xyxy + xyyx$$
$$yyzz = zzyy = zzxx = xxzz = xxyy = yyxx$$
$$yzyz = zyzy = zxzx = xzxz = xyxy = yxyx$$
$$yzzy = zyyz = zxxz = xzzx = xyyx = yxxy$$

See P. N. Butcher and D. Cotter, *The Elements of Nonlinear Optics* (Cambridge University Press, Cambridge, 1990); C. C. Shang and H. Hsu, *IEEE J. Quantum Electron.* **23**, 177 (1987).

Appendix 7
Tensor Elements of the Nuclear Third-Order Susceptibility in the Born–Oppenheimer Approximation

Since the tensor element $\chi_{ijkl}^{(3)nuc}$ of the nuclear third-order susceptibility is invariant under the permutations $(i \leftrightarrow j)$, $(k \leftrightarrow l)$, and $(ij \leftrightarrow kl)$, it can be denoted by a new compact symbol $D_{ll'}$. The relations of the cartesian indices between these two symbols are: $\begin{cases} ij, kl = xx & yy & zz & yz & xz & xy \\ l, l' = 1 & 2 & 3 & 4 & 5 & 6 \end{cases}$. The distributions of nonzero elements of $D_{ll'}$ are the same as those of the elastic compliance constant tensors of crystals. The number of the independent nonzero elements is given in parentheses.

Crystal System	Crystal Class		
Triclinic	$\begin{cases} 1 \\ \bar{1} \end{cases}$	$\begin{bmatrix} D_{11} & D_{12} & D_{13} & D_{14} & D_{15} & D_{16} \\ D_{12} & D_{22} & D_{23} & D_{24} & D_{25} & D_{26} \\ D_{13} & D_{23} & D_{33} & D_{34} & D_{35} & D_{36} \\ D_{14} & D_{24} & D_{34} & D_{44} & D_{45} & D_{46} \\ D_{15} & D_{25} & D_{35} & D_{45} & D_{55} & D_{56} \\ D_{16} & D_{26} & D_{36} & D_{46} & D_{56} & D_{66} \end{bmatrix}$	(21)
Monoclinic	$\begin{cases} 2 \\ m \\ 2/m \end{cases}$	$\begin{bmatrix} D_{11} & D_{12} & D_{13} & 0 & D_{15} & 0 \\ D_{12} & D_{22} & D_{23} & 0 & D_{25} & 0 \\ D_{13} & D_{23} & D_{33} & 0 & D_{35} & 0 \\ 0 & 0 & 0 & D_{44} & 0 & D_{46} \\ D_{15} & D_{25} & D_{35} & 0 & D_{55} & 0 \\ 0 & 0 & 0 & D_{46} & 0 & D_{66} \end{bmatrix}$	(13)
Orthorhombic	$\begin{cases} 222 \\ mm2 \\ mmm \end{cases}$	$\begin{bmatrix} D_{11} & D_{12} & D_{13} & 0 & 0 & 0 \\ D_{12} & D_{22} & D_{23} & 0 & 0 & 0 \\ D_{13} & D_{23} & D_{33} & 0 & 0 & 0 \\ 0 & 0 & 0 & D_{44} & 0 & 0 \\ 0 & 0 & 0 & 0 & D_{55} & 0 \\ 0 & 0 & 0 & 0 & 0 & D_{66} \end{bmatrix}$	(9)
Trigonal	$\begin{cases} 3 \\ \bar{3} \end{cases}$	$\begin{bmatrix} D_{11} & D_{12} & D_{13} & D_{14} & -D_{25} & 0 \\ D_{12} & D_{11} & D_{13} & -D_{14} & D_{25} & 0 \\ D_{13} & D_{13} & D_{33} & 0 & 0 & 0 \\ D_{14} & -D_{14} & 0 & D_{44} & 0 & 2D_{25} \\ -D_{25} & D_{25} & 0 & 0 & D_{44} & 2D_{14} \\ 0 & 0 & 0 & 2D_{25} & 2D_{14} & 2(D_{11}-D_{12}) \end{bmatrix}$	(7)

continued

Crystal System	Crystal Class	Matrix	
	$\begin{cases} 3m \\ \bar{3}m \\ 32 \end{cases}$	$\begin{bmatrix} D_{11} & D_{12} & D_{13} & D_{14} & 0 & 0 \\ D_{12} & D_{11} & D_{13} & -D_{14} & 0 & 0 \\ D_{13} & D_{13} & D_{33} & 0 & 0 & 0 \\ D_{14} & -D_{14} & 0 & D_{44} & 0 & 0 \\ 0 & 0 & 0 & 0 & D_{44} & 2D_{14} \\ 0 & 0 & 0 & 0 & 2D_{14} & 2(D_{11}-D_{12}) \end{bmatrix}$	(6)
Tetragonal	$\begin{cases} 4 \\ \bar{4} \\ 4/m \end{cases}$	$\begin{bmatrix} D_{11} & D_{12} & D_{13} & 0 & 0 & D_{16} \\ D_{12} & D_{11} & D_{13} & 0 & 0 & -D_{16} \\ D_{13} & D_{13} & D_{33} & 0 & 0 & 0 \\ 0 & 0 & 0 & D_{44} & 0 & 0 \\ 0 & 0 & 0 & 0 & D_{44} & 0 \\ D_{16} & -D_{16} & 0 & 0 & 0 & D_{66} \end{bmatrix}$	(7)
	$\begin{cases} 4mm \\ 422 \\ \bar{4}2m \\ 4/mmm \end{cases}$	$\begin{bmatrix} D_{11} & D_{12} & D_{13} & 0 & 0 & 0 \\ D_{12} & D_{11} & D_{13} & 0 & 0 & 0 \\ D_{13} & D_{13} & D_{33} & 0 & 0 & 0 \\ 0 & 0 & 0 & D_{44} & 0 & 0 \\ 0 & 0 & 0 & 0 & D_{44} & 0 \\ 0 & 0 & 0 & 0 & 0 & D_{66} \end{bmatrix}$	(6)
Hexagonal	$\begin{cases} 6 \\ \bar{6} \\ 6/m \\ 622 \\ 6mm \\ 6/mmm \\ \bar{6}m2 \end{cases}$	$\begin{bmatrix} D_{11} & D_{12} & D_{13} & 0 & 0 & 0 \\ D_{12} & D_{11} & D_{13} & 0 & 0 & 0 \\ D_{13} & D_{13} & D_{33} & 0 & 0 & 0 \\ 0 & 0 & 0 & D_{44} & 0 & 0 \\ 0 & 0 & 0 & 0 & D_{44} & 0 \\ 0 & 0 & 0 & 0 & 0 & 2(D_{11}-D_{12}) \end{bmatrix}$	(5)
Cubic	$\begin{cases} 23 \\ m3 \\ 432 \\ \bar{4}3m \\ m3m \end{cases}$	$\begin{bmatrix} D_{11} & D_{12} & D_{12} & 0 & 0 & 0 \\ D_{12} & D_{11} & D_{12} & 0 & 0 & 0 \\ D_{12} & D_{12} & D_{11} & 0 & 0 & 0 \\ 0 & 0 & 0 & D_{44} & 0 & 0 \\ 0 & 0 & 0 & 0 & D_{44} & 0 \\ 0 & 0 & 0 & 0 & 0 & D_{44} \end{bmatrix}$	(3)
Isotropic		$\begin{bmatrix} D_{11} & D_{12} & D_{12} & 0 & 0 & 0 \\ D_{12} & D_{11} & D_{12} & 0 & 0 & 0 \\ D_{12} & D_{12} & D_{11} & 0 & 0 & 0 \\ 0 & 0 & 0 & 2(D_{11}-D_{12}) & 0 & 0 \\ 0 & 0 & 0 & 0 & 2(D_{11}-D_{12}) & 0 \\ 0 & 0 & 0 & 0 & 0 & 2(D_{11}-D_{12}) \end{bmatrix}$	(2)

See R. W. Hellwarth, *Prog. Quantum Electron.* **5**, 1(1977); J. F. Nye, *Physical Properties of Crystals* (Oxford University Press, London, 1957).

Appendix 8
Derivation of Formulae for Self-Induced Transparency of a 2π-Pulse

A8.1 Derivation of Eq. (10.2-11)

Introducing a pair of new variables, $\tau = t - n_0 z/c$ and $z' = n_0 z/c$, and noticing that

$$\frac{\partial E_0(z',\tau)}{\partial t} = \frac{\partial E_0(z',\tau)}{\partial \tau}, \quad \frac{\partial E_0(z',\tau)}{\partial z} = -\frac{n_0}{c}\frac{\partial E_0(z',\tau)}{\partial \tau} + \frac{n_0}{c}\frac{\partial E_0(z',\tau)}{\partial z'},$$

one can rewrite the first equation of Eq. (10.2-7) as

$$\frac{\partial E_0(z',\tau)}{\partial z'} = \frac{\omega_0}{2n_0^2 \varepsilon_0} P_0(z',\tau). \tag{A8-1}$$

Multiplying by $2E_0(z',\tau)$ on the two sides of the above equation leads to

$$2E_0(z',\tau)\frac{\partial E_0(z',\tau)}{\partial z'} = \frac{\omega_0}{n_0^2 \varepsilon_0} E_0(z',\tau) P_0(z',\tau). \tag{A8-2}$$

Substituting the third part of Eq. (10.2-7) into the above equation yields

$$\frac{\partial}{\partial z'} E_0^2(z',\tau) = \frac{\hbar \omega_0}{2\pi n_0^2 \varepsilon_0} \frac{\partial N(z',\tau)}{\partial \tau}. \tag{A8-3}$$

Integrating with respect to τ and multiplying $\varepsilon_0 c n_0/2$ on both sides leads to

$$\frac{1}{2}\varepsilon_0 c n_0 \frac{\partial}{\partial z'} \int_{-\infty}^{\infty} E_0^2(z',\tau)d\tau = \frac{c\hbar \omega_0}{4\pi n_0} [N(z',\tau=\infty) - N_0]. \tag{A8-4}$$

As N is determined by the first part of Eq. (10.2-8), the above equation can be rewritten as

$$\frac{\partial S(z')}{\partial z'} = \frac{c\hbar \omega_0}{4\pi n_0} N_0[\cos\theta(z') - 1], \tag{A8-5}$$

or

$$\frac{\partial S(z)}{\partial z} = \frac{\hbar \omega_0}{4\pi} N_0[\cos\theta(z) - 1]. \tag{A8-6}$$

A8.2 Derivation of Eq. (10.2-18)

For mathematical convenience, we introduce a new function defined by

$$R\left(\tau = t - \frac{z}{\xi}\right) = \frac{\pi \tau_0 p_0}{h} E_0(\tau) = \sin\frac{\theta(\tau)}{2}. \tag{A8-7}$$

Noticing Eq. (10.2-15), we have

$$\begin{aligned}
\frac{dR(\tau)}{d\tau} &= \frac{\pi \tau_0 p_0}{h} \frac{dE_0(\tau)}{d\tau} = \frac{1}{2\tau_0} \sin\theta(\tau) \\
&= \frac{1}{\tau_0} \sin\frac{\theta(\tau)}{2} \cos\frac{\theta(\tau)}{2} \\
&= \frac{1}{\tau_0} \sin\frac{\theta(\tau)}{2} \left[1 - \sin^2\frac{\theta(\tau)}{2}\right]^{1/2} \\
&= \frac{1}{\tau_0} R[1 - R^2]^{1/2},
\end{aligned} \tag{A8-8}$$

or

$$\frac{d\tau}{\tau_0} = \frac{dR}{R[1 - R^2]^{1/2}} = d(\text{sech}^{-1} R). \tag{A8-9}$$

Then from Eqs. (A8-9) and (A8-7) we can obtain the final solution for the 2π-pulse:

$$E_0^o\left(\tau = t - \frac{z}{\xi}\right) = \frac{h}{\pi \tau_0 p_0} R(\tau) = \frac{h}{\pi \tau_0 p_0} \text{sech}\left(\frac{\tau}{\tau_0}\right). \tag{A8-10}$$

Index

A

Aberration influence, 225, 226
Aberrator, 243, 246, 256
Absorption dip, 280
Absorption saturation, 544, 529
Acoustic field, 168, 171
Acoustic wave, 139, 169
Amplified spontaneous emission (ASE), 459, 461
Anomalous dispersion, 64, 85, 516
Anti-Helmholtz configuration, 315, 316
Anti-Stokes Brillouin scattering, 170
Anti-Stokes Raman scattering, 142
Anti-Stokes ring, 154–157
Anti-Stokes shift, 131
Anti-Stokes shifted SRS, 150
Aperture angle, 366

B

Backward PCWs, 223
Backward scattering, 170
Backward stimulated scattering (BSS), 248
Band structure of solid, 490
Bandgap, 490
Beat-frequency enhanced, 125
Birefringence effect, 39
Bloch vector, 345
Blue-shifted, 121, 125
Bohr magneton, 162
Boltzmann constant, 275, 580
Born–Oppenheimer (B-O) approximation, 603
Boxcar, 472, 473, 501
Bright spatial soliton, 410, 411
Brightness, 8
Brillouin scattering, 138

C

CARS polarization spectroscopy, 308
Causality, 513, 526
Cavity build-up time, 364, 375
Cavity detuning, 362, 369, 373
Cavityless lasing, 255, 461, 463
Centrosymmetric crystal, 26
Centrosymmetry, 39
Chemical potential, 492
Chromophore, 440–442, 448
Coherent anti-Stokes Raman emission, 602
Coherent anti-Stokes Raman spectroscopy (CARS), 23, 296–301
Coherent anti-Stokes ring emission, 155–157
Coherent continuum generation, 127, 134
Coherent cross-section, 8
Coherent length of SHG, 37
Coherent length of THG, 63
Coherent population oscillation (CPO), 537, 544
Coherent Raman spectroscopy, 295–306
Coherent Stokes Raman emission, 601
Coherent Stokes Raman spectroscopy (CSRS), 297
Coherent Stokes ring emission, 155–157
Coherent transient effects, 4
Coherent-time, 8
Collinear emission, 130
Collision broadening, 275
Collision cross-section, 276
Complex conjugation symmetry, 27
Complex expressions of wave fields, 30
Complex refractive index, 291, 517

Conduction band, 162, 490
Conservation of energy and momentum, 35, 44, 48, 60
Conservation of energy, 145
Continuum generation, 74
Conventional optics, 1, 3
Co-phasal plane, 513
Coulomb gauge, 581
Coulomb interaction, 503
Counter-propagating, 227, 237, 280, 289, 293, 313–315
Coupled-wave equation, 36, 49, 62, 173
Crossover resonance, 287–289
Crossover signal, 287
Cross-section of Kerr scattering, 188
Cross-section of Raman scattering, 147
Cross-section of Raman–Kerr scattering, 191
Cylindrical lens, 129, 155

D

Dark spatial soliton, 410, 419
Debye time, 183, 193
Degenerate four-photon parametric interaction, 76
Degenerate four-wave mixing, 226–229
Density matrix, 579
Density matrix elements, 344
Density matrix equation, 336, 344, 579
Density matrix operator, 336, 344, 579
Dephasing process, 337
Diaphragm, 366
Dielectric constant, 83
Difference-frequency generation, 45, 560
Dipole-transition matrix element, 84, 89, 335
Directionality, 10

Dispersion shifted fiber (DSF), 256
Disturbing medium, 225
Divergence angle, 8, 9
Doppler broadening, 275, 347, 348
Doppler effect, 121, 168, 204, 275, 313
Doppler line shape, 279
Doppler line width, 314
Doppler profile, 278
Doppler-free saturation spectroscopy, 283–287
Doppler-free two-photon spectroscopy, 290, 293
Doublet absorption lines, 529, 540
Doublet gain lines, 529, 532, 534
Dynamic self-focusing, 114

E

Effective gyromagnetic ratio, 162
Effective interaction length, 37, 237
Effective second-order susceptibility value, 38
Effective third-order susceptibility value, 63
Eigenenergy, 145
Eigenfunction, 11, 143, 581, 603
Eigenstates, 145
Eigenvalue, 11
Electric dipole approximation, 26, 39, 55, 147, 583
Electric dipole moment operator, 147, 325
Electric displacement vector, 83
Electric field induced SHG (EFISHG), 78
Electric field strength, 1
Electric field-induced second harmonic (EFISH), 78, 567
Electric polarization, 1, 19
Electric susceptibilities, 1
Electric-dipole interaction, 583
Electric-quadrupole approximation, 55, 583
Electric-quadrupole interaction, 583

Electric-quadrupole operator, 582
Electromagnetically induced absorption (EIA), 535
Electromagnetically induced transparency (EIT), 379, 535, 540, 546
Electron acceptor, 441
Electron at a metal surface, 491
Electron donor, 441
Electron energy analyzer, 496
Electron kinetic energy, 495, 498–500
Electron-cloud distortion, 22
Electron-hole pair, 503
Electronic-transition SRS, 165
Electro-optic (EO) sampling, 559
Electro-optic effect, 559, 565
Electrostriction, 22, 95
Electrostrictive coefficient, 171
Electrostrictive force, 95, 171, 312
Energy conservation, 145
Energy-level diagram, 14, 34
Er-doped fiber amplifier (EDFA), 266
Excited state absorption (ESA), 449, 466, 536
Exciton, 503, 504
Excitonic transition, 503–504
Exponential gain, 149, 176, 192

F

Fabry–Perot (F-P) etalon, 5, 170, 179, 274, 359, 364
Far-field pattern, 207, 240, 244, 246, 247, 256, 259
Fast and slow light, 4, 513, 524, 526
Fast light effects, 529
Fast light propagation, 513, 516
Fermi energy level, 492, 493
Fermi–Dirac distribution, 492
Fiber soliton lasers, 401
Fidelity, 238
First-order susceptibility $\chi^{(1)}$, 2, 23
Five-photon photoemission (5PPE), 494
Five-photon pumped (5PP) lasing, 465

Forward PCWs, 224
Forward/backward asymmetry, 154
Fourier components, 20, 28, 581
Fourier integral, 20, 28, 581
Fourier transform, 19, 123
Four-photon absorption (4PA), 433
Four-photon parametric interaction, 14, 61, 74, 131
Four-photon photoemission (4PPE), 494
Four-photon pumped (4PP) lasing, 464
Four-wave frequency mixing (FWFM), 4, 59
Four-wave mixing (FWM), 226, 229, 563, 566
Free-space permeability (μ_0), 109, 325
Free-space permittivity (ϵ_0), 1
Frequency chirp, 121
Frequency up-conversion lasing, 458
Frequency-chirped pulse, 121
Frequency-degenerate, 76, 223
Frequency-nondegenerate, 223
Fundamental ($N = 1$) temporal soliton, 391
FWFM in SRS, 154

G

Gabor's principle, 249
Gain coefficient, 148, 176, 192
Gain factor, 176
Gain length, 154
Gaussian amplitude profile, 126
Gaussian beam, 112
Gaussian distribution, 4, 111, 348
Gaussian function, 110
Gaussian shape, 275, 373
Gaussian temporal profile, 120
Gradient magnetic field, 315
Grating, 129, 155
Group refractive index, 519
Group velocity, 386, 514, 516
Group velocity dispersion (GVD), 6, 263, 386, 522, 539

Group velocity in a gain medium, 522
Group velocity in an absorbing medium, 520

H
Hamiltonian, 143, 335
Hamiltonian operator, 143, 579
Hermite-Gaussian beam approximation, 254
High-order ($N > 1$) temporal solitons, 391, 396
Hole-burning, 278, 544
Holographic model, 229
Homogeneous broadening, 193, 324
Hyperfine splitting, 284
Hyperfine structure, 294, 286
Hypersonic wave, 177, 178, 204

I
Image charge, 492
Image potential, 491, 492
Image-potential state (IPS), 491, 495, 497
Induced acoustic motion, 22
Induced birefringence, 306
Induced Bragg grating, 203, 204
Induced dichroism, 306
Induced electric dipole moment, 19
Induced holographic grating, 229, 230
Induced phonon, 170
Induced population change, 22
Induced refractive-index change, 4, 86
Induced stationary Bragg grating, 203, 211
Inhomogeneous broadening, 193, 278, 324, 337
Inhomogeneous relaxation (dephasing) time T_2^*, 324
Intensity, 7
Intensity fluctuation, 471–473
Intensity-dependent nonlinear absorption, 203–204
Interaction Hamiltonian, 143, 146, 335, 344, 581
Interference, 123, 203, 498, 515

Intermediate state(s), 12, 33, 45, 48, 60, 66, 67, 71, 74, 142, 163, 167, 188, 290, 435, 494, 495, 565
Intracavity intensity, 360
Intramolecular motion, 22, 602
Intrinsic permutation symmetry, 595
Inverse-Fourier transform, 30, 123
Inverse-Raman spectroscopy (IRS), 303
Inverted Lamb dip, 284
Isotope shift, 286
Isotropic media, 26

K
Kerr effect, 570, 571
Kerr liquids, 93, 95, 183
Kerr scattering, 187
Kinetic energy, 489
Kinetic temperature, 314
Kleinman symmetry, 597
Kramers–Kronig relations, 299, 307, 528

L
Lamb dip, 278, 284
Lamb shift, 285
Landau levels, 162, 503
Laser cooling and trapping, 314
Laser cooing and trapping spectroscopy, 312
Laser polarization spectroscopy, 306
Light polarization, 306
Light scattering, 136
Light speed, 513
Linear absorption coefficient, 519
Linear absorption cross-section, 93
Linear electro-optic (Pockels) effect, 416
Linear optics, 3
Linear polarization, 23
Linear refractive index, 11, 85, 360
Linear susceptibility $\chi^{(1)}$, 2, 23
Liquid crystals (LC), 415

Liquid-filled hollow fiber, 185–187, 194–198
Local field correction, 590–592
Longitudinal relaxation time T_1, 324
Lorentzian function, 193
Lorentzian line shape, 281
Lorentzian profile, 149

M
Magnetic-dipole approximation, 55
Magnetic-dipole interaction, 583
Magnetic-dipole operator, 582
Magneto-optical trap (MOT), 314, 318
Manley–Rowe relations, 49, 62
Materials for SBS, 177–179
Materials for THG, 68–70
Maxwell distribution, 275
Maxwell's electromagnetic theory, 27
Maxwell's equation, 109
Midway phase conjugator, 264
Mie scattering, 139
Mode number, 9, 145
Modulation depth, 203
Molecular orientation angle, 187, 188
Molecular polarizability tensor, 189
Molecular reorientation, 22, 93
Monochromatic optical wave, 1, 513
Monochromaticity, 10
Moving foci, 116
MPA spectra, 450
MPA-based optical limiting, 466
MPA-based optical reshaping, 474
MPA-based optical stabilization, 470
MPA-induced light attenuation, 436
Multi-photon absorption (MPA), 3, 5, 432, 466, 470, 494
Multi-photon active materials, 441
Multi-photon excitation (MPE), 5, 476

Multi-photon excited fluorescence, 456, 457
Multi-photon photoconductivity (MPPC), 500, 502
Multi-photon photoelectric effects, 7
Multi-photon photoemission (MPPE), 493, 497, 500
Multi-photon processes, 4
Multi-photon pumped (MPP) lasing, 6, 458–461
Multi-photon techniques, 4
Multi-valued solutions, 359, 361

N
Natural linewidth, 281, 282, 284, 316
Near-field pattern, 129, 207, 256
Negative group velocity (GVD), 388 Negative GVD, 387
Non-centrosymmetrical crystal, 26
Non-collinear emission, 130
Nonlinear coupled wave equation(s), 27, 30, 49
Nonlinear crystals for SHG, 39, 42
Nonlinear differential equation, 29
Nonlinear dimensionless Schrödinger equation, 391
Nonlinear F-P etalon, 359
Nonlinear optics, 1, 3
Nonlinear photoelectric effects, 4, 7, 489
Nonlinear photonics, 3, 7, 14
Nonlinear polarization, 3, 18, 22
Nonlinear polarization spectroscopy, 306–312
Nonlinear polarization theory, 3
Nonlinear reflectivity, 237
Nonlinear refractive-index coefficient n_2, 86
Nonlinear spectroscopic effects, 4
Nonlinear susceptibility, 2, 4
Nonlinear transmission (NLT), 437, 445, 467–469
Nonlinear wave equation, 28

Non-resonant interaction, 26, 84
Normal dispersion, 84
Normalized line-shape function, 149, 177
Nth-order polarization component, 20
Nuclear magnetic resonance, 334, 342, 352
Numerical aperture (NA), 476

O
One-photon absorption (1PA), 2, 18, 85
One-photon resonance, 87
OPC applications, 260–270
Optical (reorientational) Kerr effect, 93, 187
Optical bistability, 4, 359
Optical bistable devices, 368, 378
Optical Bloch equation, 344, 345, 603
Optical breakdown, 23
Optical clipper, 359
Optical coherent transient effects, 325
Optical damage, 23
Optical difference-frequency generation, 45, 560
Optical differential amplification, 359
Optical discriminator, 359
Optical electrostriction, 95
Optical filament, 108
Optical free induction decay (OFID), 352–355
Optical heterodyne, 170
Optical hysteresis loop, 359, 363
Optical limiter, 359
Optical limiting, 466–470
Optical logical circuits, 6
Optical memory, 363
Optical nutation, 342, 349
Optical parametric amplification (OPA), 47
Optical parametric generator (OPG), 53, 451, 460
Optical parametric oscillation (OPO), 51
Optical phase conjugation (OPC), 4, 6, 222
Optical phase conjugator, 226

Optical Pockels effect, 99, 413
Optical rectification, 508, 559
Optical reshaping, 474–476
Optical solitons, 4
Optical spatial soliton, 410
Optical stabilization, 470–474
Optical steering, 359
Optical sum-frequency generation, 43
Optical switching, 6, 363
Optical telecommunication, 134
Optical temporal soliton, 6, 386
Optical-frequency Pockels effect, 103
Opto-thermal effect, 98, 232, 359, 376, 377, 579
Orientational Kerr effect, 23
Overall permutation symmetry, 26, 596

P
Parametric interaction, 169
Paraxial approximation, 111
Pattern plate, 156
PCW generation via backward stimulated emission (lasing), 254–260
PCW generation via backward stimulated scattering, 243–248
PCW generation via FWM, 226, 233
PCW generation via three-wave mixing, 235
Permutation symmetry, 26
Phase change function, 253
Phase conjugate reflector, 225
Phase conjugation wave (PCW), 11, 222
Phase matching of SHG, 35, 39
Phase matching of THG, 64
Phase mismatch factor, 36, 63, 175
Phase relaxation, 323
Phase shift factor, 360
Phase velocity, 386, 513
Phase-conjugator, 265
Phase-distortion function, 223
Phase-matching angle, 40
Phase-matching type I, 39
Phase-matching type II, 39
Phonon, 169, 177, 178

Photoconductivity, 491
Photoelastic effect, 173
Photoemission effect, 489
Photon annihilation, 12
Photon annihilation operator, 11
Photon creation, 12
Photon creation operator, 11
Photon degeneracy, 9, 143, 146
Photon echo, 334
Photon ensemble, 9
Photon mode, 9
Photonic crystal structure, 379, 382, 583
Photonics, 3, 14
Photorefractive (PR) crystals, 416, 427
Photorefractive effect, 416
Pinhole, 129
Plane-wave approximation, 29, 64
Pockels effect, 559, 565, 570, 571
Polarization labeling spectroscopy, 310
Polarization response function, 19, 20, 27
Polarization spectroscopy, 306
Poling process, 38
Ponderomotive force, 95, 171, 312
Population change, 22, 232
Population distribution, 279
Population redistribution, 369
Positive group velocity dispersion (GVD), 387
Power broadening, 282
Poynting vector, 11
Precession, 336, 346
Probe beam, 279, 280
Propagation period of soliton, 392
Pseudo-electric dipole moment operator, 335
Pseudo-electric field, 335
Pseudo-electric polarization, 335
Pump beam, 279, 280

Q

Quantum beat effect, 498
Quantum electrodynamics, 3, 11, 33, 142
Quantum mechanics, 10, 11
Quantum theory of radiation, 3, 11
Quantum transition, 33
Quasi-collinear four-wave mixing model, 248
Quasi-monochromatic light pulse, 514
Quasi-parallel beam, 106, 109
Quasi-periodic modulation, 131
Quasi-periodic oscillation, 355
Quasi-resonance condition, 26
Quasi-resonant interaction, 323

R

Rabi frequency, 323, 337, 344, 348, 541
Raman anisotropy, 191
Raman attenuation effect, 601
Raman cross-section, 146
Raman enhanced FWFM, 70, 73
Raman excited state, 71
Raman gain effect, 601
Raman gain spectroscopy (RGS), 303
Raman media, 151
Raman mode, 71
Raman polarizability, 191
Raman resonance, 90
Raman scattering, 138
Raman susceptibility, 300
Raman transition, 71
Raman-enhanced coherent ring emission, 72
Raman-enhanced refractive-index change, 23, 90
Raman-induced Kerr effect spectroscopy (RIKES), 301
Rayleigh scattering, 138
Rayleigh-wing scattering, 137
Recoil broadening, 277
Red-shifted, 121, 125
Refractive-index change, 96
Relaxation time of refractive-index change, 96
Reorientation work, 187
Reorientational Kerr effect, 95, 187
Resonance enhanced, 26, 65, 88, 90, 166, 167, 205, 494

Resonance enhancement of THG, 65
Resonance-enhanced refractive-index change, 88–92
Resonant absorbing medium, 34
Resonant interaction, 84, 97, 323, 432
Resonant medium, 323, 517, 524
Resonant radiation pressure, 313
Reverse saturable absorption (RSA), 466
Rise time of refractive-index change, 96
Rotational transition SRS, 168
Round-trip phase change, 365

S

Saturable absorption, 282, 306
Saturation intensity parameter, 282
Saturation spectroscopy, 278–289, 319
Scalar potential, 581
Second harmonic generation (SHG), 1, 33–39
Secondary wave, 1, 9
Second-order Doppler broadening, 277
Second-order intensity correlation, 504
Second-order nonlinear crystals, 413, 426, 504
Second-order nonlinearity, 99
Second-order polarization, 24
Second-order sum-frequency generation, 12
Second-order susceptibility $\chi^{(2)}$, 24
Self-broadening, 4, 123
Self-defocusing, 107
Self-focusing, 4, 105
Self-induced transparency, 324
Self-phase modulation, 119, 388
Self-trapping, 108
Semiclassical description of SHG, 35
Semiclassical regime, 11, 18
Semiclassical theory, 10, 21, 579

SHG in optical fibers, 79
SHG in surfaces and interfaces, 55
Single mode fiber span, 265, 266
Slit, 129
Slow light effects, 540
Slow light propagation, 516
Slowly-varying amplitude approximation, 30, 110, 174
Small aberration approximation, 252
Soliton lasers, 401–407
Spatial (transverse) coherence, 8
Spatial coherent length, 8
Spatial soliton interaction, 423
Spatial symmetry, 25
Special relativity, 513, 526
Spectra self-broadening, 123, 263
Spectral brightness, 8, 274
Spectral intensity, 7
Spectral resolution, 274
Spectral resolving power, 274
Spectral reversal, 264, 265
Spin echo, 334
Spin nutation, 342
Spin-down state, 163
Spin-flip SRS, 163
Spin-up state, 163
Spiraling spatial soliton, 417
Spontaneous scattering, 140
S-shape curve, 363
Standing-wave field, 313
Standing-wave induced grating, 203
Stark effect, 349
Stark electric field, 352, 353
Stark splitting, 294
Stationary Bragg grating, 203
Steady-state operation, 364
Steady-state self-focusing, 109
Steep anomalous dispersion, 526
Steep normal dispersion, 526
Stimulated Brillouin scattering (SBS), 5, 168
Stimulated emission, 141
Stimulated hyper-Raman scattering, 166
Stimulated Kerr scattering (SKS), 5, 142, 183–200

Stimulated Mie scattering, 5, 210–217
Stimulated Raman scattering (SRS), 5
Stimulated Raman–Kerr scattering, 191, 197
Stimulated Rayleigh–Bragg scattering (SRBS), 5
Stimulated Rayleigh–Kerr scattering, 191, 194, 198
Stimulated Rayleigh-wing scattering (SRWS), 183
Stimulated scattering, 4, 140
Stimulated thermal Rayleigh scattering, 200
Stokes Brillouin scattering, 169
Stokes Raman scattering, 146
Stokes ring, 154–157
Stokes shift, 131
Stokes-shifted SRS, 150
Sum-frequency generation (SFG), 68
Super-broadband stimulated scattering, 185
Superluminal light propagation, 513, 516
Susceptibility, 1
Susceptibility tensor, 23

T
Temperature tuning, 40
Temperature-tuning phase matching, 40
Temporal (longitudinal) coherence, 8
Terahertz (THz) radiation, 508, 559
Terahertz nonlinear photonics, 4, 559
Thermal bistability, 375
Thermally-induced refractive-index change, 98
Third-harmonic generation (THG), 62
Third-order intensity correlation, 506
Third-order polarization, 24
Third-order susceptibility $\chi^{(3)}$, 24

Three-dimensional (3D) data storage, 476, 479
Three-dimensional (3D) micro-fabrication, 476, 481
Three-photon absorption (3PA), 432
Three-photon absorption coefficient, 437
Three-photon absorption cross-section, 437
Three-photon parametric interaction, 14
Three-photon photoconductivity (3PPC), 500–505, 507
Three-photon photoemission (3PPE), 493
Three-photon pumped lasing, 462–465
Three-wave frequency mixing, 3, 33
Threshold condition, 51, 148, 177, 194, 222
Time-reversal symmetry, 27, 596
Transient and thermal optical bistability, 175
Transient time broadening, 276
Transverse (spatial) coherence length, 8
Transverse Doppler effect, 277
Transverse relaxation time T_2, 324
Trapped atom cloud, 316
Two-photon absorption (2PA), 3, 5, 18, 432
Two-photon absorption coefficient, 436
Two-photon absorption cross-section, 437
Two-photon absorption spectroscopy, 289–295
Two-photon excited fluorescence (2PEF), 447
Two-photon photoconductivity (2PPC), 500–505, 507
Two-photon photoemission (2PPE), 493
Two-photon pumped (2PP) lasing, 461
Two-photon resonance, 88
Two-photon resonance enhancement, 88, 600

U

Ultrahigh resolution spectroscopy, 274, 316
Ultrahigh vacuum, 315, 496
Uncertainty principle, 12, 143, 514, 560
Unfocused-beam approximation, 250
Unperturbed eigenfunction, 143
Unperturbed eigenstate, 144
Unperturbed Hamiltonian, 580

V

Vacuum (energy) level, 492
Valence band, 162, 490
Vector potential, 581
Virtual energy level, 8, 14, 34
Volume grating, 229

W

Walk-off effect, 40
Wavefront reconstruction, 222, 229, 249
Wave-packet, 515, 526
Wavevector, 29, 35, 60, 146
White-light continuum (WLC), 128, 445, 451–453
Work function, 489, 494

Z

Zeeman effect, 315
Zeeman shifter, 314
Zeeman slower, 316
Zeeman splitting, 315, 503
Z-scan, 446, 451, 576

2π pulse, 324, 327
$\pi/2$ pulse, 334
π pulse, 334
π-conjugated bridge, 442
3D data storage, 476–481
3D microfabrication, 481–484